Biotechnology

INDUSTRIAL MICROBIOLOGY

A TEXTBOOK

W CLARKE

CBSPD

CBS Publishers & Distributors Pvt Ltd

New Delhi • Bengaluru • Chennai • Kochi • Kolkata • Lucknow • Mumbai
Hyderabad • Jharkhand • Nagpur • Patna • Pune • Uttarakhand

Disclaimer

Science and technology are constantly changing fields. New research and experience broaden the scope of information and knowledge. The author has tried his best in giving information available to him while preparing the material for this book. Although, all efforts have been made to ensure optimum accuracy of the material, yet it is quite possible some errors might have been left uncorrected. The publisher, the printer and the author will not be held responsible for any inadvertent errors, omissions or inaccuracies.

Biotechnology

INDUSTRIAL MICROBIOLOGY

A TEXTBOOK

ISBN: 978-81-239-2864-7

First Edition: 2016
Reprint: 2024

Published by **Satish Kumar Jain** and produced by **Varun Jain** for

CBS Publishers & Distributors Pvt Ltd
4819/XI Prahlad Street, 24 Ansari Road, Daryaganj, New Delhi 110 002, India.
Ph: 011-23289259, 23266861 Website: www.cbspd.com
e-mail: delhi@cbspd.com

Corporate Office: 204 FIE, Industrial Area, Patparganj, Delhi 110 092
Ph: 011-4934 4934 Fax: 011-4934 4935 e-mail: publishing@cbspd.com;publicity@cbspd.com

Branches

- **Bengaluru:** Seema House 2975, 17th Cross, K.R. Road, Banasankari 2nd Stage, Bengaluru 560 070 Karnataka, India
  Ph: +91-80-26771678/79 Fax: +91-80-26771680 e-mail: bangalore@cbspd.com
- **Chennai:** 7, Subbaraya Street, Shenoy Nagar, Chennai 600 030, Tamil Nadu, India
  Ph: +91-44-26680620, 26681266 Fax: +91-44-42032115 e-mail: chennai@cbspd.com
- **Kochi:** 42/1325, 1326, Power House Road, Opp KSEB, Ernakulam 682 018, Kochi, Kerala, India
  Ph: +91-484-4059061-67 Fax: +91-484-4059065 e-mail: kochi@cbspd.com
- **Kolkata:** 147, Hind Ceramics Compound, 1st Floor, Nilgunj Road, Belghoria, Kolkata 700 056, West Bengal, India
  Ph: +91-33-25633055/56 e-mail: kolkata@cbspd.com
- **Lucknow:** Basement, Khushnuma Complex, 7-Meerabai Marg (Behind Jawahar Bhawan), Lucknow 226 001, UP, India
  Ph: +0552-4000032 e-mail:tiwari.lucknowi@cbspd.com
- **Mumbai:** PWD Shed. Gala no. 25/26, Ramchandra Bhatt Marg, Next to JJ Hospital Gate no. 2, Opp. Union Bank of India, Noorbaug, Mumbai 400 009, Maharashtra, India
  Ph: 022-66661880/89 e-mail: mumbai@cbspd.com

Representatives

- **Hyderabad** 0-9885175004
- **Patna** 0-9334159340
- **Jharkhand** 0-9811541605
- **Pune** 0-9664372571
- **Nagpur** 0-8692091830
- **Uttarakhand** 0-9716462459

Printed at: Rashtriya Printer, Dilshad Garden, Delhi, India

Biotechnology

INDUSTRIAL MICROBIOLOGY

A TEXTBOOK

Preface

Biotechnology is technology based on biology, agriculture, food science, and medicine and deals with the use of living organisms or their products on large scale industrial processes.

Biotechnology has applications in four major industrial areas, including health care (medical), crop production and agriculture, nonfood (industrial) uses of crops and other products (e.g. biodegradable plastics, vegetable oil biofuels), and environment.

Environmental microbiology is the study of the composition and physiology of microbial communities in the environment. The environment in this case means the soil, water, air and sediments covering the planet and can also include the animals and plants that inhabit these areas. Environmental microbiology also includes the study of micro-organisms that exist in artificial environments such as bioreactors.

Environmental biotechnology can simply be described as 'the optimal use of nature, in the form of plants, animals, bacteria, fungi and algae, to produce renewable energy, food and nutrients in a synergistic integrated cycle of profit making processes where the use of each process becomes the feedstock for another process'.

Industrial biotechnology is the application of biotechnology for industrial purposes, including manufacturing, alternative energy (or bioenergy), and biomaterials. It includes the practice of using cells or components of cells like enzymes to generate industrially useful products.

Industrial microbiology encompasses the use of micro-organisms in the manufacture of food or industrial products. The use of micro-organisms for the production of food, either human or animal, is often considered a branch of food microbiology. The micro-organisms used in industrial processes may be natural isolates, laboratory selected mutants or genetically engineered organisms.

Chapter 1 deals with basic concepts of biotechnology and microbiology. Chapter 2 introduces screening and testing of new metabolites which are intermediates and products of metabolism. Chapter 3 discusses strain development and gene technology for a cost effective process, strains with improved fermentation properties may also be needed. The fermentation medium forms the environment in which the fermentation micro-organisms live, reproduce and carry out their specific metabolic reactions to produce useful products. Considering this chapter 4 focuses on substrates for industrial fermentation. Chapter 5 deals with various fermentation methods and systems such as batch, fed-batch and continuous fermentation along with classification of fermentation processes. Chapter 6 is devoted to recovery and and purification of products. Down stream processing refers to the recovery and purification of biosynthetic products, various methods such as filtration, centrifugation, sedimentation, flocculation, chromatography and crystallisation are discussed. Chapter 7 introduces organic feedstocks produced by fermentation. Fermentation processes play a major role in the production of most organic acids which are produced microbiologically in high yields. Keeping this in mind chapter 8 focuses on organic acids and their production. Chapter 9 explains amino acids and their commercial uses. Chapter 10 concentrates on nucleosides, nucleotides, and allied compounds.

Chapter 11 focuses on production of enzymes and discusses enzyme assays, kinetics of enzymes, purification and recovery of enzymes. Micro-organisms can be used for the commercial production of vitamins such as thiamine, riboflavin, folic acid, pyridoxal, and biotin. Considering this chapter 12 is devoted to vitamins and their synthesis. Chapter 13 concentrates on antibiotics which are products of secondary metabolism which inhibit growth processes of other organisms even when used at low concentrations. Chapter 14 introduces ergot alkaloids and their synthesis with naturally occurring chemical compounds containing basic nitrogen atoms. Classification, synthesis, along with biological aspects of ergot alkaloids are discussed in detail. Chapter 15 focuses on microbial transformations. Micro-organisms have the ability to chemically modify a wide variety of organic compounds. Such changes are called biological or microbial transformations or bioconversion.

Chapter 16 discusses single cell protein (SCP). Various methods of production of SCP from alkanes, carbohydrates, sewage are discussed. Chapter 17 explains new sewage and sludge treatment processes along with production of methane. Chapter 18 is devoted to bioleaching which is the process by which metals are dissolved from ore-bearing rocks using micro-organisms. Chapter 19 deals with extracellular polysaccharides and their derivatives. Polysaccharides are used commercially to produce gels, and to thicken and stabilised foods, medicines and industrial products. Chapter 20 explains other fermentation processes which do not fit into the categories established in previous chapters.

This textbook is essential reading for all students, teachers, professionals, researchers and industrialists involved with chemical engineering, biochemical engineering, environmental science, microbiology, biotechnology and life sciences. It is also a valuable source of information for industrialists and chemists already associated with food processing, agricultural, etc. The reference textbook also caters to the requirement of the syllabus prescribed by various Indian universities for undergraduate and postgraduate students pursuing these courses. Constructive suggestions are always welcome from users of this book.

Diagrams, figures, tables and index supplement the text. All the topics have been covered in a cogent and lucid style to help the reader grasp the information quickly and easily.

W Clarke

Contents at a Glance

Contents

Chapter 1

Basic Concepts of Biotechnology and Microbiology

INTRODUCTION

Biotechnology is technology based on biology, agriculture, food science, and medicine. Modern use of the term usually refers to genetic engineering as well as cell and tissue culture technologies. However, the concept encompasses a wider range and history of procedures for modifying living things according to human purposes, going back to domestication of animals, cultivation of plants and 'improvements' to these through breeding programs that employ artificial selection and hybridisation. By comparison to biotechnology, bioengineering is generally thought of as a related field with its emphasis more on mechanical and higher systems approaches to interfacing with and exploiting living things United Nations Convention on Biological Diversity defines biotechnology as:

'Any technological application that uses biological systems, dead organisms, or derivatives thereof, to make or modify products or processes for specific use'.

Biotechnology draws on the pure biological sciences (genetics, microbiology, animal cell culture, molecular biology, biochemistry, embryology, cell biology) and in many instances is also dependent on knowledge and methods from outside the sphere of biology (chemical engineering, bioprocess engineering, information technology, biorobotics). Conversely, modern biological sciences (including even concepts such as molecular ecology) are intimately entwined and dependent on the methods developed through biotechnology and what is commonly thought of as the life sciences industry.

Biotechnology has applications in four major industrial areas, including health care (medical), crop production and agriculture, non food (industrial) uses of crops and other products (e.g. biodegradable plastics, vegetable oil, biofuels), and environmental uses.

For example, one application of biotechnology is the directed use of organisms for the manufacture of organic products (examples include beer and milk products). Another example is using naturally present bacteria by the mining industry in bioleaching. Biotechnology is also used to recycle, treat waste, clean up sites contaminated by industrial activities (bioremediation), and also to produce biological weapons.

A series of derived terms have been coined to identify several branches of biotechnology, for example:

Bioinformatics: Bioinformatics is an interdisciplinary field which addresses biological problems using computational techniques, and makes the rapid organisation and analysis of biological data possible. The field may also be referred to as computational biology, and can be defined as, 'conceptualising biology in terms of molecules and then applying informatics techniques to understand and organise the information associated with these molecules, on a large scale'. Bioinformatics plays a key role in various areas, such as functional genomics, structural genomics, and proteomics, and forms a key component in the biotechnology and pharmaceutical sector.

Blue biotechnology: Blue biotechnology is a term that has been used to describe the marine and aquatic applications of biotechnology, but its use is relatively rare.

Green biotechnology: Green biotechnology is biotechnology applied to agricultural processes. An example would be the selection and domestication of plants via micropropagation.

Red biotechnology: Red biotechnology is applied to medical processes. Some examples are the designing of organisms to produce antibiotics, and the engineering of genetic cures through genomic manipulation.

White biotechnology: White biotechnology also known as industrial biotechnology, is biotechnology applied to industrial processes. An example is the designing of an organism to produce a useful chemical. Another example is the using of enzymes as industrial catalysts to either produce valuable chemicals or destroy hazardous/polluting chemicals. White biotechnology tends to consume less in resources than traditional processes used to produce industrial goods. The investments and economic output of all of these types of applied biotechnologies form what has been described as the bioeconomy.

In medicine, modern biotechnology finds promising applications in such areas as: (i) drug production, (ii) pharmacogenomics, (iii) gene therapy, and (iv) genetic testing.

Pharmacogenomics is the study of how the genetic inheritance of an individual affects his/her body's response to drugs. It is a coined word derived from the words 'pharmacology' and 'genomics'. It is hence the study of the relationship between pharmaceuticals and genetics. The vision of pharmacogenomics is to be able to design and produce drugs that are adapted to each person's genetic makeup.

Pharmacogenomics results in the following benefits:

1. Development of tailormade medicines. Using pharmacogenomics, pharmaceutical companies can create drugs based on the proteins, enzymes and RNA molecules that are associated with specific genes and diseases. These tailormade drugs promise not only to maximise therapeutic effects but also to decrease damage to nearby healthy cells.
2. More accurate methods of determining appropriate drug dosages. Knowing a patient's genetics will enable doctors to determine how well his/her body can process and metabolise a medicine. This will maximise the value of the medicine and decrease the likelihood of overdose.
3. Improvements in the drug discovery and approval process. The discovery of potential therapies will be made easier using genome targets. Genes have been associated with numerous diseases and disorders. With modern biotechnology, these genes can be used as targets for the development of effective new therapies, which could significantly shorten the drug discovery process.
4. Better vaccines. Safer vaccines can be designed and produced by organisms transformed by means of genetic engineering. These vaccines will elicit the immune response without the attendant risks of infection. They will be inexpensive, stable, easy to store, and capable of being engineered to carry several strains of pathogen at once.

Most traditional pharmaceutical drugs are relatively simple molecules that have been found primarily through trial and error to treat the symptoms of a disease or illness. Biopharmaceuticals are large biological molecules known as proteins and these usually target the underlying mechanisms and pathways of a malady (but not always, as is the case with using insulin to treat type 1 diabetes mellitus, as that treatment merely addresses the symptoms of the disease, not the underlying cause which is autoimmunity); it is a relatively young industry. They can deal with targets in humans that may not be accessible with traditional medicines. A patient typically is dosed with a small molecule via a tablet while a large molecule is typically injected. Small molecules are manufactured by chemistry but larger molecules are created by living cells such as those found in the human body: for example, bacteria cells, yeast cells, animal or plant cells.

Modern biotechnology is often associated with the use of genetically altered micro-organisms such as *E. coli* or yeast for the production of substances like synthetic insulin or antibiotics. It can also refer to transgenic animals or transgenic plants, such as Bt corn. Genetically altered mammalian cells, such as Chinese Hamster Ovary (CHO) cells, are also used to manufacture certain pharmaceuticals. Another promising new biotechnology application is the development of plantmade pharmaceuticals.

Biotechnology is also commonly associated with landmark breakthroughs in new medical therapies to treat hepatitis B, hepatitis C, cancers, arthritis, haemophilia, bone fractures, multiple sclerosis, and cardiovascular disorders. The biotechnology industry has also been instrumental in developing molecular diagnostic devices that can be used to define the target patient population for a given biopharmaceutical.

Modern biotechnology can be used to manufacture existing medicines relatively easily and cheaply. The first genetically engineered products were medicines designed to treat human diseases. To cite one example, in 1978 Genentech developed synthetic humanised insulin by joining its gene with a plasmid vector inserted into the bacterium *Escherichia coli*. Insulin, widely used for the treatment of diabetes, was previously extracted from the pancreas of abattoir animals (cattle and/or pigs). The resulting genetically engineered bacterium enabled the production of vast quantities of synthetic human insulin at relatively low cost. Modern biotechnology has evolved, making it possible to produce more easily and relatively cheaply human growth hormone, clotting factors for hemophiliacs, fertility drugs, erythroprotein and other drugs. Most drugs today are based on about 500 molecular targets. Genomic knowledge of the genes involved in diseases, disease pathways, and drug response sites are expected to lead to the discovery of thousands more new targets.

Genetic testing: Genetic testing involves the direct examination of the DNA molecule itself. A scientist scans a patient's DNA sample for mutated sequences.

Gene therapy: Gene therapy may be used for treating, or even curing, genetic and acquired diseases like cancer and AIDS by using normal genes to supplement or replace defective genes or to bolster a normal function such as immunity. It can be used to target somatic (i.e. body) or gametes (i.e. egg and sperm) cells. In somatic gene therapy, the genome of the recipient is changed, but this change is not passed along to the next generation. In contrast, in germline gene therapy, the egg and sperm cells of the parents are changed for the purpose of passing on the changes to their offspring.

Cloning: Cloning involves the removal of the nucleus from one cell and its placement in an unfertilised egg cell whose nucleus has either been deactivated or removed.

Crop yield: Using the techniques of modern biotechnology, one or two genes may be transferred to a highly developed crop variety to impart a new character that would increase its yield. However, while increases in crop yield are the most obvious applications of modern biotechnology in agriculture, it is also the most difficult one. Current genetic engineering techniques work best for effects that are controlled by a single gene. Many of the genetic characteristics associated with yield (e.g. enhanced growth) are controlled by a large number of genes, each of which has a minimal effect on the overall yield. There is, therefore, much scientific work to be done in this area.

Proteins in foods may be modified to increase their nutritional qualities. Proteins in legumes and cereals may be transformed to provide the amino acids needed by human beings for a balanced diet.

Modern biotechnology can be used to slow down the process of spoilage so that fruit can ripen longer on the plant and then be transported to the consumer with a still reasonable shelf life. This alters the taste, texture and appearance of the fruit. More importantly, it could expand the market for farmers in developing countries due to the reduction in spoilage.

Most of the current commercial applications of modern biotechnology in agriculture are on reducing the dependence of farmers on agrochemicals. For example, *Bacillus thuringiensis* (Bt) is a soil bacterium that produces a protein with insecticidal qualities. Traditionally, a fermentation process has been used to produce an insecticidal spray from these bacteria. In this form, the Bt toxin occurs as an inactive protoxin, which requires digestion by an insect to be effective. There are several Bt toxins and each one is specific to certain target insects. Crop plants have now been engineered to contain and express the genes for Bt toxin, which they produce in its active form. When a susceptible insect ingests the transgenic crop cultivar expressing the Bt protein, it stops feeding and soon thereafter dies as a result of the Bt toxin binding to its gut wall. Bt corn is now commercially available in a number of countries to control corn borer (a lepidopteran insect), which is otherwise controlled by spraying (a more difficult process).

Biological engineering: Biotechnological engineering or biological engineering is a branch of engineering that focuses on biotechnologies and biological science. It includes different disciplines such as biochemical engineering, biomedical engineering, bioprocess engineering, biosystem engineering and so on. Because of the novelty of the field, the definition of a bioengineer is still undefined. However, in general it is an integrated approach of fundamental biological sciences and traditional engineering principles.

Bioengineers are often employed to scale up bioprocesses from the laboratory scale to the manufacturing scale. Moreover, as with most engineers, they often deal with management, economic and legal issues. Since patents and regulation are very important issues for biotech enterprises, bioengineers are often required to have knowledge related to these issues.

Bioremediation and biodegradation: Biotechnology is being used to engineer and adapt organisms especially micro-organisms in an effort to find sustainable ways to clean up contaminated environments. The elimination of a wide range of pollutants and wastes from the environment is an absolute requirement to promote a sustainable development of our society with low environmental impact. Biological processes play a major role in the removal of contaminants and biotechnology is taking advantage of the astonishing catabolic versatility of micro-organisms to degrade/convert such compounds. New methodological breakthroughs in sequencing, genomics, proteomics, bioinformatics and imaging are producing vast amounts of information. In the field of 'environmental microbiology', genome based global studies open a new era providing unprecedented *in silico* views of metabolic and regulatory networks, as well as clues to the evolution of degradation pathways and to the molecular adaptation strategies to changing environmental conditions. Functional genomic and metagenomic approaches are increasing our understanding of the relative importance of different pathways and regulatory networks to carbon flux in particular environments and for particular compounds and they will certainly accelerate the development of bioremediation technologies and biotransformation processes.

Marine environments are especially vulnerable since oil spills of coastal regions and the open sea are poorly containable and mitigation is difficult. In addition to pollution through human activities, millions of tons of petroleum enter the marine environment every year from natural seepages. Despite its toxicity, a considerable fraction of petroleum oil entering marine systems is eliminated by the hydrocarbon degrading activities of microbial communities, in particular by a remarkable recently discovered group of specialists, the so called hydrocarbonoclastic bacteria (HCCB).

MICROBIOLOGY

Microbiology is the study of micro-organisms, which are unicellular or cellcluster microscopic organisms. This includes eukaryotes such as fungi and protists, and prokaryotes. Viruses, though not strictly classed as living organisms, are also studied. In short; microbiology refers to the study of life and organisms that are too small to be seen with the naked eye.

Microbiology is a broad term which includes virology, mycology, parasitology, bacteriology and other branches. A microbiologist is a specialist in microbiology.

Microbiology is researched actively, and the field is advancing continually. We have probably only studied about one per cent of all of the microbe species on earth. Although microbes were directly observed over three hundred years ago, the field of microbiology can be said to be in its infancy relative to older biological disciplines such as zoology and botany.

The field of microbiology can be generally divided into several subdisciplines:

1. Microbial physiology: The study of how the microbial cell functions biochemically. Includes the study of microbial growth, microbial metabolism and microbial cell structure.
2. Microbial genetics: The study of how genes are organised and regulated in microbes in relation to their cellular functions. Closely related to the field of molecular biology.
3. Cellular microbiology: A discipline bridging microbiology and cell biology.
4. Medical microbiology: The study of the pathogenic microbes and the role of microbes in human illness. Includes the study of microbial pathogenesis and epidemiology and is related to the study of disease pathology and immunology.
5. Veterinary microbiology: The study of the role in microbes in veterinary medicine or animal taxonomy.
6. Environmental microbiology: The study of the function and diversity of microbes in their natural environments. Includes the study of microbial ecology, microbially mediated nutrient cycling, geomicrobiology, microbial diversity and bioremediation. Characterisation of key bacterial habitats such as the rhizosphere and phyllosphere, soil andgroundwater ecosystems, open oceans or extreme environments (extremophiles).
7. Evolutionary microbiology: The study of the evolution of microbes. Includes the study of bacterial systematics and taxonomy.
8. Industrial microbiology: The exploitation of microbes for use in industrial processes. Examples include industrial fermentation and waste-water treatment. Closely linked to the biotechnology industry. This field also includes brewing, an important application of microbiology.
9. Aeromicrobiology: The study of airborne micro-organisms.
10. Food microbiology: The study of micro-organisms causing food spoilage and foodborne illness. Using micro-organisms to produce foods, for example by fermentation.
11. Pharmaceutical microbiology: The study of micro-organisms causing pharmaceutical contamination and spoil.

Benefits: Whilst there are undoubtedly some who fear all microbes due to the association of some microbes with various human illnesses, many microbes are also responsible for numerous beneficial processes such as industrial fermentation (e.g. the production of alcohol and dairy products), antibiotic production and as vehicles for cloning in higher organisms such as plants. Scientists have also exploited their knowledge of microbes to produce biotechnologically important enzymes such as Taq polymerase, reporter genes for use in other genetic systems and novel molecular biology techniques such as the yeast two hybrid system.

Bacteria can be used for the industrial production of amino acids. *Corynebacterium glutamicum* is one of the most important bacterial species with an annual production of more than two million tons of amino acids, mainly L-glutamate and L-lysine.

A variety of biopolymers, such as polysaccharides, polyesters, and polyamides, are produced by micro-organisms. Micro-organisms are used for the biotechnological production of biopolymers with tailored properties suitable for highvalue medical application such as tissue engineering and drug delivery. Micro-organisms are used for the biosynthesis of xanthan, alginate, cellulose, cyanophycin, poly(gamma glutamic acid), levan, hyaluronic acid, organic acids, oligosaccharides and polysaccharide, and polyhydroxyalkanoates.

Micro-organisms are beneficial for microbial biodegradation or bioremediation of domestic, agricultural and industrial wastes and subsurface pollution in soils, sediments and marine environments. The ability of each micro-organism to degrade toxic waste depends on the nature of each contaminant. Since sites typically have multiple pollutant types, the most effective approach to microbial biodegradation is to use a mixture of bacterial species and strains, each specific to the biodegradation of one or more types of contaminants.

There are also various claims concerning the contributions to human and animal health by consuming probiotics (bacteria potentially beneficial to the digestive system) and/or prebiotics (substances consumed to promote the growth of probiotic micro-organisms).

Recent research has suggested that micro-organisms could be useful in the treatment of cancer. Various strains of nonpathogenic clostridia can infiltrate and replicate within solid tumors. Clostridial vectors can be safely administered and their potential to deliver therapeutic proteins has been demonstrated in a variety of preclinical models.

ENVIRONMENTAL BIOTECHNOLOGY

Environmental biotechnology is when biotechnology is applied to and used to study the natural environment. Environmental biotechnology could also imply that one try to harness biological process for commercial uses and exploitation. The International Society for Environmental Biotechnology defines environmental biotechnology as 'the development, use and regulation of biological systems for remediation of contaminated environments (land, air, water), and for environment friendly processes (green manufacturing technologies and sustainable development)'.

Environmental biotechnology can simply be described as 'the optimal use of nature, in the form of plants, animals, bacteria, fungi and algae, to produce renewable energy, food and nutrients in a synergistic integrated cycle of profit making processes where the waste of each process becomes the feedstock for another process'.

Significance Towards Industrial Biotechnology

Consider an environment in which pollution of a particular type is maximum. Let us consider the effluents of a starch industry which has mixed up with a local water body like a lake or pond. We find huge deposits of starch which are not so easily taken up for degradation by micro-organisms except for a few exemptions. We isolate a few micro-organisms from the polluted site and scan for any significant changes in their genome like mutations or evolutions. The modified genes are then identified. This is done because, the isolate would have adapted itself to degrade/utilise the starch better than other microbes of the same genus. Thus, the resultant genes are cloned onto industrially significant micro-organisms and are used for more economically significant processess like in pharmaceutical industry, fermentations etc.

Similar situations can be elucitated like in the case of oil spills in the oceans which require cleanup, microbes isolated from oil rich environments like oil wells, oil transfer pipelines etc. have been found having the potential to degrade oil or use it as an energy source. Thus they serve as a remedy to oil spills.

Still another elucidation would be in the case of microbes isolated from pesticide rich soils. These would be capable of utilising the pesticides as energy source and hence when mixed along with biofertilisers, would serve as excellent insurance against increased pesticide toxicity levels in agricultural platform. But the counter argument would be that whether these newly introduced micro-organisms would create an imbalance in the environment concerned.The mutual harmony in which the organisms in that particular environment existed may have to face alteration and we should be extremely careful so as to not disturb the mutual relationships already existing in the environment to which we are introducing the newly discovered and cloned micro-organisms. Analysis of both the benefits and the disadvantages would pave way for an improvised version of environmental biotechnology. After all it is the environment that we strive to protect.

ENVIRONMENTAL MICROBIOLOGY

Environmental microbiology is the study of the composition and physiology of microbial communities in the environment. The environment in this case means the soil, water, air and sediments covering the planet and can also include the animals and plants that inhabit these areas. Environmental microbiology also includes the study of micro-organisms that exist in artificial environments such as bioreactors.

Microbial life is amazingly diverse and micro-organisms literally cover the planet. Micro-organisms can survive in some of the most extreme environments on the planet and some, for example the Archaea, can survive high temperatures, often above 100°C, as found in geysers, black smokers, and oil wells. Some are found in very cold habitats and others in highly saline, acidic, or alkaline water.

An average gram of soil contains approximately one billion (1,000,000,000) microbes representing probably several thousand species. Micro-organisms have special impact on the whole biosphere. They are the backbone of ecosystems of the zones where light cannot approach. In such zones, chemosynthetic bacteria are present which provide energy and carbon to the other organisms there. Some microbes are decomposers which have ability to recycle the nutrients. Microbes have a special role in biogeochemical cycles. Microbes, especially bacteria, are of great importance because their symbiotic relationship (either positive or negative) have special effects on the ecosystem.

Micro-organisms are used for *in situ* microbial biodegradation or bioremediation of domestic, agricultural and industrial wastes and subsurface pollution in soils, sediments and marine environments. The ability of each micro-organism to degrade toxic waste depends on the nature of each contaminant. Since most sites typically have multiple pollutant types, the most effective approach to microbial biodegradation is to use a mixture of bacterial species and strains, each specific to the biodegradation of one or more types of contaminants. It is vital to monitor the composition of the indigenous and added bacteria in order to evaluate the activity level and to permit modifications of the nutrients and other conditions for optimising the bioremediation process.

Oil biodegradation: Petroleum oil is toxic, and pollution of the environment by oil causes major ecological concern. Oil spills of coastal regions and the open sea are poorly containable and mitigation is difficult; much of the oil can, however, be eliminated by the hydrocarbon degrading activities of microbial communities, in particular the hydrocarbonoplastic bacteria (HCB). These organisms can help remedy the ecological damage caused by oil pollution of marine habitats. HCB also have potential biotechnological applications in the areas of bioplastics and biocatalysis.

Degradation of aromatic compounds by *Acinetobacter*: *Acinetobacter* strains isolated from the environment are capable of the biodegradation of a wide range of aromatic compounds. The predominant route for the final stages of assimilation to central metabolites is through catechol or protocatechuate (3,4-dihydroxybenzoate) and the β-ketoadipate pathway and the diversity within the genus lies in the channelling of growth substrates, most of which are natural products of plant origin, into this pathway.

Analysis of waste biotreatment: Biotreatment, the processing of wastes using living organisms, is an environmentally friendly alternative to other options for treating waste material. Bioreactors have been designed to overcome the various limiting factors of biotreatment processes in highly controlled systems. This versatility in the design of bioreactors allows the treatment of a wide range of wastes under optimised conditions. It is vital to consider various micro-organisms and a great number of analyses are often required.

Environmental genomics of Cyanobacteria: The application of molecular biology and genomics to environmental microbiology has led to the discovery of a huge complexity in natural communities of microbes. Diversity surveying, community fingerprinting and functional interrogation of natural populations have become common, enabled by a range of molecular and bioinformatics techniques. Recent studies on the ecology of Cyanobacteria have covered many habitats and have demonstrated that cyanobacterial communities tend to be habitat specific and that much genetic diversity is concealed among morphologically simple types. Molecular, bioinformatics, physiological and geochemical techniques have combined in the study of natural communities of these bacteria.

Corynebacteria: Corynebacteria are a diverse group Gram positive bacteria found in a range of different ecological niches such as soil, vegetables, sewage, skin, and cheese smear. Some, such as *Corynebacterium diphtheriae*, are important pathogens while others, such as *Corynebacterium glutamicum*, are of immense industrial importance. *C. glutamicum* is one of the biotechnologically most important bacterial species with an annual production of more than two million tons of amino acids, mainly L-glutamate and L-lysine.

Legionella: Legionella is common in many environments, with at least 50 species and 70 serogroups identified. *Legionella* is commonly found in aquatic habitats where its ability to survive and to multiply within different protozoa equips the bacterium to be transmissible and pathogenic to humans.

Archaea: Originally, archaea were once thought of as extremophiles existing only in hostile environments but have since been found in all habitats and may contribute up to 20 per cent of total biomass. Archaea are particularly common in the oceans, and the archaea in plankton may be one of the most abundant groups of organisms on the planet. Archaea are subdivided into four phyla of which two, the Crenarchaeota and the Euryarchaeota, are most intensively studied.

Lactobacillus: Lactobacillus species are found in the environment mainly associated with plant material. They are also found in the gastrointestinal tract of humans, where they are symbiotic and make up a portion of the gut flora.

Metagenomics: Metagenomics can be described as the study of uncultured micro-organisms. By permitting the direct investigation of bacteria, viruses and fungi irrespective of their culturability and taxonomic identities, metagenomics has changed microbiological theory and methods and has also challenged the classical concept of species. This new field of biology has proven to be rich and comprehensive and is making important contributions in many areas including ecology, biodiversity, bioremediation, bioprospection of natural products, and in medicine.

INDUSTRIAL BIOTECHNOLOGY

Industrial biotechnology (known mainly in Europe as white biotechnology) is the application of biotechnology for industrial purposes, including manufacturing, alternative energy (or bioenergy) and biomaterials. It includes the practice of using cells or components of cells like enzymes to generate industrially useful products. The economist speculated industrial biotechnology might significantly impact the chemical industry. The economist also suggested it can enable economies to become less dependent on fossil fuels.

The industrial biotechnology community generally accepts an informal divide between industrial and pharmaceutical biotechnology. An example would be that of companies growing fungus to produce antibiotics, e.g. penicillin from the penicillium fungi. One view holds that this is industrial production; the other viewpoint is that such would not strictly lie within the domain of pure industrial production, given its inclusion within medical biotechnology.

This may be better understood in calling to mind the classification by the US biotechnology lobby group, biotechnology industry organisation (BIO) of three 'waves' of biotechnology. The first wave, 'green biotechnology', refers to agricultural biotechnology. The second wave, 'red biotechnology', refers to pharmaceutical and medical biotechnology. The third wave, white biotechnology, refers to industrial biotechnology. In actuality, each of the waves may overlap. Industrial biotechnology, particularly the development of largescale bioenergy refineries, will likely involve dedicated genetically modified crops as well as the largescale bioprocessing and fermentation as is used in some pharmaceutical production.

INDUSTRIAL MICROBIOLOGY

Industrial microbiology or microbial biotechnology encompasses the use of micro-organisms in the manufacture of food or industrial products. The use of micro-organisms for the production of food, either human or animal, is often considered a branch of food microbiology. The micro-organisms used in industrial processes may be natural isolates, laboratory selected mutants or genetically engineered organisms.

Food microbiology: Yogurt, cheese, chocolate, butter, pickles, sauerkraut, soya sauce, vitamins, amino acids, food thickeners (microbial polysaccharides), alcohol, sausages, and silage (animal food) are all produced by industrial microbiology processes. 'Good' bacteria such as probiotics are becoming increasingly important in the food industry.

Biopolymers: A huge variety of biopolymers, such as polysaccharides, polyesters, and polyamides, are produced by micro-organisms. These products range from viscous solutions to plastics. The genetic manipulation of micro-organisms has permitted the biotechnological production of biopolymers with tailored material properties suitable for highvalue medical application such as tissue engineering and drug delivery. Industrial microbiology can be used for the biosynthesis of xanthan, alginate, cellulose, cyanophycin, poly(gamma-glutamic acid), levan, hyaluronic acid, organic acids, oligosaccharides and polysaccharides, and polyhydroxyalkanoates.

Bioremediation: Microbial biodegradation of pollutants can be used to cleanup contaminated environments. These bioremediation and biotransformation methods harness naturally occurring microbes to degrade, transform or accumulate a huge range of compounds including hydrocarbons (e.g. oil), polychlorinated biphenyls (PCBs), polyaromatic hydrocarbons (PAHs), pharmaceutical substances, radionuclides and metals.

Waste biotreatment: Micro-organisms are used to treat the vast quantities of wastes generated by modern societies. Biotreatment, the processing of wastes using living organisms, is an environmentally friendly, relatively simple and cost effective alternative to physicochemical cleanup options. Confined environments, such as bioreactors can be employed in biotreatment processes.

Healthcare and medicine: Micro-organisms are used to produce human or animal biologicals such as insulin, growth hormone, and antibodies. Diagnostic assays that use monoclonal antibody, DNA probe technology or real time PCR are used as rapid tests for pathogenic organisms in the clinical laborarory.

Micro-organisms may also help in the treatment of diseases such as cancer. Research shows that clostridia can selectively target cancer cells. Various strains of nonpathogenic clostridia have been shown to infiltrate and replicate within solid tumours. Clostridia therefore have the potential to deliver therapeutic proteins to tumours. *Lactobacillus* spp. and other lactic acid bacteria possess numerous potential therapeutic properties including antiinflammatory and anticancer activities.

Archaea: Examination of microbes living in unusual environments (e.g. high temperatures, salt, low pH or temperature, high radiation) lead to discovery of microbes with new abilities that can be harnessed for industrial purposes.

Corynebacteria: Corynebacteria are a diverse group Gram positive bacteria found in a range of different ecological niches such as soil, vegetables, sewage, skin, and cheese smear. *Corynebacterium glutamicum* is of immense industrial importance and is one of the biotechnologically most important bacterial species.

Xanthomonas: The genus *Xanthomonas* consists of 20 plant associated species, many of which cause important diseases of crops and other plants. Individual species comprise multiple pathovars, characterised by distinctive host specificity or mode of infection. Bacteria of the genus *Xanthomonas* are able to produce the acidic exopolysaccharide xanthan gum. Because of its physical properties, it is widely used as a viscosifer, thickener, emulsifier or stabiliser in both food and nonfood industries.

Chapter 2

Screening and Testing of New Metabolites

INTRODUCTION

Metabolism is the set of chemical reactions that occur in living organisms to maintain life. These processes allow organisms to grow and reproduce, maintain their structures, and respond to their environments. Metabolism is usually divided into two categories. Catabolism breaks down organic matter, for example to harvest energy in cellular respiration. Anabolism, on the other hand, uses energy to construct components of cells such as proteins and nucleic acids.

The chemical reactions of metabolism are organised into metabolic pathways, in which one chemical is transformed into another by a sequence of enzymes. Enzymes are crucial to metabolism because they allow organisms to drive desirable but thermodynamically unfavourable reactions by coupling them to favourable ones, and because they act as catalysts to allow these reactions to proceed quickly and efficiently. Enzymes also allow the regulation of metabolic pathways in response to changes in the cell's environment or signals from other cells. The metabolism of an organism determines which substances it will find nutritious and which it will find poisonous. For example, some prokaryotes use hydrogen sulphide as a nutrient, yet this gas is poisonous to animals. The speed of metabolism, the metabolic rate, also influences how much food an organism will require.

Metabolites are the intermediates and products of metabolism. The term metabolite is usually restricted to small molecules. A primary metabolite is directly involved in the normal growth, development, and reproduction. A secondary metabolite is not directly involved in those processes, but usually has important ecological function. Examples include antibiotics and pigments. The metabolome forms a large network of metabolic reactions, where outputs from one enzymatic chemical reaction arc inputs to other chemical reactions. Such systems have been described as hypercycles.

Metabonomics is defined as 'the quantitative measurement of the dynamic multiparametric metabolic response of living systems to pathophysiological stimuli or genetic modification'. This approach was pioneered by Jeremy Nicholson at Imperial College London and has been used in toxicology, disease diagnosis and a number of other fields. Historically, the metabonomics approach was one of the first methods to apply the scope of systems biology to studies of metabolism.

ISOLATION AND SCREENING OF MICRO-ORGANISMS

The success of an industrial fermentation process chiefly depends on the micro-organism strain used. An ideal producer or economically important strain should have the following characteristics.

1. It should be pure, and free from phage.

2. It should be genetically stable, but amenable to genetic modification.
3. It should produce both vegetative cells and spores; species producing only mycelium are rarely used.
4. It should grow vigorously after inoculation in seed stage vessels.
5. It should produce a single valuable product, and no toxic byproducts.
6. Product should be produced in a short time, e.g. 3 days.
7. It should be amenable to long term conservation.
8. The risk of contamination should be minimal under the optimum performance conditions.

Screening of Micro-organisms for New Products

The next step after isolation of micro-organisms is their screening. A set of highly selective procedures, which allows the detection and isolation of micro-organisms producing the desired metabolite, constitutes primary screening. Ideally, primary screening should be rapid, inexpensive, predictive, specific but effective for a broad range of compounds and applicable on a large scale.

Primary screening is time consuming and labour intensive since a large number of isolates have to be screened to identify a few potential ones. However, this is possibly the most critical step since it eliminates the large bulk of unwanted useless isolates, which are either non producers or producers of known compounds.

Computer based databases play an important role by instantaneously providing detailed information about the already known microbial antibiotic compounds.

Rapid and effective screening techniques have been devised for a variety of microbial products, which utilise either a property of the product or that of its biosynthetic pathway for detection of desirable isolates. Some of the screening techniques are relatively simple, e.g. for extracellular enzymes and enzyme inhibitors.

However, for most microbial products of high value, the screening is usually complex and tedious, and often may involve two or more steps, e.g. for antimicrobials. In some cases, it may be desirable to concentrate on a group of organisms expected to yield new products.

For example, the search for new antibiotics now focusses on rare *Actinomycetes*, i.e. *Actinomycetes* other than those belonging to the genus *Streptomyces*. Suitably designed specialised screening techniques may be used to detect compounds having various pharmacological activities other than antibiotics.

TYPES OF METABOLITES

Primary Metabolite

A primary metabolite, is directly involved in normal growth, development, and reproduction. Microbial production of primary metabolites contributes significantly to the quality of life. Through fermentation, micro-organisms growing on inexpensive carbon sources can produce valuable products such as amino acids, nucleotides, organic acids, and vitamins which can be added to food to enhance its flavour or increase its nutritive value. The contribution of micro-organisms will go well beyond the food industry with the renewed interest in solvent fermentations. Micro-organisms have the potential to provide many petroleum derived products as well as the ethanol necessary for liquid fuel. The role of primary metabolites and the microbes which produce them will certainly increase in importance.

Conversely, a secondary metabolite is not directly involved in those processes, but usually has an important ecological function.

Secondary Metabolite

Secondary metabolites are organic compounds that are not directly involved in the normal growth, development, or reproduction of organisms. Unlike primary metabolites, absence of secondary metabolities does not result in immediate death, but rather in longterm impairment of the organism's survivability, fecundity, or aesthetics, or perhaps in no significant change at all. Secondary metabolites are often restricted to a narrow set of species within a phylogenetic group.

Categories

Most of the secondary metabolites of interest to humankind fit into categories which classify secondary metabolites based on their biosynthetic origin. Since secondary metabolites are often created by modified primary metabolite synthases, or borrow substrates of primary metabolite origin, these categories should not be interpreted as saying that all molecules in the category are secondary metabolites (for example the steroid category), but rather that there are secondary metabolites in these categories.

Small 'small molecules'

1. Alkaloids (usually a small, heavily derivatised amino acid):
   (a) Hyoscyamine, present in *Datura stramonium*.
   (b) Atropine, present in *Atropa belladonna*, deadly nightshade.
   (c) Cocaine, present in *Erythroxylon coca* the Coca plant.
   (d) Codeine and Morphine, present in *Papaver somniferum*, the opium poppy.
   (e) Tetrodotoxin, a microbial product in Fugu and some salamanders.
   (f) Vincristine and Vinblastine, mitotic inhibitors found in the Rosy Periwinkle.
2. Terpenoids (come from semiterpene oligomerisation):
   (a) Azadirachtin, (Neem tree).
   (b) Artemisinin, present in Artemisia annua Chinese wormwood.
   (c) Tetrahydrocannabinol, present in *Cannabis sativa*.
   (d) Steroids (terpenes with a particular ring structure)
      (i) Saponins (plant steroids, often glycosylated).
3. Glycosides (heavily modified sugar molecules):
   (a) Nojirimycin.
   (b) Glucosinolates.
4. Phenols:
   (a) Resveratrol.
5. Phenazines:
   (a) Pyocyanin.
   (b) Phenazine-1-carboxylic acid (and derivatives).

Small molecules can have a variety of biological functions, serving as cell signalling molecules, as tools in molecular biology, as drugs in medicine, and in countless other roles.

These compounds can be natural (such as secondary metabolites) or artificial (such as antiviral drugs); they may have a beneficial effect against a disease (such as drugs) or may be detrimental (such as teratogens and carcinogens). Biopolymers such as nucleic acids, proteins, and polysaccharides (such as starch or cellulose) are not small molecules, although their constituent monomers—ribo or deoxyribonucleotides, amino acids, and monosaccharides, respectively—are often considered to be. Very small oligomers are also usually considered small molecules, such as dinucleotides, peptides such as the antioxidant glutathione, and disaccharides such as sucrose.

Drugs: Most drugs are small molecules, although some drugs can be proteins, e.g. insulin. Many proteins are degraded if administered orally and most often cannot cross the cell membranes. Small molecules are more likely to be absorbed, although some of them are only absorbed after oral administration if given as prodrugs. Many dietary supplements are small molecules (but not herb extracts, such as ginkgo). For organisms to produce small molecules they need one or more specialised enzymes (to create and destroy), which as a result are not that abundant in vertebrates (recent and small + slow population size), but very common in soil bacteria (such as streptomyces) and fungi, which in particular secrete antibiotics. Plants also have several secondary metabolites, which play a role in cell signalling, pigmentation or in defence, several of which have also been used as drugs (medical and recreational).

Examples of Secondary Metabolism Genes and their Function

Polyketide

Polyketides are secondary metabolites from bacteria, fungi, plants, and animals. Polyketides are biosynthesised by the polymerisation of acetyl and propionyl subunits in a similar process to fatty acid synthesis (a Claisen condensation). They are the building blocks for a broad range of natural products or are further derivatised. Polyketides are structurally a very diverse family of natural products with diverse biological activities and pharmacological properties. Polyketide antibiotics, antifungals, cytostatics, anticholesterolemics, antiparasitics, coccidiostatics, animal growth promoters and natural insecticides are in commercial use.

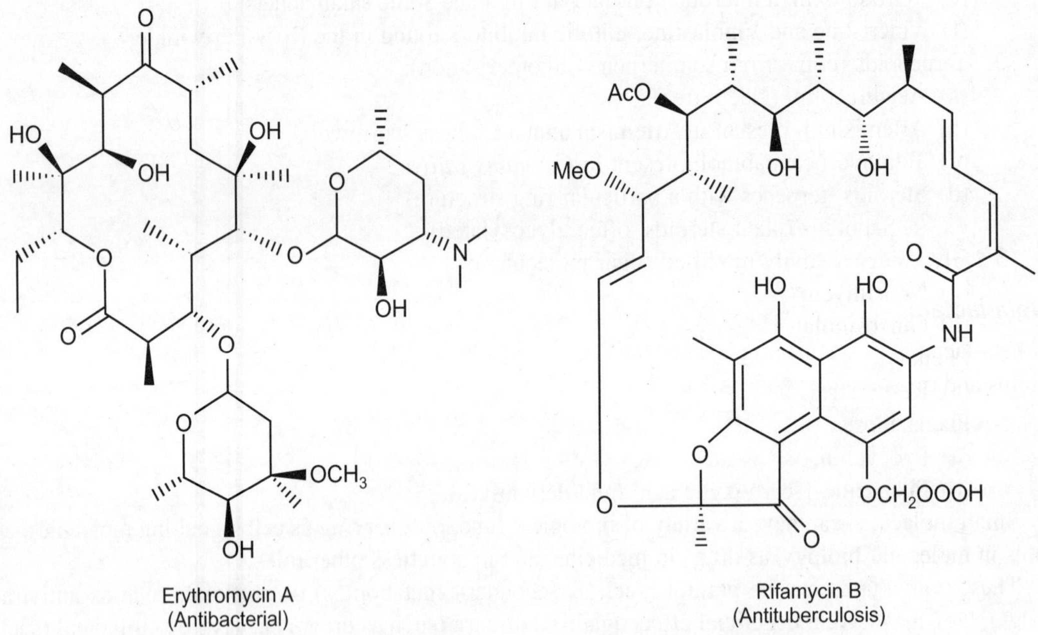

Structure of polyketide

Examples

1. Macrolides:
   (a) Picromycin, the first isolated macrolide.

   (b) The antibiotics erythromycin A, clarithromycin, and azithromycin.
   (c) The immunosuppressant tacrolimus (FK506).
2. Polyene antibiotics:
   (a) Amphotericin.
3. Tetracyclines:
   (a) The tetracycline family of antibiotics.
4. Acetogenins:
   (a) Annonacin.
   (b) Uvaricin.
5. Others:
   (a) Discodermolide.
   (b) Aflatoxin.

Biosynthesis

Polyketides are synthesised by one or more specialised and highly complex polyketide synthase (PKS) enzymes.

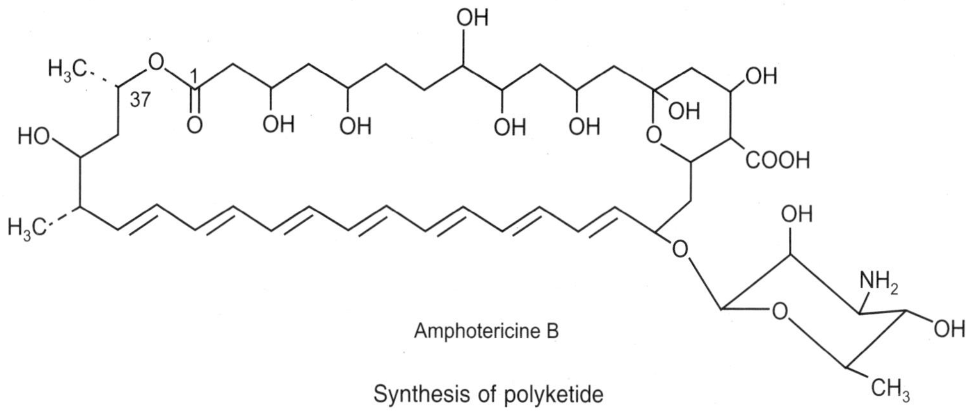

Amphotericine B

Synthesis of polyketide

Beta-lactam

A beta-lactam (β-lactam) ring is a lactam with a heteroatomic ring structure, consisting of three carbon atoms and one nitrogen atom. A lactam is a cyclic amide.

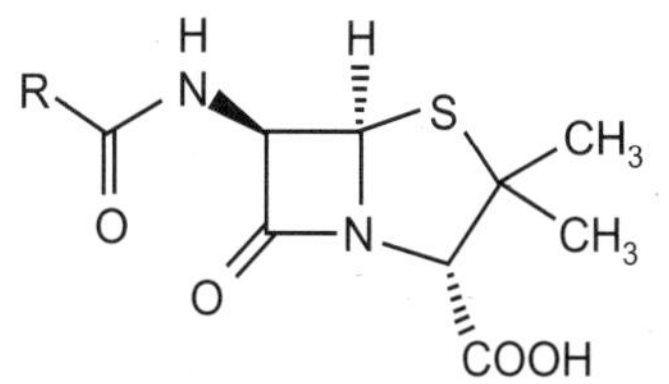

Penicillin nucleus. Beta lactam is the square at the centre.

Clinical significance

The beta-lactam ring is part of the structure of several antibiotic families, principally the penicillins, cephalosporins, carbapenems and monobactams, which are therefore also called β-lactam antibiotics. These antibiotics work by inhibiting the bacterial cell wall synthesis. This has a lethal effect on bacteria, especially on Gram positive ones. Bacteria can become resistant against β-lactam antibiotics by expressing β-lactamase.

The first synthetic β-lactam was prepared by Hermann Staudinger in 1907 by reaction of the Schiff base of aniline and benzaldehyde with diphenylketene in a [2+2]cyclo addition:

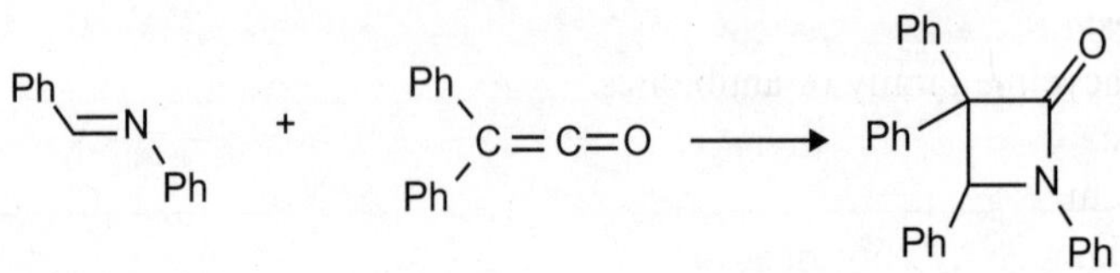

Beta-lactam resistance

Because of the popularity of β-lactam drugs, certain bacteria have been able to develop counter measures to traditional drug therapies. An enzyme called β-lactamase is present in many different types of bacteria, which serves to break the β-lactam ring, which effectively nullifies the antibiotic's effectiveness.

METHOD FOR INCREASING PRODUCTION OF MICROBIAL METABOLITES BY GENETIC ENGINEERING

Targeted gene insertion methodology can increase the production of an antimicrobial metabolite and comprises the steps of:

1. Identifying the gene in unaltered chromosomal DNA which encodes an enzyme that catalyses the rate controlling step in a biosynthetic pathway in the production of increased concentration of a precursor to a core molecule which leads to the antimicrobial metabolite.
2. Inserting into unaltered chromasomal DNA a genetic delivery vehicle (such as a vector, phage, or virus) which carries a gene which encodes the enzyme that catalyses the rate controlling step in the biosynthesis in the production of a core molecule which is a precursor to the antimicrobial metabolite into unaltered chromosomal DNA, said delivery vehicle being compatible with said unaltered chromosomal DNA, the delivery vehicle being generated by:
   (a) Inserting at least one exact or modified copy of the gene determining the rate controlling step into the compatible delivery vehicle.
   (b) Inserting the delivery vehicle containing the gene into the unaltered chromosomal DNA so that the resulting altered chromosomal DNA has increased genetic material for performing the rate controlling function.

SCREENING FOR ACTIVITIES

Microbes are exceptionally rich, diverse, and easily accessible sources of novel metabolites that can inhibit enzyme pathways related to disease targets. These metabolites vary enormously in structural complexity and biological activity. To discover therapeutically useful metabolites, it is critical not only to design suitable and sensitive assays for screening microbial extracts but also to test extracts that contain most or all of the metabolites from culture broths with a minimum of interference.

In addition to sensitive assays, novel and diverse producing micro-organisms are critical to the success of any natural products programme; implicit in this statement is that biological diversity may lead to chemical diversity.

Types of Samples to be Screened

The resting of microbial products presents several interesting challenges. For example, the question of whether to test whole broth or extracts depends on a variety of factors. Antimicrobial assays using whole organisms can tolerate whole broth; other assays, such as enzyme and receptor targets or mammalian cell based assays, are susceptible to contaminating metabolites that may be present in the broth. One way to circumvent some of these problems is to use either single phase or two phase extraction procedures that utilise methanol or ethyl acetate. These types of extracts are usually devoid of proteases and are suitable for testing in enzyme and receptor assays. In some cases, solid phase extraction protocols that make use of Sep Pak C-18 cartridges or ultrafiltration filters (with 10,000 molecular weight cutoff) result in samples that are suitable for screening in enzyme and receptor binding assays. Because of the highly coloured nature of the extracts, the effect of colour interference in some assays needs to be taken into account and corrected. Test samples can be dissolved in any solvent that is tolerated in the assay. Generally dimethyl sulphoxide (DMSO) or 50 per cent aqueous DMSO is added directly into the assay. The final concentration of DMSO ranges from 1 to 5 per cent and varies from assay to assay.

Developing and Semiautomating Screening Tests

Prior to the development of automatic pipetting stations, most assays were performed manually in volumes exceeding 1 ml and in large test tubes. The pressure to screen more samples as rapidly as possible led to the use of automatic pipetting stations as well as accelerating miniaturisation of assays to preserve assay reagents. In 1984, a shift in screening strategies occurred. This was precipitated by the advent of small volume polypropylene tube strips and the development of the Skatron cell harvester, which greatly increased the number of samples that could be tested in receptor binding assays. Since then, new technologies for high throughput screening have rapidly advanced into a fully integrated robotics system that has developed within the last 5 years. New assay technologies such as scintillation proximity assays, homogeneous time resolved fluorometric assays, FlashPlate assays, and reporter gene assays using luciferase and β-lactamase are being adapted in many productive high throughput screens. All of the screening assays that are described in this chapter were performed either with semiautomated machines such as pipetting stations and cell harvesters or with 96 well pipetters in a 96 well format unless specified otherwise.

TYPES OF SCREENS

Antihypercholesterolemia Screen

Hypercholesterolemia is an important risk factor for atherosclerosis and coronary heart disease. The microsomal enzyme 3-hydroxy-3-methylglutaryl-coenzyme A reductase (HMG-CoA reductase, EC 1.1.1.34) is the major rate limiting step in the cholesterol biosynthetic pathway. Mevinolin (lovastatin), a potent inhibitor of HMG CoA reductase, was discovered in several strains of *Aspergillus terreus* in 1978 . Mevinolin and related compounds are now the drugs of choice for treating hypercholesterolemia. Other enzymes involved in cholesterol biosynthesis can also serve as therapeutic targets. One such enzyme, squalene synthase (farnesyl diphosphate:farnesyldiphosphate farnesyltransferase, EC 2.5.1.2.1), catalyses the reductive dimerisation of farnesyl pyrophosphate to squalene and is the first committed step in the biosynthesis of cholesterol. Inhibitors of squalene synthase have the potential to be useful antihypercholesterolemia agents. The search for natural product inhibitors of this enzyme has resulted in the discovery of several potent inhibitors, namely, zaragozic acids, squalestatins and their minor components, TAN-1607A, and viridio fungins A, B, and C. Inhibitors of squalene biosynthesis have been reviewed. Below the screening assays used to detect natural product inhibitors of HMG-CoA reductase and squalene synthase are discussed.

HMG-CoA reductase

Assays are performed utilising a resin technique in glass test tubes. [^{3}H]HMG-CoA (NEN, Boston, Mass) and rat liver homogenates are used as substrate and enzyme, respectively. Radioactive enzyme products are separated by absorption of [^{3}H]HMG-CoA with Bio-Rex-5 resin suspension and unabsorbed radioactive enzyme products are counted. The assay mixture (final volume, 300 μl) contains 50 mM potassium phosphate (pH 7.0), 3 mM dithiothreitol (DTT), 1.5 mM potassium chloride, 300 μM EDTA, 10 μM HMG-CoA, and 0.09 μCi of [^{3}H]HMG-CoA. Rat liver enzyme homogenate (5 μl), prepared as described, is added with 10 μl of test sample into glass tubes. Tubes are preincubated at 37°C for 5 min., and the reaction is initiated with 10 μl of 20 μM NADPH; tubes are incubared further at 37°C for 15 min. The reaction is stopped with 70 μl of 5 N HCl, and the tubes are vortexed and incubated at 37°C for 15 min. Resin suspension (3.2 ml), prepared by suspending 250 g of Bio-Rex-5 (100 to 200 mesh) resin (Bio-Rad laboratories, Hercules, Calif.) in 1600 ml of water with three washes and repeated decantations, is added to each tube. Tubes are vortexed for 5 secs and centrifuged at 1450 × g for 10 min. Supernatants (2 ml) are pipetted into plastic counting vials, the scintillation cocktail is added, and the radioactivity in the vials is counted. Assay controls include tubes with no enzyme, tubes with enzyme and solvent, and tubes with enzyme and a known concentration of mevinolin.

Screening assay from mevalonate to squalene for squalene synthase inhibitors

To look more globally for inhibitors of cholesterol biosynthesis, a multienzyme assay has been developed. Rat liver homogenates (S20 fractions) containing at least six enzymes, namely, mevalonate kinase, 5-phosphomevalonate kinase, 5-pyrophosphatemevalonate decarboxylase, isomerase, prenyl transferase, and squalene synthase, are prepared by standard procedures. The assay mixture (final volume, 100 μl) contains 10 mM KH_2PO_4, 3 mM glucose-6-phosphate, 11 mM potassium fluoride, 3 mM DTT, 150 mM HEPES buffer at pH 7.5, 7.5 mM ATP, 7.5 mM $MgCl_2 \cdot 6H_2O$, 1 mM NADPH, and 0.2 mM sodium mevalonate plus 0.2 μCi of 5-[^{3}H]mevalonic acid. Organic microbial extracts are dissolved in 100 per cent DMSO and diluted 20 fold into the reaction mixture. Reaction products are extracted with heptane after heating with a mixture of ethanol and KOH and analysed by using a scintillation counter. Assays are also performed with an 520 fraction prepared from HepG2 cells, as described below. The final protein concentrations of rat liver homogenate and HepG2 cell extracts in the assays are 1.56 and 0.32 mg per ml, respectively. To measure selectivity of putative inhibitors, extracts are tested against various squalene synthases.

Preparation of cell extracts of HepG2 cells

HepG2 cells are grown in minimum essential medium (MEM) with nonessential amino acids, sodium pyruvate, penicillin, L-glutamate, and 10 per cent fetal bovine serum. The medium is changed twice weekly, and a confluent mo.,olayer is achieved in 1 week. Fortyeight hours prior to harvest, cells are switched from MEM with 10 per cent fetal calf serum to MEM with 10 per cent delipidated serum, as described by Cham and Knowles. Cells are harvested and washed with phosphate buffered saline. Fresh trypsin (0.25 per cent) EDTA (0.02 per cent) with Hanks' balanced salt solution (GIBCO BRL, 310-4060 AJ; Life Technologies, Gairhersburg, Md.) is then added and removed. The flasks are incubated at 37°C until the cells detach. Detached cells are resuspended in MEM centrifuged at 1000 × g for 5 min. and the cell pellets are washed once with phosphate buffered saline. Cells are resuspended in 50 mM HEPES containing 5 mM $MgCl_2$, 2 mM $MnCl_2$ and 10 mM DTT (pH 7.5) (enzyme suspension buffer), sonicated twice (Branson Sonifier, Cell Disruptor 200; Branson Sonic Power Company, Danbury, Conn.) (sonicator setting #60, pulse) on ice for 1 min. and then centrifuged for 10 min. at 1000 × g. Supernatants are transferred to clean tubes and centrifuged at 20,000 × g for 20 min. and S20 fractions are used directly in the screening assay. Some of the enzyme preparations (S20 fractions) are also further centrifuged at 1,00,000 × g for 1 hr. to obtain heavy microsomes that contain squalene synthetase. This membrane preparation is resuspended in the above described buffer and used as the source of squalene synthase.

Squalene synthase assays

Squalene synthase assays are performed as described by Lingham. The enzyme sources are heavy microsomes prepared from HepG2 cells arid rat liver. The final protein concentrations of both enzymes vary from 1.2 to 3 µg per ml in the assay mixtures. Compounds to be screened are dissolved in 100 per cent DMSO and diluted 20 fold into the assay mixture. The assay mixture (final volume, 100 µl) contains 15 mM HEPES, pH 7.5, 11 mM potassium fluoride, 3 mM DTT, 5.5 mM $MgCl_2$, 1 mM NADPH, 0.1 µg/ml of a squalene epoxidase inhibitor (L-688, 709), 0.06 µM [^{3}H)farnesyl diphosphate ([^{3}H]FPP; 740 CBq/mmol, NEN, Boston, Mass), 2.94 µM unlabelled FPP, and 1 to 3 µg/ml of enzyme preparation. Reactions are initiated by the addition of substrate, incubated at 30°C for 20 min. and stopped with 100 µl of 100 per cent ethanol. One hundred microlitres of resin (AG1-X8, chloride, 200 to 400 mesh; Bio Rad Laboratories, Hercules, Calif.) is added to each tube, and [^{3}H]squalene is extracted with 300 µl of heptane, and radioactivity is counted in a TopCount Scintillation Counter (Packard Instrument Company, Downers Grove, Ill).

Interpretation of results

In typical HMG-CoA reductase assays, the amount of radioactivity recovered without enzyme and in the enzyme reaction controls ranges between 150–200 cpm and 7000–8000 cpm, respectively. In the multienzyme pathway assays, the amount of radioactivity recovered without enzyme and in the enzyme reaction controls ranges between 150–300 cpm and 5000–8000 cpm, respectively. The per cent inhibition of test samples is calculated as shown in the equation:

$$\{[\text{cpm (control)} - \text{cpm (test sample)}]/[\text{cpm (control)} - \text{(blank)}]\} \times 100.$$

Samples are considered active when the degree of inhibition exceeds 65 per cent inhibition and the activity can be titrated. The active samples in the multienzyme assays are confirmed with squalene synthase using enzymes prepared from both rat liver HepG2 cells.

Anticancer Screen

Historically, most of the anticancer screens reported in the literature are cytotoxicity tests using various types of tumour cells. These tests, while usually informative, are time consuming and difficult to perform. In addition, the capacity of these assays is limited and not amenable to high throughput screening. In this section, we illustrate a novel mechanism based assay that takes advantage of the prominent role that the ras oncogene product, Ras (p21), plays in signal transduction and cellular growth. Among the more prevalent and pervasive cancer causing mutations are those involving a mutated form of the ras oncogene that associates with about 25 per cent of all human cancers. The incidence of mutated ras is significantly higher in colon tumours (>50 per cent) and pancreatic tumours (>90 per cent). Ras functions, in part, by transmitting signals involved in cellular proliferation and differentiation. Ras is a GTP-binding protein that couples the activation of growth factor receptor tyrosine kinases (e.g. epidermal growth factor and platelet derived growth factor) to intracellular signal transduction pathways. An intrinsic GTPase activity regulates Ras in normal cells, resulting in the formation of a Ras-GDP complex that is functionally silent. Oncogenic forms of Ras lack the GTPase activity. This results in a constitutively active, predominant GTP-bound form of Ras that continually transmits signals inside the cell. Transformation, in part, results from the unregulated stimulation of mitogenic pathways. Continued expression of oncogenic *ras* is obligatory to maintain the transformed phenotype; as such, Ras is an attractive target for therapeutic intervention.

Functionally, normal and oncogenic forms of Ras undergo a series of complex post-translational processing events prior to associating with the plasma membrane. These include farnesylation, proteolysis, methylation, and palmitoylation. The first and obligatoty step in Ras processing is farnesylation by farnesyl protein transferase (FPTase). Genetic experiments have shown that farnesylation is necessary for Ras cell transforming activity. The later processing events (proteolysis, methylation, and palmitoylation) are not necessary for Ras function. Farnesylation of Ras by FPTase directs and anchors Ras in the cell membrane. FPTase consists of two subunits (an αβ heterodimer) with molecular sizes of 48,000 and 45,000 Da, respectively.

FPTase transfers a farnesyl group from farnesyl pyrophosphate (FPP) to the cysteine residue at the carboxyl terminus CaaX box (C, Cys; a, usually an aliphatic amino acid; X, another amino acid) of Ras. Inhibition of Ras farnesylation prevents Ras membrane localisation and blocks the cell transforming activity of Ras. Several FPTase inhibitors have been reported to selectively inhibit Ras processing in cell lines, prevent tumorigenesis in nude mice, and promote regression of mammary and salivary carcinomas in Ha-ras transgenic mice.

Farnesyl protein transferase

The FPTase assay is performed as described by Lingham. Partially purified bovine FPTase and Ras peptides are prepared as reported by Schaber and Gibbs, respectively. Human FPTase is prepared as described by Omer. Compounds are dissolved in 100 per cent DMSO and diluted 20 fold directly into the assay mixtures. Bovine FPTase is assayed in a mixture (final volume, 100 μl) containing 100 mM HEPES (pH 7.4), 5 mM $MgCl_2$, 5 mM DTT, 100 mM [^{3}H]FPP (740 Cbq/mmol), 650 nM Ras-CVLS (cysteine, valine, leucine, and serine), and 10 mg/ml of FPTase at 31°C for 60 min. Reactions are initiated with FPTase and stopped with I ml of 1.0 M HCl in ethanol.

Precipitates are collected onto filtermats using a TomTec Mach II cell harvester (TOMTEC, Harnden; Conn.), washed with 100 per cent ethanol, and dried, and radioactivity is counted in an LKB β-plate 1205 Scintillation Counter (Wallac Inc., Gaithersburg, Md.). Human FPTase activity is assayed as described above with the exception that 0.1 per cent (wt./vol) PEG 20,000, 10 mM $ZnCl_2$, and 100 nM Ras-CVIM (cysteine, valine, isoleucine, and methionine) are added to the reaction mixture. After 30 min. reactions are stopped with 100 μl of 30 per cent (wt./vol) trichloroacetic acid in ethanol and processed as described above for the bovine enzyme.

Interpretation of results

Typical experimental results obtained without enzyme and in uninhibited enzyme controls range between 1300–1600 and 7300–7600 cpm, respectively. The per cent inhibition of the test sample is calculated as described above. Similarly, samples are considered active when the degree of inhibition exceeds 65 per cent and can be titrated. Test samples are first screened against bovine FPTase, and active samples are tested against human FPTase. To determine specificity, active samples are also treated against rat liver or human squalene synthase and bovine brain geranylgeranyl protein transferase.

Antihypertensive Screen

Cardiovascular disease is one of the major causes of death in modern society. One contributing factor to this process is hypertension, which is arguably caused by overwork, stress, lack of exercise, modernday lifestyles, and diet. The search for antihypertensive agents has focused on the role of the renin angiotensin system. This system plays a central role in the regulation of normal blood pressure and appears to be involved in hypertension as well as in congestive heart failure, cirrhosis, and nephrosis. Angiotensin II, the active hormone of the renin angiotensin system, is a powerful arterial vasoconstrictor that exerts its action by interacting with specific receptors located on the cell membranes of various target organs. Endothelin, a peptide secreted by endothelial cells, is a potent constrictor of vascular smooth muscle.

Elevated levels of endothelin are found in myocardial infarction, systemic hypertension, cardiac ischemia, and coronary vasospasm. Inhibitors of angiotensin converting enzyme (which converts angiotensin to angiotensin II) and antagonists of angiotensin II or endothelin receptors are potential drugs for treating hypertensive conditions. The search for natural product inhibitors has resulted in the discovery of several inhibitors, including cochinmicins, cytosporins, osteromycin, and namibione.

Angiotensin converting enzyme assay

The angiotensin converting enzyme assay is performed essentially as described by Huang. Hippuryl-L-histidyl-L-leucine and rat lung homogenate are used as substrate and enzyme, respectively. Routinely, assay volumes are 0.2 ml; the mixture consists of 1 mM substrate, 0.5 M NaCl, 0.06 M potassium phosphate buffer (pH 8.3), and test samples (10 µl of 50 per cent aqueous methanol or a 50 per cent methanolic extract). The reaction is initiated with 25 µl of diluted enzyme suspension and incubated at 37°C for 30 min. The enzyme reaction is stopped by boiling for 10 min. One millilitre of 0.2 M potassium phosphate buffer (pH 8.3) is added, followed by 0.5 ml of 3 per cent 2,4,6-trichloro-S-triazine in dioxane, after which the tube is immediately vortexed. The product is measured at a wavelength of 382 nm. This assay can easily be miniaturised by decreasing the volume of the reagents and using microtubes or microlitre plates.

Angiotensin II receptor binding

Bovine arterial membranes are prepared as described by Stevens-Miles. Membranes (about 100 µg/ml) are incubated in the presence of $[^{125}]Tyr^4$-angiotensin II (40 pM) with 100 mM Tris-HCl (pH 7.4), 5 mM $MgCl_2$, 0.2 per cent bovine serum albumin, and 0.2 mg/ml bacitracin. Assays are incubated at 37°C for 90 min, after which the mixtures are filtered using a TomTec Mach II cell harvester with GF/B filtermats.

The filtermats are dried, placed in bags with scintillation fluid, and sealed, and radioactivity is counted in an LKB β-plate 1205 Scintillation Counter.

Endothelin receptor binding

Endothelin binding assays can be performed by using membrane preparations of CHO cells expressing cloned endothelin receptors. Furthermore, wholecell binding assays with these same cells can also be developed. Briefly, cells are detached from flasks using a cell dissociation buffer (20 mM phosphate buffered saline and 2 mM EDTA) and collected by centrifugation. Cells are washed once and resuspended in assay buffer (100 mM HEPES with 5 mM EDTA and 0.1 per cent human serum albumin [pH 7.5]) at 5×10^5 cells/ml. Membranes or cells (in assay buffer) are incubated with [^{125}]endothelin (50 pM) and test samples at 37°C for 60 min. Assays are terminated by filtration using a TomTec Mach II (6 by 16 format) 96 well cell harvester with GF/B filtermats. Filtermats are dried, placed in bags with scintillation fluid, and sealed, and radioactivity is counted in an LKB β-plate 1205 Scintillation Counter.

Interpretation of results

In a typical experiment for angiotensin II binding, the number of counts recovered for the blank and uninhibited controls is 100 and 1000 cpm, respectively. Similarly, for endothelin binding, the blank and control counts are 600 and 6000 cpm, respectively. The per cent inhibition of test sample is calculated by the equation:

$$\{[\text{cpm (control)} - \text{cpm (test sample)}]/[\text{cpm (control)} - \text{cpm (blank)}]\} \times 100.$$

When the test sample shows more than 65 per cent inhibition and is dose related, the sample is considered active. Similarly, for angiotensin converting enzyme, the optical density (OD) values for blank and uninhibited enzyme are 0.1 and 0.5, respectively. Furthermore, the per cent inhibition of the test sample is calculated as shown in the equation:

$$\{[\text{OD (control)} - \text{OD (test sample)}]/[(\text{OD (control)} - \text{OD (blank)}]\} \times 100.$$

As in all the examples presented, when the test sample shows more the 65 per cent dose related inhibition, the sample is considered active.

Antiviral Screen

Historically, microbes have been the most dangerous pathogens known to humans. However, the advent of potent and specific antibiotics led to the premature and ill founded claim that there is no need to discover and develop new antibiotic agents. Viruses are, in general, more insidious infectious agents than microbes: To date, there are very few therapies that can completely eradicate viruses. Traditionally, prevention of viral infection relies upon vaccination with inactive virus or viral antigens. However, some viruses such human immunodeficiency virus (HIV-1, HIV-2) and influenza mutate rapidly. There are currently no effective vaccines available to prevent the spread of these and other viruses. One novel strategy is to target enzyme pathways that viruses use to replicate. Enzymes that lend themselves to the development of screening assays for HIV-1 and HIV-2 include integrase, reverse iranscriptase, and protease. For the purpose of this topic, the protease and reverse transcriptase activities of HIV are illustrated as *in vitro* mechanism based screens.

HIV-1 is a retrovirus that causes AIDS. The genome of HIV-1 encodes a 99 residue protease that processes the Gag, Pol, and Env polyproteins. Genetic disruptions in the HIV protease result in noninfectious viral particles. Potent protease inhibitors halt the spread of viruses in cell culture; inhibitors such as ritonavir and MK-639 have been approved for use in humans. These compounds are very efficacious and lower plasma levels of virus by about 2 logs. On the basis of results, the inhibition of HIV protease is believed to be a vital and effective mechanism for reducing and managing the HIV infection.

Another essential step in the cycle of HIV-1 is reverse transcription of the viral RNA genome to produce a double stranded DNA copy. This process is mediated by the virally encoded reverse transcriptase. Inhibition of reverse transcriptase by nucleoside analogs such as 3′-azido-3′-deoxythymidine and dideoxyinosine are clinically effective in treating HIV-1 infection, and certain nonnucleoside blockers of reverse transcriptase, such as nevirapine, lower plasma viremia by about 1.0 to 1.5 logs. However, one of the main problems with the nonnucleoside inhibitors is the rapid development of resistance, which can be attributed to mutation of the virus. Nevertheless, reverse transcriptase is still an important therapeutic target for intervention in the progress of AIDS.

With respect to influenza virus, primary transcription is a unique antiviral target. As an obligatory step in the life cycle, the eight RNA segments of negative polarity that make up the genome are transcribed into positivesense mRNAs by an associated viral transcriptase. Transcription is initiated by a novel 'mechanism in which capped and methylated (capl) RNAs are used to prime mRNA synthesis. Capped RNAs are derived from mammalian RNA polymerase II transcripts in the nuclei of infected cells by a virally encoded endonuclease. Highly selective inhibitors, such as 4-substituted 2,4-dioxobutanoic acids, target the endonuclease activity of influenza viral transcriptase and inhibit the replication of influenza A and B viruses in cell culture assays *in vitro* and in a mouse challenge model *in vivo*. These results indicate that the viral transcriptase is an effective therapeutic target against influenza viruses.

HIV-I protease

The cloning, expression, and purification of the HIV protease is performed as described by Darke. The peptide substrate Val-Ser-Gln-Asn-β-naphthylAla-Pro-Ile-Val-Gln-Gly-Arg-Arg is synthesised according to Merrifield and labelled with [^{3}H] acetic anhydride as described by Lingham. Five microlitre extracts in 100 per cent DMSO are mixed with 0.4 μM [^{3}H]acetyl-Val-Ser-Gln-Asn-β-naphthylAla-Pro-Ile-Val-Gln-GlyArg-Arg, 100 μM Val-Ser-Gln-Asn-β-naphthylAla-Pro-Ile-Val-Gln-Gly-Arg-Arg, and 2 nM HIV protease in a final volume of 100 μl containing 100 mM 2-(*N*-morpholino)ethanesulphonic acid (MES) buffer, pH 6.0, 10 mM DDT, 1 mM EDTA, and 0.1 per cent bovine serum albumin. After 60 min. at 37°C, the reaction is stopped by placing the assay tubes on ice and by the addition of 200 μl of resin (Dowex AG50W-X8, 200 to 400 mesh, H^+ form; BioRad Laboratories, Hercules, Calif.) slurry, which is prepared by repeated decantation of fine particles and with 500 g of resin in 750 ml of 0.1 N HCl. The tubes are capped, mixed for 15 min. and centrifuged for 10 min. in a Savant centrifuge using the swing bucket rotor to sediment the resin.

Aliquots of the supernatants are mixed with scintillation fluid, and radioactivity is counted in a TopCount Scintillation Counter (Packard Instrument Company, Downers Grove, Ill).

HIV-1 reverse transcriptase

The recombinant HIV-1 reverse transcriptase is prepared as described by Azwlina and Stahlhut and Olsen. Poly(rA)·oligo(dT)$_{12-18}$ (Pharmacia 27-7878-03; Uppsala, Sweden) is used as template primer, and [^{3}H]TTP (NET 221-X; NEN, Boston, Mass) is used as a tracer. All reagents are prepared with sterile doubleglass distilled water. The assay is performed in a reaction mixture of 100 μl containing 100 mM Tris-HCl (pH 8.2), 80 mm KCl, 12 mM $MgCl_2$, 2 mM DTT, 0.5 mM EGTA, 1 mg/ml of bovine serum albumin, 1.5 μg/ml of poly(rA)·ohgo(dT), 10 μM nonradioactive TTP, 0.5 μCi of [^{3}H]TTP, 0.01 per cent (wt/vol) Triton X-100, 5 μl of extracts, and 0.1 nM HIV-1 reverse transcriptase enzyme. After 45 min. at 37°C, the reaction is stopped with 100 μl of 300 per cent aqueous trichloroacetic acid in 10 mM sodium pyrophosphate. The mixtures are kept on ice for 30 min, and the precipitate is collected on a filtermat using a TomTec Mach II cell harvester. The filtermats are washed with 0.1 M HCl in 10 mM sodium pyrophosphate using the same harvester. The filtermats are baked in a microwave oven for 7 min. and placed in an LKB counting bag. Thirty millilitres of scintillant is added, and radioactivity is counted in an LKB β-plate 1205 Scintillation Counter.

Influenza a virus transcriptase

All reagents are prepared with sterile doubleglass distilled water, and gloves are worn at all times when performing the assay. Influenza virus polymerase cores are prepared and purified as described by Tomassini and Honda. The Alfalfa mosaic virus (ALMV) substrates are prepared as described by Tomassini. Briefly, ALMV segment 4 RNA containing cap 0 structures is purchased from J. Bioi, Leiden, Netherlands. The 5′ cap of ALMV RNA is methylated with nucleoside-2′-O-methyl-transferase and used directly for the assay. The transcriptions are performed in a microtiter plate using a final volume of 75 μl containing 90 mM HEPES (pH 7.3), 0.05 per cent Triton N-101, 80 mM KCl, 5 mM $MgCl_2$, 1 mM DTT, 2 μg of tRNA per ml, 20 μg of purified polymerase cores per ml, 100 μM ATP, 50 μM CTP, 50 μM GTP, 1 μM UTP, 0.3 μM [^{35}S]UTP, 5 μl of sample, and 7 nM substrate ALMV capped primer of 880 nucleotides. After 30 min. of incubation at 30°C, the reaction is stopped with 75 μl of sterile saturated sodium pyrophosphate solution and 50 μl of icecold 40 per cent trichloroacetic acid, and the plates are mixed and placed on the ice for 15 min. The precipitated RNA samples are collected using a TomTec MACH III cell harvester and Packard GF/C filter plates. Filters are dried in an oven at about 60°C for 5 min. The undersides of the plates are sealed with the white sealer sheet, and 25 μl of Microscint 20 cocktail is added. The top of the plate is then sealed with Topseal-S sealing film. Radioactivity is counted in a TopCount Scintillation counter.

Interpretation of results

In a typical experiment, the blank and uninhibited enzyme control counts recovered are as follows: HIV-1 protease, 100 and 900 cpm, respectively; HIV-1 reverse transcriptase, 800 and 10,000 cpm; and influenza virus transcription, 1000 and 20,000 cpm. As described earlier, activity of samples is calculated using the equation:

$$\{[\text{cpm (control)} - \text{cpm (test sample)}]/[\text{cpm (control)} - \text{cpm (blank)}]\} \times 100.$$

A sample is considered active when inhibition exceeds 65 per cent and is dose dependent.

CHEMICAL SCREENING: A SIMPLE APPROACH TO VISUALISING *STREPTOMYCES* DIVERSITY FOR DRUG DISCOVERY AND FURTHER RESEARCH

The morphological and biochemical characteristics of *Streptomyces* species tend to be very similar, and thus the elucidation of their diversity is both time and money consuming. Here various streptomycetes are evaluated and isolates using a chemical screening approach in order to establish their secondary metabolite patterns, thereby avoiding the possible discounting of morphologically similar strains. Results demonstrated that each isolate presented a unique pattern of secondary metabolites independently of their morphological and biochemical characteristics. We also established the enormous diversity in metabolic products among our isolates, and thus many potentially new metabolites may be studied in further research.

Our results indicate that chemical screening is a simple selection method for recognising the specific fingerprint of each isolate, highlighting the particular metabolic characteristics of each with respect to the other studied strains.

Thus, natural products lie at the heart of organic chemistry, and organic chemistry begins at the end of secondary metabolism. In addition to the isolation of unique and diverse cultures and the availability of robust assays, the isolation and characterisation of the compounds eliciting the inhibition are essential to the success of natural products discovery programmes. One of the major problems with testing natural product extracts for compounds of interest is that the screening for the activity is usually not trivial. A significant commitment on the part of the assayist is necessary to ensure that success can be achieved.

Another major issue concerns the identification of active compounds relatively early in the isolation process. One possible way to accomplish this would be to couple traditional methods of classification to more innovative technologies such as liquid chromatography/mass spectrometry. There are numerous other assays that, when coupled to high throughput technologies, can make positive contributions to the future of natural products discovery. Some of the assays presented here can be adapted by using high throughput formats.

STRAIN IMPROVEMENT METHODS

Strain improvement can generally be described as the use of any scientific techniques that allow the isolation of cultures exhibiting a desired phenotype. The technology has bean utilised for more than 50 years in conjunction with modern submerged culture fermentations and perhaps in a less systematic way for as long as fermented products have been made by humans. Most commonly, the ability of a strain to exhibit increased product accretion is the desired phenotype. However, the spectrum of improvements can include other traits, such as the elimination of toxic cometabolites or those problematic in downstream processing, the ability to degrade complex waste materials, or greater genetic stability of recombinant hosts.

The utility of strain improvement arises because of the existence of rate limiting steps within all metabolic pathways. Most of these events are not readily measurable owing to the nature of metabolic transients at reaction fluxes during the course of an industrial fermentation. The metabolic flux involved in the biosynthesis of a secondary metabolite generally includes numerous specific reactions from the biosynthetic enzymes, primary metabolism for the supply of precursors to growth and secondary metabolites, and regulatory circuits involved in cell growth and differentiation. Likewise, heterologous protein expression in bacterial or fungal systems offers a significantly complex pathway. Because the rate limiting enzymatic reactions or flux nodes are often unknown, an empirical process such as classical strain improvement is well suited to manipulation of the pathway. A screening programme can be initiated with limited knowledge of the physiology or genetics associated with production of the molecule of interest.

Classical mutagenesis and screening, also referred to as nonrecombinant strain improvement, can thus offer a significant advantage over genetic engineering approaches alone by yielding gains with minimal startup time and sustaining such gains over years despite a lack of detailed knowledge concerning the physiology of the producing micro-organism.

This empirical approach has a long history of success, as best exemplified by the improvements achieved for penicillin production in which reported penicillin titers are 50 g/l, an improvement of at least 4000 fold over the original parent.

Examples of fungal or actinomycaete cultures capable of overproducing metabolites in quantities as high as 80 g/l can be found in the literature. Thus, application of strain improvement to new fermentation processes continues to be documented in the literature despite the age of the technology. One part of this continued interest in strain improvement is the marriage of classical techniques and molecular genetics to create a synergistic effect for process improvement. Fermentation processes for products as diverse as antibiotics and human proteins have benefited from this combination of approaches. A second area of interest has been the application of new approaches or technology to strain improvement. The greater availability of userfriendly equipment and enhanced detection limits for mass spectroscopy and high pressure liquid chromatography (HPLC) have made their use more common. In addition to using design to improve media during fermentation development, statistical analysis can also lead to enhancements in screening programs, as will be discussed. The availability of userfriendly software such as JMP (JMP Statistical Discovery Software; SAS Institute Inc., Cary, NC) has allowed the wider use of such analyses by the scientist. Finally, the growing field of metabolic flux analysis or quantitative physiology will likely become a tool in directing screening work or explaining the success of such work.

Regardless of the methods of strategy, strain improvement relies on the iteration of three operations: genetic alteration, fermentation, and assay.

Genetic Alterations

Mutagenesis

The first key step is the generation of mutants. This can be accomplished by using either chemical or physical treatments to modify the genome of the target organism. It should be stressed that safety of the handler should be a consideration before starting such work. All mutagens should be considered potential carcinogens, and care should be taken to avoid exposure. Biosafety cabinets and protective equipment should be used, surfaces should be decontaminated, and used equipment should be decontaminated or disposed of by incineration.

There are many excellent reviews that list protocols for mutagenesis with a variety of different mutagens. Different mutagens are presumed to have different mechanisms of action, such as genetic alteration by base transitions or by frameshifts. During a longterm strain improvement programme, it is advisable to change mutagens periodically to take advantage of these different mechanisms of action. The detailed procedure for the isolation of mutations and the sensitivity of an organism to a particular mutagen will vary considerably from organism to organism. For example, a highly pigmented organism will show increased resistance to the killing effects of UV light exposure. Alkalophilic organisms need to be harvested and resuspended in a neutral pH buffer before treatment with chemical mutagens because of the inactivation of the mutagen at high pH.

The degree of killing versus the frequency of observed mutants will vary with different organisms. This can easily be verified by using antibiotic resistance as an indicator of mutation frequency. It is recommended that the multiplicity of mutation be determined in addition to monitoring survival rates as a more meaningful measure of efficiency.

In addition to vegetative cells and spore preparations, protoplasts can be used as starting material for mutagenesis. This is especially useful for basidiomycetes and other mycelial organisms. The nature of the protoplasting condition itself may prove to be mutagenic, and this can be enhanced by exposure of protoplasts to *N*-methyl-*N′*-nitro-*N*-nitrosoguanidine (MNNG) or UV light.

Protoplast fusion

Protoplast fusion is another tool to be used to achieve genetic alterations in the lineage of industrial micro-organisms. The technique can offer a means of combining favourable traits from two lineages or parental cultures. Fusion unfortunately does not allow the scientist to direct specific genes or DNA segments and thus, like mutagenesis, relies on empirical measurements to determine the success at combining two or more traits. The need for genetic markers thus becomes important in measuring efficiency of the approach. Phenotypic determinants such as auxotrophy, extracellular enzyme production, morphological differences, levels of antibiotic production, or antibiotic resistance can offer selectable traits. However, the use of auxotrophic markers may not be desirable for industrial micro-organisms because of the cost of supplementing medium at the production scale. Protoplast fusion also offers the advantage of not requiring significant knowledge of the genetics of a particular culture. In addition, fusion is considered natural or homologous recombination and thus can avoid the regulatory constraints of fermenting the resulting strains at largescale. A derailed description of protocols and considerations for the application of protoplast fusions to a variety of micro-organisms is available from Matsushima and Baltz. As with lignin degradation, nodulation and nitrogen fixation are determined on multiple alleles. The requirement for agricultural application is ultimately to identify strains capable of enhanced fixation but also robust enough for field application. Despite the economic importance, the application of protoplast fusion to enhance nitrogen fixation capability in soyabean rhizobia has been limited, perhaps because of the difficulty of establishing conditions as well as the slow growth of the cultures. Eisa reported the successful intraspecific fusion of *Bradyrhizobium japonicum* strains, resulting in isolates capable of forming 50 per cent more nodules than parental strains. In addition, some intraspecific and interspecific fusion products (with *Sinorhizobium fredii*) exhibited increased nitrogenase activity. A procedure for preparing and fusing such protoplasts is described below.

Protocol

Protoplast fusion

1. Grow cells in yeast extract mannitol broth for 48 hr. at 30°C (per litre: 5 g mannitol, 3 g yeast extract, 0.5 g $MgSO_4$, 0.7 g K_2HPO_4, 0.1 g KH_2PO_4, and 0.04 g $FeCl_3$).
2. Harvest cells in log phase by centrifugation.
3. Prewash the cells with 1 per cent *N*-laurylsarcosine. Wash three times with Tris-HCl, pH 7.5, containing 0.6 M $MgSO_4$.
4. Incubate cells with 5 mg/mllysozyme in the same buffer as above for 30 min.; spin out cells at 2000 × g for 10 min. and add fresh enzyme solution to the pellet.
5. Centrifuge cells again at 2000 × g for 10 min. and wash the precipitate with buffer. Resuspend the cells in the same buffer.
6. Add the suspension to a minimal medium (MM) with 7 g/l agar and overlay onto the same medium containing 15 g/l agar to determine protoplast efficiency. MM contains the following (per litre): 6.5 g HEPES, 5.5 g MES (morpholineethanesulphonic acid), 0.067 g $FeCl_3$, 1.8 g $MgSO_4 \cdot 7H_2O$, 0.13 g $CaCl_2$, 2.5 g Na_2SO_4, 3.2 g NH_4Cl, 1.25 g Na_2HPO_4, and 1 g L-(+) arabinose.

The exchange of the enzyme buffer solution midway through the reaction in step 4 was found to be critical. Regeneration of the prepared protoplasts ranged in efficiency from 3×10^{-2} to 6.4×10^{-3}. Fusion of protoplasts was accomplished by using a polyethylene glycol treatment. Regeneration of protoplasts following fusion was achieved on the order of 10^{-7}.

Natural recombination

Natural recombination mechanisms using naturally occurring conjugative plasmids have been used for strain improvement of *Lacrococcus* starter cultures in the dairy industry. Phage infection can lead to slow acid production, which can have a significant economic effect in a large cheese factory. It has been established that many naturally occurring phage resistant strains have a number of resistance mechanisms that in many cases are carried on conjugative plasmids. The phage resistance mechanisms can include abortive infection, restriction/modification, and adsorption inhibition. These approaches have been applied successfully to dairy starter cultures. As these bacteria are part of a finished food product, it is important to avoid antibiotic resistance markers or DNA from nonfood accepted organisms. Introduction of the phage resistance mechanism using conjugation is seen as exploiting a method of gene transfer used naturally by these organisms. The methods for conjugative transfer involve solid surface matings on milk agar; cells can then be harvested in 0.85 per cent saline solutions and plated on selective medium. A more rapid method for conjugation uses a direct plate technique, whereby donor and recipient cells are mixed directly on selective medium. The advantages of this technique include improvement in transfer frequency, time savings, and the detection of lowfrequency conjugal events because cells can be concentrated prior to mating. This method has been applied to the construction of, nisin producing *Lactococcus* strains.

Operational Considerations

The ultimate success of a strain improvement programme charged with developing and improving a fermentation process will be based largely on resource allocation. The key labour intensive steps in strain improvement include the segregation and isolation of individual clones, preparation and dispensing of sterile media, transfer of the isolates in order to initiate the vegetative and fermentation stages, and assay of the fermentation broth from individual flasks or other containers.

In general, the number of isolates screened will determine the success of detecting improved strains. If a manual operation is employed, the number of strains examined will be roughly proportional to the number of workers available.

Alternative approaches toward minimising the number of manipulations include bioassays or selective agents (see rational screening below). However, such methods have limited applications in a lineage and may not be effective over the longer lifetimes of some processes. Alternatively, automating the key steps in the process can drive throughput higher without adding labour.

Automated screening

Toward the goal of increasing throughput without adding significant labour, automation of the key steps is a useful approach. It is generally desirable to miniaturise where possible to reduce the cost of equipment required as well as to reduce volumes of solvents and fermentation waste streams in the laboratory, which have come under more stringent control by environmental regulations. Efforts can range from automating single steps in a process to construction of an integrated system. A schematic of an automated, high throughput system is shown in Fig. 2.1. In this industrial system, both agar and liquid media are charged into a vessel or fermenter, where they are sterilised under conditions that can simulate those at pilotscale. Media are robotically dispensed in sterile laminar flow hoods into plates or fermentation bottles. The bottles, each of which will serve as a fermentation vessel, are arrayed in fixed positions in sterilisable, cleanable modules, each containing 100 bottles. Aeration is ensured by the use of covers capable of allowing the passage of air. The modules allow the easy handling and tracking of multiple bottles. Bottles generally contain 5 to 10 ml of medium, approximately 3 to 10 fold less than that typically used in 250 ml flasks. The inoculation of the vegetative stage for the primary screen is performed by a robot. Individual colonies are detected by an optical system and plugged from agar based medium. Liquid transfer of the grown vegetative cultures is also accomplished robotically in a laminar flow hood. Extraction and HPLC analysis of the fermentation broth are also automated to match the throughput of the fermentation stage. A custom designed unit that dispenses solvent and mixes the bottle contents is employed. Commercially available equipment transfers extract to microtiter dishes, and the extract is analysed by HPLC units capable of performing isocratic separations. The use of isocratic separations of less than 5 min. is preferred to maintain throughput. One significant advantage of such a system is the high throughput, which can exceed that of a manual process by 10 fold based on comparable manpower. In addition, the automation facilitates the capture and downloading of process data, which allows statistical process control approaches to be used. Perhaps most significantly, the replication of fermentation results crucial to making informed decisions becomes more feasible. The disadvantage of such high throughput systems is the initial capital investment and continual maintenance of equipment and software. Compared to a smaller manual operation, an automated screen can also result in less flexibility in moving operations to new processes. However, the continued evolution of newer commercial equipment capable of handling microtiter plates or smaller formats could alleviate these concerns.

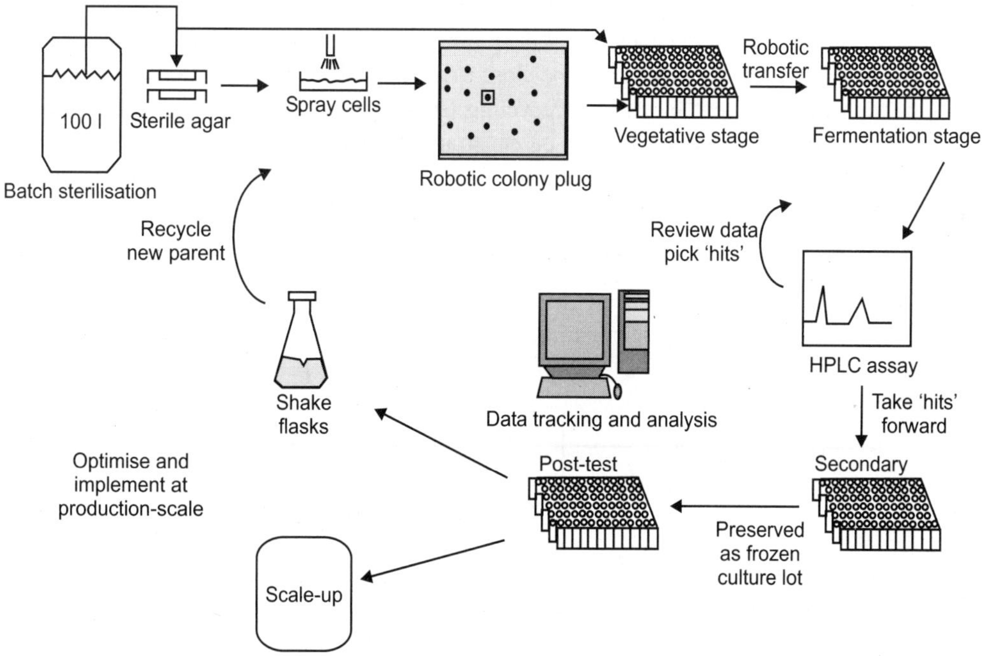

Fig. 2.1. Schematic representation of an automated screening system.

Manual screening

A manual screen had advantages in that improved yield can be obtained readily with more limited capital investment. If a manual operation is employed, the number of strains examined will roughly be proportional to the number of workers available. Thus, labour tends to be the key driver of cost. The desired target of a programme will dictate approaches. For a single protein product, it is often advisable to employ recombinant techniques as a first step. There may be multiple regulatory steps, precursor availability, export functions, and end product sensitivity that must be addressed, however, for complex antibiotic pathways. The flexibility of a manual screening system and the use of bioassays and selective agents allow for specific mutations to be built into the genealogy. When this approach is coupled with ongoing medium optimisation, yields can be increased rapidly. In a typical strain improvement project, this usually equates to examination of around 500 mutants in shake flasks per week. The expected frequency of gains could be in the order of 1 in 10,000. Thus, new strains may only be found every 10 to 20 weeks. The use of prescreens and rational selection allows for a significant increase in efficiencies for the process. With a large pathway, the frequency of detecting an improved strain from random selection is around 1 in 1000 to 2000, depending on the organism and the product and the history of the production strain. This can be accomplished in 1 to 2 months per round, depending on the fermentation cycle time.

A manual screen starts in a similar fashion as any strain improvement programme, with mutagenesis of conidia, spores, cells, or hyphal fragments (following maceration). If possible, an enrichment step is included from the rational screen, strains are selected, and working slants are produced that are then used to start the fermentation. After assay, the top few producing strains are selected for reisolation or natural selection.

A number of reisolates are then reevaluated in shake flasks. The highest producer is then selected as the parent for the next round of mutagenesis. If the producing strain has already undergone considerable mutagenesis, then the level of increase may be close to the noise level of the system. It is still possible to continue strain improvement by using a recycling technique whereby a percentage of the highest producing strains are pooled and treated as a parent strain for mutagenesis in a succeeding round. This can continue for several iterations followed by single spore or cell reisolation to examine the end productivity (Fig. 2.2).

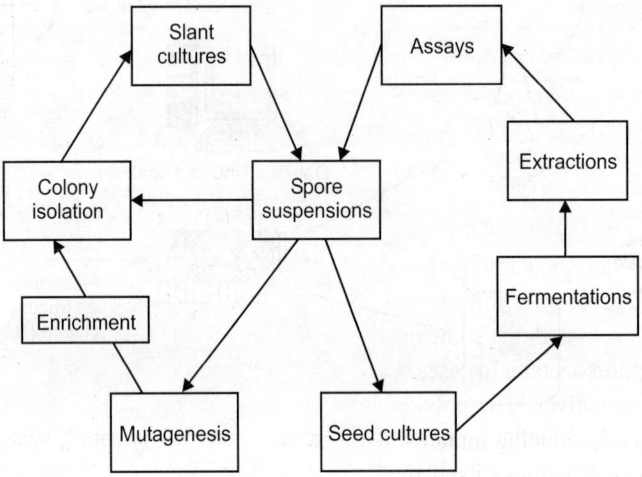

Fig. 2.2. Key steps in the progression of cultures through a manual screen.

One of the advantages of the manual screen is the trained eye of the operator. With a trained person, random selection is not quite as random as it appears. Differences in morphology, pigmentation; and growth rate can all be taken into account by the person doing the selection. In fact, selection for pelleted strains of filamentous organisms is quite commonly based on morphological criteria.

Selection and rational screening

Use of a selection strategy can greatly increase the efficiency of a strain improvement project. In random screening, a high percentage of putative mutants examined will be carried over as survivors from the mutagenesis and will exhibit the same yields or lower yields than the parent strain. However, direct selection on plates, for example, allows for a higher throughput, and only mutants are examined in shake flask experiments. An example of such an enrichment would be selection of amino acid over producers using amino acid analogs. Only mutants that either overproduce the desired amino acid or are modified in transport or degradation are selected against a background of sensitive organisms that may be on the order of 10^8. Rational screening requires some knowledge or inferred knowledge of the biosynthetic pathway to a product. A general selection scheme is outlined in Fig. 2.3. As an example, a polyketide derived secondary metabolite offers several potential targets of attack for yield improvement. Mutants that survive treatment and overcome the toxic or inhibitory analog are often designated as resistant, e.g. chloroacetate resistant mutants can overproduce polyketide products. If the carbon source in the fermentation is starch or dextrine, then overproduction of amylase may lead to increased production of available acetate as a precursor.

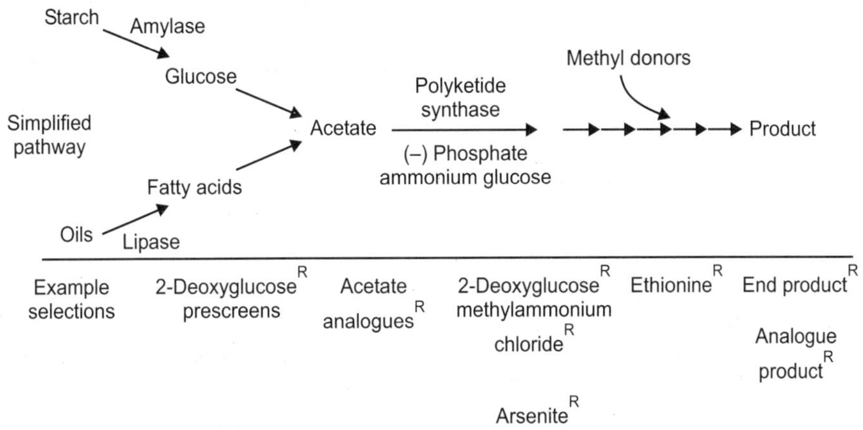

Fig. 2.3. Strategies for screening of strains in rational selections.

Similarly, if lipids are the carbon source, over production of lipase may lead to increased pools of acetyl coenzyme A. Direct selection against an acetate analog may then be useful to increase precursor production. It should be noted that some organisms are incapable of taking up free acetate analogs such as fluoroacetate or chloroacetate. In cases such as these, amides of the analog (e.g. chloroacetamide) are sometimes useful alternatives. The polyketide synthase can also be a target, using inhibitors such as cerulenin to select overproducing mutants.

The end product of a pathway itself may inhibit growth or overproduction. It is relatively easy to select mutants that are capable of growth in the presence of the end product. If stability of the compound is an issue, more stable analogs can be used, assuming a similar mechanism of action. Additional targets will depend on the physiology of the production strain. Other examples of phenotypes to overcome include repression by glucose (2-deoxyglucose resistance), ammonium ion, or phosphate. In these examples, selection strategies would encompass 2-deoxyglucose, methylammonium chloride, or arsenite resistance, respectively. If there is a specific property of a desired product, this can be incorporated into a selection scheme. For example, carotenoids have been shown to protect *Phaffia rhodozyma* against singlet oxygen damage. Combinations of rose Bengal and thymol in visible light select for carotenoid overproducing strains, thus enriching the population for pigmented organisms. Direct enrichment with this singlet oxygen system led to an increase in certain carotenoids but a decrease in astaxanthin. As with any empirical selection, mutants with many phenotypes other than that of interest might be isolated.

Chapter 3

Strain Development and Gene Technology

INTRODUCTION

Biotechnology is truly multidisciplinary in nature and it encompasses several disciplines of basic sciences and engineering. The science disciplines from which biotechnology draws heavily are: microbiology, chemistry, biochemistry, genetics, molecular biology, immunology, cell and tissue culture, and physiology. On the engineering side, it leans heavily on process, chemical and biochemical engineerings since large scale cultivation of micro-organisms and cells, their downstream processing, etc. are based on them.

This may be explained by using as example the production of a compound by micro-organisms, animal cells or plant cells. This biotechnological process can be separated into five major steps or operations: (i) strain (or culture) choice and improvement, (ii) mass culture, (iii) optimisation of cell responses, (iv) process operation, and (v) product recovery or downstream processing (Fig. 3.1).

The first step in such a biotechnological process is the identification of biological agent (micro-organism/animal cell/plant cell) capable of producing the desired compound. This would generally involve the isolation of such a micro-organism from an appropriate habitat and its improvement through suitable strain development strategies. These activities would require a knowledge of general biology and ecology of the organism to decide on what organism to isolate and from where, and then to be able to assess its ability to perform the desired functions through appropriate chemical and biochemical tests. Often it may be necessary to produce genetically engineered micro-organisms (GEMs) capable of production of the desired compound. Obviously, all these activities require major inputs from microbiology (systematics, ecology etc.), cell biology, physiology, botany, zoology, genetics and molecular genetics.

Once a suitable strain has been developed, it needs to be maintained for as long as it is needed. Such strains can be used either to produce the biomass which *per se* is the desired product, e.g. in case of single cell protein (SCP), or to recover some compounds from the biomass or the medium. In either case it is necessary to culture the strain on a large scale, the scale being much larger for biomass than that for chemical production. This operation, therefore, requires knowledge of cell physiology, genetics etc. (needed for the first step); in addition an understanding of process engineering is necessary.

The genetic machinery of cells is so geared that they perform a specified function only under specific conditions. In general, the conditions favouring rapid cell growth and biomass production are different from those conducive to the production of compound of interest, e.g. antibiotic. Therefore, in order to optimise the biochemical yields, the culture conditions have to be precisely regulated and, if needed, sequentially manipulated to fully exploit the intrinsic capabilities of cells. Obviously, this would require inputs from cell biology, physiology, genetics, etc. and process engineering.

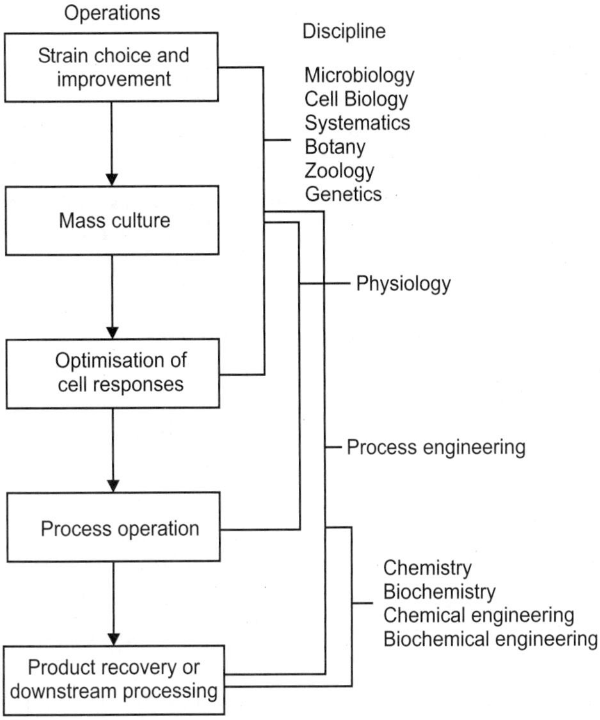

Fig. 3.1. A schematic representation of the various operations and the disciplines on which they are based In the production of a compound using micro-organisms, animal cells or plant cells (i.e. a biotechnological operation).

The various steps of a biotechnological process need to be fully optimised for safety, reproducibility, control and efficiency at all the scales of operation. In major part, this is the function of process engineering design developed with a full understanding of the biological, chemical and socioeconomic factors. The problems related to this aspect generally require fresh solutions for each new process and, often, process improvement. The practical exploitation of a biotechnological process chiefly depends on the successful implementation of the process operation.

The essence and termination of any biotechnological process is the recovery of the concerned product in a useful form. The efficiency of product recovery is directly reflected in the product cost. The mode of this operation also determines the environmental friendliness of the process. In some cases, the process of product recovery or downstream processing may be either inefficient or costly preventing the commercial exploitation of the biotechnological process, e.g. recovery of insulin, interferon, etc. from seeds of transgenic plants. This operation derives from chemistry, biochemistry, chemical engineering and biochemical engineering.

Enzyme yield is increased by careful selection of the organism, improvements in fermentation process, strain development, and refinement of downstream processing. The end use of the product and the conditions under which it is to be used will dictate the choice of organism and the product purity. The ultimate objective of strain improvement is to obtain regulatory mutants for the concerned enzyme so that its production is no more subject to the normal regulation. When immobilised cells are to be used, cell lysis must be absent and all the enzyme should be retained within the cells.

MUTATION

In biology, mutations are changes to the nucleotide sequence of the genetic material of an organism. Mutations can be caused by copying errors in the genetic material during cell division, by exposure to ultraviolet or ionising radiation, chemical mutagens, or viruses, or can be induced by the organism itself, by cellular processes such as hypermutation. In multicellular organisms with dedicated reproductive cells, mutations can be subdivided into germ line mutations, which can be passed on to descendants through the reproductive cells, and somatic mutations, which involve cells outside the dedicated reproductive group and which are not usually transmitted to descendants. If the organism can reproduce asexually through mechanisms such as cuttings or budding the distinction can become blurred. For example, plants can sometimes transmit somatic mutations to their descendants asexually or sexually where flower buds develop in somatically mutated parts of plants. A new mutation that was not inherited from either parent is called a *de novo* mutation.

The source of the mutation is unrelated to the consequence, although the consequences are related to which cells were mutated.

Mutations create variation within the gene pool. Less favourable (or deleterious) mutations can be reduced in frequency in the gene pool by natural selection, while more favourable (beneficial or advantageous) mutations may accumulate and result in adaptive evolutionary changes. For example, a butterfly may produce offspring with new mutations. The majority of these mutations will have no effect; but one might change the colour of one of the butterfly's offspring, making it harder (or easier) for predators to see.

If this colour change is advantageous, the chance of this butterfly surviving and producing its own offspring are a little better, and over time the number of butterflies with this mutation may form a larger percentage of the population.

Neutral mutations are defined as mutations whose effects do not influence the fitness of an individual. These can accumulate over time due to genetic drift. It is believed that the overwhelming majority of mutations have no significant effect on an organism's fitness. Also, DNA repair mechanisms are able to mend most changes before they become permanent mutations, and many organisms have mechanisms for eliminating otherwise permanently mutated somatic cells.

Mutation is generally accepted by the scientific community as the mechanism upon which natural selection acts, providing the advantageous new traits that survive and multiply in offspring or disadvantageous traits that die out with weaker organisms.

Classification of Mutation Types

By effect on structure

The sequence of a gene can be altered in a number of ways. Gene mutations have varying effects on health depending on where they occur and whether they alter the function of essential proteins. Mutations in the structure of genes can be classified as:

Small-scale mutations, such as those affecting a small gene in one or a few nucleotides, including:

1. Point mutations, often caused by chemicals or malfunction of DNA replication, exchange a single nucleotide for another. These changes are classified as transitions or transversions. Most common is the transition that exchanges a purine for a purine (A↔G) or a pyrimidine for a pyrimidine, (C↔T). A transition can be caused by nitrous acid, base mis-pairing, or mutagenic base analogs such as 5-bromo-2-deoxyuridine Bromodeoxyuridine (BrdU). Less common is a

transversion, which exchanges a purine for a pyrimidine or a pyrimidine for a purine (C/T↔A/G). An example of a transversion is adenine (A) being converted into a cytosine (C). A point mutation can be reversed by another point mutation, in which the nucleotide is changed back to its original state (true reversion) or by second-site reversion (a complementary mutation elsewhere that results in regained gene functionality). Point mutations that occur within the protein coding region of a gene may be classified into three kinds, depending upon what the erroneous codon codes for:

(a) Silent mutations: which code for the same amino acid.
(b) Missense mutations: which code for a different amino acid.
(c) Nonsense mutations: which code for a stop and can truncate the protein.

2. Insertions add one or more extra nucleotides into the DNA. They are usually caused by transposable elements, or errors during replication of repeating elements (e.g. AT repeats). Insertions in the coding region of a gene may alter splicing of the mRNA (splice site mutation), or cause a shift in the reading frame (frameshift), both of which can significantly alter the gene product. Insertions can be reverted by excision of the transposable element.
3. Deletions remove one or more nucleotides from the DNA. Like insertions, these mutations can alter the reading frame of the gene. They are generally irreversible: though exactly the same sequence might theoretically be restored by an insertion, transposable elements able to revert a very short deletion (say 1–2 bases) in any location are either highly unlikely to exist or do not exist at all. Note that a deletion is not the exact opposite of an insertion: the former is quite random while the latter consists of a specific sequence inserting at locations that are not entirely random or even quite narrowly defined.

Large-scale mutations in chromosomal structure, including:

1. Amplifications (orgene duplications) leading to multiple copies of all chromosomal regions, increasing the dosage of the genes located within them.
2. Deletions of large chromosomal regions, leading to loss of the genes within those regions.
3. Mutations whose effect is to juxtapose previously separate pieces of DNA, potentially bringing together separate genes to form functionally distinct fusion genes (e.g. bcr-abl). These include:
   (a) Chromosomal translocations: interchange of genetic parts from nonhomologous chromosomes.
   (b) Interstitial deletions: an intra-chromosomal deletion that removes a segment of DNA from a single chromosome, thereby apposing previously distant genes. For example, cells isolated from a human astrocytoma, a type of brain tumor, were found to have a chromosomal deletion removing sequences between the 'fused in glioblastoma' (FIG) gene and the receptor tyrosine kinase 'ROS', producing a fusion protein (FIG-ROS). The abnormal FIG-ROS fusion protein has constitutively active kinase activity that causes oncogenic transformation (a transformation from normal cells to cancer cells).
   (c) Chromosomal inversions: reversing the orientation of a chromosomal segment.
4. Loss of heterozygosity: loss of one allele, either by a deletion or recombination event, in an organism that previously had two different alleles.

Harmful Mutations

Changes in DNA caused by mutation can cause errors in protein sequence, creating partially or completely non-functional proteins. To function correctly, each cell depends on thousands of proteins to function in the right places at the right times. When a mutation alters a protein that plays a critical role in the body, a medical condition can result. A condition caused by mutations in one or more genes is called a genetic disorder. Some mutations alter a gene's DNA base sequence but do not change the function of the protein made by the gene. Studies in the fly *Drosophila melanogaster* suggest that if a mutation does change a protein, this will probably be harmful, with about 70 per cent of these mutations having damaging effects, and the remainder being either neutral or weakly beneficial.

If a mutation is present in a germ cell, it can give rise to offspring that carries the mutation in all of its cells. This is the case in hereditary diseases. On the other hand, a mutation can occur in a somatic cell of an organism. Such mutations will be present in all descendants of this cell, and certain mutations can cause the cell to become malignant, and thus cause cancer.

Often, gene mutations that could cause a genetic disorder are repaired by the DNA repair system of the cell. Each cell has a number of pathways through which enzymes recognise and repair mistakes in DNA. Because DNA can be damaged or mutated in many ways, the process of DNA repair is an important way in which the body protects itself from disease.

Beneficial Mutations

Although most mutations are deleterious, mutations may have a positive effect given certain selective pressures in a population.

For example, a specific 32 base pair deletion in human CCR5 (CCR5-Δ32) confers HIV resistance to homozygotes and delays AIDS onset in heterozygotes. The CCR5 mutation is more common in those of European descent. One theory for the etiology of the relatively high frequency of CCR5-Δ32 in the European population is that it conferred resistance to the bubonic plague in mid-14th century Europe. People who had this mutation were able to survive infection; thus, its frequency in the population increased. It could also explain why this mutation is not found in Africa where the bubonic plague never reached. Newer theory says the selective pressure on the CCR5 Δ32 mutation has been caused by smallpox instead of the bubonic plague.

STRAIN IMPROVEMENT

After an organism producing a valuable product is identified, it becomes necessary to increase the product yield from fermentation to minimise production costs. Product yields can be increased by: (i) developing a suitable medium for fermentation, (ii) refining the fermentation process and (iii) improving the productivity of the strain. Generally, major improvements arise from the last approach; all fermentation enterprises place a considerable emphasis on this activity. The techniques and approaches used to genetically modify strains to increase the production of the desired product is called strain improvement or strain development. Strain improvement is based on the following three approaches: (i) mutant selection, (ii) recombination, and (iii) recombinant DNA technology.

Mutant Selection

Large scale mutant selection programmes begin when favourable reports of clinical trials are obtained. In the early stages, selection of spontaneous mutants may be helpful, but induced mutations are the most common sources of improvements. Mutations occurring without any specific treatment are called spontaneous mutation, while those resulting due to a treatment with certain agents are known as induced mutations, such agents are referred to as mutagens. Either physical and chemical mutagens can be employed. Usually, the frequency of mutants with desirable phenotype is quite low; hence the major bottleneck is the identification and isolation of such cells from among the large number of non-mutant/

undesirable mutant cells. Many mutations (a sudden and heritable change in the traits of an organism) bring about marked changes in a biochemical character of practical interest; these are called major mutations. Some major mutations can be useful in strain improvement. For example, the original strain of *Streptomyces griseus* produced small amounts of streptomycin and large amounts of mannosidostreptomycin which has low antibiotic activity. A major mutant isolated from this strain produced negligible amounts of mannosidostreptomycin and much larger quantities of streptomycin. Similarly, a mutant strain (S-604) of *Streptomyces aureofaciens* produces 6-demethyl tetracycline in place of tetracycline; this demethylated form of tetracycline is the major commercial form of tetracycline.

In contrast, most improvements in biochemical production have been due to the stepwise accumulation of so called minor genes. These genes lead to small increases (or decreases) in the antibiotic or other biochemical production, and selection may be expected to result in a 10–15 per cent increase in yield. The selected strains are usually subjected to successive cycles of mutagenesis and selection, and after several cycles large increases is yields are likely to be obtained. Application of mutagens to induce mutations is called mutagenesis. In some cases, improvements have been obtained even without the use of mutagens. Mutants of *Penicillium chrysogenum* were selected for increased penicillin production; each cycle of selection was preceded by mutagen (chemical) treatment and resulted in only small changes in penicillin yield. But after several (about dozen) cycles of selection, a strain (E 15-1) was obtained that yielded 55 per cent more penicillin than the original strain (Fleming strain).

Selective isolation of mutants

A majority of desirable mutants, especially the 'minor gene' mutants showing increased production, are isolated by screening a large number of clones surviving the mutagen treatment; this is called secondary screening. But this approach requires a large amount of work. Therefore, efforts have increasingly focused on developing techniques for the isolation of particular classes of mutants which are likely to be over producers (Table 3.1).

Table 3.1. A summary of different approaches in utilisation of mutation and genetic recombination for strain improvement.

Approach	*Chief feature*	*Example/Remark*
Mutant selection: Types		The main approach to strain improvement; produces new alleles of existing genes
Spontaneous mutations	Occurs without any treatment with a mutagen	Used in the initial stages of strain improvement; also for maintenance of improved strains
Induced mutations	Induced by chemical (mainly) or physical mutagens	Mutagenesis followed by selection; several cycles employed
Major mutations	Affect the pattern of metabolite production	Production of 6-demethyl tetracycline in place of tetracycline by *S. griseus*
Minor mutations	Affect the rate of metabolite production	Small gains in each cycle of selection; substantial improvement after several cycles
Mutant selection: Strategies		
Auxotrophic mutants	Defective biosynthesis of a biochemical	Enhanced production of an amino acid, e.g. *phe*$^-$ mutants accumulate tyrosine

(Contd ...)

Approach	*Chief feature*	*Example/Remark*
Analogue-resistant mutants	Feed-back insensitive enzymes	Overproduction of metabolites, e.g. amino acids by *C. glutamicus*
Revertants of nonproducing mutants		Some mutants are high producers, e.g. chlortetracycline by *S. viridifaciens*
Revertants of auxotrophic mutants		Some are high producers, e.g. chlortetracycline by *S. viridifaciens*
Resistance to the antibiotic produced by the organism itself		Increased production, e.g. chlortetracyline by *S. aureofaciens*
Recombination		Produces new combinations of existing alleles
Sexual reproduction	Conjugation; fusion of gametes	Some bacteria and Actinomycetes; fungi and yeast
Heterokaryosis	Nuclear fusion followed by mitotic recombination and mitotic reduction	Fungi
Protoplast fusion	Protoplasts produced by lytic enzymes; fusion by PEG, recombinant recovery	Bacteria, Actinomycetes, fungi; quite successful

Some of the strategies are briefly summarised below; the selection for these classes of mutants is simple, easy and effective.

1. Isolation of auxotrophic mutants is the basis for commercial amino acid production in Japan from the bacterium *Corynebacterium glutamicus*. An auxotrophic mutant has a defect in one of its biosynthetic pathways so that it requires a specific biochemical for normal growth and development. For example, *phe*$^-$ mutants require phenylalanine for growth; such mutants of *C. glutamicus* accumulate tyrosine. Similarly, *tyr*$^-$ mutants accumulate phenylalanine, while phe-+ tyr- mutants accumulate tryptophan.
2. Many analogue-resistant mutants have feedback insensitive enzymes of the biosynthetic pathway the analogue of whose product was used for selection of such cells. In feedback inhibition, activity of an enzyme is inhibited by the end-product of the biosynthetic pathway in which the enzyme participate. For example, when *tyr*$^-$ mutants of *C. glutamicus* were selected for resistance to 50 mg/l *p*-fluorophenylalanine (analogue of phenylalanine), there was a nearly seven-fold increase in phenylalanine accumulation over that of the *tyr*$^-$ mutant.
3. Sometimes revertants from non producing mutants of a strain are high producers, e.g. one such reversion mutant of *Streptomyces viridifaciens* showed over 6-fold increase in chlortetracycline production over the original strain from which the nonproducing mutant was obtained. When a mutant mutates back to its original phenotype it is called reversion, and the mutant is known as revertant, e.g. nonproducer mutant mutating to back producer.
4. Reversion mutants of appropriate auxotrophs may often be high producers e.g. in case of *S. viridifaciens* reversion mutants of an auxotrophic mutant requiring homocysteine showed 28 per cent more chlortetracycline yield than the original strain.
5. In some cases, selection for resistance to the antibiotic produced by the organism itself may lead to increased yields. For example, *Streptomyces aureofaciens* mutants selected for resistance to 200–400 mg/l chlortetracycline showed a four-fold increase in the production of this antibiotic.

6. Sometimes, mutants with altered cell membrane permeability show high production of some metabolites. A mutant *E. coli* strain has defective lysine transport; it actively excretes L-lysine into the medium to 5-times as high concentration as that within its cells.
7. Mutants have been selected to produce altered metabolites, especially in case of aminogycoside antibiotics. For example, *Pseudomonas aureofaciens* produces the antibiotic pyrrolnitrin; a mutant of this fungus yields 4′-fluoropyrrolnitrin.

The above and many other approaches for selection of mutants can be most profitably used when the biosynthetic pathway for the concerned product is known, as are the precursors and the regulatory mechanisms. Mutant selection has been the most successful approach for strain improvement, but major advances are being made in the exploitation of other strategies, i.e. recombination and recombinant DNA technology.

Recombination

Recombination may be defined as formation of new gene combinations among those present in different strains. This approach has been highly successful in the improvement of animals and plants. Once several different mutants have been isolated recombination is used for both genetic analysis as well as strain improvement. It is used to combine desirable alleles present in two or more strains into one to increase product yields or to generate new products. Recombination may be based on: (i) sexual reproduction, (ii) parasexual cycle, and (iii) protoplast fusion (Table 3.1).

Sexual reproduction

Conjugation, mediated by sex-factor, occurs in many bacteria and actinomycetes, including *Streptomyces*. Conjugation leads to the formation of, usually, partial diploids in which crossing overproduces recombinant genotypes. Recombinants are recovered and used for genetic studies like linkage mapping. Similarly, yeast has two mating types; the cells of opposite mating types fuse to form diploid heterozygous cells (which are nonmating). The diploid cells undergo meiosis to produce four haploid spores which give rise to vegetative cells.

Parasexual cycle

Most industrially important fungi are asexual. However, their haploid hyphae often fuse to produce heterokaryons, i.e. cells having two distinct nuclei. Sometimes, the two nuclei of heterokaryons fuse and produce diploid nuclei. Occasionally, mitotic recombination coupled with mitotic reduction yields haploid nuclei from the diploid ones, giving rise to recombinants. In some cases, attempts have been made to use parasexuality for strain improvement, e.g. in *Penicillium chrysogenum*.

Protoplast fusion

Protoplasts of bacteria, Actinomycetes and fungi are isolated by treatment with a variety of lytic enzymes. An osmoticum is necessary for protoplast stability, and fusion is usually induced by PEG (polyethylene glycol) treatment. Protoplast fusion has been used to produce *Cephalosporium acremonium* strains which yield significantly higher cephalosporin C; this was done by fusing auxotrophic mutants. It may be noted that this organism had not responded well to mutant selection programmes. Similarly, *P. chrysogenum* strains producing low *p*-hydroxypenicillin were developed by protoplast fusion; this is important because it interferes with the formation of cephalosporin. Protoplast fusion between nonproducing strains of two species, *Streptomyces griseus* and *Streptomyces tenjimariensis*, has yielded a strain that produces indolizomycin, a new indolizine antibiotic. Usually, recombinants are recovered from protoplast fusion products, and some of them may possess desirable features.

Recombinant DNA Technology

Recombinant DNA technology involves the isolation and cloning of genes of interest, production of the necessary gene constructs using appropriate enzymes and then transfer and expression of these genes into an appropriate host organism. This approach is also called genetic engineering. This technique has been used to achieve the following two broad objectives: (i) production of recombinant proteins, and (ii) modification of the organism's metabolic pattern for production of new, modified or more quantity of metabolites.

Recombinant proteins

These are the proteins produced by the transferred gene or transgene; they themselves are of commercial value, e.g. insulin, interferon, etc. produced in bacteria, etc.

Metabolic engineering

When metabolic activities of an organism are modified by introducing into it transgenes which affect enzymatic, transport and/or regulatory functions of its cells, it is known as metabolic engineering. The various approaches of metabolic engineering are numerous, some of which are briefly summarised below (Table 3.2) to give an outline of these approaches, and their potential consequences.

Table 3.2. Summary of some of the types of changes in metabolite production induced by recombinant DNA technology or genetic engineering.

Approach	*Features*	*Example/Remarks*
Recombinant DNA technology	Genes from other organisms transferred into micro-organisms	New genes transferred; entirely new products, modified products, enhanced product yields, etc.
Recombinant proteins	Proteins encoded by the transgenes are the products of interest	–
Metabolic engineering	Metabolites catalysed by the transgene encoded enzymes are the products of interest	Existing metabolic pathways modified, extended, made more efficient or new pathways introduced
Product modification	The new enzyme modifies the product of existing biosynthetic pathway	Conversion of cephalosporin C into 7-amino-cephalosporanic acid by D. amino acid oxidase (in *A. chrysogenum*).
New substrate utilisation	Inacessible substrate converted into accessible form	Beer fermentation by yeast: cyclodextrins converted into glucose utilised by yeast
Completely new metabolite	All the genes of a new pathway transferred	*E. coli*; transfer of two genes for polyhydroxybutyrate synthesis from *Alcaligenes eutrophus*
Enhanced metabolite production	Amplification of the gene encoding that enzyme whose activity is rate limiting	Gene *cefEF* of *C. acremonium* catalysing conversion of penicillin N; increased cyclosporin yield
Enhanced growth	Enhanced substrate utilisation	*E. coli* glutamate dehydrogenase into *M. methylotrophus*; carbon conversion increased from 4 to 7 per cent

It is imperative for metabolic engineering that the biosynthetic pathways to be modified and their regulatory controls are well known, and the genes involved are identified and cloned.

1. A transgene may be added which encodes an enzyme to modify a metabolite produced by the organism to yield a new product of interest. For example, *Acremonium chrysogenum* produces cephalosporin C. The gene encoding D-amino acidoxidase from *Fusarium* was introduced into *A. chrysogenum*. This enzyme converts cephalosporin C into 7-aminocephalosporanic acid, a precursor of several semisynthetic antibiotics.
2. The enzyme encoded by transgene may enable a better utilisation of the substrate or even of previously inaccessible components of the substrate. For example, normal yeasts are unable to utilise cyclodextrins present in the malt; this increases the caloric content of beer. Transgenic yeasts capable of utilising cyclodextrins are now commercially used to produce low calorie beer with 1 per cent more alcohol content.
3. All the genes of an entirely new biosynthetic pathway may be transferred to generate new products. For example, two genes are involved in the conversion of acetyl-CoA to poly-hydroxybutyrate (PHB) which is used to produce biodegradable plastic. The two genes were transferred into *E. coli* from *Alcaligenes eutrophus*. Transgenic *E. coli*, under appropriate conditions, accumulate PHB to upto 50 per cent of their dry weight.
4. Several gene transfers have enhanced growth rates of the organisms, reduced their nutrient requirements and enabled their growth to higher cell densities. For example, transfer of the gene encoding glutamate dehydrogenase from *E. coli* to glutamate synthase (GOGAT) deficient mutants of *Methylophilus methylotrophus* increased the efficiency of carbon conversion from 4 to 7 per cent. *M. methylotrophus* is used to produce microbial biomass from methanol, which is used as animal feed. Glutamate dehydrogenase utilises ammonium to produce glutamic acid but uses little energy, while glutamate synthase pathway, (in which GOGAT is involved) uses up one additional ATP molecule for each molecule of glutamic acid produced.
5. In some cases, conversion of an intermediate product to the end product is slow due to a low activity of the rate limiting enzyme. In such cases, the activity of rate limiting enzyme can be increased by increasing its dosage. For example, in case of *C. acremonium* the enzyme (encoded by the gene *cefEF*) that converts penicillin N intermediate in the cephalosporin C biosynthesis is rate limiting. The dosage of *cefEF* was increased leading to a 15 per cent higher cephalosporin C yield. (The penicillin content was reduced by a factor of 15).

The above listed examples are only atraction of the possibilities offered by the recombinant-DNA technology. The powers of this technology shall increase with the refinements of techniques for genetic modification of industrially important micro-organisms and, the knowledge of various details of their metabolic pathways.

RECOMBINANT DNA TECHNOLOGY

Gene and Gene Function

Strain selection and development have their basis in the fact that the intrinsic capabilities of a cell are due to the genes contained by it. A gene is a part of a chromosome and is responsible for some character or trait of an organism/cell; it is made up of DNA (in some viruses, RNA functions as the genetic material). DNA is usually double-stranded, but in some viruses it is single-stranded; the latter also has a double-stranded replicative form. The two strands of a DNA double helix are complementary (A in one strand faces T in the other and G in one faces C in the other) and run anti-parallel to each other (the 3′-OH of ribose present in one strand and the 5′-P residue of the other strand are located at the same end of a double helix). DNA replication is highly faithful, is semiconservative and is catalysed by DNA polymerase.

Genes produce their phenotypic effects by specifying the amino acid sequences of specific proteins. The nucleotide sequence of one strand (the antisense strand) of a DNA double helix is used by RNA polymerase as template to generate its complementary copy of RNA; this RNA ultimately functions as messenger RNA (mRNA) and the process of its production is called transcription. Three bases beginning from a specific point of a mRNA molecule, together code for a specific amino acid; in this way the base sequence of a mRNA molecule dictates the amino acid sequence of the polypeptide specified by it. The formation of protein using mRNA as template is known as translation; it is performed together by ribosomes, tRNA, and appropriate cofactors. The polypeptide chain then folds to give a functional protein; many proteins contain two or more identical or distinct polypeptides. Most of the proteins function as enzymes each of which catalyses a specific biochemical reaction. These reactions ultimately give rise to the observable phenotypes. Thus genes determine the intrinsic capabilities of cells which form the basis of biotechnological processes.

Gene Transmission

The high fidelity semiconservative replication of DNA ensures transmission of genes from parents to progeny without change; this is the reason for stability of genetically controlled phenotypes over generations. However, a low frequency (10^{-4} to 10^{-7} per gene per generation) of changes occur in genes naturally (spontaneous mutation); these mutations are the ultimate source of all the heritable variation observed in living forms. Clearly, mutations create the variation that is exploited through selection during strain development.

Another source of genetic variation is recombination between genes. In eukaryotes, recombination occurs regularly during the meiotic cell division necessary for the formation of gametes involved in sexual reproduction; this phenomenon has been highly successfully exploited by plant and animal breeding programmes. In prokaryotes, on the other hand, recombination occurs when foreign DNA is brought into a cell during any of the following three events: (i) transformation, (ii) transduction, and (iii) conjugation. In transformation, DNA is directly taken up by the cell and a portion of it becomes integrated in the cells chromosome by a process of recombination.

But in case of transduction, the DNA is transferred from one cell into another by a virus; it may either be generalised or specialised. Conjugation, on the other hand, is a sexual process of DNA transfer; it may occur between individuals of the same or different species. During conjugation, a part or whole of the bacterial chromosome may be transferred into the recipient cell; often even large plasmids are exchanged. Segments of the transferred chromosome become integrated into the chromosome of the recipient cell by a process of recombination.

The natural processes of gene transfer vary appreciably in their range and specificity. In general they are rather imprecise which makes the recovery of desired gene combination dependent on efficient screening and selection.

In addition, their range in terms of the species involved is rather restricted depending on sexual compatibility (sexual reproduction), and virus host range (transduction). These put a serious limitation on the movement of genes across taxonomic borders.

Recombinant DNA Technology: Definitions

The above problems in gene transmission are circumvented by the recombinant DNA technology. A recombinant DNA molecule is produced by joining together two or more DNA segments usually originating from different organisms. More specifically, a recombinant DNA molecule is a vector (e.g. a plasmid, phage or virus) into which the desired DNA fragment has been inserted to enable its cloning in an appropriate host. This is achieved by using specific enzymes for cutting the DNA (restriction enzymes) into suitable fragments and then for joining together the appropriate fragments (ligation). In this manner, a recombinant DNA molecule may be produced which contains a gene from one organism joined to regulatory sequences from another organism; such a gene is called chimaeric gene. Clearly, the capability to produce recombinant DNA molecules has given man the power and opportunity to create novel gene combinations to suit specific needs.

Recombinant DNA molecules are produced with one of the following three objectives: (i) to obtain a large number of copies of specific DNA fragments, (ii) to recover large quantities of the protein produced by the concerned gene, and (iii) to integrate the gene in question into the chromosome of a target organism where it expresses itself. Even for the latter two objectives, it is essential to first obtain a large number of copies of the concerned genes. To achieve this, the DNA segments are integrated into a self-replicating DNA molecule called vector; most commonly used vectors are either bacterial plasmids or DNA viruses. All these steps concerned with piecing together DNA segments of diverse origin and placing them into a suitable vector together constitute recombinant DNA technology.

The vectors containing DNA segments to be cloned, called DNA inserts (chimaeric vectors) are then introduced into a suitable organism, usually a bacterium; this organism is called host, while the process is called transformation. The transformed most cells are selected and cloned. The vector present in such clones would replicate either in synchrony with or independent of the host cell; the gene present in the vector may or may not express itself by directing the synthesis of concerned polypeptide. The step concerned with transformation of a suitable host with a chimaeric vector, and cloning of the transformant cells is called DNA cloning or gene cloning. However, often DNA or gene cloning is taken to include both the development of chimaeric vectors as well as their cloning in a suitable host.

A clone consists of asexual progeny of a single individual or cell, while the process/technique of producing a clone is called cloning. As a result, all the individuals of a clone have the same genotype which is also identical with that of the individual from which the clone was derived. Therefore, the genomes present in members of a single clone are also identical; this applies to the recombinant DNA as well. Therefore, gene or DNA cloning produces large numbers of copies of the gene/DNA being cloned. Similarly, often the term recombinant DNA technology is used as a synonym for DNA or gene cloning used in the broader sense. A rather popular term for these activities is genetic engineering.

Restriction Endonucleases

Endonucleases are enzymes that produce internal cuts, called cleavage, in DNA molecules. Many endonucleases cleave DNA molecules at random sites. But a class of endonucleases cleaves DNA only within or near those sites which have specific base sequences; such endonucleases are known as restriction endonucleases, and the sites recognised by them are called recognition sequences or recognition sites. The recognition sequences are different and specific for the different restriction endonucleases or restriction enzymes.

Restriction enzymes were discovered due to and named after the phenomenon of host restriction of bacterial phages. When a phage (say λ C) DNA released from one bacterial strain (say, *E. coli* strain C;

such a λ phage is designated as λ C) is used to infect another strain (say, *E. coli* strain K) the efficiency of phage growth on the latter (strain K) is only a very small fraction of the efficiency on the former (strain C). Some phages that survive grow normally on the second strain, and are now referred to as λ K; these phage particles infect strain K with the same efficiency as λ C does on strain C. Thus multiplying a phage on one bacterial strain seems to restrict the growth of that phage to the strain concerned. This restriction on phage host range is due to the presence of restriction enzymes in the host cells which recognise and cleave foreign DNA introduced in the cell. The DNA of a cell is protected from its own endonucleases by methylation (usually of A and C) within their recognition sites. Thus DNA molecules having the same methylation pattern as that of a bacterial cell itself will be recognised as own DNA, while those lacking this will be regarded as foreign DNA. The presence of restriction enzymes was postulated by W. Arber during 1960s, while the first true restriction endonuclease was isolated in 1970. Restriction endonucleases are indispensable for DNA cloning and sequencing. They serve as the tools for cutting DNA molecules at predetermined sites, which is the basic requirement for gene cloning or recombinant DNA technology.

Types of restriction endonucleases

There are three distinct types of restriction endonucleases: (i) type I, (ii) type II, and (iii) type III.

Type I restriction endonucleases are complex endonucleases, and have recognition sequences of about 15 bp; they cleave the DNA about 1000 bp away from the 5′-end of the sequence 'TCA' located within the recognition site, e.g. *Eco*K, *Eco*B, etc.

Type II restriction endonucleases are remarkably stable and induce cleavage either, in most cases, within their recognition sequences or very close to them. More than 350 different type II endonucleases with over 100 different recognition sequences are known.. They require Mg^{2+} ions for cleavage. The first type II enzyme to be isolated was *Hind* II in 1970. Only type II restriction endonucleases are used for restriction mapping and gene cloning in view of their cleavage only at specific sites.

Type III restriction endonucleases are intermediate between the type I and type II enzymes; they cleave DNA in the immediate vicinity of their recognition sites, e.g. *Eco*P1, *Eco*P15, *Hinf* III, etc. Type I and Type III restriction enzymes are not used in gene cloning.

Nomenclature

The nomenclature of restriction endonucleases follows a general pattern: (i) the first letter of the name of genus in which a given enzyme is discovered is written in capital, (ii) this is followed by the first two letters of species name of the organism. These three letters are generally written in italics, e.g. *Eco* from *Escherichia coli*, *Hin* from *Haemophilus influenzae*, *Hpa* from *Haemophilus parainfluenzae*, etc. (iii) strain or type identification is depicted as subscript, e.g. Eco_{K}; if the enzyme is coded by a plasmid, the plasmid name is written as a subscript, e.g. Eco_{RI}, (iv) when an organism produces more than one enzyme, they are identified by sequential Roman numerals, e.g. the different enzymes produced by *H. influenzae* strain Rd are named Hind II, Hin_{d} III, etc. (v) all restriction enzymes are designated by the general symbol R which is prefixed to their names, e.g. $REco_{RI}$, $RHin_{d}$ III, $RBam_{H}$ I, etc. (this is to distinguish them from the corresponding methylases isolated from the same strains; the methylases are prefixed by M).

However, in practice, the following simplifications are used: (i) the subscripts (due to strain or plasmid names) are ordinarily written on line, e.g. R*Eco*RI, R*Hind* III, etc. and (ii) the prefix R is not used where the context makes it clear that the enzyme under reference is a restriction endonuclease.

Recognition sequences

The recognition sequences for Type II endonucleases form palindromes with rotational symmetry. In a palindrome, the base sequence in the second half of a DNA strand is the mirror image of the sequence in its first half; consequently, the complementary DNA strand of a double helix also shows the same situation (Fig. 3.2).

5′GAA ↑ AAG 3′

(a) Single strand

5′GAA ↑ AAG 3′
3′CTT | TTC 5′

(b) DNA double helix

Fig. 3.2. A palindromic sequence: (a) sequence in a single DNA strand, and (b) sequence in a DNA double helix. The arrow represents the axis of symmetry.

But in a palindrome with rotational symmetry, the base sequence in the first half of one strand of a DNA double helix is the mirror image of the second half of its complementary strand (Fig. 3.3). Thus in such palindromes, the base sequence in both the strands of a DNA duplex reads the same when read from the same end (either 5′ or 3′) of both the strands.

5′GAA ↑ TTC 3′
3′CTT | AAG 5′

*Eco*RI recognition site

Fig. 3.3. A palindrome with rotational symmetry. The arrow represents the axis of symmetry.

Most of the type II restriction endonucleases have recognition sites of 4, 5 or 6 bp (base pairs) which are predominantly GC-rich. Longer palindromic target sequences are also known, and so are nonpalindromic ones (specific for some enzymes). Some restriction enzymes have ambiguities in their recognition sites, e.g. *Eco*RII, so that they may recognise upto 4 different target sequences.

Cleavage patterns

Most type II restriction endonucleases cleave the DNA molecule within their specific recognition sequences, but some produce cuts immediately outside the target sequence, e.g. *Nla*III, *Sau*3A, etc. These cuts are either: (i) staggered, or (ii) even, depending on the enzyme.

Most enzymes produce staggered cuts in which the two strands of a DNA double helix are cleaved at different locations; this generates protruding (3′ or 5′) ends (Fig. 3.4), i.e. one strand of the double helix extends some bases beyond the other.

Due to the palindromic (symmetrical) nature of the target sites, the two protruding ends generated by such a cleavage by a given enzyme have complementary base sequence. As a result, they readily pair with each other; such ends are called cohesive or sticky ends.

An important consequence of this fact is that when fragments generated by a single restriction enzyme from different DNAs are mixed, they join together due to their sticky ends (Fig. 3.5). Therefore, this property of the restriction enzymes is of great value for the construction of recombinant DNAs.

Some restriction enzymes, on the other hand, cut both the strands of a DNA molecule at the same site so that the resulting termini or ends have blunt or flush ends in which the two strands end at the same point (Fig. 3.4). The blunt cut ends also can be effectively utilised for construction of recombinant DNAs following one of several strategies.

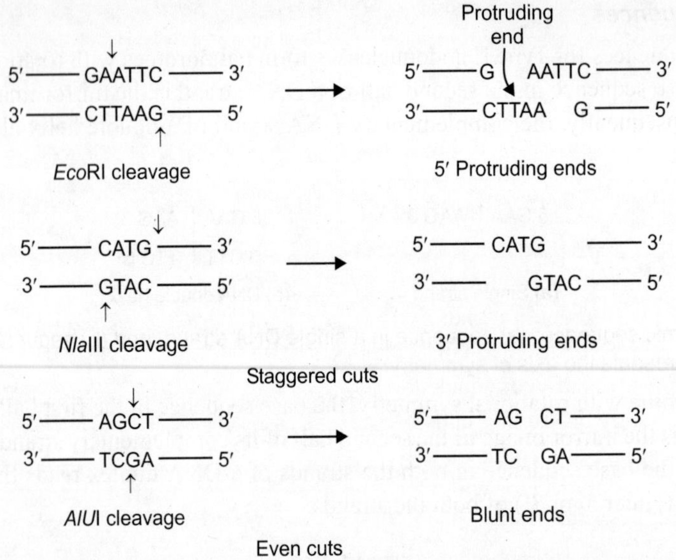

Fig. 3.4. DNA cleavage by restriction endonucleases. Staggered cuts: 5′ producing (*Eco*RI) and 3′ (*Nla*III) protruding ends, and event cuts (*Alu*I) producing blunt ends. The vertical arrows indicate the site of cut in DNA strand.

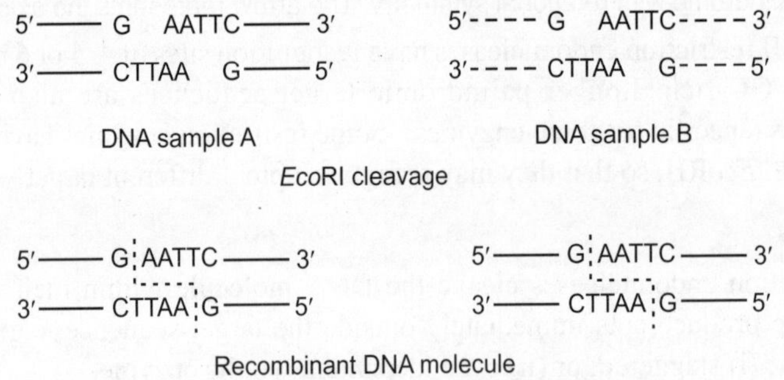

Fig. 3.5. Two distinct samples (A and B) of DNA are cleaved with the same restriction enzyme (*Eco*RI). The fragments from sample A readily join with those from sample B due to their cohesive (or sticky = complementary) protruding ends. The dotted line within the recognition sequences separates the base sequences belonging to the different fragments.

Modification of Cut Ends

The 3′ ends of DNA strands always carry a free hydroxyl (—OH) group, while their 5′ ends always bear a phosphate group. Often the ends produced by restriction enzymes have to be modified for further manipulation of the fragments; some of the modifications are summarised below.

1. Removal of the 5′ phosphate group of vector DNA by alkaline phosphatase treatment in order to prevent vector circularisation during DNA insert integration.
2. Addition of a phosphate group to a free 5′ hydroxyl group by T_4 polynucleotide kinase.
3. Removal of the protruding ends by digestion with, say, S1 nuclease; this enzyme digests both 3′ and 5′ protruding ends.

4. Filling in of the protruding ends by extending the recessed (shorter) strand with, say, Klenow fragment of *E. coli* DNA polymerase I. This strategy is preferred to S1 nuclease digestion for various reasons (bith the strategies 3 and 4 generate blunt ends which can be ligated by T_4 polynucleatide ligase.)
5. Removal of nucleotides from the 5′ ends using λ exonuclease.
6. Removal of nucleotides from the 3′ ends using *E. coli* exonuclease III. (Both the strategies 5 and 6 can vert blunt ends into protruding single-stranded ends of undefined base sequence).
7. Treatment of double-stranded DNAs with exonuclease *Bal*31 which simultaneously digests both the strands (from both the ends) of a DNA molecule; this treatment produces shortened DNA fragments with blunt ends.
8. Synthesis of single-stranded tails (protruding ends) at the 3′ ends of blunt-ended fragments by the enzyme terminal deoxynucleotidyl transferase; this is called tailing. This reaction can be used to generate protruding ends of defined sequence, e.g. poly-A tails on the 3′ ends of the DNA insert and poly-T tails on the 3′ ends of the vector (Fig. 3.6); the protruding ends of the DNA insert and the vector will, therefore, base pair under annealing conditions.
9. Linker and/or adaptor molecules can be joined to the cut ends. Linkers are short, chemically synthesised, self-complementary, double-stranded oligonucleotides which contain within them one or more restriction endonuclease sites, e.g. linker 5′ CC GAATTC GG (only one strand of the linker is shown here) contains one *Eca*RI site. Linkers are fused with blunt-ended DNA fragments; cleavage of the linker with the appropriate restriction enzyme creates suitable cohesive protruding ends (Fig. 3.7). Linkers have two applications: (i) creating cohesive ends on blunt ended DNA fragments, and (ii) on fragments having unmatched or undefined sequences in their protruding ends. In the latter situation, the DNA fragments are first made blunt-ended using either strategy 3 or 4, following which the selected linkers are ligated to them by T_4 ligase.
10. Adaptors are short, chemically synthesised DNA double strands which can be used to link the ends of two DNA molecules which have different sequences at their ends, when used in conjunction with linkers or other adaptor molecules. There are different kinds of adaptors suited for different purposes. For example, a conversion adaptor is used to join a DNA fragment or insert cut with one restriction enzyme, say, *Eco*RI, with a vector opened with another enzyme, e.g. *Bam*HI. These adaptors are prepared from two different single-stranded oligonucleotides which carry the recognition sequence of different endonucleases at their 5′ ends, and a self-complementary sequence at their 3′ ends (Fig. 3.8). For example, the oligonucleotide strand 5′ GATCCTCGAG 3′ has the recognition sequence GATC for *Bam*HI at its 5′ end, while its complementary adaptor oligonucleotide 5′ AATTCTCGAG 3′ has the recognition sequence AATT (5′ end) for *Eco*RI. Both these contain at their 3′ ends the complementary sequence 5′ CTCGAG 3′ which allows them to pair together to give rise to a double-stranded adaptor having recognition sequence for *Bam*HI at one end and that for *Eco*RI at the other (Fig. 3.8).

Steps in Gene Cloning

The entire procedure of cloning or recombinant DNA technology may be classified into the following five steps for the convenience in description and on the basis of the chief activity performed.

1. Identification and isolation of the desired gene or DNA fragment to be cloned.
2. Insertion of the isolated gene in a suitable vector.

3. Introduction of this vector into a suitable organism/cell called host (transformation).
4. Selection of the transformed host cells.
5. Multiplication/expression/integration followed by expression of the introduced gene in the host.

A brief description of these steps is given in the following sections.

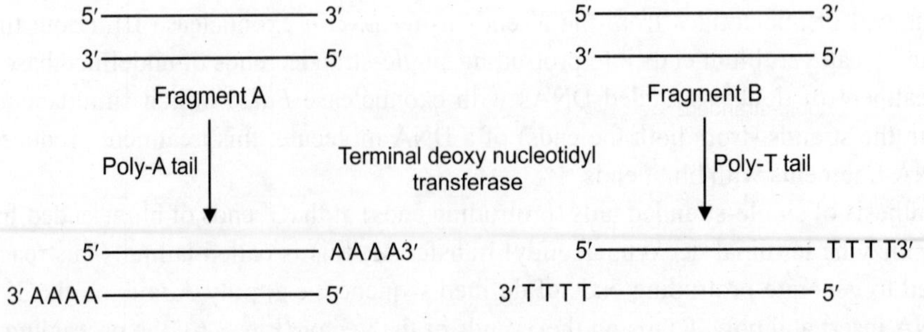

Fig. 3.6. Tailing of blunt-ended DNA fragments A and B using terminal nucleotidyl transferase. Poly-A tail is added to the 3′ ends of fragment A, while poly-T tail is added to the 3′ ends of fragment B so that the protruding ends of fragment A are complementary to those of B.

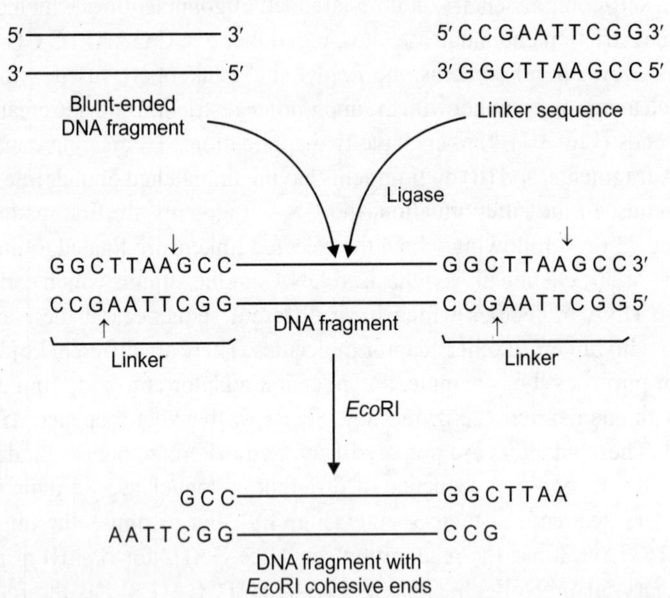

Fig. 3.7. Creation of cohesive ends on blunt-ended DNA fragments. Suitable linkers are ligated to the blunt ends by T_4 DNA ligase. The linker is then cleaved with the appropriate restriction enzyme to generate sticky ends.

Isolation of the Desired Gene

The identification and isolation of the desired gene or DNA fragment, called DNA insert, to be cloned is a critical step in gene cloning.

The desired DNA inserts can be obtained from the following: (i) cDNA libraries, (ii) genomic library, (iii) chemical (or enzymatic) synthesis, and (iv) amplification through polymerase chain reaction (PCR).

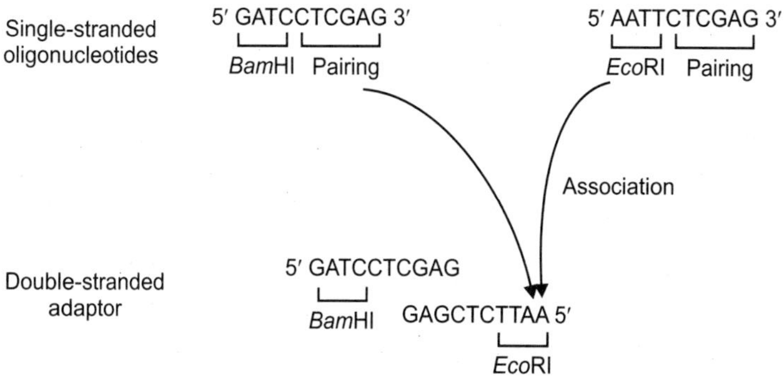

Fig. 3.8. A conversion adaptor produced by associating two oligonucleotides, each having a different recognition sequence at its 5′ end but a complementary sequence for base pairing at 3′ end.

Chemical Synthesis of Gene

The amino acid sequence of the protein (or the base sequence of mRNA) produced by a gene enables the deduction of base sequence of the concerned gene on the basis of the codons for the various amino acids. However, the degeneracy of genetic code may present some problems but a functional sequence of the gene can nonetheless be worked out. Once the base sequence of a gene is deduced, a polynucleotide of the same base sequence can be synthesised either: (i) chemically, or even (ii) enzymatically.

The chemical synthesis of a gene utilises chemical reagents for the various steps of the process. There are three distinct methods, differing mainly in the strategy of protection of OH groups of the phosphate residues: (i) phosphodiester approach, (ii) phosphotriester or phosphate triester approach, and (iii) phosphite triester or phosphoramidite approach.

Gene Amplification through Polymerase Chain Reaction

The polymerase chain reaction (PCR) technique, developed by Kary Mullis in 1985, is extremely powerful. It generates microgram (μg) quantities of DNA copies (upto billion copies) of the desired DNA (or RNA) segment, present even as a single copy in the initial preparation, in a matter of few hours. The PCR process has been completely automated and compact thermal cyclers are available in the market.

The PCR is carried out *in vitro*. It utilises: (i) a DNA preparation containing the desired segment to be amplified, (ii) two nucleotide primers (about 20 bases long) specific, i.e. complementary, to the two 3′ borders (the sequences present at the 3′ ends of the two strands) of the desired segment, and (iii) the four deoxynucleoside triphosphates, viz. TTP (thymidine triphosphate), dCTP (deoxycyclidine triphosphate), dATP (deoxyadenosine triphosphate) and dGTP (dexyguanosine triphosphate), and a heat stable DNA polymerase, e.g. *Taq* (isolated from bacterium the *Thermus acquaticus*), *Pfu* (from *Pyrococcus furiosus*) and *Vent* (from *Thermococcus litoralis*) polymerases. *Pfu* and *Vent* polymerases are more efficient than the *Taq* polymerase.

REGULATION

Enzymes are regulated by changes in their activity levels and by the amount of enzyme. There are two types of enzyme activity regulation: noncovalent and covalent. Noncovalent can often be called allosteric regulation. Allosteric regulation of activity will usually take the form of positive cooperativity where binding of the substrate leads to a higher activity form of the enzyme by a change its 3-D shape as the first molecule of substrate binds. The next molecule of substrate is now more easily bound because the enzyme has shifted into a higher activity form with a higher affinity for the substrate (i.e. a low K_m and higher V_{max} form of enzyme).

This has been illustrated with data and models for the comparison of oxygen loading to haemoglobin and myoglobin. Allosteric regulation (noncovalent regulation of enzyme activity) can also cause inhibition of enzyme activity, which was illustrated with the concept of 'feedback inhibition' of enzyme activity by a downstream product of the pathway. In general, noncovalent modification is useful for short term regulation of enzyme activity which can respond quickly to changes in cellular conditions. For example, when a lot of substrate is available for a short period of time, the enzyme shifts into a more active form to process that substrate and relaxes into its low activity form when the burst of substrate has been converted.

Longer term conversion of an enzyme into a more active state is most easily achieved by a covalent modification of the enzyme. This has been illustrated by discussing cellular conditions where the cell has been sent a hormone signal which tells it to convert to a different metabolic pattern. Thus, a cascade of modifications take place in the cell's enzymes mediated by special regulatory enzymes (such as protein kinases leading to phosphorylation of other enzymes by using ATP) which covalently modify the enzymes response for metabolism. This changes the whole pattern of the cell from one metabolic state to another. This is called a shift in 'metabolic mode' since the cell starts doing things differently when stimulated by a hormone. Once the hormone signal dissipates, the cell will return to the original state via another set of enzymes acting on the metabolic enzymes. The second set of regulatory enzymes (such as protein phosphatases which remove the phosphate esters from the phosphorylated proteins) reverse the covalent modifications put in place by the first set of regulatory enzymes. An important principle is at work here: in order for cells to response to stimuli like hormones, there must be two sets of regulatory factors or enzymes: ones to change the metabolic mode of the cell and ones to reverse these changes in order to return the cell to its original metabolic mode.

Both catabolic and anabolic processes are regulated and metabolism is generally so efficient that excess products are not formed. Strains with less efficient regulation can be selected in a screening process. Microbial metabolism is controlled by the regulation of both enzyme activity and enzyme synthesis.

Enzyme Regulation Activity

Enzyme regulation is done in two basic ways:

1. Control of enzyme activity level:
   (a) Noncovalent modifiers cause conformational change between less active and more active states of the enzyme.
   (b) Covalent modification causes interconversion between inactive and active forms of the enzyme.
2. Control the amount of the enzyme:
   (a) Isozymes—forms of the enzyme which differ in properties but catalyse the same reaction. For example, enzyme forms which differ in V_{max} and/or K_m. The isozymes can be forms found in different tissues and organs of an animal or for any eukaryotic organism, isozymes can be located in different parts of the cell. For example, different isozymes of lactate dehydrogenase are found in muscle and liver. Malate dehydrogenase occurs in different forms in the cytoplasm and the soluble matrix phase of the mitochondria.

(b) Biosynthesis of the enzyme protein can be controlled at the level of the gene via regulation of transcription (i.e. synthesis of the enzyme's mRNA). This is more of a molecular biologic type of regulation and involves molecules which bind to DNA and influence gene expression. This type of control where the amount of the enzyme is governed can also be done after the mRNA is made, but this is quite rare. In this mechanism, the mRNA is prevented from being translated and since mRNA is rather unstable, it is degraded before it is effectively used by the ribosomes to make the protein.

Allosteric regulation

Control of enzyme activity by noncovalent modifiers is usually called allosteric regulation since the modifier binds to the enzyme at a site other than the active site but alters the shape of the active site. Allosteric is a word derived from two Greek words: 'allo' meaning other and 'steric' meaning place or site; so allosteric means other site and an 'allosteric enzyme' is one with two binding sites—one for the substrate and one for the allosteric modifier molecule, which is not changed by the enzyme so it is not a substrate. The molecule binding at the allosteric site is not called an inhibitor because it does not necessarily have to cause inhibition—so they are called modifiers.

A negative allosteric modifier will cause the enzyme to have less activity, while a positive allosteric modifier will cause the enzyme to be more active.

In order for allosteric regulation to work, the enzyme must be multimeric (i.e. a dimer, trimer, tetramer, etc.). The concept is easily illustrated using a dimer as the model system, but it applies equally well to higher order multimers such as trimers and tetramers, etc.

The allosteric enzyme is a dimer with identical subunits so it is called a homodimer (A-A, where 'A' represents one of the identical subunits). Each subunit has an active site, which are shown being located at the interface of the two subunits in the model in Fig. 3.9. The binding sites for the allosteric modifier are shown as being remote from the active site. Thus, the influence of the allosteric modifier must be transmitted through the framework of the enzyme in order to influence the activity of the enzyme—either positively or negatively.

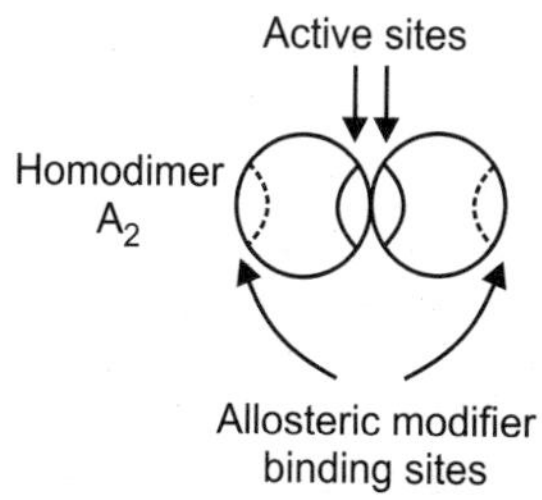

Fig. 3.9. A model of an allosteric enzyme.

First, the enzyme in absence of the positive allosteric modifier (M) has low activity and a high K_m for the substrate. Then, when the positive allosteric modifier is present, the enzyme is altered in shape, usually both subunits change shape and the dimer takes on a new conformation. In the altered form of the enzyme with M bound, it will have a higher activity and a lower K_m for substrate (Fig. 3.10).

A special case of the positive allosteric modifier is when the substrate itself is the modifier. In this case, there is no 'M' or allosteric modifier site; the binding of the substrate causes the enzyme dimer to take on a new shape and be converted to a more active form. This is called 'positive cooperative' (Fig. 3.11).

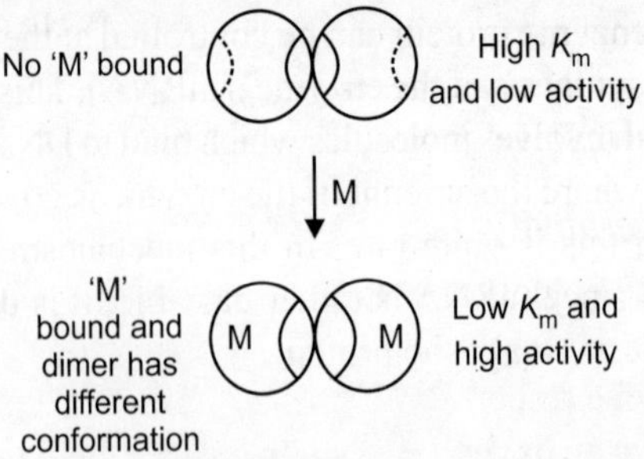

Fig. 3.10. General example of the regulation of an enzyme by its allosteric modifier. In this model, the allosteric modifier is represented by 'M'.

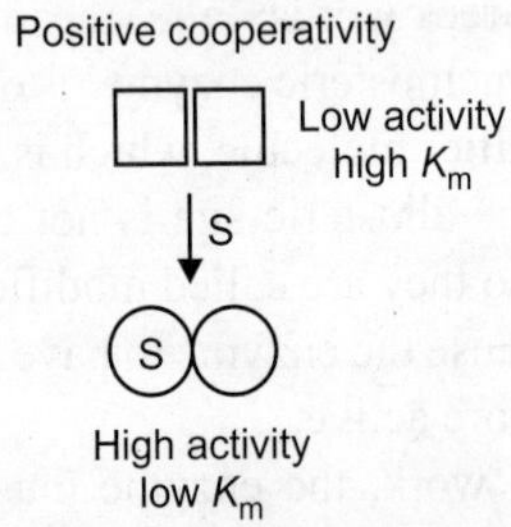

Fig. 3.11. Model of a dimer with positive cooperative when substrate (S) binds.

When the first substrate molecule binds it makes it easier for the second substrate molecule to bind. Enzymes with positive cooperative do not have Michaelis-Menten kinetics (Fig. 3.12).

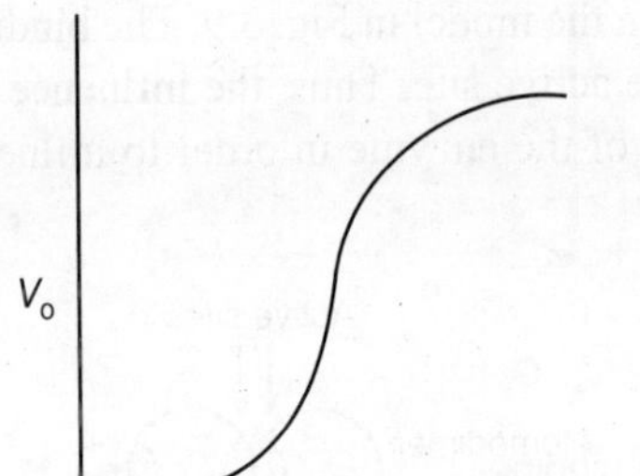

Fig. 3.12. V_o versus [S] plot for an enzyme with positive cooperative to binding of substrate.

Enzymes displaying positive cooperative have sigmoidal kinetics reflecting the very large increase in V_o with increasing [S] over a very narrow range of [S]. Note the very small change in V_o at first as more substrate is added. Then at a certain level of [S], the enzyme population changes to the more active form with a higher affinity for substrate and more activity (i.e. more production of product). But eventually, the V_{max} limit is reached since there are still a limited number of substrate binding sites, so the increase in V_o plateaus like it does for an M–M type enzyme. This is the reason to always draw a V_o versus [S] plot as your first step in analysis of an enzyme's kinetics. If you are dealing with an enzyme with sigmoidal kinetics, it will be much easier to detect using the V_o versus plot than the double reciprocal plot (plot of $1/V_o$ versus 1/[S]). In fact, the double reciprocal plot will be difficult to construct and make a good graph for because the properties of the enzyme actually change in response to adding more substrate. The advantage of sigmoidal enzyme kinetics is that a large increase in catalytic rate results from only a very small increase in [S]. Logarithmic change in V_o for a linear change in [S].

Haemoglobin—positive cooperativity

Although it is not an enzyme, haemoglobin works for oxygen transport from lungs to muscles via the principle of positive cooperative. Haemoglobin binds oxygen with positive cooperative in much the same way that the allosteric enzyme responses with positive cooperative to the binding of its substrate. The impact of positive cooperative on oxygen binding to haemoglobin is most easily seen by comparing the oxygen saturation curves for myoglobin (the muscle protein with a virtually identical 3-D shape but only one subunit) to haemoglobin with its tetrameric subunit composition. Myoglobin cannot respond cooperatively to oxygen binding since it has only one subunit and one oxygen binding site. While haemoglobin has one binding site for oxygen in each of its four subunits and can therefore have interaction among these binding sites for oxygen (Fig. 3.13).

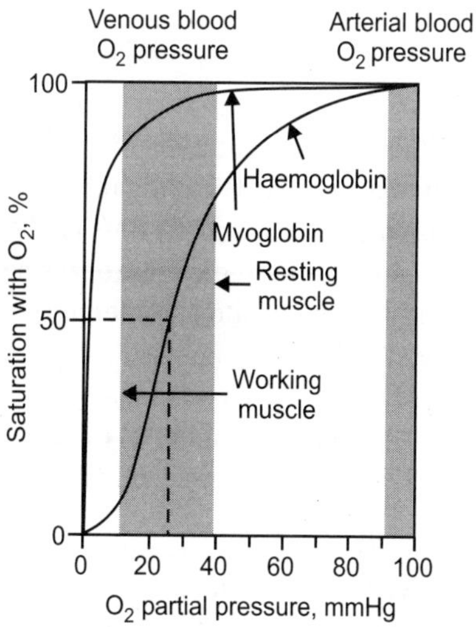

Fig. 3.13. Comparison of the oxygen binding curves for myoglobin and haemoglobin at different [oxygen].

In the lungs where oxygen is abundant and it is transferred to haemoglobin for transport to the tissue where it will be used as an electron acceptor, haemoglobin becomes nearly saturated with oxygen (shown as 'arterial pressure' for oxygen in the figure above). Then as the blood is pumped from the lungs to the more remote tissues of the body, the haemoglobin carries the oxygen into an environment low in oxygen and will release the bound oxygen (shown as 'venous pressure' in the Fig. 3.13). It can also be seen from the Fig. 3.13 that if myoglobin were used as the oxygen transport protein in blood, only about 10 per cent of the oxygen would be released from the binding sites in transporting this protein from the conditions of the lung to the outer tissues (compare the degree of oxygen saturation of myoglobin and haemoglobin under venous and arterial pressures), thus, it is the tetrameric structure of haemoglobin that accounts for this difference and the positive cooperative in loading oxygen as shown by sigmoidal curve for it as compared to the Michaelis-Menten type of oxygen loading curve for myoglobin. Thus, the myoglobin–haemoglobin comparison represents what one would find for comparison of a monomeric M-M type of enzyme and a multimeric enzyme with positive cooperatively response to substrate binding in catalysing its reaction.

Feedback regulation and covalent modification

Another type of allosteric regulation is via feedback inhibition (Fig. 3.14).

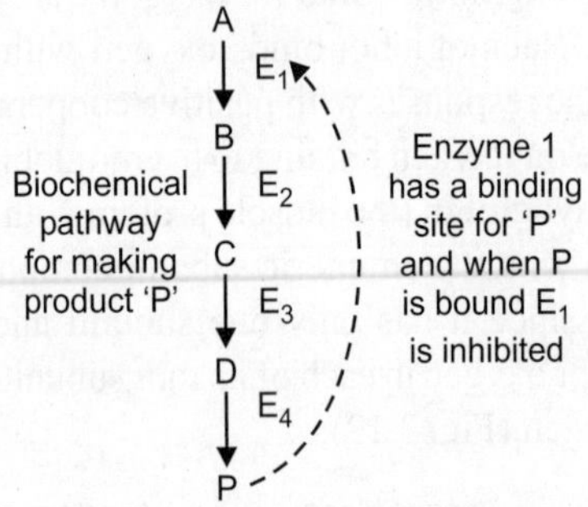

Fig. 3.14. Model of feedback inhibition using a metabolic pathway to illustrate this concept.

The biosynthesis of amino acids is often regulated by feedback inhibition (in those organisms that make amino acids since not all do). When the concentration of an amino acid is high, it inhibits its own biosynthesis until the level of the amino acid falls. Then, the inhibition is relieved by the consumption of the amino acid and its production is enhanced. Basically, this is like allosteric modifier action which increases the activity of the enzyme but in this case the modifier is an inhibitor. The allosteric site for the feedback inhibitor is separate from the active site and the inhibitor acts via the framework of the enzyme to alter the shape of the active site. This is of course related to noncompetitive inhibition of the enzyme, but we discussed only the case where the enzyme loses activity but can still bind its substrate normally, which we called a classic noncompetitive inhibitor. In reality most noncompetitive inhibitors impact both the V_{max} and K_m of the enzyme and this is generally the case for a feedback inhibitor.

Covalent modification for enzyme activity regulation

A more definitive control can be achieved by covalent modification of an enzyme and is used to switch an organism's cells from one metabolic state to another. For example, hormones like insulin can switch a liver cell from degrading and storing glucose to synthesising it and exporting it to the blood. This is done by turning on enzymes via covalent modification (Fig. 3.15).

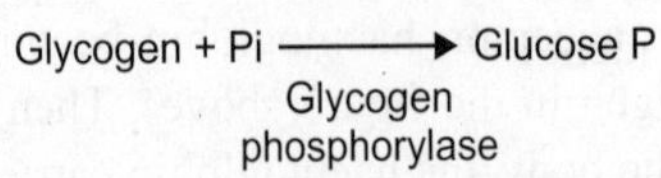

Fig. 3.15. Glycogen phosphorylase catalyses mobilisation of stored carbohydrate reserves.

Phosphorylase exists in two forms 'A' and 'B'. Form A is active phosphorylated 'A'-P. Form B is inactive and not phosphorylated. The phosphate on form A is present as a phosphate ester (Fig. 3.16).

The ser residue which is phosphorylated is remote from the active site of phosphorylase and this mechanism for controlling enzyme activity also works (in most cases) by altering the shape of the enzyme by covalent modification. This modification system is run by two enzymes:

1. 'Protein kinase' catalyses: 'B' + ATP — 'A'-P + ADP
   The protein kinase converts the inactive B form of phosphorylase into the active A form.

2. 'Protein phosphatase' catalyses: 'A'-P + H_2O — 'B' + Pi (inorganic phosphate). The protein phosphatase converts 'A'-P back into 'B' by catalysing hydrolysis of the phosphate ester.

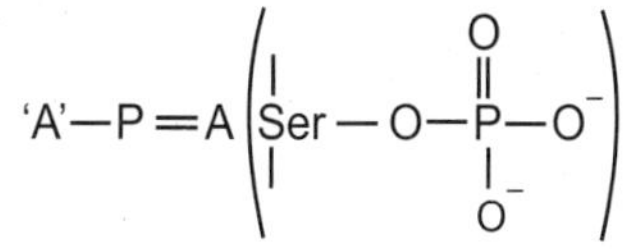

Fig. 3.16. Structure of serine-phosphate. The *ser* residue is part of the polypeptide chain of phosphorylase and is converted to the phosphorylated form after the enzyme is synthesised on the ribosomes. So this is called post-translational modification.

Thus, to change the metabolic state or mode of the cell, the protein kinase is turned on and the protein phosphatase is turned off. The hormone controls the activity of these controlling or regulatory enzymes and this involves several other proteins and receptors.

VECTORS

A vector is a DNA molecule that has the ability to replicate in an appropriate host cell, and into which the DNA fragment to be cloned (called DNA insert) is integrated for cloning. Therefore, a vector must have an origin of DNA replication (denoted as *ori*) that functions in the host cell. Any extrachromosomal small genome, e.g. plasmid, phage and virus, may be used as a vector.

Properties of a Good Vector

A good vector must have the following properties.

1. It should be able to replicate autonomously. When the objective of cloning is to obtain a large number of copies of the DNA insert, the vector replicon must be under relaxed control so that it can generate multiple copies of itself in a single host cell.
2. It should be easy to isolate and purify.
3. It should be easily introduced into the host cells, i.e. transformation of the host with the vector should be easy.
4. The vector should have suitable marker genes that allow easy detection and/or selection of the transformed host cells.
5. When the objective is gene transfer, it should have the ability to integrate either itself or the DNA insert it carries into the genome of the host cell.
6. The cells transformed with the vector molecules containing the DNA insert (recombinant or chimaeric vector) should be identifiable or selectable from those transformed by the vector molecules only.
7. A vector should contain unique target sites for as may restriction enzymes as possible into which the DNA insert can be integrated without disrupting an essential function.
8. When expression of the DNA insert is desired, the vector should contain at least suitable control elements, e.g. promoter, operator and ribosome binding sites; several other features may also be important.

It should be kept in mind that the DNA molecules used as vectors have coevolved with their specific natural host species, and hence are adapted to function well in them and in their closely related species. Therefore, the choice of vector depends largely on the host species into which the DNA insert or gene is to be cloned.

In addition, most naturally occurring vectors do not have all the required functions; therefore, useful vectors have been created by joining together segments performing specific functions (called modules) from two or more natural entities. A brief description of some of the important vectors used in different hosts is given below.

Cloning and Expression Vectors

All vectors used for propagation of DNA inserts in a suitable host are called cloning vectors. But when a vector is designed for the expression of, i.e. production of the protein specified by the DNA insert, it is termed as expression vector. As a rule, such vectors contain at least the regulatory sequences, i.e. promoters, operators, ribosomal binding sites, etc. having optimum function in the chosen host. It is desirable that all cloning vectors have relaxed replication control so that they can produce multiple copies per host cell.

When an eukaryotic gene to be expressed in a prokaryote, the eukaryotic coding sequence has to be placed after prokaryotic promoter and ribosome building site since the regulatory sequences of eukaryotic are not recognised in prokaryotes. In addition, eukaryotic genes, as a rule, contain introns (noncoding regions) present within their coding regions. These introns must be removed to enable the proper expression of eukaryotic genes since prokaryotes lack the machinery needed for their removal from the RNA transcripts. When eukaryotic genes are isolated as cDNA they are intron-free and, hence, suitable for expression in prokaryotes.

Several strategies have been attempted for the construction of expression vectors using regulatory sequences of the appropriate hosts. These approaches may be grouped into the following two broad categories.

1. Construction of vectors allowing the synthesis of fusion proteins comprising amino acids coded by a sequence in the vector and those encoded by the DNA insert (translational fusion).
2. Development of vectors permitting the synthesis of pure proteins encoded exclusively by the DNA inserts (transcriptional fusion).

Examples of the first strategy producing fusion proteins are: the expression of rat insulin, rat growth hormone, structural protein VP1 of foot and mouth disease virus, human growth hormone, etc. Some examples of the second approach producing unique proteins are: rabbit β-globin, small *t*-antigen of SV 40, human fibroblast interferon, human IGF-I protein. It may be pointed out that the undesired amino acids encoded by the vector sequence in cases of translational fusion must be removed from the fusion proteins by a suitable chemical cleavage.

Several other problems are faced when eukaryotic genes are expressed in a prokaryotic system, e.g. removal of signal sequences from precursor proteins to obtain active mature protein molecules. Various strategies are being rapidly devised to effectively overcome these problems.

E. coli Vectors

Bacteria are the hosts of choice for DNA cloning. Among them, *E. coli* occupies a prominent position since cloning and isolating DNA inserts for structural analysis is the easiest in this host. Therefore, the initial cloning experiments are generally carried out in *E. coli*. The *E. coli* strain K12 is the most commonly used; it has several substrains, e.g. C 600, RRI, HB 101, etc. each of which has some specific features important in cloning. For example, the substrain RRI has, in addition to certain other features the mutation *hsdR* which inactivates the restriction enzyme endogenous to *E. coli* K12; this minimises the degradation of recombinant DNA introduced into it.

Properties of a good host

A good host should have the following features: (i) is easy to transform, (ii) supports the replication of recombinant DNA, (iii) is free from elements that interfere with replication of recombinant DNA, (iv) lacks active restriction enzymes, e.g. *E. coli* K12 substrain HB 101, (v) does not have methylases since these enzymes would methylate the replicated recombinant DNA which, as a result, would become resistant to useful restriction enzymes, and (vi) is deficient in normal recombination function so that the DNA insert is not altered by recombination events.

E. coli supports several types of vectors, so natural, some constructed, which can be grouped as follows: (i) plasmids, (ii) bacteriophages (both natural), (iii) cosmids, (iv) plasmids, and (v) shuttle vectors (the last three constructed by man). A brief description of these vectors follows.

Plasmids

A plasmid is a DNA molecule, other than the bacterial chromosome, that is capable of independent replication and transmission. Plasmids are circular and may exist either independent of or may become integrated into the bacterial chromosome; generally they are not essential for the host cell except under specific environments. There are several types of bacterial plasmids, but the three widely studied types are: (i) *F* plasmids (responsible for conjugation), (ii) *R* plasmids (carry genes for resistance to antibiotics), and (iii) *Col* plasmids (code for colicins, the proteins that kill sensitive *E. coli* cells; they also carry genes that provide immunity to the particular colicin). The plasmids may either be conjugative or transmissible (mediate DNA transfer through conjugation, and as a result spread rapidly among the bacterial cells of a population), e.g. *F* plasmids, many *R* plasmids and some *Col* plasmids, or nonconjugative (do not mediate DNA transfer through conjugation), e.g. many *R* plasmids and most *Col* plasmids.

Stringent and relaxed replication

Each plasmid is maintained in the bacterial cell at a characteristic copy number mainly due to its replication control system. In this respect the plasmids are of two types: (i) single copy, and (ii) multicopy plasmids. The replication control of single copy plasmids is the same as that of their bacterial host cells so that they replicate and segregate with the bacterial chromosome; this is called stringent replication. In contrast, the replication control of multicopy plasmids is different from that of their bacterial host genome so that they undergo more than one replication for each replication of their host genome; this is referred to as relaxed replication.

Modular organisation

Plasmids may be visualised as constructed from modular DNA segments. A module may be regarded as a DNA segment or sequence performing a specific function; each module may contain one or more genes. The various modules present in the different plasmids are: (i) replication module (essential for all plasmids; it functions only in the natural host or closely related species), (ii) sex factors (found in conjugative plasmids; represented by *tra* genes in *F* factor and RTF, resistance transfer factor, in *R* plasmids), (iii) *R*-determinant module (specific to *R* plasmids; genes contained in it produce proteins that inactivate antibiotics), (iv) *Col* module (contains genes for colicins, antibacterial agents produced by bacteria), (v) modules specifying restriction modification systems, e.g. *Eco*RI system, and (vi) IS (insertion sequences) elements which enable transposition (movement from one DNA molecule into another) of the modules flanked by them, e.g. *R*-determinant modules of many *R* plasmids. Each plasmid must have the replication module; in addition it may or may not contain one or more of the other modules.

Plasmid Vectors

Many different *E. coli* plasmids are used as vectors. The natural plasmids have been modified, shortened, reconstructed and recombined both *in vitro* and *in vivo* to create plasmids of enhanced utility and also with specific functions; the best ones include genes (Table 3.3) that allow easy selection of the recombinant or chimaeric vectors (vectors containing DNA inserts). Some *E. coli* plasmid vectors are briefly described below.

Table 3.3. Antibiotic resistance genes found in *R* plasmids, their proteins and mechanism of antibiotic resistance.

Antibiotic (gene conferring resistance)	*Protein produced by the gene*	*Mechanism of resistance*
Ampicillin (*amp*)	Penicillinase or β-lactamase	Hydrolysis of C-N bond in β-lactam ring
Kanamycin (*kan*)	Kanamycin acetyltransferase*	*N*-acetylation of the antibiotic
Neomycin (*neo*)	Aminoglycoside phosphotransferase*	O-phosphorylation of the antibiotic
Streptomycin (*str*)	Streptomycin phosphotransferase	Phosphorylation of OH on the antibiotic
	Streptomycin adenylate synthetase	Adenylation of the OH on the antibiotic

* The antibiotics kanamycin and neomycin are related; hence *neo* product also inactivates kanamycin, and *kan* product inactivates neomycin as well.

Selection of recombinant vectors

It may be pointed out that when an experiment is performed to insert a DNA fragment into a vector, two types of vector molecules are obtained: (i) many vector molecules will contain the DNA insert (recombinant or chimaeric vector), but (ii) many others will contain only the vector sequences (unaltered vector, or simply vector). This mixture of vector molecules is used for transformation of host cells: (i) some host cells will receive the recombinant vector, (ii) some others will contain the normal unaltered vector, and while (iii) the majority of them will contain no vector, i.e. will not be transformed. In a cloning experiment it is critical to effectively select for the low frequency of cells transformed by the recombinant vector from among the cells containing the unaltered vector and the nontransformed cells.

Selection of host cells transformed by the recombinant vector is easily achieved by placing two selectable markers, e.g. antibiotic resistance genes, such as, ampicillin resistance (*amp*^{r}) and tetracycline resistance (*tet*^{r}), in the vector. The DNA insert is integrated within one of the two selectable markers. If the DNA insert is integrated within the ampicillin resistance gene, the cells containing the recombinant vector will be resistant to tetracycline but sensitive to ampicillin. In contrast, nontransformed cells will be sensitive to both the antibiotics, while those containing the unaltered vector will be resistant to both. Therefore, following transformation with the above recombinant vector, cells are plated on a tetracycline supplemented medium; this eliminates the nontransformed cells.

The remaining colonies are now replica-plated on ampicillin containing medium to identify sensitive colonies; these colonies contain the recombinant vector, and are isolated from the master plate. Further, transformed cells tend to lose the recombinant vector, since cells lacking such vectors divide much faster. The use of a vector having two selectable markers allows the maintenance of cells containing the recombinant vector on antibiotic medium (tetracycline in the above case) which eliminates the vector-free cells produced culturing culture.

Bacteriophage Vectors

Bacteriophages are viruses that attack bacteria. Most phages lyse the bacterial cells they infect (lytic phages). But many others can choose to follow either a lytic or a lysogenic cycle; in the latter situation, the phage chromosome integrates into the bacterial chromosome and multiplies with the latter as prophage (temperate or lysogenic phages). The prophage may dissociate from the bacterial chromosome and follow the lytic cycle.

Several bacteriophage are used as cloning vectors, the most commonly used *E. coli* phages being λ (lambda) and M13 phages. Plasmid vectors have to be, introduced into bacterial cells which are then cloned and selected for the recovery of recombinant vectors. In contrast, the phage vectors are directly tested on an appropriate bacterial lawn (a continuous bacterial growth on an agar plate) where each phage particle forms a plaque (a clear bacteria-free zone in the bacterial lawn). Phage vectors present two advantages over plasmid vectors: (i) they are more efficient than plasmids for cloning of large DNA fragments; the largest cloned insert size in a λ vector is just over 24 kb, while that for plasmid vectors it is less than 15 kb. In addition, (ii) it is easier to screen a large number of phage plaques than bacterial colonies for the identification of recombinant vectors.

Cosmid Vectors

Cosmids are essentially plasmids that contain a minimum of 250 bp of λ. DNA which includes: (i) the *cos* site (the sequence yielding cohesive ends), and (ii) sequences needed for binding of and cleavage by terminase so that under appropriate conditions they are packaged *in vitro* into empty λ phage particles. A typical cosmid has: (i) replication origin, (ii) unique restriction sites, and (iii) selectable markers from the plasmid (Fig. 3.17); therefore selection strategy for obtaining the recombinant vectors is based on that for the contributing plasmid. Cosmid vectors are constructed using recombinant DNA techniques.

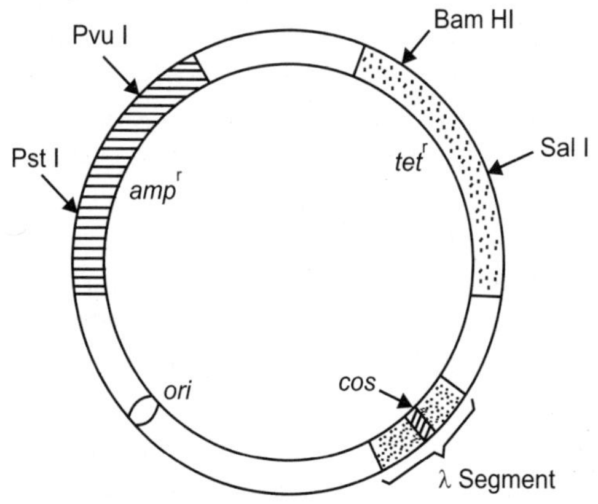

Fig. 3.17. A typical cosmid vector, containing *p*BR 322 modules and a λ segment having the *cos* site, and terminase binding and cleavage sequences.

The cosmid vectors are opened by the appropriate restriction enzyme at a unique site, are then mixed with DNA inserts prepared by using the same enzyme and annealed. Among the several types of products, long cancatemers are present which are the appropriate precursors for packaging in λ particles. This procedure selects for long DNA inserts since for packaging the distance between two *cos* sites must be between 38 and 52 kb. Cosmids can accommodate upto 45 bp long DNA inserts. Packaged cosmids infect host cells like λ particles, but once inside the host they replicate and propagate like plasmids.

The typical features of cosmids are as follows: (i) they can be used to clone DNA inserts of upto 45 kb, (ii) they can be packaged into λ particles which infect host cells, which is many-fold more efficient than plasmid transformation, (iii) selection for recombinant vector is based on the procedure applicable to the plasmid making up the cosmid, and (iv) finally, these vectors are amplified and maintained in the same manner as the contributing plasmid.

Phasmid Vectors

These vectors are shortened linear λ genomes containing DNA replication and lytic function plus the cohesive ends of the phage; their middle nonessential segment is replaced by a linearised plasmid with intact replication module. In practice, a phasmid vector contains several tandem copies of the plasmid to make it longer than 38 kb, the minimum size needed for packaging in λ particles. During construction of the recombinant DNA, one or more copies of the plasmid are deleted from and the DNA insert is integrated into the vector, but generally one copy of the plasmid is retained in the recombinant vector. Phasmids, both recombinant and unaltered, are packaged in λ particles *in vitro* and used for infection of appropriate *E. coli* cells.

If a phasmid lacks the λ gene *cI*, which produces the lysis repressor, it multiplies like a phage and produces plaques on a bacterial lawn. But if *cI* gene is present, the phasmid replicates like a plasmid. Further a phasmid may contain a mutant *cI* gene which produces a temperature sensitive *CI* protein (inactive at higher temperatures); such vectors replicate as plasmids at lower temperatures but behave like phage at higher temperatures. This feature is quite useful in some experiments.

Vectors for Other Bacteria

E. coli and its vector systems are the most widely used for obtaining large quantities of DNA inserts. But in view of medical, industrial or scientific importance of other bacteria, vectors have been developed for them as well. Vectors derived from *R* plasmids can propagate in many gram-negative bacteria. A plasmid vector has been developed for *Hemophilus influenzae*, which is naturally competent for transformation. Useful vectors for *Klebsiella pneumoniae* have been developed from the small phage P_4; these vectors call replicate either as a phage or as a plasmid depending on their environment. Both *Hemophilus* and *Klebsiella* are Gram-negative bacteria.

Among gram-positive bacteria *Bacillus* and *Streptomyces* are of special interest. *Bacillus* is widely used in commercial fermentation. More importantly, it secretes proteins into medium making the recovery of proteins encoded by DNA inserts rather easy. Both plasmid and phage vectors have been devised for these bacteria. A useful plasmid vector (5 kb) is derived from a *Staphylococcus aureus* plasmid. This vector has the gene for kanamycin resistance and a *B. subtilis* gene needed for tryptophan biosynthesis; unique restriction sites are available in both the genes. The selection strategy for recombinant vectors, therefore, is analogous to that for the *E. coli* UC series vectors. This vector has a relaxed replication control producing 50 copies per *Bacillus* cell.

Bacillus species are much easier to transform than *E. coli*, but the efficiency is markedly increased when *Bacillus* protoplasts (produced by treating cells with lysozyme) are used; cell wall is regenerated in a special nutrient medium.

Streptomyces species produce most of the antibiotics. The best developed vector system for these bacteria is based on plasmids. These vectors have either stringent or relaxed replication control, and some of them are self-transmissible by conjugation. A typical plasmid vector contains: (ii) 11 kb DNA from a plasmid isolated from *S. lividans*, and a gene for resistance to methylenomycin A from the chromosome of *S. coelicolor*. The plasmid has a growth inhibition gene which confers on the plasmid bearing bacterial cells the ability to inhibit the growth of those lacking this plasmid. Protoplasts of *Streptomyces* are used for transformation; cell walls regenerate in the standard medium. Transformed cells are placed on a lawn of plasmid-free cells; transformed clones are picked from the centre of regions in which lawn growth is inhibited.

Shuttle Vectors

These vectors have been designed to replicate in cells of two different species; therefore, they contain two origins of replication, one specific for each host species as well as those genes necessary for their replication and not provided by the host cells. These vectors are created by recombinant techniques. Some of them can be grown in two different prokaryotic species, while others can propagate in a prokaryotic species, usually *E. coli*, and a eukaryotic one, e.g. yeast, plants, animals. Since these vectors can be grown in one host and then moved into another without any extra manipulation, they are called shuttle vectors.

A shuttle vector designed to replicate in *E. coli* and *Streptomyces* has been constructed as follows: (i) the modules for DNA replication in *Streptomyces* and methylenomycin A resistance are derived from a streptomyces plasmid, and (ii) the replication module for maintenance in *E. coli* and a gene for antibiotic resistance are taken from an *E. coli* plasmid. This shuttle vector allows the initial cloning of *Streptomyces* DNA inserts in *E. coli* and their subsequent functional tests in *Streptomyces*. Shuttle vectors have been designed to specifically satisfy this need, i.e. the initial cloning of DNA inserts in *E. coli* and subsequent functional tests in the species to which the DNA inserts belong. Most of the eukaryotic vectors are, in fact shuttle vectors.

Yeast Vectors

Yeast, *Saccharomyces cerevisiae*, is an eukaryote with 34 ($2n$) chromosomes; it reproduces sexually as well as asexually by budding. In suspension cultures it grows as single cells with cell doubling time of 1.5 to 2.5 hours; but on agar plate cells produce colonies. Its haploid DNA content (1.4×10^7 bp) is only three times that of *E. coli*, and its genetics has been extensively studied. Yeast viruses are not known. Only a single yeast plasmid has been discovered which has been used to construct some useful vectors. The tough polysaccharide wall of yeast is an effective barrier to DNA molecules. Therefore, yeast cell wall is enzymatically digested to produce spheroplasts which can take up DNA following treatment with $CaCl_2$; walls regenerate in specific media.

The different vectors used in yeast may be grouped as follows: (i) plasmid vectors, (ii) ARS vectors, (iii) micro-chromosome vectors, and (iv) YAC vectors.

SELECTION OF RECOMBINANT CLONES

When recombinant vector is constructed and used for transformation of *E. coli*, cells, following types of bacterial cells are obtained: (i) majority of the cells are nontransformed, (ii) a proportion of the transformed cells contain unaltered vector, and while (iii) the remainder cells have recombinant vector. The first objective of cloning experiments is to identify and isolate those small number of cells that contain the recombinant vector from among a very large number of nontransformed cells. Since the DNA inserts are generally mixtures, particularly when cDNA preparations and genomic DNA fragments are used, the various transformed clones would contain a variety of different DNA inserts. The next step, therefore, is to identify the clone having the desired DNA insert from among the large number of clones containing the recombinant vectors. Suitable selection strategies have been devised to achieve these two critical objectives; this is the most important step in DNA cloning.

Colony Hybridisation

The most efficient and rapid strategy for identification of a clone having the desired insert uses the technique of colony hybridisation. The bacterial colonies are replica-plated or phage plaques are directly lifted on nitrocellulose filters, the cells are lysed and their DNA is denatured: the filter is incubated with the specific radioactive (^{32}P-labelled) probe under annealing conditions. After some time, the probe is washed out leaving only those probe molecules that have hybridised with the denatured DNA from bacterial cells or phage particles. The colonies/plaques with whose DNA the probe has hybridised are identified by autoradiography; these contain the desired DNA insert. These colonies/plaques are isolated from the master plate used for replica plating.

A very large number of colonies or plaques (upto 10,000 plaques) can be lifted on to a single 10 cm diameter filter. But it is essential that a specific probe for the DNA insert is available. A probe is a polynucleotide (DNA or RNA; usually small molecules of as few as 15 bases, but more often of 25–30 bases) molecule of a specific base sequence which is used to detect DNA molecules having the same base sequence by complementary base pairing. Generally, the probes are labelled with ^{32}P to enable autoradiography for an easy identification of the DNA samples that base pair with the probe. It is desirable that the probes are single-stranded to avoid pairing between the two strands of the probe itself. Either DNA or RNA can be used as probe; when RNA is used, it is of necessity single-stranded. There are several approaches for developing specific probes.

Other Approaches

When specific probes are not available, many indirect approaches may used for the identification of clones having the desired DNA insert. These procedures are not generally convenient for screening of a large number of clones. Two of such procedures, called: (i) hybrid arrested translation (HART), and (ii) hybrid selection, use *in vitro* translation systems and the identification of resulting polypeptide(s). It is, therefore, necessary that the protein product of the DNA insert being searched should be known, at least in terms of its electrophoretic mobility.

Chapter 4

Substrate for Industrial Fermentation

INTRODUCTION

The fermentation medium forms the environment in which the fermentation micro-organisms live, reproduce, and carry out their specific metabolic reactions to produce useful products. The importance of this environment cannot be overemphasised when it comes to the development of a productive fermentation process. Over the years, substantial progress has been made in developing fermentation medium design as a systematic science. However, experienced industrial microbiologists and biochemical engineers will be the first to point out that this field is as much an art as it is a science. In most industrial fermentations, where the product is something other than the cell mass itself, there are two distinct biological requirements for medium design. First, nutrients have to be supplied to establish the growth of the organism. Second, after growth is established, proper nutritional conditions have to be provided to maximise product formation. Besides these obvious biological requirements, one needs to worry about selection of nutrient components that are cost-effective, readily available, and consistent from lot to lot. In recent years, as integrated approaches to fermentation and downstream processing have been developed, it has also been recognised that the fermentation medium should not unduly hinder the downstream processing and, if possible, should even facilitate downstream processing. For new fermentation processes brought up from microbiology laboratories, considerable flexibility and latitude in medium design are possible. The process is not locked into a fixed set of raw materials (for example, due to a food and drug administration [FDA] filing), and the medium components can be freely selected for the sole purpose of maximising the product yield and minimising the cost. For an established fermentation process, the choice of medium components may be limited by such factors as FDA filing, the cost structure for the product, and the requirements of downstream processing. In spite of these limitations, continued medium development remains a necessity so that an established product retains its competitive edge in the marketplace.

This chapter focuses primarily on raw materials and medium development for microbial fermentation processes. Although general principles also apply to it, mammalian cell culture will not be emphasised. The chapter is designed to provide practicing microbiologists and biochemical engineers with a rational basis for medium development and improvement. At the start of the chapter, a chemically defined fermentation medium is considered with its pros and cons. Then various commercially available ingredients for key nutrient components of traditional complex fermentation media are described in generic terms. This is followed by a discussion of general considerations and a set of guidelines for medium development and improvement.

CHEMICALLY DEFINED FERMENTATION MEDIA

In industrial fermentation, it is very rare that a chemically defined medium is used. Media derived from complex ingredients such as flours and by-products of the brewery, meat, and corn-milling industries are common. Complex media usually give higher fermentation yields at a lower cost. There are rare cases, however, when downstream processing considerations may prohibit the use of complex raw materials. For example, when peptide products are made by cell culture processes, the use of serum in the medium may make the downstream processing more difficult. In recombinant *Escherichia coli* fermentations for production of biological proteins, the use of complex ingredients can introduce a myriad of proteins and polypeptides that can interfere with the recovery process. In many such cases, for reasons of growth and productivity, a small amount of complex nutrients such as yeast extract or bovine serum is still used. While not considered the ideal medium formulation, the completely defined medium can be designed to support most organisms to at least some extent. A completely defined medium is indeed the medium of choice for studying the metabolic pathways of a fermentation process, for example, when determining substrate uptake kinetics or when studying nutritional control and nutrient requirements for product formation. The metabolic information obtained with a defined medium can then be used to understand and further improve the industrial process involving complex ingredients. Typically, the design of a chemically defined medium is based on the elemental composition of the micro-organism being cultivated. Since the elemental compositions of organisms differ according to the type of organism and the growth conditions, excess nutrients are generally used. The cell density is then controlled by using a limiting nutrient such as glucose or ammonia. When glucose is used as the limiting nutrient, it can be assumed that the cell yield on a dry weight basis will be about 50 per cent of the glucose consumed. As a rule of thumb, to calculate the nitrogen requirement of a nitrogen-limited culture, it can be assumed that 10 per cent of the dry weight of the organism will be nitrogen. Table 4.1 illustrates the typical components and their proportions used in a glucose-limited defined fermentation medium when a cell density of 10 g (dry cell weight) per liter is desired.

Table 4.1. Components of a chemically defined fermentation medium needed to obtain about 10 g of dry cell weight per litre.

Source of	*Typical ingredient*[a]	*Concn (g/litre)*
Carbon	Glucose	20
	Sucrose	20
	Glycerol	20
Nitrogen	$(NH_4)_2SO_4$	5
	$NaNO_3$	7
	NH_4NO_3	3
	Alanine or other amino acids	7
Phosphorus	KH_2PO_4	1
	K_2HPO_4	1
Sulphur	K_2SO_4	0.4
	$MgSO_4 \cdot 7H_2O$	0.5
	Methionine	0.3

(Contd ...)

Source of	*Typical ingredient*[a]	*Concn (g/litre)*
Metals		
Mg	$MgSO_4 \cdot 7H_2O$	0.1
K	K_2SO_4	0.1
Ca	$CaCl_2$	0.05
Fe	$FeSO_4 \cdot 7H_2O$	0.001
Zn	$ZnSO_4 \cdot 7H_2O$	0.001
Cu	$CuSO_4 \cdot 5H_2O$	0.0004
Mn	$MnSO_4 \cdot H_2O$	0.0004

[a] Typically, one component from each grouping, is used based on the substrate preference of the organism being cultivated.

It should be noted that a medium formulation based on the information in Table 4.1 will not give the desired results with all cell types. In addition to these basic medium components, individual cell types may require for growth other unique medium ingredients such as amino acids, vitamins, purines, pyrimidines, additional metal ions, and chelating agents. Furthermore, the medium given in Table 4.1 does not take into account the additional nutrients and other unique components required for successful synthesis of the desired product. If the metabolic processes involved in product formation are reasonably known, the nutrients needed for product synthesis can be calculated from the stoichiometry and yield factors and included in the medium formulation.

One of the most important aspects of medium design that is often overlooked is the pH balance of the medium. In complex fermentation media, the natural amino acids, peptides, and other organic compounds provide pH buffering. In a chemically defined medium, inorganic buffering agents such as phosphates, organic acids, and carbonates have to be used or provision has to be made for external addition of acid or base to control the pH during the fermentation process.

COMPONENTS OF INDUSTRIAL FERMENTATION MEDIA

As noted above, most industrial fermentation media are complex formulations containing poorly defined ingredients. Often these ingredients contain multiple nutrients for the growth of fermentation micro-organisms. However, for the purposes of medium development, a given ingredient is thought to provide primarily a single nutrient. For example, soya flour is used primarily to supply complex nitrogen or protein for the growth of an organism. However, soya flour also contains substantial amounts of metabolisable carbohydrates and minerals. In the discussion below, the medium ingredients are classified according to their primary role in the fermentation process. On this basis, we can classify the fermentation raw materials in four broad nutrient categories: materials used primarily as sources of carbon, nitrogen, or minerals, and materials used for special purposes.

Carbon Sources

Factors influencing the choice of carbon source

It is now recognised that the rate at which the carbon source is metabolised can often influence the formation of biomass or production of primary or secondary metabolites. Fast growth due to high concentrations of rapidly metabolised sugars is often associated with low productivity of secondary metabolites. This has been demonstrated for a number of processes (Table 4.2). At one time the problem was overcome by using the less readily metabolised sugars such as lactose, but many processes now use semi-continuous or continuous feed of glucose or sucrose. Alternatively, carbon catabolite regulation might be overcome by genetic modification of the producer organism.

Table 4.2. Carbon catabolite regulation of metabolite biosynthesis.

Metabolite	*Micro-organism*	*Interfering carbon source*
Griseofulvin	*Penicillium griseofulvin*	Glucose
Penicillin	*P. chrysogenum*	Glucose
Cephalosporin	*Cephalosporium acremonium*	Glucose
Aurantin	*Bacillus aurantinus*	Glycerol
α-Amylase	*B. licheniformis*	Glucose
Bacitracin	*B. licheniformis*	Glucose
Puromycin	*Streptomyces alboniger*	Glucose
Actinomycin	*S. antibioticus*	Glucose
Cephamycin C	*S. clavuligerus*	Glycerol
Neomycin	*S. fradiae*	Glucose
Cycloserine	*S. graphalus*	Glycerol
Streptomycin	*S. griseus*	Glucose
Kanamycin	*S. kanamyceticus*	Glucose
Novobiocin	*S. niveus*	Citrate
Siomycin	*S. sioyaensis*	Glucose

The main product of a fermentation process will often determine the choice of carbon source, particularly if the product results from the direct dissimilation of it. In fermentations such as ethanol or single-cell protein production where raw materials are 60 to 77 per cent of the production cost, the selling price of the product will be determined largely by the cost of the carbon source. It is often part of a company development programme to test a range of alternative carbon sources to determine the yield of product and its influence on the process and the cost of producing biomass and/or metabolite. This enables a company to use alternative substrates, depending on price and availability in different locations and remain competitive.

The purity of the carbon source may also affect the choice of substrate. For example, metallic ions must be removed from carbohydrate sources used in some citric acid processes.

The method of media preparation, particularly sterilisation, may affect the suitability of carbohydrates for individual fermentation processes. It is often best to sterilise sugars separately because they may react with ammonium ions and amino acids to form black nitrogen containing compounds which will partially inhibit the growth of many micro-organisms. Starch suffers from the handicap that when heated in the sterilisation process it gelatinises, giving rise to very viscous liquids, so that only concentrations of up to 2 per cent can be used without modification.

The choice of substrate may also be influenced by government legislation. Within the European Economic Community (EEC), the use of beet sugar and molasses is encouraged and the minimum price controlled. The quantity of imported cane sugar and molasses is carefully monitored and their imported prices set so that they will not be competitive with beet sugar. If the world market sugar price is very low then the EEC fermentation industry will be at a disadvantage unless it receives realistic subsidies.

Refunds for a defined list of products are available in the EEC when sugar and starch are used as substrates. Legislation for recognition of new products is time consuming and manufacturers may be uncertain as to whether they would benefit from carbon substrate refunds. This uncertainty has meant that some manufacturers might prefer to site factories for new products outside the EEC.

Local laws may also dictate the substrates which may be used to make a number of beverages. There are similar laws applying to beer production in Germany. Scotch malt whisky may be made only from barley malt, water and yeast. Within France, many wines may be called by a certain name only if the producing vineyard is within a limited geographical locality.

Examples of commonly used carbon sources

Carbohydrates

It is common practice to use carbohydrates as the carbon source in microbial fermentation processes. The most widely available carbohydrates is starch obtained from maize grain. It is also obtained from other cereals, potatoes and cassava. Maize and other cereals may also be used directly in a partially ground state, e.g., maize chips. Starch may also be readily hydrolysed by dilute acids and enzymes to give a variety of glucose preparations (solids and syrups). Hydrolysed cassava starch is used as a major carbon source for glutamic acid production in Japan. Syrups produced by acid hydrolysis may also contain toxic products which may make them unsuitable for particular processes.

Barley grains may be partially germinated and heat treated to give the material known as malt, which contains a variety of sugars besides starch (Table 4.3). Malt is the main substrate for brewing beer and lager in many countries. Malt extracts may be prepared from malted grain.

Table 4.3. Carbohydrate composition of barley malt (expressed as % of dry weight of total).

Starch	58–60
Sucrose	3–5
Reducing sugars	3–4
Other sugars	2
Hemicellulose	6–8
Cellulose	5

Sucrose is obtained from sugarcane and sugarbeet. It is commonly used in fermentation media in a very impure form as beet or cane molasses (Table 4.4), which are the residues left after crystallisation of sugar solutions in sugar refining. Molasses is used in the production of high-volume/low-value products such as ethanol, SCP, organic and amino acids and some microbial gums. Molasses or sucrose also may be used for production of higher value/low-bulk products such as antibiotics, speciality enzymes, vaccines and fine chemicals. The cost of molasses will be very competitive when compared with pure carbohydrates. However, molasses contains many impurities and molasses-based fermentations will often need a more expensive and complicated extraction/purification stage to remove the impurities and effluent treatment will be more expensive because of the unutilised waste materials which are still present. Some new processes may require critical evaluation before the final decision is made to use molasses as the main carbon substrate.

Table 4.4. Analysis of beet and cane molasses (expressed as % of total w/v).

	Beet	*Cane*
Sucrose	48.5	33.4
Raffinose	1.0	0
Invert sugar	1.0	21.2

The use of lactose and crude lactose (milk whey powder) in media formulations is now extremely limited since the introduction of continuous-feeding processes utilising glucose, discussed in a later section of this chapter.

Corn steep liquor (Table 4.5) is a by-product after starch extraction from maize. Although primarily used as a nitrogen source, it does contain lactic acid, small amounts of reducing sugars and complex polysaccharides. Certain other materials of plant origin, usually included as nitrogen sources, such as soyabean meal and pharmamedia, contain small but significant amounts of carbohydrate.

Table 4.5. Partial analysis of corn-steep liquor.

Total solids	51% w/v
Acidity as lactic acid	15% w/v
Free reducing sugars	5.6% w/v
Free reducing sugars after hydrolysis	6.8% w/v
Total nitrogen	4% w/v
Amino acids as % of nitrogen	
Alanine	25
Arginine	8
Glutamic acid	8
Leucine	6
Proline	5
Isoleucine	3.5
Threonine	3.5
Valine	3.5
Phenylalanine	2.0
Methionine	1.0
Cystine	1.0
Ash	1.25% w/v
Potassium	20%
Phosphorus	1–5%
Sodium	0.3–1%
Magnesium	0.003–0.3%
Iron	0.01–0.3%
Copper	0.01–0.03%
Calcium	0.01–0.03%
Zinc	0.003–0.08%

(Contd ...)

Lead	0.001–0.003%
Silver	0.001–0.003%
Chromium	0.001–0.003%
B Vitamins	
Aneurine	41–49 μg g^{-1}
Biotin	0.34–0.38 μg g^{-1}
Calcium pantothenate	14.5–21.5 μg g^{-1}
Folic acid	0.26–0.6 μg g^{-1}
Nicotinamide	30–40 μg g^{-1}
Riboflavine	3.9–4.7 μg g^{-1}
Also niacin and pyridoxine	

Oils and Fats

Oils were first used as carriers for antifoams in antibiotic processes. Vegetable oils (olive, maize, cotton-seed, linseed, soyabean, etc.) may also be used as carbon substrates, particularly for their content of the fatty acids, oleic, linoleic and linolenic acid, because costs are competitive with those of carbohydrates. Bader discussed factors favouring the use of oils instead of carbohydrates. A typical oil contains approximately 2.4 times the energy of glucose on a per weight basis. Oils also have a volume advantage as it would take 1.24 dm^3 of soyabean oil to add 10 kcal of energy to a fermenter, whereas it would take 5 dm^3 of glucose or sucrose assuming that they are being added as 50 per cent w/w solutions. Ideally, in any fermentation process, the maximum working capacity of a vessel should be used. Oil based fed-batch fermentations permit this procedure to operate more successfully than those using carbohydrate feeds where a larger spare capacity must be catered for to allow for responses to a sudden reduction in the residual nutrient level. Oils also have antifoam properties which may make downstream processing simpler, but normally they are not used solely for this purpose.

Stowell reported the results of a Pfizer antibiotic process operated with a range of oils and fats on a laboratory scale. On a purely technical basis glycerol trioleate was the most suitable substrate. In the UK however, when both technical and economic factors are considered, soyabean oil or rapeseed oil are the preferred substrates. Glycerol trioleate is known to be used in some fermentations where substrate purity is an important consideration. Methyl oleate has been used as the sole carbon substrate in cephalosporin production.

Hydrocarbons and Their Derivatives

There has been considerable interest in hydrocarbons. Development work has been done using *n*-alkanes for production of organic acids, amino acids, vitamins and co-factors, nucleic acids, antibiotics, enzymes and proteins. Methane, methanol and *n*-alkanes have all been used as substrates for biomass production.

Drozd discussed the advantages and disadvantages of hydrocarbons and their derivatives as fermentation substrates, particularly with reference to cost, process aspects and purity. In processes where the feedstock costs are an appreciable fraction of the total manufacturing cost, cheap carbon sources are important. In the 1960s and early 1970s there was an incentive to consider using oil or natural gas derivatives as carbon substrates as costs were low and sugar prices were high. On a weight basis *n*-alkanes have approximately twice the carbon and three times the energy content of the same weight of sugar. Although petroleum-type products are initially impure they can be refined to obtain very pure products in bulk quantities which would reduce the amount of effluent treatment and downstream processing. At this time the view was also held that hydrocarbons would not be subject to the same fluctuations in cost as agriculturally derived feedstocks because it would be a stable priced commodity and might be used to provide a substrate for conversion to microbial protein (SCP) for economic animal and/or human consumption. Sharp gives a very good account of market considerations of changes in price and how this would affect the price of SCP. The SCP would have had to have been cheaper or as cheap as soya meal to be marketed as an animal feed supplement.

Drozd has made a detailed study of hydrocarbon feedstocks and concluded that the cost of hydrocarbons does not make them economically attractive bulk feedstocks for the production of established products or potential new products where feedstock costs are an appreciable fraction of manufacturing costs of low-value bulk products. In SCP production, raw materials account for three quarters of the operating or variable costs and about half of the costs of manufacture. It was considered that hydrocarbons and their derivatives might have a potential role as feedstocks in the microbial production of higher value products such as intermediates, pharmaceuticals, fine chemicals and agricultural chemicals.

Sources of Organic Nitrogen

There are principally three classes of raw materials available to supply the organic nitrogen requirement of a fermentation process: (i) those derived from agricultural products, (ii) those derived from brewery industry by-products, and (iii) those derived from meat and fish by-products. All of these products supply other important fermentation nutrients in addition to organic nitrogen.

Nitrogen sources derived from agricultural products

The sources derived from agricultural products are the workhorse ingredients of fermentation industry. They include the products of commodities such as various grains and soyabean. The soyabean flours, meals, and grits head the list of applications in antibiotic fermentations. The popularity of the soya products is based on the fact that after the soya oil is extricated from the soyabeans, the residue is about 50 per cent protein, which is readily available for cell growth. In addition, soya flour, meals, and grits contain up to 30 per cent utilisable carbohydrates. Most minerals required for microbial growth are also present in soya-based products.

In many seed medium applications, where growth is the primary consideration, all that is required in the medium is soya flour along with salts such as magnesium sulphate and potassium phosphate. A product that is processed very similarly to soya flour is cottonseed flour. In recent years, cottonseed flour has become the nitrogen source of choice in penicillin fermentations. The protein in the cottonseed flour is less readily available and thus makes a good slow-releasing nitrogen source. Corn gluten meal is another readily available product that is suitable as a slow-releasing nitrogen source. Corn steep liquor, a by-product of the corn milling industry, was very extensively used in the early years of the antibiotic fermentation industry. In recent years, though, due to the variability in the product quality, the liquid form of corn steep liquor has fallen out of favour. Spray-dried corn steep liquor is now available and is used in many antibiotic fermentations because it is less variable. Othet agricultural commodities used as nitrogen sources in fermentation industry include peanut meal, linseed meal, wheat flour, barley meal, and rice meal.

Nitrogen sources derived from brewery industry by-products

The brewing industry is an important source of fermentation raw materials. The principal product is the yeast left over after beer fermentation. The suitability of the yeast by-product for a given fermentation depends upon the method of drying. The yeast may be drum dried or spray dried. It is also sold as a paste produced by water evaporation in an industrial evaporator. All of these products have found applications in the fermentation industry as sources of nitrogen. However, the yeast is never used as the primary source of nitrogen. Instead, it is thought of as a nitrogen supplement with additional beneficial nutrients that are not available from grain-based nitrogen sources. Generally, these additional nutrients are organic phosphorus and unknown micronutrients. Brewery yeast is also refined into yeast extracts of different water solubilities, which are more expensive and used in smaller quantities. Yeast extract is often the single undefined component used in so-called semidefined fermentation media to provide micronutrients. The brewing and distilling industries supply two other by-products that are sometimes used in the fermentation industry: distillers' solubles, in the form of a concentrate or a spray-dried powder, and leftover grains from the brewing process.

Nitrogen sources derived from meat and fish by-products

Meat and fish products are very rich in protein. So are the by-products of these industries. The primary meat-based product is generically known as spray-dried lard water. This is a by-product of lard processing. The animal bones and tissues are boiled in water, sometimes in the presence of proteases, to free the fat. The resulting liquor is separated into fat and water layers. The water part is rich in proteins and peptides. This lard water, when spray dried, gives a product with a protein content of 80 per cent or greater. The lard water can be obtained with different degrees of chemical or enzymatic hydrolysis. Hydrolysed lard water products are sold as meat peptones under various brand names. A parallel line of products labelled fish meals and fish hydrolysates is derived from heat and enzymatic treatment of fish wastes. These products are generally about 70 per cent protein.

Examples of commonly used nitrogen sources

Most industrially used micro-organisms can utilise inorganic or organic sources of nitrogen. Inorganic nitrogen may be supplied as ammonia gas, ammonium salts or nitrates. Ammonia has been used for pH control and as the major nitrogen source in a defined medium for the commercial production of human serum albumin by *Saccharomyces cerivisiae*. Ammonium salts such as ammonium sulphate will usually produce acid conditions as the ammonium ion is utilised and the free acid will be liberated. On the other hand nitrates will normally cause an alkaline drift as they are metabolised. Ammonium nitrate will first cause an acid drift as the ammonium ion is utilised and nitrate assimilation is repressed. When the ammonium ion has been exhausted, there is an alkaline drift as the nitrate is used as an alternative nitrogen source. One exception to this pattern is the metabolism of *Gibberella fujikuroi*. In the presence of nitrate the assimilation of ammonia is inhibited at pH 2.8–3.0. Nitrate assimilation continues until the pH has increased enough to allow the ammonia assimilation mechanism to restart.

Organic nitrogen may be supplied as amino acid, protein or urea. In many instances growth will be faster with a supply of organic nitrogen and a few micro-organisms have an absolute requirement for amino acids. It might be thought that the main industrial need for pure amino acids would be in the deliberate addition to amino acid requiring mutants used in amino acid production. However, amino acids are more commonly added as complex organic nitrogen sources which are non-homogeneous, cheaper and readily available. In lysine production, methionine and threonine are obtained from soyabean hydrolysate since it would be too expensive to use the pure amino acids. Other proteinaceous nitrogen compounds serving as sources of amino acids include corn-steep liquor, soya meal, peanut meal, cotton-seed meal (Pharmamedia, Table 4.6 and Proflo), distiller's solubles, meal and yeast extract. In storage these products may be affected by moisture, temperature changes and ageing.

Table 4.6. The composition of Pharmamedia.

Component	*Quantity*
Total solids	99%
Carbohydrate	24.1%
Reducing sugars	1.2%
Non-reducing sugars	1.2%
Protein	57%
Amino nitrogen	4.7%
Components of amino nitrogen	
Lysine	4.5%
Leucine	6.1%
Isoleucine	3.3%
Threonine	3.3%
Valine	4.6%
Phenylalanine	5.9%
Tryptophan	1.0%
Methionine	1.5%
Cystine	1.5%
Aspartic acid	9.7%
Serine	4.6%
Proline	3.9%
Glycine	3.8%
Alanine	3.9%
Tyrosine	3.4%
Histidine	3.0%
Arginine	12.3%
Mineral components	
Calcium	2530 ppm
Chloride	685 ppm
Phosphorus	13100 ppm
Iron	94 ppm
Sulphate	18000 ppm
Magnesium	7360 ppm
Potassium	17200 ppm

(Contd ...)

Component	*Quantity*
Fat	4.5%
Vitamins	
Ascorbic acid	32.0 mg kg^{-1}
Thiamine	4.0 mg kg^{-1}
Riboflavin	4.8 mg kg^{-1}
Niacin	83.3 mg kg^{-1}
Pantothenic acid	12.4 mg kg^{-1}
Choline	3270 mg kg^{-1}
Pyidoxine	16.4 mg kg^{-1}
Biotin	1.5 mg kg^{-1}
Folic acid	1.6 mg kg^{-1}
Inositol	10,800 mg kg^{-1}

Chemically defined amino acid media devoid of protein are necessary in the production of certain vaccines when they are intended for human use.

Factors influencing the choice of nitrogen source

Control mechanisms exist by which nitrate reductase, an enzyme involved in the conversion of nitrate to ammonium ion, is repressed in the presence of ammonia. For this reason ammonia or ammonium ion is the preferred nitrogen source. In fungi that have been investigated, ammonium ion represses uptake of amino acid by general and specific amino acid permeases. In *Aspergillus nidulans*, ammonia also regulates the production of alkaline and neutral proteases. Therefore, in mixtures of nitrogen sources, individual nitrogen components may influence metabolic regulation so that there is preferential assimilation of one component until its concentration has diminished.

It has been shown that antibiotic production by many micro-organisms is influenced by the type and concentration of the nitrogen source in the culture medium. Antibiotic production may be inhibited by a rapidly utilised nitrogen source (NH_4^+, NO_3^-, certain amino acids). The antibiotic production only begins to increase in the culture broth after most of the nitrogen source has been consumed.

In shake flask media experiments, salts of weak acids (e.g. ammonium succinate) may be used to serve as a nitrogen source and eradicate the source of a strong acid pH change due to chloride or sulphate ions which would be present if ammonium chloride or sulphate were used as the nitrogen source. This procedure makes it possible to use lower concentrations of phosphate to buffer the medium. High phosphate concentrations inhibit production of many secondary metabolites.

The use of complex nitrogen sources for antibiotic production has been common practice. They are thought to help create physiological conditions in the trophophase which favour antibiotic production in the idiophase. For example, in the production of polyene antibiotics, soyabean meal has been considered a good nitrogen source because of the balance of nutrients, the low phosphorus content and slow hydrolysis. It has been suggested that this gradual breakdown prevents the accumulation of ammonium ions and repressive amino acids. These are probably some of the reasons for the selection of ideal nitrogen sources for some secondary metabolites (Table 4.7). In gibberellin production the nitrogen source has been shown to have an influence on directing the production of different gibberellins and the relative proportions of each type.

Table 4.7. Best nitrogen sources for some secondary metabolites.

Product	*Main nitrogen source(s)*
Penicillin	Corn-steep liquor
Bacitracin	Peanut granules
Riboflavin	Pancreatic digest of gelatine
Novobiocin	Distiller's solubles
Rifomycin	Pharmamedia
	Soyabean meal, $(NH_4)_2SO_4$
Gibberellins	Ammonium salt and natural plant nitrogen source
Butirosin	Dried beef blood or haemoglobin with $(NH_4)_2SO_4$
Polyenes	Soyabean meal

Other predetermined aspects of the process can also influence the choice of nitrogen source. Rhodes has shown that the optimum concentration of available nitrogen for griseofulvin production showed some variation depending on the form of inoculum and the type of fermenter being used. Obviously these factors must be borne in mind in the interpretation of results in media-development programmes.

Some of the complex nitrogenous material may not be utilised by a micro-organism and create problems in downstream processing and effluent treatment. This can be an important factor in the final choice of substrate.

Speciality Chemicals

Several types of speciality chemicals are added to large scale fermentation media. The most important of these chemicals are the defoamers. The defoamers reduce the interfacial surface tension between air and water to facilitate bubble coalescence. In the fermentation industry, silicone and polyol-based defoamers have largely replaced, vegetable oils as defoamers. The advantages of the synthetic defoamers are that they are cost-effective and very slowly metabolised and do not have appreciable metabolic side effects. The two most popular defoamers in use in the fermentation industry are polypropylene glycol and silicone emulsion. The defoamers are generally batched with the starring medium. In many fermentations, however, it is necessary to supply defoamers throughout the fermentation cycle to control foam and to control air holdup. Emulsifiers used in fermentations (such as Tween and Span) play a role opposite to that of defoamers. They are added to stabilise small droplets of oily nutrients by increasing the surface tension between oil and water. The small droplets have a dramatically increased surface area and thus allow oily substrates to be more readily utilised by the fermentation organism. Metal-chelating agents such as EDTA often included in fermentation media. The chelating agents have two diametrically opposed effects. On the one hand, they can tie up metal ions that 'are toxic to the organism. On the orher hand, they can prevent the precipitation of a required trace metal by forming a soluble complex. The availability of the metal to the fermenting organism depends upon whether the organism can effectively compete with the complexing agent for the required metal.

An important class of speciality products used in the fermentation industry is made up of various enzyme preparations. Crude preparations of enzymes such as amylase, protease, and cellulase are used to precondition the medium. Invariably, these enzymes are used at the mixing stage before medium sterilisation. A partial breakdown of the starch of medium components such as corn flour can be achieved by the addition of amylase enzyme. The cellulase complex can be used to reduce the viscosity of a medium containing a high concentration of ingredients such as soya or cottonseed flour. The protease enzymes can predigest the medium proteins before sterilisation. Enzymatic pretreatment of a fermentation medium thus allows a crude and cheaper raw material to be substituted for a more refined and expensive raw material.

Minerals

All micro-organisms require certain mineral elements for growth and metabolism. In many media, magnesium, phosphorus, potassium, sulphur, calcium and chlorine are essential components and because of the concentrations required, they must be added as distinct components. Others such as cobalt, copper, iron, manganese, molybdenum and zinc are also essential but are usually present as impurities in other major ingredients. There is obviously a need for batch analysis of media components to ensure that this assumption can be justified, otherwise there may be deficiencies or excesses in different batches of media. Tables 4.5 and 4.6 for analysis of corn steep liquor and Pharmamedia. When synthetic media are used the minor elements will have to be added deliberately. The form in which the minerals are usually supplied and the concentration ranges are given in Table 4.8. As a consequence of product composition analysis, as outlined earlier in this chapter, it is possible to estimate the amount of a specific mineral for medium design, e.g. sulphur in penicillins and cephalosporins, chlorine in chlortetracycline.

Table 4.8. The range of typical concentrations of mineral components (g dm^{-3}).

Component	*Range*
*KH_2PO_4	1.0–4.0 (part may be as buffer)
$MgSO_4 \cdot 7H_2O$	0.25–3.0
KCl	0.5–12.0
$CaCO_3$	5.0–17.0
$FeSO_4 \cdot 4H_2O$	0.01–0.1
$ZnSO_4 \cdot 8H_2O$	0.1–1.0
$MnSO_4 \cdot H_2O$	0.01–0.1
$CuSO_4 \cdot 5H_2O$	0.003–0.01
$Na_2MoO_4 \cdot 2H_2O$	0.01–0.1

*Complex media derived from plant and animal materials normally contain a considerable concentration of inorganic phosphate.

The concentration of phosphate in a medium, particularly laboratory media in shake flasks, is often much higher than that of other mineral components. Part of this phosphate is being used as a buffer to minimise pH changes when external control of the pH is not being used.

In specific processes the concentration of certain minerals may be very critical. Some secondary metabolic processes have a lower tolerance range to inorganic phosphate than vegetative growth. This phosphate should be sufficiently low as to be assimilated by the end of trophophase. Garner suggested that an important function of calcium salts in fermentation media was to precipitate excess inorganic phosphates and also the calcium indirectly improved the yield of streptomycin. The inorganic phosphate concentration also influences production of bacitracins, citric acid (surface culture), ergot, monomycin, novobiocin, oxytetracycline, polyenes, ristomycin, rifamycin Y, streptomycin, vancomycin and viomycin. However, pyrrolnitrin, bicyclomycin, thiopeptin and methylenomycin are produced in a medium containing a high concentration of phosphate. Two monomycin antibiotics are selectively produced by *Streptomyces jamaicensis* when the phosphate is 0.1 mM or 0.4 mM.

In a recent review of antibiotic biosynthesis, Liras recognised target enzymes which are: (i) repressed by phosphate, (ii) inhibited by phosphate, and (iii) repression of an enzyme occurs but phosphate repression is not clearly proved. A phosphate control sequence has also been isolated and characterised from the phosphate regulated promoter that controls biosynthesis of candicidin.

Weinberg has reviewed the nine trace elements of biological interest (atomic numbers 23–30, 42). Of these nine, the concentrations of manganese, iron and zinc are the most critical in secondary metabolism. In every secondary metabolic system in which sufficient data has been reported, the yield of the product varies linearly with the logarithmic concentration of the 'key' metal. The linear relationship does not apply at concentrations of the metal which are either insufficient or toxic, to cell growth. Some of the primary and secondary microbial products whose yields are affected by concentrations of trace metals greater than those required for maximum growth are given in Table 4.9.

Table 4.9. Trace elements influencing primary and secondary metabolism.

Product	*Trace element(s)*
Bacitracin	Mn
Protease	Mn
Gentamicin	Co
Riboflavin	Fe, Co
Mitomycin	Fe
Monensin	Fe
Actinomycin	Fe, Zn
Candicidin	Fe, Zn
Chloramphenicol	Fe, Zn
Neomycin	Fe, Zn
Patulin	Fe, Zn
Streptomycin	Fe, Zn
Citric acid	Fe, Zn, Cu
Penicillin	Fe, Zn, Cu
Griseofulvin	Zn

Chlorine does not appear to play a nutritional role in the metabolism of fungi. It is, however, required by some of the halophilic bacteria. Obviously, in those fermentations where a chlorine-containing metabolite is to be produced, the synthesis will have to be directed to ensure that the non-chloro-derivative is not formed.

The most important compounds are chlortetracycline and griseofulvin. In griseofulvin production, adequate available chloride is provided by the inclusion of at least 0.1 per cent KCl, as well as the chloride provided by the complex organic materials included as nitrogen sources. Other chlorine containing metabolites are caldriomycin, nornidulin and mollisin.

Chelators

Many media cannot be prepared or autoclaved without the formation of a visible precipitate of insoluble metal phosphates. Gaunt demonstrated that when the medium of Mandels and Weber was autoclaved, a white precipitate of metal ions formed, containing all the iron and most of the calcium, manganese and zinc present in the medium. The problem of insoluble metal phosphate(s) may be eliminated by incorporating low concentrations of chelating agents such as ethylene diamine tetraacetic acid (EDTA), citric acid, polyphosphates, etc. into the medium. These chelating agents preferentially form complexes with the metal ions in a medium. The metal ions then may be gradually utilised by the micro-organism. Gaunt showed that the precipitate was eliminated from Mandel and Weber's medium by the addition of EDTA at 25 mg dm^{-3}. It is important to check that a chelating agent does not cause inhibition of growth of the micro-organism which is being cultured.

In many media, particularly those commonly used in large scale processes, there may not be a need to add a chelating agent as complex ingredients such as yeast extracts or proteose peptones will complex with metal ions and ensure gradual release of them during growth.

Growth factors

Some micro-organisms cannot synthesise a full complement of cell components and therefore require preformed compounds called growth factors. The growth factors most commonly required are vitamins, but there may also be a need for specific amino acids, fatty acids or sterols. Many of the natural carbon and nitrogen sources used in media formulations contain all or some of the required growth factors. When there is a vitamin deficiency it can often be eliminated by careful blending of materials. It is important to remember that if only one vitamin is required it may be occasionally more economical to add the pure vitamin, instead of using a larger bulk of a cheaper multiple vitamin source. Calcium pantothenate has been used in one medium formulation for vinegar production. In processes used for the production of glutamic acid, limited concentrations of biotin must be present in the medium. Some production strains may also require thiamine.

Sources of Information on Fermentation of Raw Materials

The best source of information on a given class of fermentation raw material is the industry in which it is generated. Information about such things as the protein, fat, carbohydrate, and mineral contents of various raw materials is readily available from the supplier of the raw materials. However, this information is not necessarily generated for the use of the fermentation industry. It is generated for the benefit of the primary users, which in most cases are the animal feed and food industries. As a result, interpretation of the information for fermentation use is up to the fermentation scientist. For example, while the total nitrogen value of a grain-based product may be meaningful from the point of view of a weight gain calculation when the product is fed to a farm animal, it may not necessarily have the same meaning as the nitrogen available for the fermentation organism to grow on. For the same reason, the carbohydrate value provided by the manufacturer of one product may be higher than the value provided for a second product, and yet the second product could have more available carbon for a particular fermentation organism. The information provided by the manufacturer is a good approximation for the initial evaluation and for preliminary cost calculations. Actual fermentation experiments are necessary in all cases to justify a change of raw material.

Chapter 5

Fermentation Methods and Systems

INTRODUCTION

Fermentation is classically defined as the microbiological conversion, in the absence of oxygen, of sugars to alcohol or lactic acid. By the enzymatic conversion of sugars to alcohol, the micro-organisms gain biochemical energy (ATP). Fermentation is used in many commercial food processes, including the making of sauerkraut, yogurt, wines and beers. Certain bacteria and fungi are capable of causing fermentation to occur in the absence of oxygen.

Some industrial processes are called fermentation which are aerobic (conducted in the presence of oxygen) and would be more correctly be termed oxidative processes.

Fruits, grains, milk, and other organic substances naturally undergo fermentation during spoilage, a phenomenon which cavemen no doubt observed.

INDUSTRIAL FERMENTATION PROCESSES

There are two main types of industrial fermentation processes, batch fermentations and continuous fermentations. In batch fermentations, sterile growth medium is inoculated with the micro-organisms and no additional growth medium is added. In continuous fermentations, growth medium is added to the fermenting medium to sustain the fermentation process. Fermentation occurs by the production of cellular enzymic reactions instead of chemical reactions aided by inanimate catalysts, sometimes operating at elevated temperature and pressure.

GROWTH KINETICS

Growth kinetics, i.e. the relationship between specific growth rate and the concentration of a substrate, is one of the basic tools in microbiology. However, despite more than half a century of research, many fundamental questions about the validity and application of growth kinetics as observed in the laboratory to environmental growth conditions are still unanswered. For pure cultures growing with single substrates, enormous inconsistencies exist in the growth kinetic data reported.

Microbial growth kinetics, i.e. the relationship between the specific growth rate (μ) of a microbial population and the substrate concentration (s), is an indispensable tool in all fields of microbiology, be it physiology, genetics, ecology, or biotechnology, and therefore it is an important part of the basic teaching of microbiology.

Batch Fermentation Process

A tank of fermenter is filled with the prepared mash of raw materials to be fermented. The temperature and pH for microbial fermentation is properly adjusted, and occasionally nutritive supplements are added to the prepared mash. The mash is steam sterilised in a pure culture process. The inoculum of a pure culture is added to the fermenter, from a separate pure culture vessel.

Fermentation proceeds, and after the proper time the contents of the fermenter, are taken out for further processing. The fermenter is cleaned and the process is repeated. Thus each fermentation is a discontinuous process divided into batches.

In autecological studies, bacterial growth in batch culture can be modelled with four different phases: (i) lag phase, (ii) exponential or log phase, (iii) stationary phase, and (iv) death phase.

1. During lag phase, bacteria adapt themselves to growth conditions. It is the period where the individual bacteria are maturing and not yet able to divide. During the lag phase of the bacterial growth cycle, synthesis of RNA, enzymes and other molecules occurs.
2. Exponential phase (sometimes called the log phase) is a period characterised by cell doubling. The number of new bacteria appearing per unit time is proportional to the present population. If growth is not limited, doubling will continue at a constant rate so both the number of cells and the rate of population increase doubles with each consecutive time period. For this type of exponential growth, plotting the natural logarithm of cell number against time produces a straight line. The slope of this line is the specific growth rate of the organism, which is a measure of the number of divisions per cell per unit time. The actual rate of this growth (i.e. the slope of the line in the Fig. 5.1) depends upon the growth conditions, which affect the frequency of cell division events and the probability of both daughter cells surviving. Exponential growth cannot continue indefinitely, however, because the medium is soon depleted of nutrients and enriched with wastes.

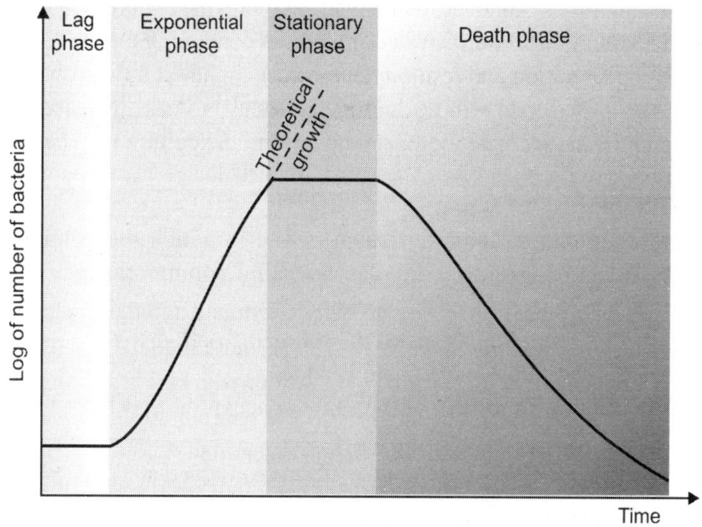

Fig. 5.1. Bacterial growth curve.

3. During stationary phase, the growth rate slows as a result of nutrient depletion and accumulation of toxic products. This phase is reached as the bacteria begin to exhaust the resources that are

available to them. This phase is a constant value as the rate of bacterial growth is equal to the rate of bacterial death.

4. At death phase, bacteria run out of nutrients and die.

Fed-Batch

A 'fed-batch' is a biotechnological batch process which is based on feeding of a growth limiting nutrient substrate to a culture. The fed-batch strategy is typically used in bioindustrial processes to reach a high cell density in the bioreactor. Mostly the feed solution is highly concentrated to avoid dilution of the bioreactor. The controlled addition of the nutrient directly affects the growth rate of the culture and allows to avoid overflow metabolism (formation of side metabolites, such as acetate for *Escherichia coli*, lactic acid in cell cultures, ethanol in *Saccharomyces cerevisiae*), oxygen limitation (anaerobiosis). In most cases the growth-limiting nutrient is glucose which is fed to the culture as a highly concentrated glucose syrup (600–850 g/l).

Two basic approaches to the fed-batch fermentation can be used: the constant volume fed-batch culture—Fixed volume fed-batch—and the variable volume fed-batch.

Fixed volume fed-batch

In this type of fed-batch, the limiting substrate is fed without diluting the culture. The culture volume can also be maintained practically constant by feeding the growth limiting substrate in undiluted form, for example, as a very concentrated liquid or gas (e.g. oxygen).

Alternatively, the substrate can be added by dialysis or, in a photosynthetic culture, radiation can be the growth limiting factor without affecting the culture volume.

A certain type of extended fed-batch—the cyclic fed-batch culture for fixed volume systems—refers to a periodic withdrawal of a portion of the culture and use of the residual culture as the starting point for a further fed-batch process. Basically, once the fermentation reaches a certain stage, (for example, when aerobic conditions cannot be maintained anymore) the culture is removed and the biomass is diluted to the original volume with sterile water or medium containing the feed substrate. The dilution decreases the biomass concentration and result in an increase in the specific growth rate. Subsequently, as feeding continues, the growth rate will decline gradually as biomass increases and approaches the maximum sustainable in the vessel once more, at which point the culture may be diluted again.

Variable volume fed-batch

As the name implies, a variable volume fed-batch is one in which the volume changes with the fermentation time due to the substrate feed. The way this volume changes is dependent on the requirements, limitations and objectives of the operator.

The feed can be provided according to one of the following options:

1. The same medium used in the batch mode is added.
2. A solution of the limiting substrate at the same concentration as that in the initial medium is added.
3. A very concentrated solution of the limiting substrate is added at a rate less than (1), (2) and (3).

This type of fed-batch can still be further classified as repeated fed-batch process or cyclic fed-batch culture, and single fed-batch process.

The former means that once the fermentation reached a certain stage after which is not effective anymore, a quantity of culture is removed from the vessel and replaced by fresh nutrient medium. The decrease in volume results in a increase in the specific growth rate, followed by a gradual decrease as the quasi-steady state is established.

The latter type refers to a type of fed-batch in which supplementary growth medium is added during the fermentation, but no culture is removed until the end of the batch. This system presents a disadvantage over the fixed volume fed-batch and the repeated fed-batch process: much of the fermentor volume is not utilised until the end of the batch and consequently, the duration of the batch is limited by the fermentor volume.

Equipment

No special piece of equipment is required over the equipment required for batch. However, some considerations should be made over the equipment used for a fed-batch fermentation.

Vessels

The vessels, particularly those used for the acid and base control, must be constructed from a nontoxic, corrosion-resistant material which is capable of withstanding repeated sterilisation cycles. Figure 5.2 illustrates two methods of assembling vessels for easy transfer of either inoculum or medium to the fermentor.

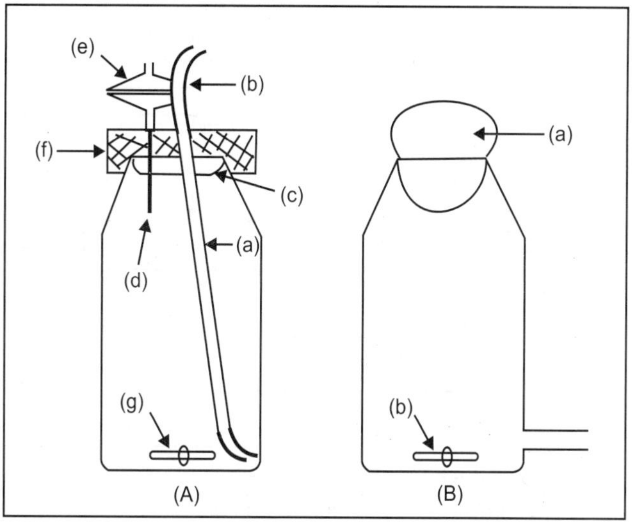

Fig. 5.2. Holding vessels. (A) Screw-neck borosilicate glass vessel with medium/inoculum addition assembly: (a) stainless steel rod, (b) silicon tubing, (c) silicon disc, (d) hypodermic needle, (e) air vent, (f) screw cap, (g) magnetic bar. (B) Aspirator-type vessel for introducing an inoculum of filamentous fungi into the fermentor: (a) cotton-wool plug, (b) magnetic stirrer bar.

Pumps

There are two types of pumps which are suitable for the aseptic pumping of small volumes of culture media: the peristaltic pump and the diaphragm-dosing pump. Other pumps are unsuitable because they are difficult to sterilise and cannot be used for pumping small volumes.

The peristaltic pump is typically constituted by a main body that comprises both the drive motor and electrics, and the rotating unit of rollers. This unit of rollers occludes the tube which, as it recovers to its original size passes to the nest roller until expelled, as the unit moves round. The flow rate can be varied by either the speed setting or by changing the diameter of the tube being used.

The diaphragm-dosing pump consists of a main body and a detachable heat-sterilisable head. The fluid is sucked in to the pump head. The suction inlet tube then closes and the pressure discharge tube opens and forces the fluid out. The suction and pressure forces in the pump head are generated by the reciprocating action of both the diaphragm plunger and the return spring.

Control techniques for fed-batch fermentation

Adaptive control is the name given to a control system in which the controller learns about the process by acquiring data from a certain process and keeps on updating a control model. A parameter estimator monitors the process and estimates the process dynamics in terms of the parameters of a previously defined mathematical model of the process. A control design algorithm is then used to generate controller coefficients from those estimates, and a controller sets up the required control signals to the devices controlling the process. An extremely important feature of an adaptive controller is the structure of the model used by the parameter estimator to analyse estimates of process dynamics. The process can be described by a set of mass balance equations, whose quantities can be measured directly or indirectly. Figure 5.3 describes schematically the concept.

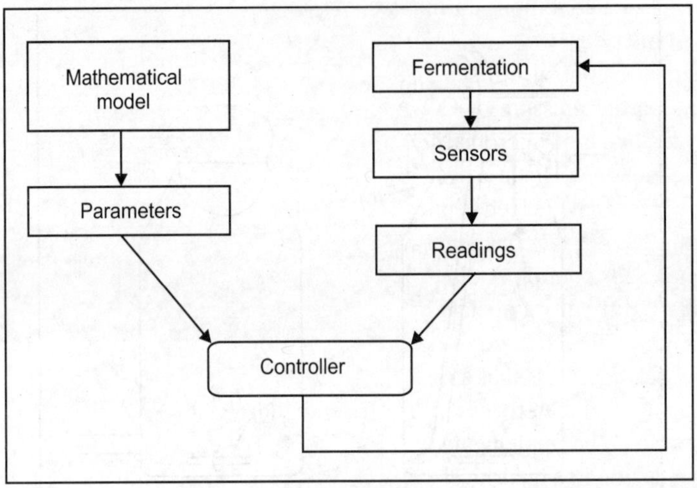

Fig. 5.3. Adaptive control: the controller compares the estimates from a mathematical model applied to the system to the readings obtained from the fermentation process. The controller then sends the signal to the device controlling the fermentation, for example, by increasing or decreasing a flow rate.

The optimal strategy for the fed-batch fermentation of most organisms is to feed the growth-limiting substrate at the same rate that the organism utilises the substrate, this is, to match the feed rate with demand for the substrate.

Four basic approaches have been used in attempts to balance substrate feed with demand (listed in order of increasing accuracy and/or complexity):

1. Open-loop control schemes in which feed is added according to historical data or predicted data.
2. Indirect control of substrate feed based on nonfeed source parameters such as pH, offgas analysis, dissolved O_2 or concentrations of organic products.

3. Indirect control schemes based on mass balance equations, the values of which are calculated from data obtained by sensors.
4. Direct control schemes based on direct, on-line measurements substrates.

Better and more flexible control may be obtained when there is direct measurement of substrate or an excreted metabolite in the medium, which can be used to influence feeding rates to the fermentation. This can be done off-line or semi-on-line, but on-line measurements are more useful because of

1. The shorter analysis required.
2. Lower personnel requirement.
3. A reduced chance of fermentor contamination.

Regardless the type of control, the design is strongly influenced by both mathematical model availabilities and measurement possibilities.

Control and optimisation of bioreactors is strongly influenced by the quality of the sensors available for crucial response variables. Of primary importance is the ratio of the dynamic parameters of the sensor to those of the process. When these variables cannot be measured easily or quickly enough, a mathematical model must be used in some way in place of feedback information.

When an exact mathematical model is at disposal, an open-loop process control can be proposed which generally proves to be insufficient. The advantage of a feedback control is that a response to unforeseen and unexpected conditions during the fermentation is achieved and the process is controlled within the desired limits.

An indirect feedback control utilises an observable parameter, such as dissolved oxygen, pH, respiratory quotient, partial pressure of CO_2, culture fluorescence or by-product formation, which is closely related to the course of microbial fermentation. As examples of fed-batch systems using this concept, one can mention the pH-stat-a system in which the feed is provided depending on the pH, and the DO-stat-a system in which the feed is provided depending on the reading of the dissolved oxygen. A direct feedback controller uses the concentration of limiting substrate in the culture medium as a feedback feed -related parameter for control. A direct feedback control can have the disadvantage of not being very feasible due to the difficulty associated with obtaining accurate on-line measurements of substrate concentrations or even by the absence of on-line sensors for the important compound to control. The advantage of a feedback control is that a response to unforeseen and unexpected conditions during the fermentation is achieved and the process is controlled within the desired limits.

A feedback control can be implemented accordingly to not only a single measurement, but also to obtain a finer control action in a dual-level system. Turner, describes a control method applied to a fed-batch culture of recombinant *Escherichia coli* in which a two-level control was preferred because it provided much greater flexibility and better control over the substrate concentration in the medium and the production of by-products.

As compared with the batch fermentation, two more parameters need to be specified to determine the operating conditions of a fed-batch fermentation: feed and initial feeding time. These parameters are usually process and/or micro-organism specific and the parameters commonly used to define them.

Continuous Fermentation Process

Growth of micro-organisms during batch fermentation confirms to the characteristic growth curve, with a lag phase followed by a logarithmic phase. This, in turn, is terminated by progressive decrements in the rate of growth until the stationary phase is reached. This is because of limitation of one or more of the essential nutrients.

In continuous fermentation, the substrate is added to the fermenter continuously at a fixed rate. This maintains the organisms in the logarithmic growth phase. The fermentation products are taken out continuously. The design and arrangements for continuous fermentation, are some what complex.

Homogeneous bioreactor systems suspension cultures

The classical system is the suspension culture using a stirred tank with different impeller types and installations, equipped with or without a spin filter. In large-scale bioreactors, slight modifications of several internal parts of bioreactors used for bacterial fermentation are made in order to adopt them for culturing animal cells. The modifications are in the agitation system. Marine type impeller, vibromixer or rotating flexible sheets replace the turbine type impeller widely used in microbial fermentation. Perfusion systems were also developed for submerged cultivation of animal cells.

This is run as either a chemostate or turbidostat. A chemostat is a bioreactor to which fresh medium is continuously added, while culture liquid is continuously removed to keep the culture volume constant. By changing the rate with which medium is added to the bioreactor the growth rate of the micro-organism can be easily controlled (Fig. 5.4).

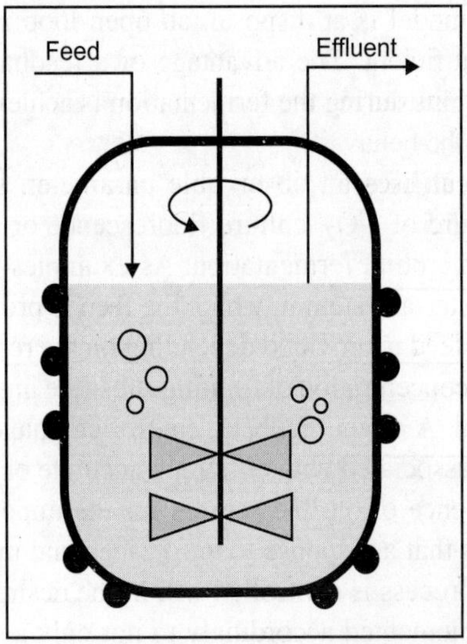

Fig. 5.4. Stirred bioreactor operated as a chemostat, with a continuous inflow (the feed) and outflow (the effluent). The rate of medium flow is controlled to keep the culture volume constant.

Steady state

One of the most important features of chemostats is that micro-organisms can be grown in a physiological steady state. In steady state, growth occurs at a constant rate and all culture parameters remain constant (culture volume, dissolved oxygen concentration, nutrient and product concentrations, pH, cell density, etc.). In addition environmental conditions can be controlled by the experimenter. Micro-organisms grown in chemostats naturally strive to steady state: if a low amount of cells are present in the bioreactor, the cells can grow at growth rates higher than the dilution rate, as growth is not limited by the addition of the limiting nutrient. The limiting nutrient is a nutrient essential for growth, present in the media at a limiting concentration (all other nutrients are usually supplied in surplus). However, if the cell concentration becomes too high, the amount of cells that are removed from the reactor cannot be replenished by growth as the addition of the limiting nutrient is insufficient. This results in an equilibrium situation (steady state), where the rate of cell growth is equal to the rate of cell removal.

Because obtaining a steady state requires at least 5 volume changes, chemostats require large nutrient and waste reservoirs.

Dilution rate

At steady state the specific growth rate (μ) of the micro-organism is equal to the dilution rate (D). The dilution rate is defined as the rate of flow of medium over the volume of culture in the bioreactor:

$$D = \frac{\text{Medium flow rate}}{\text{Culture volume}} = \frac{F}{V}$$

Maximal growth rate

Each micro-organism growing on a particular substrate has a maximum specific growth rate (μ_{max}) (the rate of growth observed if none of the nutrients are limiting). If a dilution rate is chosen that is higher than μ_{max}, the culture will not be able to sustain itself in the bioreactor, and will wash out.

Plug flow reactor model

The plug flow reactor (PFR) model is used to describe chemical reactions in continuous, flowing systems. The PFR model is used to predict the behaviour of chemical reactors, so that key reactor variables, such as the dimensions of the reactor, can be estimated. PFRs are also sometimes called continuous tubular reactors (CTRs) as shown in Fig. 5.5.

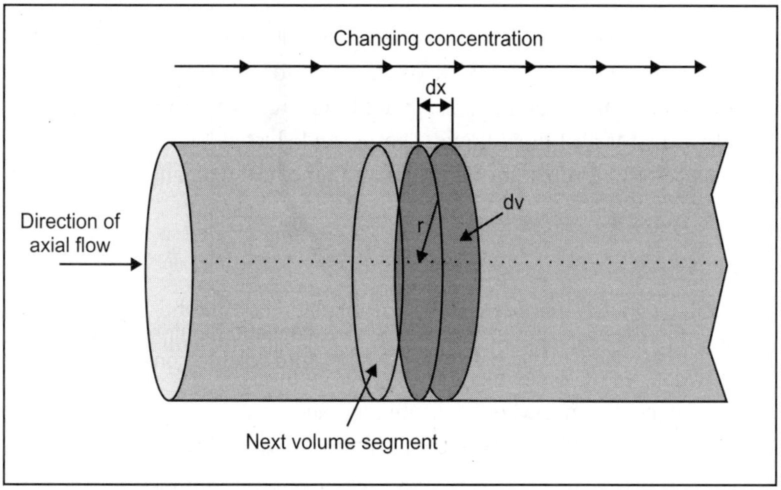

Fig. 5.5. Schematic diagram of a plug flow reactor (PFR).

Fluid going through a PFR may be modelled as flowing through the reactor as a series of infinitely thin coherent 'plugs', each with a uniform composition, travelling in the axial direction of the reactor, with each plug having a different composition from the ones before and after it. The key assumption is that as a plug flows through a PFR, the fluid is perfectly mixed in the radial direction but not in the axial direction (forwards or backwards). Each plug of differential volume is considered as a separate entity, effectively an infinitesimally small batch reactor, limiting to zero volume. As it flows down the tubular PFR, the residence time (τ) of the plug is a function of its position in the reactor. In the ideal PFR, the residence time distribution is therefore a 'Dirac delta' function with a value equal to τ.

PFR modelling

PFRs are frequently referred to as piston flow reactors, or sometimes as continuous tubular reactors. They are governed by ordinary differential equations, the solution for which can be calculated provided the appropriate boundary conditions are known.

The PFR model works well for many fluids: liquids, gases, and slurries. Although turbulent flow and axial diffusion cause a degree of mixing in the axial direction in real reactors, the PFR model is appropriate when these effects are sufficiently small that they can be ignored.

Operation and uses: PFRs are used to model the chemical transformation of compounds as they are transported in systems resembling 'pipes'. The pipe can represent a variety of engineered or natural conduits through which liquids or gases flow (e.g. rivers, pipelines, regions between two mountains, etc.).

An ideal plug flow reactor has a fixed residence time: Any fluid (plug) that enters the reactor at time t will exit the reactor at time $t + \tau$, where τ is the residence time of the reactor. The residence time distribution function is therefore a 'Dirac delta' function at τ. A real plug flow reactor has a residence time distribution that is a narrow pulse around the mean residence time distribution.

A typical plug flow reactor could be a tube packed with some solid material (frequently a catalyst). Typically these types of reactors are called packed bed reactors or PBR's. Sometimes the tube will be a tube in a shell and tube heat exchanger.

Continuous Stirred-tank Reactor

The continuous stirred-tank reactor (CSTR), also known as vat- or backmix reactor, is a common ideal reactor type in chemical engineering. A CSTR often refers to a model used to estimate the key unit operation variables when using a continuous agitated-tank reactor to reach a specified output. The mathematical model works for all fluids: liquids, gases, and slurries.

Baffle: Baffle may refer to a separator in a shell and tube heat exchanger used to support the tubes and direct fluid flow.

Reynolds Number

In fluid mechanics, the Reynolds number (Re) is a dimensionless number that gives a measure of the ratio of inertial forces ($\rho V^2/L$) to viscous forces ($\mu V/L^2$) and consequently quantifies the relative importance of these two types of forces for given flow conditions.

Reynolds numbers frequently arise when performing dimensional analysis of fluid dynamics problems, and as such can be used to determine dynamic similitude between different experimental cases. They are also used to characterise different flow regimes, such as laminar or turbulent flow: laminar flow occurs at low Reynolds numbers, where viscous forces are dominant, and is characterised by smooth, constant fluid motion, while turbulent flow occurs at high Reynolds numbers and is dominated by inertial forces, which tend to produce random eddies, vortices and other flow instabilities.

Reynolds number can be defined for a number of different situations where a fluid is in relative motion to a surface (the definition of the Reynolds number is not to be confused with the Reynolds Equation or lubrication equation). These definitions generally include the fluid properties of density and viscosity, plus a velocity and a characteristic length or characteristic dimension. This dimension is a matter of convention - for example a radius or diameter are equally valid for spheres or circles, but one is chosen by convention. For aircraft or ships, the length or width can be used. For flow in a pipe or a sphere moving in a fluid the internal diameter is generally used today. Other shapes (such as rectangular pipes or nonspherical objects) have an equivalent diameter defined. For fluids of variable density (e.g. compressible gases) or variable viscosity (non-Newtonian fluids) special rules apply. The velocity may also be a matter of convention in some circumstances, notably stirred vessels.

$$\mathrm{Re} = \frac{\rho VL}{\mu} = \frac{VL}{\nu} = \frac{QL}{\nu A}$$

where,

V is the mean fluid velocity (SI units: m/s)

L is a characteristic linear dimension, (travelled length of fluid, or hydraulic radius when dealing with river systems) (m)

μ is the dynamic viscosity of the fluid (Pa·s or N·s/m^2 or kg/m·s)

ν is the kinematic viscosity ($\nu = \mu/\rho$) (m^2/s)

ρ is the density of the fluid (kg/m^3)

Q is the volumetric flow rate (m^3/s)

A is the pipe cross-sectional area (m^2).

Note that this is equal to the ratio between $\rho V^2/L$, which is the drag (up to a numerical factor, half the drag coefficient), and $\mu V/L^2$, which is the force due to viscosity (up to a numerical factor depending on the form of the flow).

Power Number

The power number N_p (also known as Newton number) is a commonly-used dimensionless number relating the resistance force to the inertia force. The power-number has different specifications according to the field of application, e.g. for stirrers the power number is defined as:

$$N_p = \frac{P}{\rho n^3 d^5}$$

where,

P = power

ρ = fluid density

n = rotational speed

d = diameter of stirrer.

GAS EXCHANGE AND MASS TRANSFER

Mass transfer is the transfer of mass from high concentration to low concentration. The phrase is commonly used in engineering for physical processes that involve molecular and convective transport of atoms and molecules within physical systems. Mass transfer includes both fluid flow and separation unit operations.

Some common examples of mass transfer processes are the evaporation of water from a pond to the atmosphere; the diffusion of chemical impurities in lakes, rivers, and oceans from natural or artificial point sources; mass transfer is also responsible for the separation of components in an apparatus such as a distillation column. In HVAC examples of a heat and mass exchangers are cooling towers and evaporative coolers where evaporation of water cools that portion which remains as a liquid, as well as cooling and humidifying the air passing through.

The driving force for mass transfer is a difference in concentration; the random motion of molecules causes a net transfer of mass from an area of high concentration to an area of low concentration. The amount of mass transfer can be quantified through the calculation and application of mass transfer coefficients. Mass transfer finds extensive application in chemical engineering problems, where material balance on components is performed.

Henry's Law

Henry's law states that: 'At a constant temperature, the amount of a given gas dissolved in a given type and volume of liquid is directly proportional to the partial pressure of that gas in equilibrium with that liquid.' An equivalent way of stating the law is that the solubility of a gas in a liquid at a particular temperature is proportional to the pressure of that gas above the liquid. Henry's law has since been shown to apply for a wide range of dilute solutions, not merely those of gases. An everyday example of Henry's law is given by carbonated soft drinks. Before the bottle or can is opened, the gas above the drink is almost pure carbon dioxide at a pressure slightly higher than atmospheric pressure. The drink itself contains dissolved carbon dioxide. When the bottle or can is opened, some of this gas escapes, giving the characteristic hiss (or pop in the case of a champagne bottle). Because the pressure above the liquid is now lower, some of the dissolved carbon dioxide comes out of solution as bubbles. If a glass of the drink is left in the open, the concentration of carbon dioxide in solution will come into equilibrium with the carbon dioxide in the air, and the drink will go flat.

Formula and the Henry's law constant

Henry's law can be put into mathematical terms (at constant temperature) as:

$$p = k_H c$$

where, p is the partial pressure of the solute in the gas above the solution, c is the concentration of the solute and k_H is a constant with the dimensions of pressure divided by concentration. The constant, known as the Henry's law constant, depends on the solute, the solvent and the temperature.

Some values for k_H for gases dissolved in water at 298 kelvins include:

1. Oxygen (O_2) : 769.2 L·atm/mol.
2. Carbon dioxide (CO_2) : 29.4 L·atm/mol.
3. Hydrogen (H_2) : 1282.1 L·atm/mol.

There are other forms of Henry's Law, each of which defines the constant k_H differently and requires different dimensional units. In particular, the 'concentration' of the solute in solution may also be expressed as a mole fraction or as a molality.

Oxygen Transfer in Bioreactors

Oxygen is needed by cells for respiration. Oxygen used by cells in suspension must be available as dissolved oxygen. Since oxygen solubility is quite small, about 6 to 7 mg/l under normal cultivation conditions, metabolic oxygen requirement is supplied on needed basis by continuous aeration of culture medium. Actively respiring yeast requires about 0.15 g O^2 (g cell)$^{-1}$ hr. At a cell concentration of 10 g l^{-1}, medium saturated with air can support less than 30 seconds worth of metabolic oxygen. That is, a continuous supply of oxygen must be maintained in any viable aerobic manufacturing process. In this section, we will first get a quantitative appreciation for metabolic oxygen demand, followed by methods used in calculating rates at which oxygen is transfered from sparged air. We will then examine methods useful in characterising oxygen mass transfer coefficient. Finally we will evaluate bioreactor operation and design based on oxygen transfer capability.

Metabolic oxygen demand

Metabolic oxygen demand of an organism depends on the biochemical nature of the cell and cultivation conditions. Oxygen need is usually satisfied in most cells if the dissolved oxygen concentraiton in the medium is kept at about 1 mg/l. If the oxygen level is allowed to fall far below this value, oxygen consumption rate decreases with concomitant decrease in biochemical energy production, and as a result cell growth rate also decreases.

Volumetric oxygen mass transfer coefficient

In a typical aeration system, oxygen from the air bubble is transferred through the gas-liquid interface followed by liquid phase diffusion/bulk transport to the cells. Although this is a multi-step serial transport, in a well dispersed systems, the major resistance to oxygen transfer is in the liquid film surrounding the gas bubble.

Bioreactor oxygen balance

Let us now consider the case of oxygen balance within a bioreactor in which cells are growing and in the process consuming oxygen. There is a continuous inflow of air at a constant volumetric flow rate. The liquid broth is agitated by a Rushton agitator (flat blade stirrer). Let the metabolic oxygen uptake rate be q_{O_2} and cell concentration is X. Let us examine the reactor system over a sufficiently short period that we can treat X as a constant. Consider oxygen balance over the liquid phase of the bioreactor.

O_2 transfered from gas phase – O_2 consumed by cells = Accumulation

$$\left[k_{La}\left(C^*_{DO} - C_{DO}\right)\right]\cdot V - q_{O_2}XV = \frac{d(V\,C_{DO})}{dt} \qquad \text{... (5.1)}$$

For constant liquid phase volume, the above can be simplified to:

$$\frac{d(C_{DO})}{dt} = k_{La}\left(C^*_{DO} - C_{DO}\right) - q_{O_2}X \qquad \text{... (5.2)}$$

The concentration, C_{DO} is readily measured using an dissolved oxygen electrode. A later segment of the course on biosensors, will deal with principle of measurement and construction of DO electrodes.

If oxygen being supplied is in exact balance with the oxygen consumed by the cells, we expect the dissolved oxygen concentration to remain constant; that is, the derivative in Eq. 5.2 will vanish. That is,

$$q_{O_2}X = k_{La}\left(C^*_{DO} - C_{DO}\right)$$

One useful application of the above is in estimating the maximum cell concentration a particular bioreactor is capable of supporting in terms of oxygen supply.

Factors affecting k_{La}

The mass transfer coefficient is strongly affected by agitation speed and air flow rate. In general,

$$k_{La} = k\,(P_Q/V_R)^{0.4}\,(V_S)^{0.5}\,(N)^{0.5}$$

where, k is a constant, P_Q is the power required for aerated bioreactor, V_R is the bioreactor volume, V_S is air flow rate, N is agitator speed.

Note that the mass transfer coefficient increases with agitation speed and air flow rate.

Measurement of k_{La}

Most common method of measuring k_{La} is to conduct experiments in the bioreactor when cells are absent, or cell concentration is low so that consumption by cells can be neglected. The latter condition is present immediately after inoculating the bioreactor. Consider Eq. 5.2 under these conditions:

$$\frac{d(C_{DO})}{dt} = k_{La}\left(C_{DO}^{*} - C_{DO}\right)$$

If we allow steady state to occur, the dissolved oxygen concentration will reach saturation value, C_{DO}^{*} and the concentration-time profile will be flat, as shown in the Fig. 5.6.

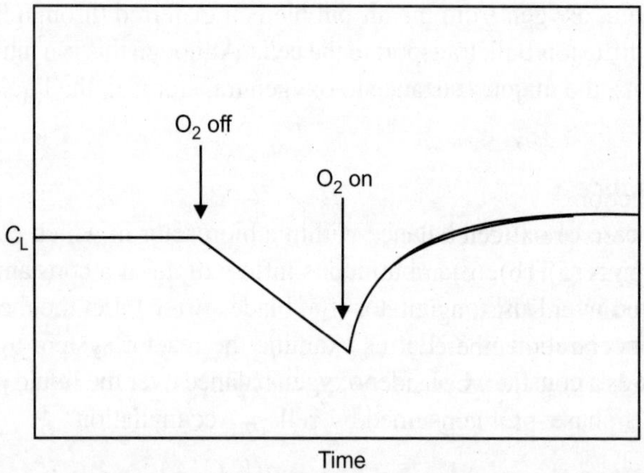

Fig 5.6. Oxygen profile during a transient. The responses will be exponential, rather than straight lines.

If the oxygen source (air) is replaced by nitrogen, the resulting response of the system is described by the above equation with the term, C_{DO}^{*} set to zero. That is,

$$\frac{d(C_{DO})}{dt} = k_{La}\left(0 - C_{DO}\right) \text{ and } C_{DO}(t=0) = C_{DO}^{*}$$

The solution to the above is:

$$C_{DO} = C_{DO}^{*} \text{ Exp } (-k_{La}t)$$

If one plots the response on a semi-log plot, the slope will equal to the negative of mass transfer coefficient. It is relatively a simple experiment and the data analysis is also easy to do. When other type of transient mass transfer experiments are conducted, the above equations should be suitably modified. For example for the case of nitrogen to air switch, we should suitably modify the solution because the initial condition is now different.

SCALE-UP AND SCALE-DOWN

Scale-up means increasing the scale of a fermentation, for example from the laboratory scale to the pilot plant scale or from the pilot plant scale to the production scale. Increase in scale means an increase in volume and the problems of process scale-up are due to the different ways in which process parameters are affected by the size of the unit. It is the task of the fermentation technologist to increase the scale of a fermentation without a decrease in yield or if a yield reduction occurs, to identify the factor which gives rise to the decrease and to rectify it. The major factors involved in scale-up are:

1. Inoculum development. An increase in scale may mean the extra stages have to be incorporated into the inoculum development program.
2. Sterilisation. Sterilisation is a scale dependent factor because the number of contaminating micro-organisms in a fermenter must be reduced to the same absolute number regardless of scale. Thus, when the scale of a process is increased the sterilisation regime must be adjusted accordingly, which may result in a change in the quality of the medium after sterilisation.
3. Environmental parameters. The increase in scale may result in a changed environment for the organism. These environmental parameters may be summarised as follows:
   (a) Nutrient availability.
   (b) pH.
   (c) Temperature.
   (d) Dissolved oxygen concentration.
   (e) Shear conditions.
   (f) Dissolved carbon dioxide concentration.
   (g) Foam production.

All the above parameters are affected by agitation and aeration, either in terms of bulk mixing or the provision of oxygen. Points (a), (b), (c) are related to bulk mixing whist (d), (e), (f) and (g) are related to airflow and oxygen transfer. Thus, agitation and aeration tends to dominate the scale-up literature. However, it should always be remembered that inoculum development and sterilisation difficulties may be the reason for a decrease in yield when a process is scaled up and that achieving the correct aeration/agitation regime is not the only problem to be addressed.

Scale-up of Aeration/Agitation Regimes in Stirred Tank Reactors

From the list of environmental parameters affected by aeration and agitation it will be appreciated that it is extremely unlikely that the conditions of the small-scale fermentation will be replicated precisely on the large scale. Thus, the most important criteria for a particular fermentation must be established and the scale-up based on reproducing those characteristics. The scale-up window represents the boundaries imposed by the environmental parameters and cost on the aeration/agitation regime and is shown in Fig. 5.7. Suitable conditions of mixing and oxygen transfer can be obtained with a range of aeration/agitation combinations.

The two axis of Fig. 5.7 are agitation and aeration and the zone within the hexagon represents suitable aeration/agitation regimes. The boundary of the hexagon is defined by the limits of oxygen supply, carbon dioxide accumulation, shear damage to the cells, cost, foam formation and bulk mixing. For example, the agitation rate must fall between a minimum and maximum value—mixing is inadequate below the minimum level and shear damage to the cells is too great above the maximum value. The limits for aeration are determined at the minimum end by oxygen limitation and carbon dioxide accumulation and at the maximum end by foam formation. The shape of the window will depend on the fermentation—for example, the supply of oxygen would be irrelevant in an anaerobic fermentation, whereas the limitation due to shear would be of major importance in the scale-up of animal cell fermentations.

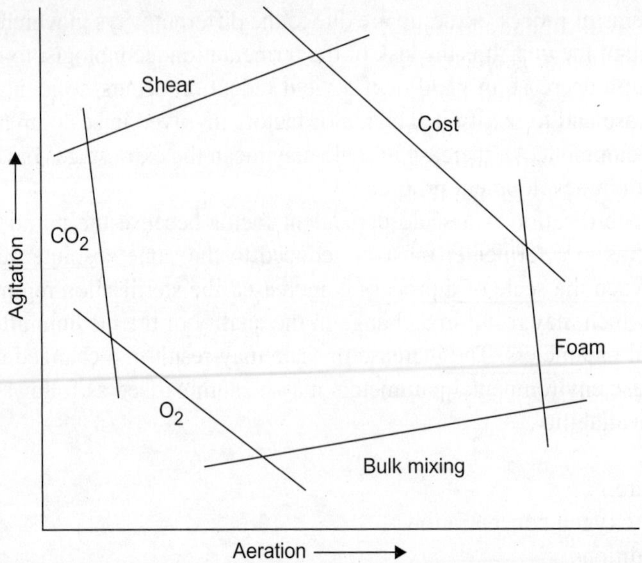

Fig. 5.7. The 'scale-up' window defining the operating boundaries for aeration and agitation in the scale-up of a fermentation.

The solution of the scale-up problem is three-fold:

1. The identification of the principal environmental domain affected by aeration and agitation in the fermentation, e.g. oxygen concentration, shear, bulk mixing.
2. The identification of a process variable (or variables) which affects the identified environmental domain.
3. The calculation of the value of the process variable to be used on the large scale which will result in the replication of the same environmental conditions on both scales.

The process variables which affect mixing and mass transfer are summarised in Table 5.1. Thus, if dissolved oxygen concentration is perceived as the overriding environmental condition then power consumption per unit volume and volumetric airflow rate per unit volume should be maintained constant on scale-up.

However, as a result, the other parameters will not be the same in the larger scale and therefore, neither will the environmental factors which they influence. This phenomenon is summarised in Table 5.2 where a 125 fold increase in scale is represented.

If power consumption per unit volume is kept constant then impeller tip speed (i.e. shear) increases and flow min^{-1} vol^{-1} (i.e. mixing) decreases. If mixing is kept constant, an enormous (and totally uneconomic) increase in power is required and shear increases 5 fold.

If impeller tip speed (shear) is kept constant then power consumption (hence, k_{La}) and mixing decrease. This analysis indicates that it is economically impossible to maintain the same degree of mixing on scale-up and therefore, a decrease in yield may be due to mixing anomalies.

The most important environmental domains affected by aeration and agitation for the majority of fermentations are oxygen concentration and shear. Thus, the most widely used scale-up criteria are the maintenance of a constant k_{La} or constant shear conditions. Constant shear may be achieved by scaling up on the basis of constant impeller tip speed. Constant k_{La} may be achieved on the basis of constant power consumption per unit volume and constant volumetric airflow rate. The operating variable dictating constant power consumption in geometrically similar vessels is the agitator speed. The agitator speed on the large scale is then calculated from the correlations between k_{La} and power consumption and between power consumption and operating variables. The accuracy of these scale-up techniques is only as good as the power and k_{La} correlations, so it is worth expending some considerable time to test the validity of potential correlations for the fermentation in question.

Table 5.1. The effect of process variables on mass transfer or mixing characteristics.

Process variable	*Mass transfer or mixing characteristic affected*
Power consumption per unit volume	Oxygen-transfer rate
Volumetric air flow rate	Oxygen-transfer rate
Impeller tip seed	Shear rate
Pumping rate	Mixing time
Reynolds number	Heat transfer

Table 5.2. The effect of the choice of scale-up criteria on operating conditions in the scaled-up vessel. Based on scale-up from 80 dm^3 to 10^4 dm^3.

Criterion used in scale-up	*Effect on the operating conditions on the large scale (large scale value/small scale value)*			
	P	*P/V*	*Flow min^{-1} vol^{-1}*	*ND_i*
P/V	125.0	1.0	0.34	1.7
Flow min^{-1} vol^{-1}	3125.0	25.0	1.0	5.0
ND_i (impeller tip speed)	25.0	0.2	0.2	1.0
Reynolds number	0.2	0.0016	0.04	0.2

Scale-up of Air-lift Reactors

Bubble columns and air-lift vessels tend to be scaled-up on the basis of geometric similarity and constant gas velocity. Under these conditions the k_{La} and shear rate in the two scales will be similar. The major difference will be the height of the vessels resulting in increased pressure at the base of the larger vessel. This would result in higher oxygen and carbon dioxide solubility which would give a higher k_{La} but might result in carbon dioxide inhibition.

The other problem in the scale-up of air-lift systems is that the organism is exposed to extremes of oxygen levels in the riser and downcomer and the effects of these conditions should be investigated on the laboratory scale.

Scale-down Methods

Scale-down is the situation where laboratory or pilot-scale experiments are conducted under conditions which mimic the industrial-scale conditions. This approach is important in both the development of a new product and the improvement of an existing full-scale fermentation. Frequently, conditions achievable on a laboratory scale are impractical on an industrial scale, which means that if inappropriate conditions have been used in the laboratory unrealistic yield objectives may be set for the scaled-up process. The aspects to consider in the design of laboratory or pilot-plant experiments in the context of scale-down may be summarised as follows:

Medium design

Media relevant to the industrial situation should be used in development experiments.

Medium sterilisation

If the medium is to be batch sterilised on the large scale its exposure time at a high temperature will be much greater than that experienced in the laboratory or pilot plant. Thus, the sterilisation times on the smaller scales should be increased to mimic the industrial situation. Alternatively, medium sterilised in the production fermenter may be used in the laboratory and pilot plant. This highlights the advantage of continuous sterilisation where little loss of medium quality occurs. Furthermore, the same continuous steriliser may be used for both full-scale and pilot scale vessels.

Inoculation procedures

Due to a range of circumstances, it may not always be possible to inoculate every production fermentation with inoculum in optimum condition. The scale-down approach can be used to predict the consequences of such events by mimicking these situations in the laboratory, for example by storing inoculum or using inocula of different ages.

Number of generations

An industrial scale fermentation requires a greater number of generations than does a laboratory one; this may place more severe stability criteria on the process strain than may have been appreciated on the small scale. The industrial situation may be modelled in the laboratory by using serial subculture to ensure that the strain is sufficiently stable. This approach is particularly pertinent in the development of recombinant fermentations.

Mixing

It is almost inevitable that the degree of mixing will decrease with an increase in scale. Thus, it is possible to model inadequate mixing in the laboratory by subjecting the organism to pulse medium feeds or fluctuating process conditions such as oxygen concentration, pH and temperature. Such scaled-down experiments then allow predictions to be made about the suitability of new strains for industrial exploitation.

Oxygen transfer rate

Far higher oxygen transfer rates can be achieved in laboratory fermenters than in industrial-scale ones. Thus, unrealistic demands may be made of a fermentation plant if the development work has been done at very high oxygen-transfer rates. Therefore, the laboratory and pilot fermenters should reflect the oxygen transfer rates achievable in the full-scale fermenters. The adoption of these simple approaches to small scale experimentation can prevent many scale-up problems before they even occur.

FILTER STERILISATION OF FERMENTATION MEDIA

Media for animal-cell culture cannot be sterilised by steam because they contain heat-labile proteins. Thus, filtration is the method of choice and fixed pore or absolute filtration is the better system to use.

An ideal filtration system for the sterilisation of animal cell culture media must fulfil the following criteria:

1. The filtered medium must be free of fungal, bacterial and mycoplasma contamination.
2. There should be minimal adsorption of protein to the filter surface.
3. The filtered medium should be free of viruses.
4. The filtered medium should be free of endotoxins.

Several filter manufacturers supply absolute filtration systems for the sterilisation of animal cell culture medium. Such systems consist of membrane cartridges which are fitted into stainless steel, steam sterilisable modules. The membranes for media filtration are constructed from steam sterilisable hydrophilic material and are treated to produce a filtrate of particular quality. For example, if minimal protein adsorption is a major criterion then a specially coated filter membrane is used. It would be very difficult to construct a single filtration membrane which would fulfil all four criteria cited above. Thus, a series of filters are used to achieve the desired result.

The example shown in Fig. 5.8 illustrates a system to produce sterile, mycoplasma free serum and consists of four filters arranged in sequence.

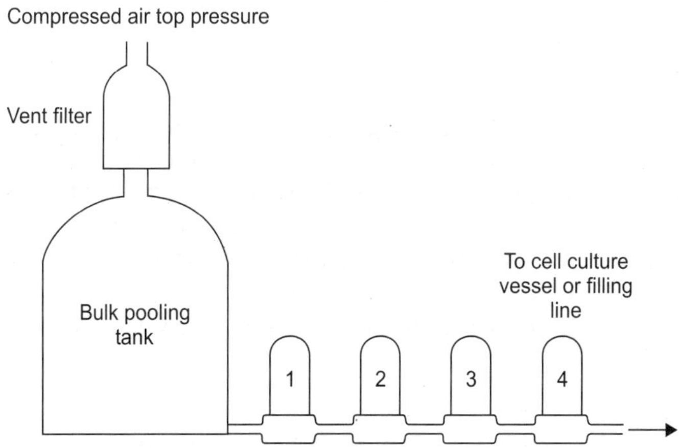

Filter 1: 5 μm absolute rated pre-filter for removal of coarse precipitates.
Filter 2: 0.5 μm absolute rated pre-filter for bulk bioburden removal.
Filter 3: 0.1. μm absolute rated single layer pre-filter for further bioburden and endotoxin removal.
Filter 4: 0.1 μm absolute rated double layer final filter for absolute sterility, mycoplasma removal and further endotoxin control.

Fig. 5.8. Filtration system for the provision of sterile, mycoplasma free serum.

The first filter is a positively charged polypropylene prefilter with an absolute rating of 5 μm for the removal of coarse precipitates, clot-like material and other gross contaminants; the second filter is also positively charged polypropylene but with an absolute rating of 0.5 μm for bulk microbial removal, deformable gels, lipid-based materials and endotoxin reduction; the third filter is a single layered, nylon/polyester positively charged filter with a 0.1 μm absolute rating for further microbial and endotoxin removal and optimum protection of the final filter; the fourth filter is similar to the third and has the same rating, but is double layered and removes mycoplasmas, gives absolute sterility and final endotoxin control. Thus, the combination of four filters gives a sequential removal of decreasingly small particles and prolongs the life of the final filter. If it is necessary to remove viral contamination then a final 0.04 μm nylon/polyester filter would be added.

Similar systems may be used in downstream processing of animal-cell products where the rating and properties of the filters would be optimised for the particular process. Figure 5.9 illustrates a system for the removal of cells and cell debris from an animal-cell fermentation broth.

The prefilter is a polypropylene 1.0 µm rated filter to remove the bulk of the cells and debris and the second filter is an hydroxyl modified nylon/polyester 0.2 µm rated filter giving absolute cell removal with minimal protein adsorption.

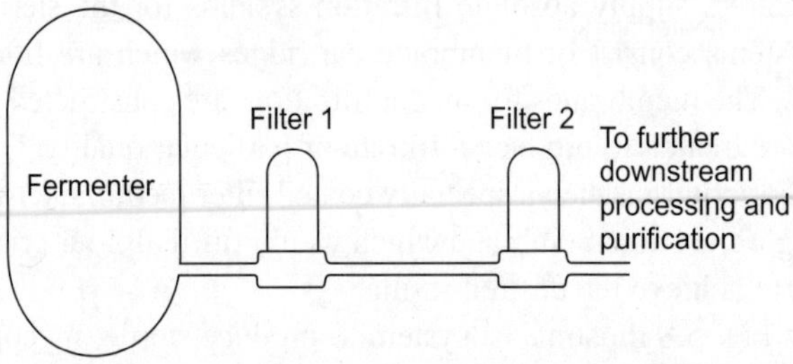

Filter 1:1.0 µm absolute rated pre-filter for bulk cell and cell debris removal.
Filter 2: 0.2 µm absolute rated single layer 'bio-inert' filter for final bioburden removal.

Fig. 5.9. Filtration system for the removal of cells and cell debris from an animal cell culture fermentation.

Filter Sterilisation of Air

Aerobic fermentations require the continuous addition of considerable quantities of sterile air. Although it is possible to sterilise air by heat treatment, the most commonly used sterilisation process is filtration. Fixed pore filters (which have an absolute rating) are very widely used in the fermentation industry and several manufacturers produce filtration systems for air sterilisation. These systems, like those for the sterilisation of liquids, consist of pleated membrane cartridges designed to be accommodated in stainless steel modules. A sectioned filter unit is shown in Fig. 5.10. The most common construction material used for the pleated membranes for air sterilisation is PTFE, which is hydrophobic and is therefore resistant to wetting.

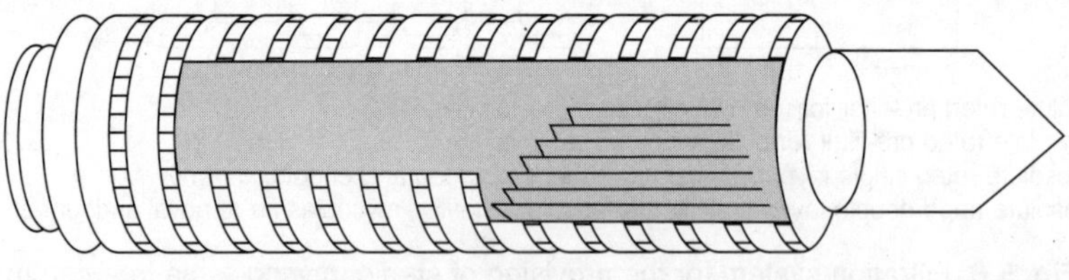

Fig. 5.10. An absolute membrane filter sectioned to show the pleated membrane structure.

Also, PTFE filters may be steam sterilised and are resistant to ammonia which may be injected into the air stream, prior to the filter, for pH control.

As was seen for the filter sterilisation of liquids it is essential that a pre-filter is incorporated upstream of the absolute filter. The prefilter traps large particles such as dust, oil and carbon (from the compressor) and pipescale and rust (from the pipework). The use of a coalescing pre-filter also ensures the removal of water from the air; entrained water is coalesced in the filter (air flow being from the inside of the filter to the outside) and is discharged via an automatic drain. Figure 5.11 illustrates the layout of such a filtration unit showing the steam sterilisation system and Fig. 5.12 is a photograph of the air sterilisation plant for an 85 m^3 fermenter.

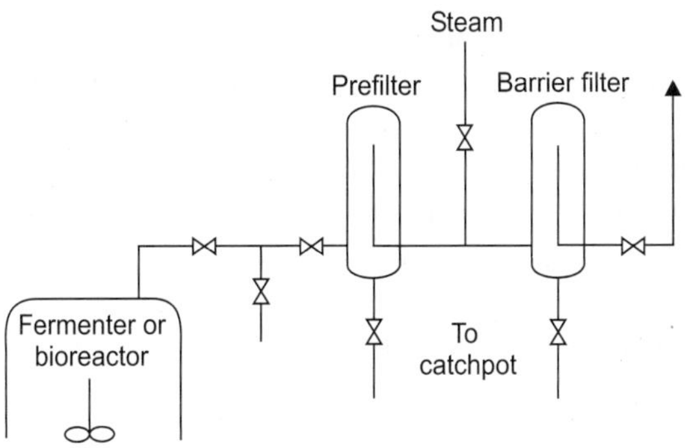

Fig. 5.11. Dual hydrophobic filter system for the sterilisation of off-gas from a fermenter.

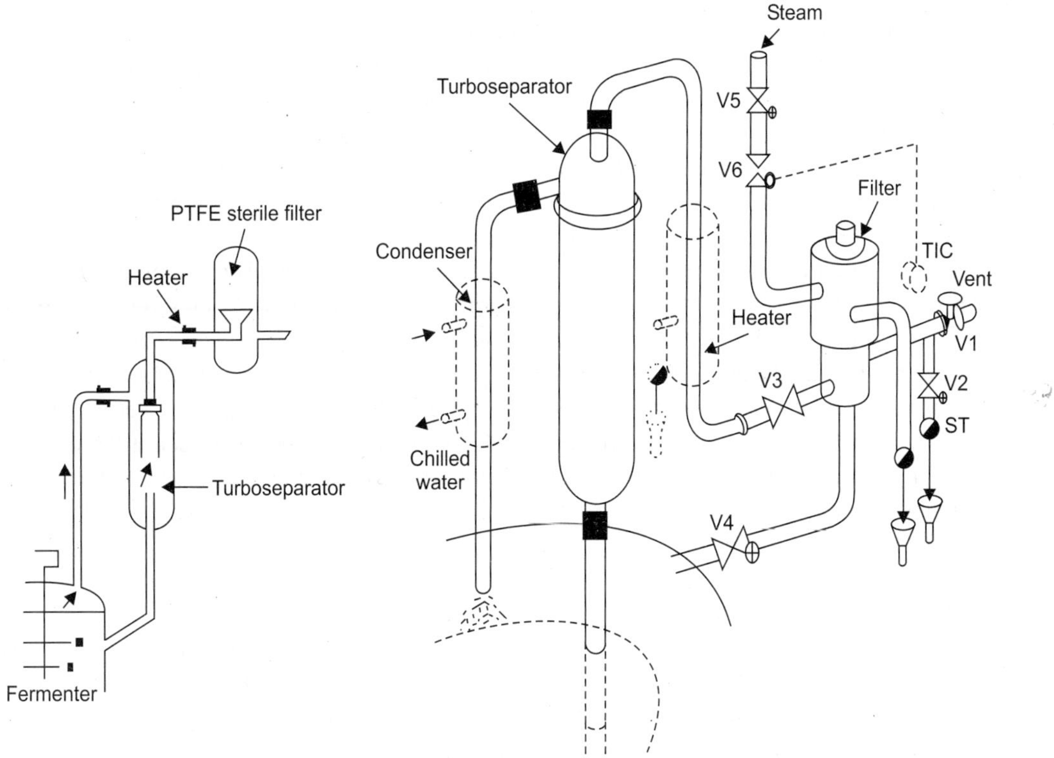

Fig. 5.12. A mechanical separator and hydrophobic filter system for the sterilisation of off-gas from a fermenter. Left. Cut-away diagram. Right, equipment arrangement, showing steam supply. V1–V6, valves; and O, steam, traps.

Sterilisation of Fermenter Exhaust Air

In many traditional fermentations the exhaust gas from the fermenter was vented without sterilisation or vented through relatively inefficient depth filters. With the advent of the use of recombinant organisms and a greater awareness of safety and emission levels of allergic compounds the containment of exhaust air is more common (and in the case of recombinant organisms, compulsory).

Fixed pore membrane modules are also used for this application but the system must be able to cope with the sterilisation of water saturated air, at a relatively high temperature and carrying a large contamination level.

Also, foam may overflow from the fermenter into the air exhaust line. Thus, some form of pretreatment of the exhaust gas is necessary before it enters the absolute filter. This pretreatment may be a hydrophobic prefilter or a mechanical separator to remove water, aerosol particles and foam. The pretreated air is then fed to a 0.2 μm hydrophobic filter. Again, it is important to appreciate that the filtration system must be steam sterilisable. Figures 5.11 and 5.12 illustrate the pre-filter and mechanical separator systems respectively.

Sterilisation of the Fermenter

The fermenter should be so designed that it may be steam sterilised under pressure. The medium may be sterilised in the vessel or separately and subsequently added aseptically. If the medium is sterilised *in situ* its temperature should be raised proper to the injection of live steam to prevent the formation of large amounts of condensate. This may be achieved by steam being introduced into the fermenter coils or jacket.

As every point of entry to and exit from the fermenter is a potential source of contamination, steam should be introduced through all the entry and exit points except the air outlet from which steam should be allowed to leave.

All pipes should be constructed as simply as possible and slope towards drainage points to make sure that steam reaches all parts of the equipment and is not excluded by siphons or pockets of condensate or mash. Each drainage point in the pipework should be fitted with a steam trap. Parker, Chain and Muller have all stressed the need to eliminate fine fissures or gaps such as flange seals which might be filled with nutrient solutions and micro-organisms. Hambleton described a high specification pilot scale fermenter with surfaces free of crevices greater than 0.05 mm depth, which is needed if the vessel was to be used for animal cells in suspension culture or on micro-carriers. For long-term aseptic operation welded joints should be used wherever possible, even though sections may have to be cut and re-welded during maintenance and repair.

Sterilisation of the Air Supply

Sterile air will be required in very large volumes in many aerobic fermentation processes. Although there are a number of ways of sterilising air, only two have found permanent application. These are heat and filtration. Heat is generally too costly for full-scale operation.

Historically, glass wool, glass fibre or mineral slag wool have been used as filter material, but currently most fermenters are fitted with cartridge-type filters. However, before the filter may be used, it too must be sterilised in association with the fermenter. Two procedures are commonly followed depending on the construction of the filter unit.

Figure 5.13 shows the simple unit. During sterilisation the main nonsterile air-inlet valve A is shut and initially the sterile air valve B is closed. Steam is applied at valve C and air is purged downwards through the filter to a bleed valve at the base. When the steam is issuing freely through the bleed valve, the valve B is opened to allow steam to pass into the fermenter as well as the filter. It is essential to adjust the bleed valve to ensure that the correct sterilisation pressure is maintained in the fermenter and filter for the remainder of the sterilisation cycle.

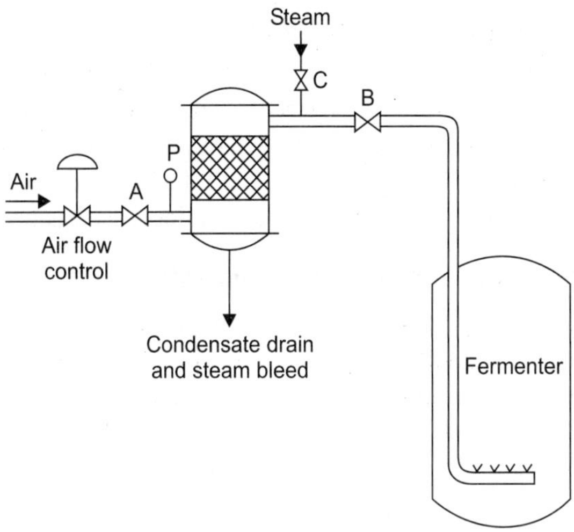

Fig. 5.13. An arrangement of packed air filter and fermenter.

An alternative approach is to use a steam-jacketed air filter (Fig. 5.14). At the beginning of a sterilisation cycle valve A will be closed and steam passed through valves B and C and bled out of D. Simultaneously steam will be passed into the steam jacket through valve F and out of G. When steam is issuing freely from valve D, valve F, may be opened and steam circulated into the fermenter. The bleed valve D will have to be adjusted to ensure that the correct pressure is maintained.

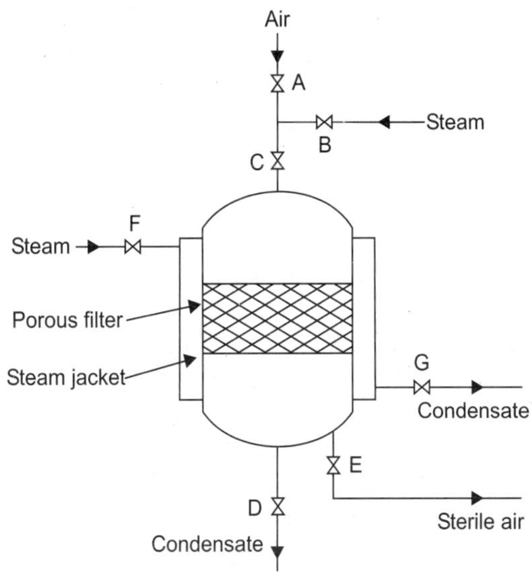

Fig. 5.14. Design for a simple steam-jacketed air-filter.

Once the sterilisation cycle is complete, valves B and E are closed and valve A is opened to allow air to pass through the heated filter and out of valve D to dry the filter. Finally the steam supply to the steam jacket is stopped. Valve D is closed and valve E opened, thus introducing sterile air into the fermenter to achieve a slight positive pressure in the vessel.

Sterilisation of the Exhaust Gas from a Fermenter

Sterilisation of the exhaust gas can be achieved by 0.2 μm filters on the outlet pipe. Under normal operation aerosol formation may occur in the fermenter and moisture and solid matter may then plug the filter. To ensure satisfactory operation a cyclone separator (for solids) and a coalescer (for liquids) would be included upstream of two filters in series. The filters should be checked regularly to ensure that no viable cells are escaping. A test procedure to ensure integrity has been described by Hesselink.

Addition of Inoculum, Nutrients and Other Supplements

To prevent contamination when operating a fermenter requiring good industrial large scale practice (GILSP), it is essential that both the addition vessel and the fermenter should be maintained at a positive pressure and that the addition port is equipped with a steam supply.

At containment levels 1 and B2, the addition of inoculum, nutrients, etc. must be carried out in such a way that release of micro-organisms is restricted. This should be done by aseptic piercing of membranes or connections with steam locks.

At Containment levels 2 and B 3/4, no micro-organisms must be released during inoculation or other additions. In order to meet these stringent requirements all connections must be screwed or clamped and all pipelines must be steam sterilisable.

DEVELOPMENT OF INOCULA FOR BACTERIAL PROCESSES

The main objective of inoculum development for traditional bacterial fermentations is to produce an active inoculum which will give as short a lag phase as possible in subsequent culture. A long lag phase is disadvantageous in that not only is time wasted but also medium is consumed in maintaining a viable culture prior to growth. The length of the lag phase is affected by the size of the inoculum and its physiological condition. As already stated, the inoculum size normally ranges between 3 and 10 per cent of the culture volume. Lincoln stressed that bacterial inocula should be transferred in the logarithmic phase of growth, when the cells are still metabolically active. The age of the inoculum is particularly important in the growth of sporulating bacteria, for sporulation is induced at the end of the logarithmic phase and the use of an inoculum containing a high percentage of spores would result in a long lag phase in a subsequent fermentation.

Keay quoted the use of a 5 per cent inoculum of a logarithmically growing culture of a thermophilic *Bacillus* for the production of proteases. Aunstrup described a two-stage inoculum development program for the production of proteases by *Bacillus subtilis*. Inoculum for a seed fermenter was grown for 1 to 2 days on a solid or liquid medium and then transferred to a seed vessel where the organism was allowed to grow for a further ten generations before transfer to the production stage. Priest and Sharp cited the use of a 5 per cent inoculum, still in the exponential phase, for the commercial production of *Bacillus* α-amylase. Underkofler emphasised that, in the production of bacterial enzymes, the lag phase in plant fermenters could be almost completely eliminated by using inoculum medium of the same composition as used in the production fermenter and employing large inocula of actively growing seed cultures. The inoculum development program for a pilot-plant scale process for the production of vitamin B_{12} from *Pseudomonas denitrificans* is shown in Fig. 5.15.

The necessity to use an inoculum in an active physiological state is taken to its extreme in the production of vinegar. The acetic-acid bacteria used in the vinegar process are extremely sensitive to oxygen starvation.

Stock culture
Lyophilised with skim milk

↓

Maintenance culture
Agar slope incubated 4 days at 28°C

↓

Seed culture—first stage
2 dm^3 flask containing 0.6 dm^3 medium inoculated with culture from one slope; incubated with shaking for 48 hours at 28°C

↓

Seed culture—second stage
40–80 dm^3 fermenter containing 25–50 dm^3 medium inoculated with 1–1.2 per cent first stage seed culture. Incubated 25–30 hours at 32°C

↓

Production culture
500 dm^3 fermenter with 300 dm^3 medium inoculated with 5 per cent second stage seed culture. Incubated at 32°C for 140–160 hours

Fig. 5.15. The inoculum development program for a vitamin B_{12} pilot scale fermentation using *Pseudomonas denitrificans.*

Therefore, to avoid disturbing the system, the cells at the end of a fermentation are used as inoculum for the next batch by removing approximately 60 per cent of the culture and restoring the original level with fresh medium. The advantage of a highly active inoculum apparently outweighs the disadvantages of possible strain degeneration and contaminant accumulation. However, strain stability is a major concern in inoculum development for fermentations employing recombinant bacteria. Sabatie and others demonstrated that plasmid stability and productivity in an *E. coli* biotin fermentation was greatly improved if stationary, rather than exponential, phase cells were used as inoculum. They postulated that the plasmid copy number may be higher in stationary cells than in exponential ones, resulting in a lower plasmid loss in the subsequent fermentation when a stationary culture is used as inoculum. A stationary phase inoculum would result in a lag phase, but this disadvantage was more than compensated for by the considerable improvement in plasmid retention and biotin production compared with that obtained using an exponential inoculum.

In the lactic-acid fermentation the producing organism may be inhibited by lactic acid. Thus, production of lactic acid in the seed fermentation may result in the generation of poor quality inoculum. Yamamoto generated high quality inoculum of *Lactococcus lactis* IO-1 on a laboratory scale using electrodialysis seed culture which reduced the lactate in the inoculum and reduced the length of the lag phase in the production fermentation.

An example of the development of inoculum for an anaerobic bacterial process is provided by the clostridial acetone-butanol fermentation. The process was out-competed by the petrochemical industry but there is still considerable interest in re-establishing the fermentation. The inoculum development program described by McNeil and Kristiansen is given in Table 5.3.

Table 5.3. The inoculum development program for the clostridial acetone-butanol fermentation.

Stage	*Cultural conditions*	*Incubation time (hours)*
1	Heat-shocked spore suspension inoculated into 150 cm^3 of potato glucose medium	12
2	Stage 1 culture used as inoculum for 500 cm^3 molasses medium	6
3	Stage 2 culture used as inoculum for 9 dm^3 molasses medium	9
4	Stage 3 culture used as inoculum for 90,000 dm^3 molasses medium	–

The stock culture is heat shocked to stimulate spore germination and to eliminate the weaker spores. The production stage is inoculated with a very low volume and this corresponds with Lurie's description of the South African acetone-butanol fermentation in which a 1,00,000 dm^3 fermenter was inoculated with only 10 dm^3 of seed.

The use of such small inocula necessitates the achievement of as near perfect conditions as possible to prevent contamination and to avoid an abnormally long lag phase.

DEVELOPMENT OF INOCULA FOR MYCELIAL PROCESSES

The preparation of inocula for fermentations employing mycelial (filamentous) organisms is more involved than that for unicellular bacterial and yeast processes. The majority of industrially important fungi and streptomycetes are capable of asexual sporulation, so it is common practice to use a spore suspension as seed during an inoculum development program.

A major advantage of a spore inoculum is that it contains far more 'propagules' than a vegetative culture. Three basic techniques have been developed to produce a high concentration of spores for use as an inoculum.

Sporulation on Solidified Media

Most fungi and streptomycetes will sporulate on suitable agar media but a large surface area must be employed to produce sufficient spores. Parker described the 'roll-bottle' technique for the production of spores of *Penicillium chrysogenum*. Quantities of medium (300 cm^3) containing 3 per cent agar were sterilised in 1 dm^3 cylindrical bottles, which were then cooled to 45°C and rotated on a roller mill so that the agar set as a cylindrical shell inside the bottle. The bottles were inoculated with a spore suspension from a sub-master slope and incubated at 24°C for 6 to 7 days. Parker claimed that although the use of the 'roll-bottle' involved some sacrifice in ease of visual examination, it provided a large surface area for cultivation of spores in a vessel of a convenient size for handling in the laboratory.

Hockenhull described the production of 10^{10} spores of *Penicillium chrysogenum* on a 300 cm^2 agar layer in a Roux bottle and El Sayed quoted the use of spore suspensions derived from agar media containing between 10^7 and 10^8 cm^{-3}.

Butterworth described the use of a Roux bottle for the production of a spore inoculum of *Streptomyces clavuligerus* for the production of clavulanic acid. The spores produced from one bottle containing 200 cm^2 agar surface could be used to inoculate a 75 dm^3 seed fermenter which, in turn, was used to inoculate a 1500 dm^3 fermenter. The clavulanic acid inoculum development program is illustrated in Fig. 5.16. Some representative solidified media for the production of streptomycete and fungal spores are given in Tables 5.4 and 5.5 respectively.

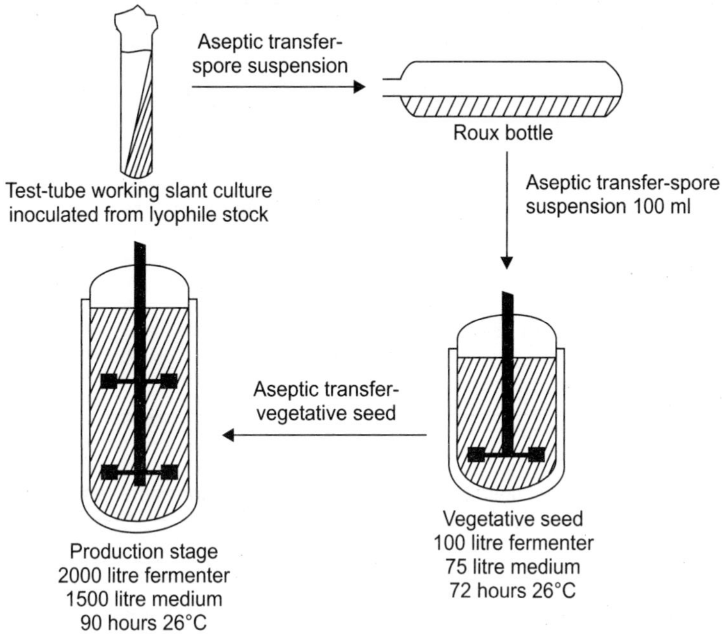

Fig. 5.16. The inoculum development program for the production of clavulanic acid from *Streptomyces clavuligerus*.

Table 5.4. Solidified media suitable for the sporulation of some representative *streptomycetes*.

Organism	*Product*	*Medium*	*Per cent*
S. aureofaciens	Tetracycline*	Malt extract (Difco)	1.0%
		Yeast extract (Difco)	0.4%
		Glucose	0.4%
S. erythreus	Erythromycin*	Beef extract (Difco)	0.1%
		Yeast extract (Difco)	0.1%
		Casamino acids (Difco)	0.2%
		Glucose	0.2%
S. vinaceus	Viomycin*	Corn-steep liquor (50% dry matter)	1.0%
		Starch	1.0%
		$(NH_4)_2SO_4$	0.3%
		NaCl	0.3%
		$CaCO_3$	0.3%
S. clavuligerus	Clavulanic acid*	Soluble starch	1.0%
		K_2HPO_4	0.1%
		$MgSO_4·7H_2O$	0.1%
		NaCl	0.1%
		$(NH_4)_2SO_4$	0.2%

(Contd ...)

Organism	*Product*	*Medium*	*Per cent*
		$CaCO_3$	0.2%
		$FeSO_4 \cdot 7H_2O$	0.001%
		$MnCl_2 \cdot 4H_2O$	0.001%
		$ZnSO_4 \cdot 7H_2O$	0.001%
S. hygroscopicus	Maridomycin	Soluble starch	1.0%
		Peptone	0.04%
		Meat extract	0.02%
		Yeast extract	0.02%
		N-Z amine (type A)	0.02%
		Agar	2.0%

*Agar content not quoted.

Table 5.5. Solidified media suitable for the sporulation of some representative fungi.

Fungus	*Medium*	
		(g dm^{-3})
Penicillium chrysogenum	Glycerol	7.5
	Cane molasses	7.5
	Curbay BG	2.5
	$MgSO_4 \cdot 7H_2O$	0.05
	KH_2PO_4	0.06
	Peptone	5.0
	NaCl	4.0
	Agar	20
Aspergillus niger	Molasses	300
	KH_2PO_4	0.5
	Agar	20

Sporulation on Solid Media

Many filamentous organisms will sporulate profusely on the surface of cereal grains from which the spores may be harvested. Substrates such as barley, hard wheat bran, ground maize and rice are all suitable for the sporulation of a wide range of fungi. The sporulation of a given fungus is particularly affected by the amount of water added to the cereal before sterilisation and the relative humidity of the atmosphere, which should be as high as possible during sporulation. Smith has described a system for the sporulation of *Aspergillus ochraces* in which a 2.8 dm^3 Fernbach flask containing 200 grams of 'pot' barley or 100 grams of moistened wheat bran produced 5×10^{11} conidia after six days at 28°C and 98 per cent relative humidity. This was 5 times the number obtainable from a Roux bottle batched with Sabouraud agar and 50 times the number obtainable from such a vessel batched with Difco Nutrient Agar, incubated for the same time period. Vezina has published a list of fungi which are capable of sporulating heavily on cereal grains. El-Sayed quoted the use of cooked rice for the production of spores of *Penicillium* and *Cephalosporium* in penicillin and cephalosporin inoculum development.

Sansing and Cieglem described the mass production of spores of several *Aspergillus* and *Penicillium* species on whole loaves of white bread and Podojil quoted the use of millet for the sporulation of *Streptomyces aureofaciens* in the development of inoculum for the chlortetracycline fermentation (Fig. 5.17).

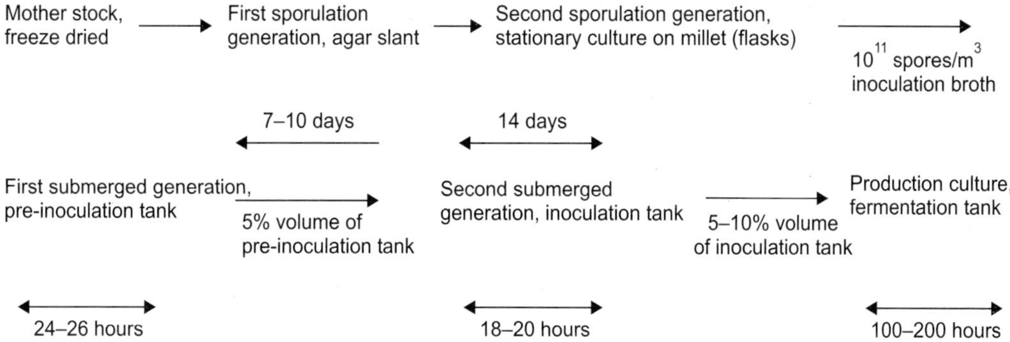

Fig. 5.17. The inoculum development program for the production of chlortetracycline by *Streptomyces aureofaciens.*

Sporulation in Submerged Culture

Many fungi will sporulate in submerged culture provided a suitable medium is employed. This technique is more convenient than the use of solid or solidified media because it is easier to operate aseptically and it may be applied on a large scale. The technique was first adopted by Foster who induced submerged sporulation in *Penicillium notatum* by including 2.5 per cent calcium chloride in a defined nitrate-sucrose medium. An example of the use of this technique for the production of inoculum for an industrial fermentation is provided by the griseofulvin process. Rhodes described the conditions necessary for the submerged sporulation of the griseofulvin-producing fungus, *Penicillium patulum* and the medium utilised is given in Table 5.6. These authors found that for prolific sporulation the nitrogen level had to be limited to between 0.05 and 0.1 per cent w/v and that good aeration had to be maintained. Also, an interaction was demonstrated between the nitrogen level and aeration in that the lower the degree of aeration the lower the concentration of nitrogen needed to induce sporulation. Submerged sporulation was induced by inoculating 600 cm^3 of the above medium, in a 2 dm^3 shake flask, with spores from a well-sporulated Czapek-Dox agar culture and incubating at 25°C for 7 days. The resulting suspension of spores was then used as a 10 per cent inoculum for a vegetative seed stage in a stirred fermenter, the seed culture subsequently providing a 10 per cent inoculum for the production fermentation.

Table 5.6. Media for the submerged sporulation of selected fungi.

Rhodes: *Penicillium patulum*		
Whey powder, to give	Lactose 3.5%	
	Nitrogen 0.05%	
KH_2PO_4		0.4%
KCl		0.05%
Corn-steep liquor solids to give approximately 0.04% N		0.38%
Foster: *Penicillium notatum*		
Sucrose		2.0%

(Contd ...)

$NaNO_3$	0.6%
KH_2PO_4	0.15%
$MgSO_4{\cdot}7H_2O$	0.05%
$CaCl_2$	2.5%
Vezina: *Aspergillus ochraceus*	
Glucose	2.5%
NaCl	2.5%
Corn-steep liquor	0.5%
Molasses	5.0%

Most actinomycetes do not sporulate in submerged culture and thus, solid or solidified media tend to be used for the production of spore inocula.

Use of the Spore Inoculum

The stage in an inoculum development program at which a large-scale spore inoculum is used varies according to the process; it appears to be common practice that the penultimate stage is so inoculated, but this will depend on the scale of the production fermentation. In the inoculum development program for the early penicillin fermentation described by Parker the penultimate stage was inoculated with a spore suspension (from a roll-bottle) and this stage may have produced either a vegetative or a submerged spore inoculum for the final fermentation. For the griseofulvin process, Rhodes stated that the spore suspension obtained from the submerged sporulation stage could either be used for direct inoculation of the production fermentation or it could be germinated in an inoculum development medium to yield a vegetative inoculum for the final fermentation. The latter course was preferred and an inoculum volume of 7–10 per cent was used. From Figs 5.16 to 5.18, it can be seen that in the clavulanic acid process the spore inoculum is used to inoculate the final seed stage, in the chlortetracycline process a vegetative stage is interspersed between the spore inoculated batch and the production fermentation and in the sagamicin process the spore inoculum is used at a very early stage followed by the vegetative growth.

When considering the production of gluconic acid by *Aspergillus niger*, Lockwood discussed the merits of inoculating the final fermentation directly with a spore suspension as compared with germinating the spores in a seed tank to give a vegetative inoculum. Direct spore inoculation would avoid the cost of installation and operation of the seed tanks whereas the use of germinated spores would reduce the fermentation time of the final stage, thus allowing a greater number of fermentations to be carried out per year. However, labour costs for the production of the vegetative inoculum could be almost as high as for the final fermentation although some of these costs may be recovered, in that gluconic acid produced in the penultimate stage would be recoverable from the final fermentation broth and would contribute to the buffering capacity throughout the fermentation. Thus, Lockwood claimed that the choice of inoculum for the production stage depends on the length of the cycle of the fermentation process, plant size and the availability and cost of labour.

Inoculum Development for Vegetative Fungi

Some fungi will not produce asexual spores and therefore, an inoculum of vegetative mycelium must be used. *Gibberella fujikuroi* is such a fungus and is used for the commercial production of gibberellin. Hansen described an inoculum development program for the gibberellin fermentation. Cultures were

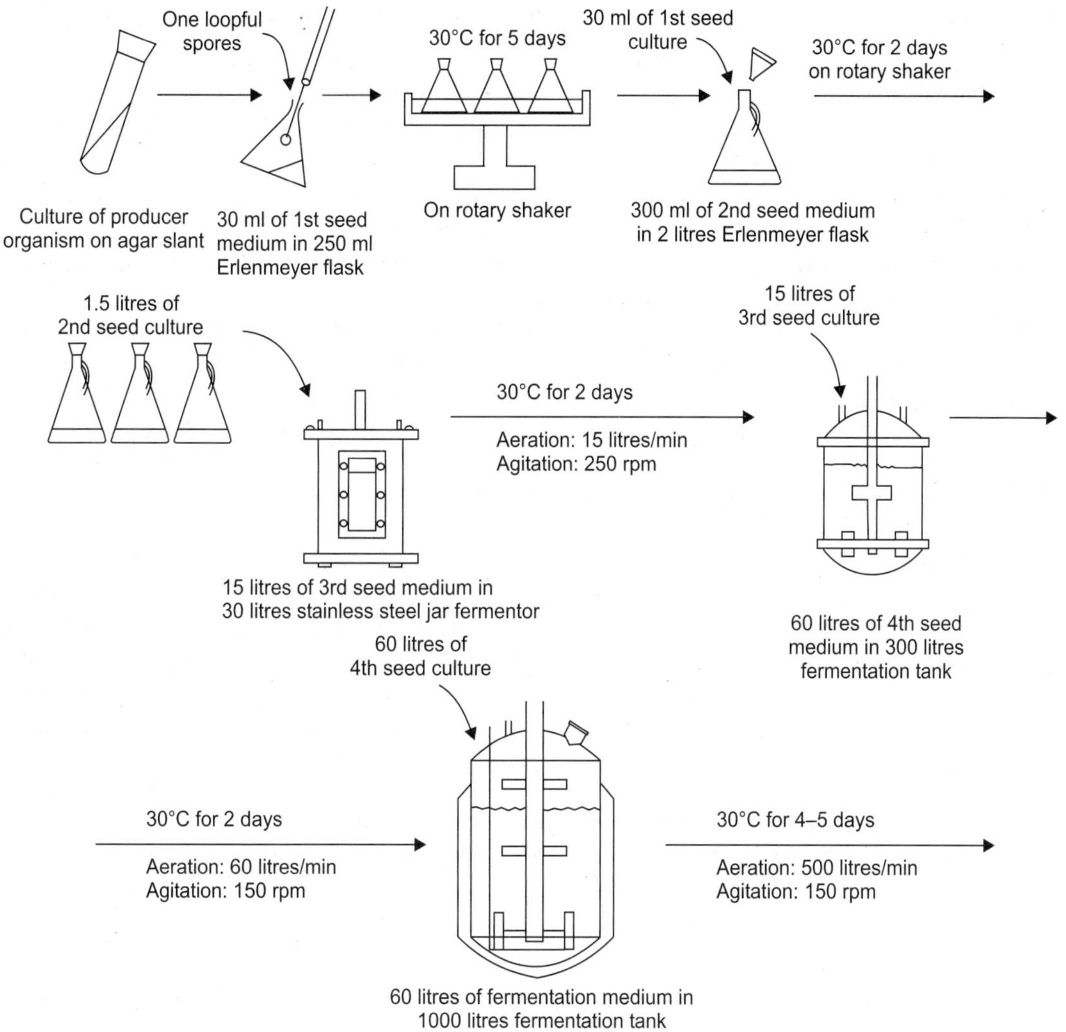

Fig. 5.18. The inoculum development program for the production of sagamicin by *Micromonospora sagamiensis.*

grown on long slants (25 × 10 mm test tubes) of potato dextrose agar for 1 week at 24°C. Growth from three slants were scraped off and transferred to a 9 dm^3 carboy containing 4 dm^3 of a liquid medium composed of 2 per cent glucose, 0.3 per cent $MgSO_4 \cdot 7H_2O$, 0.3 per cent NH_4Cl and 0.3 per cent KH_2PO_4. The medium was aerated for 75 hours at 28°C before transfer to a 100 dm^3 seed fermenter containing the same medium.

The major problem in using vegetative mycelium as initial seed is the difficulty of obtaining a uniform, standard inoculum. The procedure may be improved by fragmenting the mycelium in a homogeniser, such as a Waring blender, prior to use as inoculum. This method provides a large number of mycelial particles and therefore a large number of growing points. Worgan has given a detailed account of the use of this technique in the preparation of inocula for the submerged culture of the higher fungi.

Effect of the Inoculum on the Morphology of Filamentous Organisms in Submerged Culture

When filamentous fungi are grown in submerged culture the type of growth varies from the 'pellet' form, consisting of compact discrete masses of hyphae, to the filamentous form in which the hyphae form a homogenous suspension dispersed through the medium. The filamentous type of habit gives rise to an extremely viscous broth which may be very difficult to aerate adequately, whereas the pellet type of habit gives rise to a far less viscous, but also less homogenous, broth. In a pelleted culture there is a danger that the mycelium at the centre of the pellet may be starved of nutrients and oxygen due to diffusion limitations. Also, there is considerable evidence that the morphological form of the organism influences the productivity of the culture, but whether this is due to the phenomena already mentioned or to some form of metabolic control is far from clear. Thus, some fermentations are carried out with the fungus in a filamentous habit, whereas others are carried out with the organism growing as pellets. For example, filamentous growth has been claimed to be optimum for penicillin production by *P. chrysogenum*, whereas pelleted growth has been claimed to be optimum for citric acid production from *Aspergillus niger* and lovastatin from *Aspergillus terreus*. The necessity for filamentous growth is taken to the extreme in the ICI-Rank Hovis McDougal mycoprotein process where *Fusarium graminearium* is produced for human consumption. A highly filamentous morphology is required to produce the desired texture in the product which resembles the strength and eating texture of white and soft, red meats. Thus, in this process a median hyphal length of 400 μm is required.

The relevance of this consideration of mycelial morphology to inoculum development is that the morphology may be influenced considerably by both the concentration of spores in a spore inoculum and the inoculum development medium. Usually, a high spore inoculum will tend to produce a dispersed form of growth whilst a low one will favour pellet formation. The effect of the concentration of a spore inoculum on the morphology of *P. chrysogenum* is given in Table 5.7.

Table 5.7. The effect of spore concentration and medium on the morphology and penicillin productivity of *Penicillium chrysogenum*.

Medium	*Spore concentration in the medium*	*Morphology*
Camici:		
Corn-steep dextrin	More than 10×10^5 dm^{-3}	Filamentous
	Less than 10×10^5 dm^{-3}	Pellets
Czapek-dox	More than 3.0×10^5 dm^{-3}	Filamentous
	Less than 3.0×10^5 dm^{-3}	Pellets
Glucose, lactose and ammonium lactate	More than 2.0×10^5 dm^{-3}	Filamentous
	Less than 2.0×10^5 dm^{-3}	Pellets
Spore concentration in the inoculum (cm^{-3})	*Penicillin yield (units cm^{-3})*	*Morphology*
Calam:		
10^2	500	Dense pellets
10^3	1800	Dense pellets
2×10^3	4000	Open pellets
10^4	5000	Filamentous

Thus, in the commercial production of fungal products it is critical to grow the organism in the desired morphological form which necessitates the use of an inoculum which achieves this end. If the production fermentation is to be inoculated with a spore suspension then the spore concentration must be such as to produce the production culture in the desired morphological form; if a vegetative inoculum is to be used for the production fermentation then, again, the concentration of its spore inoculum must be such as to produce the vegetative inoculum in the desired morphological form. Although the effects of media on morphological form can be extremely varied dispersed growth is more likely in rich, complex media and pelleted growth tends to occur in chemically defined media. Thus, the medium used in the spore germination stage must be optimised in terms of the morphology of the inoculum.

An interesting series of experiments on the effects of inoculum conditions on the morphology of *Penicillium citrinum* were reported by Hosobuchi. This *Penicillium* species synthesises compound ML-236B, a precursor of pravastatin which is a cholesterol-lowering drug. Optimum productivity was achieved when the organism grew as compact pellets in the production fermentation. The vegetative inoculum for the production fermentation had to contain an optimum number of short, filamentous propagules in order to initiate pellet formation in the final culture. This was achieved by using a four-stage inoculum development program (initiated by a spore-inoculated shake flask) with very rich media in the third and fourth cultures. Thus, this system required a dispersed vegetative inoculum to generate a pelleted production fermentation.

The information available on the morphology of actinomycetes in submerged culture is very limited compared with that on fungi. However, Whitaker has reviewed the area and it is obvious that actinomycetes are capable of producing a wide range of morphological types. Also, it appears to be accepted that a dispersed mycelial morphology is desirable for most industrial actinomycete fermentations. Mycelial forms have been shown to be desirable for the production of streptomycin by *Streptomyces griseus* and turimycin by *S. hygroscopicus*, whereas the pelleted form of *S. nigrificans* was better for glucose isomerase production. As already discussed for the fungi, the concentration of spores in the inoculum has also been shown to influence the morphology of certain streptomycetes. These workers also demonstrated that medium composition and the shear forces operating during culture also affect morphological form. Thus, the principles applied to the optimisation of fungal inoculum development regimes are also relevant to actinomycete processes. Hunt and Stieber described the optimisation of the inoculum regime of a small-scale streptomycete cephamycin C fermentation. Pellet formation was observed to be detrimental to product formation and the key factor in establishing the correct form appeared to be the concentration of iron in the seed medium, a higher iron concentration giving the optimum inoculum.

BIOSENSORS IN BIOPROCESS MONITORING AND CONTROL

Biosensors belong to a subgroup of chemical sensors in which a biologically based mechanism is used for analyte detection. By definition, a biosensor is an analytical device that combines the specificity of a biological sensing element for the analyte of interest with a transducer to produce a signal proportional to target analyte concentration. This signal can result from: a change in pH, release or uptake of gases, light emission, heat emission, mass change, etc. The transducer converts the biological reaction into a measurable response such as current, potential, temperature change, modulation of light intensity, etc.

The current state of sensor technology permits the measurement and control of dissolved oxygen, dissolved carbon dioxide, pH, redox potential, temperature, agitation and the level of foam in the fermenter vessel. Recent advances in sensor technologies have led to the on-line determination of biomass through *in situ* optical density or fluorometric probes and capacitance measurements using low radio frequencies.

Oxygen uptake and/or carbon dioxide evolution during cell growth have been monitored using piezoelectric mass balances and off-gas analysis. In addition, head space analysis of volatiles such as methanol has been used to monitor and control fermentative processes.

Basic Components of Online Process Monitoring and Control

Any system for on-line monitoring and control must include three essential elements: (i) a sensor (that is a biosensor), (ii) a suitable analysis manifold employing an *in situ* or *ex situ* arrangement for contacting the fermentation liquor with the bioprobe, and (iii) a control system with the necessary hardware and control algorithms in order to employ a suitable control strategy. In order to use the analysis data efficiently, the time lag between sampling and analysis must be within the timescales of the bioprocess. Thus, in a bacterial cultivation process with short generation times, the time delay should be in the order of a very few minutes. In cultivation processes employing much slower growing mammalian cells, this time delay can be in the order of 1–2 hours. In a reactor employing an immobilised enzyme, the time delay should be of the order of seconds. The time delay is a function of the sampling, the sample handling, thc analysis and the data process.

Sensors can be interfaced to a biotechnological process in different ways. A biosensor can be used as an *in situ* probe or can be separated from the fermentation broth by a filtration unit, e.g. tangential flow units in a by-pass. An alternative and perhaps more practical way of utilising biosensors in continuous fermentation monitoring and control is to use them in conjunction with flow injection analysis (FIA). In FIA, the liquid sample to be analysed, which forms a discrete zone, is injected into a moving, non-segmented carrier stream flowing continuously past a detector. The many advantages of this technique when applied to process control include:

1. Reduced risk of contamination because the sample is not returned to the bioreactor.
2. Automatic recalibration of the sensor to counter any problems associated with drift.
3. Short response time.
4. Requirement for low sample volumes.
5. Multi-component monitoring if an array of biosensors is used as the detector.

In terms of process monitoring and control applications, the emphasis has been on the use of enzymes and in particular in an enzyme electrode configuration. [The enzyme electrode is a combination of any type of electrochemical probe with a thin layer (10–200 μm) of immobilised enzyme.] Typically, the progress of a particular enzymic reaction (which is related to the concentration of analyte) is measured by monitoring the rate of product formation or the rate of disappearance of a reactant. If either the product or reactant is electroactive, the reaction may be monitored directly using amperometry.

Biosensors based on thermal effects

These types of devices are based on the principle that enzymic reactions are exothermic in nature. This fact can be used to calorimetrically determine the amount of substrate converted to product during the enzyme-catalysed reaction.

Biosensors based on optical effects

In sensors based on optical methods of detection, the modulation of electromagnetic radiation such as UV/vis absorption, bio and chemiluminescence, reflectance, fluorescence and surface plasmon oscillation caused by the interaction of the biocatalyst with the target analyte is monitored optically. A key consideration in these types of device is the use of optical fibre waveguide technology. This involves the synthesis of two ideas: (i) the use of optical fibres to bring light of the appropriate wavelength from a spectrometer and back again, and (ii) the use of optical fibres as an immobilisation support for the biocatalyst thereby allowing the reagent to be used on a continuous rather than a 'one-off' basis.

Potentiometric biosensors

The biosensors used for process monitoring and control are an on-line fully automated system for the monitoring and control of glucose (carbon source) concentration in a fed-batch yeast fermentation system.

Amperometric biosensors

Amperometric biosensors for process monitoring and control can be traced back to the late 1980s. For example, Romette and Cooney described a biosensor for monitoring glutamine in mammalian cell cultures. Glutaminase and glutamate oxidase were insolubilised in a thin gelatine membrane that was subsequently fixed over the surface of a polarographic oxygen electrode. Correlation between oxygen consumption and glutamine concentration provided a linear response between 0.2 and 2 mM glutamine.

Thermal biosensors

The approach to process monitoring and control using thermal biosensors as the analytical device is essentially the same as that described for electrochemical sensors with the important difference that in this case, enthalpy changes are measured as opposed to electrical ones. Usage of these devices either as *in situ* devices or in an flow injection analysis (FIA) manifold is however, common.

COMPUTER APPLICATIONS IN FERMENTATION TECHNOLOGY

Since the initial use of computers in the 1960s for modelling fermentation process and in processes control for production of glutamic acid and penicillin, there have been numerous computer applications in fermentation technology. Initially, the use of large computers was restricted because of their cost but reductions in costs and the availability of cheaper small computers has widened interest in their possible applications. The availability of efficient small computers has led to their use for pilot plants and laboratory systems since the financial investment for the on-line computer amounts to a relatively insignificant part of the whole system.

There are three distinct areas of computer function, they are:

1. Logging of process data: Data logging is performed by the data acquisition system which has both hardware and software components.
2. Data analysis (reduction of logged data): Data reduction is performed by the data-analysis system, which is a computer program based on a series of selected mathematical equations.
3. Process control: Process control is also performed using a computer program. Signals from the computer are fed to pumps, valves or switches via the interface.

At this point it is necessary to be aware that there are two distinct fundamental approaches to computer control of fermenters. The first is when the fermenter is under the direct control of the computer software and is termed direct digital control (DDC). The second approach involves the use of independent controllers to manage all control functions of a fermenter and the computer communicates with the controller only to exchange information. This is termed supervisory set-point control (SSC).

It is possible to analyse data, compare it with model systems in a data store and use control programs which will lead to process optimisation. However, process optimisation by this method is not a widely used procedure in the fermentation industries at present. It is important to be aware of these different applications, since this will influence the size and type of computer system which will be appropriate for the precise role that it is intended to perform, whether in a laboratory, a pilot plant or manufacturing plant or a combination of these three.

When a computer is linked to a fermenter to operate as a control and recording system, a number of factors must be considered to ensure that all the components interact and function satisfactorily for control and data logging. A DDC system will be used as an example to explain computer controlled addition of a liquid from a reservoir to a fermenter.

The small computer itself is dedicated solely to one or more fermenters. This computer is coupled to a real-time clock, which determines how frequently readings from the sensor(s) should be taken and possibly recorded. The other ancillary equipment linked directly to the computer might include a visual display unit, a data store, a teletype, a graphic display unit, a print out, alarms and a barometer.

The small computer is often connected to a large main frame computer for random access, not on a real-time scale, but for long-term data storage and retrieval and for complex data analysis which will not be utilised subsequently in real-time control.

It is also possible to develop programs so that on-line instruments can be checked regularly and re-calibrated when necessary. Swartz and Cooney were able to routinely recalibrate a paramagnetic oxygen analyser and an infrared carbon dioxide analyser every 12 hours utilising a program which connected a gas of known composition to the analyser and subsequently monitored the analyser outputs.

Data Logging

The simplest task for a computer is data logging. Some of the following parameters can be measured by sensors which produce a signal which is compatible with the computer system: (i) temperature, (ii) pressure, (iii) flow measurement of gases and liquids, (iv) foam control, (v) microbial biomass, (vi) measurement and control of dissolved oxygen, (vii) inlet and exit gas analysis, and (viii) pH measurement and control. Programs have been developed so that by reference to the real-time clock, the signals from the appropriate sensors will be scanned sequentially in a pre-determined pattern and logged in a data store. Typically, this may be 2 to 60 seconds intervals and the data is printed out on a visual display unit. In preliminary scanning cycles the values are compared with pre-defined limit values and deviations from these values result in an error print out or if more extreme then an alarm may be activated. In the final cycle of a sequence, say every 5 to 60 minutes, the program instructs that the sensor readings are permanently recorded on a print out or in data store.

At the same time as on-line data is being recorded from sensors, analytical data for broth viscosity, microbial growth, substrate and precursor utilisation and product formation, which have to be determined separately may be logged into the data store for specific known times.

Thus, it is now possible to record data continuously for a range of parameters from a number of fermenters simultaneously using minimal manpower, provided that the capital outlay is made for fermenters with suitable instrumentation coupled with adequate computer facilities.

Data Analysis

Because a computer can undertake so many calculations very rapidly, it is possible to design programs to analyse fermentation data in a number of ways. A linked mainframe computer may be used for part of this analysis as well as the dedicated small computer. A number of the monitoring systems are given in Table 5.8. Gateway sensors are so-called because the information they yield can be processed to give further information about the fermentation.

Table 5.8. Gateway sensors.

Sensor	*Information that may be determined from the sensor signal*
pH	Acid product formation
Dissolved oxygen	Oxygen-transfer rate
Oxygen in exit gas	Oxygen-uptake rate
Gas-flow rate	Oxygen-uptake rate
Carbon dioxide in exit gas	Carbon dioxide evolution rate
Gas-flow rate	Carbon dioxide evolution rate
Oxygen-uptake rate	Respiratory quotient
Carbon dioxide evolution rate	Respiratory quotient
Sugar-level and feed rate	Yield and cell density
Carbon dioxide evolution rate	Yield and cell density

Process Control

Arminger and Moran recognised three levels of process control that might be incorporated into a system. Each higher level involves more complex programs and needs a greater overall understanding of the process. The first level of control, which is already routinely used in the chemical industries, involves sequencing operations, such as manipulating values or starting or stopping pumps, instrument recalibration, on-line maintenance and fail-safe shutdown procedures. In most of these operations the time base is at least in the order of minutes, so that high-speed manipulations are not vital. Two applications in fermentation processes are sterilisation cycles and medium batching.

The next level of computer control involves process control of temperature, pH, foam control, etc. where the sensors are directly interfaced to a computer direct digital control (DDC). When this is done separate controller units are not needed. The computer program determines the set point values and the control algorithms, such as PID, are part of the computer software package. Better control is possible as the control algorithms are mathematically stored functions rather than electrical functions. This procedure allows for greater flexibility and more precise representation of a process control policy. The system is not very expensive as separate electronic controllers are no longer needed, but computer failure can cause major problems unless there is some manual back-up facility.

The most advanced level of control is concerned with process optimisation. This will involve understanding a process, being able to monitor what is happening and being able to control it to achieve and maintain optimum conditions. Firstly, there is a need for suitable on-line sensors to monitor the process continuously.

Computer-aided Design of Integrated Biochemical Process

As a result of the advances in molecular biology and genetic engineering, the scientific community has come to realise the great potential for developing new products and systems through novel use of micro-organisms and enzymes. The challenge for the biochemical industry is now to scale-up and commercialise those products. This is a difficult task, especially for small corporations, because it is a complex problem that requires co-ordination of a large number of activities across many disciplines. Computer-aided process design tools have been successfully used in the chemical process industries for over three decades to scale-up and optimise integrated processes for the production of petrochemicals and other products. Similar benefits can be expected from the use of such tools in the biochemical industries.

Computer-aided bioprocess design tools can play an important role in the development of biochemical processes and commercialisation of biological products. At the early stages of project selection, such tools can be used to screen the large number of projects ideas from a profitability point of view and help focus development efforts on the most promising projects. During process development, such tools can be used to analyse and evaluate alternative processing schemes, reduce the impact of the whole process on the environment, interpret experimental results and help design experimental protocols. During final design and plant construction, such tools can be used to optimise the entire process from a total-systems point of view. All these benefits substantially reduce cost and development time associated with a biological product.

Chapter 6

Recovery and Purification of Products

INTRODUCTION

Downstream processing refers to the recovery and purification of biosynthetic products, particularly pharmaceuticals, from natural sources such as animal or plant tissue or fermentation broth, including the recycling of salvageable components and the proper treatment and disposal of waste. It is an essential step in the manufacture of pharmaceuticals such as antibiotics, hormones (e.g. insulin and human growth hormone), antibodies (e.g. infliximab and abciximab) and vaccines, antibodies and enzymes used in diagnostics, industrial enzymes, and natural fragrance and flavour compounds. Downstream processing is usually considered a specialised field in biochemical engineering, itself a specialisation within chemical engineering, though many of the key technologies were developed by chemists and biologists for laboratory-scale separation of biological products.

Downstream processing and analytical bioseparation both refer to the separation or purification of biological products, but at different scales of operation and for different purposes. Downstream processing implies manufacture of a purified product fit for a specific use, generally in marketable quantities, while analytical bioseparation refers to purification for the sole purpose of measuring a component or components of a mixture, and may deal with sample sizes as small as a single cell.

STAGES IN DOWNSTREAM PROCESSING

A widely recognised heuristic for categorising downstream processing operations divides them into four groups which are applied in order to bring a product from its natural state as a component of a tissue, cell or fermentation broth through progressive improvements in purity and concentration.

Removal of insolubles: Removal of insolubles is the first step and involves the capture of the product as a solute in a particulate-free liquid, for example the separation of cells, cell debris or other particulate matter from fermentation broth containing an antibiotic. Typical operations to achieve this are filtration, centrifugation, sedimentation, flocculation, electro-precipitation, and gravity settling. Additional operations such as grinding, homogenisation, or leaching, required to recover products from solid sources such as plant and animal tissues, are usually included in this group.

Product isolation: Product isolation is the removal of those components whose properties vary markedly from that of the desired product. For most products, water is the chief impurity and isolation steps are designed to remove most of it, reducing the volume of material to be handled and concentrating the product. Solvent extraction, adsorption, ultrafiltration, and precipitation are some of the unit operations involved.

Product purification: Product purification is done to separate those contaminants that resemble the product very closely in physical and chemical properties. Consequently steps in this stage are expensive to carry out and require sensitive and sophisticated equipment. This stage contributes a significant fraction of the entire downstream processing expenditure. Examples of operations include affinity, size exclusion, reversed phase chromatography, crystallisation and fractional precipitation.

Product polishing: Product polishing describes the final processing steps which end with packaging of the product in a form that is stable, easily transportable and convenient. Crystallisation, desiccation, lyophilisation and spray drying are typical unit operations. Depending on the product and its intended use, polishing may also include operations to sterilise the product and remove or deactivate trace contaminants which might compromise product safety. Such operations might include the removal of viruses or depyrogenation.

A few product recovery methods may be considered to combine two or more stages. For example, expanded bed adsorption accomplishes removal of insolubles and product isolation in a single step. Affinity chromatography often isolates and purifies in a single step.

Removal of Insoluble Components

Filtration

Filtration is a mechanical or physical operation which is used for the separation of solids from fluids (liquids or gases) by interposing a medium through which only the fluid can pass. Oversize solids in the fluid are retained, but the separation is not complete; solids will be contaminated with some fluid and filtrate will contain fine particles (depending on the pore size and filter thickness).

1. Filtration is used to separate particles and fluid in a suspension, where the fluid can be a liquid, a gas or a supercritical fluid. Depending on the application, either one or both of the components may be isolated.
2. Filtration, as a physical operation is very important in chemistry for the separation of materials of different chemical composition. A solvent is chosen which dissolves one component, while not dissolving the other. By dissolving the mixture in the chosen solvent, one component will go into the solution and pass through the filter, while the other will be retained. This is one of the most important techniques used by chemists to purify compounds.
3. Filtration is also important and widely used as one of the unit operations of chemical engineering. It may be simultaneously combined with other unit operations to process the feed stream, as in the biofilter, which is a combined filter and biological digestion device.
4. Filtration differs from sieving, where separation occurs at a single perforated layer (a sieve). In sieving, particles that are too big pass through the holes of the sieve are retained (see particle size distribution). In filtration, a multilayer lattice retains those particles that are unable to follow the tortuous channels of the filter. Oversize particles may form a cake layer on top of the filter and may also block the filter lattice, preventing the fluid phase from crossing the filter (blinding). Commercially, the term filter is applied to membranes where the separation lattice is so thin that the surface becomes the main zone of particle separation, even though these products might be described as sieves.
5. Filtration differs from adsorption, where it is not the physical size of particles that causes separation but the effects of surface charge. Some adsorption devices containing activated charcoal and ion exchange resin are commercially called filters, although filtration is not their principal function.

Liquid filtration

There are many different methods of filtration, all aim to attain the separation of substances. Separation is achieved by some form of interaction between the substance or objects to be removed and the filter. The substance that is to pass through the filter must be a fluid, i.e. a liquid or gas. Methods vary depending on the location of the targeted material, i.e. whether it is in the fluid phase or not.

Filter media

Two main types of filter media are employed in the chemical laboratory—surface filter, a solid sieve which traps the solid particles, with or without the aid of filter paper (e.g. Büchner funnel, Belt filter, rotary vacuum-drum filter, crossflow filters), and a depth filter, a bed of granular material which retains the solid particles as it passes (e.g. sand filter). The first type allows the solid particles, i.e. the residue, to be collected intact; the second type does not permit this. However, the second type is less prone to clogging due to the greater surface area where the particles can be trapped. Also, when the solid particles are very fine, it is often cheaper and easier to discard the contaminated granules than to clean the solid sieve.

Filter media can be cleaned by rinsing with solvents or detergents. Alternatively, in engineering applications, such as swimming pool water treatment plants, they may be cleaned by backwashing.

Achieving flow through the filter

Fluids flow through a filter due to a difference in pressure-fluid flows from the high pressure side to the low pressure side of the filter, leaving some material behind. The simplest method to achieve this is by gravity and can be seen in the coffee maker example. In the laboratory, pressure in the form of compressed air on the feed side (or vacuum on the filtrate side) may be applied to make the filtration process faster, though this may lead to clogging or the passage of fine particles. Alternatively, the liquid may flow through the filter by the force exerted by a pump, a method commonly used in industry when a reduced filtration time is important. In this case, the filter need not be mounted vertically.

Filter aid

Certain filter aids may be used to aid filtration. These are often incompressible diatomaceous earth or kieselguhr, which is composed primarily of silica. Also used are wood cellulose and other inert porous solids.

These filter aids can be used in two different ways. They can be used as a precoat before the slurry is filtered. This will prevent gelatinous-type solids from plugging the filter medium and also give a clearer filtrate. They can also be added to the slurry before filtration. This increases the porosity of the cake and reduces resistance of the cake during filtration. In a rotary filter, the filter aid may be applied as a precoat; subsequently, thin slices of this layer are sliced off with the cake.

The use of filter aids is usually limited to cases where the cake is discarded or where the precipitate can be separated chemically from the filter.

Alternatives

Filtration is a more efficient method for the separation of mixtures than decantation, but is much more time consuming. If very small amounts of solution are involved, most of the solution may be soaked up by the filter medium.

An alternative to filtration is centrifugation—instead of filtering the mixture of solid and liquid particles, the mixture is centrifuged to force the (usually) denser solid to the bottom, where it often forms a firm cake. The liquid above can then be decanted. This method is especially useful for separating solids which do not filter well, such as gelatinous or fine particles. These solids can clog or pass through the filter, respectively.

Centrifugation

Centrifugation is a process that involves the use of the centrifugal force for the separation of mixtures, used in industry and in laboratory settings. More-dense components of the mixture migrate away from the axis of the centrifuge, while less-dense components of the mixture migrate towards the axis. Chemists and biologists may increase the effective gravitational force on a test tube so as to more rapidly and completely cause the precipitate ('pellet') to gather on the bottom of the tube. The remaining solution is properly called the 'supernate' or 'supernatant liquid'. The supernatant liquid is then either quickly decanted from the tube without disturbing the precipitate, or withdrawn with a Pasteur pipette.

The rate of centrifugation is specified by the acceleration applied to the sample, typically measured in revolutions per minute (rpm) or g. The particles' settling velocity in centrifugation is a function of their size and shape; centrifugal acceleration, the volume fraction of solids present, the density difference between the particle and the liquid, and the viscosity.

In the chemical and food industries, special centrifuges can process a continuous stream of particle-laden liquid. It is worth noting that centrifugation is the most common method used for uranium enrichment, relying on the slight mass difference between atoms of U238 and U235 in uranium hexafluoride gas.

Centrifugation in biotechnology

Microcentrifuges and superspeed centrifuges

In microcentrifugation, centrifuges are run in batch to isolate small volumes of biological molecules or cells (prokaryotic and eukaryotic). Nuclei is also often purified via microcentrifugation. Microcentrifuge tubes generally hold 1.5–2 ml of liquid, and are spun at maximum angular speeds of 12,000–13,000 rpms. Microcentrifuges are small and have rotors that can quickly change speeds. Superspeed centrifuges work similarly to microcentrifuges, but are conducted via larger scale processes. Superspeed centrifuges are also used for purifying cells and nuclei, but in larger quantities. These centrifuges are used to purify 25–30 ml of solution within a tube. Additionally, larger centrifuges also reach higher angular velocities (around 30,000 rpm) and also use a larger rotor.

Ultracentrifugation

Ultracentrifugation makes use of high centrifugal force for studying properties of biological particles. While microcentrifugation and superspeed centrifugation are used strictly to purify cells and nuclei, ultracentrifugation can isolate much smaller particles, including ribosomes, proteins, and viruses. Ultracentrifuges can also be used in the study of membrane fractionation. This occurs because ultracentrifuges can reach maximum angular velocites in excess of 70,000 rpm. Additionally, while microcentrifuges and supercentrifuges separate particles in batch, ultracentrifuges can separate molecules in batch and continuous flow systems.

In addition to purification, analytical ultracentrifugation (AUC) can be used for determination of macromolecular properties, including the amino acid composition of a protein, the protein's current conformation, or properties of that conformation. In analytical ultracentrifuges, concentration of solute is measured using optical calibrations. For low concentrations, the Beer-Lambert law can be used to measure the concentration. Analytical ultracentrifuges can be used to simulate physiological conditions (correct pH and temperature).

In analytical ultracentrifuges, molecular properties can be modelled through sedimentation velocity analysis or sedimentation equilibrium analysis. In sedimentation velocity analysis, concentrations and solute properties are modelled continuously over time. Sedimentation velocity analysis can be used to determine the macromolecule's shape, mass, composition, and conformational properties. During sedimentation equilibrium analysis, centrifugation has stopped and particle movement is based on diffusion. This allows for modelling of the mass of the particle as well as the chemical equilibrium properties of interacting solutes.

Sedimentation

Sedimentation is the tendency for particles in suspension or molecules in solution to settle out of the fluid in which they are entrained, and come to rest against a wall. This is due to their motion through the fluid in response to the forces acting on them: these forces can be due to gravity, centrifugal acceleration or electromagnetism.

Sedimentation may pertain to objects of various sizes, ranging from large rocks in flowing water to suspensions of dust and pollen particles to cellular suspensions to solutions of single molecules such as proteins and peptides. Even small molecules such as aspirin can be sedimented, although it can be difficult to apply a sufficiently strong force to produce significant sedimentation.

The term is typically used in geology, to describe the deposition of sediment which results in the formation of sedimentary rock, and in various chemical and environmental fields to describe the motions of often-smaller particles and molecules.

Flocculation

Flocculation is, in the field of chemistry, a process where colloids come out of suspension in the form of floc or flakes. The action differs from precipitation in that, prior to flocculation, colloids are merely suspended in a liquid and not actually dissolved in a solution.

According to the IUPAC definition, flocculation is 'a process of contact and adhesion whereby the particles of a dispersion form larger-size clusters.' Flocculation is synonymous with agglomeration and coagulation.

For emulsions, flocculation describes clustering of individual dispersed droplets together, whereby the individual droplets do not lose their identity. Flocculation is thus the initial step leading to further ageing of the emulsion (droplet coalescence and the ultimate separation of the phases).

In biology, the process is used to refer to the asexual aggregation of micro-organisms, most commonly brewing yeast at the end of a brew.

Flocculation and sedimentation are widely employed in the purification of drinking water as well as sewage treatment, stormwater treatment and treatment of other industrial waste-water streams.

Flocculants, or flocculating agents, are chemicals that promote flocculation by causing colloids and other suspended particles in liquids to aggregate, forming a floc. Flocculants are used in water treatment processes to improve the sedimentation or filterability of small particles. For example, a flocculant may be used in swimming pool or drinking water filtration to aid removal of microscopic particles which would otherwise cause the water to be turbid (cloudy) and which would be difficult or impossible to remove by filtration alone.

Product Isolation

Adsorption

Adsorption is the accumulation of atoms or molecules on the surface of a material. This process creates a film of the adsorbate (the molecules or atoms being accumulated) on the adsorbent's surface. It is different from absorption, in which a substance diffuses into a liquid or solid to form a solution. The term sorption encompasses both processes, while desorption is the reverse process of 'adsorption'.

In simple terms, adsorption is 'the collection of a substance onto the surface of adsorbent solids.' It is a removal process where certain particles are bound to an adsorbent particle surface by either chemical or physical attraction. Adsorption is often confused with absorption, where the substance being collected or removed actually penetrates into the other substance.

Adsorption is present in many natural physical, biological, and chemical systems, and is widely used in industrial applications such as activated charcoal, capturing and using waste heat to provide cold water for air conditioning and other process requirements (adsorption chillers),synthetic resins, and water purification. Adsorption, ion exchange, and chromatography are sorption processes in which certain adsorbates are selectively transferred from the fluid phase to the surface of insoluble, rigid particles suspended in a vessel or packed in a column.

Similar to surface tension, adsorption is a consequence of surface energy. In a bulk material, all the bonding requirements (be they ionic, covalent, or metallic) of the constituent atoms of the material are filled by other atoms in the material. However, atoms on the surface of the adsorbent are not wholly surrounded by other adsorbent atoms and therefore can attract adsorbates. The exact nature of the bonding depends on the details of the species involved, but the adsorption process is generally classified as physisorption (characteristic of weak van der Waals forces) or chemisorption (characteristic of covalent bonding).

Precipitation

Precipitation is the formation of a solid in a solution during a chemical reaction. When the reaction occurs, the solid formed is called the precipitate, and the liquid remaining above the solid is called the supernate. Powders derived from precipitation have also historically been known as flowers.

Natural methods of precipitation include settling or sedimentation, where a solid forms over a period of time due to ambient forces like gravity or centrifugation. During chemical reactions, precipitation may also occur particularly if an insoluble substance is introduced into a solution and the density happens to be greater (otherwise the precipitate would float or form a suspension). With soluble substances, precipitation is accelerated once the solution becomes supersaturated.

An important stage of the precipitation process is the onset of nucleation. The creation of a hypothetical solid particle includes the formation of an interface, which requires some energy based on the relative surface energy of the solid and the solution. If this energy is not available, and no suitable nucleation surface is available, supersaturation occurs.

Chromatography

Chromatography is the collective term for a set of laboratory techniques for the separation of mixtures. It involves passing a mixture dissolved in a 'mobile phase' through a stationary phase, which separates the analyte to be measured from other molecules in the mixture based on differential partitioning between the mobile and stationary phases. Subtle differences in compounds partition coefficient results in differential retention on the stationary phase and thus separation.

Chromatography may be preparative or analytical. The purpose of preparative chromatography is to separate the components of a mixture for further use (and is thus a form of purification). Analytical chromatography is done normally with smaller amounts of material and is for measuring the relative proportions of analytes in a mixture. The two are not mutually exclusive.

Chromatography became developed substantially as a result of the work of Archer John Porter Martin and Richard Laurence Millington Synge during the 1940s and 1950s. They established the principles and basic techniques of partition chromatography, and their work encouraged the rapid development of several types of chromatography method: paper chromatography, gas chromatography, and what would become known as high performance liquid chromatography. Since then, the technology has advanced rapidly. Researchers found that the main principles of Tsvet's chromatography could be applied in many different ways, resulting in the different varieties of chromatography described below. Simultaneously, advances continually improved the technical performance of chromatography, allowing the separation of increasingly similar molecules.

Chromatography terms

1. The analyte is the substance that is to be separated during chromatography.
2. Analytical chromatography is used to determine the existence and possibly also the concentration of analyte(s) in a sample.
3. A bonded phase is a stationary phase that is covalently bonded to the support particles or to the inside wall of the column tubing.
4. A chromatogram is the visual output of the chromatograph. In the case of an optimal separation, different peaks or patterns on the chromatogram correspond to different components of the separated mixture.
5. A chromatograph is equipment that enables a sophisticated separation, e.g. gas chromatographic or liquid chromatographic separation.
6. Chromatography is a physical method of separation in which the components to be separated are distributed between two phases, one of which is stationary (stationary phase) while the other (the mobile phase) moves in a definite direction.
7. The effluent is the mobile phase leaving the column.
8. An immobilised phase is a stationary phase which is immobilised on the support particles, or on the inner wall of the column tubing.
9. The mobile phase is the phase which moves in a definite direction. It may be a liquid (LC and CEC), a gas (GC), or a supercritical fluid (supercritical-fluid chromatography, SFC). A better definition: The mobile phase consists of the sample being separated/analysed and the solvent that moves the sample through the column. In one case of HPLC the solvent consists of a carbonate/bicarbonate solution and the sample is the anions being separated. The mobile phase moves through the chromatography column (the stationary phase) where the sample interacts with the stationary phase and is separated.
10. Preparative chromatography is used to purify sufficient quantities of a substance for further use, rather than analysis.
11. The retention time is the characteristic time it takes for a particular analyte to pass through the system (from the column inlet to the detector) under set conditions.
12. The sample is the matter analysed in chromatography. It may consist of a single component or it may be a mixture of components. When the sample is treated in the course of an analysis, the

phase or the phases containing the analytes of interest is/are referred to as the sample whereas everything out of interest separated from the sample before or in the course of the analysis is referred to as waste.

13. The solute refers to the sample components in partition chromatography.
14. The solvent refers to any substance capable of solubilising other substance, and especially the liquid mobile phase in LC.
15. The stationary phase is the substance which is fixed in place for the chromatography procedure. Examples include the silica layer in chromatography. Thin layer chromatography.

Techniques by chromatographic bed shape

Column chromatography

Column chromatography is a separation technique in which the stationary bed is within a tube. The particles of the solid stationary phase or the support coated with a liquid stationary phase may fill the whole inside volume of the tube (packed column) or be concentrated on or along the inside tube wall leaving an open, unrestricted path for the mobile phase in the middle part of the tube (open tubular column). Differences in rates of movement through the medium are calculated to different retention times of the sample.

In 1978, W. C. Still introduced a modified version of column chromatography called flash column chromatography (flash). The technique is very similar to the traditional column chromatography, except for that the solvent is driven through the column by applying positive pressure. This allowed most separations to be performed in less than 20 minutes, with improved separations compared to the old method. Modern flash chromatography systems are sold as pre-packed plastic cartridges, and the solvent is pumped through the cartridge. Systems may also be linked with detectors and fraction collectors providing automation. The introduction of gradient pumps resulted in quicker separations and less solvent usage.

A spreadsheet that assists in the successful development of flash columns has been developed. The spreadsheet estimates the retention volume and band volume of analytes, the fraction numbers expected to contain each analyte, and the resolution between adjacent peaks. This information allows users to select optimal parameters for preparative-scale separations before the flash column itself is attempted.

In expanded bed adsorption, a fluidised bed is used, rather than a solid phase made by a packed bed. This allows omission of initial clearing steps such as centrifugation and filtration, for culture broths or slurries of broken cells.

Planar chromatography

Planar chromatography is a separation technique in which the stationary phase is present as or on a plane. The plane can be a paper, serving as such or impregnated by a substance as the stationary bed (paper chromatography) or a layer of solid particles spread on a support such as a glass plate (thin layer chromatography). Different compounds in the sample mixture travel different distances according to how strongly they interact with the stationary phase as compared to the mobile phase. The specific retardation factor (R_f) of each chemical can be used to aid in the identification of an unknown substance.

Paper chromatography

Paper chromatography is a technique that involves placing a small dot or line of sample solution onto a strip of chromatography paper. The paper is placed in a jar containing a shallow layer of solvent and sealed. As the solvent rises through the paper, it meets the sample mixture which starts to travel up the paper with the solvent. This paper is made of cellulose, a polar substance, and the compounds within the mixture travel farther if they are non-polar. More polar substances bond with the cellulose paper more quickly, and therefore do not travel as far.

Thin layer chromatography

Thin layer chromatography (TLC) is a widely-employed laboratory technique and is similar to paper chromatography. However, instead of using a stationary phase of paper, it involves a stationary phase of a thin layer of adsorbent like silica gel, alumina, or cellulose on a flat, inert substrate. Compared to paper, it has the advantage of faster runs, better separations, and the choice between different adsorbents. For even better resolution and to allow for quantitation, high-performance TLC can be used.

Displacement chromatography

The basic principle of displacement chromatography is: A molecule with a high affinity for the chromatography matrix (the displacer) will compete effectively for binding sites, and thus displace all molecules with lesser affinities. There are distinct differences between displacement and elution chromatography. In elution mode, substances typically emerge from a column in narrow, Gaussian peaks. Wide separation of peaks, preferably to baseline, is desired in order to achieve maximum purification. The speed at which any component of a mixture travels down the column in elution mode depends on many factors. But for two substances to travel at different speeds, and thereby be resolved, there must be substantial differences in some interaction between the biomolecules and the chromatography matrix. Operating parameters are adjusted to maximise the effect of this difference. In many cases, baseline separation of the peaks can be achieved only with gradient elution and low column loadings. Thus, two drawbacks to elution mode chromatography, especially at the preparative scale, are operational complexity, due to gradient solvent pumping, and low throughput, due to low column loadings. Displacement chromatography has advantages over elution chromatography in that components are resolved into consecutive zones of pure substances rather than 'peaks'. Because the process takes advantage of the nonlinearity of the isotherms, a larger column feed can be separated on a given column with the purified components recovered at significantly higher concentrations.

Techniques by physical state of mobile phase

Gas chromatography

Gas chromatography (GC), also sometimes known as gas-liquid chromatography, (GLC), is a separation technique in which the mobile phase is a gas. Gas chromatography is always carried out in a column, which is typically 'packed' or 'capillary'.

Gas chromatography (GC) is based on a partition equilibrium of analyte between a solid stationary phase (often a liquid silicone-based material) and a mobile gas (most often helium). The stationary phase is adhered to the inside of a small-diameter glass tube (a capillary column) or a solid matrix inside a larger metal tube (a packed column). It is widely used in analytical chemistry; though the high temperatures used in GC make it unsuitable for high molecular weight biopolymers or proteins (heat will denature them), frequently encountered in biochemistry, it is well suited for use in the petrochemical, environmental monitoring, and industrial chemical fields. It is also used extensively in chemistry research.

Liquid chromatography

Liquid chromatography (LC) is a separation technique in which the mobile phase is a liquid. Liquid chromatography can be carried out either in a column or a plane. Present day liquid chromatography that generally utilises very small packing particles and a relatively high pressure is referred to as high performance liquid chromatography (HPLC).

In the HPLC technique, the sample is forced through a column that is packed with irregularly or spherically shaped particles or a porous monolithic layer (stationary phase) by a liquid (mobile phase) at high pressure. HPLC is historically divided into two different sub-classes based on the polarity of the mobile and stationary phases. Technique in which the stationary phase is more polar than the mobile phase (e.g. toluene as the mobile phase, silica as the stationary phase) is called normal phase liquid chromatography (NPLC) and the opposite (e.g. water-methanol mixture as the mobile phase and C18 = octadecylsilyl as the stationary phase) is called reversed phase liquid chromatography (RPLC). Ironically the 'normal phase' has fewer applications and RPLC is therefore used considerably more.

Specific techniques which come under this broad heading are listed below. It should also be noted that the following techniques can also be considered fast protein liquid chromatography if no pressure is used to drive the mobile phase through the stationary phase.

Affinity chromatography

Affinity chromatography is based on selective non-covalent interaction between an analyte and specific molecules. It is very specific, but not very robust. It is often used in biochemistry in the purification of proteins bound to tags. These fusion proteins are labelled with compounds such as His-tags, biotin or antigens, which bind to the stationary phase specifically. After purification, some of these tags are usually removed and the pure protein is obtained.

Supercritical fluid chromatography

Supercritical fluid chromatography is a separation technique in which the mobile phase is a fluid above and relatively close to its critical temperature and pressure.

Techniques by separation mechanism

Ion exchange chromatography

Ion exchange chromatography uses ion exchange mechanism to separate analytes. It is usually performed in columns but can also be useful in planar mode. Ion exchange chromatography uses a charged stationary phase to separate charged compounds including amino acids, peptides, and proteins. In conventional methods the stationary phase is an ion exchange resin that carries charged functional groups which interact with oppositely charged groups of the compound to be retained. Ion exchange chromatography is commonly used to purify proteins using fast protein liquid chromatography (FPLC).

Size exclusion chromatography

Size exclusion chromatography (SEC) is also known as gel permeation chromatography (GPC) or gel filtration chromatography and separates molecules according to their size (or more accurately according to their hydrodynamic diameter or hydrodynamic volume). Smaller molecules are able to enter the pores of the media and, therefore, take longer to elute, whereas larger molecules are excluded from the pores and elute faster. It is generally a low resolution chromatography technique and thus it is often reserved for the final, 'polishing' step of a purification. It is also useful for determining the tertiary structure and quaternary structure of purified proteins, especially since it can be carried out under native solution conditions.

Special techniques

Reversed-phase chromatography

Reversed-phase chromatography is an elution procedure used in liquid chromatography in which the mobile phase is significantly more polar than the stationary phase.

Two-dimensional chromatography

In some cases, the chemistry within a given column can be insufficient to separate some analytes. It is possible to direct a series of unresolved peaks onto a second column with different physico-chemical (chemical classification) properties. Since the mechanism of retention on this new solid support is different from the first dimensional separation, it can be possible to separate compounds that are indistinguishable by one-dimensional chromatography.

Fast protein liquid chromatography

Fast protein liquid chromatography (FPLC) is a term applied to several chromatography techniques which are used to purify proteins. Many of these techniques are identical to those carried out under high performance liquid chromatography, however use of FPLC techniques are typically for preparing large scale batches of a purified product.

Countercurrent chromatography

Countercurrent chromatography (CCC) is a type of liquid-liquid chromatography, where both the stationary and mobile phases are liquids. It involves mixing a solution of liquids, allowing them to settle into layers and then separating the layers.

Chiral chromatography

Chiral chromatography involves the separation of stereoisomers. In the case of enantiomers, these have no chemical or physical differences apart from being three dimensional mirror images. Conventional chromatography or other separation processes are incapable of separating them. To enable chiral separations to take place, either the mobile phase or the stationary phase must themselves be made chiral, giving differing affinities between the analytes. Chiral chromatography HPLC columns (with a chiral stationary phase) in both normal and reversed phase are commercially available.

Crystallisation

Crystallisation is the (natural or artificial) process of formation of solid crystals precipitating from a solution, melt or more rarely deposited directly from a gas. Crystallisation is also a chemical solid-liquid separation technique, in which mass transfer of a solute from the liquid solution to a pure solid crystalline phase occurs.

The crystallisation process consists of two major events, nucleation and crystal growth. Nucleation is the step where the solute molecules dispersed in the solvent start to gather into clusters, on the nanometer scale (elevating solute concentration in a small region), that becomes stable under the current operating conditions. These stable clusters constitute the nuclei. However, when the clusters are not stable, they redissolve. Therefore, the clusters need to reach a critical size in order to become stable nuclei. Such critical size is dictated by the operating conditions (temperature, supersaturation, etc.). It is at the stage of nucleation that the atoms arrange in a defined and periodic manner that defines the crystal structure — note that 'crystal structure' is a special term that refers to the relative arrangement of the atoms, not the macroscopic properties of the crystal (size and shape), although those are a result of the internal crystal structure.

The crystal growth is the subsequent growth of the nuclei that succeed in achieving the critical cluster size. Nucleation and growth continue to occur simultaneously while the supersaturation exists. Supersaturation is the driving force of the crystallisation, hence the rate of nucleation and growth is driven by the existing supersaturation in the solution. Depending upon the conditions, either nucleation or growth may be predominant over the other, and as a result, crystals with different sizes and shapes are obtained (control of crystal size and shape constitutes one of the main challenges in industrial manufacturing, such as for pharmaceuticals). Once the supersaturation is exhausted, the solid-liquid system reaches equilibrium and the crystallisation is complete, unless the operating conditions are modified from equilibrium so as to supersaturate the solution again.

Many compounds have the ability to crystallise with different crystal structures, a phenomenon called polymorphism. Each polymorph is in fact a different thermodynamic solid state and crystal polymorphs of the same compound exhibit different physical properties, such as dissolution rate, shape (angles between facets and facet growth rates), melting point, etc. For this reason, polymorphism is of major importance in industrial manufacture of crystalline products.

For crystallisation to occur from a solution it must be supersaturated. This means that the solution has to contain more solute entities (molecules or ions) dissolved than it would contain under the equilibrium (saturated solution). This can be achieved by various methods, with: (i) solution cooling, (ii) addition of a second solvent to reduce the solubility of the solute (technique known as antisolvent or drown-out), (iii) chemical reaction, and (iv) change in pH being the most common methods used in industrial practice. Other methods, such as solvent evaporation, can also be used. The spherical crystallisation has some advantages (flowability, bioavailability) for the formulation of pharmaceutical drugs.

Used to improve (obtaing very pure substance) and/or verify their purity. Crystallisation separates a product from a liquid feedstream, often in extremely pure form, by cooling the feedstream or adding precipitants which lower the solubility of the desired product so that it forms crystals.

Well formed crystals are expected to be pure because each molecule or ion must fit perfectly into the lattice as it leaves the solution. Impurities would normally not fit as well in the lattice, and thus remain in solution preferentially. Hence, molecular recognition is the principle of purification in crystallisation. However, there are instances when impurities incorporate into the lattice, hence, decreasing the level of purity of the final crystal product. Also, in some cases, the solvent may incorporate into the lattice forming a solvate. In addition, the solvent may be 'trapped' (in liquid state) within the crystal formed, and this phenomenon is known as inclusion.

Chapter 7

Organic Feedstocks Produced by Fermentation and Its Utilisation

INTRODUCTION

Ethanol, also called ethyl alcohol, pure alcohol, grain alcohol, or drinking alcohol, is a volatile, flammable, colourless liquid. It is a psychoactive drug, best known as the type of alcohol found in alcoholic beverages and in modern thermometers. Ethanol is one of the oldest recreational drugs. In common usage, it is often referred to simply as alcohol or spirits.

Ethanol is a straight-chain alcohol, and its molecular formula is C_2H_5OH. Its empirical formula is C_2H_6O. An alternative notation is CH_3–CH_2–OH, which indicates that the carbon of a methyl group (CH_3–) is attached to the carbon of a methylene group (–CH_2–), which is attached to the oxygen of a hydroxyl group (–OH). It is a constitutional isomer of dimethyl ether. Ethanol is often abbreviated as EtOH, using the common organic chemistry notation of representing the ethyl group (C_2H_5) with Et.

The fermentation of sugar into ethanol is one of the earliest organic reactions employed by humanity. The intoxicating effects of ethanol consumption have been known since ancient times. In modern times, ethanol intended for industrial use is also produced from by-products of petroleum refining.

Ethanol has widespread use as a solvent of substances intended for human contact or consumption, including scents, flavourings, colourings, and medicines. In chemistry, it is both an essential solvent and a feedstock for the synthesis of other products. It has a long history as a fuel for heat and light and also as a fuel for internal combustion engines.

ETHANOL FERMENTATION

Ethanol fermentation is a biological process in which sugars such as glucose, fructose, and sucrose are converted into cellular energy and thereby produce ethanol and carbon dioxide as metabolic waste products. Because yeasts perform this process in the absence of oxygen, ethanol fermentation is classified as anaerobic. Ethanol fermentation occurs in the production of alcoholic beverages and ethanol fuel, and in the rising of bread dough. The chemical equation below summarises the fermentation of glucose. One glucose molecule is converted into two ethanol molecules and two carbon dioxide molecules:

$$C_6H_{12}O_6 \longrightarrow 2C_2H_5OH + 2CO_2$$

The process begins with a molecule of glucose being broken down by the process of glycolysis into pyruvate:

$$C_6H_{12}O_6 \longrightarrow 2CH_3COCOO^- + 2H^+$$

This reaction is accompanied by the size difference of two molecules of NAD^+ to NADH and a net of two ADP molecules converted to two ATP plus the two water molecules.

Pyruvate is then converted to acetaldehyde and carbon dioxide by an enzyme called pyruvate decarboxylase and requiring thiamine diphosphate as cofactor. The acetaldehyde is subsequently reduced to ethanol by the NADH from the previous glycolysis, which is returned to NAD^+:

$$CH_3COCOO^- + H^+ \longrightarrow CH_3CHO + CO_2$$

$$CH_3CHO + NADH \longrightarrow C_2H_5OH + NAD^+$$

Many species of yeast (*K. lactis, K. lipolytica*) will oxidise pyruvate completely to carbon dioxide and water (respiration) if oxygen is present in the environment and will ferment only in an anaerobic environment. However, the commonly used baker's yeast *S. cerevisiae* as well the yeast *S. pombe*, both prefer fermentation to respiration even in the presence of oxygen and will yield ethanol even under aerobic conditions given the right sources of nutrition.

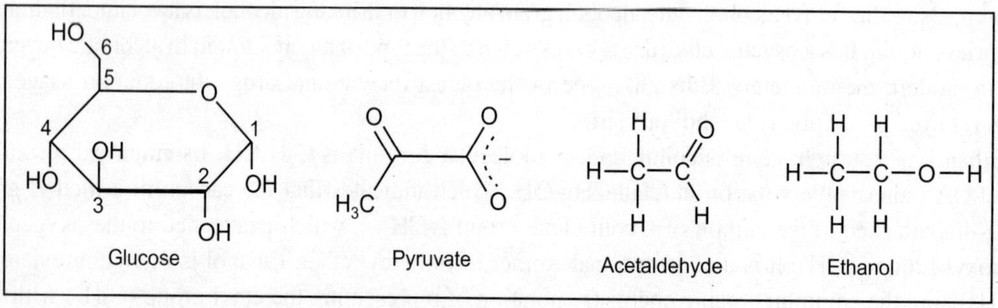

Microbes used in ethanol fermentation are yeast and *Zymomonas mobilis*

Yeasts

Yeasts are eukaryotic micro-organisms classified in the kingdom Fungi, with about 1500 species currently described; they dominate fungal diversity in the oceans. Most reproduce asexually by budding, although a few do so by binary fission.

Yeasts are unicellular, although some species with yeast forms may become multicellular through the formation of a string of connected budding cells known as *pseudohyphae*, or *false hyphae* as seen in most moulds. Yeast size can vary greatly depending on the species, typically measuring 3–4 μm in diameter, although some yeasts can reach over 40 μm.

The yeast open species *Saccharomyces cerevisiae* has been used in baking and fermenting alcoholic beverages for thousands of years. It is also extremely important as a model organism in modern cell biology research, and is one of the most thoroughly researched eukaryotic micro-organisms. Researchers have used it to gather information about the biology of the eukaryotic cell and ultimately human biology. Other species of yeast, such as *Candida albicans*, are opportunistic pathogens and can cause infections in humans. Yeasts have recently been used to generate electricity in microbial fuel cells, and produce ethanol for the biofuel industry. Yeasts do not form a specific taxonomic or phylogenetic grouping. At present it is estimated that only 1 per cent of all yeast species have been described. The term 'yeast' is often taken as a synonym for *S. cerevisiae*, but the phylogenetic diversity of yeasts is shown by their placement in both divisions Ascomycota and Basidiomycota. The budding yeasts (true yeasts) are classified in the order Saccharomycetales. Yeasts are chemoorganotrophs as they use organic compounds as a source of energy and do not require sunlight to grow. Carbon is obtained mostly from hexose sugars such as glucose and fructose, or disaccharides such as sucrose and maltose. Some species can metabolise pentose sugars like ribose, alcohols, and organic acids. Yeast species either require oxygen for aerobic cellular respiration (obligate aerobes), or are anaerobic but also have aerobic methods of energy production (facultative anaerobes). Unlike bacteria, there are no known yeast species that grow only anaerobically (obligate anaerobes). Yeasts grow best in a neutral or slightly acidic pH environment.

Yeasts will grow over a temperature range of 10°C (50°F) to 37°C (99°F), with an optimal temperature range of 30°C (86°F) to 37°C (99°F), depending on the type of species (*S. cerevisiae* works best at about 30°C (86°F). Above 37°C (99°F) yeast cells become stressed and will not divide properly. Most yeast cells die above 50°C (122°F).

If the solution reaches 105°C (221°F) the yeast will disintegrate. There is little activity in the range of 0°C (32°F)–10°C (50°F). The cells can survive freezing under certain conditions, with viability decreasing over time. Yeasts are very common in the environment, but are usually isolated from sugar-rich material. Some good examples include naturally occurring yeasts on the skins of fruits and berries (such as grapes, apples or peaches), and exudates from plants (such as plant saps or cacti). Some yeasts are found in association with soil and insects. Yeasts are generally grown in the laboratory on solid growth media or liquid broths. Common media used for the cultivation of yeasts include; potato dextrose agar (PDA) or potato dextrose broth, Wallerstein laboratories nutrient (WLN) agar, yeast peptone dextrose agar (YPD), and yeast mould agar or broth (YM). The antibiotic cycloheximide is sometimes added to yeast growth media to inhibit the growth of *Saccharomyces* yeasts and select for wild/indigenous yeast species. This will change the yeast process. Yeasts have asexual and sexual reproductive cycles, however, the most common mode of vegetative growth in yeast is asexual reproduction by budding or fission. Here a small bud, or daughter cell, is formed on the parent cell. The nucleus of the parent cell splits into a daughter nucleus and migrates into the daughter cell. The bud continues to grow until it separates from the parent cell, forming a new cell (Fig. 7.1).

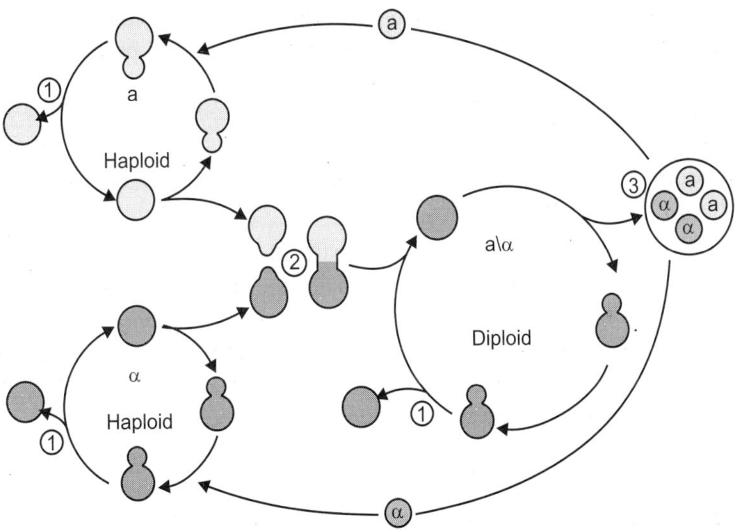

Fig. 7.1. The yeast cell's life cycle: (1) budding, (2) conjugation, and (3) spore.

Under high stress conditions haploid cells will generally die, however, under the same conditions diploid cells can undergo sporulation, entering sexual reproduction (meiosis) and producing a variety of haploid spores, which can go on to mate (conjugate), reforming the diploid. Yeast of the species *Schizosaccharomyces pombe* reproduce by binary fission instead of budding.

The useful physiological properties of yeast have led to their use in the field of biotechnology. Fermentation of sugars by yeast is the oldest and largest application of this technology. Many types of yeasts are used for making many foods: Baker's yeast in bread production, brewer's yeast in beer fermentation, yeast in wine fermentation and for xylitol production. Yeasts are also one of the most widely used model organisms for genetics and cell biology.

Strucure of biosynthesis of ethanol is given below.

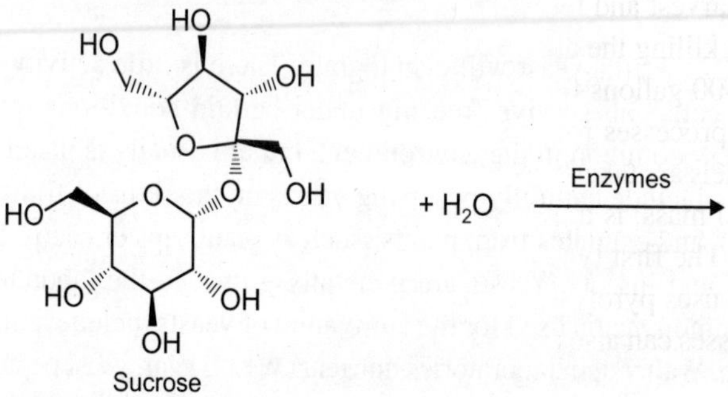

Biosynthesis of ethanol

Zymomonas mobilis

Zymomonas mobilis is a bacterium belonging to the genus *Zymomonas*. It is notable for its bioethanol-producing capabilities, which surpass yeast in some aspects. It was originally isolated from alcoholic beverages like the African palm wine, the Mexican pulque, and also as a contaminant of cider and beer in European countries. *Z. mobilis* degrades sugars to pyruvate using the Entner-Doudoroff pathway. The pyruvate is then fermentated to produce ethanol and carbon dioxide as the only products (analogous to yeast). The advantages of *Z. mobilis* over *S. cerevisiae* with respect to producing bioethanol:

1. Higher sugar uptake and ethanol yield.
2. Lower biomass production.
3. Higher ethanol tolerance.
4. Does not require controlled addition of oxygen during the fermentation.
5. Amenability to genetic manipulations.

However, it has a severe limitation compared to yeast: its utilisable substrate range is restricted to glucose, fructose, and sucrose. Using biotechnological methods, scientists are currently trying to overcome this. A variant of *Z. mobilis* that is able to use certain pentoses as a carbon source has been developed.

An interesting characteristic of *Z. mobilis* is that its plasma membrane contains hopanoids, pentacyclic compounds similar to eukaryotic sterols. This allows it to have an extraordinary tolerance to ethanol in its environment, around 13 per cent.

Ethanol Fuel

Ethanol is a renewable energy source because the energy is generated by using a resource, sunlight, which is naturally replenished.

Creation of ethanol starts with photosynthesis causing a feedstock, such as sugarcane or corn, to grow. These feedstocks are processed into ethanol.

Bioethanol is usually obtained from the conversion of carbon based feedstock. Agricultural feedstocks are considered renewable because they get energy from the sun using photosynthesis, provided that all minerals required for growth (such as nitrogen and phosphorus) are returned to the land. Ethanol can be produced from a variety of feedstocks such as sugarcane, bagasse, miscanthus, sugar beet, sorghum, grain sorghum, switchgrass, barley, hemp, kenaf, potatoes, sweet potatoes, cassava, sunflower, fruit, molasses, corn, stover, grain, wheat, straw, cotton, other biomass, as well as many types of cellulose waste and harvestings, whichever has the best well-to-wheel assessment.

An alternative process to produce bio-ethanol from algae is being developed by the company Algenol. Rather than grow algae and then harvest and ferment it the algae grow in sunlight and produce ethanol directly which is removed without killing the algae. It is claimed the process can produce 6000 gallons per acre per year compared with 400 gallons for corn production.

Currently, the first generation processes for the production of ethanol from corn use only a small part of the corn plant: the corn kernels are taken from the corn plant and only the starch, which represents about 50 per cent of the dry kernel mass, is transformed into ethanol. Two types of second generation processes are under development. The first type uses enzymes and yeast to convert the plant cellulose into ethanol while the second type uses pyrolysis to convert the whole plant to either a liquid bio-oil or a syngas. Second generation processes can also be used with plants such as grasses, wood or agricultural waste material such as straw.

Production process

The basic steps for large scale production of ethanol are: microbial (yeast) fermentation of sugars, distillation, dehydration, and denaturing (optional). Prior to fermentation, some crops require saccharification or hydrolysis of carbohydrates such as cellulose and starch into sugars. Saccharification of cellulose is called cellulolysis. Enzymes are used to convert starch into sugar (Fig. 7.2).

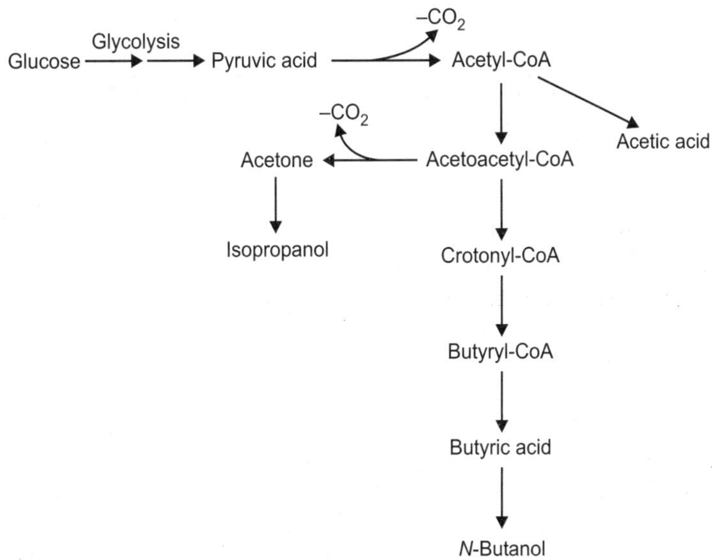

Fig. 7.2. Production of ethanol.

Fermentation

Ethanol is produced by microbial fermentation of the sugar. Microbial fermentation will currently only work directly with sugars. Two major components of plants, starch and cellulose, are both made up of sugars, and can in principle be converted to sugars for fermentation. Currently, only the sugar (e.g. sugarcane) and starch (e.g. corn) portions can be economically converted. However, there is much activity in the area of cellulosic ethanol, where the cellulose part of a plant is broken down to sugars and subsequently converted to ethanol.

Distillation

For the ethanol to be usable as a fuel, water must be removed. Most of the water is removed by distillation, but the purity is limited to 95–96 per cent due to the formation of a low-boiling water-ethanol azeotrope. The 95.6 per cent m/m (96.5 per cent v/v) ethanol, 4.4 per cent m/m (3.5 per cent v/v) water mixture may be used as a fuel alone, but unlike anhydrous ethanol, is immiscible in gasoline, so the water fraction is typically removed in further treatment in order to burn with in combination with gasoline in gasoline engines.

Dehydration

There are basically five dehydration processes to remove the water from an azeotropic ethanol/water mixture. The first process, used in many early fuel ethanol plants, is called azeotropic distillation and consists of adding benzene or cyclohexane to the mixture.

When these components are added to the mixture, it forms a heterogeneous azeotropic mixture in vapour-liquid-liquid equilibrium, which when distilled produces anhydrous ethanol in the column bottom, and a vapour mixture of water and cyclohexane/benzene. When condensed, this becomes a two-phase liquid mixture.

Another early method, called extractive distillation, consists of adding a ternary component which will increase ethanol relative volatility. When the ternary mixture is distilled, it will produce anhydrous ethanol on the top stream of the column.

With increasing attention being paid to saving energy, many methods have been proposed that avoid distillation all together for dehydration. Of these methods, a third method has emerged and has been adopted by the majority of modern ethanol plants. This new process uses molecular sieves to remove water from fuel ethanol. In this process, ethanol vapour under pressure passes through a bed of molecular sieve beads.

The bead's pores are sized to allow absorption of water while excluding ethanol. After a period of time, the bed is regenerated under vacuum to remove the absorbed water. Two beds are used so that one is available to absorb water while the other is being regenerated. This dehydration technology can account for energy saving of 3000 Btus/gallon (840 kJ/l) compared to earlier azeotropic distillation.

Technology

Ethanol-based engines

Ethanol is most commonly used to power automobiles, though it may be used to power other vehicles, such as farm tractors, boats and airplanes. Ethanol (E100) consumption in an engine is approximately 51 per cent higher than for gasoline since the energy per unit volume of ethanol is 34 per cent lower than for gasoline. However, the higher compression ratios in an ethanol-only engine allow for increased power output and better fuel economy than could be obtained with lower compression ratios. In general, ethanol-only engines are tuned to give slightly better power and torque output than gasoline-powered engines. In flexible fuel vehicles, the lower compression ratio requires tunings that give the same output when using either gasoline or hydrated ethanol. For maximum use of ethanol's benefits, a much higher compression ratio should be used, which would render that engine unsuitable for gasoline use. When ethanol fuel availability allows high-compression ethanol-only vehicles to be practical, the fuel efficiency of such engines should be equal to or greater than current gasoline engines. Current high compression ethanol-only engine designs are approximately 20–30 per cent less fuel efficient than their gasoline-only counterparts.

A study conducted by the Society of Automotive Engineers identify a method to exploit the characteristics of fuel ethanol substantially better than mixing it with gasoline. The method presents the possibility of leveraging the use of alcohol to achieve definite improvement over the cost-effectiveness of hybrid electric. The improvement consists of using dual-fuel direct-injection of pure alcohol (or the azeotrope or E85) and gasoline, in any ratio up to 100 per cent of either, in a turbocharged, high compression-ratio, small-displacement engine having performance similar to an engine having twice the displacement.

Each fuel is carried separately, with a much smaller tank for alcohol. The high-compression (which increases efficiency) engine will run on ordinary gasoline under low-power cruise conditions. Alcohol is directly injected into the cylinders (and the gasoline injection simultaneously reduced) only when necessary to suppress 'knock' such as when significantly accelerating. Direct cylinder injection raises the already high octane rating of ethanol up to an effective 130. The calculated overall reduction of gasoline use and CO_2 emission is 30 per cent. The consumer cost payback time shows a 4:1 improvement over turbo-diesel and a 5:1 improvement over hybrid. In addition, the problems of water absorption into pre-mixed gasoline (causing phase separation), supply issues of multiple mix ratios and cold-weather starting are avoided.

Ethanol's higher octane rating allows an increase of an engine's compression ratio for increased thermal efficiency. In one study, complex engine controls and increased exhaust gas recirculation allowed a compression ratio of 19.5 with fuels ranging from neat ethanol to E50. Thermal efficiency up to approximately that for a diesel was achieved. This would result in the MPG (miles per gallon) of a dedicated ethanol vehicle to be about the same as one burning gasoline.

Since 1989 there have also been ethanol engines based on the diesel principle operating in Sweden. They are used primarily in city buses, but also in distribution trucks and waste collectors. The engines, made by Scania, have a modified compression ratio, and the fuel (known as ED95) used is a mix of 93.6 per cent ethanol and 3.6 per cent ignition improver, and 2.8 per cent denaturants. The ignition improver makes it possible for the fuel to ignite in the diesel combustion cycle. It is then also possible to use the energy efficiency of the diesel principle with ethanol.

Engine cold start during the winter

High ethanol blends present a problem to achieve enough vapour pressure for the fuel to evaporate and spark the ignition during cold weather (since ethanol tends to increase fuel enthalpy of vapourisation). When vapour pressure is below 45 kPa starting a cold engine becomes difficult. In order to avoid this problem at temperatures below 11°C (59°F), and to reduce ethanol higher emissions during cold weather, both the US and the European markets adopted E85 as the maximum blend to be used in their flexible fuel vehicles, and they are optimised to run at such a blend. At places with harsh cold weather, the ethanol blend in the US has a seasonal reduction to E70 for these very cold regions, though it is still sold as E85. At places where temperatures fall below –12°C (10°F) during the winter, it is recommended to install an engine heater system, both for gasoline and E85 vehicles. Sweden has a similar seasonal reduction, but the ethanol content in the blend is reduced to E75 during the winter months.

Brazilian flex fuel vehicles can operate with ethanol mixtures up to E100, which is hydrous ethanol (with up to 4 per cent water), which causes vapour pressure to drop faster as compared to E85 vehicles. As a result, Brazilian flex vehicles are built with a small secondary gasoline reservoir located near the engine. During a cold start pure gasoline is injected to avoid starting problems at low temperatures. This provision is particularly necessary for users of Brazil's southern and central regions, where temperatures normally drop below 15°C (59°F) during the winter. An improved flex engine generation was launched in 2009 that eliminates the need for the secondary gas storage tank. In March 2009 Volkswagen do Brasil launched the Polo E-Flex, the first Brazilian flex fuel model without an auxiliary tank for cold start.

Ethanol fuel mixtures

To avoid engine stall due to 'slugs' of water in the fuel lines interrupting fuel flow, the fuel must exist as a single phase. The fraction of water that an ethanol-gasoline fuel can contain without phase separation increases with the percentage of ethanol. This shows, for example, that E30 can have up to about 2 per cent water. If there is more than about 71 per cent ethanol, the remainder can be any proportion of water or gasoline and phase separation will not occur. However, the fuel mileage declines with increased water content. The increased solubility of water with higher ethanol content permits E30 and hydrated ethanol to be put in the same tank since any combination of them always results in a single phase. Somewhat less water is tolerated at lower temperatures. For E10 it is about 0.5 per cent v/v at 70°F and decreases to about 0.23 per cent v/v at –30°F.

In many countries cars are mandated to run on mixtures of ethanol. Brazil requires cars be suitable for a 25 per cent ethanol blend, and has required various mixtures between 22 and 25 per cent ethanol, since July 2007, 25 per cent is required. The United States allows up to 10 per cent blends, and some states require this (or a smaller amount) in all gasoline sold. Other countries have adopted their own requirements. Beginning with the model year 1999, an increasing number of vehicles in the world are manufactured with engines which can run on any fuel from 0 per cent ethanol up to 100 per cent ethanol without modification. Many cars and light trucks (a class containing minivans, SUVs and pickup trucks) are designed to be flexible-fuel vehicles (also called dual-fuel vehicles). In older model years, their engine systems contained alcohol sensors in the fuel and/or oxygen sensors in the exhaust that provide input to the engine control computer to adjust the fuel injection to achieve stochiometric (no residual fuel or free oxygen in the exhaust) air-to-fuel ratio for any fuel mix. In newer models, the alcohol sensors have been removed, with the computer using only oxygen and airflow sensor feedback to estimate alcohol content. The engine control computer can also adjust (advance) the ignition timing to achieve a higher output without pre-ignition when it predicts that higher alcohol percentages are present in the fuel being burned. This method is backed up by advanced knock sensors—used in most high performance gasoline engines regardless of whether they are designed to use ethanol or not—that detect pre-ignition and detonation.

Fuel economy

In theory, all fuel-driven vehicles have a fuel economy (measured as miles per US gallon, or litres per 100 km) that is directly proportional to the fuel's energy content. In reality, there are many other variables that come in to play that affect the performance of a particular fuel in a particular engine. Ethanol contains approximately 34 per cent less energy per unit volume than gasoline, and therefore in theory, burning pure ethanol in a vehicle will result in a 34 per cent reduction in miles per US gallon, given the same fuel economy, compared to burning pure gasoline. Since ethanol has a higher octane rating, the engine can be made more efficient by raising its compression ratio. In fact using a variable turbocharger, the compression ratio can be optimised for the fuel being used, making fuel economy almost constant for any blend. For E10 (10 per cent ethanol and 90 per cent gasoline), the effect is small (~3 per cent) when compared to conventional gasoline, and even smaller (1–2 per cent) when compared to oxygenated and reformulated blends. However, for E85 (85 per cent ethanol), actual performance may vary depending on the vehicle. Based on EPA tests for all 2006 E85 models, the average fuel economy for E85 vehicles resulted 25.56 per cent lower than unleaded gasoline. The EPA-rated mileage of current USA flex-fuel vehicles should be considered when making price comparisons, but it must be noted that E85 is a high performance fuel, with an octane rating of about 104, and should be compared to premium.

Consumer production systems

While biodiesel production systems have been marketed to home and business users for many years, commercialised ethanol production systems designed for end-consumer use have lagged in the marketplace. In 2008, two different companies announced home-scale ethanol production systems. The AFS125 Advanced Fuel System from Allard Research and Development is capable of producing both ethanol and biodiesel in one machine, while the E-100 MicroFueler from E-Fuel Corporation is dedicated to ethanol only.

Environment

Energy balance

All biomass goes through at least some of these steps: it needs to be grown, collected, dried, fermented, and burned. All of these steps require resources and an infrastructure. The total amount of energy input into the process compared to the energy released by burning the resulting ethanol fuel is known as the energy balance (or 'net energy gain'). Figures compiled in a 2007 by *National Geographic Magazine* point to modest results for corn ethanol produced in the US: one unit of fossil-fuel energy is required to create 1.3 energy units from the resulting ethanol. The energy balance for sugarcane ethanol produced in Brazil is more favourable, 1:8. Energy balance estimates are not easily produced, thus numerous such reports have been generated that are contradictory. For instance, a separate survey reports that production of ethanol from sugarcane, which requires a tropical climate to grow productively, returns from 8 to 9 units of energy for each unit expended, as compared to corn which only returns about 1.34 units of fuel energy for each unit of energy expended. Table 7.1 listed the energy balance of different countries.

Table 7.1. Energy balance of various countries.

Country	*Type*	*Energy balance*
United States	Corn ethanol	1.3
Brazil	Sugarcane ethanol	8
Germany	Biodiesel	2.5
United States	Cellulosic ethanol[a]	2–36[b]

[a] Experimental, not in commercial production.
[b] Depending on production method.

Carbon dioxide, a greenhouse gas, is emitted during fermentation and combustion. However, this is cancelled out by the greater uptake of carbon dioxide by the plants as they grow to produce the biomass. When compared to gasoline, depending on the production method, ethanol releases less greenhouse gases.

Air pollution

Compared with conventional unleaded gasoline, ethanol is a particulate-free burning fuel source that combusts with oxygen to form carbon dioxide, water and aldehydes. Gasoline produces 2.44 [CO_2 equivalent] kg/l and ethanol 1.94 (this is–21 per cent CO_2). The Clean Air Act requires the addition of oxygenates to reduce carbon monoxide emissions in the United States. The additive MTBE is currently being phased out due to ground water contamination, hence ethanol becomes an attractive alternative additive. Current production methods include air pollution from the manufacturer of macronutrient fertilisers such as ammonia.

A study by atmospheric scientists at Stanford University found that E85 fuel would increase the risk of air pollution deaths relative to gasoline by 9 per cent in Los Angeles, USA: a very large, urban, car-based metropolis that is a worst case scenario. Ozone levels are significantly increased, thereby increasing photochemical smog and aggravating medical problems such as asthma.

Manufacture

In 2002, monitoring the process of ethanol production from corn revealed that they released VOCs (volatile organic compounds) at a higher rate than had previously been disclosed. The Environmental Protection Agency (EPA) subsequently reached settlement with Archer Daniels Midland and Cargill, two of the largest producers of ethanol, to reduce emission of these VOCs. VOCs are produced when fermented corn mash is dried for sale as a supplement for livestock feed. Devices known as thermal oxidisers or catalytic oxidisers can be attached to the plants to burn off the hazardous gases.

Carbon dioxide

The calculation of exactly how much carbon dioxide is produced in the manufacture of bioethanol is a complex and inexact process, and is highly dependent on the method by which the ethanol is produced and the assumptions made in the calculation. A calculation should include:

1. The cost of growing the feedstock.
2. The cost of transporting the feedstock to the factory.
3. The cost of processing the feedstock into bioethanol.

Such a calculation may or may not consider the following effects:

1. The cost of the change in land use of the area where the fuel feedstock is grown.
2. The cost of transportation of the bioethanol from the factory to its point of use.
3. The efficiency of the bioethanol compared with standard gasoline.
4. The amount of carbon dioxide produced at the tail pipe.
5. The benefits due to the production of useful by-products, such as cattle feed or electricity.

A National Geographic Magazine overview article (2007) puts the figures at 22 per cent less CO_2 emissions in production and use for corn ethanol compared to gasoline and a 56 per cent reduction for cane ethanol. Carmaker Ford reports a 70 per cent reduction in CO_2 emissions with bioethanol compared to petrol for one of their flexible-fuel vehicles.

An additional complication is that production requires tilling new soil which produces a one-off release of GHG that it can take decades or centuries of production reductions in GHG emissions to equalise. As an example, converting grass lands to corn production for ethanol takes about a century of annual savings to make up for the GHG released from the initial tilling.

Change in land use

Large-scale farming is necessary to produce agricultural alcohol and this requires substantial amounts of cultivated land. University of Minnesota researchers report that if all corn grown in the US were used to make ethanol it would displace 12 per cent of current US gasoline consumption. There are claims that land for ethanol production is acquired through deforestation, while others have observed that areas currently supporting forests are usually not suitable for growing crops.

In any case, farming may involve a decline in soil fertility due to reduction of organic matter, a decrease in water availability and quality, an increase in the use of pesticides and fertilisers, and potential dislocation of local communities. However, new technology enables farmers and processors to increasingly produce the same output using less inputs.

Cellulosic ethanol production is a new approach which may alleviate land use and related concerns. Cellulosic ethanol can be produced from any plant material, potentially doubling yields, in an effort to minimise conflict between food needs vs. fuel needs. Instead of utilising only the starch by-products from grinding wheat and other crops, cellulosic ethanol production maximises the use of all plant materials, including gluten.

This approach would have a smaller carbon footprint because the amount of energy-intensive fertilisers and fungicides remain the same for higher output of usable material. The technology for producing cellulosic ethanol is currently in the commercialisation stage.

Many analysts suggest that, whichever ethanol fuel production strategy is used, fuel conservation efforts are also needed to make a large impact on reducing petroleum fuel use.

Using ethanol for electricity

Converting biomass to electricity for charging electric vehicles may be a more 'climate-friendly' transportation option than using biomass to produce ethanol fuel. 'You make more efficient use of the land and more efficient use of the plant biomass by making electricity rather than ethanol', said Elliott Campbell, an environmental scientist at the University of California at Merced, who led the research. 'It's another reason that, rather than race to liquid biofuels, we should consider other uses of bio-resources'.

For bioenergy to become a widespread climate solution, however, technological breakthroughs are necessary, analysts say. Researchers continue to search for more cost-effective developments in both cellulosic ethanol and advanced vehicle batteries.

ACETONE/BUTANOL FERMENTATION

Processing corn by-products to create hydrogen and butanol fuels benefits the environment, reduces petrochemical dependence, and provides a potential new market for farmers. Butanol is a four carbon alcohol. It has double the amount of carbon of ethanol, which equates to a 25 per cent increase in harvestable energy (Btu's). Butanol is produced by fermentation, from corn, grass, leaves, agricultural waste and other biomass. Butanol is safer to handle with a Reid value of 0.33 psi, which is a measure of a fluid's rate of evaporation when compared to gasoline at 4.5 and ethanol at 2.0 psi.

Butanol is an alcohol that can be but does not have to be blended with fossil fuels. Butanol when consumed in an internal combustion engine yields no SO_x, NO_x or carbon monoxide all environmentally harmful by-products of combustion. CO_2 is the combustion by-product of butanol, and is considered environmentally 'green'. Butanol is far less corrosive than ethanol and can be shipped and distributed through existing pipelines and filling stations. Butanol solves the safety problems associated with the infrastructure of the hydrogen supply. Reformed butanol has four more hydrogen atoms than ethanol, resulting in a higher energy output and is used as a fuel cell fuel.

Hydrogen generated during the butanol fermentation process is easily recovered, increasing the energy yield of a bushel of corn by an additional 18 per cent over the energy yield of ethanol produced from the same quantity of corn.

The acetone butanol fermentation is one of the oldest fermentation known. The fermentation is based on culturing various strains of Clostridia in carbohydrate rich media under anaerobic conditions to yield butanol and acetone.

Clostridium acetobutylicum is the organism of choice in the production of these organic solvents. These fermentations were out of favour till very recently because of the availability of acetone and butanol from the petroleum industry.

Today there is considerable amount of interest in these fermentations. However, the concentration of end products in these fermentations is quite small and the fermentations are a type of mixed fermentation yielding a mixture of compounds such as butyric acid, butanol, acetone. etc. Attempts to increase yeilds by use of genetidally altered strains or change in fermentation conditions have been partially successful.

There is abundant biomass present in low value agricultural commodities or processing wastes requiring proper disposal to avoid our pollution problem, for example, the corn refinery industry generates more than 10 million metric tons of corn by-products that are currently of limited use and pose significant environmental problems. Similarly, there are 60 billion pounds of cheese whey generated annually in the dairy industry much of this by-product has no economical use at the present time and requires costly disposal because of its high biological oxygen demand. These various forms of biomass are inexpensive feedstocks for hydrogen, chemicals and power grade alcohol fuel (butanol) production.

Production of industrial butanol and acetone via fermentation, using *Clostridia acetobutylicum*, started in 1916, during World War I. Chime Wizemann, a student of Louis Pasture, isolated the microbe that made acetone. England approached the young microbiologist and asked for the rights to make acetone for cordite. Up until the 1920s acetone was the product sought, but for every pound of acetone fermented, two pounds of butanol were formed. A growing automotive paint industry turned the market around, and by 1927 butanol was primary and acetone became the by-product.

The production of butanol by fermentation declined from the 1940s through the 1950s, mainly because the price of petrochemicals dropped below that of starch and sugar substrates such as corn and molasses. The labour intensive batch fermentation system's overhead combined with the low yields contributed to the situation. Fermentation-derived acetone and butanol production ceased in the late 1950s.

In the 1970s the primary focus for alternative fuels was on ethanol—people were familiar with its production and did not realise that dehydration (a very energy-consuming step) was necessary in order to blend it with fossil fuels.

Nor did we realise the difficulty of distribution, since ethanol cannot be transferred through the existing pipeline infrastructure. The selection of ethanol, a lower-grade, corrosive, hard-to-purify, dangerously explosive, and very evaporative alcohol is the result. Ethanol is still subsidised by the government, since it is not profitable enough to compete with gasoline.

ABE Fermentation

Acetone butanol ethanol (ABE) fermentation by *Clostridium acetobutylicum* is one of the oldest known industrial fermentations. It was ranked second only to ethanol fermentation by yeast in its scale of production, and is one of the largest biotechnological processes ever known. The actual fermentation, however, has been quite complicated and difficult to control. ABE fermentation has declined continuously since the 1950s, and almost all butanol is now produced via petrochemical routes . Butanol is an important industrial solvent and potentially a better fuel extender than ethanol.

In a typical ABE fermentation, butyric, propionic, lactic and acetic acids are first produced by *C. acetobutylicum*, the culture pH drops and undergoes a metabolic 'butterfly' shift, and butanol, acetone, isopropanol and ethanol are formed (Fig. 7.3).

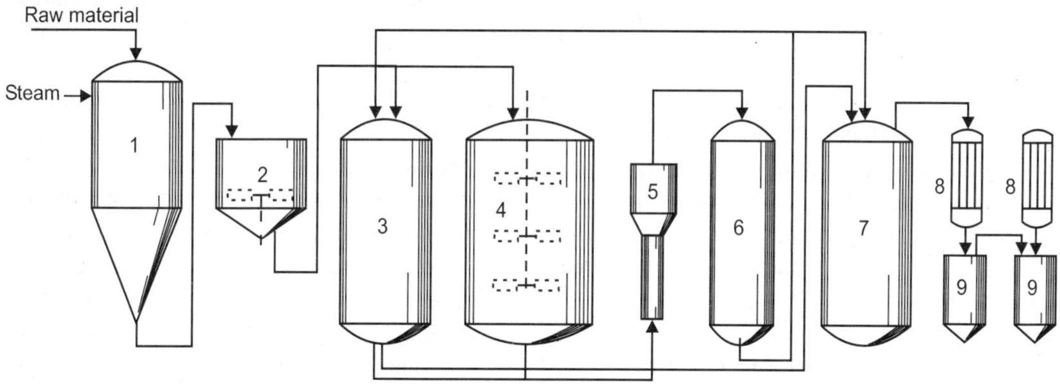

Fig. 7.3. Principal flow-sheet of the acetone-butanol fermentation. (1) Henze-Steamer, 5 m^3 (ownership of distillery), (2) mash tank, 7 m^3 (ownership of the distillery), (3) 3 m^3 substrate vessel, (4) 3.5 m^3 substrate vessel, (5) first stage fermenter, 5 litres, (6) second stage fermenter, 300 litres, (7) finishing tank, 3 m^3, (8) condenser, and (9) condensate collection tank, 50 litres.

In conventional ABE fermentations, the butanol yield from glucose is low, typically around 15 per cent and rarely exceeding 25 per cent. The production of butanol was limited by severe product inhibition. Butanol at a concentration of 1 per cent can significantly inhibit cell growth and the fermentation process. Consequently, butanol concentration in conventional ABE fermentations is usually lower than 1.3 per cent.

In the past 20+ years, there have been numerous engineering attempts to improve butanol production in ABE fermentation, including cell recycling and cell immobilisation to increase cell density and reactor productivity and using extractive fermentation to minimise product inhibition. Despite many efforts, the best results ever obtained for ABE fermentations to date are still less than 2 per cent in butanol concentration, 4.46 g/l/hr productivity, and a yield of less than 25 per cent from glucose. Optimising the ABE fermentation process has long been a goal of the industry.

With that in mind, a new process has been developed using continuous immobilised cultures of *Clostridium tyrobutyricum* and *Clostridium acetobutylicum* to produce an optimal butanol productivity of 4.64 g/l/hr and yield of 42 per cent. In simple terms, one microbe maximises the production of hydrogen and butyric acid, while the other converts butyric acid to butanol.

Compared to conventional ABE fermentation, this new process eliminates acetic, lactic and propionic acids, acetone, isopropanol and ethanol production. The fermentation only produces hydrogen, butyric acid, butanol and carbon dioxide, and doubles the yield of butanol from a bushel of corn from 1.3 to 2.5 gallons per bushel. That matches ethanol's track record—and ethanol fermentations do not yield hydrogen. Commercialisation of this new technology has the potential to reduce our nation's dependence on foreign oil, protect our fuel generation grid from sudden disruption while developing our agricultural base and reduce global warming. Butanol is a pure alcohol with an energy content similar to that of gasoline. It does not have to be stored in high pressure vessels like natural gas, and can be but does not have to be blended (10 to 100 per cent) with any fossil fuel.

Butanol can also be transported through existing pipelines for distribution. Butanol can help solve the hydrogen distribution infrastructure problems faced with fuel cell development. The employment of fuel-cell technology is held up by the safety issues associated with hydrogen distribution, but butanol can be very easily reformed for its hydrogen content and can be distributed through existing gas stations in the purity required for either fuel cells or vehicles.

Growing consumer acceptance and name recognition for butanol, incentives to agriculture and industry, falling production costs, increasing prices and taxes for fossil fuels, and the desire for cleaner-burning sources of energy should drive an increase in butanol production.

Building new, smaller, turnkey biorefineries of 5 to 30 million gallons per year for small municipalities and surrounding farming communities could introduce state-of-the-art technologies at a faster rate than has been adopted in the past. These local biorefineries would address many overwhelming problems associated with the environment, such as regional landfill burdens, and by disseminating fuel generation throughout the corn and bio-belt, any prospective disruption by terrorism is made more difficult, thus improving 'homeland security'. Cooperatively owned facilities would allow the agricultural sector to employ more people and retain profits within the local economy, bringing the resulting seven-fold multiplication. The production of butanol (15,500 Btu/lb. or 1,04,800 Btu/gallon) and hydrogen (61,000 Btu/lb.) from biomass is not constrained by technological difficulties as is the manufacturing of ethanol (12,800 Btu/lb or 84,250 Btu/gal).

New higher-value uses for coproducts of fermentation are an even more likely source of new revenues and could reduce the cost of butanol and hydrogen.

Recent advances in the fields of biotechnology and bioprocessing have resulted in a renewed interest in the fermentation production of chemicals and fuels, including *n*-butanol. With continuous fermentation technology, butanol can be produced at higher yields, concentrations and production rates.

GLYCEROL

Glycerol is an organic compound also commonly called glycerine. It is a colourless, odourless, viscous liquid that is widely used in pharmaceutical formulations. Glycerol has three hydrophilic hydroxyl groups that are responsible for its solubility in water and its hygroscopic nature. The glycerol substructure is a central component of many lipids. Glycerol is sweet-tasting and of low toxicity.

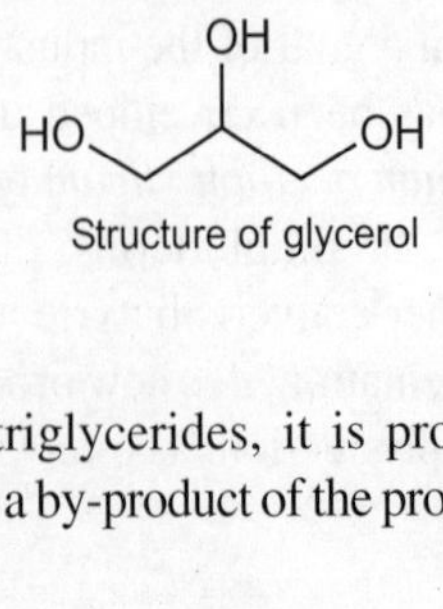

Structure of glycerol

Synthesis and Production

Since glycerol forms the backbone of triglycerides, it is produced by saponification of animal fats, e.g. a by-product of soap-making. It is also a by-product of the production of biodiesel via transesterification.

Because of the emphasis on biodiesel, the market for glycerol is depressed, and the old epichlorohydrin process for glycerol synthesis is no longer economical.

It is also produced as a by-product of refining of cooking and salad oils, and various brands are sold to the retail market as 'pure vegetable source' glycerine, 100 per cent pure, safe for ingestion. It is widely used in various industries.

Foods industry: In foods and beverages, glycerol serves as a humectant, solvent and sweetener, and may help preserve foods. It is also used as filler in commercially prepared low-fat foods (e.g. cookies), and as a thickening agent in liqueurs. Glycerol and water are used to preserve certain types of leaves. As a sugar substitute, it has approximately 27 calories per teaspoon and is 60 per cent as sweet as sucrose. Although it has about the same food energy as table sugar, it does not raise blood sugar levels, nor does it feed the bacteria that form plaques and cause dental cavities. As a food additive, glycerol is labelled as E number E422. Glycerol is also used to manufacture mono- and di-glycerides for use as emulsifiers, as well as polyglycerol esters going into shortenings and margarine.

Organic chemistry: In organic synthesis, glycerol is used as a readily available prochiral building block. Even if glycerol with no substitutions is symmetrical, and carbon atoms 1 and 3 are exchangeable, once one of them forms an ester or ether bond, the two are no longer exchangeable.

Nitroglycerine: Glycerol is used to produce nitroglycerine, or glycerol-trinitrate (GTN), which is an essential ingredient of smokeless gunpowder and various explosives such as dynamite, gelignite and propellants like cordite.

Research laboratory usage: Glycerol is a common component of solvents for enzymatic reagents stored at temperatures below zero degrees Celsius due to the depression of the freezing temperature of solutions with high concentrations of glycerol. It is also dissolved in water to reduce damage by ice crystals to laboratory organisms that are stored in frozen solutions, such as bacteria, nematodes, and fruit flies.

Pharmaceutical and personal care applications: Glycerol is used in medical and pharmaceutical and personal care preparations, mainly as a means of improving smoothness, providing lubrication and as a humectant. It is found in cough syrups, elixirs and expectorants, toothpaste, mouthwashes, skin care products, shaving cream, hair care products, soaps and water based personal lubricants. In solid dosage forms like tablets, glycerol is used as a tablet holding agent. It is also an ingredient in cigarettes that is used as a humectant. As a 10 per cent solution, glycerol prevents tannins from precipitating in ethanol extracts of plants (tinctures). It is also used as a substitute for ethanol as a solvent in preparing herbal extractions. It is less extractive and is approximately 30 per cent less able to be absorbed by the body. Fluid extract manufacturers often extract herbs in hot water before adding glycerine to make glycerites.

Topical pure or nearly pure glycerol is an effective treatment for psoriasis, burns, bites, cuts, rashes, bedsores, and calluses. It can be used orally to eliminate halitosis, as it is a contact bacterial desiccant. The same property makes it very helpful with periodontal disease; it penetrates biofilm quickly and eliminates bacterial colonies.

Alternative chemical and fuel feedstock: A great deal of research is being conducted to try to make value-added products from crude glycerol (typically containing 20 per cent water and residual esterification catalyst) obtained from biodiesel production, as an alternative to disposal by incineration.

Metabolism

Glycerol is a precursor for synthesis of triacylglycerols and of phospholipids in the liver and adipose tissue. When the body uses stored fat as a source of energy, glycerol and fatty acids are released into the bloodstream. The glycerol component can be converted to glucose by the liver and provides energy for cellular metabolism. Before glycerol can enter the pathway of glycolysis or gluconeogenesis (depending on physiological conditions), it must be converted to their intermediate glyceraldehyde 3-phosphate in the following steps:

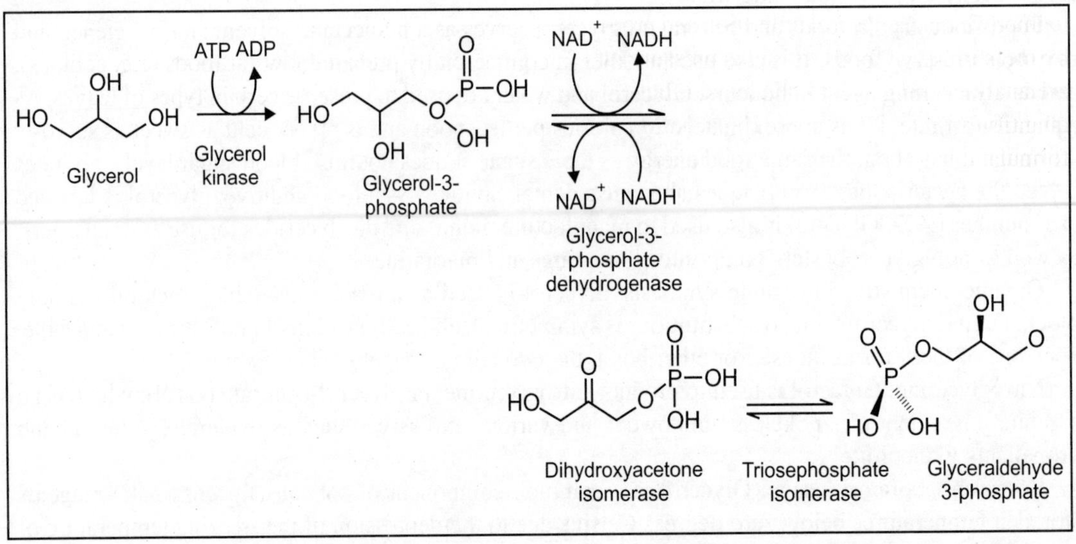

The enzyme glycerol kinase is present only in the liver. In adipose tissue, glycerol 3-phosphate is obtained from dihydroxyacetone phosphate (DHAP) with the enzyme glycerol-3-phosphate dehydrogenase.

Fermentation of Biodiesel-derived Crude Glycerol to Produce Value-added Chemicals

The rapidly expanding market for biodiesel is dramatically altering the cost and availability of glycerol, as typical biodiesel production processes generate about 10 wt per cent glycerol. As such, it is critical that new and innovative processes for managing and utilising the glycerol co-product be developed. In the longer term, as supplies continue to increase, glycerol will become a versatile building block chemical for the production of high value compounds within an integrated biorefinery.

The goal of this research is to investigate and optimise the value-added conversion of crude glycerol generated during biodiesel production using anaerobic fermentation. In addition to producing 1,3-propanediol, acetic acid, and butyric acid, clostridia have been shown to produce up to 8.8 g/l lactic acid and 17 g/l butanol during the anaerobic fermentation of glycerol. The most influential parameters affecting this fermentation are nutrient/trace metal supplementation, pH, and substrate composition and concentration. Unfortunately, none of these parameters have been thoroughly investigated or quantified with respect to production of lactic acid and butanol.

Biomass samples were collected from anaerobic digesters at both a biodiesel production facility and a municipal waste-water treatment facility. Samples were pre-treated to select for clostridia and inactivate competing organisms such as methanogens. Acclimation experiments were conducted in 500 ml batch reactor systems under conditions of excess nutrients and trace metals. Efforts have resulted in successful acclimation of an enriched culture of clostridia capable of using crude glycerol obtained from a local biodiesel facility as the sole carbon source. These cultures exhibit good glycerol utilisation and have produced notable amounts of lactic acid, 1,3-propanediol, acetic acid and butyric acid under unoptimised conditions. Continuing efforts are focused on improving lactic acid and butanol production as well as glycerol utilisation by optimising the nutrient solution/trace metal concentration and composition. Experiments are evaluating the effect of key trace metals, including Fe, Co, Ni, and Mo, to determine the appropriate concentrations for increased production of lactic acid and butanol. Additional experiments are investigating the effect of providing excess macro-nutrients, specifically nitrogen and phosphorus, to improve crude glycerol utilisation and product formation. The progress of the fermentation is monitored by measuring CO_2 production (as pressure) as well as the pH of the fermentation broth. Liquid samples are analysed for glycerol, lactic acid, butanol, and other products using HPLC. This presentation will quantify and compare product formation and glycerol utilisation over a range of nutrient solution formulations and identify those formulations that favour lactic acid and/or butanol production.

Chapter 8

Organic Acids and their Production

INTRODUCTION

An organic acid is an organic compound with acidic properties. Various organic acids are widely used as additives in food industries and also as chemical feedstocks. Fermentation processes play a major role in the production of most organic acids. Various organic acids are accumulated by several eukaryotic and prokaryotic micro-organisms. In anaerobic bacteria, their formation is usually a means by which there organisms regenerate NADH, and their accumulation therefore strictly parallels growth (e.g. lactic acid, propionic acid etc.). In aerobic bacteria and fungi, in contrast, the accumulation of organic acids is the result of incomplete substrateoxidation and is usually initiated by an imbalance in some essential nutrients, e.g. mineral ions. The most common organic acids are the carboxylic acids whose acidity is associated with their carboxyl group —COOH. Sulphonic acids, containing the group $-SO_2OH$, are relatively stronger acids. The relative stability of the conjugate base of the acid determines its acidity. Other groups can also confer acidity, usually weakly: —OH, —SH, enol group, and the phenol group. In biological systems organic compounds containing only these groups are not generally referred to as organic acids. A few common examples include: (i) lactic acid, (ii) acetic acid, (iii) formic acid, (iv) citric acid, and (v) oxalic acid.

Generally, organic acids are weak acids and do not dissociate completely in water, whereas the strong mineral acids do. Lower molecular weight organic acids such as formic and lactic acids are miscible in water, but higher molecular weight organic acids such as benzoic acid are insoluble in molecular (neutral) form.

On the other hand, most organic acids are very much soluble in organic solvents. *p*-toluenesulphonic acid is a comparatively strong acid used in organic chemistry often because it is able to dissolve in the organic reaction solvent. Exceptions to these solubility characteristics exist in the presence of other substituents which affect the polarity of the compound.

Simple organic acids like formic or acetic acids are used for oil and gas well stimulation treatments. These organic acids are much less reactive with metals than are strong mineral acids like HCl or mixtures of HCl and HF. For this reason, organic acids are used at high temperatures or when long contact times between acid and pipe are needed. The conjugate bases of organic acids such as citrate and lactate are often used in biologically-compatible buffer solutions.

Citric and oxalic acids are used as rust removal. As acids, they can dissolve the iron oxides, but without damaging the base metal like stronger mineral acids. In the dissociated form, they may be able to chelate the metal ions, helping to speed removal.

Biological systems create many and more complex organic acids such as L-lactic, citric and D-glucuronic acids that contain hydroxyl or carboxyl groups. Human blood and urine contain these plus organic acid degradation products of amino acids, neurotransmitters and intestinal bacterial action on food components. Examples of these categories are α-ketoisocaproic, vanilmandelic and D-lactic acids, derived from catabolism of L-leucine and epinephrine (adrenaline) by human tissues and catabolism of dietary carbohydrate by intestinal bacteria, respectively.

Research in the food preservation field has brought clear explanation on the mode of action of organic acids on bacteria. The key basic principle on the mode of action of organic acids on bacteria is that non-dissociated (non-ionised) organic acids can penetrate the bacteria cell wall and disrupt the normal physiology of certain types of bacteria that we call 'pH-sensitive' meaning that they cannot tolerate a wide internal and external pH gradient. Among those bacteria are *E.coli*, *Salmonella* spp., *C. perfringens*, *Listeria monocytogenes*, *Campylobacter* spp.

Upon passive diffusion of organic acids into the bacteria, where the pH is near or above neutrality, the acids will dissociate and lower the bacteria internal pH, leading to situations that will impair or stop the growth of bacteria. On the other hand, the anionic part of the organic acids that cannot escape the bacteria in its dissociated form will accumulate within the bacteria and disrupt many metabolic functions and lead to osmotic pressure increase, incompatible with the survival of the bacteria.

It has been well demonstrated that the state of the organic acids (undissociated or dissociated) is extremely important to define their capacity to inhibit the growth of bacteria, compared to undissociated acids. Lactic acid and its salts sodium lactate and potassium lactate are widely used as antimicrobials in food products, particularly meat and poultry such as ham and sausages.

Numerous trials have shown that the mode of action of organic acids also works in animal nutrition. Organic acids have been used successfully in pig production for more than 25 years and continue to be the alternative of choice. Even if less work has been done in poultry, the organic acids are very efficacious and their use is adapted to the physiology and anatomy of poultry.

Organic acids (C1–C7) are widely distributed in nature as normal constituents of plants or animal tissues. They are also formed through microbial fermentation of carbohydrates mainly in the large intestine. They are sometimes found in their sodium, potassium or calcium form.

Logically, organic acids added to feeds should be protected to avoid their dissociation in the crop and in the intestine (high pH segments) and reach far into the GIT, where the bulk of the bacteria population is located.

CITRIC ACID

Citric acid is a weak organic acid, and it is a natural preservative and is also used to add an acidic, or sour, taste to foods and soft drinks. In biochemistry, it is important as an intermediate in the citric acid cycle and therefore occurs in the metabolism of virtually all living things. It can also be used as an environmentally benign cleaning agent and acts as an antioxidant and a lubricant.

Citric acid had been synthesised from glycerol by Grimoux and Adams and later from symmetrical dichloroacetone: (i) by treating with hydrogen cyanide and hydrochloric acid to give dichloroacetonic acid, (ii) converting this into dicyano-acetonic acid, and (iii) with potassium cyanide, which on hydrolysis yields citric acid, as shown in Fig. 8.1.

Several other routes using different starting materials have since been published. All chemical methods have so far proved uncompetitive or unsuitable, mainly on economic grounds, with the starting material worth more than the end product, although poor yields due to the number of reaction steps in the synthesis and precautions necessary when handling hazardous compounds involved have contributed to the problem.

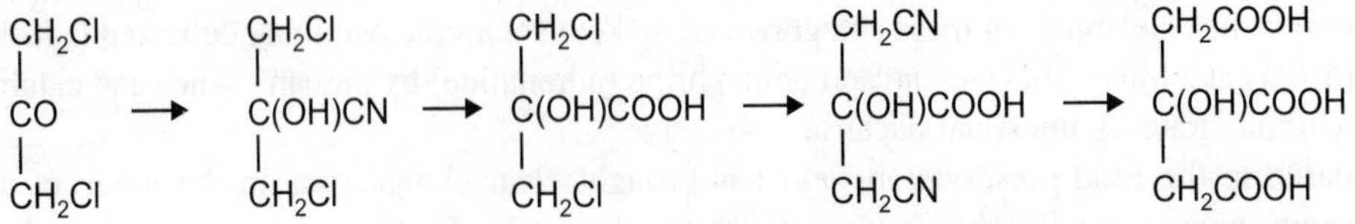

(i) (ii) (iii) (iv)

Fig. 8.1. Synthesis of citric acid.

Microbial Citric Acid

The concept of microbiological action yielding useful products followed from Pasteur's pioneering studies on fermentation and resulted in systematic investigations of fungi and bacteria. Amongst them Wehmer, showed that a *Citromyces* (now *Penicillium*) accumulated citric acid in a culture medium containing sugars and inorganic salts. This work did not lead directly to a commercial process but the subsequent search for other organisms capable of this synthesis did. Many other organisms were found to accumulate citric acid including strains of *Aspergillus niger*, *A. awamori*, *A. fonsecaeus*, *A. luchensis*, *A. phoenicus*, *A. wentii*, *A. saitoi*, *A. lanosius*, *A. flavus*, *Absidia* sp., *Acremonium* sp., *Aschochyta* sp., *Botrytis* sp., *Eupenicillium* sp., *Mucor piriformis*, *Penicillium janthinellum*, *P. restrictum*, *Talaromyces* sp., *Trichoderma viride* and *Ustulina vulgaris*.

Currie found strains of *A. niger* that produced citric acid when cultured in media with low pH values, high sugar levels and mineral salts. Prior to this *A. niger* was known to produce oxalic acid; the key difference was the low pH which, as we now know, suppressed both the production of oxalic acid, which would be toxic, and gluconic acid, which has a significantly higher production rate from sugar than citric acid. In biotechnological terms, citric acid is known as a bulk, or low value product. The market is, and always has been, very competitive, so the profit margins are small. Improvements in productivity depend on the detail of the various processes, many of which are not easily protected by patents, so that secrecy is important and understandable.

Citric Acid by the Surface Method

The general details of the original process are straightforward. The fungal mycelium is grown as a surface mat on a liquid medium in a large number of shallow trays with a capacity of 50 to 100 litres. Each tray has a surface area of about 5 m^2 and a depth of between 5 and 20 cm. The trays are manufactured from high purity aluminium or stainless steel and usually can be lifted by just two men. The trays are stacked in racks in a chamber to allow operation under relatively aseptic conditions. Various sucrose sources were used initially but cane molasses and then beet molasses soon became the norm as the sugar source. The molasses are diluted to the required concentration, usually 15 per cent and the pH adjusted to 5–7. After sterilisation, the medium is pumped into the trays and inoculation carried out directly from spores, either by adding a liquid suspension or by blowing the spores in with the air stream. Aerating the chambers is important for two purposes, oxygenation and heat removal. The air requirement depends on the stage of growth. Initially sterile air at low rates is used to prevent contamination during the germination stage, which takes about 12 hours. Later, when growth is maximal, rates of up to 10 m^3 per cubic metre medium per minute are needed to ensure heat dispersal. The heat generation is considerable, around 1 kJ hr^{-1} m^{-3} medium and the surface and medium temperatures are ideally around 28° to 30°C. This high volume air is not necessarily sterile, as contamination is normally not a problem once the pH has fallen, after about 24 hours growth. The pH falls to about 2, or slightly lower, and remains at that level until the end of the process, hence the need for high-grade materials for the construction of the trays. The incoming air is humidified to 40–60 per cent to prevent moisture loss from the high surface area of the medium. Cultivation continues for 8 to 15 days, with the objective of minimising the residence time to maximise the plant productivity. The details of time, productivity and yield are closely guarded secrets, but productivity of the order of 1 kg per square metre per day can be obtained and yield is up to 75 per cent of the initial sugar level. At the end of the process, which can be monitored by total acid production or judged by experience, the mycelial mat is removed by filtration and washed, as it contains up to 15 per cent of the total citric acid. The washings and spent medium are treated with lime (calcium hydroxide) at about 90°C to precipitate the insoluble tricalcium tetrahydrate salt of citric acid. It is not possible to crystallise the acid directly from the crude molasses medium although this can be done if pure sucrose is used as the carbon source. The precipitate of calcium citrate is washed and suspended in enough sulphuric acid to precipitate the calcium as calcium sulphate. This releases the citric acid into solution from where it can be treated further as required.

The surface process, though commercially profitable for many years, is labour intensive and inefficient in its use of space; there is a limit as to how high a large tray can be lifted. The production of citric acid by surface culture was challenged at the beginning of the 1940s by the development of submerged fermentation processes. When Shu and Johnson published their work on the effect of medium ingredients and their concentrations on citric acid production in submerged culture, the fundamental technology for submerged production was ready to be exploited on an industrial scale.

Submerged Process for Production of Citric Acid

The submerged process has become the method of choice in the industrialised countries because it is less labour intensive, gives a higher production rate, and uses less space. Several designs of reactor have been used, particularly in pilot scale systems; the stirred tank reactor is the most common design although air-lift reactors, with a higher aspect ratio than the stirred tank reactor are also used. The reactors are constructed of high-grade stainless steel, an important requirement in view of the low pH levels developed, the ability of citric acid to solubilise metal ions and the presence of manganese in stainless steels. Inferior grades of steel have caused problems in the past, both of leaching and pitting or general corrosion. Industrial rumours suggest it may still happen though not by design. The empirical process of 'conditioning' a reactor, whereby a few batches are processed before optimal production levels are achieved, may be related to this problem.

The other general requirement for reactors for citric acid production is the provision of aeration systems that can maintain a high dissolved oxygen level. With both tank and tower reactors sterile air is sparged from the base, although extra inputs are often used with tower reactors. The reactor may be held above atmospheric pressure to increase the rate of oxygen transfer into the fermentation broth. The influence of dissolved oxygen on citric acid formation has been examined and the dissolved oxygen levels are routinely monitored. The oxygen levels are also affected by the rheology of the broth.

A typical plant will consist of four areas: medium preparation, reactor section, broth separation and product recovery. The medium preparation will involve dilution of the molasses, or other raw material, addition of nutrients and other pretreatment such as ferrocyanide, and sterilisation, either in-line or in the reactor. Where in-line sterilisation is used the reactors are steam sterilised separately. It is usual to prepare an inoculum for the production reactor in a smaller reactor, in which the conditions may be modified to give rapid growth rather than product formation. Primary inoculation is by spores and the initial phase of the growth is critical.

When a separate inoculum stage is used, the correct stage for transfer, characteristically between 18 and 30 hours, is judged by pH level. Production temperature, like the inoculum temperature, is about 30°C. The process is allowed to continue until the rate of citric acid production falls below a predetermined value, which is reached many hours before the production ceases altogether.

Many reports suggest that the morphology of the mycelium is crucial to the ultimate yield; not only with respect to the shape of hyphae, but also their aggregation. Several studies suggest that hyphae should be abnormally short, bulbous and heavily branched. It is recognised that this condition is brought about by manganese deficiency or related to the addition of ferrocyanide, which is probably the same thing. The mycelium should also form small (less than 0.5 mm) pellets with a smooth, hard surface. Such pellets are produced when a number of factors are controlled, such as ferrocyanide levels, manganese levels, low iron (less than 1 ppm), low pH, control of aeration and agitation or the amount of spore inoculum.

It is clear that this morphological appearance is not in itself necessary for a successful yield, but is a result of the correct process parameters. Pellet formation is not necessary, but does give a broth with a lower energy requirement for mixing. When a change to a filamentous growth type occurs, the dissolved oxygen level may fall by 50 per cent for a fixed input. That filamentous growth can give satisfactory yields has been demonstrated and consideration of the diffusion characteristics of pellets versus filamentous mycelium would suggest that while yields may be similar, productivity should be greater without the additional diffusional constraint of pellets.

Aeration is a significant factor in the cost of the process, and although a constant aeration rate is used in many laboratory scale studies, the industrial practice is to use relatively low aeration rates initially (0.1 vvm) rising to 0.5–1 vvm as growth proceeds. Such aeration rates will lead to foaming and various devices and agents are available to minimise the problem. Although very high yields are possible, the productivity is a more important consideration on an industrial basis, and it is rare that the process is allowed to continue to the maximum yield.

The processes run today owes much to the pioneering work carried out by D. S. Clark and his co-workers at the northern regional research laboratories in Canada during the 1950s and early 1960s. Here, the technology for large-scale production of citric acid with *A. niger* using molasses was established. After the fermentation characteristics were worked out, attention was given to the controlling mechanisms of the fermentation. Numerous reports have been published on the role of metal ions on the citric acid cycle, in particular. After decades of academic discussion, there is general agreement about the factors that regulate the fermentation and give rise to the high yields obtained in industry.

Continuous and Immobilised Processes

A process for continuous production of citric acid has been described, but no commercial application of this has been made inspite of the high productivity values obtained. The process does not use the carbon source as the limiting substrate so that excess sugar will pass out of the reactor. As the carbohydrate substrate is one of the major cost factors, the continuous process will be less efficient than the batch process. This might be overcome by using several reactors in series, but this offsets any advantage from the continuous process.

Fed-batch processes have been used industrially so that the conversion of sugar concentrations greater than 15 per cent can be achieved, but the gain does not seem to be sufficient to allow the fed-batch method to become standard. The possibility of using the mycelium in an immobilised system has occurred to several workers and attempts on a small scale have been reported. Immobilisation of mycelium in alginate beads or collagen proved possible, but with very low production rates. The difficulties of avoiding oxygen limitation when preparing beads, and preventing further growth, which reduces oxygen transfer rates, have led to the immobilisation of conidia which are then grown under nitrogen limitation to the desired compact pellet. While giving a manageable system, the productivity was still too low to be of industrial interest.

Other constructs for immobilisation that have been more successful are the use of exchange filtration, and a rotating disc with an adhering mycelial film, reminiscent of sewage treatment techniques. These radical methods are unlikely to gain acceptance, even were they to give economic productivity gains, unless the engineering problems of scale-up can be overcome without making the capital costs too large.

Yeast Based Processes

From about 1965 methods using yeasts were developed, first from carbohydrate sources, then from *n*-alkanes. At this time hydrocarbons were relatively cheap and plants were built to use the method. The economics have altered since then and plants that have been built to utilise both yeast technologies have apparently switched back to carbohydrate feedstocks.

The potential advantages of using yeasts rather than filamentous fungi are the higher initial sugar concentrations that can be tolerated and the faster conversion rates possible. Further, the insensitivity to metal ions means that crude (and hence cheaper) grade molasses can be used without costly pretreatment. Since 1968, when the patent for citric acid production from molasses by eight genera of yeasts was allowed, there have been many process modifications reported. *Candida*, *Hansenula*, *Pichia*, *Debaromyces*, *Torulopsis*, *Kloekera*, *Trichosporon*, *Torula*, *Rhodotorula*, *Sporobolomyces*, *Endomyces*, *Nocardia*, *Nematospora*, *Saccharomyces*, and *Zygosaccharomyces* species are known to produce citric acid from various carbon sources. Out of these genera the *Candida* species, including *C. lipolytica*, *C. tropicalis*, *C. guillermondii*, *C. oleophila* and *C. intermedia* have been used.

The original process incorporated calcium carbonate into the medium to maintain a neutral pH, and generally a pH above 5.5 was used. Various additions have been proposed to reduce the isocitric acid contamination that afflicts yeasts even on carbohydrate media. Halogen substituted alkanoic mono- or di-substituted acids, *n*-hexadecyl citric acid or *trans*-aconitic acid, and even lead acetate have been patented, despite the possibility of toxic residues in the resulting citric acid. Many mutants have been selected for reduced isocitrate production. An osmophilic strain, which would convert sugar concentrations as high as 28 per cent without pretreatment of the molasses substrate, has been patented.

Tower reactors of fairly standard design are used, but with improved cooling systems as the rate of heat production is high. A continuous process has been described where the pH is maintained at 3.5 with ammonium hydroxide.

The industrial production of citric acid from *n*-alkanes is not now economic, although a plant was built, and operated, around 1970 at Saline, Reggio Calabria, Italy (Liquichimica). This process was based on a low aconitase mutant of *C. lipolytica* in a batch process with stirred, aerated tank reactors of 400 m^3, operating on a 72 hour cycle. The conversion from alkanes was reported to exceed 130 per cent (by weight). The theoretical yield is 250 per cent, but part of the alkanes was converted to biomass and carbon dioxide. The yeast was removed by centrifugation and the purification was traditional.

The medium used was based on the process developed for the yeast strain that had a substrate concentration of 10 per cent *n*-decane, although *n*-alkanes from 9 to 20 carbons could be used. The availability and cost of Libyan *n*-alkanes, which lead to the development of this and other plants, including the dual substrate plants, has changed over the last three decades. One unique feature of the *n*-alkane process is the insolubility of the substrate. To ensure a rapid conversion the *n*-alkane has to be thoroughly dispersed, so additives such as polyoxypropylene glycol ether, at concentrations from 20 to 200 ppm, are used to enhance this.

Koji Process

A third method for the production of citric acid is the koji process, using *Aspergillus* species. This is the solid state equivalent of the surface process described previously. It was originally developed in Japan where it uses the readily available rice bran and fruit wastes. It is confined to southeast Asia and is a relatively small-scale process. The carbohydrate source, which is principally starch and cellulose, is sterilised by steaming and the resulting semisolid paste (about 70 per cent water), at a pH of about 5.5, is inoculated by spraying on spores of *A. niger*. Additions of ferrocyanide or copper may be made. The incubation temperature is 30°C and the process takes about four to five days. Yields are low because of the difficulty of controlling trace metals and the process parameters. The fungus produces sufficient cellulases and amylases to break down the substrate, though the low yields may reflect the rate limitations of this step.

Various uses of citric acid are given in Table 8.1.

Table 8.1. Applications of citric acid.

Industry	*Property*	*Use*
Food		
Beverages	Acidulant	Flavouring
Jellies, jams, etc.	Flavouring	Acidulant
Fats and oils	Antioxidant	Metal complexing
Frozen foods	Antioxidant	
Pharmaceutical		
Effervescent	Acid	Flavour
Vitamins	Antioxidant	
Anticoagulants	Sequestering	Buffering
Iron preparations	Salt formation	
Cosmetics	Buffering	Antioxidant
Industrial		
Cleaning (metals)	Sequestering	
Detergents	Buffering	Sequestering
Photographic	Buffering	
Primer binding	Sequestering	
Polymerisations	Sequestering	

BIOCHEMISTRY OF CITRIC ACID ACCUMULATION BY *ASPERGILLUS NIGER*

The biochemical mechanism by which *Aspergillus niger* accumulates citric acid has attracted the interest of researchers since the late 1930s when the optimisation of this accumulation to give a commercial process began. In this sense, the various theories which have been proposed to explain the accumulation of citric acid in such high yields also reflect the general biochemical knowledge at the time the respective research was done. In view of the high input into this research through more than 50 years it is, therefore, rather disappointing that there is still no explanation of the biochemical basis of this process which would consistently explain all the observed factors influencing this fermentation. Reasons for this are manifold.

First, citric acid is only accumulated when several nutrient factors are present, either in excess (i.e. sugar concentration, H^+, dissolved oxygen), or at suboptimal levels (trace metals, nitrogen and phosphate), and thus is subject to multifactorial influence. Hence it is unlikely that single biochemical events are solely responsible for citric acid overflow.

Secondly, an appreciable part of the literature consists of work which has been performed using low or only moderately producing strains or by applying nutrient conditions not optimal for citric acid production, and while this may be justified for special reasons in individual cases, the respective results are not comparable to those obtained by others. Moreover, their significance for the understanding of the commercial citric acid fermentation is questionable. Thirdly, the biochemical knowledge of filamentous fungi is still significantly inferior to that of, for example, *Saccharomyces cerevisiae* or higher eukaryotes and, moreover, results from these sources cannot be uncritically transformed to filamentous fungi, which impedes a biochemically correct interpretation of results in several areas. Hence, although a considerable amount of basic biochemical research has been carried out with *A. niger*, the present state of understanding of the events relevant for citric acid accumulation (not to say production) is still a poorly resolved puzzle.

This section attempts to draw the currently recognisable picture and to aid in the further fitting together of the other scattered bits and pieces.

Glucose Catabolism in *A. niger* and Its Regulation

Citric acid biosynthetic pathway

It is well known, since the famous tracer studies by Cleland and Johnson, and Martin and Wilson, that citric acid is mainly formed via the reactions of the glycolytic pathway. Like most other fungi *Aspergillus* spp. utilise glucose and other carbohydrates for energy and cell synthesis by channelling glucose into the reactions of the glycolytic and the pentose phosphate pathway, respectively. The pentose phosphate pathway accounts for only a minor fraction of metabolised carbon during citric acid fermentation, and this decreases throughout prolonged cultivation. Legisa and Mattey speculated that this may be due to inhibition of 6-phosphogluconate dehydrogenase by citrate, but evidence for this is lacking. It should be noted that both arabitol and erythritol are accumulated as by-products until late stages of the fermentation; hence a complete blockage of the pentose phosphate pathway is obviously not taking place.

A. niger possesses a further pathway of glucose catabolism which is catalysed by glucose oxidase. This enzyme is induced by high concentrations of glucose and strong aeration in the presence of low concentrations of other nutrients, conditions which are also typical for citric acid fermentation; glucose oxidase will hence inevitably be formed during the starting phase of citric acid fermentation and convert a significant amount of glucose into gluconic acid. However, due to the extracellular location of the enzyme, it is directly influenced by the external pH and will be inactivated at pH <3.5. Because of the pK_a values for citric acid, its accumulation decreases the pH of the culture filtrate to pH 1.8 thereby inactivating glucose oxidase. It is not known if, and by which mechanism gluconic acid can be catabolised to citric acid during further fermentation.

The catabolism of glucose via glycolytic catabolism leads to 2 moles of pyruvate, and their subsequent conversion to the precursors of citrate (i.e. oxaloacetate and pyruvate). Cleland and Johnson were the first to show that *A. niger* uses 1 mole of the carbon dioxide which is released during the formation of acetyl-CoA and 1 mole of pyruvate to form 1 mole of oxaloacetate (Fig. 8.2a). This reaction is of utmost importance to high citric acid yields, because oxaloacetate could otherwise only be formed by one turn of the tricarboxylic acid cycle, which would be accompanied by the loss of two moles of CO_2 and only two thirds of the carbon of glucose could therefore accumulate as citric acid (Fig. 8.2b). The enzyme catalysing this reaction was shown to be pyruvate carboxylase, which was characterised by Feir and Suzuki and Wongchai and Jefferson. Unlike the enzyme from several other eukaryotes, the pyruvate carboxylase of *A. niger* is localised in the cytosol. Glycolytic pyruvate will therefore be converted to oxaloacetate, and further to malate by the cytosolic malate dehydrogenase isoenzyme, thereby also regenerating 50 per cent of the glycolytically produced NADH. It has been postulated that, analogous to higher eukaryotes, the cytosolic malate may serve as the cosubstrate of the mitochondrial tricarboxylic acid carrier, and that such an enhanced malate concentration may stimulate export of citrate from the mitochondrion.

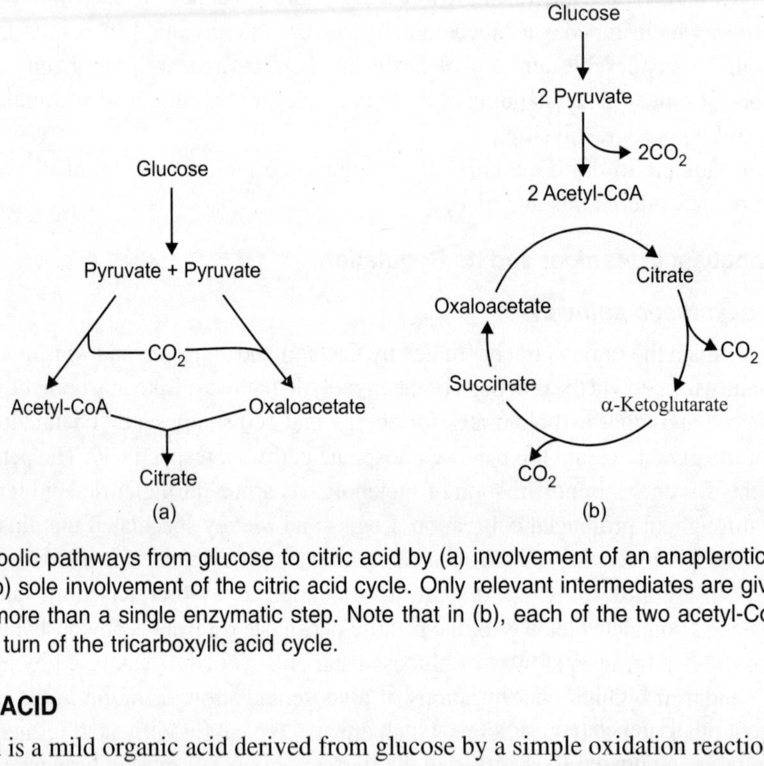

Fig. 8.2. Metabolic pathways from glucose to citric acid by (a) involvement of an anaplerotic carbon dioxide fixation, and (b) sole involvement of the citric acid cycle. Only relevant intermediates are given, and arrows may indicate more than a single enzymatic step. Note that in (b), each of the two acetyl-CoA molecules is subject to one turn of the tricarboxylic acid cycle.

GLUCONIC ACID

Gluconic acid is a mild organic acid derived from glucose by a simple oxidation reaction. The reaction is facilitated by the enzyme glucose oxidase (fungi) and glucose dehydrogenase (bacteria such as *Gluconobacter*). Microbial production of gluconic acid is the preferred method and it dates back to several decades. The most studied and widely used fermentation process involves the fungus *Aspergillus niger*. Gluconic acid and its derivatives, the principal being sodium gluconate, have wide applications in food and pharmaceutical industry. This section gives a review of microbial gluconic acid production, its properties and applications.

Gluconic acid (pentahydroxycaproic acid, Fig. 8.3) is produced from glucose through a simple dehydrogenation reaction catalysed by glucose oxidase. Oxidation of the aldehyde group on the C-1 of β-D-glucose to a carboxyl group results in the production of glucono-δ-lactone ($C_6H_{10}O_6$, Fig.8.3) and hydrogen peroxide. Glucono-δ-lactone is further hydrolysed to gluconic acid either spontaneously or by lactone hydrolysing enzyme, while hydrogen peroxide is decomposed to water and oxygen by peroxidase. The gluconate pathway is detailed in Fig. 8.4. The conversion process could be purely chemical too, but the most commonly involved method is the fermentation process. The enzymatic process could also be conducted, where the conversion takes place in the absence of cells with glucose oxidase and catalase derived from *A. niger.* Nearly 100 per cent of the glucose is converted to gluconic acid under the appropriate conditions. This method is an FDA approved process. Production of gluconic acid using the enzyme has the potential advantage that no product purification steps are required if the enzyme is immobilised, e.g. the use of a polymer membrane adjacent to anion-exchange membrane of low-density polyethylene grafted with 4-vinylpyridine. However, this approach is not yet common in the industry, and it will not be considered in this section.

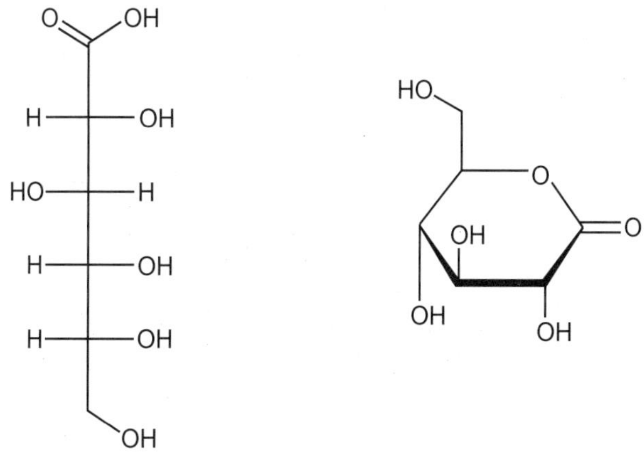

(a) (a)

Fig. 8.3. Formula of (a) gluconic acid, and (b) glucono-δ-lactone.

Gluconic acid production dates back to 1870 when Hlasiwetz and Habermann discovered gluconic acid.In 1880 Boutroux found for the first time that acetic acid bacteria are capable of producing sugar acid. In 1922 Molliard detected gluconic acid in the *Sterigmatocystis nigra*, now known as *Aspergillus niger*. Later, production of gluconic acid was demonstrated in bacterial species such as *Pseudomonas, Gluconobacter*, *Acetobacter*, and various fungal species. Studies of Bernhauer showed that *A. niger* produced high yields of gluconic acid when it was neutralised by calcium carbonate and the production was found to be highly pH dependent. However, it was found that with *Penicillium* sp., the pH dependence is not as critical when compared to *A. niger*, indicating that there was some correlation between the amount and time-dependent appearance of organic acids, such as gluconic acid, citric acid, oxalic acid, which are formed under different conditions. Gluconic acid production has been extensively studied by May Moyer, Wells, and Stubbs using *A. niger*. Using *Penicillium luteum* and *A. niger* Currie filed a patent employing submerged culture, giving yields of gluconic acid up to 90 per cent in 48–60 hrs. Later Moyer used *A. niger* in pilot plant studies and produced as high as 95 per cent of theoretical yields in glucose solution of 150 to 200 g/l in 24 hrs. Porges found that the process could be run semicontinuously, by the reuse of the mycelium for nine times repeatedly where the inoculum was recovered either by filtration or centrifugation. Findings of Moyer showed that efficiency of more than 95 per cent could be achieved by the addition of glucose at 250 g/l and boron compounds (1 per cent in solution of 250 g/l glucose) at later stages of the fungal growth with the reuse of mycelium in cycles of 24 hrs each.

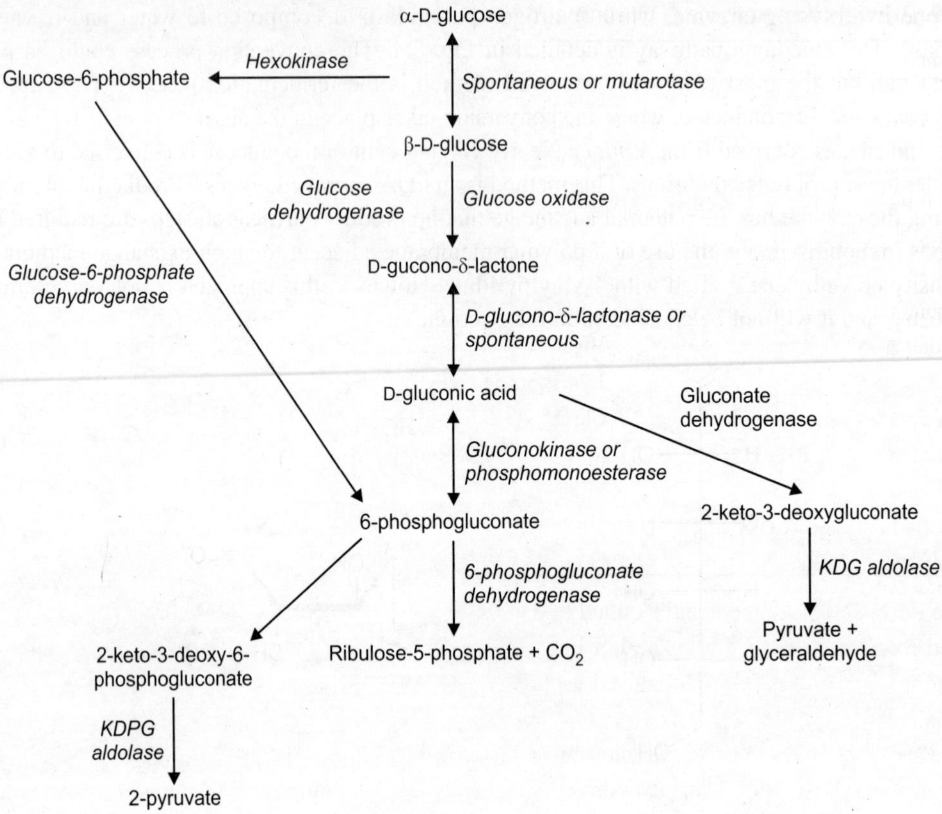

Fig. 8.4. General gluconate pathways.

Current commercial production of sodium gluconate uses submerged fermentation with *A. niger* and is based on the modified process developed by Blom. It involves fed-batch cultivation with intermittent glucose feedings and the use of sodium hydroxide as neutralising agent. pH is held at 6.0–6.5 and the temperature at about 34°C. The productivity of this process is very high, since glucose is converted at a rate of 15 g/(l·hr).

Properties

Physico-chemical behaviour

Gluconic acid is a noncorrosive, nonvolatile, nontoxic, mild organic acid. It imparts a refreshing sour taste in many food items such as wine, fruit juices, etc. Sodium gluconate has a high sequestering power. It is a good chelator at alkaline pH; its action is comparatively better than EDTA, NTA and other chelators. Aqueous solutions of sodium gluconate are resistant to oxidation and reduction at high temperatures. It is an efficient plasticiser and a highly efficient set retarder. It is easily biodegradable (98 per cent at 48 hrs). It has an interesting property of inhibiting bitterness in foodstuffs. Concentrated gluconic acid solution contains certain lactone structures (neutral cyclic ester) showing antiseptic property. The characterisitics are described in Table 8.2.

Table 8.2. General characteristics of gluconic acid.

Gluconic acid	
Nature	Noncorrosive, mildly acidic, less irritating, nonodorous, nontoxic, easily biodegradable, nonvolatile organic acid
Relative molecular mass	196.16
Chemical formula	$C_6H_{12}O_7$
Synonym	2,3,4,5,6-pentahydroxyhexanoic acid
pKa	3.7
Melting point (50% solution)	Lower than 12°C
Boiling point (50% solution)	Higher than 100°C
Density	1.24 g/ml
Appearance	Clear to brown
Solubility	Soluble in water
Sourness	Mild, soft, refreshing taste
Degree of sourness (sourness of citric acid is regarded as 100)	29–35

In the European Parliament and Council Directive No. 95/2/EC, gluconic acid is listed as a generally permitted food additive (E 574). The US FDA (Food and Drug Administration) has assigned sodium gluconate a GRAS (generally recognised as safe) status and its use in foodstuff is permitted without limitation.

There are several methods for the determination of D-gluconic acid and D-glucono-δ-lactone. Among them, isotachophoretic method and hydroxamate method are the most commonly used ones for the determination of gluconic acid. The concentration of gluconic acid is also determined by gas chromatography of their trimethylsilyl (TMS) derivatives prepared according to Laker and Mount with inositol as internal standard.

A widely used enzymatic method is based on the following principle: D-gluconic acid is phosphorylated to D-gluconate-6-phosphate by ATP in the presence of the enzyme gluconate kinase with the simultaneous formation of ADP. In the presence of NADP, D-gluconate-6-phosphate is oxidatively decarboxylated by 6-phosphogluconate dehydrogenase to ribulose-5-phosphate with the formation of reduced NADPH. The NADPH is stoichiometrically formed and its measurement allows direct determination of the amount of D-gluconic acid.

Gluconic acid is abundantly available in plants, fruits and other foodstuffs such as rice, meat, dairy products, wine (up to 0.25 per cent), honey (up to 1 per cent), and vinegar. It is produced by different micro-organisms as well, which include bacteria such as *Pseudomonas ovalis*, *Acetobacter methanolicus*, *Zymomonas mobilis*, *Acetobacter diazotrophicus*, *gluconobacter oxydans*, *gluconobacter suboxydans*, *Azospirillum brasiliense*, fungi such as *Aspergillus niger*, *Penicillium funiculosum*, *P. variabile*, *P. amagasakiense*, and various other species such as *gliocladium*, *scopulariopsis*, *gonatobotrys*, *dndomycopsis* and yeasts such as *Aureobasidium pullulans* (formerly known as *Dematium* or *Pullularia pullulans*). Ectomycorrhizal fungus *Tricholoma robustum*, which is associated with the roots of *Pinus densiflora*, was found to synthesise gluconic acid.

Applications

Gluconic acid is a mild organic acid, which finds applications in the food industry. It is a natural constituent in fruit juices and honey and is used in the pickling of foods. Its inner ester, glucono-δ-lactone imparts an initially sweet taste which later becomes slightly acidic. It is used in meat and dairy products, particularly in baked goods as a component of leavening agent for preleavened products. It is used as a flavouring agent (for example, in sherbets) and it also finds application in reducing fat absorption in doughnuts and cones. Foodstuffs containing D-glucono-δ-lactone include bean curd, yoghurt, cottage cheese, bread, confectionery and meat.

Generally speaking, gluconic acid and its salts are used in the formulation of food, pharmaceutical and hygienic products (Table 8.3). They are also used as mineral supplements to prevent the deficiency of calcium, iron, etc. and as buffer salts. Different salts of gluconic acid find various applications based on their properties. Sodium salt of gluconic acid has the outstanding property to chelate calcium and other di- and-trivalent metal ions. It is used in the bottle washing preparations, where it helps in the prevention of scale formation and its removal from glass. It is well suited for removing calcareous deposits from metals and other surfaces, including milk or beer scale on galvanised iron or stainless steel. Its property of sequestering iron over a wide range of pH is exploited in the textile industry, where it prevents the deposition of iron and for desizing polyester and polyamide fabrics. It is also used in metallurgy for alkaline derusting, as well as in the washing of painted walls and removal of metal carbonate precipitates without causing corrosion. It also finds application as an additive to cement, controlling the setting time and increasing the strength and water resistance of the cement. It helps in the manufacture of frost and crack resistant concretes. It is also used in the household cleaning compounds such as mouthwashes.

Table 8.3. Applications of gluconic acid and its derivatives.

Components	*Applications*
Gluconic acid	Prevention of milkstone in dairy industry
	Cleaning of aluminium cans
Glucono-δ-lactone	Latent acid in baking powders for use in dry cakes and instantly leavened bread mixes
	Slow acting acidulant in meat processing such as sausages
	Coagulation of soyabean protein in the manufacture of tofu
	In dairy industry for cheese curd formation and for improvement of heat stability of milk
Sodium salt of gluconic acid	Detergent in bottle washing
	Metallurgy (alkaline derusting)
	Additive in cement
	Derusting agent
	Textile (iron deposits prevention)
	Paper industry
Calcium salt of gluconic acid	Calcium therapy
	Animal nutrition
Iron salt of gluconic acid	Treatment of anaemia
	Foliar feed formulations in horticulture

Calcium gluconate is used in pharmaceutical industry as a source of calcium for treating calcium deficiency by oral or intravenous administration. It also finds a place in animal nutrition. Iron gluconate and iron phosphogluconate are used in iron therapy. Zinc gluconate is used as an ingredient for treating common cold, wound healing and various diseases caused by zinc deficiencies such as delayed sexual maturation, mental lethargy, skin changes, and susceptibility to infections.

Organic acids represent the third largest category after antibiotics and amino acids in the global market of fermentation.

The main product among the gluconic acid derivatives is the sodium gluconate due to its properties and applications. Calcium gluconate is also an important product among the derivatives of gluconic acid and it is available as tablets, powder, and liquid for dietary supplements.

Production of Gluconic Acid

There are different approaches available for the production of gluconic acid, namely, chemical, electrochemical, biochemical and bioelectrochemical. There are several different oxidising agents available, but still the process appears to be costlier and less efficient compared to the fermentation processes. Although the conversion is a simple one-step process, the chemical method is not favoured. Thus, fermentation has been one of the efficient and dominant techniques for manufacturing gluconic acid. Among various microbial fermentation processes, the method utilising the fungus *A. niger* is one of the most widely used ones. However, the process using *G. oxydans* has also gained significant importance. Irrespective of the use of fungi or bacteria, the importance lies on the product which is produced, for example, sodium gluconate or calcium gluconate, etc. As the reaction leads to an acidic product, it is required that it is neutralised by the addition of neutralising agents, otherwise the acidity inactivates the glucose oxidase, resulting in the arrest of gluconic acid production. The conditions for the fermentation processes in the production of calcium gluconate and sodium gluconate differ in many aspects such as glucose concentration (initial and final) and pH control. In the process involving calcium gluconate production, the control of pH results from the addition of calcium carbonate slurry. Another important point to be noted is about the solubility of calcium gluconate in water (4 per cent at 30°C). At high glucose concentration, above 15 per cent, supersaturation occurs, and if it exceeds the limit, the calcium salt precipitates on the mycelia and inhibits the oxygen transfer. The neutralising agent should also be sterilised separately from the glucose solution to avoid Lobry de Bruyn-van Ekenstein reaction, which alters the conformation of glucose, which results in the reduction of yield for about 30 per cent. On the contrary, the process for sodium gluconate is highly preferable as the glucose concentration of up to 350 g/l can be used without any such problems. pH is controlled by the automatic addition of NaOH solution. Sodium gluconate is readily soluble in water (39.6 per cent at 30°C).

By filamentous fungi

Glucose oxidase

The reaction involving the conversion of glucose to gluconic acid by filamentous fungi is catalysed by the enzyme glucose oxidase (β-D-glucose: oxygen-1-oxidoreductase, E.C. 1.1.3.4). The enzyme was first isolated from a press juice obtained from *Penicillium glaucum* by Müller. The enzyme was crystallised by Kusai from *P. amagasakiense*. The enzyme was previously known as notatin. Glucose oxidase is a flavoprotein which contains one very tightly but noncovalently bound FAD cofactor per monomer and is a homodimer with a molecular mass of 130–320 kDa depending on the extent of glycosylation. It catalyses the reaction where glucose is dehydrated to glucono-δ-lactone, while hydrogen is transferred to FAD. The resulting $FADH_2$ is regenerated to FAD by transmission of the hydrogen to oxygen to form hydrogen peroxide (Fig. 8.5). Glucose oxidase is a glycoprotein. The native enzyme is glycosylated, with a carbohydrate mass percentage of 16–25 per cent. The enzyme from *A. niger* contains 10.5 per cent carbohydrate, which is believed to contribute to the stability without affecting the overall mechanism.

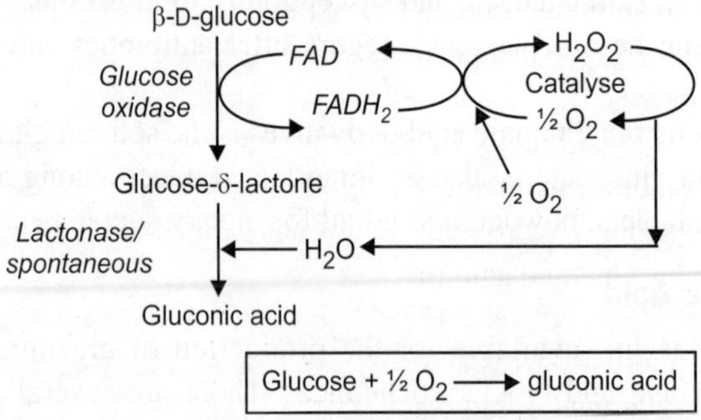

Fig. 8.5. Oxidation of glucose by *Aspergillus niger*.

The enzyme is induced in the presence of high levels of glucose in the medium, pH around 5.5 and elevated oxygen levels. The enzyme is stable between pH = 4.0 and 6.0 at 40°C for 2 hrs but is unstable above 50°C. Liu conducted a study on the effects of metallions on simultaneous production of glucose oxidase and catalase and found that calcium carbonate induced the synthesis of both enzymes. The induction of calcium carbonate was accompanied by a metabolic shift from the glycolytic pathway (EMP) to direct oxidation of glucose by the enzyme. The enzyme is found to be inhibited by hydrogen peroxide, the by-product of gluconic acid production. A study on glucose oxidase inactivation showed that only the reduced form of glucose oxidase is highly sensitive to hydrogen peroxide.

The enzyme is used in various fields such as food, clinical analysis, mainly as glucose sensor, in the quantitative determination of glucose in body fluids and urine. It is used in food processing in the removal of glucose prior to the preparation of products such as dried eggs to reduce the nonenzymatic browning. It is also used in removing residual oxygen from fruit juices, beer, and wine and also from dehydrated packaged foods.

Reports on glucose oxidase localisation are ambiguous. Van Dijken and Veenhuis, and Witteveen reported that the enzyme of *A. niger* is intracellular and found in peroxisomes, whereas Mischak reported it as extracellular. There are also reports which have stated that it is intracellular prior to fungal autolysis. These varying reports on its location in the cell could be attributed to the differences of parameters and conditions adopted for the growth or due to the age of the fungal cultures. Very little is known about the mechanisms of glucose oxidase export. Zetelaki associated export with autolysis of the fungus, whereas Mischak reported that the glucose oxidase of *A. niger* was excreted after synthesis.

Aspergillus niger

A. niger produces all the enzymes required for the conversion of glucose into gluconic acid, which include glucose oxidase, catalase, lactonase and mutarotase. Although crystalline glucose monohydrate, which is in the alpha form, is converted spontaneously into beta form in the solution, *A. niger* produces the enzyme mutarotase, which serves to accelerate the reaction. During the process of glucose conversion, glucose oxidase present in *A. niger* undergoes self-reduction by the removal of two hydrogens. The reduced form of the enzyme is further oxidised by the molecular oxygen, which results in the formation of hydrogen peroxide, a by-product in the reaction. *A. niger* produces catalase which acts on hydrogen peroxide releasing water and oxygen. Hydrolysis of glucono-δ-lactone to gluconic acid is facilitated by lactonase. The reaction can be carried out spontaneously as the cleavage of lactone occurs rapidly at pH near neutral, which are brought about by the addition of calcium carbonate, or sodium hydroxide. Removal of lactone from the medium is recommended as its accumulation in the media has a negative effect on the rate of glucose oxidation and the production of gluconic acid and its salt. There are reports stating that the enzyme gluconolactonase is also present in *A. niger*, which increases the rate of conversion of glucono-δ-lactone to gluconic acid.

Production of gluconic acid is directly linked with the glucose oxidase activity. Depending on the application, the fermentation broths containing sodium gluconate or calcium gluconate are produced by the addition of solutions of sodium hydroxide or calcium carbonate respectively, for neutralisation. The general optimal condition for gluconic acid production is as follows:

1. Glucose at concentrations between 110–250 g/l.
2. Nitrogen and phosphorus sources at a very low concentration (20 mM).
3. pH value of medium around 4.5 to 6.5.
4. Very high aeration rate by the application of elevated air pressure (4 bar).

There are two key parameters which influence the gluconic acid production. These are oxygen availability and pH of the culture medium. Oxygen is one of the key substrates in the oxidation of glucose as glucose oxidase uses molecular oxygen in the bioconversion of glucose. The concentration of oxygen gradient and the volumetric oxygen transfer coefficient are the critical factors, which monitor the availability of oxygen in the medium. These two factors highly influence the rate of the transfer of oxygen from gaseous to aqueous phase. Several reports are available on this particular aspect. The aeration rate and the speed of agitation are the two parameters which affect the availability of the oxygen in the medium. Gluconic acid production is an extremely oxygen-consuming process with a high oxygen demand for the bioconversion reaction, which is strongly influenced by the dissolved oxygen concentration. Oxygen is generally supplied in the form of atmospheric air; however, in some studies high-pressure pure oxygen has also been provided. For example, Sakurai and other supplied high-pressure oxygen at approx. 6 bar and maintained dissolved oxygen at 150 ppm. They found that immobilised mycelium of *A. niger* grown using pure oxygen produced high titres of gluconic acid in comparison with mycelium grown in air. Kapat found that at an agitation speed of 420 rpm and aeration of 0.25 vvm, the dissolved oxygen concentration was optimal for glucose oxidase production. The K_m value of glucose oxidase for oxygen lies in the range of air saturation in water. Lee obtained high volumetric productivity of gluconic acid using relatively high pressure (2–6 bar), resulting in an increase in dissolved oxygen up to 150 mg/l. Generally, during the course of fungal growth, the distribution of oxygen becomes uneven, as the size of gas bubbles increases, resulting in insufficient oxygen supply. The oxygen absorption rate is also influenced by the viscosity of the culture. A rapid decrease is observed in the absorption rate of oxygen with an increase in mycelial concentration.

pH is another important parameter that influences the gluconic acid production. *A. niger* produces weak organic acids such as citric acid, gluconic acid and oxalic acid, and their accumulation depends on the pH of the nutritive medium. pH below 3.5 triggers the TCA cycle and facilitates the citric acid formation. The pH range of the fungi for the production of gluconic acid is around 4.5 to 7.0. pH = 5.5 is generally considered as optimum for *Aspergillus niger*. Franke collected some data concerning the relative activity of glucose oxidase at different pH levels and reported 5 and 35 per cent activity at pH = 2.0 and 3.0, respectively, based on 100 per cent activity at pH = 5.6. Report by Heinrich and Rehm states that gluconic acid production occurs even at pH = 2.5 in the presence of manganese in fixed bed and stirred bed reactors, possibly because of the difference in intracellular and extracellular pH.

Cheaper raw materials as substrates

Glucose is generally used as carbon source for microbial production of gluconic acid. However, hydrolysates of various raw materials such as agro-industrial waste have also been used as substrate. Smith obtained a high yield of gluconic acid in media containing glucose or starch hydrolysate as the sole carbon source. Vassilev used hydrol (corn starch hydrolysate) as the fermentable sugar to produce gluconic acid by immobilised *A. niger*. Bailey used cane molasses as a source of glucose. The cane molasses was subjected to different pre-treatments such as acid treatment, potassium ferrocyanide treatment, salt treatment, etc. Potassium ferrocyanide treatment gave a promising result. Gluconic acid synthesis was influenced by various metal ions such as copper, zinc, magnesium, calcium, iron, etc. clive used deproteinised whey as a nutritive medium for gluconic acid production. Lactose was used as a substrate and 92 g of gluconic acid was produced from 1 litre of whey containing 0.5 per cent glucose and 9.5 per cent lactose by *A. niger* immobilised on polyurethane foam. Ikeda used saccharified solution of waste paper with glucose concentration adjusted to 50–100 g/l for bioconversion with *A. niger*. The yields were 92 per cent in Erlenmeyer flasks and 60 per cent in repeated batch cultures in the turbine blade reactor with 800 ml of working volume. Another striking feature in the study was when xylose and cellobiose were used as the sole carbon sources, yields of gluconic acid obtained were 83 and 56 per cent, respectively.

Smith and others observed that grape must and banana must resulted in significant levels of gluconic acid production, i.e. 63 and 55 g/l respectively. The purification of grape and banana must leads to a 20–21 per cent increase in gluconic acid yield. They also used molasses, where the gluconate production was 12 g/l, but a significant increase in production of 60 g/l with a yield of 61 per cent was observed following treatment of the molasses with hexacyanoferrate. Rectified grape must appeared to be the best suited substrate, which after 144 hrs resulted in 73 g/l of gluconic acid with 81 per cent yield when compared to the value of 72 per cent obtained from the rectified banana must. Buzzini and others also used grape must and rectified grape must and they found that the latter substrate was better, with a production of 67 g/l and a yield of 96 per cent in 72 hrs. Citric acid was also observed as a by-product.

Use of solid-state fermentation (SSF)

SSF has been widely described for the production of industrial enzymes and organic acids. However, for the production of gluconic acid, there are only a few reports using SSF. Roukas reported the production of gluconic acid by solid-state fermentation on figs. The maximal gluconic acid concentration was 490 g/kg of dry fig with 63 per cent yield. The addition of 6 per cent methanol into the substrate helped to increase the production of gluconic acid from 490 to 685 g/kg. Smith performed SSF by using HCl pretreated sugarcane bagasse and the highest level of gluconic acid (107 g/l) with 95 per cent yield was obtained. In comparison with the submerged culture, the degree of conversion was higher in SSF. The increased rate of product formation might be due to the variations of osmotic pressure, water content and dissolved oxygen. A study by Moksia used a two-step process, the first being the production of spores of *A. niger* by SSF on buckwheat seeds, and the second step, the bioconversion of glucose to gluconic acid by the spores recovered from the SSF medium. The interesting aspect about this work was that the spores were not allowed to germinate as the bioconversion medium did not contain any nitrogen source. The spores acted as a biocatalyst, producing 200 g/l of gluconic acid with a yield of 1.06 g per mass of glucose, very close to the stoichiometric value.

Production of gluconic acid by bacteria

Acetic acid bacteria and *Pseudomonas savastanoi* were the cultures initially observed to produce gluconic acid. Unlike in fungi, in bacteria the reaction is carried out by glucose dehydrogenase (GDH, E.C. 1.1.99.17) that oxidises glucose to gluconic acid, which is further oxidised to 2-ketogluconate by gluconic acid dehydrogenase (GADH). The final oxidation step to 2,5-diketogluconic acid (DKG) is mediated by 2-ketogluconate dehydrogenase (KGDH). The reaction steps are shown in Fig. 8.6. All three enzymes are localised in the membranes of the cells and are induced by high glucose concentrations (>15 mM). GDH is an extracellular protein and has PQQ (pyrroloquinoline quinine) as a coenzyme. Also, there is an intracellular enzyme, an $NADP^+$-dependent glucose dehydrogenase, which is less involved in the gluconic acid formation when compared to the extracellular enzyme. Gluconic acid produced is exported to the cell and further catabolised *via* the reactions in pentose phosphate pathway. When the glucose concentration in the medium is greater than 15 mM, pentose phosphate pathway is repressed and thus gluconic acid accumulation takes place.

D-glucose
↓ Glucose dehydrogenase
Gluconic acid
↓ Gluconate dehydrogenase
2-ketogluconic acid
↓ 2-ketogluconic acid dehydrogenase
2,5-diketogluconic acid

Fig. 8.6. Specific pathway for oxidation of glucose by *Gluconobacter.*

Gluconobacter oxydans: It is an obligate aerobic bacterium that oxidises glucose via two alternative pathways. The first pathway requires an initial phosphorylation followed by oxidation via the pentose phosphate pathway. The second is the »direct glucose oxidation« pathway, which results in the formation of gluconic acid and ketogluconic acid. *G. oxydans* converts D-glucose into 2,5-diketogluconic acid by the action of three membrane-bound $NADP^+$-independent dehydrogenases as mentioned in Fig. 8.6. The acidotolerant acetic acid bacterium, *Acetobacter diazotrophicus*, exhibited high rates of gluconic acid formation. Glucose oxidation by the organism was less sensitive to low pH values than glucose oxidation by *G. oxydans.* Both the phosphorylative and direct oxidative pathways of glucose metabolism appeared to be operative. In addition to a pyridine nucleotide (strictly NAD^+)-dependent glucose dehydrogenase, *A. diazotrophicus* contained a PQQ-dependent glucose dehydrogenase, which was primarily responsible for gluconic acid formation. Bacterial gluconic acid production has limited success at industrial scale, as the oxidation proceeds with the secondary reactions leading to oxogluconic acids. The ability of *Pseudomonas* and *Gluconobacter* spp. to produce gluconolactone and gluconic acid has been exploited and the process is used commercially mainly in the production of lactone.

Acetobacter methanolicus: It is also used to catalyse the conversion of glucose into gluconic acid. The key advantage of using this facultatively methylotrophic micro-organism as catalyst is that the gluconic acid formed is a metabolic dead-end product, and unlike in other bacterial fermentation processes, organism uses methanol, a cheap raw material as a substrate. Further in the process glucose is not assimilated or consumed for growth, so consequently the maximuim theoretical yield coefficient is achieved. A patent was filed by Currie and Carter in which the medium containing 200 g/l of glucose with other nutrients and a neutralising agent was allowed to flow through a tower packed with wood shavings or coke, which had been inoculated with *Acetobacter suboxydans*, while air was passed upwards through the packing. Tsao and Kempe, working with *Pseudomonas ovalis* found that a particular strain could convert glucose to gluconic acid with a yield of 99 per cent, and the rate was directly related to the efficiency of aeration.

Yeast

Research carried out by several authors utilised *Aureobasidium pullulans,* a yeastlike form of the dimorphic fungi, for the production of gluconic acid. Various process parameters for the continuous and discontinous production of gluconic acid such as pH, oxygen, temperature and medium composition, air saturation, etc. were studied. The highest glucose conversion of 94 per cent and product yield of 87.1 per cent was achieved at an optimum pH of 6.5. At pH = 4.5, the product selectivity and yield were very poor, reaching 67.8 and 20.7 per cent, respectively. Temperature range of 29° to 31°C was found to be suitable for thc production of gluconic acid by the yeast. Increase of temperature by 1°C, namely to 32°C, dramatically influenced the reduction in steady state concentration of biomass and product.

Immobilisation

Immobilisation techniques are involved where the biomass is immobilised onto the support and, in some cases, the enzyme isolated from the culture is immobilised. It enables repetitive use of the high biomass to carry out biochemical reactions rapidly leading to process economy and stability. Immobilisation seems to be an attractive method for accomplishing high cell densities in order to achieve rapid carbohydrate conversion to organic acids. Matrix immobilisation is a simple and easy technique by which mycelia are retained on a matrix by mycelial entanglement. The type of support, cell retention, stabilisation of enzyme or the mycelia and the quantum of biomass, etc. play important roles.

In the past, there were several investigations related to the production of gluconic acid with immobilised cells of *A. niger*. There are also reports of the immobilisation of *A. niger* pellets by flocculation with polyelectrolytes, calcium alginate, glycidyl ester copolymers and entrapment in gels. Glass rings were used to immobilise *A. niger* for the production of gluconic acid by Heinrich and Rehm. Sakurai adopted a novel method for the immobilisation of *A. niger* using a support of nonwoven fabric. Smith reported the immobilisation of the same filamentous fungi on polyurethane foam. Different carriers such as calcium alginate agar, polyurethane sponge, pearlite, and activated carbon were used for the immobilisation of *Penicillium* variabile by Petruccioli.

Free gluconic acid was continuously produced in an aerated tubular immobilised cell bioreactor using *G. oxydans* for at least 6 months, with a volumetric productivity of at least 5 g/(l·hr) per 100 g/l of glucose substrate and the concentration of produced gluconic acid of about 80 g/l). Spores of *A. niger* were immobilised on sintered glass, pumice stones and polyurethane foams, and mycelia which developed on the pumice stone carrier produced high extracellular glucose oxidase (80 per cent) when compared to the enzyme activity on free cells.

An attempt was made by Karze to study the bioconversion of glucose to gluconic acid using *A. niger* immobilised on cellulosic fabric as a support matrix. Glucose solution (100 g/l) was made to flow through capillaries of a vertical fabric support, used for immobilisation, and was oxidised to gluconic acid at the interface. The system was found to run continuously for a period of 61 days utilising the entire available glucose. The emerging broth contained a product concentration of 120–140 g/l of gluconic acid, which was higher than expected (maximum of 109 g/100 g of glucose), as a result of evaporative concentration during the downward flow. Smith found that the optimum biomass requirement on a porous cellulose support was 0.234 mg/cm^2 for efficient bioconversion. Increasing the quantum of biomass beyond this value resulted in an overgrown biofilm which affected productivity adversely. Morphological characteristics of immobilised *A. niger* have also been investigated.

Recovery

The recovery process depends on the method followed for broth neutralisation and the nature of carbon sources used. Generally, the downstream process is similar for the fermentation processes using fungal and bacterial species. Gluconic acid, glucono-δ-lactone, calcium gluconate, and sodium gluconate are some of the important products and their extraction process is briefly mentioned below.

For the recovery of free gluconic acid from calcium gluconate the broth is clarified, decolourised, concentrated and exposed to –10°C in the presence or absence of alcohol. Thus the calcium salt of gluconic acid crystallises, then it is recovered and further purified. Gluconic acid can also be obtained by precipitating the calcium gluconate from hypersaturated solutions in the cold and released subsequently by adding sulphuric acid stoichiometrically, removing the calcium as calcium sulphate. Another method of passing the solution through a column containing a strong cation exchanger is also practised where the calcium ions are absorbed.

For obtaining calcium gluconate as a product, calcium hydroxide or calcium carbonate is used as the neutralising agent. They are added to the nutritive broth accompanied by heating and vigorous stirring. The broth is concentrated to a hot supersaturated solution of calcium gluconate, followed by cooling at 20°C, and adding water miscible solvents, which crystallises the compound. A treatment with activated carbon facilitates the crystallisation process. Finally they are centrifuged, washed several times and dried at 80°C.

Sodium gluconate, the principal manufactured form of gluconic acid, is prepared by ion exchange. In the process developed by Blom the sodium gluconate from the filtered fermented broth is concentrated to 45 per cent (mass per volume), followed by the addition of sodium hydroxide solution raising the pH to 7.5, and drum drying. Carbon treatment of the hot solution before drying process is practised for obtaining a refined product. Glucono-δ-lactone recovery is a very simple process. Aqueous solutions of gluconic acid are an equilibrium mixture of glucono-δ-lactone, glucono-λ-lactone and gluconic acid. At temperature between 30°–70°C the crystal which is separated from the supersaturated solution is glucono-δ-lactone. At temperature below 30°C, gluconic acid results even above 70°C, and the resulting product would be glucono-λ-lactone.

Molecular biology

The molecular genetics of gluconic acid overproduction is not very well investigated. It is well known that the enzyme is actively induced by glucose concentration and high aeration and pH above 4.0. The gene encoding glucose oxidase of *A. niger* (*gox A*) has been cloned, and its amplification resulted in a 2–3-fold increase in activities. *A. niger* secretes multiple forms of catalases to shield itself against the arising hydrogen peroxide, among which one has been cloned and characterised. Swart described nine different complementation groups of glucose oxidase overproduction mutants. *Gox B*, *gox C*, and *gox F* belong to linkage group 11, *gox* 1 to linkage group 111, *gox D* and *gox G* to linkage group V, *gox A* and *gox E* to linkage group VII, and the linkage of *gox H* is unknown. Their study also indicates that *gox A* overproduction is regulated by the carbon source and oxygen in an independent manner. Knowledge about gene encoding lactonase is very narrow.

The gox-encoding gene of *Penicillium* variabile P16 was isolated and characterised to identify the molecular bases of its high level of expression and in view of improving enzyme production by developing a process based on heterologous expression.

There are some works carried out on the bacterial enzyme. A Tn*5*-induced glucose dehydrogenase (GDH) deficient mutant of *Gluconobacter oxydans* IFO 3293 was characterised. DNA sequencing showed that the insertion site occurred in an open reading frame with homology to the *pqqE* gene. It was shown that acid production could be restored by addition of the coenzyme PQQ to the medium. The *pqq* cluster of *G. oxydans* ATCC 9937 was cloned and sequenced. It has five genes, *pqqA-E*. The cluster could complement the Tn*5*-induced mutation in IFO 3293. Pulsed-field gel electrophoresis suggested that the *pqq* genes are not closely linked to the *ribF* gene that produces the riboflavin cofactor for the gluconic acid dehydrogenase.

Although the production of gluconic acid is a simple oxidation process that can be carried out by electrochemical, biochemical or bioelectrochemical methods, production by fermentation process involving fungi and bacteria is well established commercially. Considerable progress has been made in understanding the mechanism of fermentation process by different micro-organisms, and highly efficient production process, which dates back to five decades, has been developed. However, development of novel, more economical process for the conversion of glucose to gluconic acid with longer shelf life would be promising. These requirements could be met by enzymatic system. Another way of improvement is to use cheap substrates, such as methanol instead of glucose.

GLUCONOLACTONE

A lactone is an oxidised derivative of glucose. Gluconolactone is a poly hydroxy acid (PHA) that is capable of chelating metals and may also function by scavenging free radicals, there by protecting skin from some of the damaging effects of UV radiation. It is in the form of white crystals, m.p. 155°C, molecular weight 178.14 readily soluble in water, slightly soluble in alcohol. The formula of gluconolactone is $C_6H_{10}O_6$. The structure of gluconolactone is given below:

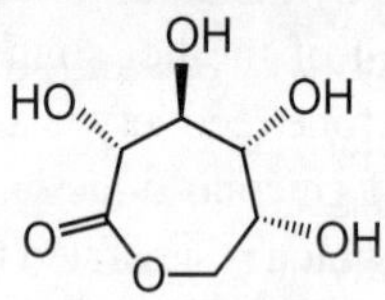

Gluconolactone

Gluconolactone is composed of multiple water-attracting hydroxyl groups, which hydrate the skin, resulting in enhanced degrees of moisturisation. Gluconolactone provided up to 50 per cent protection against UV radiation and UV radiation-induced elastin promotor activation. Contrary to some scientific fears, it does not significantly increase sunburn cells in human skin.

Although more research is being done on gluconolactone's effects on photoageing, in one study, signs of photoageing were significantly reduced after six and twelve weeks of twice daily use, including firmness, sallowness, fine lines, wrinkles, roughness, hyperpigmentation and pore size. It may also show improvements in patients with rosacea, and is gentle on sensitive skin.

There are no warnings regarding the use of gluconolactone, and because of its PHA composition, it may be safer than products containing AHA or BHA. Pregnant women should still consult a doctor before using products with gluconolactone.

Novel Pathway for Alcoholic Fermentation Δ-Gluconolactone in the Yeast *Saccharomyces*

Under anaerobic conditions, the yeast *Saccharomyces bulderi* rapidly ferments Δ-gluconolactone to ethanol and carbon dioxide. A novel pathway for Δ-gluconolactone fermentation operates in this yeast. In this pathway, Δ-gluconolactone is first reduced to glucose via an NADPH-dependent glucose dehydrogenase (EC 1.1.1.47). After phosphorylation, half of the glucose is metabolised via the pentose phosphate pathway, yielding the NADPH required for the glucose-dehydrogenase reaction. The remaining half of the glucose is dissimilated via glycolysis. Involvement of this novel pathway in Δ-gluconolactone fermentation in *S. bulderi* is supported by several experimental observations: (i) fermentation of Δ-gluconolactone and gluconate occurred only at low pH values, at which a substantial fraction of the substrate is present as Δ-gluconolactone. Unlike gluconate, the latter compound is a substrate for glucose dehydrogenase, (ii) high activities of an $NADP^{(+)}$ dependent glucose dehydrogenase were detected in cell extracts of anaerobic, Δ-gluconolactone-grown cultures, but activity of this enzyme was not detected in glucose-grown cells. Gluconate kinase activity in cell extracts was negligible, (iii) during anaerobic growth on Δ-gluconolactone, CO_2 production exceeded ethanol production by 35 per cent, indicating that pyruvate decarboxylation was not the sole source of CO_2, and (iv) levels of the pentose phosphate pathway enzymes were 10-fold higher in Δ-gluconolactone-grown anaerobic cultures than in glucose-grown cultures, consistent with the proposed involvement of this pathway as a primary dissimilatory route in Δ-gluconolactone metabolism.

GLUCOSE OXIDASE

The glucose oxidase enzyme (GO_X) (EC 1.1.3.4) binds to β-D-glucopyranose (a hemiacetal form of the six-carbon sugar glucose) and aids in breaking the sugar down into its metabolites. GOx is a dimeric protein that catalyses the oxidation of β-D-glucose into D-glucono-1,5-lactone, which then hydrolyses to gluconic acid.

The CAS number for this type of the enzyme is [9001–37–0]. In order to work as a catalyst, GO_X requires a cofactor, flavin adenine dinucleotide (FAD). FAD is a common component in biological oxidation-reduction (redox reactions). Redox reactions involve a gain or loss of electrons from a molecule. In the GO_X-catalysed redox reaction, FAD works as the initial electron acceptor and is reduced to $FADH_2$. Then $FADH_2$ is oxidised by the final electron acceptor, molecular oxygen (O_2), which can do so because it has a higher reduction potential. O_2 is then reduced to hydrogen peroxide (H_2O_2).

The glucose oxidase enzyme is commonly used in biosensors to detect levels of glucose by keeping track of the number of electrons passed through the enzyme by connecting it to an electrode and measuring the resulting charge. When produced commercially for this application, it is often extracted from *Aspergillus niger*. This has a possible application in the world of nanotechnology when used in conjunction with tiny electrodes as glucose sensors for diabetics.

Glucose oxidase is found in honey and acts as a natural preservative. GO_X at the surface of the honey reduces atmospheric O_2 to hydrogen peroxide (H_2O_2), which acts as an antimicrobial barrier.

Glucose oxidase (GO_X) catalyses the oxidation of β-D-glucose to D-glucono-β-lactone and hydrogen peroxide. It is highly specific for β-D-glucose and does not act on α-D-glucose.

A major use of glucose oxidase has been in the determination of free glucose in bodyfluids, food and agricultural products. However, it has been gaining increasing attention in the baking industry; its oxidising effects result in a stronger dough. In some applications, it can be used to replace oxidants such as bromate and L-ascorbic acid. Other uses of glucose oxidase include the removal of oxygen from food packaging and removal of D-glucose from egg white to prevent browning.

Glucose oxidase catalyses the oxidation of β-D-glucose to D-glucono-δ-lactone with the concurrent release of hydrogen peroxide. In the presence of peroxidase (POD) this hydrogen peroxide (H_2O_2) enters into a second reaction involving *p*-hydroxybenzoic acid and 4-aminoantipyrine with the quantitative formation of a quinoneimine dye complex which is measured at 510 nm.

The reactions involved are:

1. $\beta\text{-D-Glucose} + O_2 + H_2O \xrightarrow{(GOx)} \text{D-glucono-}\delta\text{-lactone} + H_2O_2$

2. $2H_2O_2 + p\text{-hydroxybenzoic acid} + \text{4-aminoantipyrine} \xrightarrow{(POD)} \text{quinoneimine dye} + 4\,H_2O$

The reagents used in the determination of glucose oxidase are not hazardous materials in the sense of the Hazardous substances Regulations. However, the buffer concentrate contains sodium azide (0.02 per cent w/v) as a preservative. The general safety measures that apply to all chemical substances should be adhered to.

Fermentation of a Yeast Producing *A. niger* Glucose Oxidase

Recently a fermentation process has been developed to produce up to three grams per litre of active, secreted glucose oxidase from a recombinant *Saccharomyces cerevisiae*. Real-time size-exclusion HPLC analysis is used to monitor enzyme production during fermentation, and purification to more than 95 per cent is obtained using only filtration methods. The recombinant enzyme is stable to higher temperatures and a wider pH range than the native *Aspergillus niger* enzyme, and is free of contaminating amylase, cellulase and catalase.

Biosynthesis of Glucose Oxidase

Glucose oxidase biosynthesis in relation to biochemical mutations in Aspergillus niger

The production of extracellular glucose oxidase in a submerged culture by a number of auxotrophic, 2-deoxy-D-glucose resistant and protease-less mutants of *Aspergillus niger* was evaluated. Among the auxotrophic strains, no evident dependence was found between the kind of the nutritional requirements and the level of the glucose oxidase activity. However, the majority of auxotrophs, requiring serine or niacin, showed a higher enzyme activity (from 16 to 680 per cent) than the parent strain. The dynamics of the glucose oxidase synthesis by the free and immobilised mycelium of the most active niacin-mutant of *A. niger* was also investigated.

Biosynthesis of glucose oxidase by producing strain Z-I-C in batch and chemostat culture

The biosynthesis of glucose oxidase (GOD) by *Penicillium*-producing strain Z-I-C in batch and chemostat culture was studied. In a glucose-limited chemostat, the experimental results showed that the maintenance coefficient of this strain was 0.04 g glucose/g dry cell wt/h; yield factor of growth was 0.714 g dry cell wt/g glucose; maximum specific growth rate was 0.385 h and saturation constant Ks was 4.76 g/l. Theoretical optimum dilution rate was 0.260 h^{-1}.

Maximum specific enzyme activity was increased from 1.51 × 10(3) mu/mg in batch culture to 2.16 × 10(3) mu/mg in chemostat culture and to 3.11 × 10(3) mu/mg in chemostat culture supplemented in feed medium with 0.02 per cent inducer, α-methyl-D-glucose a 43 per cent or 106 per cent increase, respectively, compared with that of batch culture.

Glucose oxidase biosynthesis using immobilised mycelium of Aspergillus niger

A simple method for the immobilisation of *Aspergillus niger* GIV-10 which produces an extracellular glucose oxidase. *A. niger* conidia were immobilised on sintered glass Raschig rings, pumice stones or polyurethane foam. Mycella growing out from the spores produced extracellular glucose oxidase: the highest production was with the pumice stone carriers. This technique facilitates the growth of the filamentous cultures in the spongy structure of a support with continuous accumulation of biomass. After 24 to 36 hrs, a culture liquid with 2.7 to 3.1 U of glucose oxidase/ml was obtained. This procedure also made possible repeated batch enzyme production and as many as 25 subsequent 24 hrs batches could be fermented by using the same carrier with only a small loss of glucose oxidase activity.

Purification of glucose oxidase from complex fermentation medium using tandem chromatography

A fast and efficient purification method for recombinant glucose oxidase (rGO_X) for flask fermentation scale (up to 2 l) was designed for the purposes of characterisation of rGO_X mutants during directed protein evolution. The *Aspergillus niger* GO_X was cloned into a pYES2-αMF-GO_X construct and expressed extracellularly in yeast *Saccharomyces cerevisiae*. Hydrophobic interaction (HIC)/size exclusion (SEC)-tandem chromatographic system was designed for direct purification of rGO_X from a conditioned complex expression medium with minimum preceding sample preparation (only adjustments to conductivity, pH and coarse filtering). HIC on Butyl 650s (50 mM ammonium acetate pH 5.5 and 1.5 M ammonium sulphate) absorbs GOX from the medium and later it is eluted by 100 per cent stepwise gradient with salt free buffer directly into SEC column (Sephadex 200) for desalting and final polishing separation. The electrophoretic and UV-vis spectrophotometric analyses have proven enzyme purity after purification.

ACETIC ACID

Acetic acid, CH_3COOH, also known as ethanoic acid, is an organic acid which gives vinegar its sour taste and pungent smell. Pure, water-free acetic acid (glacial acetic acid) is a colourless liquid that absorbs water from the environment (hygroscopy), and freezes at 16.7°C (62°F) to a colourless crystalline solid. It is a weak acid, in that it is only partially dissociated acid in aqueous solution.

Acetic acid is one of the simplest carboxylic acids. It is an important chemical reagent and industrial chemical, used in the production of polyethylene terephthalate mainly used in soft drink bottles; cellulose acetate, mainly for photographic film; and polyvinyl acetate for wood glue, as well as synthetic fibres and fabrics. In households, diluted acetic acid is often used in descaling agents. In the food industry acetic acid is used under the food additive code E260 as an acidity regulator.

Acetic acid bacteria are bacteria that derive their energy from the oxidation of ethanol to acetic acid during respiration. They are Gram-negative, aerobic, rod-shaped bacteria.The acetic acid bacteria are found in nature where ethanol is being formed as a result of yeast fermentation of sugars and plant carbohydrates. They can be isolated from the nectar of flowers and from damaged fruit. Other good sources are fresh apple cider and unpasteurised beer which has not been filter sterilised. In these liquids the acetic acid bacteria grow as a surface film due to their aerobic nature and active motility. Vinegar is produced when acetic acid bacteria act on alcoholic beverages such as wine. Some genera, such as *Acetobacter*, can eventually oxidise acetic acid to carbon dioxide and water using Krebs cycle enzymes. Other genera, such as *Gluconobacter*, do not further oxidise acetic acid, as they do not have a full set of

Krebs cycle enzymes. Some acetic acid bacteria, notably *Acetobacter xylinum*, are known to synthesise cellulose, something normally only done by plants. As these bacteria produce acid, they are unusually acid tolerant, growing well below pH 5.0, although the pH optimum for growth is 5.4–6.3.

Acetobacter is a genus of acetic acid bacteria characterised by the ability to convert alcohol (ethanol) to acetic acid in the presence of oxygen. There are several species within this genus, and there are other bacteria capable of forming acetic acid under various conditions; but all of the *Acetobacter* are known by this characteristic ability.

Acetobacter are of particular importance commercially, because: (i) they are used in the production of vinegar (intentionally converting the ethanol in the wine to acetic acid), (ii) they can destroy wine which they infect by producing excessive amounts of acetic acid or ethyl acetate, both of which can render the wine unpalatable, (iii) they are used to intentionally acidify beer during long maturation periods in the production of traditional Flemish Sour Ales, and (iv) *A. xylinus* is the main source of microbial cellulose.

The growth of Acetobacter in wine can be suppressed through effective sanitation, by complete exclusion of air from wine in storage, and by the use of moderate amounts of sulphur dioxide in the wine as a preservative.

Acetobacter can be easily distinguished in the laboratory by their growth of colonies on a medium containing about 7 per cent ethanol, and enough calcium carbonate to render the medium partially opaque. When Acetobacter colonies form enough acetic acid from the ethanol, the calcium carbonate around the colonies dissolves, forming a very distinct clear zone.

Oxidative Fermentation

For most of human history, acetic acid, in the form of vinegar, has been made by acetic acid bacteria of the genus Acetobacter. Given sufficient oxygen, these bacteria can produce vinegar from a variety of alcoholic foodstuffs. Commonly used feeds include apple cider, wine, and fermented grain, malt, rice, or potato mashes. The overall chemical reaction facilitated by these bacteria is:

$$C_2H_5OH + O_2 \longrightarrow CH_3COOH + H_2O$$

A dilute alcohol solution inoculated with Acetobacter and kept in a warm, airy place will become vinegar over the course of a few months. Industrial vinegar-making methods accelerate this process by improving the supply of oxygen to the bacteria.

The first batches of vinegar produced by fermentation probably followed errors in the winemaking process. If must is fermented at too high a temperature, acetobacter will overwhelm the yeast naturally occurring on the grapes. As the demand for vinegar for culinary, medical, and sanitary purposes increased, vintners quickly learned to use other organic materials to produce vinegar in the hot summer months before the grapes were ripe and ready for processing into wine. This method was slow, however, and not always successful, as the vintners did not understand the process.

One of the first modern commercial processes was the 'fast method' or 'German method', first practiced in Germany in 1823. In this process, fermentation takes place in a tower packed with wood shavings or charcoal. The alcohol-containing feed is trickled into the top of the tower, and fresh air supplied from the bottom by either natural or forced convection. The improved air supply in this process cut the time to prepare vinegar from months to weeks.

Most vinegar today is made in submerged tank culture, first described in 1949 by Otto Hromatka and Heinrich Ebner. In this method, alcohol is fermented to vinegar in a continuously stirred tank, and oxygen is supplied by bubbling air through the solution. Using modern applications of this method, vinegar of 15 per cent acetic acid can be prepared in only 24 hours in batch process, even 20 per cent in 60 hour fed-batch process.

Anaerobic Fermentation

Species of anaerobic bacteria, including members of the genus *Clostridium*, can convert sugars to acetic acid directly, without using ethanol as an intermediate. The overall chemical reaction conducted by these bacteria may be represented as:

$$C_6H_{12}O_6 \longrightarrow 3CH_3COOH$$

More interestingly from the point of view of an industrial chemist, these acetogenic bacteria can produce acetic acid from one-carbon compounds, including methanol, carbon monoxide, or a mixture of carbon dioxide and hydrogen:

$$2CO_2 + 4H_2 \longrightarrow CH_3COOH + 2H_2O$$

This ability of Clostridium to utilise sugars directly, or to produce acetic acid from less costly inputs, means that these bacteria could potentially produce acetic acid more efficiently than ethanol-oxidisers like Acetobacter. However, Clostridium bacteria are less acid-tolerant than Acetobacter. Even the most acid-tolerant Clostridium strains can produce vinegar of only a few per cent acetic acid, compared to *Acetobacter* strains that can produce vinegar of up to 20 per cent acetic acid. At present, it remains more cost-effective to produce vinegar using Acetobacter than to produce it using Clostridium and then concentrating it. As a result, although acetogenic bacteria have been known since 1940, their industrial use remains confined to a few niche applications.

Vinegar

Vinegar is an alcoholic liquid that has been allowed to sour. It is primarily used to flavour and preserve foods and as an ingredient in salad dressings and marinades. Vinegar is also used as a cleaning agent. The word is from the French *vin* (wine) and *aigre* (sour).

The use of vinegar to flavour food is centuries old. It has also been used as a medicine, a corrosive agent, and as a preservative. In the Middle Ages, alchemists poured vinegar onto lead in order to create lead acetate. Called 'sugar of lead,' it was added to sour cider until it became clear that ingesting the sweetened cider proved deadly.

By the Renaissance era, vinegar-making was a lucrative business in France. Flavoured with pepper, clovers, roses, fennel, and raspberries, the country was producing close to 150 scented and flavoured vinegars. Production of vinegar was also burgeoning in Great Britain. It became so profitable that a 1673 Act of Parliament established a tax on so-called vinegar-beer. In the early days of the United States, the production of cider vinegar was a cornerstone of farm and domestic economy, bringing three times the price of traditional hard cider.

The transformation of wine or fruit juice to vinegar is a chemical process in which ethyl alcohol undergoes partial oxidation that results in the formation of acetaldehyde. In the third stage, the acetaldehyde is converted into acetic acid. The chemical reaction is as follows:

$$CH_3CH_2OH = 2HCH_3CHO = CH_3COOH.$$

Historically, several processes have been employed to make vinegar. In the slow, or natural, process, vats of cider are allowed to sit open at room temperature. During a period of several months, the fruit juices ferment into alcohol and then oxidise into acetic acid.

The French Orleans process is also called the continuous method. Fruit juice is periodically added to small batches of vinegar and stored in wooden barrels. As the fresh juice sours, it is skimmed off the top. Both the slow and continuous methods require several months to produce vinegar. In the modern commercial production of vinegar, the generator method and the submerged fermentation method are employed. These methods are based on the goal of infusing as much oxygen as possible into the alcohol product.

Raw materials

Vinegar is made from a variety of diluted alcohol products, the most common being wine, beer, and rice. Balsamic vinegar is made from the Trebbiano and Lambrusco grapes of Italy's Emilia-Romagna region. Some distilled vinegars are made from wood products such as beech.

Acetobacters are microscopic bacteria that live on oxygen bubbles. Whereas the fermentation of grapes or hops to make wine or beer occurs in the absence of oxygen, the process of making vinegars relies on its presence. In the natural processes, the acetobacters are allowed to grow over time. In the vinegar factory, this process is induced by feeding acetozym nutrients into the tanks of alcohol.

Mother of vinegar is the gooey film that appears on the surface of the alcohol product as it is converted to vinegar. It is a natural carbohydrate called cellulose. This film holds the highest concentration of acetobacters. It is skimmed off the top and added to subsequent batches of alcohol to speed the formation of vinegar. Acetozym nutrients are manmade mother of vinegar in a powdered form.

Herbs and fruits are often used to flavour vinegar. Commonly used herbs include tarragon, garlic, and basil. Popular fruits include raspberries, cherries, and lemons.

Design

The design step of making vinegar is essentially a recipe. Depending on the type of vinegar to be bottled at the production plant—wine vinegar, cider vinegar, or distilled vinegar—food scientists in the test kitchens and laboratories create recipes for the various vinegars. Specifications include the amount of mother of vinegar and/or acetozym nutrients added per gallon of alcohol product. For flavoured vinegars, ingredients such as herbs and fruits are macerated in vinegar for varying periods to determine the best taste results.

Manufacturing process

Orleans method

1. Wooden barrels are laid on their sides. Bungholes are drilled into the top side and plugged with stoppers. Holes are also drilled into the ends of the barrels.
2. The alcohol is poured into the barrel via long-necked funnels inserted into the bungholes. Mother of vinegar is added at this point. The barrel is filled to a level just below the holes on the ends. Netting or screens are placed over the holes to prevent insects from getting into the barrels.
3. The filled barrels are allowed to settle for several months. The room temperature is kept at approximately 85°F (29°C). Samples are taken periodically by inserting a spigot into the side holes and drawing liquid off. When the alcohol has converted to vinegar, it is drawn off through the spigot. About 15 per cent of the liquid is left in the barrel to blend with the next batch.

Submerged fermentation method

1. The submerged fermentation method is commonly used in the production of wine vinegars. Production plants are filled with large stainless steel tanks called acetators. The acetators are

fitted with centrifugal pumps in the bottom that pump air bubbles into the tank in much the same way that an aquarium pump does.

2. As the pump stirs the alcohol, acetozym nutrients are piped into the tank. The nutrients spur the growth of acetobacters on the oxygen bubbles. A heater in the tank keeps the temperature between 80 and 100°F (26°–38°C).
3. Within a matter of hours, the alcohol product has been converted into vinegar. The vinegar is piped from the acetators to a plate-and-frame filtering machine. The stainless steel plates press the alcohol through paper filters to remove any sediment, usually about 3 per cent of the total product. The sediment is flushed into a drain while the filtered vinegar moves to the dilution station.

Generator method

1. Distilled and industrial vinegars are often produced via the generator method. Tall oak vats are filled with vinegar-moistened beechwood shavings, charcoal, or grape pulp. The alcohol product is poured into the top of the vat and slowly drips down through the fillings.
2. Oxygen is allowed into the vats in two ways. One is through bungholes that have been punched into the sides of the vats. The second is through the perforated bottoms of the vats. An air compressor blows air through the holes.
3. When the alcohol product reaches the bottom of the vat, usually within in a span of several days to several weeks, it has converted to vinegar. It is poured off from the bottom of the vat into storage tanks. The vinegar produced in this method has a very high acetic acid content, often as high as 14 per cent, and must be diluted with water to bring its acetic acid content to a range of 5–6 per cent.
4. To produce distilled vinegar, the diluted liquid is poured into a boiler and brought to its boiling point. A vapour rises from the liquid and is collected in a condenser. It then cools and becomes liquid again. This liquid is then bottled as distilled vinegar.

Balsamic vinegar

1. The production of balsamic vinegar most closely resembles the production of fine wine. In order to bear the name balsamic, the vinegar must be made from the juices of the Trebbiano and Lambrusco grapes. The juice is blended and boiled over a fire. It is then poured into barrels of oak, chestnut, cherry, mulberry, and ash.
2. The juice is allowed to age, ferment, and condense for five years. At the beginning of each year, the ageing liquid is mixed with younger vinegars and placed in a series of smaller barrels. The finished product absorbs aroma from the oak and colour from the chestnut.

Quality control

The growing of acetobacters, the bacteria that creates vinegar, requires vigilance. In the orleans method, bungholes must be checked routinely to ensure that insects have not penetrated the netting. In the generator method, great care is taken to keep the temperature inside the tanks in the 80°–100°F range (26°–38°C). Workers routinely check the thermostats on the tanks. Because a loss of electricity could kill the acetobacters within seconds, many vinegar plants have backup systems to produce electrical power in the event of a blackout.

By-products/waste

Vinegar production results in very little by-products or waste. In fact, the alcohol product is often the by-product of other processes such as winemaking and baker's yeast.

Some sediment will result from the submerged fermentation method. This sediment is biodegradable and can be flushed down a drain for disposal.

LACTIC ACID

Lactic acid (IUPAC systematic name: 2-hydroxypropanoic acid), also known as milk acid, is a chemical compound that plays a role in several biochemical processes. It was first isolated in 1780 by a Swedish chemist, Carl Wilhelm Scheele, and is a carboxylic acid with a chemical formula of $C_3H_6O_3$. It has a hydroxyl group adjacent to the carboxyl group, making it an alpha hydroxy acid (AHA). In solution, it can lose a proton from the acidic group, producing the lactate ion $CH_3CH(OH)COO^-$. It is miscible with water or ethanol, and is hygroscopic.

Lactic acid is chiral and has two optical isomers. One is known as L-(+)-lactic acid or (*S*)-lactic acid and the other, its mirror image, is D-(–)-lactic acid or (*R*)-lactic acid. L-(+)-Lactic acid is the biologically important isomer.

In animals, L-lactate is constantly produced from pyruvate via the enzyme lactate dehydrogenase (LDH) in a process of fermentation during normal metabolism and exercise. It does not increase in concentration until the rate of lactate production exceeds the rate of lactate removal which is governed by a number of factors including: monocarboxylate transporters, concentration and isoform of LDH and oxidative capacity of tissues. The concentration of blood lactate is usually 1–2 mmol/l at rest, but can rise to over 20 mmol/l during intense exertion.

Industrially, lactic acid fermentation is performed by *Lactobacillus* bacteria, among others. These bacteria can operate in the mouth; the acid they produce is responsible for the tooth decay known as caries.

In medicine, lactate is one of the main components of Ringer's lactate or lactated Ringer's solution (compound sodium lactate or Hartmann's solution in the UK). This intravenous fluid consists of sodium and potassium cations, with lactate and chloride anions, in solution with distilled water in concentration so as to be isotonic compared to human blood. It is most commonly used for fluid resuscitation after blood loss due to trauma, surgery, or a burn injury.

Lactic acid bacteria are a group of related bacteria that produce lactic acid as a result of carbohydrate fermentation. These microbes are broadly used by us in the production of fermented food products, such as yogurt (*Streptococcus* spp. and *Lactobacillus* spp.), cheeses (*Lactococcus* spp.), sauerkraut (*Leuconostoc* spp.) and sausage.

These organisms are heterotrophic and generally have complex nutritional requirements because they lack many biosynthetic capabilities. Most species have multiple requirements for amino acids and vitamins. Because of this, lactic acid bacteria are generally abundant only in communities where these requirements can be provided. They are often associated with animal oral cavities and intestines (e.g. *Enterococcus faecalis*), plant leaves (*Lactobacillus, Leuconostoc*) as well as decaying plant or animal matter such as rotting vegetables, fecal matter, compost, etc.

Lactic acid bacteria are used in the food industry for several reasons. Their growth lowers both the carbohydrate content of the foods that they ferment, and the pH due to lactic acid production. It is this acidification process which is one of the most desirable side-effects of their growth. The pH may drop to as low as 4.0, low enough to inhibit the growth of most other micro-organisms including the most common human pathogens, thus allowing these foods prolonged shelf life. The acidity also changes the texture of the foods due to precipitation of some proteins, and the biochemical conversions involved in growth enhance the flavour. The fermentation (and growth of the bacteria) is self-limiting due to the sensitivity of lactic acid bacteria to such acidic pH.

Lactic Acid Fermentation

Lactic acid fermentation is a biological process by which sugars such as glucose, fructose, and sucrose, are converted into cellular energy and the metabolic product lactic acid. It is the anaerobic form of respiration that occurs in some bacteria and animal cells in the absence of oxygen. During homolactic acid fermentation, one molecule of glucose is ultimately converted to two molecules of lactic acid. In heterolactic acid fermentation, sometimes referred to as the phosphoketolase pathway, the products of fermentation are one molecule of carbon dioxide, one molecule of ethanol, and one molecule of lactic acid.

Process

Glycolysis produces 2 molecules of ATP, reduces 2 molecules of NAD+ to NADH, and creates 2 three-carbon molecules of pyruvate. Most of the chemical energy (about 95 per cent) of the glucose is still trapped in pyruvate. The complete breakdown of glucose to carbon dioxide requires the oxidation of pyruvate thorough the Krebs cycle and electron transport system (ETS). When the Krebs cycle and ETS are working at capacity (this action requires oxygen), further local ATP needs can be achieved by increasing glycolysis. The resulting pyruvate is converted to lactic acid through lactic acid fermentation.

The conversion of pyruvate to lactate regenerates NAD^+, which allows glycolysis to continue. Lactate diffuses out of the cell and into the blood. The lactate in the bloodstream is converted back into pyruvate in the liver, for use when oxygen is once again present.

Certain cells, such as cardiac muscle cells, are highly permeable to lactate. Lactate is converted into pyruvate and metabolised normally (i.e. via the Krebs cycle). Since these cells are highly oxygenated, it is unlikely that lactate would accumulate (as is the case in oxygen-starved muscle cells). This also allows circulating glucose to be available to muscle cells.

Any excess lactate is taken up by the liver, converted into pyruvate and then into glucose. This, along with the production of lactate from glucose in muscle cells constitutes the Cori cycle.

Phosphofructokinase (PFK) is inhibited by a low pH and this prevents the formation of excess lactate and/or lactic acidosis (sudden drop in blood pH). PFK catalyses an irreversible step in glycolysis.

KOJIC ACID

Kojic acid ($C_6H_6O_4$; 5-hydroxy-2-(hydroxymethyl)-4-pyrone) is a chelation agent produced by several species of fungi, especially *Aspergillus oryzae*, which has the Japanese common name *koji*. Kojic acid is a by-product in the fermentation process of malting rice, for use in the manufacturing of sake, the Japanese rice wine. It is a mild inhibitor of the formation of pigment in plant and animal tissues, and is used in food and cosmetics to preserve or change colours of substances. It is used on cut fruits to prevent oxidative browning, in seafood to preserve pink and red colours, and in cosmetics to lighten skin. Kojic acid also has antibacterial and antifungal properties.

Conversion of Glucose to Kojic Acid

1. The conversion of glucose to kojic acid by two species of *Aspergilli* has been studied using both [$1–^{14}C$] and $3{:}4–^{14}C_3$] glucose. Between 70 and 90 per cent of the total radioactivity of the isolated kojic acid was found in the carbon atoms corresponding to those labelled in the glucose, indicating that the major pathway of kojic acid formation is a direct conversion of glucose without splitting of the carbon chain. Symmetrical C_3 or C_4 compounds are not intermediates in the biosynthesis.

2. A minor pathway of kojic acid biosynthesis, involving condensation of 'small' molecules, probably triosephosphates, leads to incorporation of ^{14}C into position 6 of kojic acid, when [1–^{14}C]-glucose is used as substrate.
3. The biosynthesis of kojic acid from [2–^{14}C]-dihydroxyacetone, alone or in the presence of unlabelled glucose, has also been studied. In the lstter case, ^{14}C is incorporated predominantly into kojic acid.

Kojic Acid Fermentation by *Aspergillus Flavus*

The influence of pH on kojic acid fermentation by *Aspergillus flavus* Link 44–1 in submerged fermentation and resuspended cell system was investigated separately. Although the highest growth in submerged fermentation was obtained at initial culture pH between 6 to 7, the optimum kojic acid production was achieved at initial culture pH 3. At initial culture pH 2, growth was greatly inhibited and kojic acid was not produced. In resuspended cell system using buffered glucose solution, in which the production of cell material was carried out at initial culture pH 3, the optimum kojic acid production was also obtained at pH 3. Kojic acid production in fermentation using 50 litres stirred tank fermenter with pH control strategy (fermentation was started with pH 3 and without pH control during growth phase and culture pH was controlled at 3 during the production phase) was improved by about 20 per cent as compared to fermentation without pH control. The maximum kojic acid concentration in fermentation with pH control strategy was 62 g/l and this gave yield and productivity of 0.516 g/g and 0.22 g/l.h, respectively. Kojic acid can be also be fermented by *Aspergillus flavus* link 544–1 using sucrose as a carbon source under different conditions.

ITACONIC ACID

Itaconic acid, or methylenesuccinic acid, is an organic compound that is one of the three acids obtained by the distillation of citric acid. Itaconic acid is a white crystalline powder that is soluble in water, ethanol and acetone. It has formula $C_5H_6O_4$, its structural formula is given below:

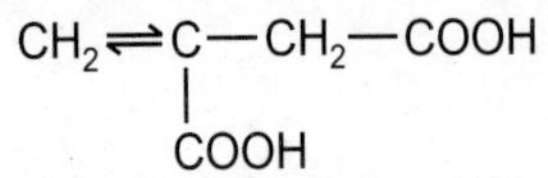

In nature, itaconic acid is being produced by *Aspergillus terreus*. Scientists are developing plants to produce itaconic acid. This may allow replacement of petrochemical products by plant based, thus more sustainable, raw material. Itaconic acid is used as a resin in detergents. The transformation of citric acid by *Aspergillus terreus* can be used for commercial production of itaconic acid. Fermentation process involves a well-aerated molasses-mineral salts medium at a pH, below 2.2. At higher pH this microbe degrades itaconic acid. Like citric acid, low levels of trace metals must be used to achieve acceptable product yields.

Fermentation of Itaconic Acid by *A. Terreus* NRRL 1960

The effect of interrupting aeration on itaconic acid fermentation by A. terreus NRRL 1960 has been studied. Under the conditions used, stopping aeration for 5 minute led to a complete cessation of itaconic acid production, which was only slowly restored after 24 hours when aeration was resumed. After a 5 minutes break in aeration and in the presence of 0.1 mM cycloheximide no itaconic acid was formed even after 3 days. It seems that, upon oxygen shortage, a rapid destruction of the itaconic-acid-producing mechanism takes place, which is restored only aerobically in a slow process involving protein synthesis. Itaconic acid fermentation is also effectively stopped by metabolic inhibitors of ATP formation, pointing to the need for biochemical energy in maintaining the fermentation. ATP is possibly needed to maintain a proper physiological (i.e. near neutral) pH inside the cells, counteracting the acid produced in the fermentation process and the low external pH (below 2.0). Inhibitors of plasma membrane ATPase have no effect on itaconic acid fermentation. This indicates that the plasma membrane might be impermeable to H^+ and that ATP might rather be involved in the transport of itaconic acid out of the cell. It is suggested that insufficient aeration may lead to insufficient production of ATP which, in turn, leads to damage of the metabolic machinery by acid produced in the fermentation process.

Itaconic Acid Fermentation by a Yeast Belonging to the Genus *Candida*

Many yeasts were isolated from natural sources in the tropics and subtropics by enrichment culture technique, using medium which contained a surfactant. The medium was acidified with citric acid. A strain S-10 belonging to the genus *Candida* was found to produce itaconic acid. Under suitable conditions in shake culture, a mutant derived from this strain produced the acid at about 35 per cent yield on the basis of glucose supplied.

Chapter 9

Amino Acids and their Commerical Uses

INTRODUCTION

Amino acids are molecules containing both amine and carboxyl functional groups. These molecules are particularly important in biochemistry, where this term refers to α-amino acids with the general formula $H_2NCHRCOOH$, where R is an organic substituent. In the α-amino acids, the amino and carboxylate groups are attached to the same carbon atom, which is called the α-carbon. The various α-amino acids differ in which side chain (R group) is attached to their α-carbon. They can vary in size from just a hydrogen atom in glycine through a methyl group in alanine to a large heterocyclic group in tryptophan.

Amino acids are critical to life, and have a variety of roles in metabolism. One particularly important function is as the building blocks of proteins, which are linear chains of amino acids. Every protein is chemically defined by this primary structure, its unique sequence of amino acid residues, which in turn define the three-dimensional structure of the protein. Just as the letters of the alphabet can be combined to form an almost endless variety of words, amino acids can be linked together in varying sequences to form a vast variety of proteins. Amino acids are also important in many other biological molecules, such as forming parts of coenzymes, as in S-adenosylmethionine, or as precursors for the biosynthesis of molecules such as heme. Due to this central role in biochemistry, amino acids are very important in nutrition. Amino acids are commonly used in food technology and industry. For example, monosodium glutamate is a common flavor enhancer that gives foods the taste called *umami*. Beyond the amino acids that are found in all forms of life, amino acids are also used in industry. Applications include the production of biodegradable plastics, drugs and chiral catalysts. Amino acids are produced using a range of technologies including direct fermentation, biotransformation of precursors using cells or enzymes, extraction of protein hydrolysates and chemical synthesis.

Amino acids are used for a variety of purposes. The food industry requires L-glutamate as a flavour enhancer and glycine as a sweetener in juices, for instance (Table 9.1). The chemical industry requires amino acids as building blocks for a diversity of compounds. The pharmaceutical industry requires the amino acids–or in special dietary food. And last but not least, a large market for amino acids is their use as animal feed additive. The reason is that typical feedstuffs, such as soyabean meal for pigs, are poor in some essential amino acids, like methionine, for instance. Methionine is added for this reason and considerably increases the effectiveness of the feed. The addition of as little as 10 kg methionine per tonne increases the protein quality of the feed just as effectively as adding 160 kg soyabean meal or 56 kg fish meal. The first limiting amino acid in feed based on crops and oil seed is usually L-methionine, followed by L-lysine and L-threonine. Another aspect of feed supplementation is that with a balanced amino acid content the manure contains less nitrogen thus reducing environmental pollution. While fermentation or biotransformation processes have been developed for production of all amino acids except glucine, L-cysteine and L-cystine, not all of these processes are commercially viable. L-Asparagine, L-leucine, L-tyrosine, L-cysteine and L-cystine are produced by purification of protein hydrolysates. Chemical synthesis is more economical for production of optically-inactive racemic mixtures of D and L-isomers and D,L-alanine, D,L-methionine, D,L-tryptophan and glycine are produced in this way. Processes involving amino acylase enzymes may be used to resolve these racemic mixtures.

Table 9.1. Amino acids, their production methods and applications.

Amino acid	*Preferred production method*	*Main use*
L-Glutamic acid	Fermentation	Flavour enhancer
L-Lysine	Fermentation	Food additive
D,L-Methionine	Chemical synthesis	Food additive
L-Aspartate	Enzymatic catalysis	Aspartame
L-Phenylalanine	Fermentation	Aspartame
L-Threonine	Fermentation	Food additive
Glycine	Chemical synthesis	Food additive, sweetener
L-Cysteine	Reduction of cystine	Food additive, pharmaceutical
L-Arginine	Fermentation, extraction	Pharmaceutical
L-Leucine	Fermentation, extraction	Pharmaceutical
L-Valine	Fermentation, extraction	Pesticides, pharmaceutical
L-Tryptophan	Whole cell process	Pharmaceutical
L-Isoleucine	Fermentation, extraction	Pharmaceutical

Microbial strains from the genera *Corynebacterium* and *Brevibacterium* have assumed major importance in the production of amino acids by fermentation. Natural isolates of these strains can excrete large quantities of glutamic acid. Because of cell metabolic regulatory mechanisms, particularly end-product repression and inhibition, substantial levels of amino acids are rarely excreted by wild-type isolates. Production of commercial quantities of the amino acids has been dependent on the successful development of deregulated mutants. The two most important methods involve use of auxotrophic and regulatory mutants or a combination of the two. Auxotrophic mutants, which lack the enzyme needed to form the regulatory effector metabolite (often the end-product), may accumulate and excrete the metabolic intermediate which is the substrate for the eliminated enzyme. A lysine auxotroph, for example, lacks an enzyme in the pathway necessary for lysine synthesis and requires lysine or a metabolic precursor which can be converted to lysine, for growth.

End-product inhibition by the amino acid product of an unbranched biosynthetic pathway may be avoided by the development of regulatory mutants, having an altered feedback-insensitive key enzyme, thus allowing accumulation of the particular amino acid. Analogues of the end-product, which are also capable of inhibition of the sensitive key enzyme, may be used in screening methods for selection of analogue-resistant or regulatory mutants. Revertants may be selected from auxotrophic mutants (apparently lacking the key regulatory enzyme) which produce a modified deregulated enzyme. Table 9.2 illustrates the genetic features of mutants of *Brevibacterium* spp. and *Corynebacterium* spp. and some published yields of amino acids over-produced from glucose.

Table 9.2. Genetic characteristics of some amino acid-producing strains of *Brevibacterium flavum* and *Corynebacterium glutamicum*.

Microbial strain	*Amino acid*	*Genetic characteristics*	*Yield* ($g\ l^{-1}$)
Brevibacterium flavum	L-Arginine	$Gua^- TA^r$	35
	L-Histidine	$TA^r SM^r Eth^r ABT^r$	10
	L-Isoleucine	$AHV^r OMT^r$	15
	L-Lysine	AEC^r	57
	L-Proline	$Ile^-\ SG^r DHP^r$	29
	L-Threonine	$Met^-\ AHV^r$	18
Corynebacterium glutamicum	L-Glutamate	Wild type	>100
	L-Glutamine	Wild type	40
	L-Lysine	$Hom^-\ Leu^-\ AEC^r$	39
	L-Phenylalanine	$Tyr^-\ PFP^r PAP^r$	9
	L-Tryptophan	$Phe^-\ Tyr^-\ 5MT^r TrpHx^r 6FT^r$ $4MT^r PFP^r PAP^r TyrHx^r PheHx^r$	12
	L-Tyrosine	$Phe^-\ PFP^r PAP^r PAT^r TyrHx^r$	18

Resistance abbreviations: r, resistant; ABT, 2-aminobenzthiazole; AEC, S-(β-aminoethyl)-L-cysteine; AHV, α-amino-β-hydroxyvaleric acid; DHP, 3,4-dehydroproline; Eth, ethionine; 6FT, 6-flurortryptophan; 4MT, 4-methyltryptophan; 5MT, 5-methyltryptophan; OMT, *o*-methylthreonine; PAP, *p*-aminophenylalanine; PheHx, phenylalanine hydroxamate; SG, sulphaguanidine; TA, 2-thiazolalanine; TyrHx, tyrosine hydroxamate; TrpHx; tryptophan hydroxamate.
Auxotroph abbreviations: –, auxotroph; Ile, isoleucine; Met, methionine; Hom, homoserine; Leu, leucine; Phe, phenylalanine; Tyr, tyrosine; Gua, guanine.

PRODUCTION METHODS OF AMINO ACIDS

Some amino acids are chemically synthesised, such as glycine, which has no stereochemical centre or D, L-methionine. This latter sulphur-containing amino acid can be added to feed as a racemic mixture, since animals contain a D-amino acid oxidase which, together with a transaminase activity, converts D-methionine to the nutritively effective L-form. The classical procedure of amino acid isolation from acid hydrolysates of proteins is still in use for selected amino acids with a low market volume, e.g. L-cysteine (Table 9.1). Other methods in use are those of precursor conversion with bacteria or enzymatic synthesis. However, for L-amino acids required in large volumes, fermentation production with bacteria is the method of choice.

Classical Strain Development

However, bacteria do not normally excrete amino acids in significant amounts because regulatory mechanisms control the amino acid synthesis in an economical way. Therefore, mutants have to be generated which over-synthesise the respective amino acid. A great number of amino-acid-producing bacteria have been derived by mutagenesis and screening programmes. This has involved the consecutive application of:

1. Undirected mutagenesis.
2. Selection for a specific phenotype.
3. Selection of the mutant with the best amino acid accumulation.

Taking the best resulting strain, the entire procedure was repeated over several additional rounds to increase the productivity each time and eventually, resulted in an industrial producer. Due to this optimisation over several decades, together with the accompanying process adaptation, excellent high-performance strains are now available. They certainly carry a variety of unknown mutations also decisive for their production properties, as will become evident from the examples described below.

Application of Recombinant Techniques

In conjunction with this classical technique for strain development, recombinant DNA techniques are also applied. They serve

1. To rapidly develop new producers by increasing limiting enzyme activities.
2. To analyse mechanisms of flux control.
3. To combine this knowledge with classically obtained strains for their further development.

Intracellular Flux Analysis

An exciting new approach in strain development combining both the genetic and classical procedure is the reliable quantification of the carbon fluxes in the living cell. A great deal of progress has been made here recently in developing to a high level of sophistication the old isotope labelling technique. In particular, with ^{13}C-NMR spectroscopy the intracellular fluxes were quantified to extreme high resolution. For instance, in *C. glutamicum* it has even been possible to quantify the exchange flux rates as are present in the pentose phosphate pathway. Such flux identifications are of major assistance in selecting the reactions in the central metabolism to be modified by genetic engineering.

Functional Genomics

Another tool whose potential is only now being exploited is the genome analysis of producer strains. The availability of the entire sequence of the chromosomes from *C. glutamicum* and *E. coli* opens up exciting possibilities to compare mutants and to uncover new mutations essential for high overproduction of metabolites. For instance, RNA analysis using chip technology will make it possible to detect whether a specific gene is altered in its expression for producers of different efficiency. New mutations and genes might thus be discovered which are not directly concerned with carbon fluxes, but rather with total cell control or are involved in energy metabolism. Chip technology will also make it possible to use genome analysis as a tool to qualify individual fermentations, thus resulting in still further improvements and consolidations of the production processes.

L-GLUTAMATE (L-GLUTAMIC ACID)

L-Glutamate was the first amino acid to be produced. The very successful production still exclusively uses the original bacterium *C. glutamicum*. As metabolic pathways *C. glutamicum* uses glycolysis, the pentose phosphate pathway and the citric acid cycle to generate precursor metabolites and reduced pyridine nucleotides. However, this bacterium displays a special feature in the anaplerotic reactions of the citric acid cycle (Fig. 9.1). Since L-glutamate is directly derived from α-ketoglutarate, a high capability for replenishing the citric acid cycle is, of course, a prerequisite for high glutamate production. It was originally assumed that only the phospho*enol*pyruvate carboxylase is present as a carboxylating enzyme within the anaplerotic reactions. However, molecular research in close conjunction with ^{13}C-labelling studies and flux analysis showed that an additional carboxylating reaction must be present. The pursuit of this enzyme activity resulted in the detection of pyruvate carboxylase activity, PyrC and the cloning of its gene. This carboxylase was not detected by the original enzyme measurements since it is very unstable in crude extracts. Its detection requires an *in situ* enzyme assay using carefully permeabilised cells. Therefore, *C. glutamicum* has the pyruvate dehydrogenase (PyrDH) shuffling acetyl-CoA into the citric acid cycle but two enzymes supplying oxaloacetate: pyruvate carboxylase (PyrC) together with a phospho*enol*pyruvate carboxylase (PEPC) (Fig. 9.1). The successful cloning of both genes together with mutant studies showed that both carboxylases can basically replace each other to ensure conversion of glucose-derived C3-units to oxaloacetate. This is different from *E. coli*, which has exclusively the phospho*enol*pyruvate carboxylase serving this purpose, or *Bacillus subtilis*, where only the pyruvate carboxylase is present. Since *C. glutamicum* possesses both enzymes, it has an enormous flexibility for replenishing citric acid cycle intermediates upon their withdrawal.

The reductive amination of α-ketoglutarate to yield L-glutamate is catalysed by glutamate dehydrogenase. The enzyme is a multimer, each subunit having a molecular weight of 49100. It has a high specific activity of 1.8 mmol min^{-1}·mg protein and L-glutamate is present in the cell in a rather high concentration of about 150 mM. In the case of other amino acids, in contrast, the intracellular concentrations are usually below 10 mM. The high concentration serves to ensure the supply of L-glutamate directly required for cell synthesis and also for the supply of amino groups via transaminase reactions for a variety of cellular reactions. As much as 70 per cent of the amino groups in cell material stems from L-glutamate.

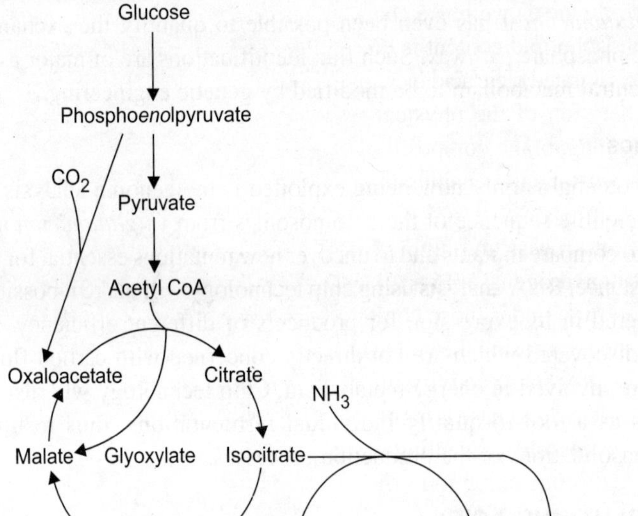

Fig. 9.1. Metabolic pathway for production of glutamic acid from glucose.

Production Strains

For the biotechnological production of L-glutamate the intracellularly synthesised amino acid must be released from the cell. This is, of course, usually not the case since the charged L-glutamate is retained by the cytoplasmic membrane, otherwise the cell would not be viable. However, as shown by the special circumstances in discovering *C. glutamicum*, L-glutamate is already excreted when biotin is limiting. This striking fact is based on two essential characteristics:

1. A carrier is present mediating the active excretion of L-glutamate.
2. The lipid environment of this carrier triggers its activity.

A specific carrier is required since otherwise, in addition to the charged L-glutamate, other metabolites and ions would also leak from the cell. Moreover, only an active export enables the energy-dependent uphill transport of L-glutamate from inside the cell (0.15 M) towards the very high concentrations obtained in fermentation broths (more than 1 M). However, for practical purpose, the triggering of active export by the appropriate molecular environment of the cytoplasmic membrane is important. The switches for tuning this environment and thus eliciting glutamate export are surprisingly diverse: (i) growth under biotin limitation, (ii) addition of local anaesthetics, (iii) addition of penicillin, (iv) addition of surfactants, (v) use of oleic acid auxotrophs, and (vi) use of glycerol auxotrophs. All of these means trigger L-glutamate excretion. Although, overall, there are as yet no completely conclusive ideas on the molecular changes thus caused, nevertheless in the classical biotin effect part of the causal link to glutamate excretion is well understood. Biotin is a cofactor of the acetyl-CoA carboxylase. With limited supply, the activity of this enzyme is thus decreased and consequently the fatty acid synthesis is diminished. This leads to a decreased availability of phospholipids and a greatly decreased lipid to protein ratio in the membrane as well as a change in the degree of saturation of the fatty acids. Under biotin limitation the phospholipid content is drastically decreased from 32 to 17 nmol mg^{-1} dry weight and the content of the unsaturated oleic acid increased relative the saturated palmitic acid by 45 per cent. This represents a severe alteration of the physical state of the membrane which thus dramatically alters L-glutamate efflux. The membrane composition is also affected in oleic acid and glycerol auxotrophic mutants. The use of such mutants enables the production of monosodium glutamate from substrate which may be rich in biotin as well.

Apart from the export process and high glutamate dehydrogenase activity, another key reaction is that of α-ketoglutarate dehydrogenase (Fig. 9.1). This enzyme has a weak activity in *C. glutamicum* and it is also unstable. Therefore, under those conditions that result in glutamate efflux, the activity of this enzyme in also diminished. Exposing the cell to either penicillin, surfactants or biotin-limitation reduces the α-ketoglutarate dehydrogenase activity up to a residual activity of only 10 per cent, whereas the activity of the glutamate dehydrogenase is hardly affected. The competing α-ketoglutarate dehydrogenase activity is therefore lowered, thus preventing an excess conversion of α-ketoglutarate to succinyl-CoA and therefore favouring its conversion to L-glutamate.

Production Process

The most relevant factors influencing L-glutamate formation are the ammonium concentration, the dissolved O_2 concentration and the pH. Although, in total, a large amount of ammonium is necessary for sugar conversion to L-glutamate, a high concentration is inhibitory to growth as well to the production of L-glutamate. Therefore, ammonium is added in a low concentration at the beginning of the fermentation and is then added continuously during the course of the fermentation.

The oxygen concentration is controlled, since under conditions of insufficient oxygen, the production of L-glutamate is poor and lactic acid as well as succinic acid accumulates, whereas with an excess oxygen supply the amount of α-ketoglutarate as a by-product accumulates. A flow diagram of the process is shown in Fig. 9.2.

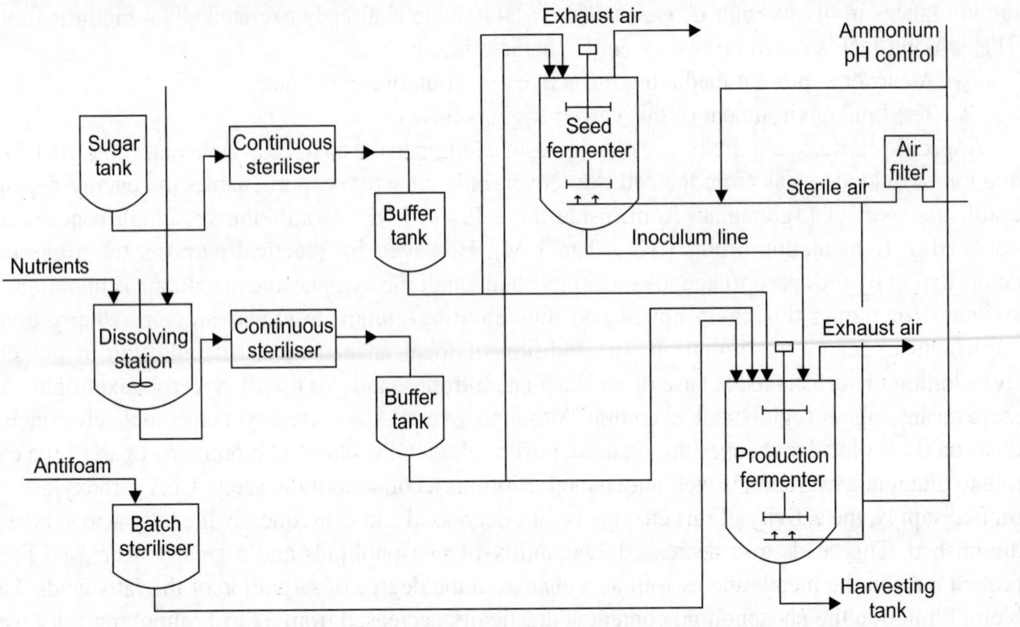

Fig. 9.2. A scheme of the material flow in an L-glutamate production plant.

For the actual fermentation the production strains are grown in fermenters as large as 500 m^3. After precultivation, the onset of L-glutamate excretion is controlled by the addition of surfactants like polyoxyethylene sorbitan monopalmitate (Tween 40). Yields of 60–70 per cent L-glutamate, based on the glucose used, have been reported. At the end of the fermentation the broth contains L-glutamate in the form of its ammonium salt. In a typical downstream process, the cells are separated and the broth is passed through a basic anion exchange resin. L-Glutamate anions will be bound to the resin and ammonia will be released. This ammonia can be recovered via distillation and reused in the fermentation. Elution is performed with NaOH to directly form MSG in the solution and to regenerate the basic anion exchanger. From the eluates, MSG may be crystallised directly followed by further conditioning steps like decolourisation and sieving to yield a food-grade quality.

L-LYSINE

The second amino acid made exclusively with *C. glutamicum*, or its subspecies *lactofermentum* and *flavum*, is L-lysine. The carbons of L-lysine are derived in the central metabolism from pyruvate and oxaloacetate (Fig. 9.3). In contrast to the special situation with L-glutamate, where practically only a single reaction represents the synthesis pathway, L-lysine is synthesised via a long pathway. Moreover the first two steps of L-lysine synthesis are shared with that of the other members of the aspartate family of amino acids: L-methionine, L-threonine and L-isoleucine.

Kinase Initiating Lysine Synthesis Feedback-Inhibited by Lysine Plus Threonine

The first reaction initiating L-lysine synthesis is catalysed by aspartate kinase. As is typical of an enzyme at the start of a lengthy synthesis pathway, aspartate kinase is controlled in its catalytic activity. The enzyme is inactive when L-lysine plus L-threonine together are present in excess, thus providing a feedback signal concerning the availability of these two major metabolites of the aspartate family of amino acids. The kinase has an interesting structure. It consists of two α-subunits of 421 amino acid residues each, and two β-subunits of 171 amino acid residues.

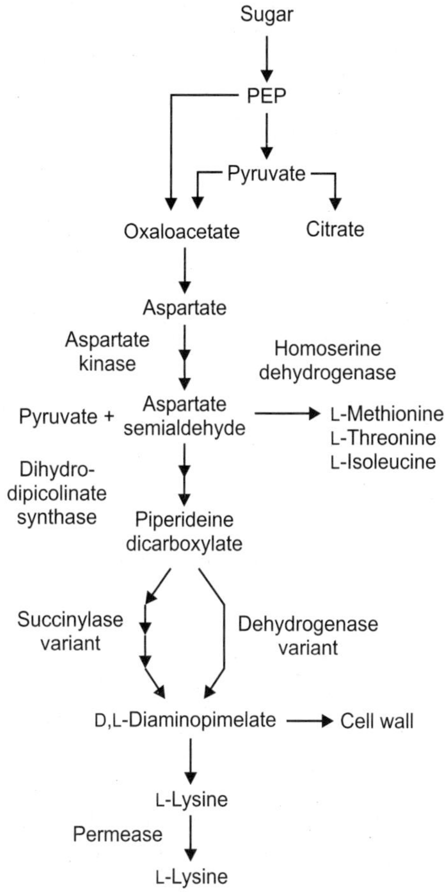

Fig. 9.3. L-Lysine biosynthesis in *C.glutamicum* witht he reactions supplying oxaloacetate and pyruvate as precursors. PEP, phospho*enol*pyruvate.

An exciting discovery was that the amino acid sequence of the β-subunit is identical to that in the carboxyterminal part of the α-subunit. The molecular basis is that the gene for the smaller β-subunit, *lysC*β, is an in-frame constituent part of the larger α-subunit (Fig. 9.4). Thus two promoters are present at this locus: one driving *lysC*α expression together with that of the downstream gene, *asd*, and one driving *lysC*β and asd expression. The regulatory features of the kinase reside in the β-subunit. Thus specifically altering the β-subunit structure, or those of both subunits together in their carboxy terminal part. results in a kinase which is no longer feedback regulated (by L-lysine plus L-threonine. With such an insensitive kinase, *C. glutamicum* already excretes some L-lysine, showing the rather simple type of flux control in this organism.

Synthase Limits Flux

A further important step afflux control within lysine biosynthesis is at the level of aspartate semialdehyde distribution. The dihydrodipicolinate synthase activity competes with the homoserine dehydrogenase for the aspartate semialdehyde (Fig. 9.3). In *C. glutamicum*, the synthase is not regulated in its catalytic activity as is the corresponding enzyme in *E. coli*, for example. Instead, in *C. glutamicum* it is the amount of the protein which directly controls the flux. This is thus different from the kinase where the catalytic activity is regulated by L-lysine and thereby controls the flux at a constant amount of protein. Graded over-expression of the synthase gene, *dapA*, has shown that with an increasing amount of synthase a graded flux increase towards L-lysine is the result. Surprisingly, *dapA* overexpression also has a second consequence: the flux of aspartate semialdehyde into the branch leading to homoserine is already diminished withjust two *dapA* copies. Due to the shortage ofthe homoserine-derived amino acids, this results in a weak: growth limitation which is advantageous for L-lysine formation, since now more intermediates of the central metabolism are used for lysine synthesis instead for cell proliferation.

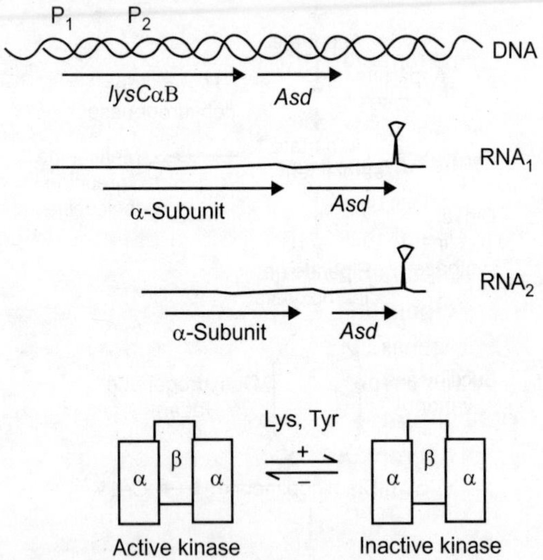

Fig. 9.4. The *lysCasd* operon of *C. glutamicum* and allosteric control of the kinase. The second promoter within *lysC* results in formation of the β subunit constituting the regulatory subunit of the kinase protein of $\alpha_2\beta_2$ structure.

Lysine Synthesis is Split which Ensures Proper Cell Wall Formation

A remarkable feature of *C. glutamicum* is its split pathway of L-lysine synthesis. At the level of piperideine-2,6-dicarboxylate, flux is possible either via the 4-step succinylase variant or the 1-step dehydrogenase variant (Fig. 9.3). In contrast, *E. coli*, for example, has only the succinylase variant and *Bacillus macerans* only the dehydrogenase variant. The flux distribution via both pathways has been quantified in a study using NMR spectroscopy and [1-^{13}C)glucose as the substrate. Surprisingly, the flux distribution is variable (Fig. 9.5). Whereas at the start of the cultivation about three-quarters of the L-lysine is made via the dehydrogenase variant, at the end the newly synthesised L-lysine is almost exclusively made via the succinylase route. There is a mechanistic reason for this. As kinetic characterisations have shown, the dehydrogenase has a weak affinity towards its substrate, ammonium, with a K_m of 28 mM. Thus at low ammonium concentrations, as are present at the end of the fermentation, the dehYdrogenase cannot contribute to L-lysine formation. Instead, flux via the succinylase variant is favoured where after succinylation of piperideine-2,6-dicarboxylate, a transaminase incorporates the second amino group into the final L-lysine molecule.

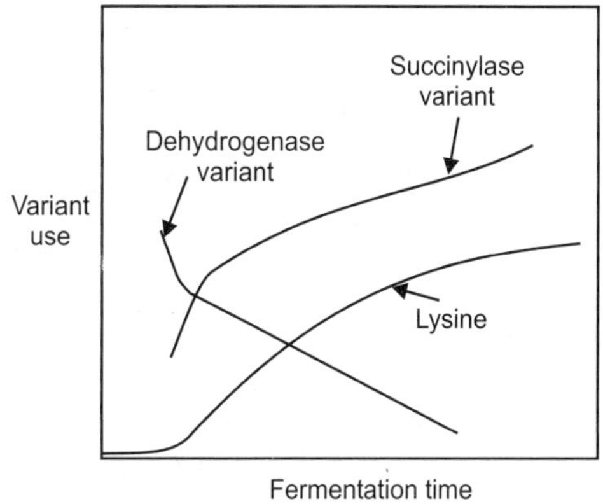

Fig. 9.5. At the beginning of the L-lysine fermentation use prevails of the dehydrogenase variant over that of the sucinylase variant, whereas at the end the succinylase variant is used almost exclusively. Variant use is in per cent.

The key to understanding this luxurious pathway construction is provided by the amino acid D,L-diaminopimelate. This amino acid is required for the synthesis of the activated muramyl peptide L-Ala-γ, D-Glu-D, L-Dap, which is one of the linking units in the peptidoglycan of the cell wall. Upon inactivation of the succinylase variant, a radical change to the cell morphology becomes apparent with low nitrogen supply. The cells are elongated, and furthermore less resistant to mechanical stress. If either the succinylase or the dehydrogenase variant is inactivated, L-lysine accumulation is reduced to 40 per cent. Thus both variants together ensure the proper supply of the crucial linking unit D,L-diaminopimelate, as well as a high throughput for L-lysine formation. The split pathway in *C. glutamicum* is an example of an important principle in microbial physiology: pathway variants are generally not redundant but evolved to provide key metabolites under different environmental conditions.

Export of L-Lysine

Amino acid transport has long been investigated in bacteria but, principally, this is only their import. In contrast, the molecular basis for amino acid export was completely unknown until 1996 since a specific export process appeared nonsensical. The breakthrough was achieved by the cloning of the lysine export carrier from *C. glutamicum*, which at one blow enabled amasing discoveries concerning the nature and relevance of a new type of exporter. The L-lysine carrier, LysE, is a comparatively small membrane protein of 25.4 Da. It has the transmembrane spanning helices typical of carriers, but only five of them (Fig. 9.6). A sixth hydrophobic segment is located between helix one and three and may dip into the membrane or be surface localised. Several distinct steps are involved in the translocation mechanism, which probably requires the dimerisation of LysE. These are: (i) the loading of the negatively charged carrier with its substrate L-lysine together with two hydroxyl ions, (ii) substrate translocation via the membrane, (iii) the release of L-lysine and the accompanying ions at the outside of the membrane, and finally, (iv) the reorientation of the carrier. The driving force for the entire translocation process is the membrane potential, $\Delta\Psi$, required for the reorientation of the carrier.

Access to the lysine-exporter gene, lysE, has also made it possible to solve the puzzle as to why *C. glutamicum* has such an exporter at all. In a *lysE* deletion mutant supplied with glucose and 1 mm of the dipeptide, lysyl-alanine, an extraordinarily high intracellular L-lysine concentration of more than 1 m accumulates, abolishing growth of the mutant. Thus, the exporter serves as a valve to excrete any excess intracellular L-lysine that may arise in the natural environment in the presence of peptides. As in the case of other bacteria, too, *C. glutamicum* has active peptide-uptake systems as well as hydrolysing enzymes giving access to the amino acids as valuable building blocks.

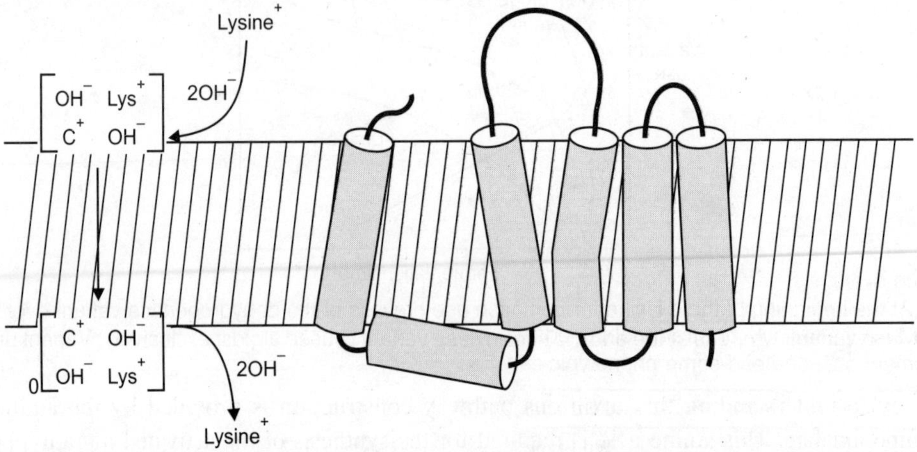

Fig. 9.6. Topology of the L-lysine exporter showing its five membrane spanning helices and the additional hydrophobic segment. The formally distinct steps of the translocation process driven by the membrane potential are included.

However, *C. glutamicum* has no L-lysine-degrading activities and therefore must prevent any piling up of L-lysine. This also happens in the lysine producer strains where the biosynthesis pathway is mutated. As genome projects have now shown, homologous structures of the L-lysine carrier LysE are present in various Gram-negative and Gram-positive bacteria. Therefore, this type of intracellular amino acid control by an exporter is expected to be present in other bacteria, too. Since the LysE structure is not shared with other translocators. LysE also represents a new superfamily of trans locators, which is probably related to its new function.

Production Strains

L-Lysine producer strains have been derived over the decades by mutagenesis to give strains excreting more than 170 g L-lysine per litre. It is clear that these strains carry a long list of phenotypic characters to achieve this massive flux directioning (Table 9.3). Typically. the strains are resistant or sensitive to some analogue of lysine. A typical feature of some L-lysine producers is their resistance to the lysine analogue S-(2-aminoethyl)-L-cysteine (Fig. 9.7). In these mutants, the aspartate kinase (Fig. 9.3) is mutated so that it is no longer inhibited by L-lysine. Dozens of other chemicals structurally related to L-lysine, such as γ-methyl-L-lysine or α-chlorocaprolactam. have been used in screenings to obtain improved producers. Fluoropyruvate has also been used to identify strains that are sensitive to it as these have decreased pyruvate dehydrogenase activities resulting in a diminished oxidation of pyruvate via the citric acid cycle (Fig. 9.3). This is also the case in strains with decreased citrate synthase activity. In another lineage of strains, over-producers were derived from mutants with diminished homoserine dehydrogenase activity to lower the availability of L-threonine inside the cell (Fig. 9.3). In this way, inhibition of the kinase activitywas abolished and, at the same time, a favourable growth limitation was introduced.

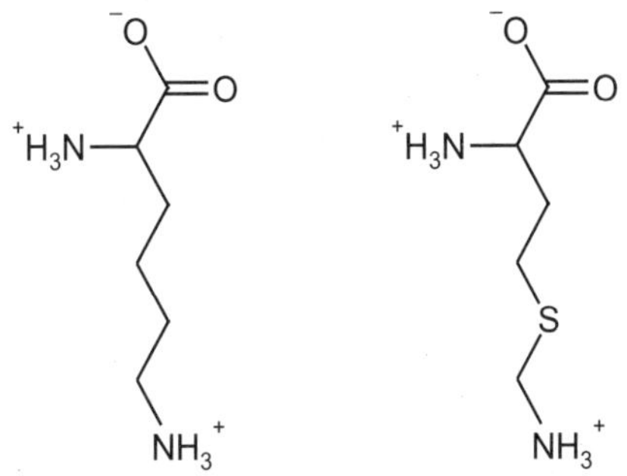

Lysine Aminoethylcysteine

Fig. 9.7. Aminoethyl cysteine is a sulphur-containing analogue of L-lysine for generating mutants deregulated in L-lysine synthesis.

Table 9.3. A genealogy of strains obtained by classical mutagenesis and screening, showing the yield improvement obtained and some phenotypic characters known.

Strain	*Character*	*Yield of L-lysine (%)*
AJ 1511	Wild type	0
AJ 3445	AEC^r	16
AJ 3424	AEC^r Ala^-	33
AJ 3796	AEcr Ala^- CCL^r	39
AJ3990	AEcr Ala^- CCL^r ML^r	43
AJ 1204	AEC^r Ala^- CCl^r ML^r FP^s	50

Notes: AEC^r: Resistant to S-(β-aminoethyl)-L-cysteine; Ala^-: L-alanine-requiring; CCL^r: resistant to α-chlorocaprolactam; ML^r: resistant to γ-methyl-L-lysine; FP^s: sensitive to β-fluoropyruvate.

Production Process

The most common carbon sources for L-lysine fermentation and also other amino acids are molasses (cane or sugar beet molasses), high test molasses (inverted cane molasses) or sucrose and starch hydrolysates. In contrast to *E. coli*, the wild type of *C. glutamicum* can utilise both glucose and sucrose. There are also production technologies available based on acetic acid or ethanol as feedstocks. In the past, molasses was mostly used for production since it is a relatively cheap carbon source. However, the utilisation of molasses has severe disadvantages:

1. Waste is exported from the sugar company to the fermentation plant and causes additional costs there.
2. The seasonal availability of molasses causes ageing effects in its quality during storage.

Therefore, there is a clear tendency away from molasses towards refined carbon sources such as hydrolysed starches. Profitable nitrogen sources are ammonium sulphate and ammonia (gaseous or ammonia water). The growth factors required are provided from plant protein hydrolysates, cornsteep liquor or by the addition of the defined compounds. A typical lysine fermentation is shown in Fig. 9.8. After consumption of the initial sugar, the substrates are added continuously and L-lysine accumulates up to 170 g l^{-1}. Ammonium sulphate provides the counterion to neutralise the accumulating basic amino acid. Therefore, L-lysine is present in the fermentation broth as its sulphate. As a convention in the literature, lysine is usually given as lysine HCl. Due to the high sugar cost, the conversion yield is a very important criterion for the entire production process. Technical processes have been published with a yield of 45–50 g Lys.HCl per 100 g carbon source.

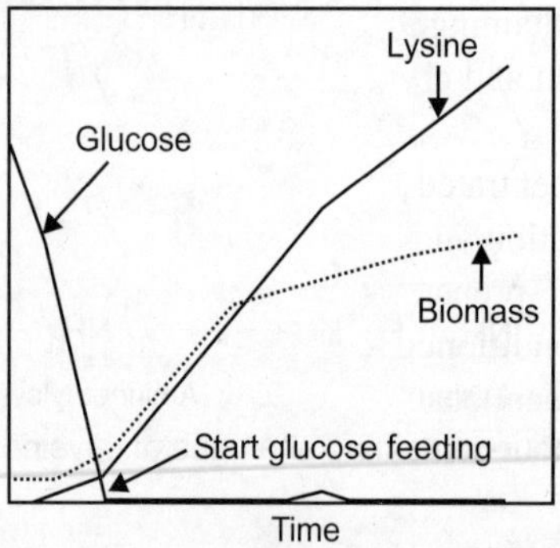

Fig. 9.8. Time cource of L-lysine accumulation in a production plant. There are three phases of growth and L-lysine accumulation.

The development of a new multi-step biotechnological process requires three steps, comprising:

1. Identification and characterisation of a suitable biological system (micro-organism, biocatalyst).
2. Increase of bioreactor productivity by systematic media optimisation and adaptation of fermentation technology to a developing process (process development and fermentation technology) (Fig. 9.9).
3. Downstream process (cell separation by centrifugation or ultrafiltration, separation, evaporation and drying).

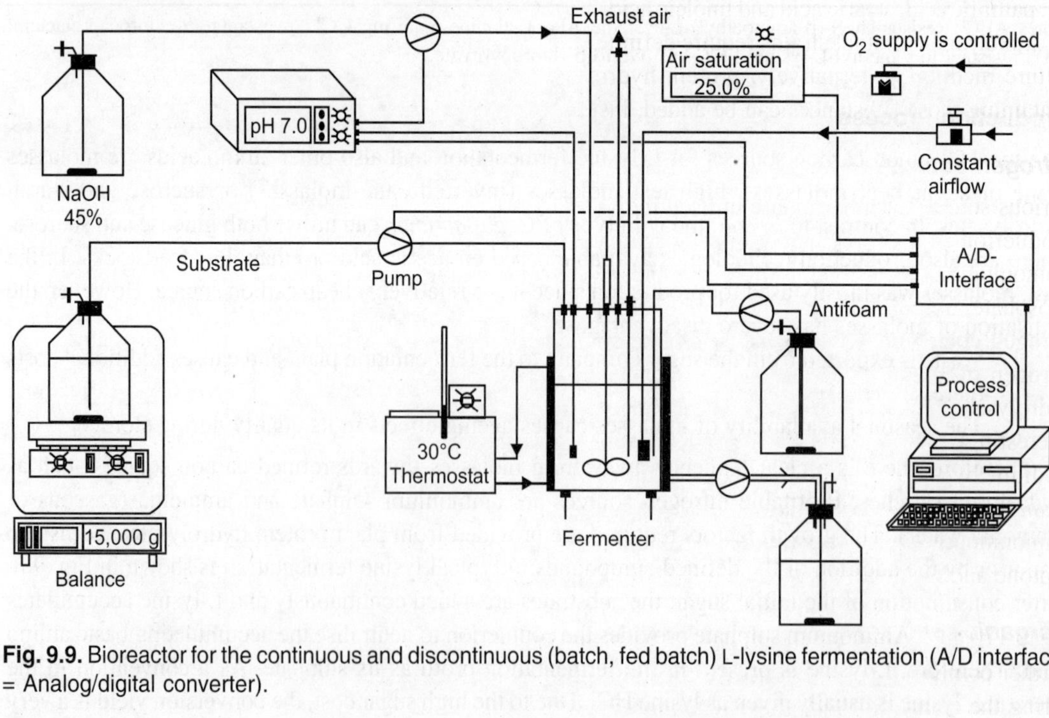

Fig. 9.9. Bioreactor for the continuous and discontinuous (batch, fed batch) L-lysine fermentation (A/D interface = Analog/digital converter).

For the recovery of L-lysine, several basically different processes have been developed. Three processes are currently in use to supply L-lysine in a form suitable for feed purposes:

1. A crystalline preparation containing 98.5 per cent L-lysine HCl. It can be made by ion exchange chromatography, evaporation and crystallisation. Also direct spray-drying of the ion exchange eluate is possible.
2. An alkaline solution of concentrated L-lysine containing 50 per cent L-lysine. It is obtained by biomass separation, evaporation and filtration.
3. A granulated lysine sulphate preparation consisting of 47 per cent L-lysine. It consists of the entire fermentation broth conditioned by spray-dryinging and granulation.

These processes differ in investment costs, losses during downstreaming, amount of waste volume, and user friendliness. All this, together with the fermentation itself, decides the success of the entire production process.

Carbon source

Mutants of *Corynebacterium* and related micro-organisms enable the inexpensive production of amino acids from cheap renewable carbon sources by direct fermentation. Various carbohydrates are utilised individually or as a mixture for the production of L-lysine such as glucose, fructose, sucrose, molasses (sucrose, glucose, fructose etc.), maltose, blackstrap molasses, starch hydrolysate (glucose, oligo-saccharides), lactose, maltose, starch and starch hydrolysates, cellulose, cellulose hydrolysate, organic acids such as acetic acid, propionic acid, benzoic acid, formic acid, malic acid, citric acid and fumaric acid, alcohols such as ethanol, propanol, inositol and glycerol and certainly hydrocarbons, oils and fats such as soyabean oil, sunflower oil, groundnut oil and coconut oil as well as fatty acids such as, e.g. palmitic acid, stearic acid and linoleic acid. Those substances may be used individually or as mixtures.

Auxotrophic mutant strains require definite substance for their growth, which should be added to the culture medium. Alternatively, protein hydrolysate, corn steep liquor, meat extract or yeast extract containing those substances can be added, instead.

Nitrogen source

Various sources of nitrogen are utilised individually or as mixtures for the commercial and pilot scale production of L-lysine, including inorganic compounds such as gaseous and aqueous ammonia, ammonium salts of inorganic or organic acids such as ammonium sulphate, ammonium nitrate, ammonium phosphate, ammonium chloride, ammonium acetate and ammonium carbonate. Alternatively, natural nitrogen containing organic materials like soyabean-hydrolysate, soyaprotein HCl-hydrolysate (total nitrogen of about 7 per cent), soyabean meal, soyabean cake hydrolysate, corn steep liquor, casein hydrolysate, yeast extract, meat extract, malt extract, urea, peptones and amino acids may also be utilised. Interestingly, Inuzuka and Hamada claimed enhanced L-lysine yields by enriching L-lysine fermentation medium by the culture liquor (2–150 ml/l) of an L-leucine-producing micro-organism. *Nocardia alkanoglutinousa* utilises various nitrogen sources including ammonium acetate, ammonium sulphate, ammonium chloride, ammonium nitrate, ammonia water, amino acids, amino acid mixtures, yeast extract, peptone and meat extract.

Inorganic salts, trace elements and growth factors

Further components are necessarily added to fermentation media at the initiation and/or intermittently during the course of L-lysine fermentation, such as inorganic salts of various metals, like magnesium (e.g. magnesium sulphate), calcium, potassium, sodium, iron (e.g. iron sulphate), manganese, and zinc or traces of other metals. Phosphoric acid, potassium dihydrogen phosphate (KH_2PO_4) or dipotassium hydrogen phosphate (K_2HPO_4) or the corresponding sodium salts are commonly used as source of phosphorus for the production of L-lysine. Essential growth factors such as amino acids and vitamins (e.g. vitamin B1) and suitable precursors are also added to the culture in addition to the substances of batch or intermittently during the cultivation. Table 9.4 illustrates the composition of a typical production medium for *C. alkanoglutinousa*.

Table 9.4. Composition of L-lysine fermentation medium of *C. alkanoglutinousa*.

Compound	*Shake flask*	*Bioreactor*	*Unit*
n-Alkane (C_{14}–C_{18})	50.0	100.0	g/l
Ammonium sulphate	30.0	35.0	g/l
$CaCO_3$	30.0	1.0	g/l
K_2HPO_4	0.5	1.0	g/l
KH_2PO_4	0.5	1.0	g/l
$MgSO_4{\cdot}7H_2O$	0.5	0.5	g/l
NaCl	1.0	1.0	g/l
$FeSO_4{\cdot}7H_2O$	20.0	100.0	mg/l
$ZnSO_4{\cdot}7H_2O$	10.0	20.0	mg/l
$MnSO_4{\cdot}4H_2O$	10.0	20.0	mg/l
Tap water	940.0	870	ml
pH	7.0	7.0	–
Steaming	15	15	Min.

The effect of various amino acids (1–3 g/l) on L-lysine production was investigated by Smith. The addition of certain amino acids such as arginine, aspartic acid, isoleucine or valine enhanced L-lysine production, in contrary to leucine. Furthermore, the degree of inhibition on bacterial growth caused by AEC (S-2-aminoethyl-cysteine) was reduced by the addition of 1 g/l of arginine to the culture medium.

Influence of oxygen

L-Lysine fermentation is an aerobic process demanding large amounts of oxygen and strongly influenced by the air saturation in bioreactor. Lactic acid is formed as a by-product under anaerobic conditions, which is reconsumed after the establishment of aerobic conditions.

Aerobic conditions are maintained by aseptically adding to the culture oxygen containing gaseous mixtures, e.g. atmospheric air or pure oxygen. Cultivation of L-lysine producing micro-organisms is carried out with shaking of shake flasks (250–300 rpm) or by the aeration (0.5–1.5 vvm) of stirring bioreactors. Enormous effects of air saturation (100 per cent air saturation corresponds to saturation at 1 vvm aeration rate at 30°C and 600 rpm agitation rate) on continuous L-lysine production by *B. lactofermentum* have been found in chemostat process development experiments.

Influence of temperature

Nocardia alkanoglutinousa grows at temperatures between 10° and 40°C producing L-lysine at 20°–40°C, optimally at 27°–37°C. Incubation of *E. coli* during L-lysine fermentation is conducted in submerged-aerial stirring, shaking or stationary culture at temperatures between 30° to 42°C, preferably at about 37°C for 4 to 24 hours or aerobically from 16 to 72 hr at temperatures between 25° and 45°C.

Influence of pH

The pH is a very important factor strongly influencing microbial fermentations. Basic compounds such as sodium hydroxide, potassium hydroxide, ammonium hydroxide, calcium carbonate, urea, ammonia and gaseous ammonia, or inorganic acid compounds such as phosphoric or sulphuric acid and organic acids are utilised for controlling pH in L-lysine cultures at a pH ranging from 5 to 9 discloses L-lysine production by mutant strains at a pH spectrum between 5 and 8.5.

Nocardia alkanoglutinousa grows at pH between 6 and 9. The *E. coli* L-lysine fermentation process operates at pH between 5 and 8 using inorganic, acidic or alkaline or ammonia gas.

Antifoaming

Foaming occurring during fermentation is controlled by the addition of antifoams such as fatty acid polyglycol esters or silicone and polypropylene.

Plasmid protection

Suitable selectively acting substances, e.g. antibiotics are added into the medium in order to maintain the stability of plasmids.

Biochemistry and regulation of L-lysine fermentation

Entry into lysine pathway begins with L-aspartate, which is synthesised by the transamination of oxaloacetate. *C. glutamicum* has the ability of converting the L-lysine intermediate piperidine 2,6-dicarboxylate to diaminopimelate through two different routes, by reactions involving succinylated intermediates or by the single reaction of diaminopimelate dehydrogenase (Fig. 9.10).

Downstream processes of L-lysine

Product separation and purification is a very important factor enormously affecting fermentation process effectiveness and production costs, steadily requesting improvements in the recovery process of amino acids, especially L-lysine. L-Lysine is recovered from fermentation broth in various ways. For many years, L-lysine-HCl solid has been produced following various steps such as fermentation, separation, purification, crystallisation and drying. L-Lysine of resultant culture broth can be recovered by known conventional methods such as using ion exchange resins, or by directly crystallisation of L-lysine from culture broths. After cell separation by cell filtration or centrifugation, L-lysine may be recovered from fermentation broth by an ion exchange step and thereafter concentrated by evaporation and spray drying.

L-THREONINE

The commercial production of L-threonine is possible with either *E. coli* or *C. glutamicum* mutants. However, the production figures of selected *E. coli* strains are superior. The synthesis of L-threonine proceeds via a short pathway comprising only five steps (Fig. 9.11). As already mentioned, the first steps are shared with that of L-lysine and L-methionine synthesis. Furthermore, L-threonine is also an intermediate in the L-isoleucine synthesis.

This naturally requires special metabolic regulation. In *C. glutamicum* this was solved in such a way that the sole aspartate kinase present was only inhibited by the joint presence of L-lysine and L-threonine, one by L-lysine and one by L-methionine. There are furthermore two homoserine dehydrogenase activities: one is inhibited by L-threonine and one by L-methionine. Additionally, the corresponding genes are grouped into transcriptional units, thereby ensuring a balanced synthesis of the appropriate amino acid at the level of gene expression. The relevant operon for L-threonine synthesis in *E. coli* is *thrABC*.

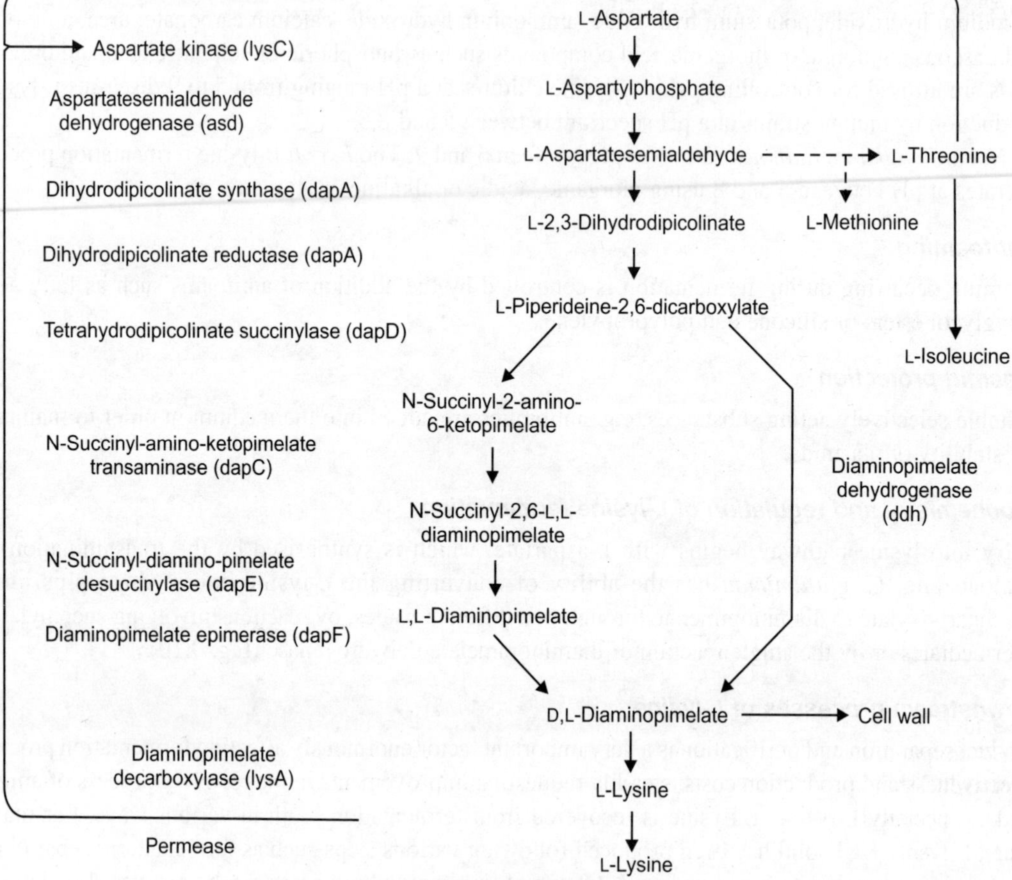

Fig. 9.10. Flow scheme of L-lysine formation pathway.

It encodes three polypeptides, with *thrA* encoding an apparently fused polypeptide with kinase plus dehydrogenase activity. Therefore, four enzyme activities of the five steps required to convert L-aspartate to L-threonine are encoded by *thrABC*. A strong expression control of this operon is provided by a transcription attenuation mechanism. The corresponding leader peptide at the beginning of the transcription unit is Thr-Thr-Ile-Thr-Thr-Thr-Ile-Thr-Ile-Thr-Thr, serving to sense the availability of L-threonine and L-isoleucine. When the corresponding tRNAs are uncharged, the leader peptide formation does not occur and transcription of the operon is increased at least ten-fold.

Production Strains

Based on this regulation there is a clear focus on two major targets for the design of a production strain: the prevention of L-isoleucine formation and stable high-level expression of *thrABC*. Therefore, in one of the first steps of strain development, chromosomal mutations were introduced to give an isoleucine leaky strain (Fig. 9.12). The isoleucine mutation is a very specific and important one. L-Isoleucine is required for growth only a low L-threonine concentrations but, at thigh concentrations of L-threonine, growth is independent of L-isoleucine. The mutation therefore has several advantageous consequences. In the first place, it prevents an excess formation of the undesired by-product L-isoleucine. Additionally, it prevents the L-isoleucine-dependent premature termination of the *thrABC* transcription due to limiting tRNA^{Ile}. A high transcription rate is, of course, required to have high specific enzyme activities.

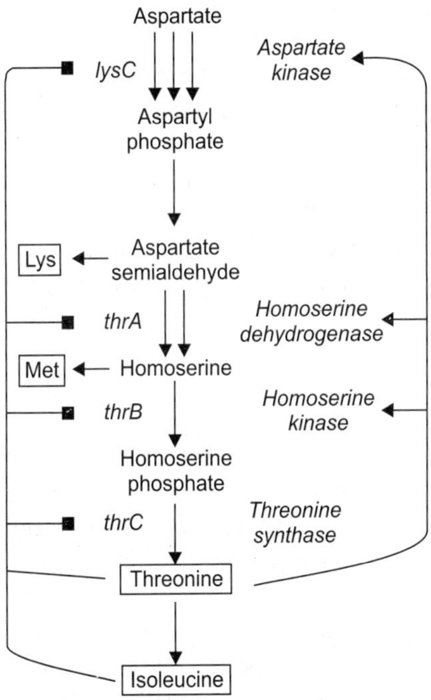

Fig. 9.11. L-Threonine synthesis in *E. coli.* The thin arrows indicate individual enzyme activities and the genes *thrABC* constitute an operon. Only the regulation of genes and enzymes by L-threonine and L-isoleucine is shown, where the square ends indicate gene repression and the arrowhead ends enzyme activity inhibition.

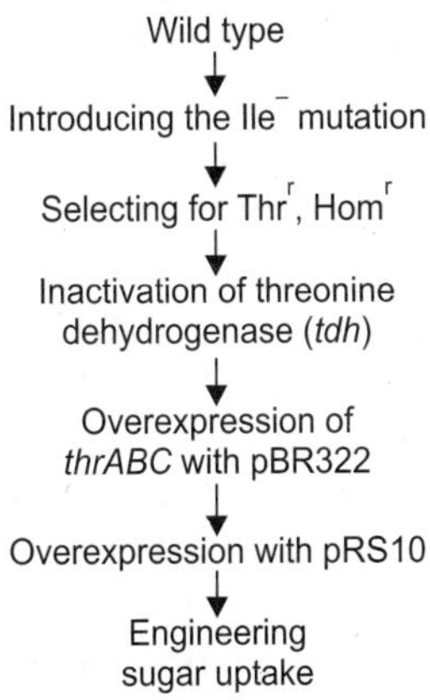

Fig. 9.12. Relevant steps in the development of an *E. coli* strain suitable for L-threonine production involving undirected mutagenesis, gene inactivation and use of different plasmids.

Another consequence of the isoleucine mutation is more subtle. It relates to the stability of the plasmid-containing producer strain in the various precultivation steps. Starting from a single clone, a preculture is inoculated for each production run and is then enlarged in several stages. This means that the clone is fermented for about 25 generations so that there is a great danger of the plasmid containing the *thrABC* operon being lost. This would of course be a complete disaster if it happened in the final production stage. In the presence of the isoleucine leaky mutation, however, cells that have lost the plasmid now are clearly disadvantaged when not supplied with L-isoleucine. Their further proliferation is halted, thereby stabilising a culture where almost all the cells that are growing contain the plasmid. Further engineering during strain evolution involved the introduction of resistance to L-threonine and L-homoserine. Subsequently, *tdh*, which encodes threonine dehydrogenase, was inactivated thus preventing threonine degradation. To obtain very high activities of the *thrABC*-encoding enzymes, the operon was cloned from a strain whose kinase and dehydrogenase activities are resistant to L-threonine inhibition. In addition, the transcription attenuator region was deleted. In fermentations the operon engineered in this way was successfully used with pBR322 as a vector, but a further improvement was obtained by replacing this plasmid by a pRS1010 derivative, resulting in an even more stable high-level expression.

Substrate uptake

Since the cost of the sugar source has a decisive influence on the price of the amino acid produced it is essential to be able to switch between glucose and sucrose as substrates. However, only a few of the *E. coli* strains can use sucrose. Two different sucrose-utilising systems of *E. coli* are available to engineer sugar utilisation in L-threonine producing strains (Fig. 9.13). One of them is represented by the *scr* regulon, where the actual translocator consists of a phospho*enol*pyruvate: sugar phosphotransferase system (PTS). Introduction of the *scr* genes into a glucose-utilising *E. coli* strain results in the uptake and phosphorylation of sucrose. Due to subsequent hydrolase and fructokinase activities the sugar is then channelled into the central metabolism. An alternative sucrose utilisation system is provided by the *csc* regulon of some *E. coli* strains. In this case, sucrose is translocated by the *cscB* encoded translocator is symport with protons. Using transposition the sucrose-utilisation capability of the *csc* regulon was introduced into a glucose-utilising strain. Although originally without uptake of sucrose, this strain now imported sucrose at a rate of 9 pmol min^{-1}·mg dry wt. With the plasmid-encoded regulon the rate obtained was 43 pmol min^{-1}·mg cell dry wt, which was almost identical to that of the strain from which the *csc* regulon had been isolated.

Production Process

The fermentation of the engineered L-threonine producer is in a simple mineral salts medium with either glucose or sucrose as the substrate with addition of a small amount of a complex medium component like yeast extract. After the inoculation and consumption of the initially provided sugar, continuous feeding of sugar begins. Additionally, ammonia has to be fed in the form of gas or as NH_4OH which is regulated via pH control. Thus the feeding strategy in the case of L-threonine fermentation is quite easy compared to L-lysine fermentation where the accumulation of the basic product requires the feeding of sulphate as the counter-ion. At the end of the fermentation, L-threonine is present in concentrations of about 85 g l^{-1} with a conversion yield of up to 60 per cent based on the carbon source used. Such fermentations with high yields show quite low by-product levels. This is an advantage for downstream processing. Crystallisation of L-threonine is easy due to its low solubility (about 90 g l^{-1} in water) and the low salt concentration present. A process is described where the cells are initially coagulated by a heat-or pH-treatment step, followed by filtration. Subsequently, the broth is concentrated and crystallisation initiated by cooling. The separation and drying of the crystals leads to an isolation yield of 80 to 90 per cent with the L-threonine having a purity of more than 90 per cent. A recrystallisation step may be required for high-purity L-threonine.

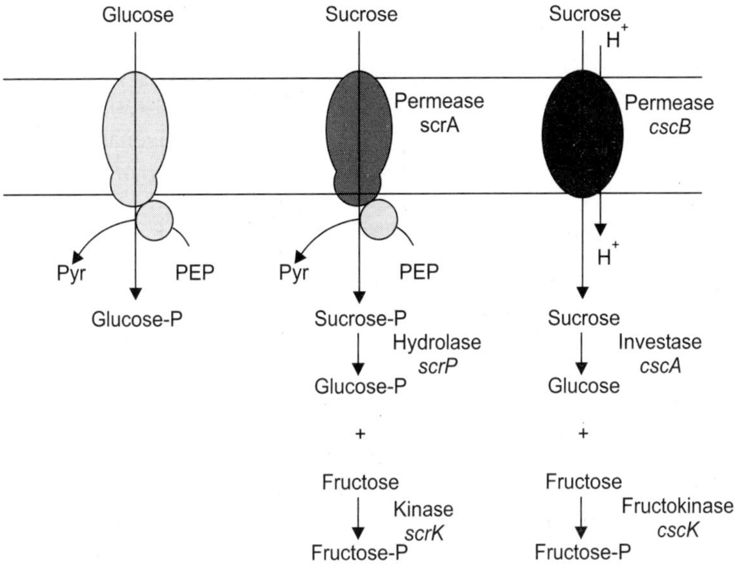

Fig. 9.13. Mechanisms of sugar uptake and phosphorylation in *E. coli*. Translocation is coupled by phosphorylation, as is the case for the phosphotransferase system (left and middle) or occurs in symport with protons without phosphorylation (right). The phosphotransferase translocating sucrose (middle) shares one of the phosphoryl transfer domains with a component of the phosphotransferase translocating glucose. Pyr, pyruvate; PEP, phosphoenolpyruvate.

L-PHENYLALANINE

L-Phenylalanine can be produced with *E. coli* or *C. glutamicum*. The pathway for L-phenylalanine synthesis is shared in part with that of L-tyrosine and L-tryptophan. These three aromatic amino acids have in common the condensation of erythrose 4-phosphate and phospho*enol*pyruvate to deoxy*arabino*heptulosonate phosphate (DAHP) with further conversion in six steps up to chorismate. L-Phenylalanine is then finally made in three further steps (Fig. 9.14). There are three DAHP synthase enzymes in *E. coli* encoded by *aroF, aroG* and *aroH*.

These enzymes play a key role in flux control. Their regulation of catalytic activity, in each case by one of the three aromatic amino acids, recalls the specific regulation of aspartate kinase in the synthesis of threonine. About 80 per cent of the total DAHP-synthase activity is contributed by the *aroG*-encoded enzyme. Increased flux towards L-phenylalanine can be obtained by over-expression of either *aroF* or *aroG* encoding feedback-resistant enzymes.

Furthermore, *pheA* overexpression is essential. This gene encodes the bifunctional corismate mutase-prephenate dehydratase. A second chorismate activity is present as a bifunctional chorismate mutase-prephenate dehydrogenase. The *pheA*-encoded enzyme activities are inhibited by L-phenylalanine and *pheA* expression is dependent on the level of $tRNA^{Phe}$.

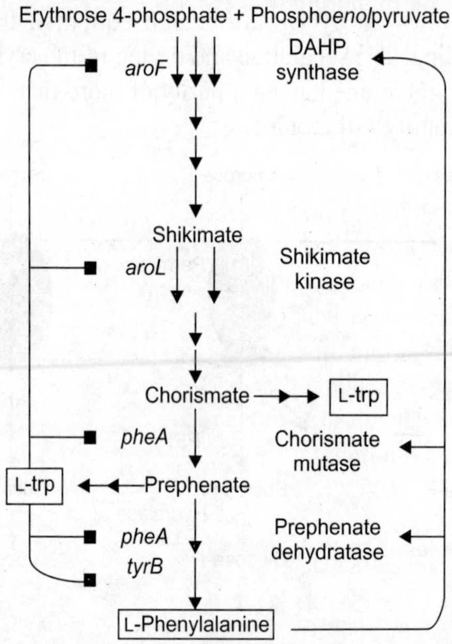

Fig. 9.14. Simplified pathway of L-phenylalanine synthesis and the relevant regulation by L-phenylalanine and L-tyrosine (L-tyr) with feedback control of enzyme activity (arrowhead ends) and gene repression (square ends).

Production Strains

Producer strains have a DAHP activity that is resistant to feedback inhibition and which is encoded either by *aroF* or *aroG* and a feedback-resistant corismate mutase-prephenate dehydratase. As a rule, the producers are L-tyrosine auxotrophic mutants. There are very good reasons for this, one of which is that the enzymes of the common pathway from DAHP to prephenate are no longer regulated by L-tyrosine and enzyme activities are no longer feedback-inhibited. Another reason is that in this way tyrosine accumulation is prevented, which would otherwise undoubtedly result as a by-product since there are only two additional steps from prephenate to L-tyrosine. An essential aspect is that due to the auxotrophy, a beneficial growth limitation is possible by appropriate tyrosine feeding. In some *E. coli* strains, the temperature-sensitive cI_{857} repressor of bacteriophage λ has been used together with the λP_L promoter to enable inducible expression of key genes *pheA* and *aroF*. This enables extremely high enzyme activities to be adjusted solely in the actual production runs thus eliminating the inherent problems of strain stability due to the resulting high metabolite concentrations or side activities of the enzymes. It enables the precultivation steps up to the seed fermenter to be performed with low expression of the key genes but in the actual large production fermenter the genes are now induced to a high level of expression.

Production Process

As with the other amino acids, effective L-phenylalanine production is the joint result of engineering the cellular metabolism and control of the production process. Control is necessary for two reasons. First, the carbon flux has to be optimally distributed between the four major products of glucose conversion, which are L-phenylalanine, biomass, acetic acid and CO_2. The second reason is that the cellular physiology is not constant during the course of fermentation, which correspondingly requires an adaptation of fermentation control during the process. Figure 9.15 shows the typical time curve of L-phenylalanine production. The major problem is that *E. coli* tends to produce acetic acid which has a strong negative effect on process efficiency. To prevent this, researchers have developed an ingenious sugar-feeding strategy, which first collects on-line data and fluxes such as oxygen concentration, sugar consumption and biomass concentrations. These are then counterbalanced during the process to control the optimum sugar concentration. The feeding of sugar starts when the cells enter Stage 2 of the fermentation where the glucose initially provided has almost been consumed. The trick is to prevent too high a glucose concentration occurring since this would result in acetic acid formation and at the same time, to prevent too low a glucose concentration since this would result in an excess of CO_2 evolution. Thus the feeding rate is a compromise where the process is run at the highest possible feeding rate which still provides a sufficiently strong limitation to prevent acetic acid excretion. When the L-tyrosine initially present has been consumed, the cells proceed to Stage 3. As already mentioned, almost all L-phenylalanine producers cannot synthesise tyrosine. The L-tyrosine concentration selected at the start of the culture therefore fixes the minimum amount of biomass necessary to efficiently metabolise the predetermined amount of glucose. In Stage 3, the metabolic capacity of the cells decreases which brings about a consequent decrease of the glucose feeding rate. At the end of Stage 3, acetic acid excretion begins and the cells enter Stage 4 where no further L-phenylalanine accumulation occurs and the process is eventually terminated. This example of amino acid production shows that by the sophisticated application of feeding strategies with adaptive control a very high L-phenylalanine concentration can be achieved with a high yield within 3 days. Values of 50.8 gram phenylalanine per litre with a yield of 27.5 per cent of carbon used have been reported.

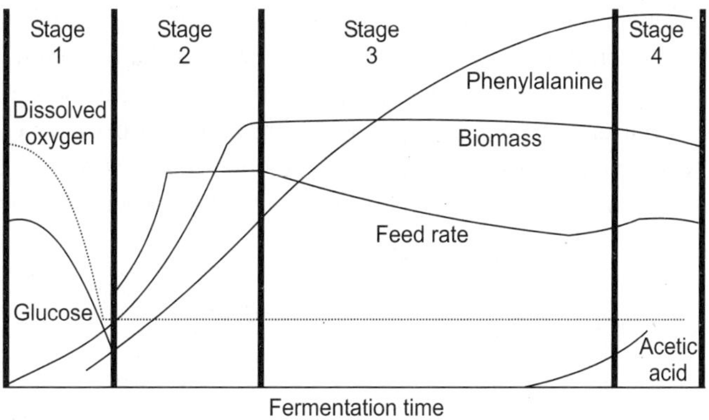

Fig. 9.15. The four stages of L-phenylalanine production characterised by different physiology requiring different process control regimes to give the highest yields in shortest times.

L-TRYPTOPHAN

L-Tryptophan is a high-price amino acid which still has a rather low market volume. Effective production processes are available with mutants of different bacteria, including *Bacillus subtilis*. However, cellular synthesis is no longer performed due to originally not realised impurities in the final product used for medical purposes. These impurities arose during the isolation of L-tryptophan from a chemical reaction with traces of acetaldehyde at low pH. An alternative process is the enzymatic synthesis of L-tryptophan from precursors. The current enzymatic production process uses the activity of the biosynthetic tryptophan synthase (Fig. 9.16).

This enzyme catalyses the last step in the tryptophan synthesis, which consists in fact of two partial reactions:

Indole 3-glycerol phosphate → indole + glyceraldehyde 3-phosphate Indole + L-serine→ L-tryptophan + H_2O

These separate reactions are catalysed by separate subunits of the enzyme: α and β. The enzyme of *E. coli* is an $\alpha_2\beta_2$ tetramer, which can be dissociated into two α subunits and a β_2 subunit. The α subunit catalyses the cleavage of indole 3-glycerol phosphate, whereas the β_2 subunit catalyses the condensation of L-serine with indole to form L-tryptophan. Each β subunit contains one molecule of covalently bound pyridoxal phosphate, forming a Schiff's base with L-serine. This enzyme-bound aminoacrylate is attacked when indole is provided from the a subunit. But how does indole get to the β subunit? The problem is that indole is very hydrophobic so that with free diffusion it can pass through the cell membrane and be lost. The crystal structure of the synthase revealed the ingenious solution for solving this problem. To prevent a loss of indole it is channelled within the enzyme protein. There is a 25 Å long tunnel from the α subunit, where indole is formed, to the β subunit where, as the enzyme-bound aminoacrylate, L-serine is ready to accept the indole. Furthermore, within the native tetramer both partial reactions are coordinated. Only when L-serine, as aminoacrylate, is ready to accept the indole, does indole 3-glycerol phosphate conversion occur at the β subunit. Tryptophan synthase is thus an example of how an enzyme complex is used as sophisticated device to handle a reactive and diffusible intermediate within the cell.

Production from Precursors

The process of L-tryptophan production with this enzyme is based on *E. coli* cells which have a high tryptophan synthase activity. The α and β subunits encoding genes *trypA* and *trpB*, respectively, are located on the *trpEDCBA* operon which is regulated by repression and attenuation. In the *E. coli* mutant used, the repressor of that operon has been deleted as is part of the attenuator region together with the first structural genes of the operon. In the resulting strain, about 10 per cent of the total protein is tryptophan synthase with an excess of the β subunit. Although indole is not the true substrate of the enzyme (Fig. 9.16), with a sufficiently high concentration the enzyme will react with it.

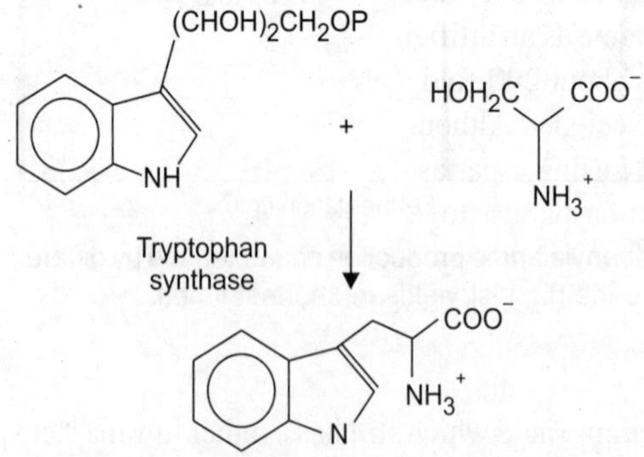

Fig. 9.16. The tryptophan synthase uses *in vivo* indole 3-glycerol phosphate plus L-serine and in the production process indole plus L-serine.

Indole is available from the petrochemical industry as a comparably cheap educt, whereas the second educt, L-serine, is recovered from molasses during sugar refinement using ion exclusion chromatography and further purification steps (Fig.9.17). The resulting L-serine is fed to the previously cultivated *E. coli* and indole is added continuously at a concentration adjusted to 10 mM, which is controlled on-line. This type of process ensures an almost quantitative conversion of indole to yield L-tryptophan with a space-time yield of about 75 g per litre day. Further processing of the L-tryptophan solution can be taken from Fig. 9.17 leading to a pyrogen-free pharmaceutical product of the highest quality.

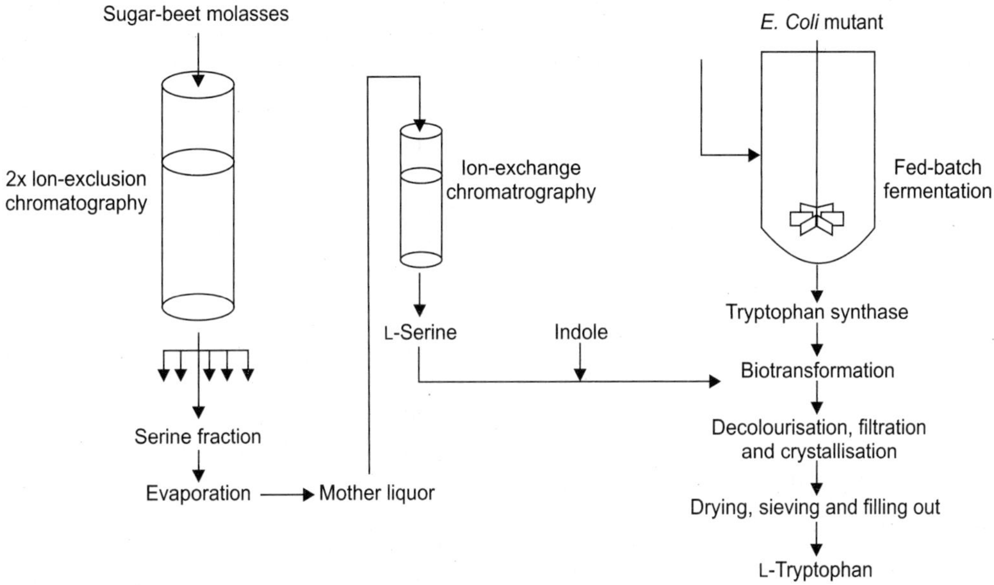

Fig. 9.17. Production plant to fractionate molasses by ion-exclusion chromatography, with isolation of L-serine. An *E. coli* mutant overexpressing tryptophan synthase is pregrown and subsequently mixed with L-serine plus indole to convert these substrate to L-tryptophan.

L-ASPARTATE

L-Aspartic acid is widely used as a food additive and in pharmaceuticals. Demand increased rapidly with the introduction of aspartame as an artificial sweetener. This is a dipeptide consisting of L-aspartate and L-phenylalanine which is about 200-fold sweeter than sugar and was successfully introduced into the market as a low-calorie sweetener. Although L-aspartate as originally produced fermentatively, it is currently produced exclusively using aspartase due to the high productivities and the cost effectiveness of the process. In fact, the use of aspartase to make L-aspartate represents one of the highest productivity known for an enzyme used in biotechnology. The method developed allows reuse of the enzyme to the extent that over 2,20,000 kg of product can be produced per kg of enzyme.

Aspartase catalyses the interconversion between L-aspartate and fumarate plus ammonia (Fig. 9.18). The reaction favours the amination reaction. The enzyme of *E. coli* is tetramer with a molecular weight of 1,96,000 which has an absolute requirement for divalent metal ions. A severe disadvantage at the beginning of the work by the Tanabe Seiyaku company, which now successfully uses aspartase, was the instability of the enzyme. After incubation of the enzyme in solution for just half an hour at 50°C, activity was no longer detectable. Nevertheless, a residual activity of 10 per cent is present when the enzyme is immobilised in polyacrylamide. Such a physical confinement of cells in space turned out to be the method of choice. Table 9.4 shows that with the natural polymer *k*-carrageenan, resulting from a screening of different polymers and use of appropriate cross-lining exceptional improvements are obtained in the relative productivity as well as in the stability of the catalyst. The final material has a half-life of almost two years. This represents almost unimaginable progress in comparison to the initial situation where enzyme in free solution only had a half-life measured in minutes. An initial disadvantage of the original cells used, was their fumarase activity which results in the partial conversion of fumarate to L-malic acid. To solve this problem a heat treatment step of the cells is used which eliminates the fumarase activity almost completely. Using such conditioned cells and starting with 1 M ammonium fumarate, the final product solution contains 987 mM L-aspartate, 10.7 mM non-reacted fumarate and only trace quantities of L-malic acid of 1.9 mM.

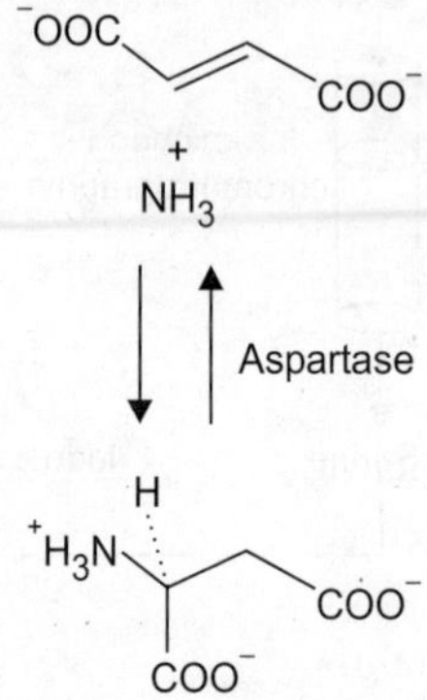

Fig. 9.18. Fumarate and ammonium serve as substrates for the aspartase.

Table 9.5. Comparison of immobilised *E. coli* cells used for production of L-aspartate.

Immobilisation method	*Aspartase activity (U/g cells)*	*Half-life (days)*	*Relative productivity (%)*[a]
Polyacrylamide	18850	120	100
Carrageenan	56340	70	174
Carrageenan (GA)[b]	37460	240	397
Carrageenan (GA + HA)[b]	49400	680	1498

Notes: [a]Considers the activity, decay constant and operation period. [b]GA = glutaraldehyde, HA = hexamethylene diamine.

For the production process the immobilised cells are packed into a column designed as a multistage system. The stages introduced, consisting of horizontal tubes, serve two purposes. On the one hand, they allow effective cooling to prevent decay of the catalytic activity since the aspartase reaction is exergonic.

About 6 kcal heat mol^{-1} substrate evolves in the actual large-scale production process which is very close to that calculated from the standard free energy change of the aspartase reaction of 4 kcal mol^{-1}. On the other hand, the flow properties of the column are increased. Any compacting of the bed over time is prevented and the preferred plug-flow characteristics are obtained. With such a column, flow rates of two column volumes per hour are possible. The continuous process enables full automation and control to achieve an optimum throughput with the highest product quality. Yet another advantage as such a controlled continuous process is its reduced waste production. A typical volumetric activity is about 200 mmol hr^{-1} g cells. Assuming a 1000 litre column, the yield of L-aspartate is 3.4 tonnes per day which is 100 tonnes per month. The final product is eventually purified by crystallisation.

Chapter 10

Nucleosides, Nucleotides and Allied Compounds

INTRODUCTION

Nucleosides are structural subunits of nucleic acids, the macromolecules that convey genetic information in living cells. They consist of a nitrogen-containing base bonded to a five-carbon (pentose) sugar. Nucleosides are the biochemical precursors of nucleotides, the molecular building blocks of the nucleic acids DNA and RNA. Nucleotides also are important in cell metabolism (ATP is the energy currency of the cell) and as co-enzymes. Nucleotides are formed by the addition of one or more phosphate groups to the nucleoside. Some nucleosides have important clinical applications; for example, puromycin and certain other antibiotics are nucleosides produced by moulds or fungi.

Nucleotides are molecules that, when joined together, make up the structural units of RNA and DNA. In addition, nucleotides play central roles in metabolism. In that capacity, they serve as sources of chemical energy (adenosine triphosphate and guanosine triphosphate), participate in cellular signaling (cyclic guanosine monophosphate and cyclic adenosine monophosphate), and are incorporated into important cofactors of enzymatic reactions (coenzyme A, flavin adenine dinucleotide, flavin mononucleotide, and nicotinamide adenine dinucleotide phosphate).

NUCLEOSIDES

Nucleosides are glycosylamines consisting of a nucleobase (often referred to simply base) bound to a ribose or deoxyribose sugar. Examples of these include cytidine, uridine, adenosine, guanosine, thymidine and inosine. Nucleosides can be phosphorylated by specific kinases in the cell on the sugar's primary alcohol group (—CH_2·OH), producing nucleotides, which are the molecular building blocks of DNA and RNA. Nucleosides can be produced by *de novo* synthesis pathways, particularly in the liver, but they are more abundantly supplied via ingestion and digestion of nucleic acids in the diet, whereby nucleotidases break down nucleotides (such as the thymine nucleotide) into nucleosides (such as thymidine) and phosphate. The nucleosides, in turn, are subsequently broken down:

1. In the lumen of the digestive system by nucleosidases into nucleobases and ribose or deoxyribose.
2. Inside the cell into nitrogenous bases, and ribose-1-phosphate or deoxyribose-1-phosphate.

In medicine several nucleoside analogues are used as antiviral or anticancer agents. The viral polymerase incorporates these compounds with non-canonical bases. These compounds are activated in the cells by being converted into nucleotides, they are administered as nucleosides since charged nucleotides cannot easily cross cell membranes. In molecular biology several analogues of the sugar back bone exist. Due to the low stability of RNA, which is prone to hydrolysis, several more stable alternative nucleoside/nucleotide analogues are used which correctly bind to RNA. This is achieved by using a different backbone sugar. These analogues include LNA, morpholino, PNA.

In sequencing dideoxynucleotides are used. These nucleotides possess the non-canon sugar dideoxyribose, which lacks 3′-hydroxyl group (which accepts the phosphate) and therefore cannot bond with the next base, terminating the chain as DNA polymerases mistake it for a regular deoxyribonucleotide. Structure of some common nuclosides is given in Fig. 10.1.

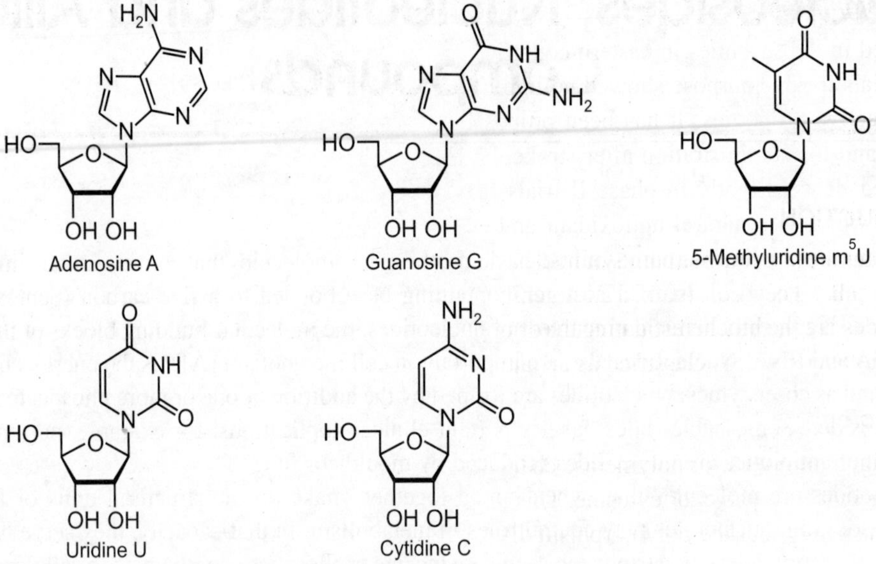

Fig. 10.1. Structure of some common nuclosides.

Inosine

Inosine is a nucleoside that is formed when hypoxanthine is attached to a ribose ring (also known as a ribofuranose) via a β-N_9-glycosidic bond.

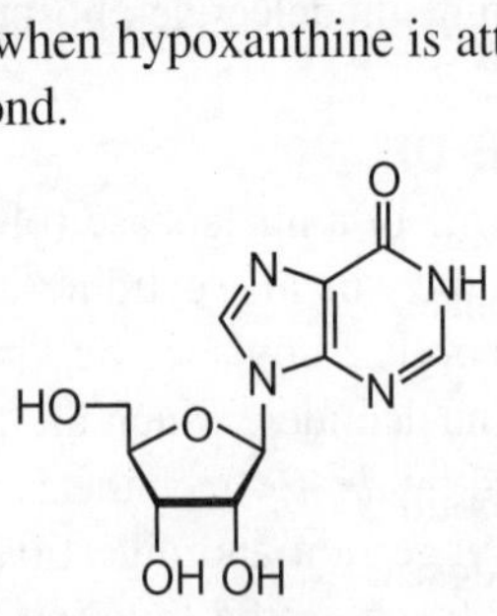

Inosine

Inosine is commonly found in tRNAs and is essential for proper translation of the genetic code in wobble base pairs. Knowledge of inosine metabolism has led to advances in immunotherapy in recent decades. Inosine monophosphate is oxidised by the enzyme inosine monophosphate dehydrogenase yielding xanthosine monophosphate, a key precursor in purine metabolism. Mycophenolate mofetil is an anti-metabolite, anti-proliferative drug, used in the treatment of a variety of autoimmune diseases including Wegener's granulomatosis. The uptake of purine by actively dividing B cells can exceed eight times that of normal body cells and therefore this set of white cells (which cannot operate purine salvage pathways) is selectively targeted by the purine deficiency resulting from IMD inhibition.

Reactions

Adenine is converted to adenosine or inosine monophosphate (IMP), either of which in turn is converted into inosine (I), which pairs with adenine (A), cytosine (C) and uracil (U). Purine nucleoside phosphorylase intraconverts inosine and hypoxanthine. Inosine is also an intermediate in a chain of purine nucleotides reactions required for muscle movements.

Clinical significance

It was tried in the seventies in eastern countries for improving athletic performance. Nevertheless the clinical trials for this purpose showed no improvement. Nowadays, it has been shown that inosine has neuroprotective properties. It has been proposed for spinal cord injury; because it improves axonal rewiring, and for administration after stroke, because observation has shown that axonal re-wiring is encouraged. It is currently in phase II trials for multiple sclerosis (MS). It produces uric acid after ingestion, which is a natural antioxidant and a peroxynitrite scavenger, which can suggest possible benefit in multiple sclerosis (peroxynitrite has been correlated with the axons degeneration).

Alseres pharmaceuticals (named Boston Life Sciences when patent was granted) patented the treatment for stroke and is currently investigating the drug in the MS setting. In the anatomical therapeutic chemical classification system, it is classified as an antiviral.

Biotechnology

When designing primers for polymerase chain reaction, inosine is useful in that it will indiscriminately pair with adenine, thymine, or cytosine. This allows for design of primers that span a single nucleotide polymorphism, without the polymorphism disrupting the primer's annealing efficiency.

Fitness

Despite a lack of clinical evidence that it improves muscle development, inosine remains an ingredient in some fitness supplements.

STRUCTURE AND NOMENCLATURE OF NUCLEOTIDES

A nucleotide is composed of a nucleobase (nitrogenous base), a five-carbon sugar (either ribose or 2′-deoxyribose), and one to three phosphate groups. Together, the nucleobase and sugar comprise a nucleoside.

The phosphate groups form bonds with either the 2, 3, or 5-carbon of the sugar, with the 5-carbon site most common. Cyclic nucleotides form when the phosphate group is bound to two of the sugar's hydroxyl groups. Ribonucleotides are nucleotides where the sugar is ribose, and deoxyribonucleotides contain the sugar deoxyribose. Nucleotides can contain either a purine or pyrimidine base. Nucleic acids are polymeric macromolecules made from nucleotide monomers. In DNA, the purine bases are adenine and guanine, while the pyrimidines are thymine and cytosine. RNA uses uracil in place of thymine.

Nucleotides can be synthesised by a variety of means both *in vitro* and *in vivo*. *In vivo*, nucleotides can be synthesised *de novo* or recycled through salvage pathways. Nucleotides undergo breakdown such that useful parts can be reused in synthesis reactions to create new nucleotides. *In vitro*, protecting groups may be used during laboratory production of nucleotides. A purified nucleoside is protected to create a phosphoramidite, which can then be used to obtain analogues not found in nature and/or to synthesise an oligonucleotide. The term 'nucleotide' originates from the name of nucleic acids, originally found in nuclei, and which are polymers of nucleotides. A nucleotide has the general structure

phosphate — pentose sugar — base.

A nucleoside has the structure

pentose sugar — base.

Thus, AMP and the corresponding nucleoside, adenosine, have the structures shown below:

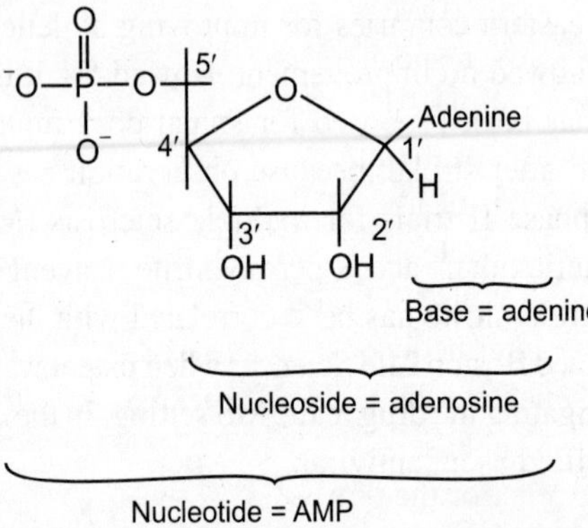

Strictly speaking, the AMP shown here should be written as 5′ AMP. The prime (′) indicates that the number refers to the position on the ribose sugar ring, to which the phosphate is attached, rather than to the numbering of atoms in the adenine ring.

However, it is a common practice to assume that the phosphate is 5′ AMP unless specified, since this is the most usual position. Thus 5′ AMP is often called AMP, whereas if the phosphate is on the carbon atom 3 of the ribose, this is always specified as 3′ AMP. Be careful to note that AMP is chemically and biochemically different from cyclic AMP.

Sugar Component of Nucleotides

The sugar component of a nucleotide is always a pentose, ribose, or 2′-deoxyribose, which are always in the D-configuration, never the L-form.

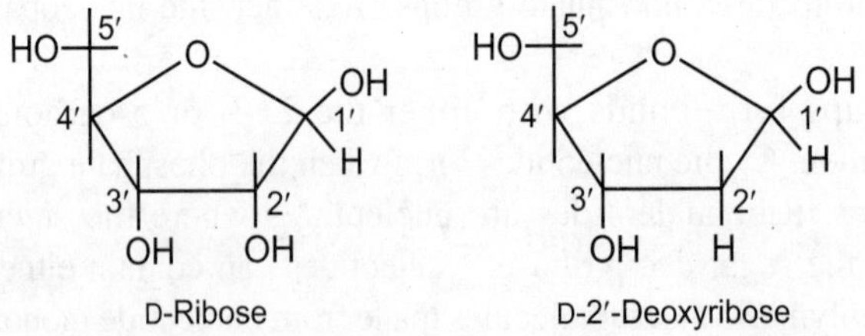

In RNA the sugar is always ribose (hence the name, ribonucleic acid) and in DNA, deoxyribose (hence, deoxyribonucleic acid).

A nucleotide containing ribose is a ribonucleotide but this is not usually specified; unless otherwise stated, a named nucleotide such as AMP is taken to be a ribonucleotide. A deoxyribonucleotide is always specified; for example, deoxyadenosine monophosphate or dAMP, etc. (with the one occasional exception mentioned below).

Base Component of Nucleotides

Nomenclature

We are primarily concerned with five different bases—adenine, guanine, cytosine, uracil, and thymine, all often abbreviated to their initial letter.

A, G, C, and U are found in RNA.

A, G, C, and T are found in DNA.

The ribonucleotides are AMP, GMP, CMP, and UMP, but older and still used terms are adenylic, guanylic, cytidylic, and uridylic acids, respectively (or adenylate, guanylate, cytidylate, and uridylate for the ionised forms at physiological pH). The deoxyribonucleotides are dAMP, dGMP, dCMP, and dTMP. The latter often is called TMP, or thymidylate, without the d-prefix because T is found only in deoxynucleotides (with rare exceptions of no significance here).

When the intention is to indicate a nucleotide without specifying the base, the abbreviations NMP or 5′ NMP are often used, or dNMP or 5′ dNMP for deoxynucleotides.

Deoxy UMP exists only as an intermediate in the formation of dTMP; it does not occur in DNA (except as a result of chemical damage to the DNA).

The corresponding ribonucleosides (base—sugar) are, respectively, adenosine, guanosine, cytidine, and uridine, and deoxyadenosine, etc. for the deoxyribose compounds. The names for the nucleosides, cytidine and uridine, sound like those of free purine bases (cf. adenine and guanine), while the name of the free base cytosine sounds like that of a nucleoside (cf. adenosine).

Other so-called 'minor' bases exist and are found in transfer RNA. Hypoxanthine is one of these. It is also the first base produced by the pathway of purine biosynthesis as hypoxanthine ribotide or inosine monophosphate (IMP; see below); hypoxanthine riboside is called inosine (a somewhat confusing nomenclature).

Structure of the bases

The first point is that:

A and G are purines.

C, U, and T are pyrimidines.

These names originate from their being structurally related to purine and pyrimidine, respectively (neither of which occur in nature).

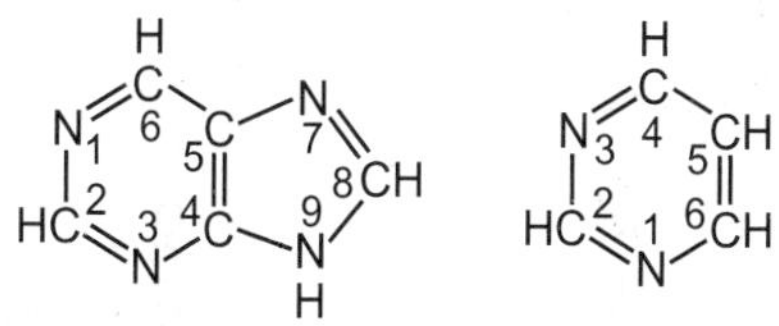

Purine Pyrimidine

These rings in future structures will be represented by the simplified forms below.

Purine Pyrimidine

The structures of the nucleotide bases are represented in Fig. 10.2. Of especial importance, note that T is simply a methylated U. It will be useful to fix in your mind that T is essentially the same as U except that it is 'tagged' by a methyl group. T is found only in DNA; U only in RNA. The significance of this will be apparent later.

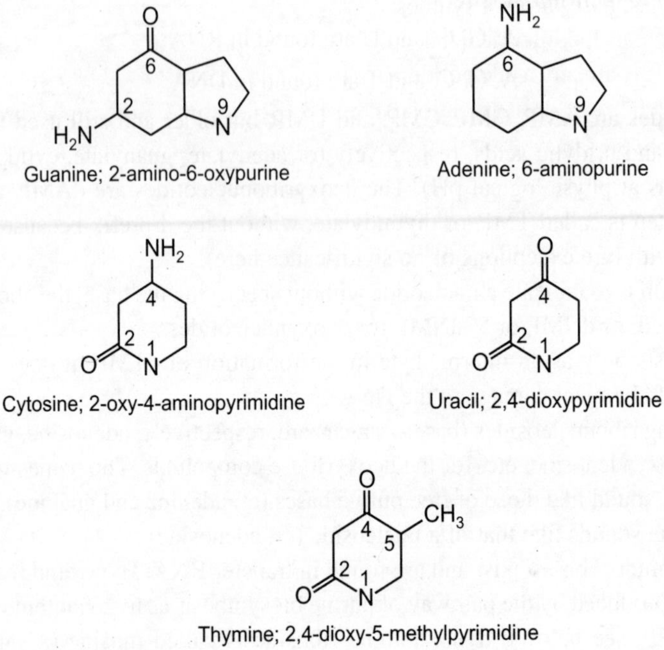

Fig. 10.2. Diagrammatic representation of structures of purine and pyrimidine bases found in nucleic acids.

Attachment of the bases in nucleotides

The bases are attached to the pentose sugar moieties of nucleotides at the N-9 position of purines and the N-1 position of pyrimidines.

The glycosidic bond is in the β-configuration, that is, it is above the plane of the pentose ring. The structures of AMP and CMP are given in Fig. 10.3.

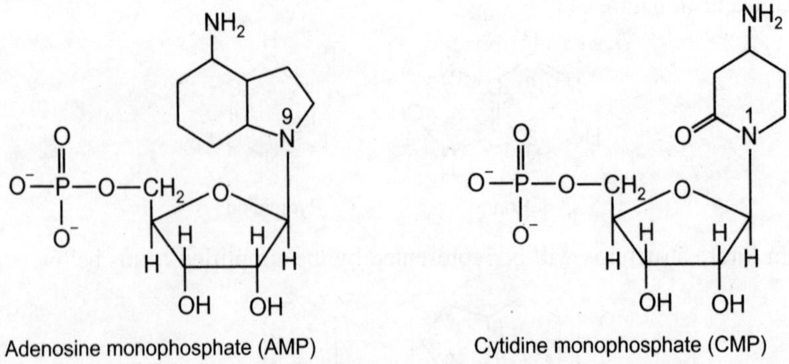

Fig. 10.3. Structures of a purine and a pyrimidine nucleotide.

Synthesis of Purine and Pyrimidine Nucleotides

Purine nucleotides

Most cells can synthesise purine bases *de novo* from smaller precursor molecules. In the *de novo* synthesis of purine nucleotides, bases are not synthesised in the free form but rather the purine ring is assembled piece by piece with all the intermediates attached to ribose-5-phosphate so that, by the time a purine ring is assembled, it is already a nucleotide. (This refers only to the *de novo* synthesis of purines because free purine bases released by degradation of nucleotides are utilised for nucleotide synthesis by the separate salvage pathway to be described later.) The mechanism of ribotidation (the addition of ribose-5-phosphate) is the same in both pathways, as well as in pyrimidine nucleotide synthesis (but the compound that is ribotidated differs in each case, as described below). This brings us to PRPP, the metabolite that is involved in all ribotidation.

PRPP—the ribotidation agent

PRPP is 5-phosphoribosyl-l-pyrophosphate. It is formed from ribose-5-phosphate (produced by the pentose phosphate pathway) by the transfer of a pyrophosphate group from ATP by the enzyme PRPP synthetase.

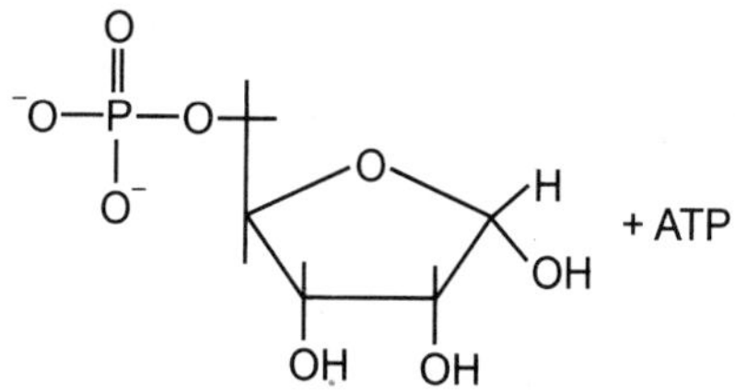

Ribose-5-phosphate

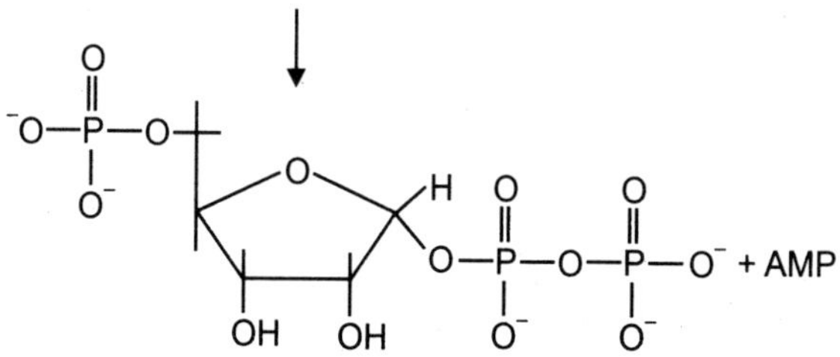

PRPP

The PRPP is an 'activated' form of ribose-5-phosphate; appropriate enzymes can donate the latter to a base forming a nucleotide, and splitting out —P—P. Hydrolysis of the latter to 2Pi drives the reaction thermodynamically.

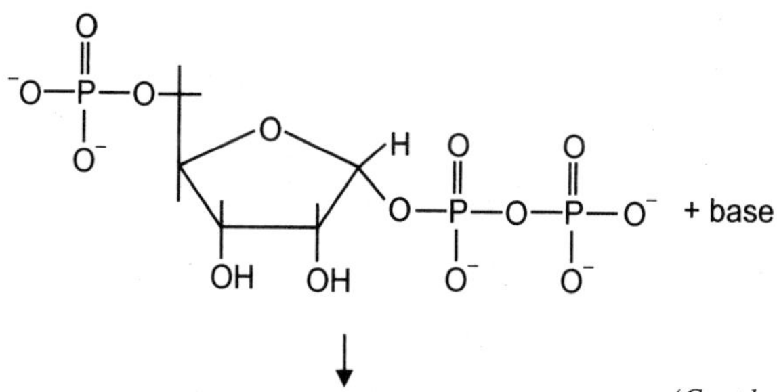

(Contd ...)

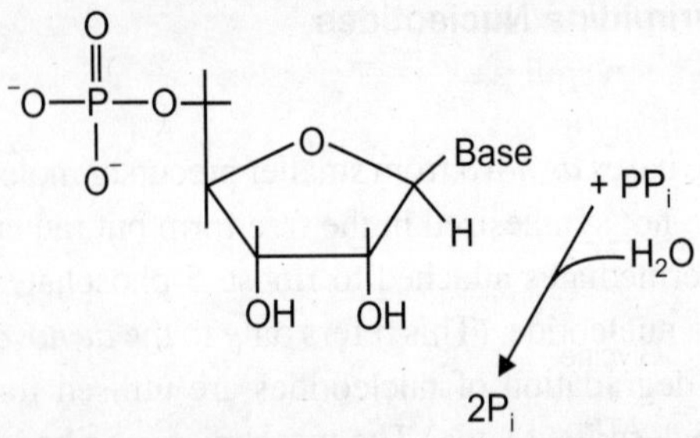

In this reaction the configuration at carbon atom 1 is inverted so that the base is in the required β position. Note that in the *de novo* pathway (see reaction 1 of this pathway in Fig. 10.5) the base is simply —NH_2, derived from glutamine and the product, 5-phosphoribosylamine; in the purine salvage pathway it is a purine. We usually associate the term nucleotide with a purine or pyrimidine base but it can be applied to any base attached in the appropriate manner to the sugar phosphate.

The de novo purine nucleotide synthesis pathway

To return from this general point to the purine *de novo* pathway in particular, after the formation of 5-phosphoribosylamine, there follows a series of nine reactions resulting in the assembly of the first purine nucleotide in which hypoxanthine is the base (Fig. 10.4). This nucleotide is IMP or inosinic acid.

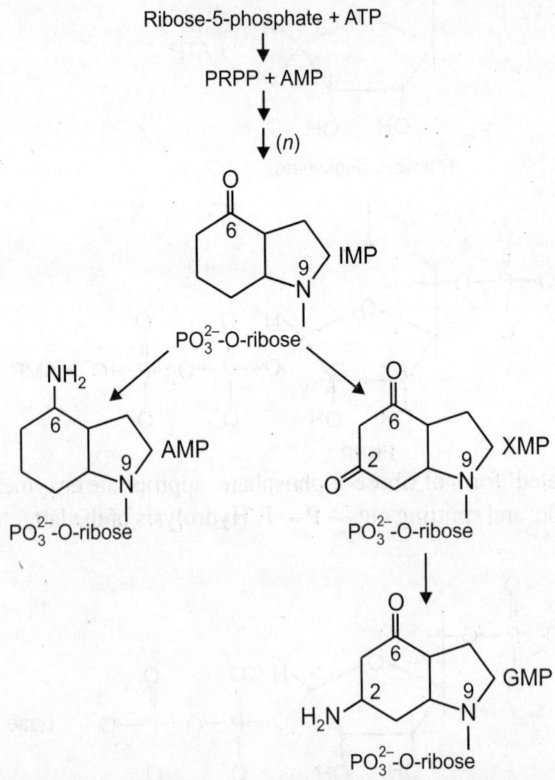

Fig. 10.4. Diagram of the purine *de novo* pathway of GMP, AMP, and XMP synthesis. The base in IMP or inosine monophosphate is hypoxanthine. The base in XMP or xanthosine monophosphate is xanthine. The complete pathway of AMP and GMP synthesis can be seen in Fig. 10.5 and 10.6.

IMP is a branch-point since its hypoxanthine base is converted either to adenine or guanine yielding AMP and GMP, respectively. The overall pathway is summarised in Fig. 10.4, but you should also look at all of the reactions of the pathway as set out in Figs 10.5 and 10.6.

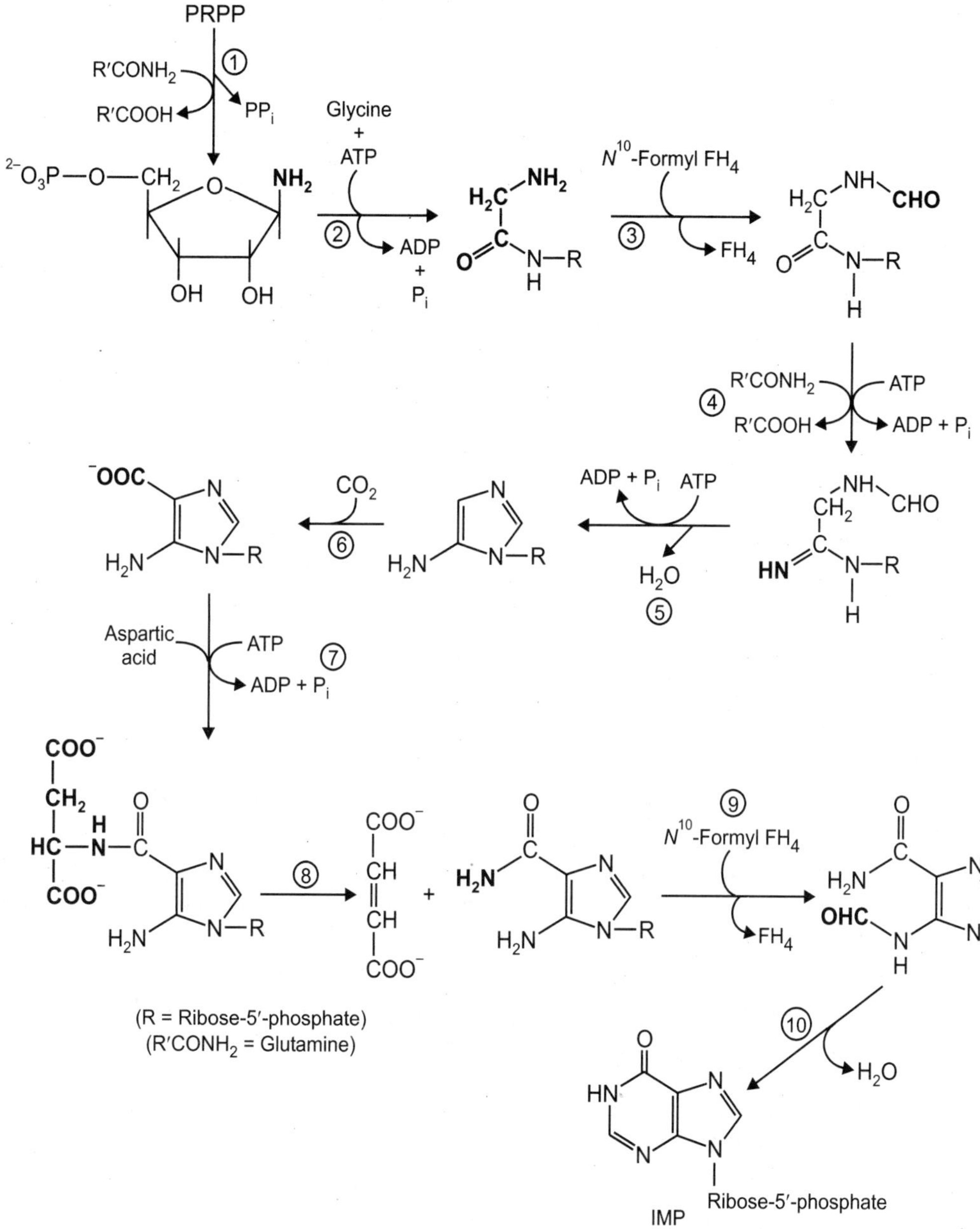

Fig. 10.5. Details of the pathway for the *de novo* synthesis of the purine ring from PRPP to inosinic acid (IMP), given for reference purposes. (The circled reaction numbers are referred to in the text) Bold indicates the structural change resulting from the latest reaction. N^{10}-formyl FH_4, N^{10}-formyltetrahydrofolate.

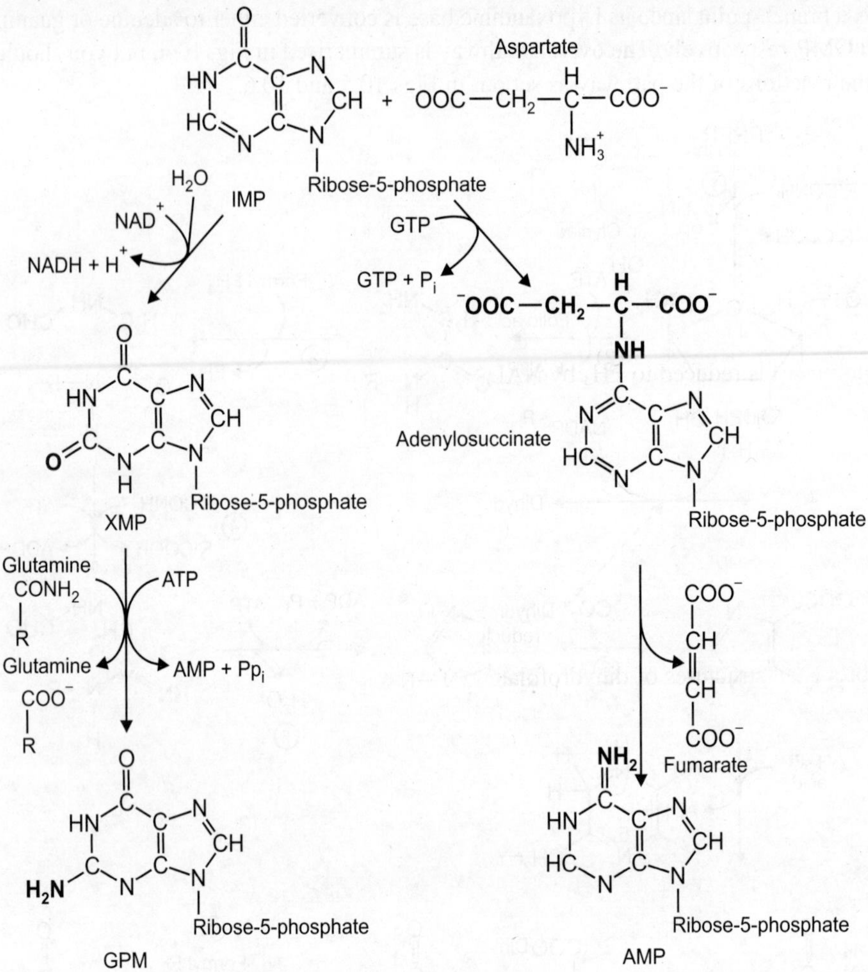

Fig. 10.6. Details of the pathways for the synthesis of GMP and AMP from inosinic acid (IMP), given for reference purposes. Bold shows the change resulting from each reaction. The pathways are summarised in Fig. 10.4. XMP, xanthosine monophosphate.

The daunting *de novo* pathway in Fig. 10.5 is given not for detailed study but so that we can refer to some reactions of specific interest. It will also give you an impression of its remarkable complexity. You will see that six molecules of ATP are consumed in the synthesis of one purine nucleotide molecule. The ATP utilisation refers only to —Ⓟ groups; there is no loss of the adenine nucleotide of ATP so that the pathway results in a net synthesis of AMP.

Reactions 3 and 9 of this pathway are an important class of reaction and have a general interest and we need to divert to deal with these in some detail.

One-carbon transfer reaction in purine nucleotide synthesis

Reactions 3 and 9 of the pathway involve the addition of a formyl (HCO—) group to intermediates in the pathway (Fig. 10.5). The donor molecule in both cases is N^{10}-formyltetrahydrofolate, a molecule not mentioned before. Tetrahydrofolate (FH_4, or sometimes THF) is the carrier in the cell of formyl groups. It is a coenzyme derived from the vitamin folic acid (F) or pteroylglutamic acid. We suggest that you need to learn only the relevant part of this and related structures below (not the whole molecules).

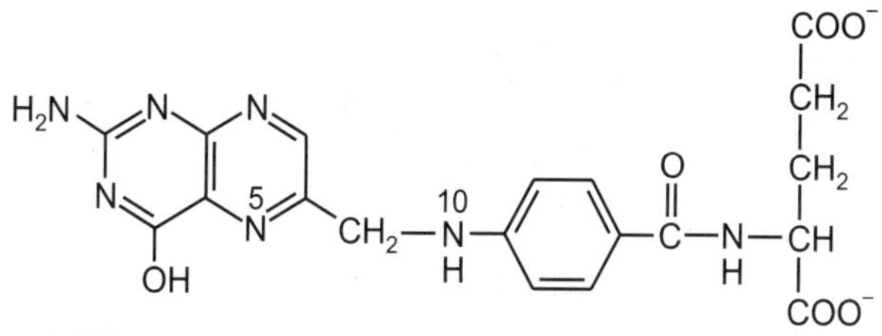

Folic acid (pteroylglutamic acid or F)

The vitamin (F) is reduced to FH_4 by NADPH in two stages.

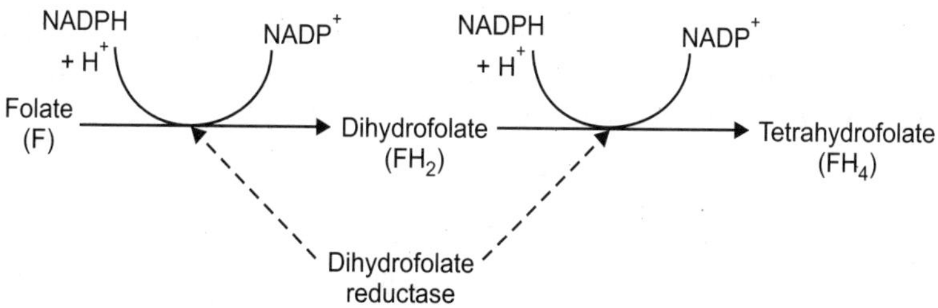

The abbreviated structures of dihydrofolate (FH_2, or sometime DHF) and FH_4 are given below.

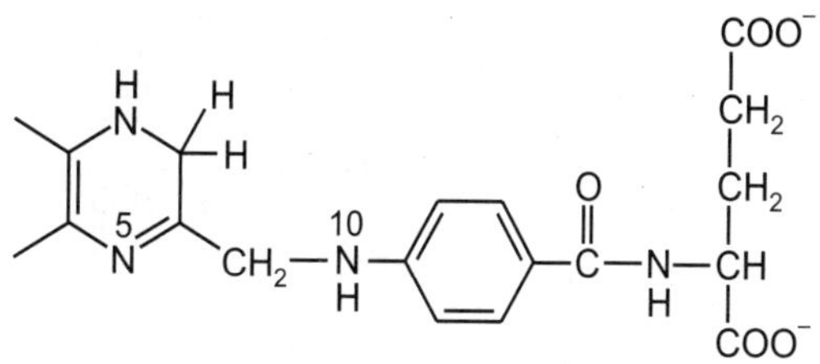

Dihydrofolate (FH_2)

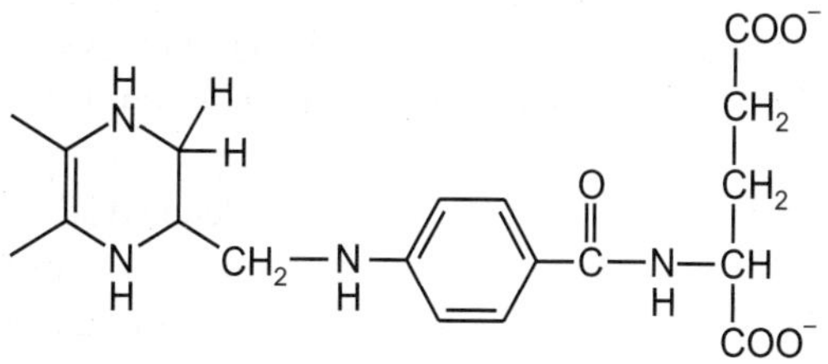

Tetrahydrofolate (FH_2)

Have a look at the structure of FH_4 and notice that the N^5 and N^{10} atoms are placed such that a single carbon atom can neatly bridge the gap between them. For our present purposes we can therefore represent FH_4 as:

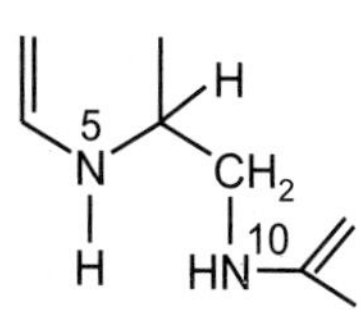

and N^{10}-formyl-FH_4 as:

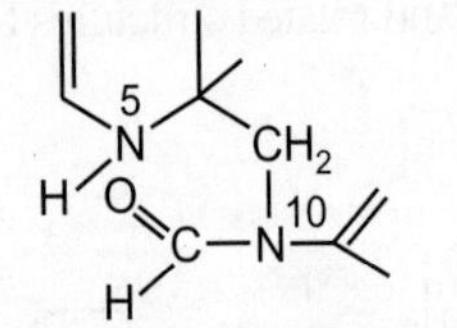

N^{10}-Formyltetrahydrofolate
(N^{10}-formyl FH_4)

This is the donor of the formyl group in reactions 3 and 9 of the purine biosynthesis pathway; specific formyl transferase enzymes catalyse the reactions.

Where does the formyl group in N^{10}-formyl FH_4 come from?

The answer is the amino acid serine (which is readily synthesised from the glycolytic intermediate 3-phosphoglycerate). An enzyme, serine hydroxymethylase, transfers the hydroxymethyl group ($—CH_2OH$) to FH_4 leaving glycine and forming N^5, N^{10}-methylene FH_4.

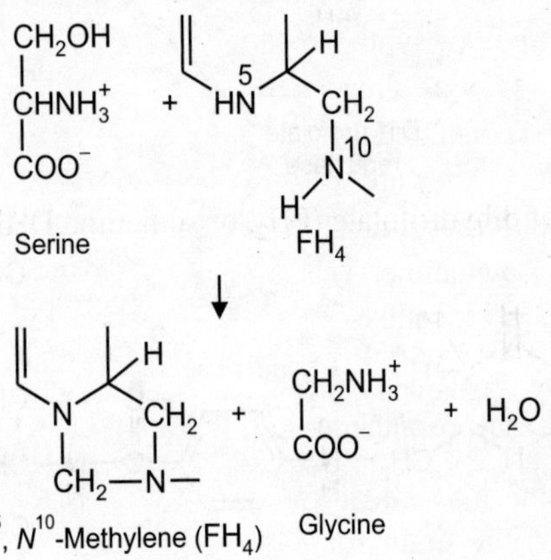

The product, N^5, N^{10}-methylene FH_4 is not quite what we want for formylation because the $—CH_2—$ group is more reduced than a formyl group. It is therefore oxidised by an $NADP^+$-requiring enzyme, forming the methenyl derivative, which is hydrolysed to N^{10}-formyl-FH_4, the formyl group donor.

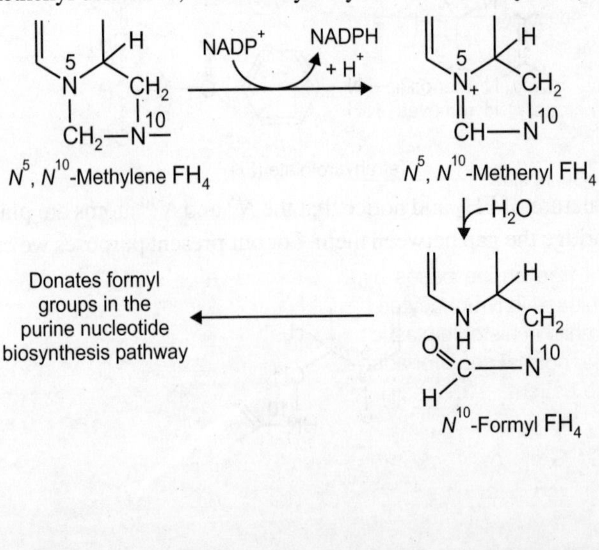

How are ATP and GTP produced from AMP and GMP?

Most of the synthetic reactions of the cell involve nucleoside triphosphates. As discussed later, these are needed for nucleic acid synthesis. It is a simple but especially important concept that enzymes (kinases) exist in the cell to transfer —Ⓟ groups between nucleotides at the high-energy level. There is little free-energy change involved so that —Ⓟ groups can be shuffled around from nucleotide to nucleotide with ease. The main source of —Ⓟ is, of course, ATP, for remember that the energy-generating metabolism constantly regenerates ATP from ADP and P_i. Newly formed AMP and GMP are phosphorylated by kinase enzymes as shown.

$$\text{AMP} + \text{ATP} \rightleftharpoons 2\ \text{ADP} \qquad \text{Adenylate kinase}$$

$$\text{GMP} + \text{ATP} \rightleftharpoons \text{GDP} + \text{ADP} \qquad \text{Guanylate kinase}$$

$$\text{GDP} + \text{ATP} \rightleftharpoons \text{GTP} + \text{ADP} \qquad \text{Nucleoside diphosphate kinase}$$

The nucleoside diphosphate kinase has a wide specificity and can use any pair of nucleoside di- and triphosphates.

Purine salvage pathway

We have emphasised that the *de novo* synthesis of purines does not involve free purine bases—purine nucleotides are produced. However, as already indicated, there is a separate route of purine nucleotide synthesis in which free bases are converted to nucleotides by reaction with PRPP. The free bases originate from degradation of nucleotides—they are salvaged and hence the name of the pathway. Two enzymes are involved—these are phosphoribosyltransferases, one of which forms nucleotides from adenine and the other from hypoxanthine or guanine. The latter enzyme, known as HGPRT (for hypoxanthine-guanine phosphoribosyltransferase), catalyses the reaction

$$\begin{matrix}\text{Guanine or}\\ \text{hypoxanthine}\end{matrix} + \text{PRPP} \longrightarrow \begin{matrix}\text{GMP}\\ \text{or}\\ \text{AMP}\end{matrix} + \text{PP}_i.$$

The enzyme salvaging adenine may be of lesser importance than that dealing with guanine and hypoxanthine in humans, for the main routes of nucleotide breakdown produce the free bases hypoxanthine (from AMP) and guanine (from GMP) as shown in Fig. 10.7.

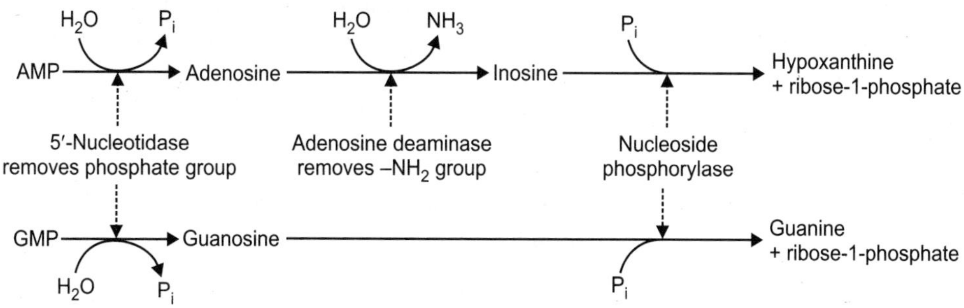

Fig. 10.7. Production of free purine bases hypoxanthine and guanine by nucleotide breakdown. Patients lacking adenosine deaminase in lymphocytes have an immune deficiency that formerly could be treated only by keeping the affected child in a sterile plastic bubble. The disease was the first to be successfully treated by gene therapy in which the normal gene for adenosine deaminase was inserted *in vitro* into bone marrow stem cells and returned to the patient.

Physiological role of the purine salvage pathway

Since purines are energetically expensive to make, a mechanism for re-utilising free purine bases is economical since it can reduce the amount of *de novo* synthesis a cell has to carry out. Moreover, certain cells such as erythrocytes have no *de novo* purine synthesis pathway and must rely on the salvage pathway.

The physiological importance of purine salvage is underlined by the rare genetic disease of infants called the Lesch-Nyhan syndrome, in which the enzyme HGPRT is missing. This results in neurological problems including mental retardation and self-mutilation. Brain possesses the *de novo* pathway only at low levels, so purine nucleotide synthesis is very sensitive to the salvage defect. Lack of the salvage reaction leads to a hepatic overproduction of purine nucleotides by the *de novo* pathway in these patients because the level of PRPP rises (due to lack of utilisation by the salvage reaction) and stimulates the *de novo* pathway. This explains why excessive uric acid production occurs as in gout, which may result in kidney failure caused by the urate crystals. The connection between the biochemical defect and the neurological symptoms is not clear in the Lesch-Nyhan patients. While uric acid overproduction is treatable with allopurinol, this does not relieve the neurological problems. Nor do patients with gout develop the neurological symptoms.

The sources of free purine bases for salvage are probably several. Although the diet contains purines in the form of nucleic acids, most of them are destroyed by epithelial cells of the intestine and not absorbed. By contrast, purines injected into the bloodstream are utilised by cells. The liver is a major site of purine synthesis and some evidence exists that it releases the bases into the blood for use by other cells, such as reticulocytes, that do not have the complete *de novo* pathway. The salvage pathway recycles, within cells, purine bases released from breakdown of nucleic acids.

The recycling of preformed purine bases has the obvious advantage of energy-saving provided, of course, that the *de novo* pathway synthesis is correspondingly reduced. This is achieved in two ways: (i) salvage reduces the level of PRPP and hence of the pathway, and (ii) the AMP and GMP produced by salvage exert feedback inhibition on the pathway (see below).

Formation of uric acid from purines

Nucleotide degradation leads to the production of free hypoxanthine and guanine. Part of this is salvaged back to nucleotides but part is oxidised to produce uric acid (Fig. 10.8). The enzyme, xanthine oxidase, that produces uric acid is present mainly in the liver and intestinal mucosa. Gout is due to a raised level of urate in the blood, leading to the deposition of crystals in tissues. Although gout is traditionally associated with rich living, the main source of uric acid is excess *de novo* production of purine nucleotides due, in some patients, to a high level of PRPP synthetase activity. Also, as described above, deficiency of the HGPRT enzyme leads to the overproduction of purine nucleotides. The drug allopurinol, used in the treatment of gout, mimics the structure of hypoxanthine. It is a potent xanthine oxidase inhibitor. This inhibition results in xanthine and hypoxanthine formation rather than that of uric acid (Fig. 10.8). These products are more water soluble than uric acid and more readily excreted, thus preventing the deposition of insoluble uric acid crystals in tissues that results in the clinical symptoms of gout.

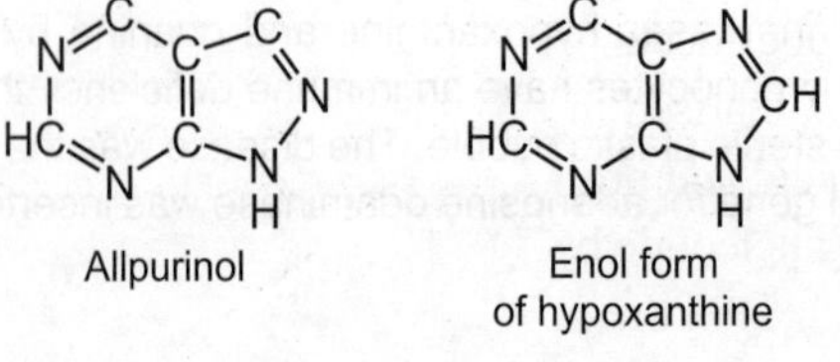

Allpurinol

Enol form of hypoxanthine

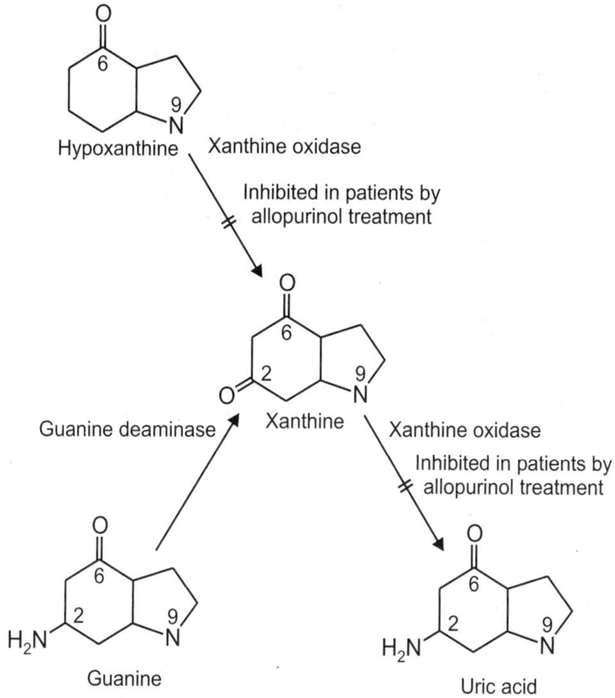

Fig. 10.8. Conversion of hypoxanthine and guanine to uric acid. The drug allopurinol is closely related in structure to hypoxanthine.

Control of purine nucleotide synthesis

As with all metabolic pathways, there must be regulation or chemical anarchy would prevail. The *de novo* pathway is a classical example of allosteric feedback control. The first step of a pathway is a logical place for control. In the *de novo* pathway this is the PRPP synthetase. This enzyme is negatively controlled by AMP, ADP, GMP, and GDP. The next enzyme, which catalyses the first committed step to synthesis of purine nucleotides (reaction 2, Fig. 10.5), is inhibited also, as shown in Fig. 10.9. However, this is not quite the end of the story, because the *de novo* pathway produces IMP, and then the IMP goes in two directions—to AMP and GMP. These latter feedback-control their own production. The regulatory loops serve to ensure a balanced production of ATP and GTP since both are required for nucleic acid synthesis.

Synthesis of pyrimidine nucleotides

Most cells of the body synthesise pyrimidine nucleotides *de novo* but, unlike bacteria, mammals do not appear to have significant pyrimidine salvage pathways for free bases, analogous to those for purines. The nucleoside thymidine, however, is readily phosphorylated to TMP by thymidine kinase and, in that sense, salvage of this nucleoside does occur.

The pyrimidine pathway is summarised in Fig. 10.10 and given in full (for reference purposes) in Fig. 10.11. It starts with aspartic acid and produces a ring structure compound, orotic acid. Orotic acid is converted to the corresponding nucleotide by the PRPP reaction and this is converted to UMP. UTP is produced by kinase enzymes much as in the purine pathway. CTP is produced by amination of UTP.

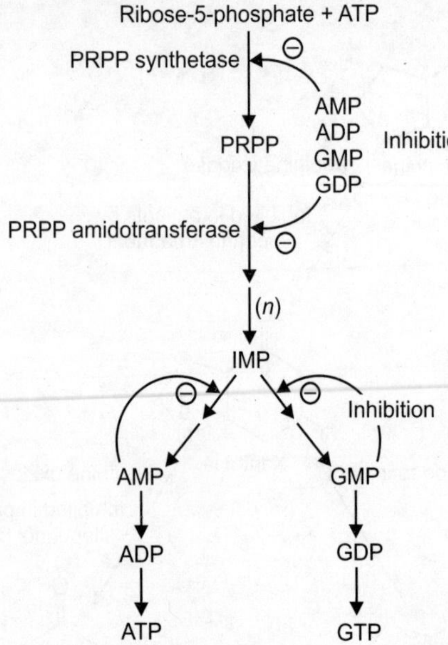

Fig. 10.9. Simplified scheme of the control of the purine nucleotide biosynthesis pathway.

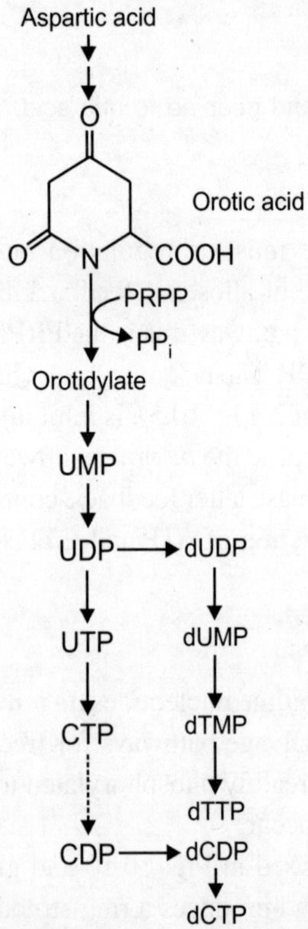

Fig. 10.10. Summary of pyrimidine nucleotide synthesis. (It is assumed here that CTP is converted to COP). The complete pathway is given in Fig. 10.11.

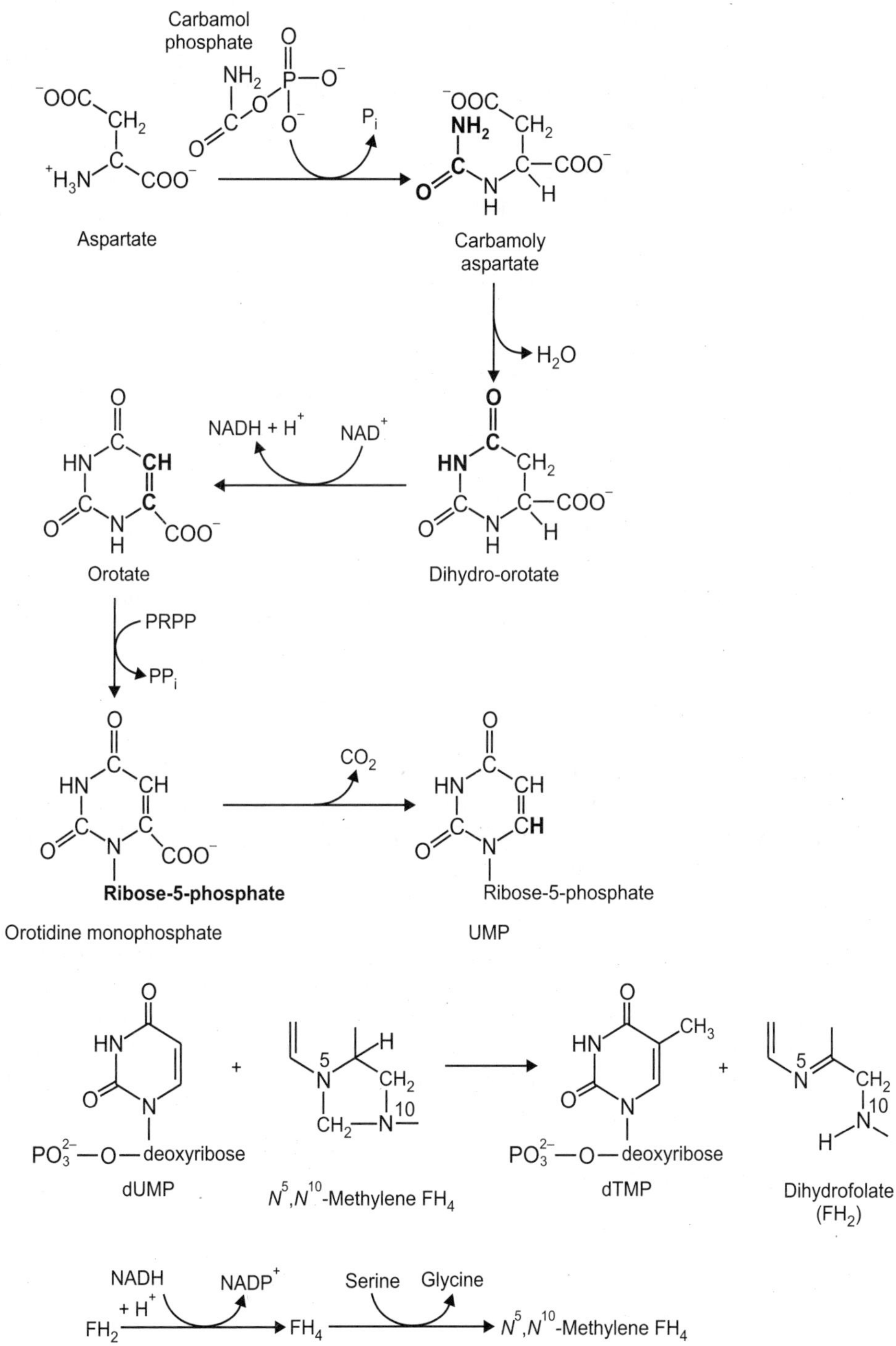

Fig. 10.11. Details of the pathway for the *de novo* synthesis of pyrimidine nucleotides, given for reference purposes Bold indicates the structural change resulting from the latest reaction.

In *E. coli*, control of pyrimidine nucleotide synthesis is mainly at the aspartate transcarbamoylase step. In mammals the pathway is controlled at the carbamoyl phosphate synthase step; it is inhibited by pyrimidine nucleotides and activated by purine nucleotides. The latter control serves to keep the supply of all the nucleotides required for nucleic acid synthesis in balance.

Formation of deoxyribonucleotides

For DNA synthesis dATP, dGTP, dCTP, and dTTP are required. The reduction of ribonucleotides to deoxy compounds occurs at the diphosphate level with NADPH as the reductant (the electrons being transported to the reductase by a complex pathway, not described here).

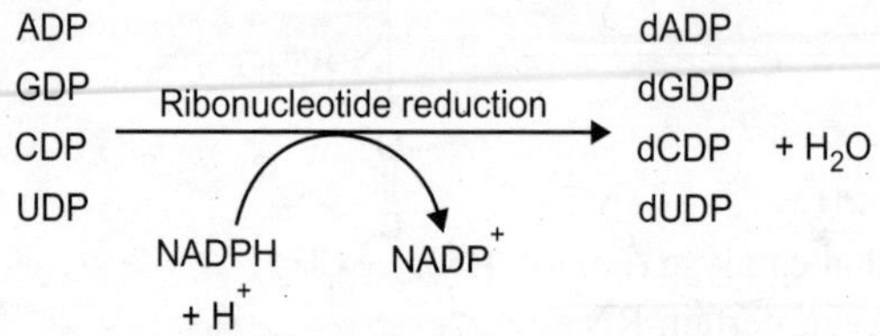

The resultant dADP, dGDP, dCDP, and dUDP are converted to the triphosphates by phosphoryl transfer from ATP.

However, dUTP is not used for DNA synthesis since DNA, you will recall, has thymine (the methylated uracil) as one of its four bases, but never uracil. The dUTP is converted to dTTP. This is done in three steps:

1. dUTP is hydrolysed to dUMP.

$$\text{dUTP} + H_2O \longrightarrow \text{dUMP} + PP_i.$$

2. The dUMP is converted to dTMP.
3. And then dTMP is converted to dTTP by phosphoryl transfer from ATP.

An appropriate system of allosteric feedback controls exists to keep the production of the four deoxynucleotide triphosphates in balance.

Thymidylate synthesis—conversion of dUMP to dTMP

The methylation of dUMP is of especial interest. The enzyme involved is called thymidylate synthase and it utilises N^5, N^{10} methylene FH_4. In purine synthesis, the methylene group of the latter is oxidised to produce a formyl group. In thymidylate synthesis, the methylene group is transferred and, at the same time, reduced to the methyl group of thymine. The reducing equivalents for the reduction come from FH_4 itself, leaving it as FH_2 (note how versatile this coenzyme is). In the Fig. 10.11, only the relevant part of N^5, N^{10}-methylene FH_4 is shown.

The FH_2 produced in this reaction is reconverted to FH_4 by dihydrofolate reductase. The FH_4 is reconverted to methylene FH_4 by reaction with serine.

Chapter 11

Production of Enzymes

INTRODUCTION

Enzymes are biomolecules that catalyse (i.e. increase the rates of) chemical reactions. Nearly all known enzymes are proteins. However, certain RNA molecules can be effective biocatalysts too. These RNA molecules have come to be known as ribozymes. In enzymatic reactions, the molecules at the beginning of the process are called substrates, and the enzyme converts them into different molecules, called the products. Almost all processes in a biological cell need enzymes to occur at significant rates. Since enzymes are selective for their substrates and speed up only a few reactions from among many possibilities, the set of enzymes made in a cell determines which metabolic pathways occur in that cell.

Like all catalysts, enzymes work by lowering the activation energy (E_a or $\Delta G^{\ddagger}$) for a reaction, thus dramatically increasing the rate of the reaction. Most enzyme reaction rates are millions of times faster than those of comparable uncatalysed reactions. As with all catalysts, enzymes are not consumed by the reactions they catalyse, nor do they alter the equilibrium of these reactions. However, enzymes do differ from most other catalysts by being much more specific. Enzymes are known to catalyse about 4000 biochemical reactions. A few RNA molecules called ribozymes catalyse reactions, with an important example being some parts of the ribosome. Synthetic molecules called artificial enzymes also display enzyme-like catalysis.

Enzyme activity can be affected by other molecules. Inhibitors are molecules that decrease enzyme activity; activators are molecules that increase activity. Many drugs and poisons are enzyme inhibitors. Activity is also affected by temperature, chemical environment (e.g. pH), and the concentration of substrate. Some enzymes are used commercially, for example, in the synthesis of antibiotics. In addition, some household products use enzymes to speed up biochemical reactions (e.g. enzymes in biological washing powders break down protein or fat stains on clothes; enzymes in meat tenderisers break down proteins, making the meat easier to chew).

Pectinase is a general term for enzymes that break down pectin, a polysaccharide substrate that is found in the cell walls of plants. One of the most studied and widely used commercial pectinases is polygalacturonase. It is useful because pectin is the jelly-like matrix which helps cement plant cells together and in which other cell wall components, such as cellulose fibrils, are embedded. Therefore pectinase enzymes are commonly used in processes involving the degradation of plant materials, such as speeding up the extraction of fruit juice from fruit, including apples and sapota. Pectinases have also been used in wine production since the 1960s.

They can be extracted from fungi such as *Aspergillus niger*. The fungus produces these enzymes to break down the middle lamella in plants so that it can extract nutrients from the plant tissues and insert fungal hyphae. If pectinase is boiled it is denatured (distorted) making it harder to connect with the pectin at the active site, and produce as much juice.

Pectinases are also used for retting. Addition of chelating agents or pretreatment of the plant material with acid enhance the effect of the enzyme.

As they are enzymes, pectinases have an optimum temperature and pH at which they are most active. For example, a commercial pectinase might typically be activated at 45° to 55°C and work well at a pH of 4.0 to 5.

ENZYME ASSAY

Enzyme assays are laboratory methods for measuring enzymatic activity. They are vital for the study of enzyme kinetics and enzyme inhibition.

Enzyme units: Amounts of enzymes can either be expressed as molar amounts, as with any other chemical, or measured in terms of activity, in enzyme units.

Enzyme activity: Enzyme activity = moles of substrate converted per unit time = rate × reaction volume. Enzyme activity is a measure of the quantity of active enzyme present and is thus dependent on conditions, which should be specified. The SI unit is the katal, 1 katal = 1 mol s^{-1}, but this is an excessively large unit. A more practical and commonly-used value is 1 enzyme unit (EU) = 1 μmol min^{-1} (μ = micro, × 10^{-6}). 1 U corresponds to 16.67 nanokatals.

Specific activity: The specific activity of an enzyme is another common unit. This is the activity of an enzyme per milligram of total protein (expressed in μmol $min^{-1}mg^{-1}$). Specific activity gives a measurement of the purity of the enzyme. It is the amount of product formed by an enzyme in a given amount of time under given conditions per milligram of enzyme. Specific activity is equal to the rate of reaction multiplied by the volume of reaction divided by the mass of enzyme. The SI unit is katal kg^{-1}, but a more practical unit is μmol mg^{-1} min^{-1}. Specific activity is a measure of enzyme processivity, usually constant for a pure enzyme.

Related terminology: The rate of a reaction is the concentration of substrate disappearing (or product produced) per unit time (mol $L^{-1} s^{-1}$). The per cent purity is 100 per cent × (specific activity of enzyme sample/specific activity of pure enzyme). The impure sample has lower specific activity because some of the mass is not actually enzyme. If the specific activity of 100 per cent pure enzyme is known, then an impure sample will have a lower specific activity, allowing purity to be calculated.

Types of Assay

All enzyme assays measure either the consumption of substrate or production of product over time. A large number of different methods of measuring the concentrations of substrates and products exist and many enzymes can be assayed in several different ways. Biochemists usually study enzyme-catalysed reactions using four types of experiments:

1. Initial rate experiments: When an enzyme is mixed with a large excess of the substrate, the enzyme-substrate intermediate builds up in a fast initial transient. Then the reaction achieves a steady-state kinetics in which enzyme substrate intermediates remains approximately constant over time and the reaction rate changes relatively slowly. Rates are measured for a short period after the attainment of the quasi-steady state, typically by monitoring the accumulation of product with time. Because the measurements are carried out for a very short period and because of the large excess of substrate, the approximation free substrate is approximately equal to the initial substrate can be made. The initial rate experiment is the simplest to perform and analyse, being

relatively free from complications such as back-reaction and enzyme degradation. It is therefore by far the most commonly used type of experiment in enzyme kinetics.

2. Progress curve experiments: In these experiments, the kinetic parameters are determined from expressions for the species concentrations as a function of time. The concentration of the substrate or product is recorded in time after the initial fast transient and for a sufficiently long period to allow the reaction to approach equilibrium. We note in passing that, while they are less common now, progress curve experiments were widely used in the early period of enzyme kinetics.
3. Transient kinetics experiments: In these experiments, reaction behaviour is tracked during the initial fast transient as the intermediate reaches the steady-state kinetics period. These experiments are more difficult to perform than either of the above two classes because they require rapid mixing and observation techniques.
4. Relaxation experiments: In these experiments, an equilibrium mixture of enzyme, substrate and product is perturbed, for instance by a temperature, pressure or pH jump, and the return to equilibrium is monitored. The analysis of these experiments requires consideration of the fully reversible reaction. Moreover, relaxation experiments are relatively insensitive to mechanistic details and are thus not typically used for mechanism identification, although they can be under appropriate conditions.

Enzyme assays can be split into two groups according to their sampling method: continuous assays, where the assay gives a continuous reading of activity, and discontinuous assays, where samples are taken, the reaction stopped and then the concentration of substrates/products determined.

Continuous assays

Continuous assays are most convenient, with one assay giving the rate of reaction with no further work necessary. There are many different types of continuous assays.

Spectrophotometric

In spectrophotometric assays, you follow the course of the reaction by measuring a change in how much light the assay solution absorbs. If this light is in the visible region you can actually see a change in the colour of the assay, these are called colourimetric assays. The MTT assay, a redox assay using a tetrazolium dye as substrate is an example of a colourimetric assay.

UV light is often used, since the common coenzymes NADH and NADPH absorb UV light in their reduced forms, but do not in their oxidised forms. An oxidoreductase using NADH as a substrate could therefore be assayed by following the decrease in UV absorbance at a wavelength of 340 nm as it consumes the coenzyme.

Direct versus coupled assays

Even when the enzyme reaction does not result in a change in the absorbance of light, it can still be possible to use a spectrophotometric assay for the enzyme by using a coupled assay. Here, the product of one reaction is used as the substrate of another, easily-detectable reaction.

Fluorometric

Fluorescence is when a molecule emits light of one wavelength after absorbing light of a different wavelength. Fluorometric assays use a difference in the fluorescence of substrate from product to measure the enzyme reaction. These assays are in general much more sensitive than spectrophotometric assays, but can suffer from interference caused by impurities and the instability of many fluorescent compounds when exposed to light.

An example of these assays is again the use of the nucleotide coenzymes NADH and NADPH. Here, the reduced forms are fluorescent and the oxidised forms non-fluorescent. Oxidation reactions can therefore be followed by a decrease in fluorescence and reduction reactions by an increase. Synthetic substrates that release a fluorescent dye in an enzyme-catalysed reaction are also available, such as 4-methylumbelliferyl-β-D-galactoside for assaying β-galactosidase.

Calorimetric

Calorimetry is the measurement of the heat released or absorbed by chemical reactions. These assays are very general, since many reactions involve some change in heat and with use of a microcalorimeter, not much enzyme or substrate is required. These assays can be used to measure reactions that are impossible to assay in any other way.

Chemiluminescent

Chemiluminescence is the emission of light by a chemical reaction. Some enzyme reactions produce light and this can be measured to detect product formation. These types of assay can be extremely sensitive, since the light produced can be captured by photographic film over days or weeks, but can be hard to quantify, because not all the light released by a reaction will be detected.

The detection of horseradish peroxidase by enzymatic chemiluminescence (ECL) is a common method of detecting antibodies in western blotting. Another example is the enzyme luciferase, this is found in fireflies and naturally produces light from its substrate luciferin.

Light scattering

Static light scattering measures the product of weight-averaged molar mass and concentration of macromolecules in solution. Given a fixed total concentration of one or more species over the measurement time, the scattering signal is a direct measure of the weight-averaged molar mass of the solution, which will vary as complexes form or dissociate. Hence the measurement quantifies the stoichiometry of the complexes as well as kinetics. Light scattering assays of protein kinetics is a very general technique that does not require an enzyme.

Discontinuous assays

Discontinuous assays are when samples are taken from an enzyme reaction at intervals and the amount of product production or substrate consumption is measured in these samples.

Radiometric

Radiometric assays measure the incorporation of radioactivity into substrates or its release from substrates. The radioactive isotopes most frequently used in these assays are ^{14}C, ^{32}P, ^{35}S and ^{125}I. Since radioactive isotopes can allow the specific labelling of a single atom of a substrate, these assays are both extremely sensitive and specific. They are frequently used in biochemistry and are often the only way of measuring a specific reaction in crude extracts (the complex mixtures of enzymes produced when you lyse cells). Radioactivity is usually measured in these procedures using a scintillation counter.

Chromatographic

Chromatographic assays measure product formation by separating the reaction mixture into its components by chromatography. This is usually done by high-performance liquid chromatography (HPLC), but can also use the simpler technique of thin layer chromatography. Although this approach can need a lot of material, its sensitivity can be increased by labelling the substrates/products with a radioactive or fluorescent tag. Assay sensitivity has also been increased by switching protocols to improved chromatographic instruments (e.g. ultra-high pressure liquid chromatography) that operate at pump pressure a few-fold higher than HPLC instruments.

Factors to Control in Assays

1. Salt concentration: Most enzymes cannot tolerate extremely high salt concentrations. The ions interfere with the weak ionic bonds of proteins. Typical enzymes are active in salt concentrations of 1–500 mm. As usual there are exceptions such as the halophilic (salt loving) algae and bacteria.
2. Effects of temperature: All enzymes work within a range of temperature specific to the organism. Increases in temperature generally lead to increases in reaction rates. There is a limit to the increase because higher temperatures lead to a sharp decrease in reaction rates. This is due to the denaturating (alteration) of protein structure resulting from the breakdown of the weak ionic and hydrogen bonding that stabilise the three dimensional structure of the enzyme. The 'optimum' temperature for human enzymes is usually between 35°C and 40°C. The average temperature for humans is 37°C. Human enzymes start to denature quickly at temperatures above 40°C. Enzymes from thermophilic archaea found in the hot springs are stable up to 100°C. However, the idea of an 'optimum' rate of an enzyme reaction is misleading, as the rate observed at any temperature is the product of two rates, the reaction rate and the denaturation rate. If you were to use an assay measuring activity for one second, it would give high activity at high temperatures, however, if you were to use an assay measuring product formation over an hour, it would give you low activity at these temperatures.

Effects of pH

Most enzymes are sensitive to pH and have specific ranges of activity. All have an optimum pH. The pH can stop enzyme activity by denaturating (altering) the three dimensional shape of the enzyme by breaking ionic, and hydrogen bonds. Most enzymes function between a pH of 6 and 8; however, pepsin in the stomach works best at a pH of 2 and trypsin at a pH of 8.

Substrate saturation

Increasing the substrate concentration increases the rate of reaction (enzyme activity). However, enzyme saturation limits reaction rates. An enzyme is saturated when the active sites of all the molecules are occupied most of the time. At the saturation point, the reaction will not speed up, no matter how much additional substrate is added.

Level of crowding

Large amounts of macromolecules in a solution will alter the rates and equilibrium constants of enzyme reactions, through an effect called macromolecular crowding.

LARGE SCALE PRODUCTION:FERMENTATION

Sources of Enzymes

Industrial enzymes are produced from plants, animals, and micro-organisms, but manufacture from the first two groups is limited for several reasons. Cultivation of plants is restricted to areas where climate is suitable. It is generally seasonal, impeding steady enzyme production. As the concentration of enzymes in plant tissues is generally low, processing of large amounts of plant material is necessary. Enzymes of animal origin are by-products of the meat industry and for this reason limited in supply. Moreover, they often compete with other end-users for the supply of suitable glands.

In contrast, microbial enzymes can be produced in amounts meeting all demands of the market. Seasonal fluctuations of raw material do not count and there are possibilities for genetic and environmental manipulation of bacteria and fungi to give increased yields of desired enzymes in a way not possible with higher organisms. Moreover, the diversity of enzymes available from micro-organisms is very great. Lastly, microbial enzymes present a wide spectrum of characteristics that makes them utilisable for quite specific applications.

Selection of Micro-organisms

The first step in the manufacture of an enzyme involves the selection of an organism suitable to produce the desired enzyme in amounts as large as possible. The general aspects of this procedure can be outlined as follows:

1. Extracellular enzymes are preferred, because difficult and costly methods of cell disruption are not necessary. As compared with intracellular enzymes, they are present in a relatively pure form in the culture liquor. Intracellular enzymes are industrially used to a lesser extent because of difficult procedures of cell disruption and separation of contaminating cell components.
2. High yields of enzymes should be obtained with an economical time required for culture production.
3. The strain must be stable with respect to productivity, requirement for culture conditions, and sporulation.
4. The organism should be able to grow on cheap substrates.
5. Synthetic activity should be as far as possible in the direction of the desired enzyme. Formation of interfering by-products should be low.
6. Clarification of the culture liquor or extract should be possible without difficulties.
7. The strain must not produce toxic substances and should be free of antibiotic activities. It should not belong to related strains that synthesise toxins.

Mostly, enzymes with particular properties, e.g. with respect to stability and activity, are desired. This requires special screening programs. As has been demonstrated, the technique of screening is also influenced by the method of cultivation of the organisms at the industrial production stage. This is particularly valid for fungi. If the submerged culture technique is employed, an appropriate selection of strains should be done at an early stage of the program. It is worth mentioning that shaken cultures behave differently from deep cultures. For this reason shaken culture screens only permit a first rough selection. Further selection in deep culture is necessary.

Detection of mutants with increased productivity is difficult. Methods based on checking for halo formation are possible with extracellular enzymes but often worthless for industrial purposes. Therefore, attempts have been undertaken to find out correlations between production of a particular enzyme and one, or several, distinct physiological or morphological characteristics. For example, Nasuno correlated formation of *Aspergillus oryzae*-type alkaline protease with smooth conidia and production of *Aspergillus sojae*-type alkaline protease with echinulate or tuberculate conidia. This correlation was valid for a large number of *Aspergillus* species tested.

Mechanisms of Enzyme Biosynthesis

Metabolism is principally regulated by a change of the rate of enzyme reactions. Therefore, regulation of metabolism is mainly a problem of kinetics. As outlined previously, the rate of most enzyme reactions can be described by the Michaelis-Menten equation:

$$v = \frac{k_{-2}[E_o][S]}{K_M - [S]}$$

where, [S] (substrate concentration), [E_o] (total enzyme concentration), k_{-2} (rate constant), and K_M (Michaelis-Menten constant) are independent variables. The enzyme concentration is varied by two mechanisms, namely, by controlled protein synthesis and controlled protein degradation.

For control of enzyme synthesis the microbial cells bring into action the mechanisms of induction and repression. Since Jacob and Monod outlined their theory of enzyme regulation, these mechanisms have become increasingly understandable, and practical applications have led to sometimes drastic increases of enzyme production by environmental and genetic manipulations.

Inducible enzymes

There are only few enzymes synthesised in substantial concentration under all conditions of growth. These 'constitutive' enzymes include, for example, the enzymes of the hexose-monophosphate pathway. Many of the enzymes used commercially fall into the inducible group. Their biosynthesis requires the presence of substrate in the medium. For example, starch acts as an inducer for amylase. Dextrin, a degradation product of starch, was found to give 16 per cent higher amylase production than did starch in *Bacillus polymyxa*, whereas maltose induced the synthesis of only 50 per cent of the activity induced by starch. Often inducers are analogs or derivatives of the substrate, e.g. isopropyl-β-D-thiogalactoside for β-galactosidase. In other cases compounds structurally similar to the substrate may serve as inducer, e.g. sophorose for cellulase. For polymer substrates which cannot enter the microbial cell, it has been shown that the dimer is the true inducer, such as cellobiose for cellulase. However, the dimers are active as inducers only when they are present in very low concentrations. At higher levels catabolite repression occurs. Thus, it appears that the inductive effects of the polymers result from their slow hydrolysis to dimers which are consumed by the organism as rapidly as they are formed. The same result can be achieved when slowly metabolisable derivatives of the dimers are used, e.g. sucrose monopalmitate for invertase synthesis in yeasts and moulds. Table 11.1 is a representation of inducers active toward commercial enzymes.

Table 11.1. Inducible enzymes.

Enzyme	*Organism*	*Inducer*
α-Amylase	*Bacillus* spp.	Starch, dextrin, maltose
Catalase	*Aspergillus niger*	H_2O_2, O_2
	Candida tropicalis	H_2O_2, O_2, hydrocarbons
Cellulase	*Pestalotiopsis westerdijkii*	Cellulose, cellobiose, cellobiose octoacetate
	Trichoderma lignorum	Lactose, sucrose monopalmitate
	Trichoderma viride	Cellulose, cellobiose, cellobiose tripalmitate, sophorose
Glucose isomerase	*Bacillus coagulans*	D-Xylose
Glucose oxidase	*Aspergillus niger*	Glucose, sucrose
	Penicillium vitale	Sucrose
Invertase	*Pullularia pullulans*	Sucrose, sucrose monopalmitate
Lactase	*Aspergillus nidulans*	Lactose
	Escherichia coli	Lactose, isopropyl-β-D-thiogalactoside
Lipase	*Candida cylindracea*	Tripalmitin
	Candida lipolytica	Sorbitan monooleate
	Candida paralipolytica	Cholesterol

(Contd ...)

Enzyme	*Organism*	*Inducer*
Pectinmethylesterase	*Rhizopus stolonifer*	Pectin
Pectin *trans*-eliminase	*Rhizoctonia solani*	Citrus pectin
Polygalacturonase	*Aspergillus niger*	Pectin
	Rhizoctonia solani	Na-polypectlaite
Pullulanase	*Aerobacter* sp.	Maltose

Repression mechanisms

Feedback repression: Repression mechanisms are of the feedback or of the catabolite type. Feedback repression means that biosynthesis of an enzyme is inhibited when end products of a pathway are accumulated or added to the growth medium. For example, protease production in many bacilli is repressed in certain amino acid containing media, and protease formation by *Aspergillus niger* is apparently sensitive to repression by sulphur-containing amino acids.

Catabolite repression

Carbon catabolite repression, or simply catabolite repression, is an important part of global control system of various bacteria and other micro-organisms. Catabolite repression allows bacteria to adapt quickly to a preferred (rapidly metabolisable) carbon and energy source first. This is usually achieved through inhibition of synthesis of enzymes involved in catabolism of carbon sources other than the preferred one. The catabolite repression was first shown to be intiated by glucose and therefore sometimes referred to as the glucose effect. However, the term 'glucose effect' is actually a misnomer since other carbon sources are known to induce catabolite repression. Catabolite repression was extensively studied in *Escherichia coli*. *E. coli* grows faster on glucose than on any other carbon source. For example, if *E. coli* is placed on agar plate containing only glucose and lactose, the bacteria will use glucose first and lactose second. When glucose is available in the environment, the synthesis of β-galactosidase is under repression due to the effect of catabolite repression caused by glucose. The catabolite repression in this case is achieved through the utilisation of phosphotransferase system.

An important enzyme from the phosphotranferase system called Enzyme II A plays a central role in this mechanism. Enzyme II A is specific for glucose transport only. When glucose levels are high inside the bacteria, Enzyme II A mostly dwells in its unphosphorylated form. This leads to inhibition of adenylyl cyclase and lactose permease, therefore cAMP levels are low and lactose cannot be transported inside the bacteria. After some time, the glucose is all used up and the second preferred carbon source (i.e. lactose) has to be used by bacteria. Absence of glucose will 'turn off' catabolite repression.

Furthermore, when glucose levels are low the phosphorylated form of Enzyme II A accumulates and consequently activates the enzyme adenylyl cyclase, which will produce high levels of cAMP. cAMP binds to catabolite activator protein (CAP) and together they will bind to a promoter sequence on the lac operon. However, this is not enough for the lactose genes to be transcribed. Lactose must be present inside the cell to remove the lactose repressor from the operator sequence (transcriptional regulation). When these two conditions are satisfied, it means for the bacteria that glucose is absent and lactose is available. Next, bacteria start to transcribe lactose gene and produce β-galactosidase enzymes for lactose metabolism. The example above is a simplification of a complex process. Catabolite repression is considered to be a part of global control system and therefore it affects more genes rather than just lactose gene transcription. Some examples of repressible enzymes are shown in Table 11.2.

Table 11.2. Catabolite repression-sensitive enzymes.

Enzyme	*Organism*	*Repressor*
α-Amylase	*Bacillus stearothermophilus*	Fructose
Amyloglucosidase	*Endomycopsis bispora*	Glucose, maltose, starch, glycerol
Catalase	*Rhodotorula mucilaginosa*	Glucose
Cellulase	*Trichoderma viride*	Glucose, cellobiose, starch, glycerol
C_1-Cellulase	*Trichoderma* sp.	Glucose, sucrose
C_x-Cellulase	*Rhizoctonia solani*	Glucose, cellobiose
Glucose isomerase	*Streptomyces phaeochromogenes*	Glucose
Invertase	*Aspergillus nidulans*	Glucose
Lactase	*Escherichia coli*	Glucose
Pectin *trans*-eliminase	*Penicillium expansum*	Glucose, arabinose, mannose, galactose, sucrose, raffinose, galacturonic acid
Polygalacturonic acid *trans*-eliminase	*Aeromonas liquefaciens*	Glucose, polygalacturonic acid
Polygalacturonase	*Aspergillus niger*, *Penicillium expansum*,	Glucose, galacturonic acid
	Rhizoctonia solani	Glucose, cellobiose
Protease	*Bacillus megaterium*, *B. subtilis*, *Candida lipolytica*	Glucose
Alkaline protease	*Aspergillus nidulans*, *Neurospora crossa*	Low molecular wt. sources of C, N, and S
Neutral protease	*Aspergillus nidulans*, *Neurospora crossa*	Low molecular wt. sources of C, N, and S

Cyclic AMP

Investigations by Perlman and Pastan indicate that inhibition of cyclic 3′,5′-adenosine monophosphate formation holds a key position in catabolite repression. In *Escherichia coli*, its intracellular concentration is depressed 1000-fold by growth on glucose, whereas the addition of this nucleotide reverses catabolite repression of many enzymes.

Manipulation of Enzyme Biosynthesis

A number of methods are available to overcome anyone of the control mechanisms which may exert an inhibiting effect on the production of large amounts of a given enzyme. These techniques can be divided into two main categories: manipulation of the genetic function of the organism, and manipulation of the environment of the organism.

The methods of genetic manipulation include the classical techniques of mutant formation and a class of novel techniques which is often termed 'genetic engineering.' While mutant formation is quite usual in enzyme manufacture, genetic engineering techniques are restricted to research laboratories, unless easier handling permits their introduction into industrial practice.

There are two ways in which mutations can cause overproduction of enzymes. The first one is concerned with an alteration in the regulation mechanisms. Such mutational events effect removal of inducer requirement, resistance to end product repression, and resistance to catabolite repression. The second group of mutations leads to an increase in copies of the gene responsible for the production of the enzyme.

Genetic engineering means transfer of genes from one strain to another. Terms such as 'plasmid transfer,' 'phage escape synthesis,' etc. may be cited to characterise the methods employed.

Manipulations of the environment enable the biochemical engineer to overcome inhibition of enzyme biosynthesis as caused by regulatory mechanisms, by selection of suitable medium composition, or culture conditions. In the following, methods of genetic and environmental manipulations, which lead to substantial increases in enzyme production, are enumerated.

For inducible enzymes the application of two methods is possible: (i) mutation to constitutivity, or (ii) incorporation of inducers into the medium. Often the most potent inducers are nonmetabolisable substrate analogs. For industrial practice it is important for a particularly expensive or not readily available inducer to be successfully replaced by compounds which can be converted by the organism to the required inducer (Table 11.1).

End product repression of enzyme biosynthesis can be avoided by several means:

1. Avoiding presence of end products as medium constituents. For instance, protease production by *Aspergillus niger* is derepressed under conditions of sulphate limitation.
2. Limitation of end product accumulation is generally possible by adding an inhibitor of the pathway to the medium. External accumulation can further be limited by using mixed cultures with a second organism that metabolises the repressive substance and by applying dialysis or ultrafiltration techniques in the fermenter system. Internal build-up of end product co-repressors can be limited by starving an auxotroph mutant of the end product required for growth. There are several means: limited feeding of the end product, using slowly utilised derivatives of the required end product, growing partial auxotrophs ('leaky mutants') in the absence of their end product requirement.
3. Selection of regulatory mutants which are not repressed by end products (constitutive mutants). A commonly used method is selection for resistance to a toxic analog of the end product, but other methods are also available.

As mentioned, catabolite repression is of great importance since many enzymes produced in industry are subject to this type of regulation. Catabolite repression can be avoided by the following means:

1. Avoidance of the use of repressing carbon sources in the medium. For example, replacement of fructose by glycerol increases α-amylase production of *Bacillus stearothermophilus* more than 25-fold.
2. Derepression of the enzyme synthesis by growth limitation. Suitable means are slow feeding of the repressive substrate and use of slowly metabolisable analogs or derivatives of the substrate. For example, application of sucrose monopalmitate instead of sucrose was found to increase invertase production 80-fold: Cultivation at lower temperatures, addition of toxic substances to the medium, etc. are further methods of growth restriction.
3. Mutation to resistance against catabolite repression may increase productivity considerably. For example, a yeast mutant has been obtained that produced 2 per cent of its cellular protein in the form of invertase.

Kinetics of Enzyme Biosynthesis

Rates of fermentation processes are desirable to know for both engineering and fundamental scientific reasons. Bioengineers are concerned with microbial dynamics from the point of view of process design and optimisation, building on the experience in the classical chemical industry. Bioscientists, on the other hand, are interested in the dynamic response of micro-organisms as a tool for gaining insight into the mechanisms of microbial physiology.

However, biochemical processes are extremely complex and sensitive to a number of factors and rendering their mathematical modelling is most difficult. The available models for predicting the causal effects of changes of control variables are not simple and not accurate. This is the reason why commercial fermentation processes have not been significantly optimised in an engineering sense.

Excellent contributions to the development of kinetic models of enzyme formation stem from Terui and his associates. The presented models refer to commercially produced hydrolases. They are based on the assumptions that the rate-limiting ability of the enzyme-forming system (EFS) corresponds to mRNA and that the specific rate of enzyme production (ε, units $\cdot$ mg^{-1} $\cdot$ hr^{-1}) is proportional to the quantity per cell of mRNA (r, quantity' mg^{-1}), i.e. $\varepsilon \propto a\mu$.

For growth-associated enzyme production ($\varepsilon = a\mu$) the following hypothetical relationship was proposed:

$$\frac{d\varepsilon}{dt} = a\mu - b\frac{d\mu}{dt} - k$$

where, μ is the specific growth rate (hr^{-1}), k the monomolecular decay rate constant of the specific mRNA (hr^{-1}), and a and b are the system constants. The second term on the right-hand side of the equation is based on the negative correlation of the change of E with that of μ. It expresses the rate of growth-associated repression exerted at the level of transcription. The third term represents the decay rate of mRNA or EFS.

The preceding model has been shown to be in accord with the actual fermentation processes for the production of α-amylase by *Bacillus subtilis*, and amyloglucosidase, acid protease, and polygalacturonase by *Aspergillus niger*. In other cases, such as production of amyloglucosidase, acid protease, polygalacturonase, and C_x-cellulase by *Aspergillus niger* and C_x-cellulase by *Penicillium variabile*, where enzyme formation in the stationary growth phase plays a major role in enzyme accumulation, another model was proposed. It concerns remaining mRNA formed in the preceding growing phase and turnover of RNA to mRNA in the nongrowing phase. This model has the form

$$\varepsilon = \varepsilon_m e^{-k(t-t_m)} + K_1(e^{-\lambda(t-t_m)} - e^{-k(t-t_m)})$$

where, ε_m is the maximum rate of enzyme production at time t_m, when growth has just ceased; λ is the monomolecular decay constant for cell RNA; and K_1 is a system constant. To comment upon the terms on the right-hand side of the equation: The first term represents ε due to the mRNA carried over from the growing phase and the second term is the change of ε due to the turnover synthesis and degradation of the mRNA.

Cultivation Techniques

Solid substrate cultivation

This method plays an important role in commercial enzyme production from fungal sources, especially in Japan. Advantages and disadvantages of this method, as compared with the submerged culture technique, have often been analysed. Considering modern deep bed processes, the following advantages can be stated as a matter of fact: (i) enzyme yield per unit volume of incubator is high, (ii) power requirement is low, (iii) minimum control is necessary, (iv) extraction yields highly concentrated enzyme solutions, (v) only small equipment for enzyme recovery is required due to the small amounts of extracts obtained, and (vi) scaling-up is easy.

Problems which can be solved, if need be, may be listed as follows: (i) continuous operation is possible, (ii) feeding substrates during cultivation is possible, and (iii) defined media can be applied by using suitable inert carriers. Caused by the nature of the complex media used, the extracts contain considerable amounts of fungal pigments, the removal of which is difficult and costly. Their formation might be avoided by using the previously mentioned 'solidified' synthetic media.

The methods of solid substrate cultivation can be divided into two groups: thin layer and deep bed processes.

The thin layer techniques, also called tray processes, work with substrate layers of 2 to 4 cm height spread on wooden or metallic trays. These are incubated in air-conditioned rooms or cabinets (hence cabinet method). Figure 11.1 shows a flow sheet of the tray cultivation. Usually the heat produced by the growing culture is removed by moistened cool air which passes over the surfaces of the trays or is pressed through the bran mass. Kalashnlkov recommended a water-cooling system for the trays.

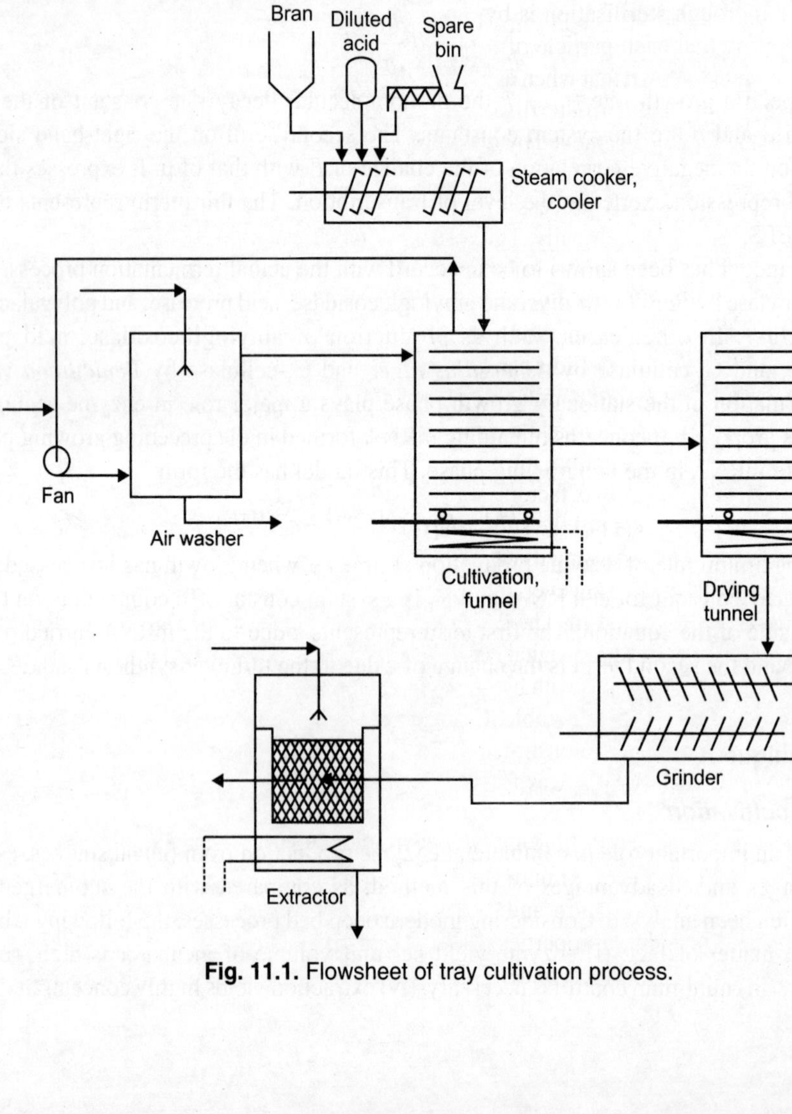

Fig. 11.1. Flowsheet of tray cultivation process.

The deep bed process, developed by Terui and co-workers in order to meet the enormous demand for enzymes needed for the traditional soyabean fermentations, uses substrate layers usually as deep as 0.6 m (2 ft), but beds as deep as 1.5–1.8 m (5–6 ft) have also been reported. The dimension of rectangular beds is of the order of 5.5 × 61 m (18 × 200 ft). Circular beds are also operated. In general, the equipment used in deep bed processes is quite similar to that known in the malting industry. As mentioned above, the deep bed plants are fully automated.

Media used in solid substrate fermentations are mainly based on wheat bran. This material is particularly suitable because of its high content of nutrients and its large surface. Other basic substrates are rice bran and soyabean or sweet potato flakes, as well as grains or soyabeans, etc. Kernels should be selected for optimal size or cracked to give particles of the desired size. On bran, superficial growth of the fungus is sufficient for utilisation of nutrients. Kernels, however, must permit the inoculum to penetrate. This is not difficult with polished rice, but corn or soyabeans must be cracked or freed of the hulls. The amount of water needed for moistening the substrates is in the range of 40 to 70 per cent. In the case of grains and soyabeans the optimum, content of moisture is about 30 per cent.

Pressure sterilisation of large masses of wet bran in bulk presents serious problems. A convenient method to ensure thorough sterilisation is by direct steam injection into the bran. During this process the mass is agitated so that each particle of the moist bran is in constant direct contact with the steam. Experience, however, has shown that when acidic solutions are employed in place of water for moistening the bran it was quite sufficient to sterilise the bran at 95°C for 15 to 30 minutes. In some cases decontamination of the bran was achieved by means of bactericides, e.g. formaldehyde or β-propiolactone.

Inoculation of the sterilised medium is carried out by use of spores in a dry or suspended form. The amount of inoculum varies from process to process depending on a number of factors. The actual amount must be determined empirically. Underkofler found that even as low an inoculation ratio as 0.04 per cent of dry spore culture was satisfactory in fungal amylase production. Attempts had been undertaken to use an inoculum in the mycelial form which can be easily produced in large quantities by submerged culture. However, this method is not convenient, resulting in nonuniform growth of the fungus throughout the bran mass. Conidia can be produced in large quantities in special cultures, whereby the method of fermentation is quite similar to that applied in enzyme production. In order to promote spore formation it may be beneficial to add a balanced solution of trace elements (Fe^{3+}, Zn^{++}, C^{++}, Mn^{++}).

Among the conditions of incubation, moisture and temperature present the most serious problems. It was soon recognised that these two factors are dependent variables. Growing cultures produce heat which tends to evaporate the water of the substrate. This effect is intensified by warming the air when it passes the culture and thereby enlarging its capacity to dissolve water vapour. These processes inevitably cause a drying out of the culture. However, the water lost is partially replaced by water that is formed by the metabolic activity of the organism. The difference must be made up by application of sprayed water (this method, of course, is unfit for nonagitated cultures). The microbiological approach to this problem is the selection of slowly growing strains with high productivity. Short-time fermentations may also be helpful.

Heat production is, indeed, considerable. Its magnitude can be readily appraised by the loss in dry weight of the culture. It has been demonstrated that almost all this loss is due to the oxidation of organic compounds to CO_2. In many cases it was found that approximately half of the dry weight of the bran disappears.

Kalashnikov observed with *Aspergillus niger* that under industrial conditions up to 380 $J \cdot h^{-1} \cdot kg^{-1}$ of fermented bran were liberated at maximum heat production. Removal of this amount of heat required 20 m^3 air with 20°–28°C temperature and 100 per cent moisture. The problem of heat removal is complicated by the heat-insulating properties of the bran. Attempts to minimise this problem by diluting the bran through incorporation of inert materials, such as grain husks, only cause other problems, e.g. a greater requirement of space. Regarding the moisture content of the medium, it has been found in many cultures that careful observation of the tolerated limits has a decisive influence on the production of the enzyme. In this respect it is very important to keep in mind the previously mentioned formation of water by metabolic processes. Initial pH and the course of the pH during the development of the culture also play a great role. Several means are known to influence the pH development. For instance, this can be done by incorporation of suitable inorganic or organic salts into the medium.

Submerged cultivation techniques

Treatment of this subject may be limited to some considerations of enzyme production. The fermentation equipment used for the large-scale production of enzymes is the same as that used in the production of other microbial metabolites. As far as is known, processes are batch-operated; however, attempts have been made to introduce continuous processes to enzyme production. Continuous fermentation may be employed if the optimal conditions for the process are known. For inducible enzymes, however, optimal conditions for cell growth are often different from those required for induction and synthesis of the enzymes. In such cases a 2-stage, continuous culture system can be used efficiently for production of the enzyme. By using this method, the growth stage can be operated with an optimal set of culture conditions. It may be useful to insert a washing of the cells between the first and the second stage in order to decrease the requirement of inducer and to produce less impure preparations. Two-stage, continuous culture processes can also be used when reduction or elimination of catabolite repression due to a high medium concentration is desired.

Addition of a substrate to a batch process under controlled conditions is a well-known technique in fermentation technology for prolonging growth and increasing product accumulation. Since it permits prolonged maintenance of a constant environment with respect to the added substrate, the system is more similar to a continuous process than to a batch fermentation. In many cases drastic increases in enzyme production result from the use of extended culture.

Quite similar to solid substrate fermentations, the pH also plays an important role in submerged cultivations. Generally, the culture starts with a certain pH which depends on the strain employed and the enzyme desired. The course of pH during the fermentation is often manipulated by addition of suitable agents. In some processes the pH is 'fixed' at a certain level, within limits. This is the case in the production of amyloglucosidase from *Aerobacter aerogenes*. In a number of batch processes it is convenient to change the pH because of differences in optimum pH for growth and enzyme formation. For example, in the case of neutral metalloproteinase from a *Bacillus* species, the pH is allowed to change 'naturally.' The course of pH may, however, be programmed. As an example, the acid stable α-amylase of *Aspergillus niger* may be mentioned.

ENZYME RECOVERY

In enzyme production there is a very unfavourable ratio between input of raw material and output of product. This requires the installation of concentration procedures. For economic reasons of enzyme application a concentration up to 10-fold is usually satisfactory for industrial enzyme preparations. For example, enzyme products employed in detergents contain about 5–10 per cent protease while amylase preparations for use in flour treatment contain only about 0.1 per cent pure α-amylase. However, in applications where high purity enzymes are required, e.g. in enzymie analysis, 1000-fold purification is quite common.

In some applications, such as baking and dextrose manufacture, the presence of contaminating enzymes must be very low or rigidly controlled. Moreover, the raw enzyme solutions obtained from microbial cultures contain—independent of their source—different types of by-products. Separation of all these substances may be necessary because of the possibility of undesired effects. Considering enzyme stability there is another reason for treatment of crude enzyme preparations. Since the trend in enzyme applications is toward use of liquid preparations, stabilisation is an important procedure.

Figure 11.2 is a presentation of some treatments which are used in the preparation of enzymes on a commercial scale. Techniques for the large-scale isolation and (partial) purification of enzymes from microbial sources make use mainly of traditional procedures. Most of the equipment can be found in food-processing plants. Large-scale equipment specific for enzyme isolation is not marketed.

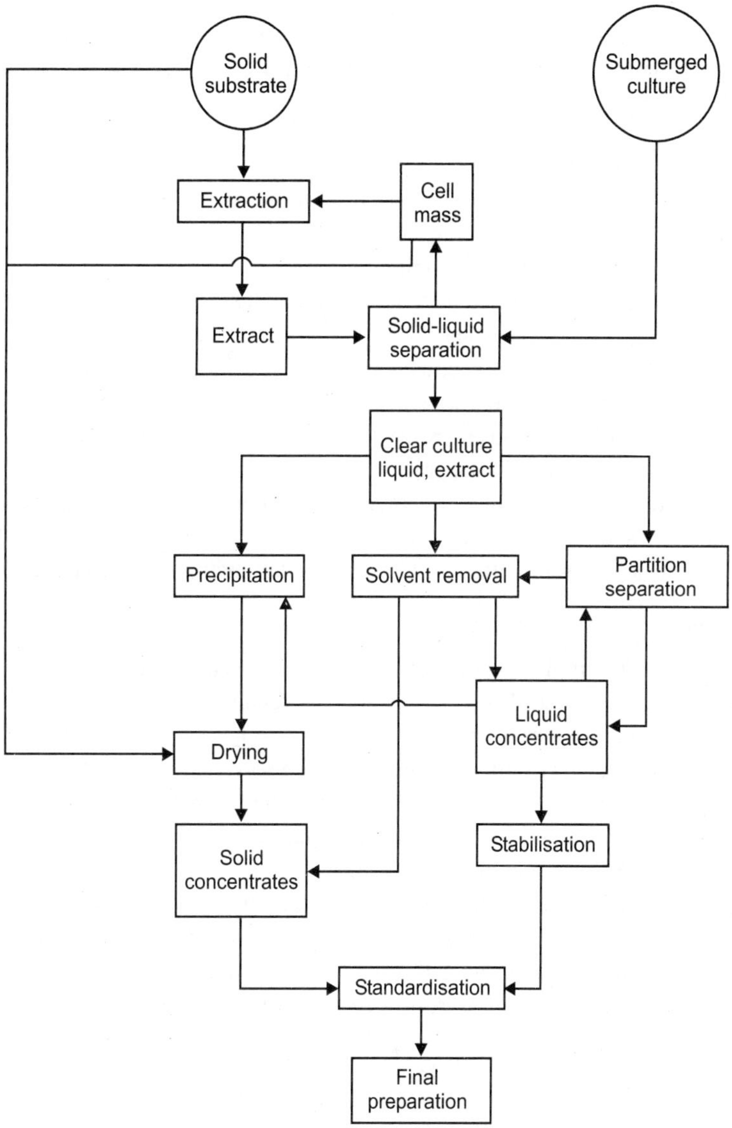

Fig. 11.2. Concentration and purification of enzymes.

Nearly all process operations are carried out at low temperatures (preferably 0°–10°C), with the exception of drying. Separation processes are usually conducted in batches rather than continuously. However, the scale-up of batch operations inherently causes extended processing times which for many enzymes result in increased losses of activity due to denaturation of the enzyme protein. For this reason the application of continuous operations seems to be useful, but the necessity for highly reliable machines and ingenious process control delays introduction of continuous methods. In addition the value of continuous processing is lost when a single process step is conducted batchwise, perhaps during precipitation.

Extraction Methods

The first step in the isolation of enzymes is their extraction. Techniques that fall into this group are employed either to separate enzymes from solid substrate culture or to release enzymes from the interior of microbial cells.

Extraction of solid substrate cultures

Enzymes produced by solid substrate cultivation used to be of the extracellular type. It is therefore easily conceived that extraction of mould brans is rather a washing out process. Countercurrent techniques of percolation are the most frequently used unit operation.

In many cases the mould bran is dried prior to extraction. This is convenient when the utilisation of the particular enzyme preparation is seasonal. The cultures can be produced in relatively small equipment all the year round, while the extraction is conducted in times of enzyme demand. On the other hand, it is easily seen that extraction from dried bran will yield solutions with higher enzyme concentrations. And last, drying avoids interference caused by the activity of living cells of fresh cultures. This argument, however may not apply in continuously operated culture plants.

In all cases the extractant is water which, however, may contain acids (inorganic or organic), salts, buffer, or other substances to facilitate solubilisation of the enzyme or to improve its stability in solution, or to exclude or minimise undesired effects caused by contaminating by-products or micro-organisms.

Extraction of cells

The decision on whether to employ whole cells for a biochemical process or to use isolated enzymes depends on many factors. Technical difficulties and the related cost of large-scale isolation play an important role. There are a number of methods for cell disruption, as reviewed by Hughes. Chemical and biochemical methods, such as autolysis, treatment with solvents, detergents, or lytic enzymes, have the disadvantage of being in principle batch operations. Their conduct is difficult to standardise and optimise. More recommendable are mechanical techniques.

At present, the APV-Manton-Gaulin homogeniser seems to be the most versatile type for cell disintegration. In this machine the cell suspension passes a homogenising valve at the selected operating pressure and impinges on an impact ring. The strong shearing forces combined with the sudden decompression lead to a disruption of the cell wall. Dunnill and Lilly, who examined the disruption of yeast, found that release of protein can be described by a first order rate equation:

$$\log \frac{R_m}{R_m - R} = KNP^{2.9}$$

where R is the amount of soluble protein released in g per kg cell mass, R_m the maximum amount of soluble protein released, K a temperature-dependent rate constant, N the number of times the cell suspension has passed the homogeniser, and P the operating pressure. With industrial models, between 50 and 9000 litres of bacterial suspension per hr can be treated, depending on the size of the machine. Ball mills available on the market have a volume capacity of 0.6–250 litres.

PURIFICATION OF ENZYMES

Preliminary Purification Procedures

As mentioned already in this chapter, a preliminary purification step is usually included in the procedure used to extract an enzyme. For example, some degree of cell fractionation is often carried out before extraction, so that only enzymes in the same subcellular compartment as the relevant enzyme are extracted. Also, the composition of the extraction medium is chosen so that the enzyme of interest is soluble but many others are not.

Extracted nucleic acids may be precipitated by treatment with basic substances such as streptomycin or protamine, or with $MnCl_2$ or $MgCl_2$.

All precipitates and cell debris are then removed by centrifugation and discarded. Polysaccharides may also be removed by high-speed centrifugation, although the results are not always satisfactory. An alternative way to dispose of polysaccharides is to break them down into small units with enzymes such as α-amylase; nucleic acids may similarly be broken down by the use of nuclease enzymes.

The next stage of purification is usually to precipitate the enzyme of interest from solution, thus separating it from mono- and oligosaccharides, nucleotides, free amino acids, etc. and from many other proteins which remain in solution. This may be achieved by altering the pH and the organic or salt concentrations of the medium. For example, the pH may be adjusted to the isoelectric point of the enzyme. Alternatively a high salt concentration can be introduced: ammonium sulphate is often used for this purpose, because it is extremely soluble in water. The salt concentration is usually increased in stages, the aim being to precipitate other proteins (which are then discarded) before precipitating the fraction containing the relevant enzyme. This is then redissolved in water, and residual ammonium sulphate removed by dialysis.

Although a considerable degree of purification can be achieved by these procedures, many other proteins will still be present because of overlap of solubility ranges. One possible way to remove some of these is to raise the temperature of the medium for a few minutes to a value where the enzymes being purified is known to be stable, but others might be denatured and precipitate from solution. Enzymes are often particularly stable in the presence of their substrates, so the relevant substrate can be added to increase the effectiveness of this procedure.

Further Purification Procedures

The crude extract, partially purified, may be treated in a variety of ways to increase the purification of the relevant enzyme.

Adsorbent gels

Adsorbent gels (e.g. zinc hydroxide) have been used to remove pigments from enzyme preparations. A mixture of enzymes may also be adsorbed by a suitable gel (e.g. aluminium hydroxide) and then fractionated by elution with buffers of increasing ionic strength. Partition chromatography has similarly been used to separate mixtures of enzymes: for example, ribonuclease was purified by Martin and Porter on columns packed with kieselguhr, using a two-phase system consisting of ammonium sulphate, ethyl cellosolve and water; the sample was applied to a column which had been pre-equilibrated with the organic phase, and then eluted by the aqueous phase, separation of components being on the basis of relative solubility in the two phases. Such techniques played an important part in the development of enzymology, but have now been superseded by others.

Electrophoresis

Electrophoresis is mainly an analytical procedure, since it is ideally suited to the separation of small amounts of material, but it has also been used for the purification of proteins. The rate and direction of migration of a protein in an electric field depends on its net-charge at the pH used and also on the size of the molecule, since this imposes some resistance to movement. In zone electrophoresis, a mixture of proteins is introduced at a common point (the origin) and allowed to move in an electric field, usually in a horizontal direction, each molecule travelling in the same zone as others of the same charge and size. To minimise diffusion the whole process is usually carried out in a solid but porous support medium (e.g. starch gel) through which the buffer and proteins permeate. When separation has been achieved, as indicated by cutting a narrow longitudinal section from the support medium and staining for proteins (e.g. with amido black) then the rest of the support medium may be cut into strips across the direction of travel and the contents of each extracted by elution. Alternatively, zone electrophoresis may be carried out in a vertical direction in a closed column with, for example, powdered cellulose as the support medium; after electrophoresis has been completed, a tap at the bottom of the column is opened to enable the liquid contents to be run out and collected in fractions.

Some forms of electrophoresis capable of giving extremely high resolution, e.g. isotachophoresis and isoelectric focusing, which are employed at present mainly for analytical purposes, are likely to find increasing use in preparative work.

Cation exchange chromatography

Cation exchange chromatography on columns packed with carboxylated polystyrene has been used in the successful purification of stable, low molecular weight basic proteins such as lysozyme and ribonuclease. However, cellulosic ion exchange chromatography has proved of more general application to the separation of proteins. Such resins, in contrast to most other ion exchange resins, are relatively hydrophilic and have an open structure readily penetrated by large molecules such as proteins. The most frequently used are diethylaminoethyl (DEAE)-cellulose which is an anion exchanger, and carboxymethyl (CM) cellulose, a cation exchanger. The DEAE group, $-OC_2H_5NH(C_2H_5)_2$, is highly positively charged at pH 6–8, so DEAE-cellulose is most useful for the chromatography of proteins which are negatively charged in this range; similarly, CM-cellulose, cellulose–$OCH_2CO_2^-$, is most applicable for the separation of proteins which are positively charged at around pH 4.5.

Elution of proteins from the columns may be brought about by changes in either salt concentration or pH: as the concentration of salt (e.g. NaCl) increases, protein is displaced from DEAE-cellulose by the anion (Cl^-) and from CM-cellulose by the cation (Na^+); if the pH is altered over the relatively narrow working range, proteins are eluted as their isoelectric point is reached, since they then have no net charge with which to bind to the resin. Ion exchange chromatography has been used in the purification of membrane proteins in non-ionic or zwitterionic detergents as well as for the purification of soluble enzymes. The buffering capacity of ion exchangers may be exploited to create a linear pH gradient in the column and separate proteins by chromatofocusing. This technique has features in common with isoelectric focusing.

Gel filtration

Gel filtration (size exclusion or molecular-sieve chromatography), another important technique in enzyme purification, is carried out with columns packed with swollen gels which separate components of a sample on the basis of molecular size: molecules too large to enter the pores of the gels will pass quickly through the columns; smaller molecules will pass through the column more slowly, the actual speed for each component being dependent on the ease with which its molecules can pass into the gels and be thus retarded. Gels commonly used include cross-linked dextrans (Sephadex), cross-linked agarose (Sepharose) and cross-linked polyacrylamide (Biogel). These are graded according to pore size, and thus to the size (and hence molecular weight) of protein molecule which can enter. For a successful separation, the enzyme of interest must not be too large to penetrate the gel, because it would then pass straight through the column together with all other proteins of similar and greater molecular size.

Eluting buffers should be of high ionic strength (e.g. 20 mM) to counteract the few charges which may be present on the gel: apart from that, the only criterion for the buffer is that the proteins are stable in it: no pH or salt gradient is employed, since the separation is by size alone. However, it should be noted that some of these gels are available with attached ion exchange groups (an example is DEAE-Sephadex) in which case the principles of elutions are as for cellulosic ion exchange chromatography.

In the case of membrane proteins solubilised in detergents, gel filtration is widely used to separate protein-detergent micelles from excess detergent: this is particularly easy to achieve if detergent molecules are kept in the monomeric form, e.g. by ensuring that the detergent concentration is less than the CMC, or by adding a small amount of a bile salt.

Affinity chromatography

Affinity chromatography is a bio-specific process and thus ideally suited for the separation of one protein from all others, including those which resemble it so closely in physical characteristics that separation by any other procedure is extremely difficult. The column is packed with an inert matrix (e.g. agarose) to which ligands for the required enzyme have been attached. These immobilised ligands may be, for example, substrate analogues, and must be covalently linked to the matrix in such a way that they can still bind to their enzyme, if this is present: this usually means that a spacer arm (e.g. a hydrocarbon chain) must be included between matrix and ligand, joining to the ligand at a point which does not interact with the enzyme, if steric hindrance to the formation of an enzyme-ligand complex is to be avoided. The attachment of the spacer arms to –OH groups of the matrix is often achieved by the use of CNBr reagent.

The protein mixture is applied to the column, and the relevant enzyme is trapped by the immobilised ligands while all other proteins pass through and are discarded. The enzyme is then liberated from the column either by eluting with a deforming buffer at a pH which changes the characteristics of the enzyme and no longer allows it to bind to the immobilised ligand, or by the use of a competitive counter-ligand, which displaces the immobilised ligand on the enzyme.

In both cases, the enzyme passes through the column and can be collected, now free of other proteins. An example of an enzyme extracted by the use of such a technique is β-galactosidase with the substrate-analogue *p*-aminophenyl β-D-galactoside as ligand, the deforming buffer being 0.1 M borate at pH 10.

Immunoaffinity chromatography

A closely related procedure is immunoaffinity chromatography, where the immobilised ligand is an antibody specific for the enzyme which is being purified. The use of an immobilised monoclonal antibody as ligand gives even greater specificity than use of a more conventional antibody.

The bio-specific nature of these techniques means that a high degree of purification can be achieved; on the other hand, it also means that a polymer-ligand complex prepared for the purification of one enzyme is limited to that particular application.

Another rapidly developing form of affinity chromatography involves the use of reactive triazine-based dyes, e.g. Cibacron Blue F3G-A and dyes of the Procion series, as ligand. Under alkaline conditions the chlorotriazine dye is linked, usually directly, to a matrix such as agarose by a triazine bond; the reaction involves a hydroxyl group on the matrix and a chloride on the dye. Such immobilised dyes bind a wide range of NAD- and NADP-dependent dehydrogenases and other enzymes. They are, of course, less specific than biological group-specific ligands, and their specificity is difficult to predict, but they have many advantages: they have a greater protein-binding capacity, they are extremely resistant to chemical and enzymic degradation, they can be used and reused in a variety of applications, and the triazine bond is more stable than the isouronium linkage introduced by the CNBr activation of agarose. Elution of enzyme from the ligand is usually achieved by changing pH or increasing salt concentration.

Even less specific, though still useful, is hydrophobic interaction chromatography. This technique depends on the fact that although amino acids with polar side chains predominate at the surface of enzymes, some non-polar ones will be present which can form hydrophobic interactions with column packings such as phenyl-Sepharose, particularly when the ionic strength is high. Elution can be brought about by decreasing the ionic strength or introducing an organic solvent.

A major advance over the past few years has been the application of HPLC technology to protein purification. The use of stainless steel columns and high-quality robust packing material of particle size 10 μm or less has allowed separations to be obtained similar to those for conventional low-pressure chromatography but in minutes rather than hours. HPLC columns for preparative work are wider than those for analytical use (at least 20 mm internal diameter), which allows milligram or even gram amounts of protein to be applied; a fraction collector is usually included in the system (e.g. the Waters Delta Prep 3000 system) after the detector. High-performance size-exclusion chromatography (HPSEC) is based on the same principle as conventional gel filtration, but Sephadex, Sepharose and other non-rigid or semi-rigid gels can only withstand low pressures. Instead, HPSEC often utilises rigid spherical beads of porous silica with bonded hydrophilic polar groups. High-performance ion exchange chromatography (HPIEC) utilises amines as anion exchangers and sulphonic or carboxylic acids as cation exchangers, each bonded to some rigid support such as silica.

Proteins may also be separated by reverse-phase HPLC on alkylsilica columns, the eluting solvents being buffered aqueous and organic mixtures.

When after the use of one or more of these procedures it is considered that the relevant enzyme has been separated completely from other proteins mineral salts and any small molecules present may be removed by dialysis, if this is desired.

The purified enzyme preparation is likely to be quite dilute, so it might require concentrating before being used for the purpose for which it was prepared. One convenient method for doing this is lyophilisation, followed by the redissolving of the enzyme in a small volume of liquid. Another suitable procedure is ultrafiltration: this involves forcing the solvent molecules through a membrane of chosen porosity (e.g. by the application of nitrogen gas pressure) and thus separating them from the protein molecules, these being too large to pass through the pores.

The ideal way to complete a purification procedure is to crystallise the enzyme, if this is possible. Sometimes it may be achieved by adding not quite enough ammonium sulphate to cause precipitation of the enzyme and leaving this solution in a coldroom for several days. In general, successful conditions have to be determined empirically by attempting crystallisation at a variety of salt and organic solvent concentrations and pH values, and by showing much patience.

Membrane enzymes may be inactive after purification unless re-introduced into a phospholipid environment. For example, phospholipid may be added to purified protein-detergent micelles and concentrations adjusted so that the phospholipid replaces detergent in the micelles; released detergent, in the monomeric form, may then be removed by gel filtration, dialysis or ultrafiltration. An example of a membrane enzyme that has been characterised is Na^+–K^+ ATPase.

Criteria of Purity

At each stage of a purification procedure, an assay for the enzyme being purified should be performed on all fractions and its specific activity in each fraction determined. This will show, irrespective of preconceived ideas, precisely which fractions contain the important enzyme and will enable the degree of purification to be calculated. The total activity of this enzyme will be unchanged as a result of the purification step, unless some has been lost during purification; under no circumstances can it be increased. However, some of the contaminating proteins which were originally present in the same preparation as the enzyme should now be in different fractions. Hence, if the fractions containing the relevant enzyme are combined, the total activity of the enzyme in the combined fraction should be the same as that started with, whereas the total amount of protein in this fraction will be less than that originally present. The specific activity of the enzyme in the combined fraction should, therefore, be greater than that in the preparation before purification, and the increase in specific activity will be a measure of the purification achieved. With each successive purification step, the specific activity of the fraction (or fractions) containing the enzyme should be greater than before until complete purification is achieved and the specific activity reaches a limiting value.

However, the finding of the same specific activity value before and after a purification step does not necessarily mean that the enzyme preparation is completely pure: it could simply mean that contaminating proteins have passed through the procedure in the same fraction as the enzyme. Similarly, crystallisation cannot be taken as proof that only one protein is present, for many mixed protein crystals have been found. Hence other criteria of purity have to be considered. If a particular enzyme is known to be a possible contaminant, then a logical step is to carry out an assay for that enzyme and demonstrate its absence in that way. However, that would not exclude the possibility that other contaminants were present.

Conversion to Storage Form

Storability of an enzyme requires the preparation of a suitable storage form. Commercial enzyme products are available either in solution or in solid state. Generally, users prefer solutions because of their easier handling, but enzymes are usually very unstable in aqueous solution. For this reason stabilisation of dissolved enzymes is a very important step in the manufacture of liquid enzyme preparations. The storage stability is affected by the following two factors: microbial deterioration of the enzyme solution and denaturation of the enzyme protein. These two problems seem to be closely related to each other.

Many treatments have been tried in order to prevent growth of micro-organisms. The methods include, for instance, incorporation of chemical preservatives, pasteurisation, addition of salts and polyhydric alcohols, and irradiation. But some of those treatments are undesirable due to legal aspects. Therefore, the most suitable method to repress microbial growth is to dissolve the enzyme in a highly concentrated solution of salts and sugars.

With liquid preparations, storage at low temperatures and at suitable pH is essentially inevitable. It is well known that substrates almost invariably protect the corresponding enzyme against physical, chemical, or physico-chemical agents. This can be attributed to either conformational stabilisation or steric or competitive protection. From a number of publications it can be seen that almost any effect on enzyme stability may *a priori* be an allosteric one due to attack at sites other than the active site of the enzyme. For example, in thermolysin, a bacterial protease, Ca-ions stabilise the enzyme molecule, while Zn-ions are required for activity. The enormous value of Ca-ions in stabilising bacterial α-amylase has long been known.

A number of techniques are available for stabilisation. Some of them are presented in the following list, immobilisation methods excluded:

1. Conformational or charge stabilisation and/or protection from dilution-dissociation by using buffers, glycerol, substrates, or inhibitors.
2. Protection of active site thiol via disulphide exchange by thiols, redox dyes, oxygen-binding agents, or chelating agents.
3. Miscellaneous methods include, e.g. inhibition or removal of proteolytic enzymes; protection from light by photosensitive dyes; lowering activity of water by viscosity effectors, salts, or sugars; lowering surface energy by antifoams; cooling and crystallisation protection by antifreeze; removal of harmful agents; and sterilisation for protection against microbial attack.

Commercially available solid enzyme preparations are dried mould brans, dried precipitates, or dried solutions. Spray drying is the preferred method for removal of water from enzyme solutions due to economic reasons. However, it is only applicable to enzymes sufficiently resistant to the temperatore conditions of this process. On the other hand, freeze drying is most preserving, but its use is limited by cost considerations as well as by the fact that unless the salt concentrations of the enzyme solution are sufficiently reduced, eutectic mixtures may be formed. This may lead to incomplete drying or to severe foaming and protein denaturation. A specific method of drying sometimes used is granulation in a fluidised bed with milk sugar or maltodextrin as carrier. In this case, of course, sufficiently high specific activity of the enzyme is required in order to ensure satisfactory activity of the commercial preparation.

EXTREMOZYMES

Many enzymes function optimally under extreme conditions of temperature, pH etc. they are called extremozymes. For example, about 30 enzymes have been isolated from hyperthermophile (optimal growth at or above 100°C), sulphur-reducing bacteria; their optimal temperatures for catalysis are above 100°C and some of them have half-life of several days at the optimum temperature. Such enzymes are called high temperature enzymes. A hyperthermophile organism may be defined as an organism that grows at 90°C and above with optimum growth at 80°C or above. Only two of the 20 odd hyperthermophile genera known so far are conventional bacteria, the remaining being classified as Archaea (formerly Archaebacteria). In the 1970s, Archaea were recognised as a distinct kingdom of lif. Hyperthermophiles are the most ancient forms of life. Almost all hyperthermophiles are strict anaerobes and strict organotrophs that use complex organic mixtures as sources of C and N. Most of them must reduce elemental sulphur to hydrogen sulphide for optimal growth.

Several enzymes have been isolated from hyperthermophiles, e.g. protease, amylase, α-glucosidase, hydrogenase, glutamate dehydrogenase, DNA polymerase etc. The interest in these enzymes arises because they may, be used for catalysis under high temperatures which in some cases may offer unique advantages. An example is provided by hyperthermostable DNA polymerases; these are essential for polymerase chain seac (PCR). So far, three high temperature DNA polymerases are commercially available. The enzyme obtained from *Pyrococcus furiosus* takes 20 hrs at 95°C to lose 50 per cent of catalytic activity, while that from *Thermococcus litoralis* takes 7 hrs. This very high thermal stability of these DNA polymerases is very useful for PCR work since such enzymes need to be added only at the stare of PCR, and there is no decline in the enzyme activity with the progression of PCR cycles.

NONTRADITIONAL ENZYMES

A great majority of the enzymes are highly evolved extremely specific proteins. However, two other forms of biological catalysts are now known; these are: (i) abzymes, and (ii) ribozymes.

Abzymes

Abzymes are antibodies that catalyse specific chemical reactions, i.e. function as enzymes. Antibodies, by definition, have evolved to recognise and bind to the ground states of the molecules they are specific to. In contrast, enzymes have binding sites that preferentially bind to the transition state of their substrate molecules. For this reason, abzymes are not produced naturally. A catalytic antibody is produced in response to molecules that have a structure similar to the proposed/expected transition state of the reaction to catalyse which the antibody is sought. The fact that such catalytic antibodies are in fact produced is the strongest evidence in support of the transition theory of enzyme action.

Ribozymes

RNA molecules that have the capability of catalysing chemical reactions are called ribozymes. Ribozymes are true catalysts in that they enhance the rate of chemical reactions without any net change to themselves. They are capable of turnover, i.e. recycling, and show kinetics typical of enzymes. Ribozymes are so far known to catalyse only two reactions: (i) cleavage of RNA, and (ii) cleavage of DNA. Ribozymes are able to cut and splice themselves into a form that can catalyse the cleavage of other RNA/DNA molecules. The catalytic ability of ribozymes arises due to their three dimensional structures, which is able to generate in them the substrate-specific binding site. This is similar to enzymes in which the substrate specific binding site is produced by their three-dimensional structure. It may be emphasised that all biological catalysts must possess the ability to bind to their specific substrates.

Ribozymes offer exciting possibilities of varied applications, e.g. in human, animal and plant disease control. It has been suggested that RNA was the first macromolecule on earth which catalysed its own replication, and it subsequently developed other catalytic capabilities. Much later, RNA acquired the ability to catalyse protein synthesis. In due course of time proteins became the primary biological catalysts since they were capable of catalysing a much larger variety of reactions more efficiently than ribozymes. The first DNA molecules are thought to have originated by reverse transcription of RNA molecules. Ultimately, DNA became the primary genetic material in view of its greater stability than RNA. Thus, according to this hypothesis, RNA has been dethroned from its princely status by the same two macromolecules, i.e. proteins and DNA, the synthesis of which it either catalysed (proteins) or provided the primary information for (DNA by reverse transcription); RNA is now generally relegated to a mere secondary (although essential) role as the link between DNA and proteins.

IMMOBILISATION OF ENZYMES

Enzyme immobilisation may be defined as confining the enzyme molecules to a distinct phase from the one in which the substrates and the products are present; this may be achieved by fixing the enzyme molecules to or within some suitable material. It is critical that the substrates and the products move freely in and out of the phase to which the enzyme molecules are confined. Immobilisation of enzyme molecules does not necessarily render them immobile; in some methods of immobilisation, e.g. entrapment and membrane confinement, the enzyme molecules move freely within their phase, while in cases of adsorption and covalent bonding they are, in fact, immobile.

The materials used for immobilisation of enzymes, called carrier matrices, are usually inert polymers or inorganic materials. The ideal carrier matrix has the following properties: (i) low cost, (ii) inertness, (iii) physical strength, (iv) stability, (v) regenerability after the useful lifetime of the immobilised enzyme, (vi) enhancement of enzyme specificity, (vii) reduction in product inhibition, (viii) a shift in the pH optimum for enzyme action to the desired value for the process, and (ix) reduction in microbial contamination and non-specific adsorption. Clearly, most matrices possess only some of the above features. Therefore, carrier matrix for the immobilisation of an enzyme must be chosen with care keeping in view the properties and limitations of various matrices.

Methods of Immobilisation

The various methods used for immobilisation of enzymes may be grouped into the following four types: (i) adsorption, (ii) covalent bonding, (iii) entrapment, and (iv) membrane confinement.

Adsorption

In case of adsorption, the enzyme molecules adhere to the surface of carrier matrix due to a combination of hydrophobic effects and the formation of several salt links per enzyme molecule. The binding of enzyme molecules to the carrier matrix is usually very strong, but it may be weakened during use by many factors, e.g. addition of substrate, pH or ionic strength. Therefore, the matrix should be carefully chosen keeping these factors in mind. Some of the commonly used matrices are, ion exchange matrices, porous carbon, clays, hydrous metal oxides, glasses, and polymeric aromatic resins. Ion exchange matrices are costly, but they can be readily regenerated (at the end of active life of the absorbed enzyme) by a simple operation, e.g. washing-off the adsorbed enzyme with a concentrated salt solution.

Adsorption of enzymes to the matrices is very easy and widely used. The enzyme is mixed with a suitable adsorbent under appropriate conditions of pH and ionic strength. After incubation for a sufficient period of time, the carrier is washed to remove unadsorbed enzyme molecules, and the immobilised enzyme is ready for use. This method usually produces a high loading (about 1 g enzyme/g matrix) of the enzyme.

Covalent binding

In this system the enzyme molecules are attached to the carrier matrix by formation of covalent bonds. As a result, the strength of binding is very strong, and there is no enzyme loss during use. The covalent bond formation occurs with the side chains of amino acids of the enzyme, their degree of reactivity being dependent on their charged status. The following relation is observed in reactivity:

$$-S^- > -SH > -O^- > -NH_2 > -COO^- > -OH >> -NH_3^+$$

Lysine residues are the most useful in covalent binding of enzymes since they are usually exposed on the surface, are highly reactive and only very rarely occur at active sites of enzymes. Enzyme loading is quite low (ca. 0.02 g/g matrix); only in exceptional cases it may be as high as 0.3 g/g matrix.

The most commonly employed matrices are agarose, celluloses and polyacrylamides. Sepharose, an agarose, is available commercially as beads, is highly hydrophilic and is generally inert to microbial attack. On a research scale, it is activated by a treatment with cyanogen bromide which forms highly reactive intermediates with the –OH groups of sepharose; the enzyme molecules bind to these activated groups. However, cyanogen bromide is highly toxic so that it is not used commercially. Alternatively, sepharose is activated by treating it with chloroformates, carbodiimides, glutaraldehyde or other compounds. Treatment with trialkoxysilanes enables the use of such inert matrices as glass for covalent binding of enzymes.

Gluteraldehyde is a bifunctional reagent. It exists as an equilibrium mixture of monomer and oligomers It can be used to covalently bind the enzyme molecules or, alternatively, it can be used to cross-link enzyme molecules. In cross-linking, each bifunctional reagent molecule binds to two enzyme molecules; ultimately, a network of enzyme molecules linked together is produced. Glutaraldehyde is particularly useful for producing immobilised enzyme membranes for use in biosensors; this is achieved by cross-linking the enzyme plus a noncatalytic protein, used for dilution within a porous sheet; e.g. lens tissue paper or nylon net fabric.

Immobilisation may lead to a loss in enzyme activity due to the involvement of active site in immobilisation, or immobilisation of the enzyme in an orientation which either distorts the active site or renders it unavailable (Fig. 11.3). This can be markedly reduced as follows. The enzyme is immobilised in the presence of saturating concentration of its substrate or a competitive inhibitor (such inhibitors bind to the active site). In such a situation, the active site is occupied and kept in the correct conformation while immobilisation, takes place; this minimises the involvement of active site as well as incorrect orientation of the enzyme molecules during immobilisation.

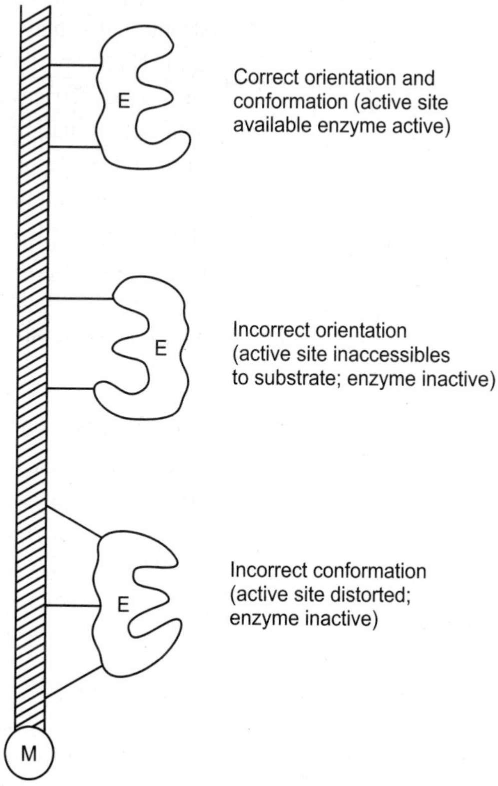

Fig. 11.3. The possible effects of enzyme immobilisation by covalent binding on enzyme activity. E = enzyme, M = matrix.

Entrapment

In this approach, enzyme molecules are held or entrapped within suitable gels or fibres and there mayor may not be covalent bond formation between the enzyme molecules and the matrix. A non-covalent entrapment may be viewed as putting the enzyme molecule in a molecular cage just as a caged bird/animal. When covalent binding is also to be generated, the enzyme molecules are usually treated with a suitable reagent. For example, acryloyl chloride is used to prepare lysine residues for binding by forming acryloyl amides; these are then copolymerised and cross-linked with acrylamide and bisacrylanide to form a gel containing the entrapped enzyme which may be used to form small beads or a film on a solid support.

Enzymes may be entrapped within cellulose acetate fibres as follows. An emulsion of the enzyme and cellulose acetate is prepared in methylene chloride. The emulsion is extruded as fibres into a solution of an aqueous precipitant. Entrapment in calcium alginate is the most widely used method for entrapment of microbial, animal and plant cells. Enzyme loading is very high (1 g/g gel or fibre). However, diffusion of the substrate to, the enzyme and of the product away from the enzyme creates difficulties.

Membrane confinement

Enzyme molecules, usually in an aqueous solution, may be confined within a semipermeable membrane which, ideally, allows a free movement in either direction to the substrates and products but does not permit the enzyme molecules to escape. A number of strategies are employed for this purpose which are briefly outlined below:

1. The reaction vessel may be partitioned into two chambers by a semipermeable membrane; one chamber contains the enzyme while the other has the substrate and the product.
2. Hollow-fibre membrane units contain the enzyme in their lumen or hollow space, and themselves are submerged in the substrate. This strategy provides an extremely large surface area per unit volume, e.g. > 20 m^2/l, but is useful for only such substrates that are much smaller than the enzyme molecules. Hollow-fibres are costly, and can be used with a variety of enzymes including coenzyme regenerating systems.
3. Enzymes may be packed in microcapsules formed by a polymerisation reaction, e.g. by using 1, 6-diaminohexane. In addition, they may be enclosed within liposomes which are small spheres made up of concentric lipid membranes.

Each immobilisation strategy has some strong and some weak points which are summarised in Table 11.3.

Table 11.3. Comparison of the various features of the different immobilisation systems.

Feature	*Adsorption*	*Immobilisation system* *Covalent binding*	*Entrapment*	*Membrane confinement*
Matrices (examples)	Ion-exchange matrices, clays, glasses etc.	Sepharose, cellulose, acrylamide, etc.	Acrylamide, cellulose acetate etc.	Semipermeable membranes, e.g. hollow-fibres, lipsomes etc.
Preparation	Simple	Difficult	Difficult	Simple
Immobilisation mechanism	Hydrophobic effects, salt links	Covalent bonds	Trapping in gel/ fibre; even covalent bonds	Confinement in a semipermeable membrane

(Contd ...)

Feature	Adsorption	*Immobilisation system* Covalent binding	Entrapment	Membrane confinement
Binding force	Variable	Strong	Weak	Strong
Enzyme loading	High (ca. 1g/g matrix)	Small (ca 0.02 g/g matrix)	–	–
Enzyme leakage during use	Yes	No	Yes	No
Applicability	Wide	Selective	Wide	Very wide
Problems during operation	High	Low	High	High
Matrix effects on enzyme	Yes	Yes	Yes	No
Diffusional barriers to substrate and product molecules	Absent	Absent	Large	Large
Protection from microbial attack	No	No	Yes	Yes
Cost	Low	High	Moderate	High

Effects of Immobilisation on Enzyme

Often kinetic behaviour of an immobilised enzyme may differ significantly from that of its free molecules. Different enzymes respond differently to the same immobilisation protocol. Therefore, a suitable immobilisation protocol has to be worked out for a given enzyme. The effects on enzyme kinetics (i.e. activity) may be due to the influence of matrix *per se* or due to conformational changes in the enzyme molecules induced by the procedure of immobilisation.

Immobilisation protocol may increase or decrease enzyme stability. When immobilisation produces a strain in the enzyme molecules, they become more prone to inactivation by higher temperatures, pH etc. In contrast, binding of an enzyme molecule at several points without creating any strain in the enzyme molecule may lead to substantial stabilisation. This is primarily due to the physical prevention (due to multipoint binding) of large changes in the conformation of enzyme molecules which is essential for their inactivation. However, enzymes, whether free or immobilised, loose activity with time due to denaturation.

Advantages of Immobilisation

1. Enzymes are costly items, and can be used repeatedly only if they can be recovered from the reaction mixtures. Immobilisation permits their repeated use since such enzyme preparations can be easily separated from the reaction system.
2. The product is readily freed from the enzyme. This saves on the cost of downstream processing of the product.
3. Immobilised enzymes can be used in nonaqueous systems as well, which may be highly desirable in some cases.
4. Continuous production systems can be used, which is not possible with free enzymes.

5. Thermostability of some enzymes may be increased. For example, glucose isomerase denatures at 45°C in solution, but is stable for about 1 year even at 65°C when suitably immobilised.
6. Recovery of enzyme may also reduce effluent handling problems.
7. Enzymes can be used at much higher concentrations than free enzyme.

Disadvantages of Immobilisation

1. Immobilisation means additional cost. Therefore, it should be used only when there is a sound economic, safety or process advantages over soluble enzymes.
2. Immobilisation often adversely affects the stability and/or activity of the enzymes. In such cases suitable immobilisation protocols should be developed.
3. This approach can not be used when one of the substrates is insoluble.
4. Some immobilisation strategies present large problems in diffusion of the substrate to reach the enzyme.

ENZYME UTILISATION IN INDUSTRIAL PROCESSES

From the standpoint of chemical process engineering, enzymes have some properties which make them ideal catalysts. They are able to carry out chemical reactions at high rates and with high specificity at ambient temperatures and normal pressure in aqueous solutions. Thus enzymes provide a means of eliminating many of the operational difficulties of high temperatures and pressures often encountered in chemical catalytic proesses. Practical application of bulk enzymes, however, is rather limited to industries which process natural materials. The reasons for this comparatively mediocre utilisation are complex, but the following ones may be mentioned:

1. Until recently, the development of enzyme applications was founded on an empirical rather than a rational basis.
2. Enzymes are almost exclusively used as processing materials in other industries. Consequently, the growth of the enzyme market is determined by the growth in the market of the final product.
3. Strong government regulations, required to protect health and safety, tend to inhibit development and introduction of potential enzyme applications in many industries.

Economic considerations of the use of enzymes are governed by a number of variables and are not always simple to determine. Justification of utilisation is often calculated on a plant-by-plant basis rather than on an industry-wide basis.

Fields of Application

At present, the bulk enzyme industry still relies almost entirely on the production of relatively simple enzymes, primarily for application in the food and other industries. By far the majority of these enzymes are hydrolases, such as amylases, cellulases, pectinases, proteases, etc. They are used mostly as additives or processing aids in the baking, dairy, fruit juice, and other industries. In contrast to the bulk enzymes, the market for the individual enzymes glucose isomerase, glucose oxidase, etc. is expanding rapidly. This fits also for highly purified enzymes used in the pharmaceutical industry, therapeutic application, and clinical and chemical analysis. A brief abstract of applications is presented in Table 11.4.

Use of microbial enzymes in food processing falls roughly into three categories:

1. Those in which enzymes form an essential part of a process, e.g. production of cheese, beer, spirits.
2. Those in which enzymes are used to improve the economics of a process, e.g. extraction of fruit juices and essential oils.

3. Those in which enzymes are used to improve product quality, e.g. meat tenderisation, loaf volume, flexibility of glucose syrups and dextrins, etc.

Application of immobilised enzymes in food industries is restricted to a very few cases. For instance, aminoacylase (for resolution of racemic mixtures of amino acids) and glucose isomerase (for converting glucose to fructose) may be mentioned.

Table 11.4. Commercial applications of microbial enzymes.

Industry	*Application*	*Enzyme*	*Source*
Baking and milling	Reduction of dough viscosity, acceleration of fermentation process, increase in loaf volume, improvement of crumb score and softness, maintenance of freshness and softness	Amylase	Fungal
	Improvement of dough texture, reduction of mixing time, increase in loaf volume	Protease	Fungal/bacterial
Beer	Mashing	Amylase	Fungal/bacterial
	Chillproofing	Protease	Fungal/bacterial
	Improvement of fine filtration	β-Glucanase	Fungal/bacterial
Cereals	Precooked baby foods, breakfast foods	Amylase	Fungal
	Condiments	Protease	Fungal/bacterial
Chocolate, cocoa	Manufacture of syrups	Amylase	Fungal/bacterial
Coffee	Coffee bean fermentation	Pectinase	Fungal
	Preparation of coffee concentrates	Pectinase, hemicellulase	Fungal
Confectionery, candy	Manufacture of soft centre candies and fondants	Invertase, pectinase	Fungal/yeast
	Sugar recovery from scrap candy	Amylase	Fungal
Corn syrup	Manufacture of high-maltose syrups	Amylase	Fungal
	Production of low D.E. syrups	Amylase	Bacterial
	Production of glucose from corn syrup	Amyloglucosidase	Fungal
	Converting corn syrup to a sweeter fructose-containing product	Glucose isomerase	Bacterial
Dairy	Residual H_2O_2 removal from milk (subsequent to sterilisation by H_2O_2)	Catalase	Fungal
	Manufacture of protein hydrolysates	Protease	Fungal/bacterial
	Stabilisation of evaporated milk	Protease	Fungal
	Production of whole milk concentrates, whey concentrates, and ice cream and frozen desserts	Lactase	Yeast
	Curdling milk	Protease	Fungal/bacterial
Distilled beverages	Mashing	Amylase	Fungal/bacterial
Eggs, dried	Glucose removal	Glucose oxidase	Fungal

(Contd ...)

Industry	*Application*	*Enzyme*	*Source*
Feeds, animal	Pig starter rations	Amylase, protease	Fungal
Flavours	Clarification (starch removal)	Amylase	Fungal
	Oxygen removal	Glucose oxidase	Fungal
Fruit juices	Clarification, preventing gelling of concentrates, improvement of juice extraction yield	Pectinases	Fungal
	Oxygen removal	Glucose oxidase	Fungal
Laundry	Detergents	Protease	Bacterial
Leather	Dehairing, bating	Protease	Fungal/bacterial
Meat	Tenderisation	Protease	Fungal
	Preparation of fish protein concentrates	Protease	Fungal/bacterial
Pharmaceutical and clinical	Digestive aids	Amylase, protease	Fungal
	Injection for bruises, inflammation, etc.	Streptokinase	Bacterial
	Various clinical tests	Numerous	Fungal/bacterial
Photography	Recovery of silver from spent film	Protease	Bacterial
Protein hydrolysates	Preparation of protein hydrolysates	Protease	Fungal/bacterial
Soft drinks	Stabilisation of citrus terpenes from light-catarased oxidation	Glucose oxidase and catalase	Fungal
Textiles	Desising of fabrics	Amylase	Bacterial
Vegetables	Preparation of hydrolysates	Pectinase, cellulase	Fungal
	Liquefying purees an soups	Amylase	Fungal
Wine	Clarification of must	Pectinase	Fungal

Legal Aspects

Large amounts of enzymes are presently used in the food industry. Therefore, manufacturers must conform to food laws which, however, differ from country to country. The Codex Committee of Food Additives has published a list of enzymes derived from plant, animal, and microbial sources, limiting their application in food products. This has to be in accordance with the general principles regulating the use of 'food additives'. In the United States the following enzymes are generally recognised as safe (GRAS), when manufactured with sound processing practice: proteases and carbohydrases from *Aspergillus oryzae*; carbohydrases, cellulases, glucose oxidase, pectic enzymes, and lactase from *Aspergillus niger*; invertase from *Saccharomyces cerevisiae*; lactase from *Saccharomyces fragilis*; and carbohydrases and proteases from *Bacillus subtilis*. The use of enzymes from other microbial sources is limited by the 'Food Additive Amendment.' This is true, for example, for carbohydrases from *Rhizopus oryzae*; milk clotting enzymes from *Mucor miehei*, *Mucor pusillus*, *Bacillus cereus*, and *Endothia parasitica*; and for catalase from *Micrococcus lysodeikticus*.

Care must be taken that the organisms used for enzymes that will be applied in the food industry are known or shown to be nonpathogenic and do not produce toxic substances, β-nitropropionic acid included. New strains must be screened also for variability; it must be certain that the properties of the organism show no difference from those originally tested and approved. Safety, furthermore, is dependent on freedom from the organism used as well as on the materials used in the concentration and purification processes. Where carryover of materials or residues might persist, these must be of food grade. Fillers and additives must conform to normal standards.

Harmful effects on process workers must also be considered. Suitable means for circumventing these effects are careful handling and change of the physical form of the enzyme preparation. Thus, enzymes are now usually in granular or, less often, encapsulated form.

PARTICULAR TECHNICAL ENZYME PREPARATIONS

Of the 1000 and more enzymes isolated, fewer than 20 are now commercially used on a scale that has significant impact on either the enzyme industry or the user industries. Only the more important will be treated here. It is worth remembering that commercial enzyme preparations represent mixtures in which the principal enzyme activity is only one among many others. In enzyme applications this fact may imply that desired effects are interfered with by contaminating enzymes, or that effects observed trace back to such activities. The situation becomes even more complex when one considers the varying qualities and quantities of accompanying enzymes depending on the microbial source, the cultivation method, and the recovery process. This is one of the reasons for variable suitability of different products of the market or even of different prices for the same product. In some applications, it is invariably necessary to eliminate the interfering activity unless a mutant unable to produce this activity has been found. The success of such a procedure can decisively influence the development of a special enzyme market.

Amylolytic Enzymes

Amylolytic enzymes represent a group of catalytic proteins of great importance to the food industry. They were also one of the first enzymes to be produced commercially by micro-organisms. Since industrial starch processing requires certain physical properties or a specific carbohydrate composition of the final starch hydrolysate, great efforts have been made to find microbial enzymes with the specific characteristics required. Consequently, a remarkable number of new starch-degrading enzymes have been discovered, most of which have characteristics and properties that clearly distinguish them from all amylolytic enzymes previously known.

Starch-degrading enzymes can be divided into two main groups, α-1,4-glucanases and α-1,6-glucanases (debranching enzymes). The more important members of the class are shown in Fig. 11.4.

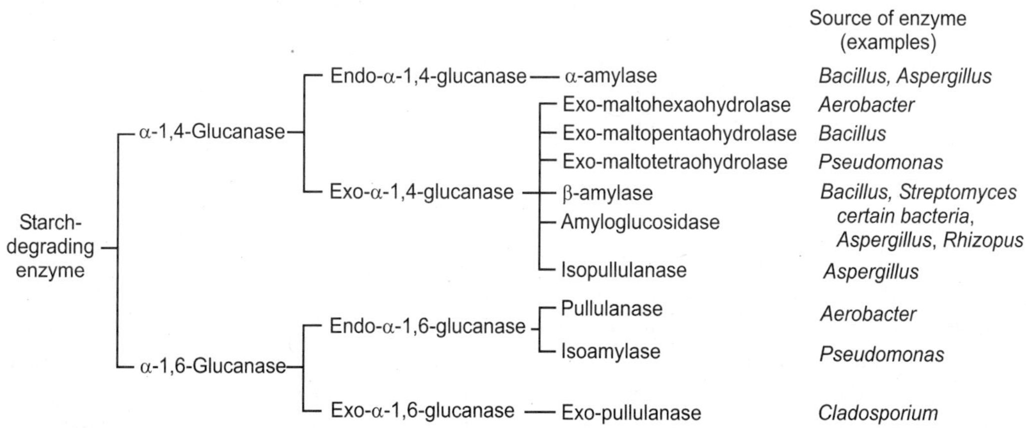

Fig. 11.4. Classification of starch-degrading enzymes of industrial interest.

α-Amylase

This enzyme (α-l,4-glucan-glucanohydrolase, EC 3.2.1.1) acts on starch components, which contain at least three α-1,4-linked glucose units, as an endoase, i.e. in an essentially random manner, with the production of reducing sugars. Mode of action, properties, and products of hydrolysis differ somewhat, depending on the source of the enzyme. Two types of microbial α-amylases have been recognised, termed 'liquefying' and 'saccharifying' α-amylases. The main difference between them is that the saccharifying enzyme produces a higher yield of reducing sugars than the liquefying enzyme.

Bacterial α-amylase might be produced by a number of *Bacillus* species, *Pseudomonas saccharophila*, and *Clostridium* species, but on an industrial scale specially selected strains of *Bacillus subtilis* seem to be preferred. Fungal α-amylases for commercial purposes are derived from *Aspergillus oryzae*. Certain strains of *Aspergillus niger* and particularly of *Aspergillus oryzae* produce large amounts of the saccharifying enzyme.

In industrial fermentation the cultivation of bacterial α-amylase producers is principally conducted by the submerged culture technique. The media employed are generally based on the use of natural raw materials, because of their cheapness and stimulatory effect. The latter effect is attributed to enzyme inducers or certain growth factors, such as trace elements, vitamins, or suitable combinations of amino acids. A careful balance of carbohydrate and nitrogen ingredients of the medium is most important. From reports in the literature it appears that the concentration of the N source should normally be higher than necessary for meeting full growth of the organism. Mostly, organic N is supplied, but inorganic N can also serve as the major nitrogenous component: it is, however, usually combined with small amounts of proteinaceous compounds.

Fungal α-amylase was originally and is still produced in significant amounts in solid substrate culture. Wheat bran serves as the basic component of the medium. Most publications on this procedure mention weakly acidic solutions as a means of moistening the bran, but acid has been shown to be harmful to amylase production. The moisture content of the culture depends on the height of the culture and on incubation temperature. The incubation time required to reach maximum yield of enzyme varies between 'overnight' and about 4 days.

While Dunn succeeded in selecting strains capable of producing good yields of α-amylase under submerged culture conditions, fungal α-amylase is also produced in deep tank fermentation. Usually flour or starch serves as raw material, supplemented with inorganic salts. Addition of stillage, corn steep liquor, or yeast extract contributes stimulating agents for some strains. In most cases of fungal α-amylase fermentation, the enzyme accumulates largely in the stationary phase, and quite often during the phase of autolysis. The amount of α-amylase that is formed during these growth phases is very strongly dependent on environmental and internal factors.

Published work indicates the existence of common features in both bacterial and fungal α-amylase fermentations. It has been shown quite generally that in both cases a proper course of pH change toward alkalinity during cultivation is of great importance. Salts of organic acids, e.g. citrate, gluconate, or acetate, can serve as pH regulators acting in the desired sense. The necessary amounts depend on the carbohydrate concentration or on the kind of concentration of organic and inorganic N sources. The same effect can be achieved by alkalising nitrogenous compounds, such as nitrates, urea, and proteinaceous matter. When ammonium salts are used as the N source, organic salts have to be given preference due to their capability of maintaining the pH of the culture at neutral or of resulting in a rising pH, as shown by Horvath and Inczefi. Proteinaceous matter, e.g. peptone, exerts the same effect. Fully synthetic media must not contain ammonium salts, when lacking salts of organic acids, but should contain nitrates or urea. In contrast to the acidifying effect of inorganic NH_4^+ salts, ammonium acetate tends to maintain the pH of the culture at neutral or to result in a rising pH. This can be attributed to the fact that the organisms produce nonutilisable organic acids.

The question of the type of mechanism that regulates the formation of α-amylase cannot be answered clearly or generally. While Coleman stated that in *Bacillus subtilis* α-amylase is constitutive and controlled by the size of the pool of nucleic acid precursors, Schaeffer and Meers found in bacilli that biosynthesis of this enzyme is governed by catabolite repression. Using a carbon-limited medium it was demonstrated that the presence of an inducer was not necessary for α-amylase biosynthesis.

Dunn found that addition of Ca phytate to natural and synthetic media increased the yield of dextrinogenic α-amylase from a *Bacillus* strain. Similarly, Yamada observed an increase in production of both dextrinogenic and saccharogenic amylase as a response to the incorporation of 0.01–0.05 per cent phytic acid in the medium used for *Aspergillus oryzae* or *Aspergillus awamori*.

Commerical preparations of bacterial α-amylase are commonly produced with a minimum of purification. Highly active or purified preparations are obtained by precipitation and/or adsorption techniques. For some applications the absence of other enzymes, especially proteinase, is essential. For this purpose several methods are available, e.g. adsorption procedures, fractional precipitation, and selective inactivation.

Bacterial α-amylase has a molecular size on the order of 50,000, each molecule containing 1 gram atom of Ca^{++}. In the presence of zinc, a dimer is formed containing 1 atom of zinc. The calcium maintains the enzyme molecule in the optimum conformation for maximum activity and stability; it does not participate directly in the reaction. The addition of Ca salts is generally recommended to achieve maximum heat stability of the enzyme. The maximum activities of α-amylase are in the slightly acidic region between pH 4.5 and 7.0, with differences depending on the enzyme source. The α-amylases from *Bacillus subtilis*, and especially *Bacillus stearothermophilus*, are particularly heat stable. In contrast, thermal stability of fungal α-amylase is relatively low.

Bacterial α-amylases preparations are mainly used in the continuous process for desizing of textile fabrics. Other applications include modification of starches suitable for preparation of adhesives, sizes and coatings for the paper industry, as well as manufacture of glucose and glucose syrups and brewing processes. Fungal α-amylases are extensively used in flour treatment for supplementing the diastatic activity of flour.

Amyloglucosidase

The enzyme amyloglucosidase (glucoamylase, α-1,4-glucan glucohydrolase. EC 3.2.1.3) acts as an exoase that liberates the α–1,4–linked glucose units consecutively from the nonreducing ends of the starch chains. Terminal α-1,6-bonds are also cleaved, but much more slowly than α-1,4-linkages. Maltose is only slowly, attacked; increasing the chain length up to 5 or 6 glucose units gives faster attack.

Amyloglucosidase occurs in many micro-organisms, particularly in starch degrading moulds (*Aspergillus*, *Mucor*, *Rhizopus*, *Endomyces*) and certain bacteria (*Aerobacter*, *Clostridium*). The mould species commonly used for large-scale production is *Aspergillus niger*. However, this species also produces transglucosidase, which transfers glucose and higher saccharides to oligosaccharides, resulting in synthesis of polysaccharide. These products cause serious reduction in glucose yield and impede its crystallisation. Strains are chosen which produce low levels of transglucosidase. Some organisms, e.g. members of the *Mucor*, *Rhizopus* and *Aspergillus phoenicis* groups, usually produce amyloglucosidase without simultaneous formation of transglucosidase.

When *Aspergillus niger* is used as the production strain, the submerged fermentation is the cultivation method of choice. *Mucor* and *Rhizopus* strains are used in solid substrate fermentation because they are obviously unsuitable for submerged culture. The reason is that when the nonseptate hyphae of these species are damaged by the strong shearing action of high-speed impellers at one single point the protoplasm of the whole filament will be extruded.

Media employed for amyloglucodsidase production in submerged fermentation contain usually high solid contents of organic matter (of the order of 12–20 per cent). Starchy material, such as maize, wheat, barley, rye, and sorghum, are the raw materials of choice. These cereals provide also the N source, the content of which is usually sufficient. Inorganic nitrogen can be utilised, ammonium salts being preferred. Stimulation of amyloglucosidase formation may be obtained by incorporation of yeast extract or stillage into the medium. An acid reaction of the mash seems to be necessary for maximal production of amyloglucosidase. In *Aspergillus niger* fermentations the pH of the culture decreases to values as low as 2.8 at the end of the cultivation, caused by the excretion of organic acids.

Removal of transglucosidase activity is an important step in the processing of amyloglucosidase. One of the oldest methods to free the culture filtrate from transglucosidase takes advantage of the fact that amyloglucosidase is highly acid resistant. Adsorption techniques using synthetic water-insoluble hydrous magnesium silicate or clay minerals (e.g. attapulgite and bentonite) are said to be either unreliable or not sufficiently selective. Barton claimed coprecipitation with both maleic anhydride copolymers and heteropolyacids to be commercially practicable for removal of transglucosidase. Another claim of Sternberg suggests the precipitation of transglucosidase by chloroform, but the conditions of the precipitation process are very critical.

Amyloglucosidase has been bound covalently to a variety of matrices and also has been physically adsorbed on or bound to various materials. Further, ultrafiltration reactors have been used for the continuous conversion of starch to glucose but obviously there is little industry to adopt processes based on immobilised amyloglucosidases. Amyloglucosidase has largely replaced acid hydrolysis in glucose production. Following the solubilisation of starch by acid or, preferably by a heat stable bacterial α-amylase, further degradation is achieved using amyloglucosidase.

Pullulanase

The enzyme pullulanase (amylopectin 6-glucanohydrolase, EC 3.2.1.41) splits α-1,6-glucosidic linkages at a branch point as well as in a linear chain of a number of oligo- and polysaccharides. It degrades pullulan completely to maltotriose and is active on amylopectin, glycogen, and their β-dextrins. The minimum chain which pullulanase liberates is maltose; a three unit chain (maltotriose) is optimal.

Pullulanase has been found in *Aerobacter aerogenes* and other bacteria, *Streptomyces* included. Production of the enzyme is inducible by growth on α-glucans ranging from maltose to glycogen. Yokobayashi found that starch or liquefied starch with a low dextrose equivalent (D.E. 5–10) was particularly inducive. Highest yields occurred when this low D.E. syrup was used in all steps of inoculum build-up as well as in the production stage. Further, for optimal synthesis, a comparatively high pH is required. Ueda found that in *Streptomyces* the pH of the culture rose up to 8.5 and the enzyme was abundantly produced in the autolytic phase.

Isoamylase

This enzyme (amylopectin 6-glucanohydrolase, EC 3.2.1.68) is able to completely debranch glycogen, but unable to act on pullulan. It differs from-pullulanase in that it acts more readily on the native polymers. The α-1,6-linkages are only split when present at branch points in oligo-and polysaccharides. Isoamylase requires a minimum of three glucose units on the A-chain of amylopectin, while pullulanase requires only two.

Isoamylase occurs in yeast and bacteria. It is produced commercially from *Cytophaga* and *Pseudomonas*, and is an inducible enzyme.

Other starch-degrading enzymes

Among these, the following are worthy of mention: An exo-amylase from *Pseudomonas stutzeri* which splits off maltotetraose from the ends of starch chains; an exo-amylase releasing maltohexaose, which is produced by *Aerobacter aerogenes*; an enzyme from *Bacillus licheniformis* that forms maltopentaose as the principal product; isopullulanase produced by certain species of *Aspergillus niger*. This enzyme degrades pullulan to isopanose. It shows no action on amylopectin but is able to hydrolyse terminal isomaltosyl groups as produced by amyloglucosidase.

Cellulase

The name 'cellulase' is given to all enzymes which cleave β-l,4-glucosidic linkages in cellulose and chemically or physically modified cellulose, in cellodextrin, and in cellobiose.

It is well established that cellulase is a multi-enzyme complex, the different components of which bring about the complete degradation of cellulose to monosaccharide residues. A classification scheme of cellulolytic enzymes is given in Fig. 11.5. The terms C_1-cellulase and C_x-cellulase were originally proposed by Reese for two components of the cellulase complex that differed in their substrate specificities against cotton fibre. According to this concept, C_1 can attack native cellulose of higher crystallinity (e.g. cotton fibre), while C_x cannot attack such cellulose but can split in turn the cellulose fragments which have been produced by the action of C_1. The mode of action of the C_1-enzyme has long been questioned, but now many investigators tend to identify it with β-1,4-glucan cellobiohydrolase, which liberates cellobiose units from the nonreducing end of the cellulose chain.

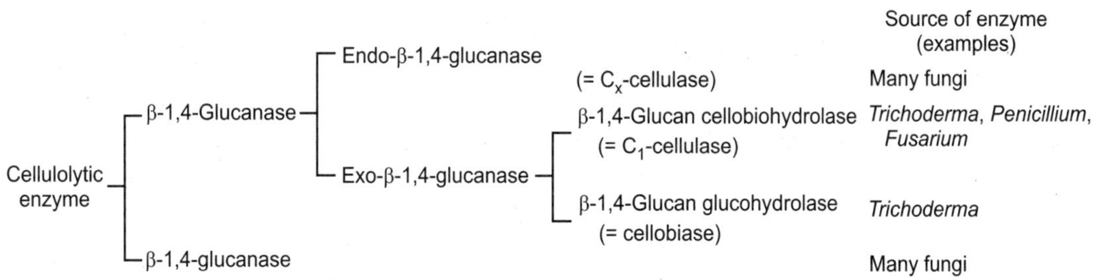

Fig. 11.5. Classification of cellulolytic enzymes.

Many bacteria and fungi are cellulolytic, but preparations marketed for industrial applications are derived only from *Aspergillus niger*, *Trichoderma viride*, *Neurospora*, and some other organisms. The *Aspergillus* enzyme exerts good activity on carboxymethylcellulose (CMC), but fails to attack solid cellulose because it lacks C_1-cellulase. In contrast, *Trichoderma viride* produces and enzyme complex with high levels of C_1-cellulase, which extensively degrades insoluble cellulose.

Cellulase in fungi is an inducible enzyme. It is only produced when the cells are grown on cellulose, on glucans of mixed linkages including the β-1,4 bond, and on a few oligosaccharides. The 'inducing effect of cellulose is due to soluble hydrolysis products of the cellulose, in particular cellobiose. Lactose, a β-1,4-galactoside, and sophorose, a β-1,2-glucoside, are the only known cellulase inducers that do not have a β-1,4-glucoside bond. The inductive action of sophorose is limited to *Trichoderma viride*. However, in spite of the impressive inductive power of this rare sugar, the levels of enzyme produced are not equal to those on cellulose.

Cellobiose plays a complex role: in low concentrations (0.1 per cent) it serves as an inducer of cellulase; in high concentrations (0.5–1.0 per cent) it represses cellulase formation, and, in addition, it can also act as an inhibitor of cellulase action.

Cellulase yields can be increased by various additives to the medium. Reese and Maguire observed that Tween 80 and Tween 40 doubled the cellulase yield in *Trichoderma* cultures. The mechanism of the action of these surfactants is not understood but may be related to increased permeability of the cell membrane. Nevertheless, Tween 80 has proven useful in the fermentation industry and is routinely incorporated into the culture medium. Further, enhancement of enzyme production can be achieved by supplying peptone at one-tenth the cellulose concentration. This leads to a decrease in the lag of growth and cellulase synthesis.

In actual large-scale fermentation *Aspergillus niger* is mostly cultured by the wheat bran-tray method. This process has no problems and leads to high yields of cellulase. The extraction of cellulase from solid substrate cultures is performed by percolation of the dried mould bran with 0.02 to 0.1 M lactic acid.

Neurospora and *Trichoderma* are grown by submerged culture. For continuous culture it is advantageous that *Trichoderma viride* produces a suspension of short mycelial threads, rarely forming pellets. Mitra and Wilke proposed a two-stage operation in continuous stirred tank reactors. The first stage utilises glucose for biomass production and the second stage utilises pure spruce wood cellulose for enzyme formation. A significant increase in enzyme productivity was obtained.

Bran, straw, and other plant materials pretreated with alkali serve as cellulose-containing raw materials for submerged cultivations. Ammonium ions can be used as suitable N source. A correct pH profile is necessary to give optimum enzyme yields in batch culture. It has been shown that a drop to the range of pH 3.5–3.0 is optimal for *Trichoderma viride*. This is usually achieved by an empirical procedure which involves medium composition and initial pH.

Length of cultivation affects the relative amount of the various cellulase components present in the medium. This has been demorstrated for *Myrothecium verrucaria* with CMC and with swollen substrates. In both cases C_1 appeared prior to C_x.

Concentration and purification of the enzyme is carried out by precipitation, adsorption, or gel filtration techniques. Granulated preparations can be obtained by mixing with salt hydrates (e.g. $Na_2SO_4 \cdot H_2O$) and subsequent vacuum drying.

Current use of cellulase is limited to improving texture and palatability of poor quality vegetables. It is also useful for accelerating drying of vegetables. A potential application of cellulase is the conversion of cellulosic materials to glucose and other sugars which in turn can be used as microbial substrates to produce single cell protein or a variety of fermentation chemicals (alcohol, etc.).

Pectolytic Enzymes

Many plant pathogenic bacteria and fungi have long been known to produce pectolytic enzymes and it is widely accepted that the production of these enzymes is a major means by which micro-organisms invade the host tissue. Moreover, pectolytic enzymes are essential in the decay of dead plant material by nonpathogenic micro-organisms and thus assist in recycling carbon compounds in the biosphere. Lastly, these enzymes play a decisive role in the microbial spoilage of fruits and vegetables.

Several types of enzymes are involved in the degradation of pectic materials. They are divided into two main groups, depolymerising enzymes and saponifying enzymes or pectinesterases. According to the scheme of Neukom, the depolymerising pectolytic enzymes are further classified by applying the following three criteria: preference for pectic acid or pectin as substrate, hydrolytic or transeliminative cleavage of the glycosidic link ages, and endo- or exo-types of the action mechanism. By various combinations of these characteristics, eight groups of depolymerising enzymes can be listed, but the existence of the exo-polymethylgalacturonase and exo-polygalacturonate lyase types is doubtful.

Occurrence of pectolytic enzymes has been reported in a large number of bacteria and fungi. Commercial enzymes are generally obtained from fungal sources since the pH optima of these enzymes are in the range found naturally in materials to be processed. Most potent strains are selected from *Aspergillus niger*. Japanese enzyme manufacturers also use *Sclerotinia libertiana* and *Coniothyrium diplodiella* as producers of pectolytic enzymes.

In the majority of cases micro-organisms produce a variety of pectolytic enzymes and, for this reason, commercial preparations are mixtures of these enzymes. The relative amount of the single components varies considerably with the particular strain employed, medium composition, and culture conditions. Careful observation of the factors responsible for the promoted synthesis of certain enzyme fractions or limited formation of others enables the manufacturer to 'control' the composition of the preparation and to meet the need for specific formulations.

Biosynthesis of pectolytic enzymes is constitutive or controlled by the mechanisms of induction or catabolite repression. No uniformity exists among the various organisms and the various components of the enzyme complex. Phaff found pectolytic enzymes to be adaptive in *Penicillium chrysogenum*, but constitutive in *Aspergillus foetidus*. Saito showed that in *Aspergillus niger* endopolygalacturonase (endo-PGase) is adaptive. For *Clostridium felsineum*, a plant retting organism, Osman demonstrated pectinmethylesterase (PME) to be adaptive and PGase constitutive.

On an industrial scale pectolytic enzymes are produced by the solid substrate method as well as by submerged culture. Wheat bran or defatted rice bran have been recognised as satisfactory basic substrates in solid substrate cultures. It is well known that some by-products of the food industry, such as beet pulp, apple pulp, or grape pulp, exert a promoting effect on enzyme formation. Other ingredients, e.g. nutrient salts, acid, or buffers, are also incorporated to regulate the pH during the growth of fungi. The time of cultivation can extend up to 7 days, but when *Aspergillus niger* is used, the desired enzyme level is normally reached within 36 to 72 hrs. After fermentation the mould bran is dried and can be used as such. For obtaining concentrates the dried mould bran is extracted with suitable aqueous solutions and concentrated under vacuum or by ultrafiltration. Crude or refined solid concentrates are obtained by spray drying or precipitation with neutral salts or solvents.

Submerged cultures, in contrast to solid substrate cultures, seem to have the disadvantages of poor yields and undesirable composition. Brooks and Reid, for example, found that *Aspergillus foetidus* produced endo-PGase and exo-PGase in surface culture, but only endo-PGase in submerged culture.

The production of pectolytic enzymes by submerged fermentation has been described by Nyiri. As an example, the method reported for *Aspergillus alliaceus* can be cited. The fungus is grown in a liquid medium composed of 2 per cent wheat bran, 2 per cent $(NH_4)_2SO_4$, 0.25 per cent KH_2PO_4, 0.25 per cent yeast extract, 0.1 per cent pectin (degree of esterification = 59 per cent). The initial pH is 3.8, adjusted with HCl. Traces of silicon serve as antifoam. After inoculation with conidia, the medium is agitated and aerated, with the pressure inside the fermentor maintained at 141.8 kPa (1.4 atm). The fermentation is completed within 72 hrs and the mycelium separated by filtration. The filtrate is cooled to 0°–1°C and the enzyme precipitated by addition of $(NH_4)_2SO_4$ during a period of 4 hrs. After standing for 12 hrs the precipitate is washed and dried to give a solid concentrate.

Initial pH of the medium and pH development in the growing culture play an important role with regard to both enzyme composition and yield of the enzyme fractions. Low pH values, in particular a decreasing pH during cultivation, are favourable for the production of pectinases used in the fruit juice industry, according to Hauptmann, pH 2–3 at culture maturity of *Aspergillus niger* being desirable. In contrast, Tuttobello and Mill allowed the pH to rise up to 4 at the end of the culture following a drop to about 3 on the fifth day of cultivation.

Tuttobello and Mill found an aqueous extract of non defatted peanut flour strongly stimulated the production of pectolytic enzymes. They also observed a strong influence of inoculum on enzyme production. The kind of inoculum and, particularly, its size have to be standardised carefully. With *Aspergillus niger* Tuttobello and Mill found that the production of pectolytic enzymes was strongly influenced by inoculum size in the range of 10^4 to 2×10^5 conidia per ml, while mycelium formation remained unchanged.

From many reports in the literature it can be concluded that there is a strong variation in relative activity of the various components of the pectolytic system in the course of fermentation, indicating their sequential production.

Extensive use of pectolytic enzymes is made in processing fruit juices for increasing juice yields on pressing, as aid in clarification of juices, and for depectinising in order to obtain high density fruit juice concentrates. Fungal enzymes are widely used in producing apple juice, grape juice, and wine. In the production of coffee beans the residual mucilaginous coating surrounding the bean can be liquefied by commercial pectolytic enzyme preparations, thus offering an alternative to the usually used fermentation process. The curing or fermentation of cocoa, tea, and tobacco also can involve pectolytic enzymes. One of the oldest applications of these enzymes is the process of retting, in which textile fibres, such as flax, hemp, and jute, are loosened from their plant stems. The enzyme system of *Clostridium felsineum*, an organism that is involved in aerobic retting, contains endopolygalacturonate *trans*-eliminase, but not pectinesterase. Recently, pectolytic enzymes have been proposed as a means to make commercial softwoods, such as Sitka and Norway spruce, more permeable to preservatives. It has been demonstrated that treatment with enzyme preparations as well as with the specific bacteria that produce them is possible.

Hemicellulase

Plant cell wall polysaccharides other than cellulose and pectic substances are referred to as hemicelluloses. They are complex compounds and very few of their chemical structures have been clarified. During recent years many enzymes have been recognised which specifically act on different types of hemicelluloses. By far the best known group is that of the xylanases. Other groups of hemicellulases are, for example, mannanases, galactanases, etc.

Many strains of bacteria and fungi are known to produce hemicellulases inducibly or constitutively, but on an industrial scale only fungal strains seem to be used as enzyme sources. Even in these cases hemicellulases are mostly obtained as side activities in the production of other enzymes such as cellulase (for commercial enzyme preparations containing hemicellulases). Therefore, as a rule, the hemicellulase activity of the commercially available enzyme systems is low. This fact reflects either an inherent instability of these enzymes or lack of knowledge of how to produce them. The potential application for hemicellulases is great. For example, β-glucanase is used in the brewing industry to degrade the barley β-glucans for solving pumping and filtering problems.

Invertase

Sucrase and invertase are two of the older names of the enzyme β-fructofuranosidase (EC 3.2.1.26). It catalyses the hydrolysis of the terminal nonreducing β-fructofuranoside residues in β-fructofuranosides. The name 'invertase' was derived from the action in splitting sucrose, which is optically dextrorotatory, to form glucose and fructose, a mixture that is levorotatory:

$$\underset{[\alpha]_D = +66.5^\circ}{\text{Sucrose}} + H_2O \xrightarrow{\text{Invertase}} \underbrace{\underset{[\alpha]_D = +52.5^\circ}{\text{D (+)-glucose}} + \underset{[\alpha]_D = -92^\circ}{\text{D(–)-fructose}}}_{[\alpha]_D = -20^\circ}$$

This reaction can also be carried out by α-D-glucosidase (so-called glucosidoinvertase), but this enzyme is unable to split off fructose from the trisaccharide raffinose as is β-fructosidase (so-called fructosidoinvertase).

β-Fructosidase can be prepared from a variety of microbial sources, but only the enzymes from *Saccharomyces cerevisiae* and *Saccharomyces carlsbergensis* have industrial importance.

Biosynthesis of invertase is controlled by a catabolite repression mechanism in *Saccharomyces fragilis* and by repression through unknown effectors in *Saccharomyces cerevisiae*. For a long time invertase was considered to be totally an intracellular enzyme. It has, however, been established that in derepressed cells only a small proportion of the invertase is located inside the cytoplasmic membrane most of it being retained externally within the cell wall or between the wall and the cell membrane. In fully repressed cells all the enzyme is intracellular.

The release of the invertase from yeast is achieved by destruction of the structures responsible for the retention of the enzyme. There are various ways in which the separation from the cells can be accomplished. One method is autolysis with chloroform, toluene, or ethylacetate at 30°C for not over 3 hrs. Following extraction from yeast, comparatively high purification of invertase is necessary for its application in foods because the enzyme preparation usually has an undesirable, irritating taste originating from yeast. These procedures include common methods for purification of enzymes such as ultrafiltration, precipitation, and adsorption techniques. The commercial preparation of invertase usually starts with an accumulation step. For this purpose pressed bottom yeast is suspended in a 20-fold amount of nutrient broth containing 4 parts $(NH_4)_2 \cdot HPO_4$, 4 parts KH_2PO_4, 1 part $Mg(NO_3)_2$, and 1 part KNO_3. The mixture is aerated for 3–8 hrs, while the temperature is maintained at 28° to 30°C and at a pH of 4.5.

During the same period, 3 to 20 per cent sucrose in solution is added continuously, a procedure that ensures reduced catabolite repression. At the end of the process the invertase activity of the yeast is increased up to 15-fold. The intracellular yeast invertase has a molecular weight of 1,35,000 and is free of carbohydrate, whereas the external enzyme has a molecular weight of 2,70,000. Approximately half of the external β-fructofuranoside consists of mannan.

The pH activity curve of invertase is rather broad between pH 3.5 and 5.5, with an optimum between 4 and 4.5. This is the same range within which the enzyme exhibits its highest stability. Yeast invertase is strongly inhibited by heavy metal ions (especially Ag^+); they combine with the histidine side chains of the enzyme molecule, not with its thiol groups.

Invertase has a number of interesting uses in the confectionery industry for soft centre candies, fondant, and chocolate coatings. Its use in the preparation of invert sugar by hydrolysing sucrose has been restricted by glucose isomerase which permits production of invert sugar from cheaper sources.

Lactase

Lactose or milk sugar is enzymically split to glucose and galactose by the action of enzymes called β-galactosidases or, more commonly, lactases (β-D-galactoside galactohydrolase, EC 3.2.1.23). Lactase is very specific for the galactose residue but much less specific for the aglycone moiety of β-galactosides. The enzyme is also responsible for transfer activities which occur with the formation of oligosaccharides.

Lactase is widely distributed in micro-organisms. Some strains of *Escherichia coli* are very potent producers, but are not suitable for food purposes. Available commercial preparations are derived from lactose fermenting yeasts such as *Saccharomyces fragilis*, *Zygosaccharomyces lactis*, and *Candida pseudotropicalis* or from fungi like *Aspergillus niger*, and particularly a mutant strain of *Aspergillus foetidus*.

The biosynthesis of β-galactosidase has been extensively investigated, largely in *Escherichia coli* in connection with studies on the biosynthesis of proteins and its genetic control. The enzyme is of the inducible type, with lactose serving as an inducer. In *Fusarium oxysporum* and *Verticillium alboatrum* the lactase can be induced by D-galacturonic acid and, to a lesser extent, by D-galactose.

On an industrial scale the enzyme is obtained, for example, by growing yeast on a lactose medium or on whey. The separated yeast is autolysed or extracted, and a cell-free extract is obtained by centrifugation or filtration.

The enzyme may then be further processed by salt or solvent precipitation. Another procedure, described by Stimpson, involved spray drying of the washed yeast at temperatures which destroy any residual alcoholic fermentation activity, thus leading to crude products which can immediately be used as lactase preparations.

The lactases from various microbial sources differ in properties such as pH optima, etc. For example, the pH optimum of the bacterial enzymes is around 7.0; that of the fungal preparations near 5.0; and that of the yeast enzymes near 6.0; the lactase from *Corticium rolfsii* is distinguished by its unusual maximum activity and stability at pH 1.8–2.0.

Yeast lactase is activated by potassium and ammonium ions and is inhibited by certain metals such as copper and iron. Metal-chelating agents do not stimulate the enzyme, indicating little sensitivity to trace heavy metals. The addition of reducing compounds, e.g. cysteine, sodium sulphide, or potassium metasulphite, is able to overcome the effect of metal inactivators and to activate the enzyme.

Hereditary intolerance to lactose precludes use of milk as a valuable protein source in large areas of Asia and Africa. In addition, lactose causes a number of problems in the dairy and allied industry because of its poor solubility, resulting in crystallisation in concentrated dairy products. Enzymic hydrolysis of the milk sugar is helpful in overcoming these problems. Moreover, lactase is used in the production of sweet syrups from sources of lactose such as cheese whey and in making waste whey a better substrate for growing micro-organisms for single cell protein.

When treating milk it is preferred to employ lactase from yeast because of legal reasons, although this enzyme is less stable than the bacterial one. During past years the use of skim milk powders in bread has been considerably reduced. This has reduced the interest in lactase for the formation of fermentable sugars in baking.

Proteases

The microbial proteases which are of interest for application in the food industry are all of the endopeptidase type and are all extracellular enzymes. There are many different types of proteases produced by an extraordinarily large number of micro-organisms, but in actual practice the enzymes prepared commercially are of a very limited number of types and they are derived from very few organisms (Table 11.5).

The proteolytic enzymes from micro-organisms are classified into four main groups according to the scheme of Hartley and based on the mechanism of their action: serine proteinases, thiol proteinases, metalloproteinases, and acid proteinases. Further subgroupings refer to the side-chain, specificity of the proteinases and to the properties of their active centres.

Table 11.5. Organisms currently used for protease production.

Organism	*Enzyme type*	*Product*[1]
Bacillus subtilis	Metallo, serine	Montase, Milezyme
	Serine	Alcalase, Maxatase
B. thermoproteolyticus	Metallo	Thermoase
Streptomyces griseus	Metallo, serine	Pronase
Aspergillus oryzae	Acid	Rhozyme A-4
A. saitoi	Acid	Molsin
Mucor pusillus	Acid	Microbial Rennet

[1]Trade names, except for 'Microbial Rennet'.

Industrial production of microbial proteases is carried out on a large scale by a number of companies in Europe, Japan, and the United States. For cultivation of the micro-organisms the submerged fermentation is the preferred method; with bacteria it is the exclusively used process. However, fungi usually give higher yields when cultured on solid media so this method continues to play a role. As in most fermentations there is a trend to use highly concentrated media. The reason for this is that one can expect higher enzyme yields per unit volume with a larger cell concentration, although there is no direct correlation between growth and protease production. With regard to serine and metalloproteinases it seems that low concentrations of purely carbonaceous substrates and high concentrations of proteinaceous N sources stimulate production.

Many of the organisms excrete more than one kind of protease. The type of proteolytic enzyme formed may depend on the composition of the medium. For example, *Bacillus* NRRL B-3411 produces the preferable neutral protease when grown on a grain medium, but mainly alkaline protease when cultured on a fishmeal-enzose-cerelose medium. The biosynthesis of proteases is often correlated with particular growth phases of the microbial culture. Under most growth conditions, *Bacillus* species produce extracellular protease during the postexponential growth phase. Mandelstam attributed this behaviour to an increased need for turnover of cell proteins at the slower growth rate. Other bacilli synthesise proteases during the exponential growth phase. However, these kinetics depend on the composition of the medium.

For all protease preparations the degree of purification depends on the intended use. A number of purification procedures are in existence, which follow the general description of recovery. Of course, various combinations are possible. Particular care is necessary during the drying process in order to avoid the formation of dust. For this reason protease preparations are pelleted or coated with some suitable material.

Bacterial proteases are used on a large scale in enzyme-containing washing, powders but they are not widely used in food processing. Minor uses are in the chillproofing of beer, in the production of protein hydrolysates, in the production of condensed fish solubles, and as feed supplement. In contrast to bacterial preparations fungal proteases are the more interesting group for the food industry. They are used, for example, for the modification of wheat proteins in bread doughs, in meat tenderising, and in several less important applications.

Serine proteinases

These proteases are widespread in bacteria and fungi. They show maximum activity at neutral to alkaline pH and are inhibited by diisopropyl fluorophosphate (DEP) or phenylalanine sulphonylfluoride (PMSF). They can be classified into at least five groups: trypsinlike proteinases, alkaline proteinases, *Myxobacter* α-lytic proteinase, staphylococcal proteinases, and serine neutral proteinases. Here only the serine alkaline proteinases will be treated becauses of their superior economic importance.

Proteases of this type are most active at pH 9.5–10.5; they are sensitive to DFP and potato inhibitor, but not to tosyl-L-lysine chloromethyl ketone. Their specificity is similar to that of an α-chymotrypsin but somewhat broader. All alkaline serine proteases show specificity toward aromatic or hydrophobic amino acid residues, such as tyrosine, phenylalanine, or leucine, at the carboxyl side of the cleavage point. The molecular weights are 26,000–34,000, slightly below the range of neutral metalloproteases. The isoelectric points are about pH 9. Most of the alkaline proteases are stable from pH 5 to 10 at low temperatures, but show rapid loss of activity at 65°C. certain strains of *Bacillus*, showing alkalophilic properties, synthesise a serine a alkaline proteinase that is most active pH 11–12.

Serine alkaline proteinase is produced by numerous species of bacteria and fungi. The best known representatives of this type are the subtilisins, which are produced by *Bacillus subtilis* and related species. Due to their great economic importance, the biosynthesis of *Bacillus* alkaline proteases has been well investigated. Keay and Moser have proposed that alkaline serine proteinases produced by different bacilli or different strains of *Bacillus subtilis* can be divided into two groups: subtilisin Carlsberg and subtilisin Novo. These enzymes are quite distinct from each other, but possess many similar properties. It may be mentioned that a similar situation has also been observed with various alkaline serine proteinases from the genus *Aspergillus*.

The synthesis of these enzymes is linked to particular phases of development of the microbial culture. Some strains, e.g. those of *Bacillus megaterium*, produce the proteases during the log phase of growth, while others, like those of *Bacillus subtilis* and *Bacillus cereus*, produce it in the stationary phase. However, as mentioned above, the relationship between growth cycle and enzyme formation depends on the ingredients of the substrate. It is generally valid that the time of biosynthesis is genetically determined and can be changed or extended by selecting proper mutants.

In most species production can be inhibited by certain components in the growth media, such as free ferric ions, amino acids, carbon sources, or several of these. Catabolite repression and availability of nucleic acid precursors are also thought to play a role in alkaline protease synthesis. The concentration of purely carbonaceous medium components should normally be kept on a low level. This can be achieved by incremental feeding of the C source, e.g. glucose, keeping its concentration at a range of 0.4 to 1 per cent. High concentrations of C sources yield excess organic acids leading to a decrease in pH, which is accompanied by a decrease in alkaline protease production. On the other hand, Keay have shown that the enzyme yield can be largely enhanced when the synthesis of protease is accompanied by an increase in pH of the culture. Such an increase in pH can be reached, for example, by using an organic acid (or its salts) as major C source. Niwa observed good yields of protease under this condition, and the results of Kline Dion, and Maxwell confirm this assumption.

Bacterial alkaline proteases are produced exclusively by the submerged culture methods. Amounts of more than 1 g protease per litre culture liquor are quite usual. With specially selected strains markedly higher yields are possible; e.g. *Bacillus subtilis* strain AJ 3266 can produce more than 10 g enzyme per litre.

Continuous fermentation techniques do not seem to have been employed in the industrial production of alkaline protease, but there are several publications describing continuous culture on the laboratory scale. Heineken and O'Connor observed that a continuous fermentation of *Bacillus subtilis* yielded mutants with lower protease productivity.

Fungal alkaline proteases are mainly produced from *Aspergillus* species, in both solid substrate and deep tank fermentations. Solid substrate cultures, extensively used in Japan, are carried out with wheat or rice bran or whole grains as the basic substrate. It has been shown that NH_4^+ ions strongly inhibit production of the enzyme, while nitrates and Na salts of aspartic and glutamic acids promote its formation. Na salts of organic acids had the same effect. Probably the effect of all these compounds on protease synthesis is produced through their influence on pH development during cultivation, as described previously for α-amylase. The processing of culture filtrates or clarified extracts follows the general description of recovery.

Serine alkaline proteases of bacterial origin are used in large amounts in laundering and to a lesser extent in leather tanning and the food industry.

Metalloproteinases

Enzymes of this type play a less important role in commercial applications than the serine and acid proteinases. This is mainly due to their relatively poor stability. Metalloproteinases exhibit maximum activity at pH 7 to 8. In the majority of cases they contain a Zn atom in their active centre. They are inhibited by metal chelating agents such as ethylenediaminetetraacetate (EDTA) or *o*-phenanthroline (OP), but not by DFP or thiol reagents.

Regarding their pH activity metalloproteinases are divided into neutral and alkaline types. The neutral enzymes all have pH optima around pH 7. Their molecular weights are in the range of 35,000–45,000. The isoelectric points of the proteinases from *Bacillus subtilis*, *Pseudomonas aeruginosa*, and *Streptomyces* have been determined to be at pH 9.0, 5.9, and 4.2, respectively. Neutral metalloproteinases of bacterial and fungal origin are specific toward hydrophobic or bulky amino acid residues on the amino side of the cleavage point. In general, these enzymes are the least stable of the microbial proteases. The enzyme from *Bacillus subtilis* retains only about 10 per cent of its activity after treatment at 60°C and pH 7 for 15 min. *Bacillus thermoproteolyticus* produces a very stable neutral protease (thermolysin) that retains 50 per cent of its activity after 60 min. at 80°C. Neutral proteases tend to undergo very rapid autolysis, which makes their recovery and application difficult.

Neutral proteases are widespread in micro-organisms, both fungi and bacteria. Strains used for industrial production belong to the genera *Aspergillus*, *Bacillus*, and *Streptomyces*. Many of the organisms used for commercial production of neutral proteases also produce alkaline or acid proteases. Only few strains have been found which synthesise neutral protease free of accompanying serine and acid proteases. Such strains are, for example, *Bacillus cereus* ATCC 14579 and NCTC 945, *Bacillus megaterium* ATCC 14 581 and MA, and *Bacillus polymyxa* ATCC 842.

The formation of neutral proteases by bacteria does not seem to be correlated with sporulation. In *Bacillus subtilis* enzyme synthesis is subject to catabolite repression. Fogarty and Griffin found that the enzyme was produced irrespective of the C source used, but that the nature of the peptone had a marked effect on protease accumulation. Without pH adjustment during the fermentation, the culture produced the neutral protease parallel to growth, and enzyme formation reached its maximum toward the end of the log phase. The yield was 15 times that obtained when the culture was run with a 'fixed' pH of 6.8. According to Kalunyants the best medium pH (6.9) can be achieved by separate sterilisation of carbohydrate and N-containing compounds of the medium. Variable temperature during fermentation (45°C at the beginning and lowering it to 43°C, 40°C, and 37°C, respectively during the 1st, 2nd, and 3rd 2-hr period) was preferable to a constant temperature. Zn^{++}, Ca^{++}, and Mn^{++} exert a beneficial effect on the level of the metalloproteinase, as described by several authors.

In cultures of *Aspergillus* species, biosynthesis and extemalisation of neutral protease are repressed by low molecular weight sources of C, N, and S. Protease production and release occur when the medium is deficient for any of these elements. For *Aspergillus terricola* it has been shown that the enzyme accumulation in the medium was maximal when the N/C ratio was 0.5.

Due to their high instability, processing of metalloproteinases may lead to high activity losses. Therefore, the main problem in the concentration and purification of the enzyme is its stabilisation. This can be achieved by strictly observing the tolerated rage of pH, by the presence of metal ions (Zn^{++} for activity, Ca^{++} for stability), and by elimination of alkaline protease activity. One has also to take into consideration that the pigment complex produced by an organism can act as an inhibitor of the neutral protease of this organism, as shown for *Bacillus mesentericus* by Velcheva and Kolev.

The application of neutral metalloproteinases is very limited because of the mentioned instability of these enzymes. Actual and potential uses are: treatment of beer, application in bakeries, and reduction of dental plaque in humans.

Acid proteinases

These enzymes are without doubt the most interesting group of proteases with respect to use in the food industry. They are characterised by maximum activity and stability at pH 2.0–5.0. The molecular weight is around 35,000. Acid proteinases are low in basic amino acid content and have low isoelectric points. They are insensitive to SH-reagents, metal chelators, heavy metals, and DFP and are generally stable in the acid pH range (pH 2–6), but are rapidly inactivated at higher pH values. The acid proteases exhibit limited esterase activity, but split a wide range of peptide bonds.

Acid proteinases of commercial importance are prepared exclusively from fungal sources and are tentatively divided into two subgroups by their physiological characteristics: pepsin-like acid proteinases and rennin-like proteinases.

Pepsin-like acid proteinases have usually been reported in the group of black aspergilli, such as *Aspergillus niger*, *Aspergillus awamori*, *Aspergillus usamii*, and *Aspergillus saitoi*, but also occur in species of *Penicillium*, *Rhizopus*, and others. To a large extent they are produced in solid substrate cultures. Biosynthesis of these enzymes is favoured by high C/N ratios. Inorganic N sources show an inhibiting effect on the production of acid proteinases, whereas peptone-was found highly effective in inducing this enzyme in a strain of *Aspergillus niger*. The inducing action of peptone was much more, remarkable when this material was added to a culture during the growth phase than in the stationary phase. From this finding it is evident that the increase in acid proteinase activity observed when growth has ceased represents *de novo* synthesis. It has been shown to be due to exhaustion of adenine-group growth substances. Adipic and glutaric acids were also highly effective in supporting enzyme formation, but amino acids tested and some dipeptides were less effective than peptone.

Acid proteinases of the pepsin type play an important role in the production of fermented foods by moulds from soyabeans, rice, and other cereals. They are further used in the baking industry for the modification of wheat proteins in bread doughs.

Rennin-like acid proteinases are produced by strains of *Mucor miehei*, *Mucor pusillus*, *Endothia parasitica*, and *Trametes sanguinea*. The enzyme from the *Mucor* species has been, and is now, produced by the solid substrate culture method. However, Aunstrup isolated strain of *Mucor miehei*, which he succeeded in growing in submerged culture for rennin production.

Microbial rennet substitutes must be freed of lipase to avoid rancidity of the cheese. This can be achieved by controlled heating or by adjusting to a low pH. Unspecific proteolytic enzymes, which may cause bitter taste of the cheese, must also be removed. Their separation is obtained by adsorption on aluminosilicates. These adsorbents are particularly advantageous because they can be mixed into the culture liquid at the end of the fermentation, even in the presence of the medium, with a good separation effect and without any loss of milk clotting activity. Bentonite, permutite, and attapulgite are also suitable.

Because of their particular properties, rennet-like microbial proteases are used for clotting of milk in cheese manufacture. The process is based on the coagulation of casein under the influence of the rennet-like protease. It is known that the casein in milk is mainly composed of α_s-, β-, and κ-casein. In particular, κ-casein plays an important role in the coagulation process, because it keeps the casein micelles present in milk in solution and protects them against flocculation by calcium ions. The clotting effect of rennins consists of the destabilisation of the casein complex. Two phases can be distinguished:

1. The primary or enzymatic phase, in which the protective colloid (κ-casein) of the casein micelle is broken down and a glycomacropeptide is split off as follows:

$$\kappa\text{-casein} \xrightarrow{\text{Rennin}} \underset{\text{insobule}}{\text{para-}\kappa\text{-casein}}\text{-}\underset{\text{soluble}}{\text{glycomacropeptide}}$$

2. The secondary or nonenzymatic phase, in which the coagulum is formed under the influence of calcium ions.

The primary phase has a temperature coefficient, Q_{10}, of about two (like most enzyme reactions), whereas the secondary phase has a Q_{10} of about 15. Therefore, it is reasonable to develop a system for continuous clotting of milk employing immobilised enzymes. Consequently, in a two-stage enzyme reactor the enzymatic phase is conducted in the first stage at low temperatures in order to inhibit the nonenzymatic phase. In the second stage subsequent warming clots the milk by the action of calcium ions.

Lipases

The enzyme lipase catalyses the reaction:

$$\text{Triacylglycerol-}H_2O \xrightarrow{\text{Lipase}} \text{Diacylglycerol} + \text{fatty acid anion}$$

This reaction goes to completion, i.e. until glycerol and free fatty acids are formed.

Many, perhaps most, bacteria and fungi produce lipase. Potent producers are among the fat-producing micro-organisms. However, no distinct relationship between capacity of fat production and lipase production has been found. The enzyme from *Candida cylindracea* is commercially available. Other producers of lipase are, e.g. *Geotrichum candidum, Rhizopus arrhizus*, and *Aspergillus niger*. *Geotrichum candidum* lipase is unique with respect to its specificity properties. The enzyme from the thermophile, *Humicola lanuginosa*, exhibits better thermostability.

The production of lipase is markedly affected by many factors. Generally, synthetic media produce Iower yields of lipase then complex media. Lipase formation is highly dependent on nutrient and physical conditions. In a number of case, addition of lipid material or fatty acids to the culture medium was found to enhance lipase production. In contrast, Smith and Alford observed inhibition of lipase formation in media supplied with lard, sodium oleate, or salts of other unsaturated fatty acids. These authors also reported that the inhibition was prevented, but not reversed, for example, by some divalent cations and Tweens. Glucose is unsuitable as the C source and ammonium ions seem to be unsuitable as the N source. Incorporation of $CaCO_3$ acts differently, depending on the strain used. In some cases promotion of lipase production was observed, whereas in other cases investigators found an inhibitory effect using $CaCO_3$. Sometimes lipase preparations prove to be unstable. This can be due to the presence of proteases, the removal of which will lead to stable products.

Despite the fact that there is a considerable industry based on fats there is little industrial application of lipases. The main use is as digestive aid.

Glucose Oxidase

Glucose oxidase (β-D-glucose:oxygen oxidoreductase, EC 1.1.3.4), also known as notatin, acts in the presence of molecular oxygen to convert glucose to gluconic acid and hydrogen peroxide:

$$C_6H_{12}O_6 + O_2 + H_2 \xrightarrow{\text{Glucose oxidase}} C_6H_{12}O_7 + H_2O_2$$

It is highly specific for β-D-glucose, although slight activily is found with 2-deoxyglucose. At present, glucose oxidase is commercially prepared from *Aspergillus niger* and *Penicillium amagasakiense* in submerged culture. It has also been reported that *Penicillium notatum* and *Penicillium chrysogenum* synthesise glucose oxidase on liquid media in surface culture, but not in submerged culture.

During the growth of the fungal culture, the enzyme occurs in the phase following the lag phase. By feeding glucose this phase can be extended and thus the enzyme yield enhanced. The special culture conditions, however, depend markedly on microbial strains used. For example, beet molasses has proved to be a suitable carbon source in *Penicillium purpurogenum*, but not suitable in *Penicillium chrysogenum*; high aeration rates supported enzyme synthesis in *Penicillium purpurogenum*, but did not in *Penicillium chrysogenum*. For the purpose of concentration and purification, glucose oxidase must be separated from cells by extraction. The crude solutions also contain catalase which may interfere with glucose oxidase in some applications. In these cases separation is conducted by adsorption of the catalase on alumina or kaolin. For preparation of solid products, glucose oxidase can be precipitated by neutral salts or solvents, but liquid preparations are preferred.

Glucose oxidase is a good glycoprotein. The enzyme from *Aspergillus niger* contains 10.5 per cent carbohydrate, which is believed to contribute to the stability and not to affect the overall structure. Two FAD molecules per molecule of enzyme act as the prosthetic group. The molecular weight of the *Aspergillus niger* enzyme is 1,86,000, that of *Penicillium notatum* is 1,52,000. The optimum pH of glucose oxidase is about 5.5. The enzyme is fully stable between pH 4 and 6 at 40°C for 2 hrs. Specially stabilised preparations for use at pH 2.5 are available. Use above pH 8.0 may be possible, but requires a high, glucose concentration. Glucose oxidase is very unstable above 50°C, although glucose has some protective effect. Normal increase in activity caused by increased temperatures is counteracted by decrease in dissolved oxygen concentration at higher temperatures.

According to the reaction equation, glucose oxidase can be used in order to remove glucose or oxygen or to form hydrogen peroxide or gluconic acid. Indeed, in food processing glucose oxidase finds application for removal of residual glucose prior to the preparation of dried eggs or to remove it from other products in order to reduce nonenzymatic browning. It is highly effective in removing residual oxygen from beer, wine, fruit juices, high fat products (mayonnaise), or packaged dehydrated foods. In the treatment of flour, when the formation of peroxide is desired, only catalase-free preparations can be used.

Catalase

Catalase (EC 1.11.1.6) splits hydrogen peroxide to water and oxygen:

$$2H_2O_2 \xrightarrow{\text{Catalase}} 2H_2O + O_2$$

The enzyme is widely distributed in micro-organisms. Its biological role has been studied by a number of investigators. In methanol-utilising yeasts, it is generally accepted that catalase must be involved in the metabolism of methanol, since hydrogen peroxide is liberated during methanol oxidation by alcohol dehydrogenase. This suggestion is supported by the fact that catalase is markedly induced when the yeast cells are grown on methanol. Commercially, catalase is prepared from *Aspergillus niger*, *Penicillium vitale*, or *Micrococcus lysodeikticus*. It is a hemo-protein containing 4 ferri-protoporphyrin prosthetic groups per molecule of enzyme, with a molecular weight of 2,50,000. The optimum pH of *Aspergillus niger* catalase is at pH 6.0; 75 per cent of its optimum activity occurs between pH 3.0 and 9.0. The enzyme is inactivated by cyanides, phenol, alkali, urea, freezing, and by sunlight under aerobic conditions. Production of catalase is conducted in deep tank cultivation. Biosynthesis occurs simultaneously with glucose oxidase formation. The ratio of these enzymes is controlled by quality and quantity of the inoculum, the composition of the medium, and by aeration conditions.

With *Penicillium vitale*, Nikolskaya found that greater amounts of catalase were accumulated when the C:N ratio in the medium was as high as 12:1. Optimum pH was 4.5–5.5 during the first 48–72 hr of growth at 26°–27°C. L-Cysteine and DL-methionine promoted catalase production by the same fungus. The stimulating effect observed with $CaCO_3$ was demonstrated to be not the result of neutralisation. Ca^{++} was believed to facilitate the transport of the enzyme from the mycelium into the medium.

Catalase produced by the directed biosynthesis can be separated selectively from extracts containing catalase and glucose oxidase. The recovery of catalase from *Micrococcus lysodeikticus* starts with lysing of the cells in a solution of sodium chloride (0.5 to 2 per cent). The next step is fractionation of the lysate by centrifugation of a mixture of the lysate, an organic solvent (ethanol in an end concentration of 40 to 50 per cent v/v), and a salt (sodium or potassium chloride adjusted to a concentration of 1 to 2 per cent either before or after addition of the solvent). The dissolved catalase can then be precipitated from solution by adding ethanol to a final concentration of 75 per cent. Separation of catalase from solution can also be conducted by adsorption methods using alumina or kaolin as adsorbents. For commercial application, liquid preparations are preferred. Catalase finds application wherever the removal of hydrogen peroxide is required or the controlled release of oxygen from hydrogen peroxide is desired. Therefore, in the food industry catalase is employed to remove the excess of hydrogen peroxide used for cold sterilisation in milk and cheese processing. Catalase may also be employed in cake baking as well as in irradiated foods in the process of which hydrogen peroxide is formed.

Glucose Isomerase

This enzyme converts D-glucose to D-fructose. The main substrate of this enzyme, however, is xylose and, indeed, the glucose isomerising enzyme is a D-xyloseketoisomerase (EC 5.3.1.5) with side activities to D-glucose and D-ribose. A large number of genera of bacteria and some yeasts have been found to produce a glucose isomerising enzyme.

But the strains most widely used as sources for commercial production are members of the genus *Streptomyces*. Outtrup found thermophilic atypical variants of *Bacillus coagulans* to be particularly suited as the enzyme source due to the properties of its glucose isomerase. The isomerase in *Streptomyces* is an inducible enzyme which requires the presence of D-xylose in the culture medium for its production. Sanchez and Quinto selected a double mutant strain of *Streptomyces phaeochromogenes* which was able to produce glucose isomerase constitutively and was insensitive to catabolite repression. Diers, who worked with *Bacillus coagulans* in chemostat cultures, found that glucose isomerase production of this strain was regulated mainly by catabolic repression. The latter occurred when inorganic compounds limited growth, whereas carbon limitation and particularly carbon-oxygen limitation were advantageous.

Media for the commercial production of glucose isomerase are based on xylan or xylan-containing raw materials such as wheat bran, maize husks, sulphite liquor, etc. As the enzyme is of the intracellular type, it can be used in the form of whole cells. Takasaki described the immobilisation of Streptomyces cells by heating them to over 60°C for about 10 min. This procedure prevents autolysis of the cells and 'fixes' the glucose isomerase. Purified preparations with higher activities can be obtained by application of the usual methods of cell rupture and solubilising of the enzyme. After discarding the cellular material, the glucose isomerase is then adsorbed on DEAE-cellulose or a similar material, recovered, and washed.

Co and Mg ions are well-known activators of glucose isomerase and essential for obtaining maximum activity, whereas copper, nickel, and zinc are strong inhibitors. The other characteristics such as pH and temperature activity and stability vary with the enzyme source and depend on whether the enzyme is in the native or an immobilised form. In order to prevent alkaline conversion of fructose produced during the isomerisation process it is necessary that the employed glucose isomerase be highly active at pH 6.5.

Principal producers and users of glucose isomerase are found in the corn wet milling industry, where the enzyme is used to convert glucose in corn syrups to fructose. Crystalline D-glucose can be used as a substrate, for D-fructose production. However, in industrial practice, high D.E. starch hydrolysates are a more economical source of D-glucose. Such a hydrolysate can be prepared by the combined action of bacterial α-amylase, amyloglucosidase, and isoamylase on starch to yield a glucose syrup of about 95–98 D.E. Subsequent isomerisation is carried out in agitated vessels at 65°C and pH 7 for 18 to 24 hr using, for example, immobilised cells. Following conversion to D-fructose, the immobilised enzyme is removed and the liquor further treated to give an invert sugar syrup of about 45 per cent fructose and 55 per cent glucose. The process is illustrated in Fig. 11.6. It will probably replace invertase in the manufacture of invert sugar.

PENICILLIN ACYLASE

Penicillin acylase (penicillin amidohydrolase, EC 3.5.1.11) is widely distributed among micro-organisms, including bacteria, yeast and filamentous fungi. It is used on an industrial scale for the production of 6-aminopenicillanic acid, the starting material for the synthesis of semi-synthetic penicillins. Its *in vivo* role remains unclear, however, and the observation that expression of the *Escherichia coli* enzyme *in vivo* is regulated by both temperature and phenylacetic acid has prompted speculation that the enzyme could be involved in the assimilation of aromatic compounds as carbon sources in the organism's free-living mode. The mature *E. coli* enzyme is a periplasmic 80K heterodimer of A and B chains synthesised as a single cytoplasmic precursor containing a 26-amino-acid signal sequence to direct export to the cytoplasm and a 54-amino-acid spacer between the A and B chains which may influence the final folding of the chains. The N-terminal serine of the B chain reacts with phenylmethylsulphonyl fluoride, which is consistent with a catalytic role for the serine hydroxyl group. Modifying this serine to a cystein inactivates the enzyme, whereas threonine, arginine or glycine substitution prevents *in vivo* processing of the enzyme, indicating that this must be an important recognition site for cleavage. Here we report the crystal structure of penicillin acylase at 1.9 Å resolution. On analysis it shows that the environment of the catalytically active *N*-terminal serine of the B chain contains no adjacent histidine equivalent to that found in the serine proteases. The nearest base to the hydroxyl of this serine is its own α-amino group, which may act by a new mechanism to endow the enzyme with its catalytic properties.

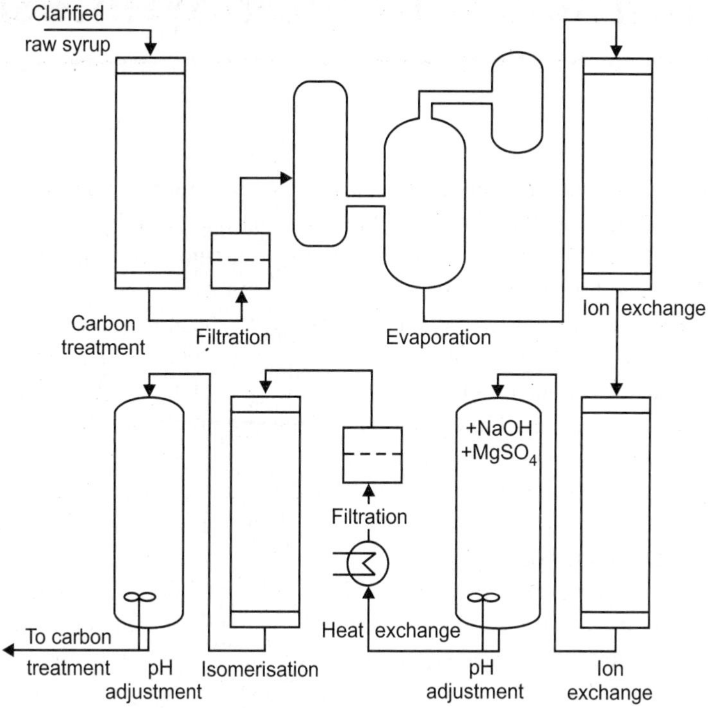

Fig. 11.6. Flowsheet of glucose isomerisation.

Chapter 12

Vitamins and their Synthesis

INTRODUCTION

A vitamin is an organic compound required as a nutrient in tiny amounts by an organism. A compound is called a vitamin when it cannot be synthesised in sufficient quantities by an organism, and must be obtained from the diet. Thus, the term is conditional both on the circumstances and the particular organism. For example, ascorbic acid functions as vitamin C for some animals but not others, and vitamins D and K are required in the human diet only in certain circumstances. The term vitamin does not include other essential nutrients such as dietary minerals, essential fatty acids, or essential amino acids, nor does it encompass the large number of other nutrients that promote health but are otherwise required less often.

Vitamins are classified by their biological and chemical activity, not their structure. Thus, each vitamin may refer to several vitamer compounds that all show the biological activity associated with a particular vitamin. Such a set of chemicals are grouped under an alphabetised vitamin 'generic descriptor' title, such as 'vitamin A', which includes the compounds retinal, retinol, and many carotenoids. Vitamers are often inter-converted in the body.

Vitamins have diverse biochemical functions, including function as hormones (e.g. vitamin D), antioxidants (e.g. vitamin E), and mediators of cell signaling and regulators of cell and tissue growth and differentiation (e.g. vitamin A). The largest number of vitamins (e.g. B complex vitamins) function as precursors for enzyme cofactor bio-molecules (coenzymes), that help act as catalysts and substrates in metabolism.

When acting as part of a catalyst, vitamins are bound to enzymes and are called prosthetic groups. For example, biotin is part of enzymes involved in making fatty acids. Vitamins also act as coenzymes to carry chemical groups between enzymes. For example, folic acid carries various forms of carbon group—methyl, formyl and methylene—in the cell. Although these roles in assisting enzyme reactions are vitamins' best-known function, the other vitamin functions are equally important.

Until the 1900s, vitamins were obtained solely through food intake, and changes in diet (which, for example, could occur during a particular growing season) can alter the types and amounts of vitamins ingested. Vitamins have been produced as commodity chemicals and made widely available as inexpensive pills for several decades, allowing supplementation of the dietary intake.

CLASSIFICATION OF VITAMINS

Vitamins are classified as either water-soluble or fat soluble. In humans there are 13 vitamins: 4 fat-soluble (A, D, E and K) and 9 water-soluble (8 B vitamins and vitamin C).

Water-soluble Vitamins

Water-soluble vitamins dissolve easily in water, and in general, are readily excreted from the body, to the degree that urinary output is a strong predictor of vitamin consumption. Because they are not readily stored, consistent daily intake is important. Many types of water-soluble vitamins are synthesised by bacteria.

Fat-soluble Vitamins

Fat-soluble vitamins are absorbed through the intestinal tract with the help of lipids (fats). Because they are more likely to accumulate in the body, they are more likely to lead to hypervitaminosis than are water-soluble vitamins. Fat-soluble vitamin regulation is of particular significance in cystic fibrosis.

NECESSITY OF VITAMINS

In Nutrition and Diseases

Vitamins are essential for the normal growth and development of a multicellular organism. Using the genetic blueprint inherited from its parents, a fetus begins to develop, at the moment of conception, from the nutrients it absorbs. It requires certain vitamins and minerals to be present at certain times. These nutrients facilitate the chemical reactions that produce among other things, skin, bone, and muscle. If there is serious deficiency in one or more of these nutrients, a child may develop a deficiency disease. Even minor deficiencies may cause permanent damage.

For the most part, vitamins are obtained with food, but a few are obtained by other means. For example, micro-organisms in the intestine—commonly known as 'gut flora'—produce vitamin K and biotin, while one form of vitamin D is synthesised in the skin with the help of the natural ultraviolet wavelength of sunlight. Humans can produce some vitamins from precursors they consume. Examples include vitamin A, produced from β-carotene, and niacin, from the amino acid tryptophan.

Once growth and development are completed, vitamins remain essential nutrients for the healthy maintenance of the cells, tissues, and organs that make up a multicellular organism; they also enable a multicellular life form to efficiently use chemical energy provided by food it eats, and to help process the proteins, carbohydrates, and fats required for respiration.

Deficiencies

Deficiencies of vitamins are classified as either primary or secondary. A primary deficiency occurs when an organism does not get enough of the vitamin in its food. A secondary deficiency may be due to an underlying disorder that prevents or limits the absorption or use of the vitamin, due to a 'lifestyle factor', such as smoking, excessive alcohol consumption, or the use of medications that interfere with the absorption or use of the vitamin. People who eat a varied diet are unlikely to develop a severe primary vitamin deficiency. In contrast, restrictive diets have the potential to cause prolonged vitamin deficits, which may result in often painful and potentially deadly diseases.

Because human bodies do not store most vitamins, humans must consume them regularly to avoid deficiency. Human bodily stores for different vitamins vary widely; vitamins A, D, and B_{12} are stored in significant amounts in the human body, mainly in the liver, and an adult human's diet may be deficient in vitamins A and B_{12} for many months before developing a deficiency condition. Vitamin B_3 is not stored in the human body in significant amounts, so stores may only last a couple of weeks.

Well-known human vitamin deficiencies involve thiamine (beriberi), niacin (pellagra), vitamin C (scurvy) and vitamin D (rickets). In much of the developed world, such deficiencies are rare; this is due to an adequate supply of food; and the addition of vitamins and minerals to common foods, often called fortification. Some evidence also suggests that there is a link between vitamin deficiency and mental disorders. There is some research suggesting appropriate levels of vitamin D may reduce cancer risk.

Side Effects and Overdose

In large doses, some vitamins have documented side effects that tend to be more severe with a larger dosage. The likelihood of consuming too much of any vitamin from food is remote, but overdosing from vitamin supplementation does occur. At high enough dosages some vitamins cause side effects such as nausea, diarrhea, and vomiting. When side effects emerge, recovery is often accomplished by reducing the dosage. The concentrations of vitamins an individual can tolerate vary widely, and appear to be related to age and state of health.

Supplements

Dietary supplements, often containing vitamins, are used to ensure that adequate amounts of nutrients are obtained on a daily basis, if optimal amounts of the nutrients cannot be obtained through a varied diet. Scientific evidence supporting the benefits of some dietary supplements is well established for certain health conditions, but others need further study.

VITAMINS FEED WE NEED

Our body is one powerful machine, capable of doing all sorts of things by itself. But one thing it cannot do is make vitamins. That's where food comes in. Your body is able to get the vitamins it needs from the foods we eat because different foods contain different vitamins. The key is to eat different foods to get an assortment of vitamins. Though some kids take a daily vitamin, most kids don't need one if they're eating a variety of healthy foods. Now, let's look more closely at vitamins—from A to K.

Vitamin A

This vitamin plays a really big part in eyesight. Vitamin A helps us in colour, too, from the brightest yellow to the darkest purple. In addition, it helps us grow properly and aids in healthy skin.

The foods rich in vitamin A are: (i) milk fortified with vitamin A, (ii) liver, (iiii) orange fruits and vegetables (like cantaloupe, carrots, sweet potatoes), and (iv) dark green leafy vegetables (like kale, collards, spinach).

Vitamin B

There's more than one B vitamin. Here's the list: B_1, B_2, B_6, B_{12}, niacin, folic acid, biotin, and pantothenic acid. The B vitamins are important in metabolic activity—this means that they help make energy and set it free when your body needs it. So the next time you're running to third base, thank those B vitamins. This group of vitamins is also involved in making red blood cells, which carry oxygen throughout your body. Every part of your body needs oxygen to work properly, so these B vitamins have a really important job. The foods rich in vitamin B are: (i) whole grains, such as wheat and oats, (ii) fish and seafood, (iii) poultry and meats, (iv) eggs, (v) dairy products, like milk and yogurt, (vi) leafy green vegetables, and (vii) beans and peas.

Vitamin C

This vitamin is important for keeping body tissues, such as gums and muscles in good shape. C is also key if you get a cut or wound because it helps you heal. This vitamin also helps your body resist infection. This means that even though you can't always avoid getting sick, vitamin C makes it a little harder for your body to become infected with an illness. The foods rich in vitamin C are: (i) citrus fruits, like oranges, (ii) cantaloupe, (iii) strawberries, (iv) tomatoes, (v) broccoli, (vi) cabbage, (vii) kiwi fruit, and (viii) sweet red peppers.

Vitamin D

Vitamin D is the vitamin you need for strong bones. It is also great for forming strong teeth. Vitamin D even lends a hand to an important mineral—it helps your body absorb the amount of calcium it needs.

The foods rich in vitamin D are: (i) milk fortified with vitamin D, (ii) fish, (iii) egg yolks, (iv) liver, and (v) fortified cereal.

Vitamin E

Everybody needs E. This hard-working vitamin maintains a lot of your body's tissues, like the ones in eyes, skin, and liver. It protects lungs from becoming damaged by polluted air. And it is important for the formation of red blood cells. The foods rich in vitamin E are: (i) whole grains, such as wheat and oats, (ii) wheat germ, (iii) leafy green vegetables, (iv) sardines, (v) egg yolks, and (vi) nuts and seeds.

Vitamin K

Vitamin K is the clotmaster. When we get cut our blood did something special called clotting. This is when certain cells in our blood act like glue and stick together at the surface of the cut to help stop the bleeding. The foods rich in vitamin K are: (i) leafy green vegetables, (ii) dairy products, like milk and yogurt, (iii) broccoli, (vi) soyabean oil.

VITAMIN B_{12}

Vitamin B_{12} is a water soluble vitamin with a key role in the normal functioning of the brain and nervous system, and for the formation of blood. It is one of the eight B vitamins. It is normally involved in the metabolism of every cell of the body, especially affecting DNA synthesis and regulation, but also fatty acid synthesis and energy production.

Vitamin B_{12} is the name for a class of chemically-related compounds, all of which have vitamin activity. It is structurally the most complicated vitamin and it contains the biochemically rare element cobalt. Biosynthesis of the basic structure of the vitamin can only be accomplished by bacteria, but conversion between different forms of the vitamin can be accomplished in the human body. A common synthetic form of the vitamin, cyanocobalamin, does not occur in nature, but is used in many pharmaceuticals and supplements, and as a food additive, due to its stability and lower cost. In the body it is converted to the physiological forms, methylcobalamin and adenosylcobalamin, leaving behind the cyanide, albeit in minimal concentration. More recently, hydroxocobalamin, methylcobalamin and, adenosylcobalamin can also be found in more expensive pharmacological products and food supplements. The utility of these is presently debated.

Historically, vitamin B_{12} was discovered from its relationship to the disease pernicious anemia, which is an autoimmune disease that destroys parietal cells in the stomach that secrete intrinsic factor. Intrinsic factor is crucial for the normal absorption of B_{12}, therefore, a lack of intrinsic factor, as seen in pernicious anemia, causes a vitamin B_{12} deficiency. Many other subtler kinds of vitamin B_{12} deficiency, and their biochemical effects, have since been elucidated.

Structure

Vitamin B_{12} is a collection of cobalt and corrin ring molecules which are defined by their particular vitamin function in the body. All of the substrate cobalt-corrin molecules from which B_{12} is made must be synthesised by bacteria. However, after this synthesis is complete, the body has a limited power to convert any form of B_{12} to another, by means of enzymatically removing certain prosthetic chemical groups from the cobalt atom. Structure of vitamin B_{12} is given in Fig. 12.1.

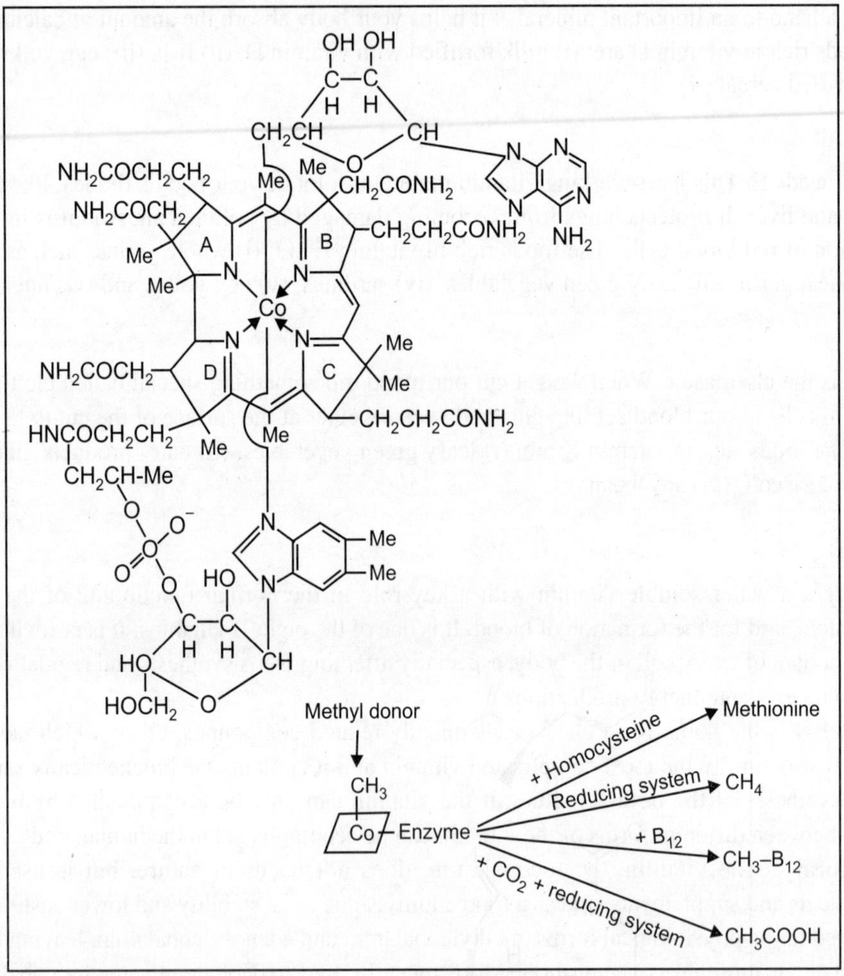

Fig. 12.1. Structure of vitamin B_{12}.

Cyanocobalamin is one such compound that is a vitamin in this B complex, because it can be metabolised in the body to an active co-enzyme form. However, the cyanocobalamin form of B_{12} does not occur in nature normally, but is a by-product of the fact that other forms of B_{12} are avid binders of cyanide (—CN) which they pick up in the process of activated charcoal purification of the vitamin after it is made by bacteria in the commercial process. Since the cyanocobalamin form of B_{12} is deeply red coloured, easy to crystallise, and is not sensitive to air-oxidation, it is typically used as a form of B_{12} for food additives and in many common multivitamins. However, this form is not perfectly synonymous with B_{12}, inasmuch as a number of substances (vitamers) have B_{12} vitamin activity and can properly be labelled vitamin B_{12}, and cyanocobalamin is but one of them. (Thus, all cyanocobalamin is vitamin B_{12}, but not all vitamin B_{12} is cyanocobalamin). B_{12} is the most chemically complex of all the vitamins. The structure of B_{12} is based on a corrin ring, which is similar to the porphyrin ring found in heme, chlorophyll, and cytochrome. The central metal ion is cobalt. Four of the six coordination sites are provided by the corrin ring, and a fifth by a dimethylbenzimidazole group. The sixth coordination site, the centre of reactivity, is variable, being a cyano group (—CN), a hydroxyl group (—OH), a methyl group (—CH_3) or a 5′-deoxyadenosyl group (here the C5′ atom of the deoxyribose forms the covalent bond with Co), respectively, to yield the four B_{12} forms mentioned above. Historically, the covalent C–Co bond is one of first examples of carbon-metal bonds to be discovered in biology. The hydrogenases and, by necessity, enzymes associated with cobalt utilisation, involve metal-carbon bonds.

Synthesis

Vitamin B_{12} cannot be made by plants or animals as only bacteria have the enzymes required for its synthesis. The total synthesis of B_{12} was reported by Robert Burns Woodward and Albert Eschenmoser, and remains one of the classic feats of organic synthesis. Figure 12.2 shows the synthesis of vitamin B_{12}.

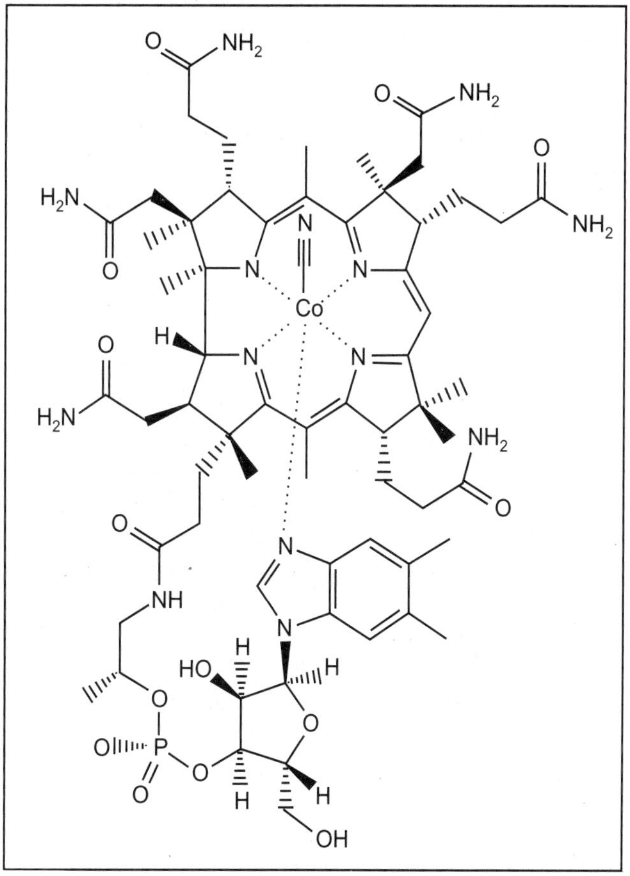

Fig. 12.2. Synthesis of vitamin B_{12}.

Species from the following genera are known to synthesise B_{12}: *Aerobacter*, *Agrobacterium*, *Alcaligenes*, *Azotobacter*, *Bacillus*, *Clostridium*, *Corynebacterium*, *Flavobacterium*, *Micromonospora*, *Mycobacterium*, *Nocardia*, *Propionibacterium*, *Protaminobacter*, *Proteus*, *Pseudomonas*, *Rhizobium*, *Salmonella*, *Serratia*, *Streptomyces*, *Streptococcus* and *Xanthomonas*. Industrial production of B_{12} is through fermentation of selected micro-organisms. *Streptomyces griseus*, a bacterium once thought to be a yeast, was the commercial source of vitamin B_{12} for many years. The species *Pseudomonas denitrificans* and *Propionibacterium shermanii* are more commonly used today. These are frequently grown under special conditions to enhance yield, and at least one company, Rhône-Poulenc of France, at one point used genetically engineered versions of one or both of these species. It is not clear whether Sanofi-Aventis, the company which the pharmaceutical division of Rhône-Poulenc merged into, has continued the use of genetically modified organisms.

Functions

Vitamin B_{12} is normally involved in the metabolism of every cell of the body, especially affecting the DNA synthesis and regulation but also fatty acid synthesis and energy production. However, many (though not all) of the effects of functions of B_{12} can be replaced by sufficient quantities of folic acid (another B vitamin), since B_{12} is used to regenerate folate in the body. Most 'B_{12} deficient symptoms' are actually folate deficient symptoms, since they include all the effects of pernicious anemia and megaloblastosis, which are due to poor synthesis of DNA when the body does not have a proper supply of folic acid for the production of thymine. When sufficient folic acid is available, all known B_{12} related deficiency syndromes normalise, save those narrowly connected with the B_{12} dependent enzymes methylmalonyl coenzyme A mutase (MUT), and 5-methyltetrahydrofolate-homocysteine methyltransferase (MTR), also known as methionine synthase; and the buildup of their respective substrates (methylmalonic acid, MMA) and homocysteine.

Human Absorption and Distribution

The human physiology of vitamin B_{12} is complex, and therefore is prone to mishaps leading to vitamin B_{12} deficiency. Unlike most nutrients, absorption of vitamin B_{12} actually begins in the mouth where small amounts of unbound crystalline B_{12} can be absorbed through the mucosa membrane. Food protein bound vitamin B_{12} is digested in the stomach by proteolytic gastric enzymes, which require an acid pH (Even small amounts of B_{12} taken in supplements bypasses these steps, and thus, any need for gastric acid, which may be blocked by antacid drugs). Once the B_{12} is freed from the proteins in food, R-proteins, such as haptocorrins and cobalaphilins, are secreted, which bind to free vitamin B_{12} to form a B_{12}-R complex. Also in the stomach, IF, a protein synthesised by gastric parietal cells, is secreted in response to histamine, gastrin and pentagastrin, as well as the presence of food. If this step fails due to gastric parietal cell atrophy (the problem in pernicious anemia), sufficient B_{12} is not absorbed later on, unless administered orally in relatively massive doses (500 to 1000 mcg/day). Due to the complexity of B_{12} absorption, geriatric patients, many of whom are hypoacidic due to reduced parietal cell function, have an increased risk of B_{12} deficiency.

Symptoms and Damage from Deficiency

Vitamin B_{12} deficiency can potentially cause severe and irreversible damage, especially to the brain and nervous system. At levels only slightly lower than normal, a range of symptoms such as fatigue, depression, and poor memory may be experienced.

However, these symptoms by themselves are too nonspecific to diagnose deficiency of the vitamin.

1. Vitamin B_{12} deficiency can also cause symptoms of mania and psychosis.
2. Vitamin B_{12} deficiency has the following pathomorphology and symptoms.

Clinical symptoms: The main syndrome of vitamin B_{12} deficiency is Biermer's disease (pernicious anemia). It is characterised by a triad of symptoms:

1. Anemia with bone marrow promegaloblastosis (megaloblastic anemia).
2. Gastrointestinal symptoms.
3. Neurological symptoms.

Each of those symptoms can occur either alone or along with others. The neurological complex, defined as *Myelosis funicularis*, consists of the following symptoms:

1. Impaired perception of deep touch, pressure and vibration, abolishment of sense of touch, very annoying and persistent paresthesias.
2. Ataxia of dorsal cord type.
3. Decrease or abolishment of deep muscle-tendon reflexes.
4. Pathological reflexes—Babinski, Rossolimo and others, also severe paresis.

During the course of disease, mental disorders can occur which include: irritability, focus/concentration problems, depressive state with suicidal tendencies, paraphrenia complex. These symptoms may not reverse after correction of hematological abnormalities, and the chance of complete reversal decreases with the length of time the neurological symptoms have been present.

Sources

Foods

Vitamin B_{12} is naturally found in meat (especially liver and shellfish), milk and eggs. Animals, in turn, must obtain it directly or indirectly from bacteria, and these bacteria may inhabit a section of the gut which is posterior to the section where B_{12} is absorbed. Thus, herbivorous animals must either obtain B_{12} from bacteria in their rumens, or (if fermenting plant material in the hindgut) by reingestion of cecotrope faeces. Eggs are often mentioned as a good B_{12} source, but they also contain a factor that blocks absorption. Certain insects such as termites contain B_{12} produced by their gut bacteria, in a manner analogous to ruminant animals.

Natural food sources of B_{12}

Vitamin B_{12} is found in foods that come from animals, including fish, meat, poultry, eggs, milk, and milk products. One half chicken breast provides some 0.3 μg per serving or 6.0 per cent of one's daily value (DV), 3 ounces of beef, 2.4 μg, or 40 per cent of one's DV, one slice of liver 47.9 μg or 780 per cent of DV, and 3 ounces of molluscs 84.1 μg, or 1400 per cent of DV, while one egg provides 0.6 μg or 10 per cent of DV.

Potential sources

The mushroom *Agaricus bisporus* could contain vitamin B_{12}. This can be seen on some nutrition sites. Some sources say it is a form that is not usable as a vitamin B_{12} source.

Legumes have root nodules that contain Rhizobia. These bacteria can create vitamin B_{12}. It may not be a source of vitamin B_{12} for nutrition purposes. Salt rising bread contains *Clostridium perfringens* which is known to create vitamin B_{12}. It may not be available as nutrition.

Supplements

Vitamin B_{12} is provided as a supplement in many processed foods, and is also available in vitamin pill form, including multi-vitamins. Vitamin B_{12} can be supplemented in healthy subjects also by liquid, transdermal patch, nasal spray, or injection and is available singly or in combination with other supplements.

Injection and patches are sometimes used in cases where digestive absorption is impaired, but there is evidence that this course of action may not be necessary with modern high potency oral supplements (such as 500 to 1000 μg or more). Even pernicious anemia can be treated entirely by the oral route. These supplements carry such large doses of the vitamin that 1 to 5 per cent of high oral doses of free crystalline B_{12} is absorbed along the entire intestine by passive diffusion.

Recommendations

The dietary reference intake for an adult ranges from 2 to 3 μg (micrograms) per day. Vitamin B_{12} is believed to be safe when used orally in amounts that do not exceed the recommended dietary allowance (RDA). The RDA for vitamin B_{12} in pregnant women is 2.6 μg per day and 2.8 μg during lactation periods. There is insufficient reliable information available about the safety of consuming greater amounts of vitamin B_{12} during pregnancy.

Allergies

Vitamin B_{12} supplements in theory should be avoided in people sensitive or allergic to cobalamin, cobalt, or any other product ingredients. However, direct allergy to a vitamin or nutrient is extremely rare, and if reported, other causes should be sought.

Side Effects, Contraindications and Warnings

1. Dermatologic: Itching, rash, transitory exanthema, and urticaria have been reported.
2. Gastrointestinal: Diarrhea has been reported.
3. Hematologic: Peripheral vascular thrombosis has been reported. Treatment of vitamin B_{12} deficiency can unmask polycythemia vera, which is characterised by an increase in blood volume and the number of red blood cells.
4. Leber's disease: Vitamin B_{12} in the form of cyanocobalamin is contraindicated in early Leber's disease, which is hereditary optic nerve atrophy. Cyanocobalamin can cause severe and swift optic atrophy, but other forms of vitamin B_{12} are available.

VITAMIN B_{12} AND ANTIBIOTIC

A new streptomycete isolated from soil has been found to produce an antibiotic, to produce vitamin B_{12}, and to oxidise progesterone to 16 α-hydroxyprogesterone. The nitrogen and energy sources in the media used for growth of the culture markedly influenced antibiotic formation, had less effect on vitamin B_{12} biosynthesis, and had hardly any effect on steroid oxidising ability of the mycelium formed.

Methods

Fermention methods

The streptomycete used in these experiments was grown on yeast extract-beef extract agar for four to seven days until sporulation had occurred. These spores were preserved by lyophilisation (with milk as suspending agent) and stored under nitrogen in small vials. These were used as the inoculum source for all experiments mentioned in this report. Fermentation media were inoculated with approximately 5 per cent (by volume) of the mycelial mass obtained during a three day incubation period while the culture was grown under submerged aeration.

In all experiments the culture was passed through two vegetative transfers before use as inoculum for the fermentations. The first inoculum pasage was made on a soyabean meal-glucose medium and the second on a medium containing the same nitrogen and energy sources used in the fermentation phase of the experiments.

The actinomycete was grown in cottonplugged 500 ml Erleumeyer flasks containing 100 ml of medium. After inoculation the flasks were placed on a reciprocating shaker (120 two inch cycles per minute) located in a 25°C constant temperature room. The water soluble components of the media used were prepared in bulk, and aliquots were added to the flasks before autoclaving at 123°C for 30 minutes. When lipids or steroids were components of the media, they were dissolved in chloroform and added to the flasks containing the other medium ingredients prior to autoclaving. At least two replicate fermentations were run for each medium combination. At the end of the incubation period, 10 to 15 ml samples were removed from the fermentation flasks for analysis. All values in Tables 12.1–12.3 represent averages of analyses on two or more samples.

Table 12.1. Effect of energy source on antibiotic and vitamin B_{12} production by *Streptomycete* sp. no. MDS418.

		Fermentation data											
*Energy sources**		*pH chages*			*Antibiotic production (Diluion units/ml)*			*Vitamin B12 production (μg/ml)*			*Carbohydrate utilised (%)*		
	g/l	*2 days*	*4 days*	*6 days*	*2 days*	*4 days*	*6 days*	*2 days*	*4 days*	*6 days*	*2 days*	*4 days*	*6 days*
Glucose	20	7.0	7.1	8.0	28	200	185	0.26	0.43	0.46	18	83	95
	40	7.0	7.1	1.8	45	285	210	0.30	0.53	0.55	15	65	90
Maltose	20	7.1	7.3	8.1	42	285	200	0.25	0.53	0.63	27	80	98
	40	7.3	7.9	8.2	27	265	300	0.31	0.63	0.60	27	68	90
Sucrose	20	7.8	8.1	8.3	<25	<25	<25	0.15	0.35	0.32	5	7	10
Starch	40	7.1	7.6	8.2	27	180	325	0.40	0.53	0.60	35	80	95
Lard oil	8.8	5.9	7.6	8.2	48	248	300	0.35	0.65	0.75	–	–	–
	17.6	6.1	7.3	8.1	40	215	285	0.25	0.55	0.70	–	–	–
Mannitol	20	7.5	7.8	8.3	35	175	260	0.31	0.55	0.60	–	–	–

* Other ingredients of media (g/l): soyabean meal, 30 g; $Co(NO_3)_2{\cdot}6H_2O$, 0.005 g; distilled water,1 litre; pH adjusted with NaOH to 7.0 before and after autoclaving at 123°C for 30 minutes.

Analytical methods

Antibiotic content of fermentation samples was measured by a modification of the tube-dilution bioassay method described by Donovick using *Klebsiella pneumoniae* as test organism. The assays are reported in dilution units, and under the conditions of the assay method one dilution unit is equivalent to 1.25 μg of streptomycin.

Preliminary experiments showed that substantial quantities of the antibiotic were adsorbed on the fermentation medium solids. This material was released by treatment of the solids with sulphuric acid as described by Rake prior to bioassay.

Table 12.2. Effect of nitrogen source on antibiotic and vitamin B_{12} production by *Streptomycete* sp. no. MD2428.

*Energy sources**		*Fermentation data*								
		pH chages			*Antibiotic production (Diluion units/ml)*			*Vitamin B_{12} production (μg/ml)*		
	g/l	*2 days*	*4 days*	*6 days*	*2 days*	*4 days*	*6 days*	*2 days*	*4 days*	*6 days*
2 days	*4 days*	*6 days*								
Soyabean meal	15	7.1	7.5	8.1	28	185	200	0.15	0.27	0.45
	30	7.3	7.8	8.1	45	285	210	0.30	0.53	0.55
	45	7.4	7.9	8.3	27	180	110	0.18	0.65	0.85
Cornsteep liquor	30	7.1	6.8	8.1	17	160	180	0.18	0.35	0.55
	60	7.3	7.5	8.3	28	253	200	0.20	0.41	0.63
	90	7.9	8.3	8.5	17	45	35	0.23	0.63	0.85
Dried yeast	15	7.1	7.9	8.3	16	151	90	0.22	0.40	0.55
	30	7.3	8.3	8.4	28	45	25	0.25	0.30	0.55
	45	7.7	7.9	8.4	16	23	14	0.23	0.41	0.80
Synthetic medium[a]	–	7.2	8.0	8.3	23	80	60	0.31	0.61	0.80

*Other ingredients of all media (g/l): glucose, 20 g; $Co(NO_3)_2{\cdot}6H_2O$, 0.005 g; distilled water, 1 litre; pH adjusted with NaOH to 7.0 before and after autoclaving for 30 minutes at 123°C.

[a]Composition of synthetic medium (g/l): glucose, 20 g, glycine, 2.6 g; sodium acid glutamate, 2.2 g; $K_2HPO_4{\cdot}3H_2O$, 0.5 g; $MgSO_4{\cdot}7H_2O$, 0.5 g; $CuSO_4{\cdot}5H_2O$, 0.015 g; $MnSO_4{\cdot}4H_2O$, 0.016 g; $ZnSO_4{\cdot}7H_2O$, 0.03 g; $FeSO_4{\cdot}7H_2O$, 0.025 g; $CaCl_2{\cdot}2H_2O$, 0.05 g; $Co(NO_3)_2{\cdot}6H_2O$, 0.005 g; distilled water, 1 litre; pH adjusted to 7.0 with NaOH before and after autoclaving for 5 minutes at 123°C.

Table 12.3. Effect of inorganic phosphate on antibiotic and vitamin B_{12} production by *Streptomycete* sp. no. MD2428.

*Medium compostion**		*Fermentation data*											
Basal medium	*$K_2HPO_4{\cdot}3H_2O$ added*	*pH chages*			*Antibiotic production (Diluion units/ml)*			*Vitamin B12 production (μg/ml)*			*Carbohydrate utilised (%)*		
	g/l	*2 days*	*4 days*	*6 days*	*2 days*	*4 days*	*6 days*	*2 days*	*4 days*	*6 days*	*2 days*	*4 days*	*6 days*
Soyabean meal-glucose	0	7.1	7.6	8.1	20	135	115	0.27	0.64	0.79	40	75	100
	0.5	6.8	7.6	8.5	30	148	15	0.31	0.51	0.79	53	80	98
	1.5	7.0	7.7	8.4	27	160	20	0.27	0.80	1.10	78	90	100
	10.0	7.0	7.6	8.3	19	226	20	0.30	0.56	0.68	83	96	100
Synthetic	0.5	6.8	7.8	8.1	35	160	215	0.28	0.55	0.60	25	68	93
	1.5	7.1	8.0	8.3	18	110	60	0.21	0.60	0.78	38	85	98
	10.0	8.5	8.1	8.3	23	63	47	0.11	0.55	0.60	43	87	98

*Composition of media (g/l): Soyabean meal-glucose: soyabean meal, 30 g; glucose, 20 g; $Co(NO_3)_2{\cdot}6H_2O$, 0.005 g; distilled water, 1 litre; pH adjusted with NaOH to 7.0 before and after autoclaving for 30 minutes at 123°C. Synthetic: Glucose, 20 g; glycine, 2.6 g; sodium acid glutamate, 2.2 g; $K_2HPO_4{\cdot}3H_2O$, 0.5 g; $MgSO_4{\cdot}7H_2O$, 0.5 g; $CuSO_4{\cdot}5H_2O$, 0.015 g; $ZnSO_4{\cdot}7H_2O$, 0.03 g; $FeSO_4{\cdot}7H_2O$, 0.025 g; $MnSO_4{\cdot}4H_2O$, 0.016 g; $CaCl_2{\cdot}2H_2O$, 0.05 g; distilled water, 1 litre; pH adjusted to 7.0 with NaOH before and after autoclaving for 5 minutes at 123°C.

The vitamin B_{12} content of the acid treated fermentation samples was determined by a modification of the bioassay method described by Brownlee and Lapedes using *Lactobacillus leichmannii* as the test organism and a sample of crystalline cyanocobalamin as the standard. Study by counter-current liquid-extraction methods showed that the vitamin B_{12} produced by the streptomycete when grown on media containing cobalt salts was mainly hydroxocobalamin. In this modified bioassay method where a sterile aliquot of the fermentation sample (obtained by filtration through an asbestos pad) was added to the previously sterilised assay media, hydroxocobalamin had growth promoting activity equal to that of cyanocobalamin. Repeated tests showed that the antibiotic content of these diluted samples was below the quantity inhibitory to *L. leichmannii.*

The residual carbohydrate content of the fermentation samples was determined by the Shaffer-Somogyi method using reagent 50 containing 5 grams of KI per litre with glucose as a standard. Samples containing sucrose, maltose, lactose, or starch were subjected to acid hydrolysis prior to analysis.

Oxidation of progesterone added to fermentation media was determined by the filter paper chromatographic methods of Zaffaroni and Burton. Samples were removed from three replicate fermentations and pooled. A 40 ml aliquot was centrifuged, and the supernatant fluid was extracted thrice with 15 ml of chloroform. The collected solids were extracted with 30 ml of chloroform. The chloroform extracts were pooled, and the steroids present were examined by the filter paper partition chromatographic technique using a propylene glycol-toluene system and a 1.5 to 2.5 hours development period. After the solvents were removed by drying the papers at room temperature for 4–6 hours, the Zimmerman test was used to locate the positions of progesterone and derivatives on the paper strips. The locations on duplicate strips were excised, the steroid eluted with 95 per cent ethanol, and the approximate quantity present in the eluates determined spectrophotometrically by measuring absorption at 240 mμ. Examination of the chromatograms showed that 16-α-hydroxyprogesterone was the major oxidation product of progesterone in these fermentations. It had a mobility relative to progesterone of about 0.3 in this chromatographic system and also reacted with an iodine solution to give a blue colour. While this chromatographic technique was found to be valuable as a qualitative tool, the reproducibility of the quantitative measurements was not as high as desired (±10 per cent) as other unknown substances found in the paper or samples often had appreciable absorption at 240 mμ.

Experimental Results and Discussion

Identification of the streptomycete

The micro-organism, carried as streptomycete MD2428 in collection, when grown on agar has mature vegetative hyphae varying from 0.9 to 1.2 mμ in diameter. The aerial mycelium is hyaline under the microscope, is generally branched, not forming loops or spirals. Individual filaments are rarely septate. The colour of colonies when viewed on agar without magnification is white to light gull gray. Mature spores range from about 1.0 to 1.2 mμ in diameter and from 1.0 to 1.2 mμ in length. Individual spores are colourless at maturity, but in mass appear white to gray.

The micro-organism liquefies gelatine, peptonises litmus milk, reduces nitrate to nitrite, and produces hydrogen sulphide when grown on Kligler agar. It grows on media containing ammonium sulphate, sodium nitrate, asparagine, or tryptophan as the sole source of nitrogen (basal medium: KH_2PO_4, 2.38 g; K_2HPO_4, 5.65 g; $MgSO_4{\cdot}7H_2O$, 1.0 g; $CuSO_4{\cdot}5H_2O$, 0.0064 g; $FeSO_4{\cdot}7H_2O$, 0.0011 g; $MnSO_4{\cdot}4H_2O$, 0.0079 g; $ZnSO_4{\cdot}7H_20$, 0.0015 g; agar, 15 g; glucose, 10 g; distilled water q.s. litre; pH adjusted with KOH to 6.8). Addition of the following sugars (10 g per litre) supports growth: arabinose, rhamnose, glucose, galactose, fructose, mannose, lactose, maltose, dextrine, starch, glycerol, mannitol, salicin. No growth is noted when sucrose, raffinose, sorbose, sorbitol, or inositol is added to the basal medium as the sole carbon source. Sodium acetate and sodium citrate support growth, but sodium tartrate does not.

According to the characteristics described, the culture appears to resemble a strain of *Streptomyces lavendukae* studied by Pridham and Gottlieb. However, the culture produces an antibiotic which is extractable from aqueous solution with butanol or organic solvents and differs in other respects from streptothricin, streptolin, and other antibiotics produced by identified strains of *S. lavendulae*.

Production of antibiotic subsances: Preliminary experiments showed that this streptomycete when grown on yeast extract-beef extract agar produced an antibiotic substance (or substances) inhibiting the following: *Esherichia coli*; *Micrococcu pyogenes* var. *aureus*; *Aerobacillus polymyxa*; *Shigella dysenteriae*; *Kibsiella pneumoniae*; *Salmonella schottmülleri*; a strain of *E. coli* known to be resistant to penicillin, streptomycin, streptothricin, neomycin, and a strain of *E. coli* resistant to chlortetracycline. Growth of *Saccharomyces pastorianus* was not inhibited, and growth of a streptomycin-dependent strain of *E. coli* was not stimulated. Other preliminary experiments showed that the antibiotic produced when the streptomycete was grown on soyabean meal-glucose medium could be extracted from the neutral aqueous solution by butanol or amyl acetate, or adsorbed by the cationic exchange resin IRC-50. It was quite stable to heating at 100°C in aqueous solution at pH 2 but not at pH 7 or pH 9. These biological and chemical characteristics showed that the antibiotic material was basic in nature and differed from streptothricin and related substances both chemically and microbiologically. *In vivo* tests with concentrates of the antibiotic substance prepared by butanol extraction of the fermented medium showed that the minimal toxic dose of the materials was 36 dilution units for mice and 17 dilution units for eggs. When administered to mice at sublethal doses these concentrates gave slight protection against a *Streptococu pyogenes* (strain C203) infection. They did not inhibit the growth of a number of viruses in eggs.

A study of the effect of medium composition on the production of antibiotic activity by this streptomycete is summarised in Tables 12.1 and 12.2. As shown in Table 12.1, relatively high antibiotic activity was obtained when the streptomycete was grown on media containing any of a number of carbohydrates and lipids as energy sources with soyabean meal as nitrogen source. The antibiotic activity was found to be partially bound to the mycelium and was released by treatment with acid or by addition of ionisable salts to the mycelial suspensions. Examination of samples of fermented media by filter paper chromatography using a butanol-water-acetic acid system (5:4:1) showed the presence of only one antibiotic substance when inhibition of *M. pyogenes* was used to determine the position of the antibiotic on the chromatogram. Replacement of the soyabean meal component of the medium with other seed meals, cornsteep liquor, or a whole yeast hydrolysate did not change the antibiotic production markedly as shown in Table 12.2.

Antibiotic production was significantly reduced when the streptomycete was grown on a synthetic medium or when it was grown under conditions such that the pH of the medium rose above pH 8. Several experiments of this type are summarised in Table 12.3. Addition of phosphate buffer to the medium reduced the pH rise as shown in Table 12.3 and resulted in somewhat higher rates of utilisation of glucose without increasing antibiotic yields markedly.

Production of vitamin B_{12}

Inclusion of cobalt salts in media used for growth of this streptomycete resulted in the production of substantial amounts of vitamin B_{12}. The vitamin was found to be partially adsorbed on the fermentation medium solids and was released by treatment with acid or inorganic salts. Variation of medium composition by changes in type or quantity of energy source included in the medium had little effect on production of vitamin B_{12} as shown in Table 12.1. Increasing the soyabean meal content of the medium resulted in a significant increase in vitamin B_{12} production while antibiotic production was somewhat reduced, as shown in Table 12.2. The same yields of vitamin B_{12} were obtained when the streptomycete was grown on a synthetic medium as when grown on the soyabean meal media, as shown in Table 12.3. These studies showed that vitamin B_{12} production was to a large degree independent of antibiotic production, a conclusion reached by Nelson in studies of antibiotic production by *Streptomyces fradiae*.

Oxidation of progesterone

In a preliminary communication from these laboratories the conversion of progesterone by this streptomycete to 16-α-hydroxyprogesterone, dihydroxyprogesterone, and pregnanol-16 adione-3,20 was reported. Data collected in experiments studying the influence of medium composition on this conversion showed that medium composition had no significant effect on steroid oxidation when progesterone was added to the media used in experiments summarised in Table 12.2. The major product of the action of the microbial enzymes on the steroid was found to be 16 α-hydroxyprogesterone with yields ranging from 30 to 40 per cent of the steroid added. Although the accuracy of techniques used to measure the quantities formed was not as high as desired, the order of magnitude of the yields would not be significantly altered by use of other fermentation methods. Extension of the incubation period resulted in formation of somewhat larger quantities of the dihydroxyprogesterone, presumably from the hydroxylated progesterone.

Conditions favouring steroid oxidation did not appear to be related to antibiotic or vitamin B_{12} production, and often low yields of the latter products were obtained coincident with high yields of the former. Pregnenolone was also found to be oxidised by this streptomycete with the formation of progesterone, 16-α-hydroxyprogesterone, and possibly other hydroxylated derivatives as shown by filter paper chromatography.

No good correlation apparently exists between antibiotic formation, production of vitamin B_{12}, and oxidative degradation of steroids by this streptomycete. Thus, it is not surprising to find that several of these activities proceed satisfactorily under conditions where other syntheses or degradations are only incompletely carried out. These experiments serve to emphasise the relationship of medium composition to metabolic activities of the streptomycetes.

RIBOFLAVIN

Riboflavin (E101), also known as vitamin B_2, is an easily absorbed micronutrient with a key role in maintaining health in humans and animals. It is the central component of the cofactors flavin adenine dinucleotide (FAD) and flavin mononucleotide (FMN), and is therefore required by all flavoproteins. As such, vitamin B_2 is required for a wide variety of cellular processes. Like the other B vitamins, it plays a key role in energy metabolism, and is required for the metabolism of fats, ketone bodies, carbohydrates, and proteins.

Milk, cheese, leafy green vegetables, liver, kidneys, legumes such as mature soyabeans, yeast, mushrooms and almonds are good sources of vitamin B_2, but exposure to light destroys riboflavin.

Toxicity

Riboflavin is not toxic when taken orally, as its low solubility keeps it from being absorbed in dangerous amounts from the gut. Although toxic doses can be administered by injection, any excess at nutritionally relevant doses is excreted in the urine, imparting a bright yellow colour when in large quantities. In humans, there is no evidence for riboflavin toxicity produced by excessive intakes.

Even when 400 mg/d of riboflavin was given orally to subjects in one study for three months to investigate the efficacy of riboflavin in the prevention of migraine headache, no short-term side effects were reported.

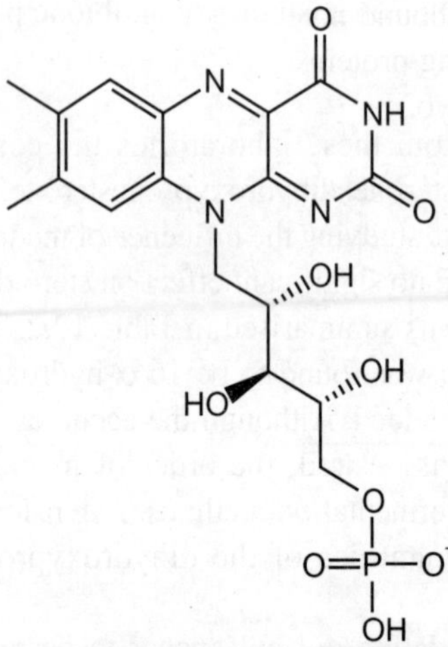

Structure of riboflavin

Industrial Synthesis

Various biotechnological processes have been developed for industrial scale riboflavin biosynthesis using different micro-organisms, including filamentous fungi such as *Ashbya gossypii*, *Candida famata* and *Candida flaveri* as well as the bacteria *Corynebacterium ammoniagenes* and *Bacillus subtilis*. The latter organism has been genetically modified to both increase the bacteria's production of riboflavin and to introduce an antibiotic (ampicillin) resistance marker, and is now successfully employed at a commercial scale to produce riboflavin for feed and food fortification purposes. The chemical company BASF has installed a plant in South Korea, which is specialised on riboflavin production using *Ashbya gossypii*. The concentrations of riboflavin in their modified strain are so high, that the mycelium has a reddish/brownish colour and accumulates riboflavin crystals in the vacuoles, which will eventually burst the mycelium.

Riboflavin in Food: Occurrence, Sources and Stability

Riboflavin is yellow or yellow-orange in colour and in addition to being used as a food colouring, it is also used to fortify some foods. It is used in baby foods, breakfast cereals, pastas, sauces, processed cheese, fruit drinks, vitamin-enriched milk products, and some energy drinks. Regarding occurrence and sources of vitamin B_2, yeast extract is considered to be exceptionally rich in vitamin B_2, and liver and kidney are also rich sources. Wheat bran, eggs, meat, milk, and cheese are important sources in diets containing these foods. Cereals grains contain relatively low concentrations of flavins, but are important sources in those parts of the world where cereals constitute the staple diet. The milling of cereals results in considerable loss (up to 60 per cent) of vitamin B_2, so white flour is enriched by addition of the vitamin. The enrichment of bread and ready-to-eat breakfast cereals contributes significantly to the dietary supply of vitamin B_2. Polished rice is not usually enriched, because the vitamin's yellow colour would make the rice visually unacceptable to the major rice-consumption populations. However, most of the flavins content of the whole brown rice is retained if the rice is steamed prior to milling. This process drives the flavins in the germ and aleurone layers into the endosperm. Free riboflavin is naturally present in foods along with protein-bound FMN and FAD. Bovine milk contains mainly free riboflavin, with a minor contribution from FMN and FAD. In whole milk, 14 per cent of the flavins are bound noncovalently to specific proteins. Egg white and egg yolk contain specialised riboflavin-binding proteins, which are required for storage of free riboflavin in the egg for use by the developing embryo.

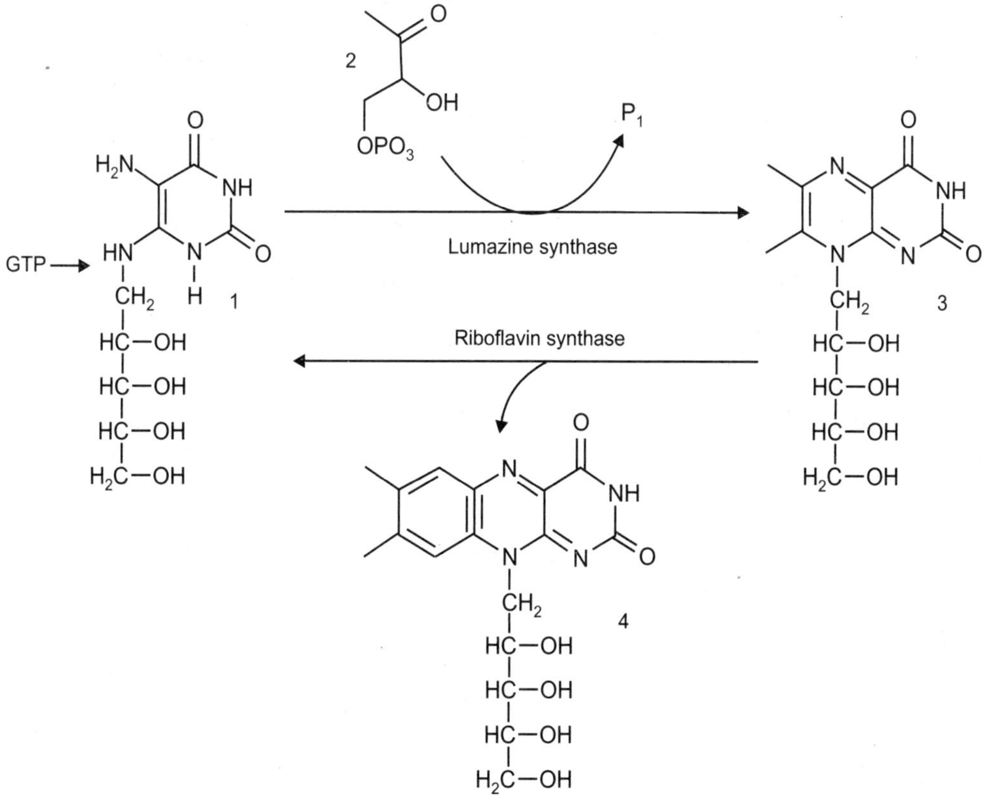

Synthesis of riboflavin

It is difficult to incorporate riboflavin into many liquid products because it has poor solubility in water. Hence the requirement for riboflavin-5′-phosphate (E101a), a more expensive but more soluble form of riboflavin. Riboflavin is generally stable during the heat processing and normal cooking of foods if light is excluded. The alkaline conditions in which riboflavin is unstable are rarely encountered in foodstuffs. Riboflavin degradation in milk can occur slowly in dark during storage in the refrigerator.

Nutrition-recommended Dietary Allowance

Recommended dietary allowance (RDA)

The latest RDA recommendation for vitamin B_2 are similar to the 1989 RDA, which for adults, suggested a minimum intake of 1.2 mg for persons whose caloric intake may be >2000 Kcal. The current RDAs for riboflavin for adult men and women are 1.3 mg/day and 1.1 mg/day, respectively; the estimated average requirement for adult men and women are 1.1 mg and 0.9 mg, respectively. Recommendations for daily riboflavin intake increase with pregnancy and lactation to 1.4 mg and 1.6 mg, respectively (1 in advanced). For infants the RDA is 0.3–0.4 mg/day and for children it is 0.6–0.9 mg/day.

Riboflavin deficiency

Riboflavin is continuously excreted in the urine of healthy individuals, making deficiency relatively common when dietary intake is insufficient. However, riboflavin deficiency is always accompanied by deficiency of other vitamins. A deficiency of riboflavin can be primary—poor vitamin sources in one's daily diet or secondary, which may be a result of conditions that affect absorption in the intestine, the body not being able to use the vitamin, or an increase in the excretion of the vitamin from the body.

In humans, signs and symptoms of riboflavin deficiency (ariboflavinosis) include cracked and red lips, inflammation of the lining of mouth and tongue, mouth ulcers, cracks at the corners of the mouth (angular cheilitis), and a sore throat. A deficiency may also cause dry and scaling skin, fluid in the mucous membranes, and iron-deficiency anemia. The eyes may also become bloodshot, itchy, watery and sensitive to bright light.

Assessment of riboflavin status

Biochemical tests are essential for confirming clinical cases of riboflavin deficiency and for establishing subclinical deficiencies.

Function, Mechanism of Action and Clinical Uses

Function and mechanism of action FMN and FAD function as coenzymes for a wide variety of oxidative enzymes and remain bound to the enzymes during the oxidation-reduction reactions. Flavins can act as oxidising agents because of their ability to accept a pair of hydrogen atoms. Reduction of isoalloxazine ring (FAD, FMN oxidised form) yields the reduced forms of the flavoproteins ($FMNH_2$ and $FADH_2$). Flavoproteins exhibit a wide range of redox potential and therefore can play a wide variety of roles in intermediary metabolism. Some of these roles are:

1. Flavoproteins play very important roles in the electron transport chain.
2. Decarboxylation of pyruvate and α-ketoglutarate requires FAD.
3. Fatty acyl CoA dehydrogenase requires FAD in fatty acid oxidation.
4. FAD is required to the production of pyridoxic acid from pyridoxal (vitamin B_6).
5. The primary coenzyme form of vitamin B_6 (Pyridoxal phosphate) is FMN dependent.
6. FAD is required to convert retinal (vitamin A) to retinoic acid.
7. Synthesis of an active form of folate (5-methyl THF) is $FADH_2$ dependent.
8. FAD is required to convert tryptophan to niacin (vitamin B_3).
9. Reduction of the oxidised form of glutathione (GSSG) to its reduced form (GSH) is also FAD dependent.

Riboflavin has been used in several clinical and therapeutic situations. Riboflavin supplements have been used as part of the phototherapy treatment of neonatal jaundice. The light used to irradiate the infants breaks down not only the toxin causing the jaundice, but the naturally occurring riboflavin within the infant's blood as well. More recently there has been growing evidence that supplemental riboflavin may be a useful additive along with beta-blockers in the prevention of migraine headaches.

Development is underway to use riboflavin to improve the safety of transfused blood by reducing pathogens found in collected blood. Riboflavin attaches itself to the nucleic acids (DNA and RNA) in cells, and when light is applied, the nucleic acids are broken, effectively killing those cells. The technology has been shown to be effective for inactivating pathogens in all three major blood components: (platelets, red blood cells, and plasma). It has been shown to inactivate a broad spectrum of pathogens, including known and emerging viruses, bacteria, and parasites.

Recently riboflavin has been used in a new treatment to slow or stop the progression of the corneal disorder keratoconus. This is called corneal collagen cross-linking (CXL). In corneal cross-linking, riboflavin drops are applied to the patient's corneal surface. Once the riboflavin has penetrated through the cornea, Ultraviolet A light therapy is applied. This induces collagen cross-linking, which increases the tensile strength of the cornea. The treatment has been shown in several studies to stabilise keratoconus.

Industrial Uses

Because riboflavin is fluorescent under UV light, dilute solutions (0.015–0.025 per cent w/w) are often used to detect leaks or to demonstrate coverage in an industrial system such a chemical blend tank or bioreactor.

Good sources

Riboflavin is found naturally in asparagus, bananas, persimmons, okra, chard, cottage cheese, milk, yogurt, meat, eggs and fish, each of which contain at least 0.1 mg of the vitamin per 3–10.5 oz (85–300 g) serving.

FERMENTATIVE PARAMETERS AND KINETIC STUDIES OF RIBOFLAVIN PRODUCTION BY LOCAL ISOLATE OF *ASPERGILLUS TERREUS*

Riboflavin, a yellow and water soluble solid, is widely distributed in plants and animals. It plays an important role in the growth and normal health of many living organisms as well as man. Lack of riboflavin causes ariboflavinosis.

Furthermore, the vitamin is used as animal feed supplements. Riboflavin is the precursor of flavin mononucleotide (riboflavin 5′-monophosphate) and flavin adenine dinucleotide, which function as coenzymes for a wide variety of metabolic enzymes. In addition, its role as to producing antioxidant agent by scavenging damaging particles in the body known as free radicals. At present, the amount of riboflavin production is speculated to be more than 300 ton/year. Riboflavin is produced commercially by chemical synthesis or biological synthesis, the latter being preferred because it costs less. Moreover, the benefit of microbial production of riboflavin is to reduce some waste products which are used as resources for fermentation procedures.

Many *Aspergillus* spp. had been used for riboflavin production using a variety of cheap by-products containing media. In the present communications, the effect of different production media and cultivation conditions on the growth and riboflavin formation by a local isolate of *A. terreus* were studied.

Materials and Methods

Micro-organism

A. terreus used in this work was obtained from the Center of Cultures of the Microbial and Natural Products Chemistry Department, National Research Center (NRC), Cairo, Egypt.

Maintenance of the micro-organism

The tested fungus was maintained on agar slants at 4°C using potato dextrose agar medium. Subcultures were carried out every 4 weeks.

Media

1. Medium 1: Consisted of (g/l) glucose, 10, peptone, 5; and corn steep liquor, 10.
2. Medium 2: Consisted of (g/l) glucose, 20, peptone, 5; yeast extract, 2; KH_2PO_4, 1.5; $MgSO_4{\cdot}7H_2O$, 0.5, and corn steep liquor, 10.
3. Medium 3: It has the same composition of medium 2, 3 but supplemented with FeCl , 0.005 mg/l, oleic acid 1.5 g/l and glycine 0.2g/l.
4. Medium 4: Consisted of (g/l) $(NH_4)_2$ SO_4, 3.75; $NH_4H_2PO_4$, 3.75; KH_2PO_4, 2.5, K_2HPO_4, 2,5; $MgSO_4$ $\cdot 7H_2O$, 0.5; and solar 5 per cent (v/v).

Cultivation

The tested medium was initially adjusted to pH 6, 50 ml portions of the medium were dispensed in 250 ml Erlenmeyer flasks. The flasks were sterilised by autoclaving at 121°C for 15 min., inoculated with 2 ml spore suspension of 72 hrs old culture and incubated at 30°C on a rotary shaker for the desired fermentation period.

Cultivation using the laboratory bioreactor

The fermentation process was performed in biofloferementer which adopted bench-scale cultivation, under the following conditions: liquid volume, 3 litres, inoculum size 2 per cent (v/v), agitation rate (200 and 400), aeration 1 v/v/min, the temperature was controlled at 30°C and the pH of the medium was automatically controlled at 6.5 by addition of (0.1N) NaOH or (0.1N) HCl. Samples were daily collected for analysis.

Dry weight estimation

The fungus growth was separated by centrifugation, washed and dried at 60°C until constant weight. The culture filtrates were then analysed for their contents of riboflavin.

Riboflavin estimation

0.8 ml of the culture filtrate was mixed with 0.2 and of 1 M NaOH. A volume of 0.4 ml of the resulting solution was neutralised with 1ml of 0.1M potassium phosphate buffer (pH 6) and the absorbance of the produced colour was measured at λ 444 nm was measured. The amount of riboflavin produced was calculated from a previously prepared standard curve. In addition, the fungus biomass (obtained at the end of the incubation period) was heated for 30 min. at 75°C in order to liberate the vitamin bound to the cells and the vitamin yield was then collectively estimated.

RIBOFLAVIN BIOSYNTHESIS IN *SACCHAROMYCES CEREVISIAE*

Saccharomyces cerevisiae has a monofunctional riboflavin synthase that catalyses the formation of riboflavin from 6,7-dimethyl-8-ribityllumazine. By isolated the gene encoding this enzyme from a yeast genomic library by functional complementation of a mutant, rib5-10, lacking riboflavin synthase activity. Deletion of the chromosomal copy of *RIB5* led to riboflavin auxotrophy and loss of enzyme activity. Intragenic complementation between point and deletion mutant alleles suggested that the encoded protein (Rib5p) assembles into a multimeric complex and predicted the existence of a discrete functional domain located at the N terminus. Nucleotide sequencing revealed a 714-base pair open reading frame encoding a 25-kDa protein. Rib5p was purified to apparent homogeneity by a simple procedure. The specific activity of the enzyme was enriched 8500-fold. The *N*-terminal sequence of the purified enzyme was identical to the sequence predicted from the nucleotide sequence of the *RIB5* gene. Initial structural characterisation of riboflavin synthase by gel filtration chromatography and both nondenaturing pore limit and SDS-polyacrylamide gel electrophoresis showed that the enzyme forms a trimer of identical 25-kDa subunits. The derived amino acid sequence of *RIB5* shows extensive homology to the sequences of the subunits of riboflavin synthase from *Bacillus subtilis* and other prokaryotes. In addition, the sequence also shows internal homology between the N-terminal and the C-terminal halves of the protein. Taken together, these results suggest that the Rib5p subunit contains two structurally related (substrate-binding) but catalytically different (acceptor and donator) domains.

Ariboflavinosis

Ariboflavinosis is the medical condition caused by deficiency of riboflavin (vitamin B_2). Ariboflavinosis is most often seen in association with protein-energy malnutrition, and also in cases of alcoholism.

It was originally known as *pellagra sin pellagra*, as it exhibits certain similarities to the niacin deficiency pellagra. The US Recommended Daily Allowance (RDA) for riboflavin ranges from 1.1 to 1.3 milligrams per day for healthy adults to as high as 1.6 mg/day for pregnant or nursing women.

Causes

The most common cause of riboflavin deficiency is an inadequate diet; thus, it occurs most frequently in populations consuming limited quantities of riboflavin-containing foods such as meats, eggs, milk, cheese, yogurt, leafy green vegetables and whole grains. Riboflavin deficiency can also occur in those with impaired liver function, which prevents proper utilisation of the vitamin. Borderline riboflavin deficiency as a consequence of certain anti-retroviral medications has also been known to cause acute lactic acidosis.

Presentation

The signs and symptoms of riboflavin deficiency typically include sore throat with redness and swelling of the mouth and throat mucosa, cheilosis and angular stomatitis (cracking of the lips and corners of the mouth), glossitis (magenta tongue with atrophy), seborrheic dermatitis or pseudo-syphilis (moist, scaly skin particularly affecting the scrotum or labia majora and the nasolabial folds), and a decreased red blood cell count with normal cell size and hemoglobin content (normochromic normocytic anemia).

Riboflavin deficiency is usually found together with other nutrient deficiencies, particularly of the other water-soluble vitamins. Phototherapy to treat jaundice in infants can cause increased degradation of riboflavin, leading to deficiency if not monitored closely. Persons with chronic alcoholism can have impaired absorption of riboflavin and other vitamins such as thiamine.

Studies of the Turkoman people of Iran, who have a significantly increased incidence of esophageal cancer, have shown some relationship between chronic riboflavin deficiency and the onset of esophageal malignancies. One study of pregnant women has found that riboflavin-deficient women were 4.7 times more likely to develop preeclampsia, though the mechanism for this is not known.

β-CAROTENE

β-Carotene is an organic compound—a terpenoid, a red-orange pigment abundant in plants and fruits. As a carotene with β-rings at both ends, it is the most common form of carotene. It is a precursor (inactive form) of vitamin A. Being highly conjugated, it is deeply coloured, and as a hydrocarbon lacking functional groups, it is very lipophilic. The structure was deduced by Karrer. In nature, β-carotene is a precursor to vitamin A via the action of β-carotene 15,15′-monooxygenase. β-Carotene is also the substance in carrots that colours them orange. β-Carotene is biosynthesised from geranylgeranyl pyrophosphate.

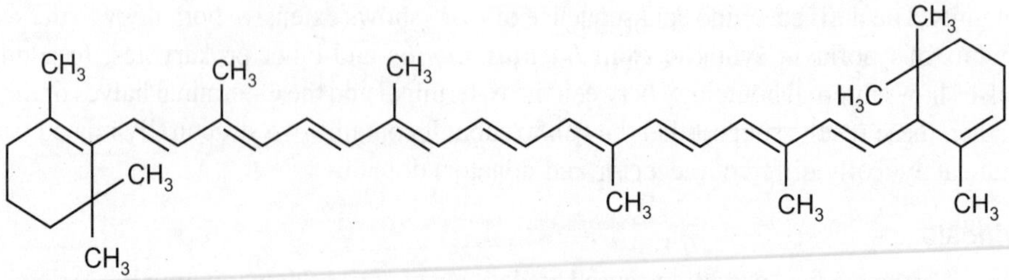

Structure of β-carotene

Pro-vitamin A activity

Plant carotenoids are the primary dietary source of pro-vitamin A worldwide, with β-carotene as the most well-known pro-vitamin A carotenoid. Others inlcude α-carotene and β-cryptoxanthin. Carotenoids are absorbed into the small intestine by passive diffusion. One molecule of β-carotene can be cleaved by a specific intestinal enzyme into two molecules of vitamin A.

Absorption efficiency is estimated to be between 9–22 per cent. The absorption and conversion of carotenoids may depend on the form that the β-carotene is in (cooked vs. raw vegetables, in a supplement), intake of fats and oils at the same time, and the current levels of vitamin A and β-carotene.

The following mnemonic has been developed by researchers to list the factors that determine the pro-vitamin A activity of carotenoids.

S = Species of carotenoid.
L = Molecular linkage.
A = Amount in the meal.
M = Matrix properties.
E = Effectors.
N = Nutrient status.
G = Genetics.
H = Host specificity.
I = Interactions between factors.

Sources in the Diet

β-Carotene contributes to the orange colour of many different fruits and vegetables. Vietnamese gac (Momordica Cochinchinensis Spreng) and crude palm oil are particularly rich sources, as are yellow and orange fruits, such as mangoes and papayas, orange root vegetables such as carrots and yams and in green leafy vegetables such as spinach, kale, sweet potato leaves, and sweet gourd leaves. Vietnam gac and crude palm oil have by far the highest content of β-carotene of any known fruit or vegetable, 10 times higher than carrots for example.

Side Effects

The most common side effect of excessive β-carotene consumption is carotenodermia, a harmless condition that presents as a conspicuous orange skin tint arising from deposition of the carotenoid in the outermost layer of the epidermis. Chronic, high doses of β-carotene have been associated with increased rate of lung cancer among those who smoke.

Biosynthesis of β-Carotene Derivatives and the Creation of Vitamin A

The plastid is involved in the synthesis of many vitamins and compounds in the plant. This section of plant biology advice takes a look at the derivatives that are created from β-carotene.

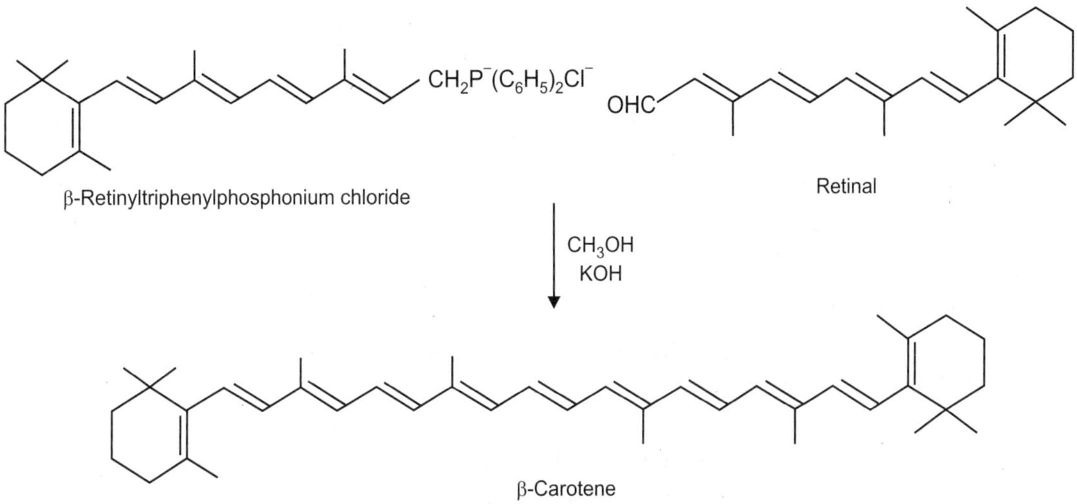

Synthesis of β-carotene

Creation of β-carotene and zeaxanthin

As mentioned previously once lycopene has been produced it can take two different routes, the β,β branch is responsible for β-carotene and its derivatives, and involves the addition of two beta rings to the cyclic end groups of lycopene.

Lycopene β-cyclases have been identified in both the *Arabidopsis* and tomato plants. In tomato the gene has been shown to have a role in fruit ripening and the pigmentation of fruit. It is also thought that the pepper enzyme capsanthin-capsorubin synthase is very closly related to these lycopene β-cyclases.

Plant *Xanthiophylls* such as zeaxanthin have hydroxyl moieties, these are located on the third carbon of the cyclic β-ionine end group. There are many β-ring hydroxylases that are involved in the addition of these end groups from carotene to produce zeaxanthin in higher plants.

Production and use of vitamin A (Retinaldehyde)

Vitamin A is essential in the human diet, and lack of it may lead to blindness and other health consequences. Because of this much work has been carried out on vitamin A research. This has led to plant breeding that has led to increased β-carotene levels in sweet potatoes and in rice.

Vitamin A itself is a C20 product of the carotenoid pathway. It has been found that any carotenoid that has an unmodified β-ionine ring can have provitamin A activity, and is therefore able to produce retinaldehyde. Vitamin A produced is used in the body as both retinoic acids and retinals.

Study on biosynthesis of β-carotene

The growth and β-carotene production of two opposite sex filamentous fungal strains of *B. trispora*, plus mating type WH1(+) and minus mating type WH2(–) were investigated in shaking flask culture.

Two fermentation media such as basal minimal medium (BM) containing glucose, L-asparagine, mineral salts, soyabean oil, thiamine and fermentation medium (FM) containing corn starch, soyabean meal, corn steep liquor, mineral salts, soyabean oil, thiamine were used in the research. The results showed that plus mating strain *B. trispora* WH1 (73.33–76.71 mg/l, equivalent 1.26–1.30 mg/g dry cell mass) produced β-carotene more than minus mating strain *B. trispora* WH2 (40.94-42.49 mg/l, equivalent 0.81–0.84 mg/g dry cell mass) in the same fermentation media shaking at 250 rpm and 28°C for 6 days. No significant differences in the morphology of mycelia, zygophores, zygospores of two opposite sex strains were observed by microscopy when they grew on potato-dextrose-agar medium (PDA) and on fermentation media. Mixed culture of opposite sex *B. trispora* WH1 and *B. trispora* WH2 producd β-carotene 2–3 times (84.78–154.76 mg/l) higher than each strain alone (41–77 mg/l). The effect of various carbon and nitrogen sources on cell growth and β-carotene production of *B. trispora* Wh1 based on BM was studied. Sucrose, maltose, glycerol, peptone, yeast extract, soyabean meal and inogranic nitrogenous compounds did not enhance the production of β-carotene. The highest growth and β-carotene production were obtained at concentration of 50 g/l of glucose and 50 g/l of corn steep liquor. The preliminary result of HPLC analysis showed that *B. trispora* WH1 might produce β-carotene and other carotenoids.

PRODUCTION OF β-CAROTENE AND ERGOSTEROL BY RED YEASTS UNDER PHYSIOLOGICAL STRESS

Huge commercial demand for natural carotenoids has focused attention on developing of suitable biotechnological techniques for their production. To increase the yield of these pigments, physiological regulation of the fermentation process has been employed.

In this work, protein and metabolic profiles of some carotenogenic yeasts grown in optimal and stress conditions were compared. Different 11 carotenogenic yeast strains were cultivated under osmotic, oxidative, metal and nutrition stress. Produced carotenoids and ergosterol were measured using HPLC/MS. Proteins were analysed by 1D microfluidic system and by 2D-gel electrophoresis. Individual spots were identified by LC/MS/MS. Presence of exogenous stress led to important overproduction of pigments as well as ergosterol. Production of carotenoids by *R. glutinis* cells was about 5–6*x* higher under oxidative stress, while production of ergosterol increased more than 10-*x*. Salt and metal stress led also to slight increase of carotenoid production. Combination of stress factors in cultivation media induced significant increase of β-carotene formation mainly in *S. roseus* (230 mg/g in medium with 2 per cent NaCl and 5 mM H_2O_2) and in salt stressed *R. glutinis* (200 mg/g). Production of carotenoids under exogenous stress changed simultaneously with ergosterol production.

Yeast proteome is quite complex, so, interpretation of 2D results is very difficult. Under all stress conditions expression of some protein fractions was changed. Osmotic stress led to overexpression of some specific protein fractions in most of studied yeast strains. These proteomic profiles were different from protein profiles obtained at oxidative as well as metal stress. Individual strains differ in their response to exogenous stress type. In *R.glutinis* and *S.roseus* some 2D-protein spots were analysed and proteins were identified according to databases. Further analysis is needed for detailed characterisation of metabolic changes occurring during controlled carotenoid production in red yeast cells.

NEW MUTANTS OF *PHYCOMYCES BLAKESLEEANUS* FOR β-CAROTENE PRODUCTION

The accumulation of β-carotene by the zygomycete *Phycomyces blakesleeanus* is increased by mutations in the *carS* gene. The treatment of spores of *carS* mutants with *N*-methyl-*N*(prm 1)-nitro-*N*-nitrosoguanidine led to the isolation, at very low frequencies, of mutants that produced higher levels of β-carotene. Strain S556 produced about 9 mg of β-carotene per g of dry mass when it was grown on minimal agar. Crosses involving strain S556 separated the original *carS* mutation from a new, unlinked mutation, *carF*. The *carF* segregants produced approximately as much carotene as did *carS* mutants, but they were unique in their ability to produce zygospores on mating and in their response to agents that increase carotenogenesis in the wild type. The carotene contents of *carF* segregants and *carF carS* double mutants were increased by sexual interaction and by dimethyl phthalate but were not increased by light or retinol. Mixed opposite-sex cultures of *carF carS* mutants contained up to 33 mg of bcarotene per g of dry mass. Another strain, S444, produced more β-carotene than did S556 but was marred by slow growth, defective morphology, and bizarre genetic behaviour. In all the strains tested, the carotene concentration was minimal during the early growth phase and became higher and constant for several days in older mycelia.

RETINOID

The retinoids are a class of chemical compounds that are related chemically to vitamin A. Retinoids are used in medicine, primarily due to the way they regulate epithelial cell growth.

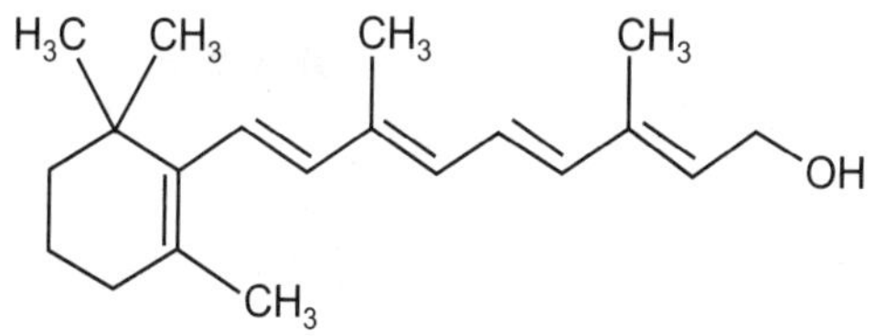

Structure of retinoid

Retinoids have many important and diverse functions throughout the body including roles in vision, regulation of cell proliferation and differentiation, growth of bone tissue, immune function, and activation of tumor suppressor genes.

Research is also being done into their ability to treat skin cancers. Currently 9-cis retinoic acid may be used topically to help treat skin lesions from Kaposi's sarcoma.

Types of Retinoids

There are three generations of retinoids:

1. First generation retinoids: which include retinol, retinal, tretinoin (retinoic acid, Retin-A), isotretinoin and alitretinoin.
2. Second generation retinoids: which include etretinate and its metabolite acitretin.
3. Third generation retinoids: which include tazarotene, bexarotene and adapalene.

Structure

The basic structure of the retinoid molecule consist of a cyclic end group, a polyene side chain and a polar end group. The conjugated system formed by alternating C=C double bonds in the polyene side chain are responsible for the colour of retinoids (typically yellow, orange, or red). Hence, many retinoids are chromophores. Alternation of side chains and end groups creates the various classes of retinoids.

First and second generation retinoids are able to bind with several retinoid receptors due to the flexibility imparted by their alternating single and double bonds. Third generation retinoids are less flexible than first and second generation retinoids and therefore, interact with fewer retinoid receptors.

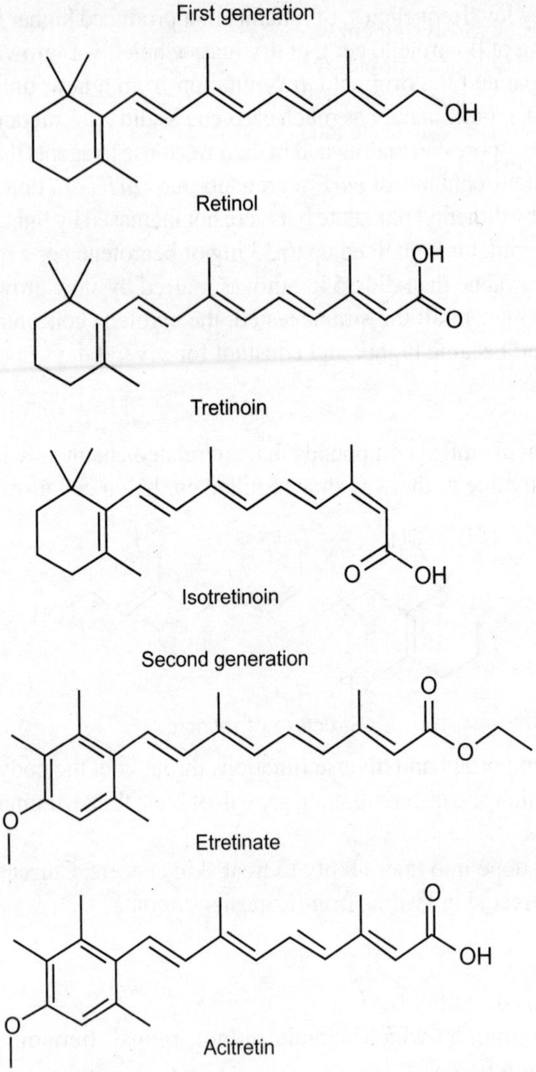

First- and second-generation retinoid compounds

Absorption

The major source of retinoids from the diet are retinyl esters derived from animal sources. Retinyl esters are hydrolysed in the intestinal lumen to yield free retinol and the corresponding fatty acid (i.e. palmitate or stearate). After hydrolysis, retinol is taken up by the enterocytes. Retinyl ester hydrolysis requires the presence of bile salts that serve to solubilise the retinyl esters in mixed micelles and to activate the hydrolysing enzymes

Several enzymes that are present in the intestinal lumen may be involved in the hydrolysis of dietary retinyl esters. Cholesterol esterase is secreted into the intestinal lumen from the pancreas and has been *shown in vitro to display retinyl ester hydrolase activity. In addition, a retinyl ester hydrolase that is* intrinsic to the brush-border membrane of the small intestine has been characterised in the rat as well as in the human. The different hydrolysing enzymes are activated by different types of bile salts and have distinct substrate specificites. For example, whereas the pancreatic estrase is selective for short-chain retinyl esters, the brush-border membrane enzyme preferentially hydrolyses retinyl esters containing a long-chain fatty acid such as palmitate or stearate. Retinol enters the absorptive cells of the small intestine, preferentially in the all-*trans*-retinol form.

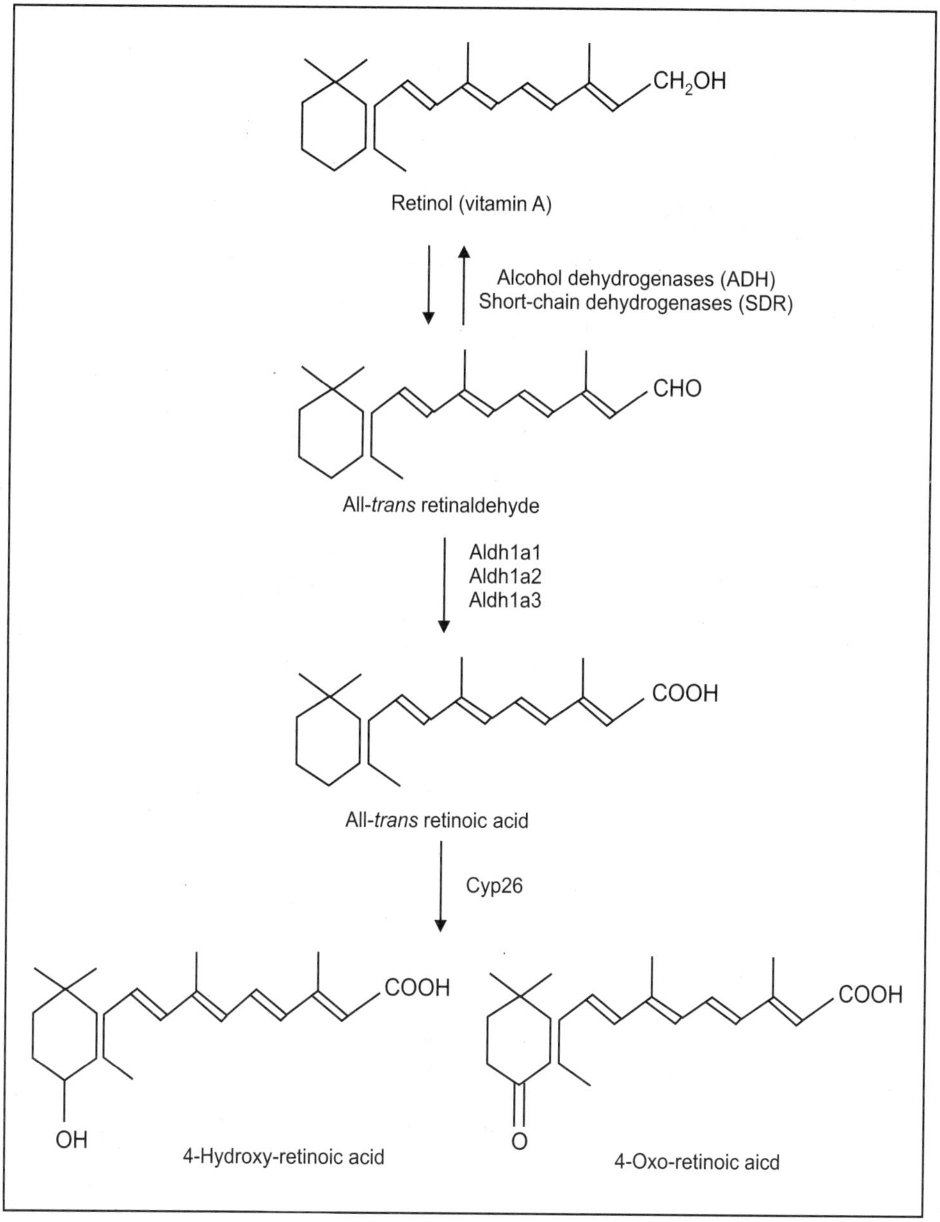

Synthesis of retinol

Uses

Retinoids are used in the treatment of many diverse diseases and are effective in the treatment of a number of dermatological conditions such as inflammatory skin disorders, skin cancers, disorders of increased cell turnover (e.g. psoriasis), and photoaging. Common skin conditions treated by retinoids include acne and psoriasis.

Toxicity

Toxic effects occur with prolonged high intake. The specific toxicity is related to exposure time and the exposure concentration. A medical sign of chronic poisoning is the presence of painful tender swellings on the long bones. Anorexia, skin lesions, hair loss, hepatosplenomegaly, papilloedema, bleeding, general malaise, pseudotumor cerebri, and death may also occur.

Chronic overdose also causes an increased liability of biological membranes and of the outer layer of the skin to peel. Recent research has suggested a role for retinoids in cutaneous adverse effects for a variety of drugs including the antimalarial drug proguanil. It is proposed that drugs such as proguanil act to disrupt retinoid homeostasis. Systemic retinoids (isotretinoin, etretinate) are contraindicated during pregnancy as they may cause CNS, cranio-facial, cardiovascular and other defects.

Chapter 13

Antibiotics

INTRODUCTION

In common usage, an antibiotic is a substance or compound that kills or inhibits the growth of bacteria. Antibiotics belong to the broader group of antimicrobial compounds, used to treat infections caused by micro-organisms, including fungi and protozoa. The term 'antibiotic' was coined by Selman Waksman in 1942 to describe any substance produced by a micro-organism that is antagonistic to the growth of other micro-organisms in high dilution. This original definition excluded naturally occurring substances that kill bacteria but are not produced by micro-organisms (such as gastric juice and hydrogen peroxide) and also excluded synthetic antibacterial compounds such as the sulphonamides. Many antibiotics are relatively small molecules with a molecular weight less than 2000 Da.

With advances in medicinal chemistry, most antibiotics are now semisynthetic—modified chemically from original compounds found in nature, as is the case with β-lactams (which include the penicillins, produced by fungi in the genus Penicillium, the cephalosporins, and the carbapenems). Some antibiotics are still produced and isolated from living organisms, such as the aminoglycosides, and others have been created through purely synthetic means: the sulphonamides, the quinolones, and the oxazolidinones. In addition to this origin-based classification into natural, semisynthetic, and synthetic, antibiotics may be divided into two broad groups according to their effect on micro-organisms: those that kill bacteria are bactericidal agents, while those that only impair bacterial growth are known as bacteriostatic agents.

Fundamentally, antibiotics are classified as either having lethal or bactericidal action against bacteria or are bacteriostatic, preventing bacterial growth. The bactericidal activity of antibiotics may be growth phase dependent and in most but not all cases the action of many bactericidal antibiotics requires ongoing cell activity and cell division for the drugs' killing activity. These classifications are based on laboratory behaviour; in practice, both of these are capable of ending a bacterial infection. *In vitro* characterisation of the action of antibiotics to evaluate activity measure the minimum inhibitory concentration and minimum bactericidal concentration of an antimicrobial and are excellent indicators of antimicrobial potency. However, in clinical practice, these measurements alone are insufficient to predict clinical outcome. By combining the pharmacokinetic profile of an antibiotic with the antimicrobial activity, several pharmacological parameters appear to be significant markers of drug efficacy.

Most anti-bacterial antibiotics do not have activity against viruses, fungi, or other microbes. Anti-bacterial antibiotics can be categorised based on their target specificity: 'narrow-spectrum' antibiotics target particular types of bacteria, such as Gram-negative or Gram-positive bacteria, while broad-spectrum antibiotics affect a wide range of bacteria. Antibiotics which target the bacterial cell wall (penicillins, cephalosporins), or cell membrane (polymixins), or interfere with essential bacterial enzymes (quinolones, sulphonamides) usually are bactericidal in nature. Those which target protein synthesis, such as the aminoglycosides, macrolides and tetracyclines, are usually bacteriostatic.

Although antibiotics are generally considered safe and well tolerated, they have been associated with a wide range of adverse effects. Side effects are many, varied and can be very serious depending on the antibiotics used and the microbial organisms targeted. The safety profiles of newer medications may not be as well established as those that have been in use for many years. Adverse effects can range from fever and nausea to major allergic reactions including photodermatitis. One of the more common side effects is diarrhea, sometimes caused by the anaerobic bacterium *Clostridium difficile*, which results from the antibiotic disrupting the normal balance of the intestinal flora. Such overgrowth of pathogenic bacteria may be alleviated by ingesting probiotics during a course of antibiotics. An antibiotic-induced disruption of the population of the bacteria normally present as constituents of the normal vaginal flora may also occur, and may lead to overgrowth of yeast species of the genus *Candida* in the vulvo-vaginal area. Other side effects can result from interaction with other drugs, such as elevated risk of tendon damage from administration of a quinolone antibiotic with a systemic corticosteroid.

The emergence of antibiotic resistance is an evolutionary process that is based on selection for organisms that have enhanced ability to survive doses of antibiotics that would have previously been lethal. Antibiotics like penicillin and erythromycin which used to be one-time miracle cures are now less effective because bacteria have become more resistant. Antibiotics themselves act as a selective pressure which allows the growth of resistant bacteria within a population and inhibits susceptible bacteria. Survival of bacteria often results from an inheritable resistance. Any antibiotic resistance may impose a biological cost and the spread of antibiotic resistant bacteria may be hampered by the reduced fitness associated with the resistance which proves disadvantageous for survival of the bacteria when antibiotic is not present. Additional mutations, however, may compensate for this fitness cost and aids the survival of these bacteria. Bacteriocins are also a growing alternative to the classic small-molecule antibiotics. Different classes of bacteriocins have different potential as therapeutic agents. Small molecule bacteriocins (microcins, for example, and lantibiotics) may be similar to the classic antibiotics; colicin-like bacteriocins are more likely to be narrow-spectrum, demanding new molecular diagnostics prior to therapy but also not raising the spectre of resistance to the same degree.

One drawback to the large molecule antibiotics is that they will have relative difficulty crossing membranes and travelling systemically throughout the body. For this reason, they are most often proposed for application topically or gastrointestinally. Because bacteriocins are peptides, they are more readily engineered than small molecules. This may permit the generation of cocktails and dynamically improved antibiotics that are modified to overcome resistance. Probiotics are another alternative that goes beyond traditional antibiotics by employing a live culture which may in theory establish itself as a symbiont, competing, inhibiting, or simply interfering with colonisation by pathogens.

The discovery of major antibiotics, such as penicillin, cephalosporin, streptomycin, tetracycline and erythromycins and their subsequent development, have been well documented. Their commercial development over the past 80 years serves as an excellent example of how the applied research has contributed to producing low cost commodities that support therapeutic products. Examples of important antibiotics produced by fermentation for pharmaceutical use are given in Table 13.1.

Antibiotics sold today are made either by total chemical synthesis or by a combination of microbial fermentation and subsequent chemical modification. The choice is one of the simple economics. The microbial fermentation produces modification, the therapeutic effects of the molecule can be increased, e.g. by increasing stability to low pH or temperature, widening the spectrum of activity, altering tissue distribution, increasing absorption and decreasing excretion.

Table 13.1. Examples of important antibiotic types produced by fermentation for pharmaceutical use.

Antibiotic group type	*Example*	*Producing organism*	*Activity spectrum*[1]	*Other comments*
β-Lactam	Penicillin G	*Penicillium thrysogenum*	G^+	Low toxicity, acid labile, β-lactamase sensitive
	Ampicillin	*Penicillium chrysogenum*	G^+G^-	β-Lactamase sensitive, acid-stable
	Cephalosporin C	*Cephalosperium acremonium*	G^+G^-	Low toxicity, penicillinase-resistant but inactivated by β-lactamases produced by some G^- bacteria
Peptide	Bacteracin	*Bacillus licheniformis*	G^+G^-	Use confined to topical application because of toxicity
Aminoglycoside	Streptomycin	*Streptomyces griseus*	G^-	Mainly used to treat tuberculosis
Macrolide	Erythromycin	*Streptomyces griseus*	G^+	Particularly effective against *Staphylococcus* and diphtheroids. Low toxicity
Polyene macrolide	Candidin	*Streptomyces viridoflavus*	F	Widely used for topical anti-funga application
Tetracycline	Chlortetracycline	*Streptomyces aureofaciens*	G^+G^-	
Aromatic	Griseofulvin	*Penicillium patulum*	F	

[1] G^+, Gram-positive; G^-, Gram-negative; F, fungi.

This chapter will discuss the biotechnology involved with the manufacture important antibiotics. In the development of all these antibiotics there are many common approaches. These will be summarised first. Specific examples associated individual antibiotics will be addressed in later sections.

BIOSYNTHESIS

Knowledge of the biosynthetic pathway of antibiotic is not necessary for the early empirical development of the fermentation process. However, in order to progress in a rational approach some knowledge of the biosynthesis is essential. This is particularly important for investigations into the genetic and enzymic regulation. For most of the important antibiotics the synthetic pathways are known, together with their relevant enzymes and gene locations and this detailed knowledge has had a significant impact on the development of improved strains and the optimisation of productive fermentations.

STRAIN IMPROVEMENT

The increase in the production of a specific microbial product such as an antibiotic has several important consequences. Higher concentrations of the antibiotic increase the volumetric productivity (output per fermenter), increase the extraction efficiency, decrease the proportion of unwanted products and make purification easier and most importantly, reduce the cost of the product.

Strain improvement programmes involve the forced creation of mutations in the DNA material of the micro-organism generally using ultraviolet radiation or a chemical mutagen such as nitrosoguanidine (NTG). The latter gives better results as it has a higher mutagenic effect compared to kill rate, however, it has to be handled carefully due its carcinogenic nature. With either treatment, the protocols are optimised to produced a kill range of 60–90 per cent, which gives the highest percentage chance of single point mutations.

Initially, improved strains can be selected empirically by choosing surviving colonies with minor morphological changes or altered colour production, however, well-established strain improvement programmes use many selective approaches designed around the known biochemistry of the biosynthesis of the antibiotic and the metabolism of the micro-organism (Table 13.2).

Table 13.2. Selection environments for strain improvement.

Resistance	*Possible effect*
Analogues of amino acids, sugars involved in biosynthesis	Remove feedback control
Antifungal agents, e.g. nystatin	Altered cell wall composition increased, permeability
Toxic metals, e.g. Cu, Cd, Hg	Increase in thiols, glutathione
Toxic metals Fe, Mn	Improved sporulation
Selenomethionine, ethionine	Increase in sulphate metabolism
Selenide, methyl selenide	Improved cysteine synthesis
Deoxyglucose	Reduced glucose regulated feedback
Carbon dioxide	Tolerate high levels CO_2 in fermentation
High phosphate	Reduced phosphate regulation
High salt	Tolerate high-salt raw materials
Nitrophenol, azide	Improved oxidative phosphorylation
Polypropylene glycol	Tolerate high levels of antifoam
Water miscible solvents	Improved downstream processing
Peroxide	Higher catalase activity
Others	
Improved tolerance to low oxygen	Improved metabolism at low dissolved oxygen
Sensitivity to chromate, selenate	Increased sulphate uptake
Selection of auxotrophs	Redirection of metabolism

The highest percentage gains occurred in the earlier years and now current strain selection is subject to the law of diminishing returns. However, even though increases of less than 5 per cent are difficult to achieve and detect analytically, they are desirable in terms of cost reduction and volumetric capacity increase.

To recognise improvements of 5 per cent or less needs the careful design of shake-flask fermentations, with the appropriate controls and replicates and the exacting skills of the analyst in sample preparation and analyte measurement.

Today's screening programmes rely on the use of miniaturisation, automation and high troughput screening to process the large numbers necessary for the recognition of superior strains. Automation also decreases procedural variabilities seen in sample preparation and dilution. Replication at the shake-flask, retest stage is necessary to minimise the inherent variability of the biological process to provide the confidence in recognising mutants with only small percentage increases in titres. Replication decreases the number of different cultures that can be handled by the system, however, good screening programmes usually have an excess of analytical capacity to meet these challenges.

Rapid recycling of the mutation cycles is often used to compensate for the selection of new strains with only minor improvements. Thus several minor changes can be built up into a culture at the laboratory level before further evaluation in the pilot plant. Good cultures typically show their superiority across different media culture conditions and through scale-up (Fig. 13.1).

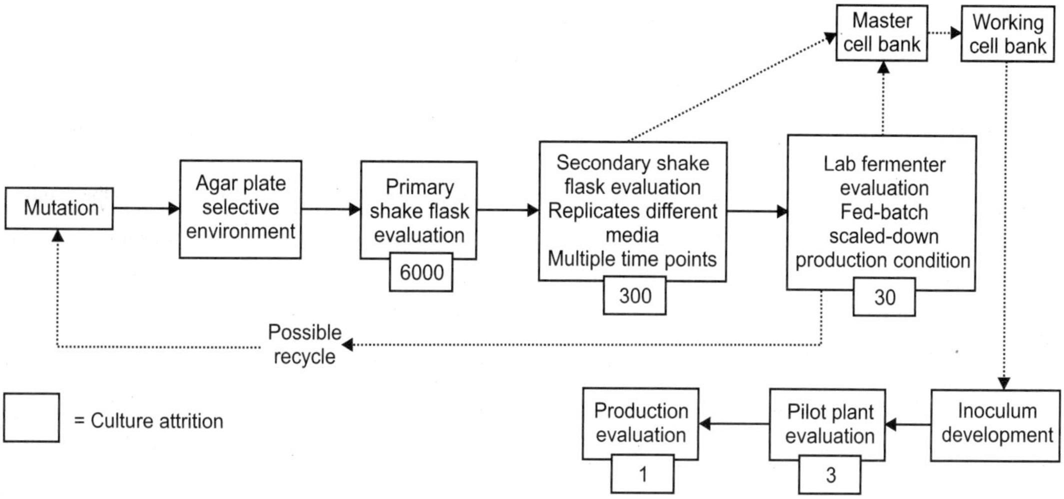

Fig. 13.1. Strain improvement scale-up.

Strain improvement programmes have been carried out as in-house projects, which have the benefit of control and rapid integration into scale-up and downstream process. However, there is an increased trend now-a-days to contract such work to third parties: companies that are specialised in the multiple skills of culture mutation, selection, automated fermentation and analysis using high throughput screening. Such companies are always a cost-effective option.

SCALE-UP

Shake-flask media and conditions are selected to provide environments as close as possible to the stirred-tank large-scale fermentations. This is not always possible and many compromises have to be taken (Table 13.3).

Table 13.3. Differences between shake-flask and stirred tank fermentations.

Shake-flask	*Stirred tank*
10–50 ml in 125–500 ml flasks	10–1,00,000 litre vessels
Batch only	Batch and feed possible
Limited controls: temperature	Continuous controls: pH, temperature, dissolved oxygen, pressure
Slow metabolising carbohydrate: lactose, starch	Readily metabolised carbohydrate: glucose
High initial salts: ammonium, precursors, stimulators added in large shots	Ammonium salts or ammonia added precursors, stimulators added continuously
Ambient pressure	Pressure at two atmospheres possible

(Contd ...)

Shake-flask	*Stirred tank*
Buffers needed to control pH: phosphate or calcium carbonate	No buffer necessary
In-process sampling difficult	In-process sampling easy and often necessary for feedback control
Volume decrease by evaporation	Volume increase by sugar feed
Solid growth on side walls	Very uniform growth
Antifoam not needed	Antifoam often required
Agitation limited: shaker speed, radial throw, baffled flasks	Wide variety of impellers and baffles to optimise mixing
No control of dissolved oxygen	Dissolved oxygen controlled by aeration, pressure, water addition

A good relationship between shake-flask performance, pilot plant and large-scale fermentations can only be established after years of careful comparison. Potential titre increases of 5 per cent or less are not only difficult to assess in shake-flask experiments but also difficult to assess at the pilot plant stage where resources are limited and evaluations expensive. It is always desirable to have new cultures that easily fit into the existing fermentation protocols without further development work. However, new cultures often have properties that need further development to express their full potential. Here the interdisciplinary skills of the bioengineers, microbiologists and biochemists can prove to be rewarding.

FERMENTATION

Large-scale antibiotic fermentations are optimised for fast culture growth, early production rates and maximum productivity. High productivity plants are characterised by their ability to maximise the use of all vessels. A train of vessels of increasing size allows for the rapid build-up of cell mass, each typically 1–3 day fermentations, with inoculum transfers at 5–10 per cent (v/v) into the next stage. The final stage is run in conditions to maximise productivity in terms of amount of antibiotic per unit fermenter volume per time period (Fig. 13.2). Down time (or turnaround time) for production fermenters is usually kept at a minimum by use of separate continuous media sterilisation, rapid harvesting and tank cleaning, sterilisation and inoculation (Fig. 13.3).

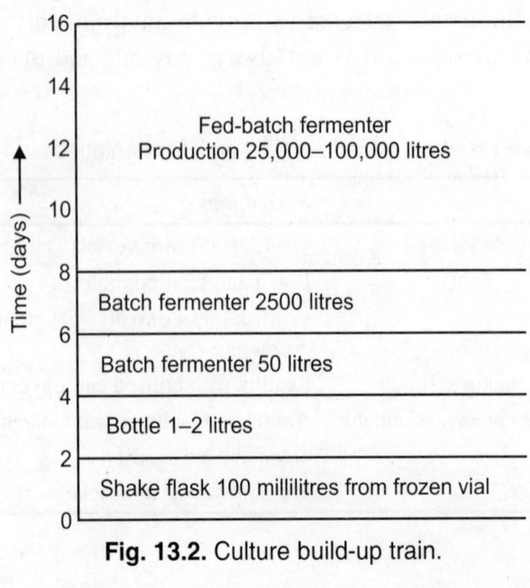

Fig. 13.2. Culture build-up train.

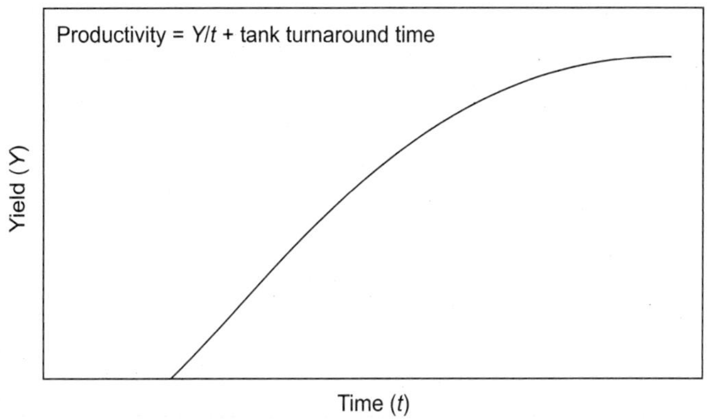

Fig. 13.3. Stirred tank productivities.

Media for cell mass build-up are designed to provide fast growth in a batch mode with minimal changes in pH. Individual medium components do not have to be greater than 3–5 per cent and provide a readily available carbohydrate, such as glucose or sucrose and a soluble form of nitrogen such as corn steep liquor or yeast extract.

Calcium carbonate or phosphates can be added if buffering is required which is often the case due to organic acids that can be produced by the rapid metabolism of sugars. Ammonium sulphate can be used to provide additional nitrogen (Table 13.4).

Table 13.4. Media for cell growth and final production.

Stage	*Components*	*Range% (w/w)*
Shake-flask/tank seed	Glucose/sucrose/starch	3.0–5.0
	Corn steep liquor	3.0–5.0
	Calcium carbonate	0.5–1.0
	Phosphate	0.1–0.5
	Ammonium sulphate	0.1–1.0
	Urea	0.1–0.5
	Oil	0.1–0.5
Shake-flask production	Glucose	0.2–1.0
	Starch	0.5–5.0
	Lactose	5.0–8.0
	Corn steep liquor	5.0–8.0
	Pharmamedia	1.0–5.0
	Soya flour	1.0–5.0
	Oil	0.5–5.0
	Ammonium sulphate	0.5–1.0
	Calcium carbonate	0.5–1.0
	Phosphate	0.1–1.0
	MOPS/MES buffers	0.1–1.0

(Contd ...)

Stage	*Components*	*Range% (w/w)*
Tank production	Glucose	0.5–1.0
	Starch	0.5–5.0
	Corn steep liquor	5.0–8.0
	Pharmamedia	1.0–5.0
	Soya flour	1.0–5.0
	Ammonium sulphate	0.5–1.0 and fed periodically
	Phosphate	0.1–1.0
	Oil	1.0 and fed continuously
	Corn syrup	Fed continuously

Media for the production stage are proprietary and have been developed and fine-tuned over the years. They are a compromise between cost and performance. The most suitable media are those that use inexpensive raw materials in combinations that can maximise productivity. Final stage fermentations are fed-batch, which gives the bioengineer the ability to optimise the fermentation to provide the fine balance between controlled cell growth and maximum biosynthesis. A fed-batch fermenter can be controlled in a number of ways: physically, e.g. by temperature, aeration, agitation pH; or biochemically, e.g. by the addition of nutrients, precursors, inducers.

Raw materials for use in the initial batch have to provide both immediate utilisable soluble nutrients as well as longer lasting and therefore less soluble sources. Initial carbon sources are the least critical as they are easily added in a soluble form during the fermentation. Nitrogen sources are more critical as they serve as a main nutrient source throughout the fermentation. Ideal nitrogen sources are derived from agricultural sources, however, questions of quality and variability can arise both with seasons and between seasons (Table 13.5). This presents an on-going concern for maintaining reproducible fermentations. To alleviate this situation several different raw materials can be used to prevent excess variation.

Table 13.5. Complex nitrogen source raw materials.

Beef blood
Casein hydrolysate
Cotton seed flour (Pharmamedia)
Cottonseed meal
Corn germ meal
Corn gluten meal
Corn steep liquor
Corn steep liquor (solid)
Distillers solubles
Fish meal
Fish solubles
Lard water solids
Linseed meal

(Contd ...)

Meat and bone meal
Peanut meal
Rape seed meal
Soyabean meal
Soyabean flour
Soyabean protein concentrate
Whey solids
Whey permeate
Whole yeast, brewer's
Whole yeast, torula
Yeast extract

In some of today's highly productive fermentations there is no clear separation of the primary (tropophasic) and secondary (idiophasic) stages. This lack of division generally depends upon the state of the fermentation technology. In batch type fermentations, clear primary and secondary stages can be seen, however, with the use of continuous feed these differences are not always apparent (Fig. 13.4). To obtain maximum production rates, conditions are created that can provide rapid, early, antibiotic production with continued cell growth.

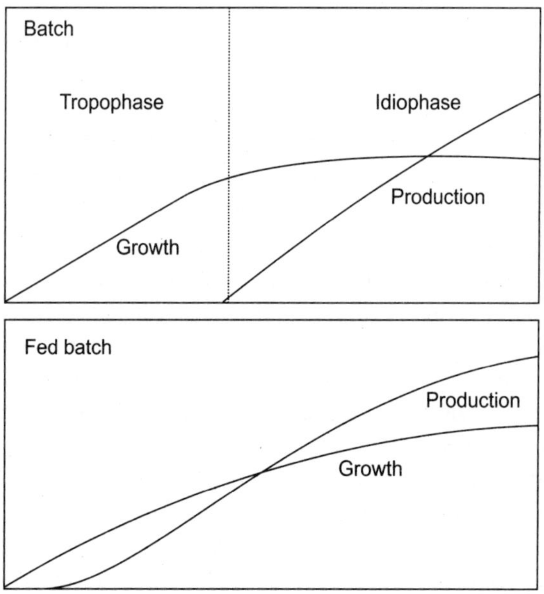

Fig. 13.4. Comparison of batch and fed-batch.

Supplemented raw materials are soluble and rapidly utilised. Suitable carbohydrates are sucrose, glucose or enzyme-hydrolysed corn syrups. Other carbon sources can be used (Table 13.6). If necessary they can be supplemented with soluble nitrogen from corn steep liquor. The diligent feeding of a soluble, readily utilised carbohydrate such as glucose can prevent catabolic repression, as the concentration of the sugar will always be very low.

Table 13.6. Carbon source raw materials.

Beet and cane molasses
Glucose
Citric acid
Corn syrup incompletely hydrolysed
Corn syrup fully hydrolysed
Dextrins
Ethanol
Glycerol
Maltose syrup
Methanol
Starch
Lactose
Cottonseed oil
Lard oil
Methyl oleate
Palm oil
Palm kernel oil
Peanut oil
Rape oil (Canola)
Soya oil
Tallow

Oil, as triacylglycerol, can be lard oil, soya oil, palm oil, peanut oil or rape seed oil, the final choice often dictated by local availability. Oil addition has the additional benefit of controlling excessive foaming and air hold-up. Antifoams, such as silicone-based products or polypropylene glycol, can be used to supplement or replace oil feeding. It is important that antifoam addition is available on an as needed basis and not simply batched into the starting medium due to the toxic nature of some antifoams. The metabolism of the proteinaceous nutrients from the complex raw materials can create foaming, often at unpredicted times, thus it is important to have automated feed-back control for effective antifoam addition to provide sufficient control without excess usage of these agents. Excess use can cause processing difficulties on downstream recovery. Control of foaming and the minimisation of air hold-up are important factors in obtaining the maximum volumetric output from a fermenter. Typically, the final harvest volumes should be in the range 80–85 per cent of the total fermenter capacity.

The added volume of soluble nutrient feed can vary depending upon its concentration (typically 30–65 per cent). At lower sugar concentrations early partial harvests may be necessary to decrease the increase in broth volume caused by the higher volume of feed addition. This addition of dilute solutions has the added benefit of lowering the viscosity of the broth, typically a problem with filamentous cultures. Early, partial harvests, produce large volumes of dilute antibiotic for product recovery. With correct handling however, such protocols can be very productive as the maximum production rate of the fermentation can be maintained for long periods.

The pH of the broth can be controlled to within 0.1 pH units by the addition of acid (sulphuric) or base (ammonia or caustic). Often ammonia gas can be added through the air input. The pH can also be controlled by using the culture's own metabolism of sugar. Excess feeding of sugar in some conditions will produce acetic acid, which will lower the pH. Conversely, a cutback in the sugar feed-rate can raise the pH.

Dissolved O_2 DO levels are critical for maintaining the maximum rate of antibiotic production and culture viability. As O_2 supplementation is too costly, as well as a safety concern, ambient air is used as the source of O_2. A fine balance has to be established between aeration and the agitation necessary to distribute the O_2 into the liquid phase, the back pressure in the tank to increase oxygen solubility, the volume expansion of the fermentation broth and the compounding of several of these effects on the dissolved CO_2 levels.

Dissolved O_2 levels should be maintained higher than 20 per cent saturation at 1.5–2 atmospheres pressure throughout the fermentation, at air flow rates high enough to sweep out as much CO_2 as possible, build up of which can be detrimental to some fermentations (Fig. 13.5).

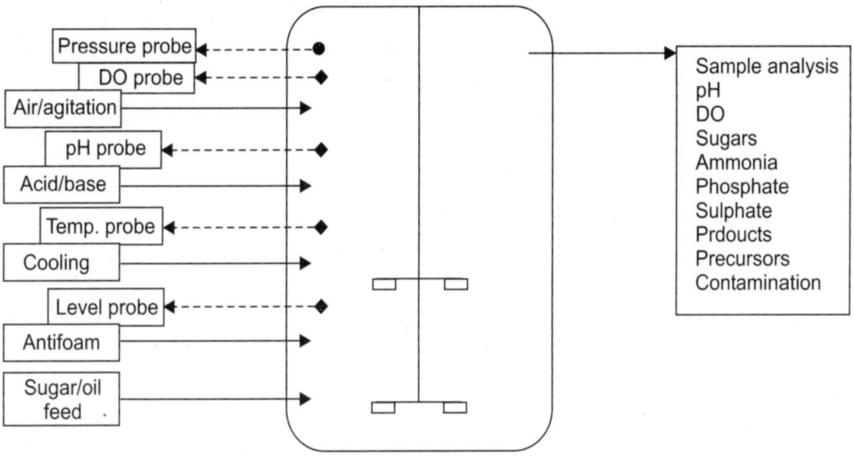

Fig. 13.5. Fermentation control parameters.

Monitoring of the fermentation can be performed by the use of multiple pH and dissolved O_2 probes, pressure and temperature measurements. However, for more detailed analysis, broth samples are taken at convenient intervals depending on the length of the fermentation and critical nature of the measurement. Samples are used to check the correct performance of pH probes and for a variety of chemical measurements, e.g. concentration of the product, ammonia, sulphate and other factors that may be considered important to control the fermentation, i.e. sugar, amino acids, organic acids, degradation products or intermediates. These analyses can be performed rapidly by shift workers and the results can be available within several hours if necessary. Additional samples are taken for microbiological analysis to asses the aseptic nature of the fermentation.

β-LACTAM ANTIBIOTIC

β-Lactam antibiotics are a broad class of antibiotics that include penicillin derivatives (penams), cephalosporins (cephems), monobactams, and carbapenems, that is, any antibiotic agent that contains a β-lactam nucleus in its molecular structure. They are the most widely-used group of antibiotics. While not true antibiotics, the β-lactamase inhibitors are often included in this group.

β-Lactam antibiotics are indicated for the prophylaxis and treatment of bacterial infections caused by susceptible organisms. At first, β-lactam antibiotics were mainly active only against Gram-positive bacteria, yet the recent development of broad-spectrum β-lactam antibiotics active against various Gram-negative organisms has increased their usefulness.

β-Lactam antibiotics are bactericidal, and act by inhibiting the synthesis of the peptidoglycan layer of bacterial cell walls. The peptidoglycan layer is important for cell wall structural integrity, especially in Gram-positive organisms. The final transpeptidation step in the synthesis of the peptidoglycan is facilitated by transpeptidases known as penicillin-binding proteins (PBPs).

β-Lactam antibiotics not only block the division of bacteria, including cyanobacteria, but also the division of cyanelles, the photosynthetic organelles of the Glaucophytes, and the division of chloroplasts of bryophytes. In contrast, they have no effect on the plastids of the highly developed vascular plants. This is supporting the endosymbiotic theory and indicates an evolution of plastid division in land plants.

β-Lactam antibiotics are analogues of D-alanyl-D-alanine—the terminal amino acid residues on the precursor NAM/NAG-peptide subunits of the nascent peptidoglycan layer. The structural similarity between β-lactam antibiotics and D-alanyl-D-alanine facilitates their binding to the active site of penicillin-binding proteins (PBPs). The β-lactam nucleus of the molecule irreversibly binds to (acylates) the Ser_{403} residue of the PBP active site. This irreversible inhibition of the PBPs prevents the final crosslinking (transpeptidation) of the nascent peptidoglycan layer, disrupting cell wall synthesis.

Under normal circumstances peptidoglycan precursors signal a reorganisation of the bacterial cell wall and, as a consequence, trigger the activation of autolytic cell wall hydrolases. Inhibition of cross-linkage by β-lactams causes a build-up of peptidoglycan precursors, which triggers the digestion of existing peptidoglycan by autolytic hydrolases without the production of new peptidoglycan. As a result, the bactericidal action of β-lactam antibiotics is further enhanced.

Modes of Resistance

By definition, all β-lactam antibiotics have a β-lactam ring in their structure. The effectiveness of these antibiotics relies on their ability to reach the PBP intact and their ability to bind to the PBP. Hence, there are two main modes of bacterial resistance to β-lactams, as discussed below.

The first mode of β-lactam resistance is due to enzymatic hydrolysis of the β-lactam ring. If the bacteria produces the enzymes β-lactamase or penicillinase, these enzymes will break open the β-lactam ring of the antibiotic, rendering the antibiotic ineffective. The genes encoding these enzymes may be inherently present on the bacterial chromosome or may be acquired via plasmid transfer, and β-lactamase gene expression may be induced by exposure to β-lactams. The production of a β-lactamase by a bacterium does not necessarily rule out all treatment options with β-lactam antibiotics. In some instances, β-lactam antibiotics may be co-administered with a β-lactamase inhibitor. However, in all cases where infection with β-lactamase-producing bacteria is suspected, the choice of a suitable β-lactam antibiotic should be carefully considered prior to treatment. In particular, choosing appropriate β-lactam antibiotic therapy is of utmost importance against organisms with inducible β-lactamase expression. If β-lactamase production is inducible, then failure to use the most appropriate β-lactam antibiotic therapy at the onset of treatment will result in induction of β-lactamase production, thereby making further efforts with other β-lactam antibiotics more difficult.

The second mode of β-lactam resistance is due to possession of altered penicillin-binding proteins. β-Lactams cannot bind as effectively to these altered PBPs, and, as a result, the β-lactams are less effective at disrupting cell wall synthesis. Notable examples of this mode of resistance include methicillin-resistant *Staphylococcus aureus* (MRSA) and penicillin-resistant *Streptococcus pneumoniae*. Altered PBPs do not necessarily rule out all treatment options with β-lactam antibiotics.

Nomenclature

β-Lactams are classified according to their core ring structures.

1. β-Lactams fused to saturated five-membered rings:
   (a) β-Lactams containing thiazolidine rings are named penams.
   (b) β-Lactams containing pyrrolidine rings are named carbapenams.
   (c) β-Lactams fused to oxazolidine rings are named oxapenams or clavams.
2. β-Lactams fused to unsaturated five-membered rings:
   (a) β-Lactams containing 2,3-dihydro thiazole rings are named penems.
   (b) β-Lactams containing 2,3-dihydro-1H-pyrrole rings are named carbapenems.
3. β-Lactams fused to unsaturated six-membered rings:
   (a) β-Lactams containing 3,6-dihydro-2H-1,3-thiazine rings are named cephems.
   (b) β-Lactams containing 1,2,3,4-tetrahydropyridine rings are named carbacephems.
   (c) β-Lactams containing 3,6-dihydro-2H-1,3-oxazine rings are named oxacephems.
4. β-Lactams not fused to any other ring are named monobactams.

Common β-lactam antibiotics

Penicillins (penams)

Semisynthetic penicillins are prepared starting from the penicillin nucleus 6-APA.

Narrow-spectrum

1. β-Lactamase sensitive:
   (a) Benzathine penicillin.
   (b) Benzylpenicillin (penicillin G).
   (c) Phenoxymethylpenicillin (penicillin V).
   (d) Procaine penicillin
2. Penicillinase-resistant penicillins:
   (a) Methicillin.
   (b) Oxacillin.
   (c) Nafcillin.
   (d) Cloxacillin.
   (e) Dicloxacillin.
   (f) Flucloxacillin.
3. β-Lactamase-resistant penicillins:
   (a) Temocillin.

Moderate-spectrum

1. Amoxycillin.
2. Ampicillin.

Broad-spectrum

1 Co-amoxiclav (amoxicillin + clavulanic acid).

Extended-spectrum

1. Azlocillin.
2. Carbenicillin.

3. Ticarcillin.
4. Mezlocillin.
5. Piperacillin.

Cephalosporins (cephems)

First generation

Moderate spectrum.

1. Cephalexin.
2. Cephalothin.
3. Cefazolin.

Second generation

Moderate spectrum with anti-*Haemophilus* activity.

1. Cefaclor.
2. Cefuroxime.
3. Cefamandole.

Second generation cephamycins

Moderate spectrum with anti-anaerobic activity.

1. Cefotetan.
2. Cefoxitin.

Third generation

Broad spectrum.

1. Ceftriaxone.
2. Cefotaxime.
3. Cefpodoxime.

Broad spectrum with anti-*Pseudomonas* activity.

1. Ceftazidime.

Fourth generation

Broad spectrum with enhanced activity against Gram-positive bacteria and β-lactamase stability.

1. Cefepime.
2. Cefpirome.

Carbapenems and penems

Broadest spectrum of β-lactam antibiotics.

1. Imipenem (with cilastatin).
2. Meropenem.
3. Ertapenem.
4. Faropenem.
5. Doripenem.

Monobactams

Unlike other β-lactams, the monobactam contains a nucleus with no fused ring attached. Thus, there is less probability of cross-sensitivity reactions.

1. Aztreonam (Azactam).
2. Tigemonam.
3. Nocardicin A.
4. Tabtoxinine-β-lactam.

β-Lactamase inhibitors

Although they exhibit negligible antimicrobial activity, they contain the β-lactam ring. Their sole purpose is to prevent the inactivation of β-lactam antibiotics by binding the β-lactamases, and, as such, they are co-administered with β-lactam antibiotics such as: (i) clavulanic acid, (ii) tazobactam, and (iii) sulbactam.

Adverse effects

Adverse drug reactions

Common adverse drug reactions (ADRs) for the β-lactam antibiotics include diarrhea, nausea, rash, urticaria, superinfection (including candidiasis).

Infrequent ADRs include fever, vomiting, erythema, dermatitis, angioedema, pseudomembranous colitis. Pain and inflammation at the injection site is also common for parenterally-administered β-lactam antibiotics.

Allergy/hypersensitivity

Immunologically-mediated adverse reactions to any β-lactam antibiotic may occur in up to 10 per cent of patients receiving that agent (a small fraction of which are truly IgE-mediated allergic reactions). Anaphylaxis will occur in approximately 0.01 per cent of patients. There is perhaps a 5–10 per cent cross-sensitivity between penicillin-derivatives, cephalosporins, and carbapenems; but this figure has been challenged by various investigators.

Nevertheless, the risk of cross-reactivity is sufficient to warrant the contraindication of all β-lactam antibiotics in patients with a history of severe allergic reactions (urticaria, anaphylaxis, interstitial nephritis) to any β-lactam antibiotic.

NEW β-LACTAM TECHNOLOGIES

There is interest in developing alternative ways to make the cephalosporin intermediates, 7-ADCA and 7-ACA, using the *P. chrysogenum* fermentation. The availability of biosynthetic genes has been used to this purpose to design new biosynthetic pathways.

The expandase enzyme have a strict substrate preference for penicillin N-like molecules and will not expand penicillin G-like molecules. It has been demonstrated, however, the adipic acid can serve as the precursor to adipyl-penicillin in *P. chrysogenum*. On the insertion of the gene, *cef*EF (expandase from *Streptomyces clavuligerus*), into *P. chrysogenum*, the transformants produced adipyl-6-APA and adipyl-7-ADCA. Transformants with the genes *cef*EF and *cef*G (acetyltransferase) produced adipyl-7-ACA in addition to the above (Fig. 13.6).

The adipyl derivatives do have the advantage of being solvent-extractable and their hydrolysis has been demonstrated using glutaryl amidases from *Pseudomonas* sp., enzymes known to have some affinity for the adipyl side chain.

Similarly directed synthesis has been carried out using carboxymethylthiopropionate, a molecule of similar structure to adipic acid.

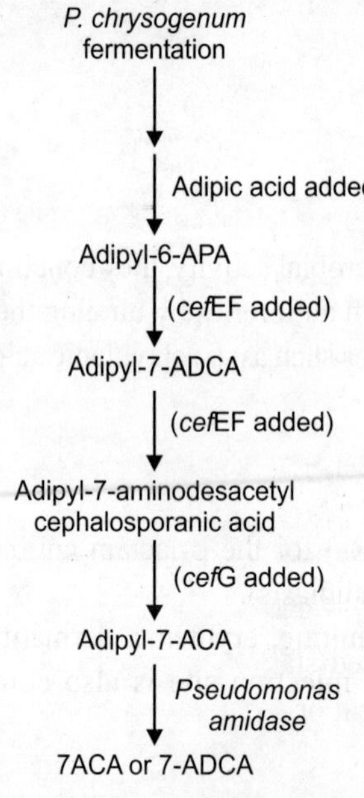

Fig. 13.6. Adipic acid precursored β-lactams.

The direct fermentation of 7-ACA has already been demonstrated by the insertion of genes for D-aminoacid oxidase from *Fusarium solani* and for glutaryl amidase from *Pseudomonas diminuta* into *A. chrysogenum*. A more challenging objective would be to produce the 7-ADCA nucleus directly by the expandase working directly on 6-APA or isopenicillin N. Again, strict substrate preferences do not permit this with the current natural enzymes.

Direct 6-APA fermentations have been developed. The instability, however, of this product, especially in the presence of CO_2 and its poor extractability were major barriers to commercialisation. On the other hand, 7-ADCA is stable in solution, does not react with CO_2 and has very low solubility at pH 4. This could conceivably be an easier compound to recover.

PENICILLINS

The basic structure of the penicillins is 6-aminopenicillanic acid (6-APA) made up of a thiazolidine ring fused with a β-lactam rung (Fig. 13.7). The 6-amino position carries a variety of acyl substituents. In the absence of addition of side-chain precursors to the fermentation medium a mixture of natural penicillins is produced but only benzylpenicillin (Pen G) and phenoxymethylpenicillin (Pen V) are therapeutically important. Both have a similar target spectrum (Gram-positive bacteria) but Pen G is acid-labile and must be administered parenterally whereas the acid stable Pen V may be taken orally. While Pen G is produced naturally, better control of the fermentation process is achieved, resulting in higher yields of Pen G requiring simpler down-stream processing, by addition of the phenylacetic acid precursor to the medium.

By incorporation of precursors phenoxyacetic acid and allylmeracaptoacetic acid into the medium, phenoxymethylpenicillin (Pen V) and the less allergenic allylmercapto-methylpenicillin (Pen O) may be produced.

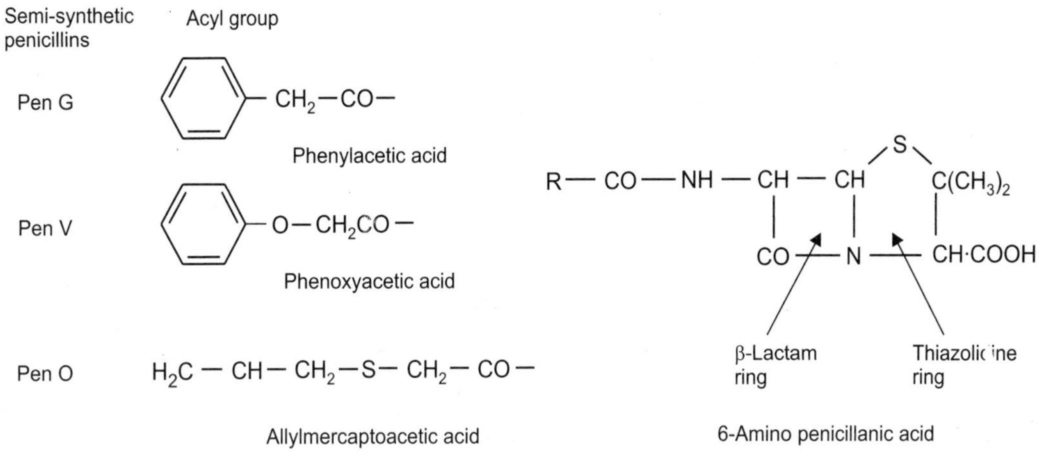

Fig. 13.7. Structure of 6-aminopenicillanic acid and acyl groups of some semi-synthetic penicillins.

Penicillin derivatives, having improved stability and antimicrobial activity, may be produced by semi-synthetic process, following chemical or enzymatic hydrolysis of Pen G to 6-APA.

Penicillin Biosynthetic Pathway

The β-lactam thiazolidine is synthesised from L-α-aminoadipate, L-cystine and L-valine by formation and cyclisation of a peptide to produce isopenicillanic acid. Benzylpenicillin is then produced by transacetylation. Lysine and penicillin share a common anabolic pathway to L-α-aminoadipic acid and lysine is an inhibitor of penicillin synthesis.

The metabolic pathway and some of the possible regulatory mechanisms involved in benzylpenicillin production are illustrated in Fig. 13.8. In addition, glucose causes catabolite repression of penicillin biosynthesis and penicillin appears to regulate its own synthesis. Penicillin production is also affected by phosphate concentration.

Strain Development

Initial yields obtained with Fleming's *P. notatum* strain, were two International Units IUs/ml or 1.2 ml/l. Isolation of *P. chrysogenum* NRRL-1951, a strain which was more suitable for submerged culture than the original *P. notatum* strain, raised the yield to 120IU/ml. Mutation of this strain produced the famous Wisconsin strain Wis Q 176, which yielded 900 IU/ml.

Mutation/selection techniques involving X-rays, short-wave UV radiation and chemical mutagens, such as methyl-*bis*-(β-chloroethyl)amine, nitrosoguanidine, alkylating agents and nitrite, were used until the early 1970s.

Studies with blocked mutants led to an understanding of the biosynthetic pathways which in turn suggested appropriate selection techniques.

The discovery of the parasexual cycle in *P. chrysogenum* led to strain improvement using parasexual breeding and protoplast fusion techniques. Through continued strain-improvement programmes, combined with fermentation optimisation, typical current penicillin yields of 85000 IU/ml or 5 g/l have been achieved.

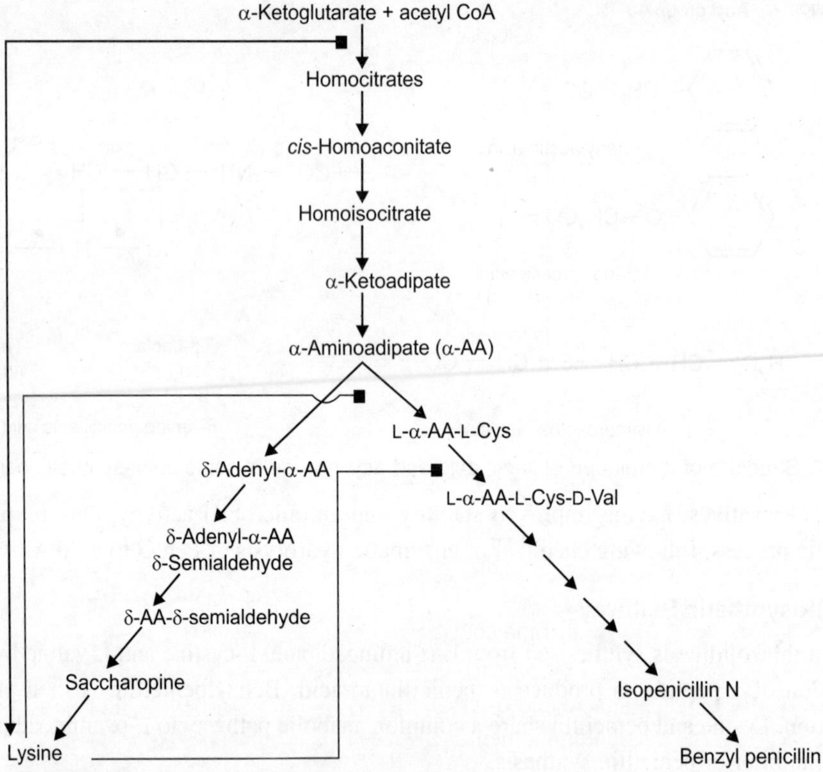

Fig. 13.8. Outline pathway of penicillin biosynthesis indicating possible points of regulation by lysine.

Production of Penicillin

Optimised inoculum spore concentration and formation of pellets in loose rather than compact form in the vegetative growth stages is essential for attainment of high penicillin yields. Biomass doubling time is usually about 6 hrs. Strain stability problems exist and careful strain maintenance is required. Typical penicillin production conditions involve use of a fed-batch culture fermentation with a medium containing corn steep, ammonia, salts and a carbon source such as glucose, lactose or molasses. Most fermentation processes included corn steep as organic nitrogen source because it improved penicillin yield due to its content of side-chain precursors. Introduction to the medium of specific side-chain precursors enabled other nitrogen sources to replace corn steep. The continuous maintenance of ammonia in the medium supports respiration, prevents mycelial lysis and is important for penicillin synthesis. pH is controlled at 6.5 and phenylacetic acid or phenoxyacetic acid is continuously fed as precursor. The rate of sugar utilisation and the rate of oxygen supply are important fermentation parameters. The oxygen supply rate is especially critical because the increasing viscosity of the broth hinders oxygen transfer. The process requires an oxygen uptake rate of 0.4–1.0 mmol per litre per minute and an RQ (moles CO_2 formed/ moles O_2 consumed) of about 0.95. Figure 13.9 illustrates an example of a fermentation outline including the patterns of substrate utilisation and product formation. The industrial fermentation is typically characterised by a high growth rate for about two days. Then growth rate declines and the rate of the penicillin formation increases and continues production for a further period of 6–8 days provided the appropriate substrate feeds are maintained.

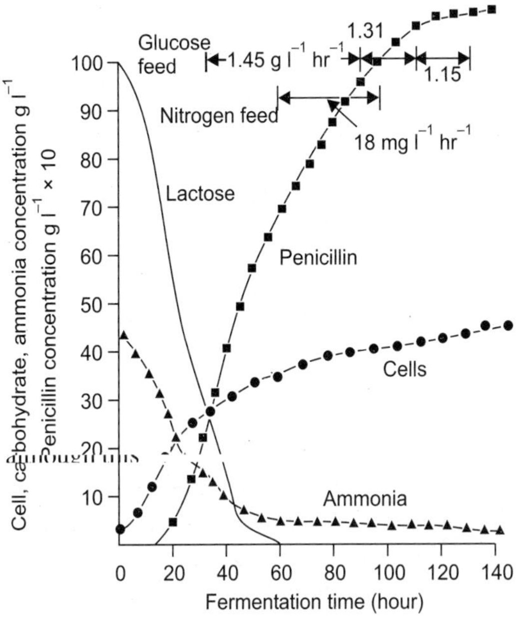

Fig. 13.9. Time course of a penicillin fermentation.

6-APA is produced from Pen G by passage through a immobilised penicillin acylase column which converts Pen G to phenylacetic acid and 6-APA. The column pH is maintained at neutrality by NaOH addition and the 6-APA in the column eluent is recovered by precipitation at pH 4.0. In production of semi-synthetic penicillins, the resultant 6-APA is chemically acylated with the appropriate side-chain standard methods.

Benefits of enzyme hydrolysis of penicillin are shown in Table 13.7.

Table 13.7. Benefits of enzyme hydrolysis of penicillin.

Elimination of chlorinated and alcohol solvents
Elimination of solvent recovery and solvent/odour release
Elimination of hazardous chemical reagents
Elimination of liquid nitrogen for cooling
Elimination of hazardous and toxic waste products and their disposal
All aqueous reactions, neutral pH and ambient temperatures
Good control and monitoring of reaction through pH measurement and adjustment
Quick removal of soluble reaction products from immobilised catalyst
Re-use of immobilised enzyme catalyst
Easy recovery of side chain for re-use
Pleasant working environment for all personnel
Improved product quality, less impurities
Improved yields and manufacturing capacity
Decreased cost of manufacture

CEPHALOSPORIN

The cephalosporins (IPA) are a class of β-lactam antibiotics originally derived from Acremonium, which was previously known as Cephalosporium. Together with cephamycins they constitute a subgroup of β-lactam antibiotics called cephems.

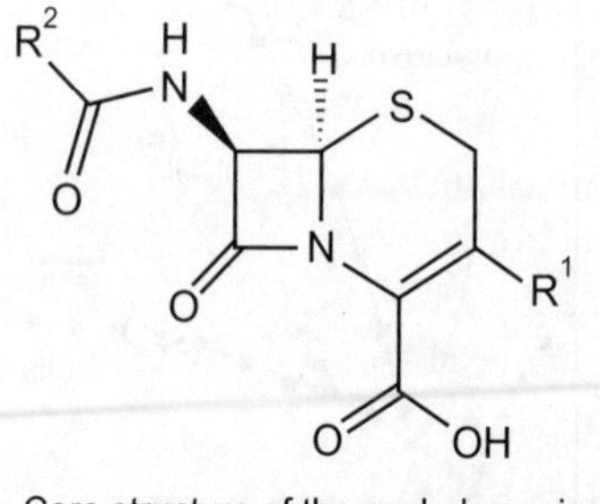

Core structure of the cephalosporins

The cephalosporin antibiotics are comprised of several different classes of compounds with dissimilar spectrums of activity and pharmacokinetic profiles. All 'true' cephalosporins are derived from cephalosporin C which is produced from *Cephalosporium acremonium*.

Cephalosporins are usually bactericidal against susceptible bacteria and act by inhibiting mucopeptide synthesis in the cell wall resulting in a defective barrier and an osmotically unstable spheroplast. The exact mechanism for this effect has not been definitively determined, but β-lactam antibiotics have been shown to bind to several enzymes (carboxypeptidases, transpeptidases, endopeptidases) within the bacterial cytoplasmic membrane that are involved with cell wall synthesis. The different affinities that various β-lactam antibiotics have for these enzymes (also known as penicillin-binding proteins; PBPs) help explain the differences in spectrums of activity of these drugs that are not explained by the influence of β-lactamases. Like other β-lactam antibiotics, cephalosporins are generally considered to be more effective against actively growing bacteria.

The cephalosporin class of antibiotics is usually divided into three classifications or generations. The so-called first generation of cephalosporins include (routes of administration in parentheses): cephalothin (IM/IV), cefazolin (IM/IV), cephapirin (IM/IV/Intramammary), cephradine (IM/IV/PO), cephalexin (PO) and cefadroxil (PO). While there may be differences in MIC's for individual first generation cephalosporins, their spectrums of activity are quite similar. They possess generally excellent coverage against most Gram-positive pathogens and variable to poor coverage against most Gram-negative pathogens. These drugs are very active *in vitro* against groups A β-hemolytic and B *Streptococci*, non-enterococcal group D *Streptococci* (*S. bovis*), *Staphylococcus intermedius* and *aureas*, *Proteus mirabilis* and some strains of *E. coli*, *Klebsiella* sp., *Actinobacillus*, *Pasturella*, *Haemophilus equigenitalis*, *Shigella* and *Salmonella*. With the exception of *Bacteroides fragilis*, most anaerobes are very susceptible to the first generation agents. Most species of *Corynebacteria* are susceptible, but *C. equi* (*Rhodococcus*) is usually resistant. Strains of *Staphylococcus epidermidis* are usually sensitive to the parenterally administered 1st generation drugs, but may have variable susceptibilities to the oral drugs. The following bacteria are regularly resistant to the 1st generation agents: Group D *streptococci/enterococci* (*S. faecalis*, S. *faecium*), Methicillin-resistant *Staphylococci*, indole-positive *Proteus* sp., *Pseudomonas* sp., *Enterobacter* sp., *Serratia* sp. and *Citrobacter* sp.

The second generation cephalosporins include: cefaclor (PO), cefamandole (IM/IV), cefonicid (IM/IV), ceforanide (IM/IV) and cefuroxime (PO/IM/IV). Although not true cephalosporins (they are actually cephamycins), cefoxitin (IM/IV) and cefotetan (IM/IV) are usually included in this group, although some references categorise cefotetan as a 3rd generation agent. In addition to the Gram-positive coverage of the 1st generation agents, these agents have expanded Gram-negative coverage. Cefoxitin and cefotetan also have good activity against *Bacteroides fragilis*. Enough variation exists between these agents in regard to their spectrums of activity against most species of Gram-negative bacteria, that susceptibility testing is generally required to determine sensitivity. The second generation agents have not found widespread use in most veterinary practices, although cefoxitin has been used somewhat.

The third generation cephalosporins retain the Gram-positive activity of the first and second generation agents, but in comparison, have much expanded Gram-negative activity. Included in this group are: cefotaxime (IM/IV), moxalactam (actually a 1-oxa-β-lacatam; IM/IV), cefoperazone (IM/IV), ceftizoxime (IM/IV), ceftazidime (IM/IV), ceftriaxone (IM/IV), ceftiofur (IM) and cefixime (PO). As with the 2nd generation agents, enough variability exists with individual bacterial sensitivities that susceptibility testing is necessary for most bacteria. Usually only ceftazidime and cefoperazone are active against most strains of *Pseudomonas aeruginosa*. Because of the excellent Gram-negative coverage of these agents and when compared to the aminoglyco-sides, their significantly less toxic potential, they have been used on an increasing basis in veterinary medicine. Ceftiofur is approved for use in beef cattle, but its use in other species is hindered by a lack of data on its spectrum of activity or availability of pharmacokinetic profiles.

Uses/Indications: Cephalosporins have been used for a wide range of infections in various species. FDA-approved indications/species, as well as non-approved uses are listed in the Uses/Indications and Dosage sections for each individual drug.

Pharmacokinetics (General): Until recently, only some first generation cephalosporins were absorbed appreciably after oral administration, but this has changed with the availability of cefuroxime axetil (2nd generation) and cefixime (3rd generation). Depending on the drug, absorption may be delayed, unaltered, or increased if administered with food. There are reported species variations in the oral bioavailability of some cephalosporins which are detailed under each individual drug's monograph.

Cephalosporins are widely distributed to most tissues and fluids, including bone, pleural fluid, pericardial fluid and synovial fluid. Higher levels are found in inflamed than in normal bone. Very high levels are found in the urine, but they penetrate poorly into prostatic tissue and aqueous humor. Bile levels can reach therapeutic concentrations with several of the agents as long as biliary obstruction is not present. With the exception of cefuroxime, no first or second generation cephalosporin enters the CSF (even with inflamed meninges) in therapeutically effective levels. Therapeutic concentrations of cefotaxime, moxalactam, cefuroxime, ceftizoxime, ceftazidime and ceftriaxone can be found in the CSF after parenteral dosing in patients with inflamed meninges. Cephalosporins cross the placenta and fetal serum concentrations can be 10 per cent or more of those found in maternal serum. Cephalosporins enter milk in low concentrations. Protein binding of the drugs is widely variable and species specific. Cephalosporins tend to bind to equine and canine plasma proteins less so then to human plasma proteins.

Cephalosporins and their metabolites (if any) are excreted by the kidneys, via tubular secretion and/or glomerular filtration. Some cephalosporins (e.g. cefotaxime, cefazolin, and cephapirin) are partially metabolised by the liver to desacetyl compounds that may have some antibacterial activity.

Contraindications/Precautions/Reproductive safety: Cephalosporins are contraindicated in patients who have a history of hypersensitivity to them. Because there may be cross-reactivity, use cephalosporins cautiously in patients who are documented hypersensitive to other β-lactam antibiotics (e.g. penicillins, cefamycins, carbapenems).

Oral systemic antibiotics should not be administered in patients with septicemia, shock or other grave illnesses as absorption of the medication from the GI tract may be significantly delayed or diminished. Parenteral routes (preferably IV) should be used for these cases.

Cephalosporins have been shown to cross the placenta and safe use of them during pregnancy have not been firmly established, but neither have there been any documented teratogenic problems associated with these drugs. However, use only when the potential benefits outweigh the risks.

Adverse effects/warnings: Adverse effects with the cephalosporins are usually not serious and have a relatively low frequency of occurrence.

Hypersensitivity reactions unrelated to dose can occur with these agents and can be manifested as rashes, fever, eosinophilia, lymphadenopathy, or full-blown anaphylaxis. The use of cephalosporins in patients documented to be hypersensitive to penicillin-class antibiotics is controversial. In humans, it is estimated that up to 15 per cent of patients hypersensitive to penicillins will also be hypersensitive to cephalosporins. The incidence of cross-reactivity in veterinary patients is unknown.

Cephalosporins can cause pain at the injection site when administered intramuscularly, although this effect is less so with cefazolin than other agents. Sterile abscesses or other severe local tissue reactions are also possible but are much less common. Thrombophlebitis is also possible after IV administration of these drugs. When given orally, cephalosporins may cause GI effects (anorexia, vomiting, diarrhea). Administering the drug with a small meal may help alleviate these symptoms. Because the cephalosporins may also alter gut flora, antibiotic-associated diarrhea can occur as well as the selection out of resistant bacteria maintaining residence in the colon of the animal. While it has been demonstrated that the cephalosporins (particularly cephalothin) have the potential for causing nephrotoxicity, at clinically used doses in patients with normal renal function, risks for this adverse effect occurring appear minimal.

High doses or very prolonged use has been associated with neurotoxicity, neutropenia, agranulocytosis, thrombocytopenia, hepatitis, positive Comb's test, interstitial nephritis, and tubular necrosis. Except for tubular necrosis and neurotoxicity, these effects have an immunologic component.

Some cephalosporins (cefamandole, cefoperazone, moxalactam) that contain a thiomethyltetrazole side chain have been implicated in causing bleeding problems in humans. These drugs are infrequently used in veterinary species at the present time, so any veterinary ramifications of this effect are unclear.

Overdosage/Acute toxicity: Acute oral cephalosporin overdoses are unlikely to cause significant problems other than GI distress, but other effects are possible.

Drug interactions: The concurrent use of parenteral aminoglycosides or other nephrotoxic drugs (e.g. amphotericin B) with cephalosporins is controversial. Potentially, cephalosporins could cause additive nephrotoxicity when used with these drugs, but this interaction has only been well documented with cephaloridine (no longer marketed). Nevertheless, they should be used together cautiously. *In vitro* studies have demonstrated that cephalosporins can have synergistic or additive activity against certain bacteria when used with aminoglycosides, penicillins, or chloramphenicol.

However, some clinicians do not recommend using cephalosporins concurrently with bacteriostatic antibiotics (e.g. chloramphenicol), particularly in acute infections where the organism is proliferating rapidly. Probenecid competitively blocks the tubular secretion of most cephalosporins, thereby increasing serum levels and serum half-lives. A disulphiram-like reaction (anorexia, nausea, vomiting) has been reported in humans who have ingested alcohol with 48–72 hrs of receiving β-lactam antibiotics (e.g. cefamandole, cefoperazone, moxalactam, cefotetan) with a thiomethyltetrazole side-chain. Because these antibiotics have been associated with bleeding, they should be used cautiously in patients receiving oral anticoagulants.

Drug/Laboratory interactions: Except for cefotaxime, cephalosporins may cause false-positive urine glucose determinations when using cupric sulphate solution. Tests utilising glucose oxidase are not affected by cephalosporins.

When using the Jaffe reaction to measure serum or urine creatinine, cephalosporins (not ceftazidime or cefotaxime) in high dosages may falsely cause elevated values.

In humans, particularly with azotemia, cephalosporins have caused a false-positive direct Combs' test. Cephalosporins may also cause falsely elevated 17-ketosteroid values in urine.

Monitoring parameters: Because cephalosporins usually have minimal toxicity associated with their use, monitoring for efficacy is usually all that is required. Patients with diminished renal function, may require intensified renal monitoring. Serum levels and therapeutic drug monitoring are not routinely done with these agents.

Biosynthesis and Fermentation

The early part of the biosynthesis pathway is shared with penicillin. The cephalosporin molecule is derived from penicillin by a ring expansion of penicillin N. This enzyme, expandase and other subsequent synthesis enzymes are not found in *P. chrysogenum*. As in the case of penicillins, yield improvements have been sought by the insertion of extra rate-limiting enzymes. Genes for cyclase (*pcb*C), expandase and hydroxylase (*cef*EF) and acetyltransferase (*cef*G) have been enriched in *A. chrysogenum* in attempts to both increase the production of cephalosporin C and reduce the production of unwanted intermediates, e.g. penicillin N and desacetoxycephalosporin C (DAOC).

High-producing cultures of *A. chrysogenum* are fermented using corn steep liquor and soya flour-based media with continuous feeding of both corn syrup and triacylglycerols (soya, rape or lard oil). Methionine is used both as a source of sulphur and as in inducer of morphological changes. Cephalosporin C is unstable and degrades chemically to desacetylcephalosporin C (DAC) and thiazole-4-carboxylate. Cephalosporin C is also hydrolysed to DAC by esterases released by the fungus. These, however, can be inhibited by the use of phosphates.

Production of 7-Aminocephalosporanic Acid

Cephalosporin C is recovered from broth filtrates by a variety of hydrophobic and ion exchange resins. The column chromatography is designed to separate cephalosporin C from related intermediates and breakdown products. The rich fractions are either treated with zinc acetate to precipitate the low solubility zinc salt or with sodium or potassium acetate followed by a water-miscible solvent to precipitate the salt complex.

Isolated cephalosporin C is efficiently hydrolysed chemically to 7-ACA. Unfortunately, the process uses similar reactants and solvents used in the chemical hydrolysis process for penicillin, with the familiar drawbacks of hazardous material handling, solvent use and negative environmental issues. The switch to enzyme hydrolysis has proved to be difficult due to the inability to identify enzymes to directly hydrolyse off the side chain, the unnatural D-amino acid D-α-aminoadipate. Indirect enzyme systems, though, have been developed which rely on the sequential use of two enzymes (Fig. 13.10).

The first enzyme, a D-aminoacid oxidase, removes the chirality of the side chain by oxidative deamination to produce a keto acid which, in the presence of the co-produced peroxide, is conveniently decarboxylated to the glutaryl side chain.

The yeast, *Trigonopsis variabilis*, is a suitable source of this enzyme. The second enzyme was discovered in *Psuedomonas* sp. and can directly hydrolyse the glutaryl side chain to produce 7-ACA. In a similar manner to penicillin hydrolysis these two enzymes are now available from recombinant sources and have been immobilised. Most of the major industrial producers of 7-ACA are now switching over to the enzyme process.

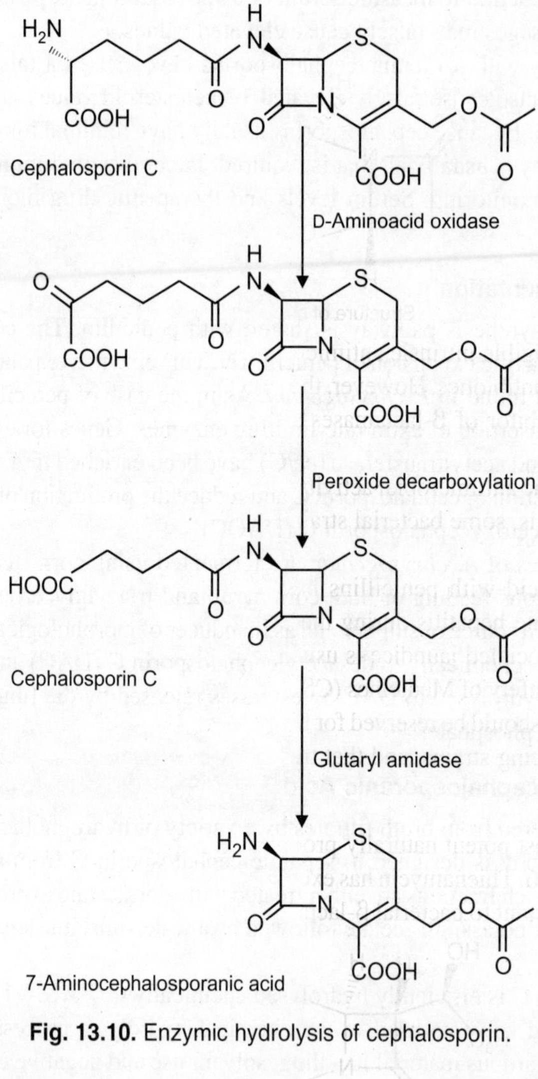

Fig. 13.10. Enzymic hydrolysis of cephalosporin.

New β-lactam ring systems

These have been discovered by the introduction of screening methods such as using as test organisms β-lactam hypersensitive strains or using enzymatic tests to detect β-lactam inhibitors.

Clavulanic acid

The name is derived from the *Streptomyces clavuligerus*, which produces clavulanic acid. Clavulanic acid is a β-lactamase inhibitor combined with penicillin group antibiotics to overcome certain types of antibiotic resistance. It is used to overcome resistance in bacteria that secrete β-lactamase, which otherwise inactivates most penicillins. In its most common form, the potassium salt potassium clavulanate is combined with amoxicillin (co-amoxiclav or the veterinary formulation Synulox from Pfizer) or ticarcillin.

Clavulanic acid is biosynthetically generated from the amino acid arginine and the sugar glyceraldehyde 3-phosphate.

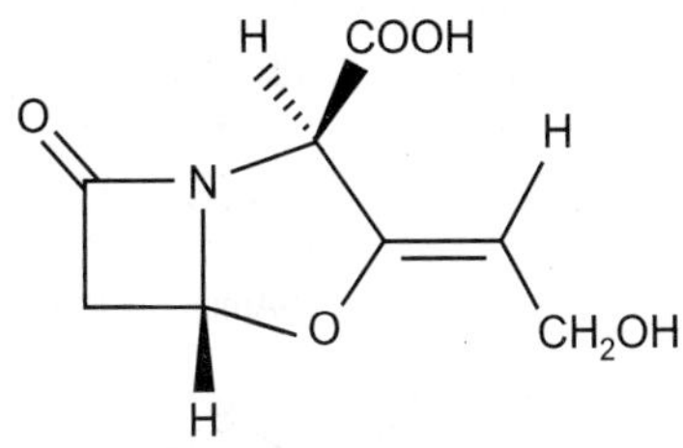

Structure of clavulanic acid

Clavulanic acid has negligible intrinsic antimicrobial activity, despite sharing the β-lactam ring that is characteristic of β-lactam antibiotics. However, the similarity in chemical structure allows the molecule to act as a competitive inhibitor of β-lactamases secreted by certain bacteria to confer resistance to β-lactam antibiotics.

This inhibition restores the antimicrobial activity of β-lactam antibiotics against lactamase-secreting-resistant bacteria. Despite this, some bacterial strains that are even resistant to such combinations have emerged.

The use of clavulanic acid with penicillins has been associated with an increased incidence of cholestatic jaundice and acute hepatitis during therapy or shortly after, particularly in men and those aged over 65 years. The associated jaundice is usually self-limiting and very rarely fatal.

The UK Committee on Safety of Medicines (CSM) recommends that treatments such as amoxicillin/clavulanic acid preparations should be reserved for bacterial infections likely to be caused by amoxicillin-resistant β-lactamase-producing strains, and that treatment should not normally exceed 14 days.

Thienamycin

Thienamycin, one of the most potent naturally-produced antibiotics known thus far, was discovered in *Streptomyces cattleya* in 1976. Thienamycin has excellent activity against both Gram-positive and Gram-negative bacteria and is resistant to bacterial β-lactamase enzymes. Thienamycin is a zwitterion at pH 7.

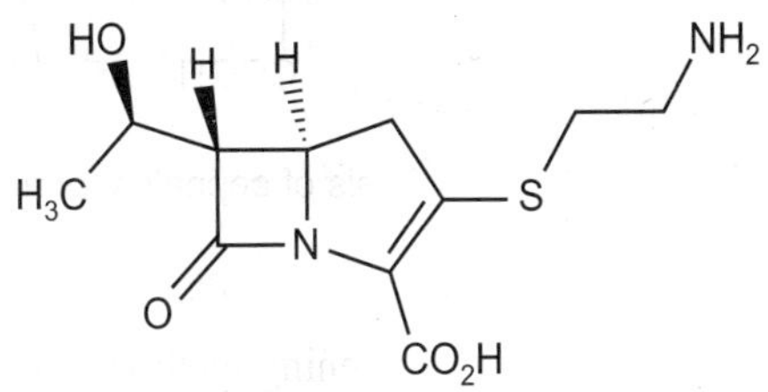

Structure of thienamycin

In 1976, fermentation broths obtained from the soil bacteria *Streptomyces cattleya* were found to be active in screens for inhibitors of peptidoglycan biosynthesis. Initial attempts to isolate the active species proved difficult due to the chemical instability of that component. After many attempts and extensive purification, the material was finally isolated in >90 per cent purity, allowing for the structural elucidation of thienamycin in 1979.

Thienamycin was the first among the naturally-occurring class of carbapenem antibiotics to be discovered and isolated. Carbapenems are similar in structure to their antibiotic 'cousins' the penicillins. Like penicillins, carbapenems contain a β-lactam ring (cyclic amide) fused to a five-membered ring. Carbapenems differ in structure from penicillins in that within the five-membered ring a sulphur is replaced by a carbon atom (C1) and an unsaturation is present between C2 and C3 in the five-membered ring. *In vitro*, thienamycin employs a similar mode of action as penicillins through disrupting the cell wall synthesis (peptidoglycan biosynthesis) of various Gram-positive and Gram-negative bacteria (*Staphylococcus aureus,Staphylococcus epidermidis*, *Pseudomonas aeruginosa* to name a few). In a study carried out by Spratt and others, they found that, although thienamycin binds to all of the penicillin-binding proteins (PBP) in *Escherichia coli*, it preferentially binds to PBP-1 and PBP-2, which are both associated with the elongation of the cell wall.

Unlike penicillins, which are rendered ineffective through rapid hydrolysis by the β-lactamase enzyme present in some strains of bacteria, thienamycin remains antimicrobially active. Neu found that thienamycin displayed high activity against bacteria that were resistant to other β-lactamase stable compounds (cephalosporins), highlighting the superiority of thienamycin as an antibiotic among β-lactams.

Biosynthesis

The formation of thienamycin is thought to occur through a different pathway from classic β-lactams (penicillins, cephalosporins). Production of classic β-lactams in both fungi and bacteria occur through two steps: First, the condensation of L-cysteine, L-valine, and L-α-amino adipic acid by ACV synthetase (ACVS, a nonribsomal peptide synthetase) and then cyclisation of this formed tripeptide by isopenicillin N synthetase (IPNS).

The gene cluster (thn) for the biosynthesis of thienamycin by *S. cattleya* was identified and sequenced in 2003, lending insight into the biosynthetic mechanism for thienamycin formation.

The biosynthesis is thought to share features with the biosynthesis of the simple carbapenems, beginning with the condensation of malonyl-CoA with glutamate semialdehyde to form the pyrroline ring. The β-lactam is then formed by a β-lactam synthetase, which makes use of ATP, providing a carbapenam. At some later point, oxidation to the carbapenem and ring inversions must occur.

The hydroxyethyl side chain of thienamycin is thought to be a result of two separate methyl transfers from S-adenosyl methionine. According to the proposed gene functions, ThnK, ThnL, and ThnP could catalyse these methyl-transfer steps. A β-lactam synthetase (ThnM) is thought to catalyse the formation of the β-lactam ring fused to the five-membered ring. How the cysteaminyl side-chain is incorporated is largely unknown, although ThnT, ThnR, and ThnH are involved in the processing of CoA to cysteamine for use in the pathway. Various oxidations complete the biosynthesis.

Due to low titre and to difficulties in isolating and purifying thienamycin produced via fermentation, total synthesis is the preferred method for commercial production.

Thienamycin itself is extremely unstable and decomposes in aqueous solution. Consequently, thienamycin is impractical for the clinical treatment of bacterial infections. For this reason, stable derivatives of thienamycin were created for medicinal consumption. One such derivative–imipenem was formulated in 1985.

Imipenem, an *N*-formimidoyl derivative of thienamycin, is rapidly metabolised by a renal dihpeptidase enzyme found in the human body. To prevent its rapid degradation, imipenem is normally co-administered with cilastatin, an inhibitor of this enzyme.

Nocardicins

Nocardicins are monocyclic β-lactams. Monocyclic, bacterially produced or semisynthetic β-lactum antibiotics and can be easily synthesised.

Monobactams

Monobactams are β-lactam compounds wherein the β-lactam ring is alone and not fused to another ring (in contrast to most other β-lactams, which have at least two rings). They work only against Gram-negative bacteria. The only commercially available monobactam antibiotic is aztreonam. Other examples of monobactams are tigemonam, nocardicin A, and tabtoxin.

AMINO ACID AND PEPTIDE ANTIBIOTICS

This group of antibiotics are amino acid derivatives such as cycloserine and azaserine, most of the β-lactam antibiotics which are also discussed in this chapter.

Cycloserine

Cycloserine is an antibiotic effective against *Mycobacterium tuberculosis*. For the treatment of tuberculosis, it is classified as a second line drug, i.e. its use is only considered if one or more first line drugs cannot be used.

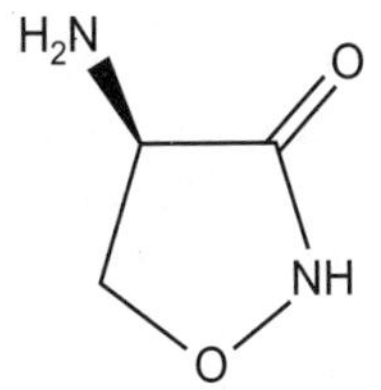

Structure of cycloserine

Although in principle active against other bacteria as well, cycloserine is not commonly used in the treatment of infections other than tuberculosis.

The terminal two amino acid residues of the murein precursor lipid II consist of D-alanine, which is produced by the enzyme alanine racemase; the two residues are joined by D-alanine ligase. Both enzymes are competitively inhibited by cycloserine. It is also being trialed for treatment of phobias as well as an adjuvant to conventional treatments for depression, obsessive-compulsive disorder and schizophrenia. It has been experimentally used for treatment of Gaucher's disease.

Recent research suggests that D-cycloserine may be effective in treating chronic pain. The side effects are mainly central nervous system (CNS) manifestations, i.e. headache, irritability, depression, psychosis convulsions.

Co-administration of pyridoxine can reduce the incidence of some of the CNS side effects (e.g. convulsions). These psychotropic responses are related to D-cycloserine's action as a partial agonist of the neuronal NMDA receptor for glutamate and have been examined in implications with sensory-related fear extinction in the amygdala.

Depsipeptide

A depsipeptide is a peptide in which one or more of the amide (–CONHR–) bonds are replaced by ester (COOR) bonds.

Depsipeptides have often been used in research to probe the importance of hydrogen bond networks in protein folding kinetics and thermodynamics. They are also found in nature as natural products. An infamous example is the L-Lys-D-Ala-D-Lac motif found in vancomycin resistant bacteria's cell wall building blocks. The amide to ester mutation disrupts its hydrogen bonding network with vancomycin, which is key to the antibiotic's activity.

'Depsipeptide' is also used to refer to (NSC 630176) which is a member of the bicyclic peptide class of histone deacetylase (HDAC) inhibitors and was first isolated as a fermentation product from *Chromobacterium violaceum*. It is also known as FK-228. It is being used in the treatment of some cancers where it is thought to reactivate silenced genes. It was first observed to have a positive role on gene expression in 1990. Another natural depsipeptide HDAC inhibitor is Spiruchostatin A.

Antibiotics of the virginiamycin family also have a depsipeptide structure. They consist of a mixture of structurally different macrocyclic lactone rings. Virginiamycin is a streptogramin antibiotic similar to pristinamycin and quinupristin/dalfopristin. It is a combination of pristinamycin IIA (virginiamycin M1) and virginiamycin S1. Virginiamycin is used in the fuel ethanol industry to prevent microbial contamination.

Linear and Cyclic Peptide Antibiotics

Cyclic peptides (or cyclic proteins) are polypeptide chains whose amino and carboxyl termini are themselves linked together with a peptide bond, forming a circular chain. A number of cyclic peptides have been discovered in nature and they can range anywhere from just a few amino acids in length to hundreds. The processes by which cyclic peptides are formed in cells are not yet fully understood. One interesting property of cyclic peptides is that they tend to be extremely resistant to digestion, allowing them to survive intact in the human digestive tract. This trait makes cyclic peptides attractive to protein based drug designers for use as scaffolds which, in theory, could be engineered to incorporate any arbitrary protein domain of medicinal value, in order to allow those components to be delivered orally.

The majority of the antibiotics formed by species of the genus *Bacillus* are peptides; other peptide antibiotics are produced by streptomycetes. Due to their toxicity the terapeutic action of peptide antibiotics is limited. Polymyxins are used for infections caused by Gram-negative pathogens (including *Pseudomonas*), and viomycin and capreomycin are used in tuberculosis therapy.

Bacitracin

Bacitracin is a mixture of related cyclic polypeptides produced by organisms of the licheniformis group of *Bacillus subtilis* var *Tracy*, isolation of which was first reported in 1945.

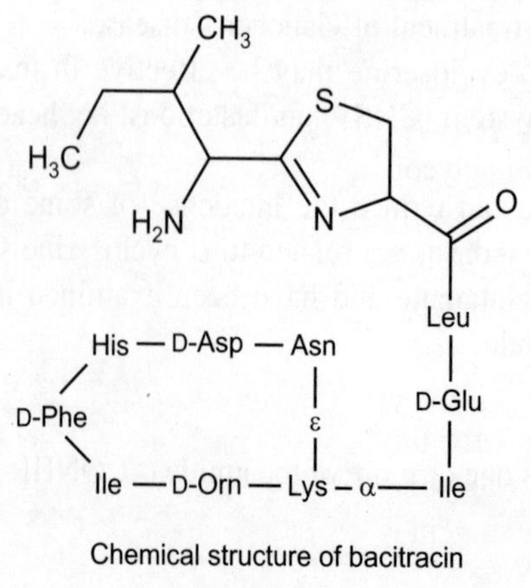

Chemical structure of bacitracin

As a toxic and difficult-to-use antibiotic, bacitracin doesn't work well orally. However, it is very effective topically. Its action is on Gram-positive cell walls. Bacitracin is synthesised via the so-called nonribosomal peptide synthetases (NRPSs), which means that ribosomes are not involved in its synthesis.

Bacitracin interferes with the dephosphorylation of the C_{55}-isoprenyl pyrophosphate, a molecule which carries the building blocks of the peptidoglycan bacterial cell wall outside of the inner membrane. Bacitracin is a protein disulphide isomerase inhibitor.

Bacitracin is used in human medicine as a polypeptide antibiotic and is 'approved by the US Food and Drug Administration (FDA) for use in chickens and turkeys'.

As bacitracin zinc salt, and in combination with other topical antibiotics (usually polymyxin B and neomycin), it is used in ointment form for topical treatment of a variety of localised skin and eye infections, as well as for the prevention of wound infections. In the United States a popular brand name Neosporin contains Bacitracin as one of its antibiotic agents along with neomycin and polymyxin B. Bacitracin can also be bought in pure form for those with allergies.

It is also commonly used as an aftercare antibiotic on tattoos. It is preferred over Neosporin because of its fewer ingredients, which lowers chances of an allergic reaction.

In infants, it is sometimes administered intramuscularly for the treatment of pneumonias,but in most cases, it has been replaced by other antibiotics. This formulation is sold under the brand name Baciim.

Bleomycin: Bleomycin is a glycopeptide antibiotic produced by the bacterium *Streptomyces verticillus*. Bleomycin refers to a family of structurally related compounds. When used as an anti-cancer agent, the chemotherapeutical forms are primarily bleomycin A_2 and B_2. The drug is used in the treatment of Hodgkin lymphoma (as a component of the ABVD regimen), squamous cell carcinomas, and testicular cancer, as well as in the treatment of pleurodesis and plantar warts.

Bleomycin acts by induction of DNA strand breaks. Some studies suggest that bleomycin also inhibits incorporation of thymidine into DNA strands. DNA cleavage by bleomycin depends on oxygen and metal ions, at least *in vitro*. It is believed that bleomycin chelates metal ions (primarily iron) producing a pseudoenzyme that reacts with oxygen to produce superoxide and hydroxide free radicals that cleave DNA. In addition, these complexes also mediate lipid peroxidation and oxidation of other cellular molecules. Bleomycin is a nonribosomal peptide that is a hybrid peptide-polyketide natural product. The peptide/polyketide/peptide backbone of the bleomycin aglycon is assembled by the bleomycin megasynthetase, which is made of both nonribosomal peptide synthetase (NRPS) and polyketide synthase (PKS) modules. Nonribosomal peptides and polyketides are synthesised from amino acids and short carboxylic acids by NRPSs and PKSs, respectively. These NRPSs and PKSs use similar strategies for the assembly of these two distinct classes of natural products. Both NRPs and type I PKSs are organised into modules. The structural variations of the resulting peptide and polyketide products are determined by the number and order of modules on each NRPS and PKS protein.

The biosynthesis of the bleomycin aglycon can be easily visualised in three stages:

1. NRPS-mediated formation of P-3A from Ser, Asn, His, and Ala.
2. PKS-mediated elongation of P-3A by malonyl CoA and AdoMet to yield P-4.
3. NRPS-mediated elongation of P-4 by Thr to P-5 that is further elongated by β-Ala, Cys, and Cys to get P-6m.

On the basis of the bleomycin structure and the deduced functions of individual NRPS and PKS domains and modules, a linear model for the bleomycin megasynthetase-templated assembly of the bleomycin peptide/polyketide/peptide aglycon was proposed from nine amino acids and one acetate. The most serious complication of bleomycin is pulmonary fibrosis and impaired lung function.

CARBOHYDRATE ANTIBIOTICS

Carbohydrate antibiotics include the therapeutically important aminoglycosides, the ortosomycins and various sugar derivatives.

Glycosides and Sugar Derivatives

In chemistry, glycosides are molecules in which a sugar is bound to a non-carbohydrate moiety, usually a small organic molecule. Glycosides play numerous important roles in living organisms. Many plants store chemicals in the form of inactive glycosides; which can be activated by enzyme hydrolysis. This causes the sugar part to be broken off, making the chemical available for use. Many such plant glycosides are used as medications. In animals (including humans), poisons are often bound to sugar molecules as part of their elimination from the body.

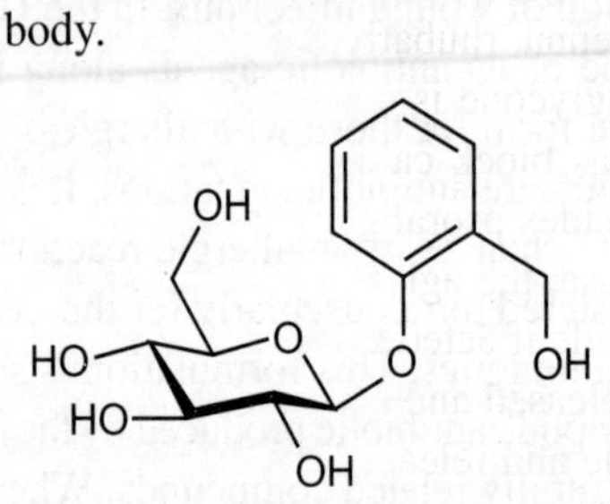

Salicin, a glycoside related to aspirin

Formally, a glycoside is any molecule in which a sugar group is bonded through its anomeric carbon to another group via a glycosidic bond. Glycosides can be linked by an O-(an O-glycoside), N-(a glycosylamine), S-(a thioglycoside) or C-(a C-glycosyl) glycosidic bond. The given definition is the one used by IUPAC. Many authors require in addition that the sugar be bonded to a non-sugar for the molecule to qualify as a glycoside, thus excluding polysaccharides. The sugar group is then known as the glycone and the non-sugar group as the aglycone or genin part of the glycoside. The glycone can consist of a single sugar group (monosaccharide) or several sugar groups (oligosaccharide). Molecules containing an N-glycosidic bond are known as glycosylamines.

The glycone and aglycone portions can be chemically separated by hydrolysis in the presence of acid. There are also numerous enzymes that can form and break glycosidic bonds. The most important cleavage enzymes are the glycoside hydrolases, and the most important synthetic enzymes in nature are glycosyltransferases. Genetically altered enzymes termed glycosynthases have been developed that can form glycosidic bonds in excellent yield.

There are a great many ways to chemically synthesise glycosidic bonds. Fischer glycosidation refers to the synthesis of glycosides by the reaction of unprotected monosaccharides with alcohols (usually as solvent) in the presence of a strong acid catalyst. The Koenigs-Knorr reaction is the condensation of glycosyl halides and alcohols in the presence of metal salts such as silver carbonate or mercuric oxide.

Classification

We can classify glycosides by the glycone, by the type of glycosidic bond, and by the aglycone.

By glycone: If the glycone group of a glycoside is glucose, then the molecule is a glucoside; if it is fructose, then the molecule is a fructoside; if it is glucuronic acid, then the molecule is a glucuronide, etc. In the body, toxic substances are often bonded to glucuronic acid to increase their water solubility; the resulting glucuronides are then excreted.

By type of glycosidic bond: Depending on whether the glycosidic bond lies below or above the plane of the cyclic sugar molecule, glycosides are classified as α-glycosides or β-glycosides. Some enzymes such as α-amylase can only hydrolyse α-linkages; others, such as emulsin, can only affect β-linkages.

By aglycone: Glycosides are also classified according to the chemical nature of the aglycone. For purposes of biochemistry and pharmacology, this is the most useful classification.

Alcoholic glycosides: An example of an alcoholic glycoside is salicin which is found in the genus *Salix*. Salicin is converted in the body into salicylic acid, which is closely related to aspirin and has analgesic, antipyretic and antiinflammatory effects.

Anthraquinone glycosides: These glycosides contain an aglycone group that is a derivative of anthraquinone. They are present in senna, rhubarb and aloes; they have a laxative effect.

Coumarin glycosides: Here the aglycone is coumarin. An example is apterin which is reported to dilate the coronary arteries as well as block calcium channels. Those obtained from dried leaves of Psoralia corylifolia have main glycosides psoralin and corylifolin.

Cyanogenic glycosides: In this case, the aglycone contains a cyanide group, and the glycoside can release the poisonous hydrogen cyanide if acted upon by some enzyme. They are stored in the vacuole but if the plant is attacked they are released and become activated by enzymes in the cytoplasm. These remove the sugar part of the molecule and release toxic hydrogen cyanide.

Storing them in inactive forms in the cytoplasm prevents them from damaging the plant under normal conditions. An example of these is amygdalin from almonds. They can also be found in the fruits (and wilting leaves) of the rose family (including cherries, apples, plums, almonds, peaches, apricots, raspberries, and crabapples).

Flavonoid glycosides: Here the aglycone is a flavonoid. This is a large group of flavonoid glycosides. Examples include:

1. Hesperidin (aglycone: Hesperetin, glycone: Rutinose).
2. Naringin (aglycone: Naringenin, glycone: Rutinose).
3. Rutin (aglycone: Quercetin, glycone: Rutinose).
4. Quercitrin (aglycone: Quercetin, glycone: Rhamnose).

Among the important effects of flavonoids are their antioxidant effect. They are also known to decrease capillary fragility.

Phenolic glycosides (simple): Here the aglycone is a simple phenolic structure. An example is arbutin found in the common Bearberry *Arctostaphylos uva-ursi*. It has a urinary antiseptic effect. Rutin found in rooibos tea.

Saponins: These compounds give a permanent froth when shaken with water. They also cause hemolysis of red blood cells. Saponin glycosides are found in liquorice. Their medicinal value is due to their expectorant effect.

Steroidal glycosides or cardiac glycosides: Here the aglycone part is a steroidal nucleus. These glycosides are found in the plant genera *Digitalis*, *Scilla*, and *Strophanthus*. They are used in the treatment of heart diseases, e.g. congestive heart failure (historically as now recognised does not improve survivability; other agents are now preferred) and arrhythmia.

Steviol glycosides: These sweet glycosides found in the stevia plant *Stevia rebaudiana* Bertoni have 40–300 times the sweetness of sucrose. The two primary glycosides, stevioside and rebaudioside A, are used as natural sweeteners in many countries. These glycosides have steviol as the aglycone part. Glucose or rhamnose-glucose combinations are bound to the ends of the aglycone to form the different compounds.

Nojirimycin: Nojirimycin is antibiotic produced by streptomyces strains, inhibits glucosidases and prevents normal glycosylation of proteins by interfering with the early pruning down to the core carbohydrate that is normally followed by addition of specific sugar residues.

Vancomycin: Vancomycin is a glycopeptide antibiotic used in the prophylaxis and treatment of infections caused by Gram-positive bacteria. It has traditionally been reserved as a drug of last resort, used only after treatment with other antibiotics had failed, although the emergence of vancomycin-resistant organisms means that it is increasingly being displaced from this role by linezolid and daptomycin.

The organism that produced it was eventually named *Amycolatopsis orientalis*. The original indication for vancomycin was for the treatment of penicillin-resistant *Staphylococcus aureus*.

Vancomycin never became first line treatment for *Staphylococcus aureus* for several reasons:

1. The drug must be given intravenously, because it is not absorbed orally.
2. β-Lactamase-resistant semi-synthetic penicillins such as methicillin (and its successors, nafcillin and cloxacillin) were subsequently developed.
3. Early trials used early impure forms of vancomycin (Mississippi mud) which were found to be toxic to the ears and to the kidneys; these findings led to vancomycin being relegated to the position of a drug of last resort.

It is a branched tricyclic glycosylated nonribosomal peptide produced by the fermentation of the Actinobacteria species *Amycolatopsis orientalis* (formerly designated *Nocardia orientalis*).

Vancomycin acts by inhibiting proper cell wall synthesis in Gram-positive bacteria. The mechanism inhibited, and various factors related to entering the outer membrane of Gram-negative organisms mean that vancomycin is not active against Gram-negative bacteria (except some non-gonococcal species of Neisseria). Specifically, vancomycin prevents incorporation of *N*-acetylmuramic acid (NAM)- and *N*-acetylglucosamine (NAG)-peptide subunits into the peptidoglycan matrix; which forms the major structural component of Gram-positive cell walls.

The large hydrophilic molecule is able to form hydrogen bond interactions with the terminal D-alanyl-D-alanine moieties of the NAM/NAG-peptides. Normally this is a five-point interaction. This binding of vancomycin to the D-Ala-D-Ala prevents the incorporation of the NAM/NAG-peptide subunits into the peptidoglycan matrix.

Vancomycin exhibits atropisomerism—it has two chemically distinct rotamers owing to the rotational restriction of the chlorotyrosine residue (on the right hand side of the figure). The form present in the drug is the thermodynamically more stable conformer and, importantly, has more potent activity.

Vancomycin is indicated for the treatment of serious, life-threatening infections by Gram-positive bacteria which are unresponsive to other less toxic antibiotics. In particular, vancomycin should not be used to treat methicillin-sensitive *Staphylococcus aureus* because it is inferior to penicillins such as nafcillin.

Damage to the kidneys and to the hearing were a side effect of the early impure versions of vancomycin, and these were prominent in the clinical trials conducted in the mid-1950s. Later trials using purer forms of vancomycin found that nephrotoxicity is an infrequent adverse effect (0.1–1 per cent of patients), but that this is accentuated in the presence of aminoglycosides.

Lately it has been emphasised that vancomycin can induce platelet-reactive antibodies in the patient, leading to severe thrombocytopenia and bleeding with florid petechial hemorrhages, ecchymoses, and wet purpura.

Lincomycin: Lincomycin is a lincosamide antibiotic that comes from the actinomyces *Streptomyces lincolnensis*. It has been structurally modified by thionyl chloride to its more commonly known 7-chloro-7-deoxy derivative, clindamycin.

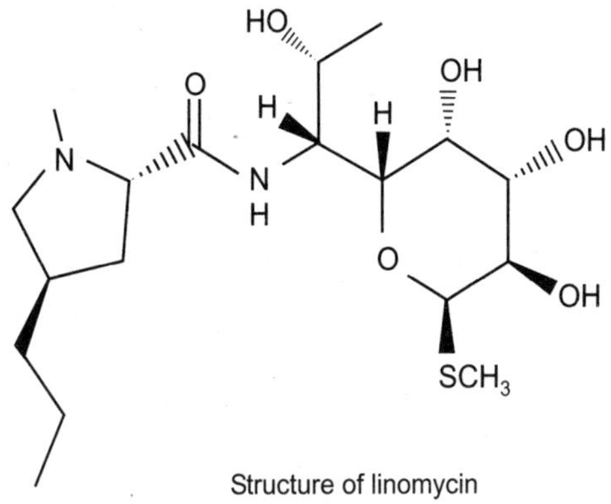

Structure of linomycin

Although similar in structure, antibacterial spectrum, and in mechanism of action to macrolides they are also effective against other species as well i.e. actinomycetes, mycoplasma, and some species of *Plasmodium*. However, because of its adverse effects and toxicity, it is rarely used today and reserved for patients who are either allergic to penicillin or where bacteria has developed resistance.

Intramuscular administration of a single dose of 600 mg of Lincomycin produces average peak serum levels of 11.6 micrograms/ml at 60 min., and maintains therapeutic levels for 17 to 20 hrs, for most susceptible Gram-positive organisms. Urinary excretion after this dose ranges from 1.8 to 24.8 per cent (mean: 17.3 per cent).

Mdenomycins: Mdenomycins are any range of phosphoglycolipid antibiotic, of complex structure, that inhibits the growth of a broad spectrum of Gram-positive bacteria.

Aminoglycoside: An aminoglycoside is a molecule composed of a sugar group and an amino group. Several aminoglycosides function as antibiotics that are effective against certain types of bacteria. They include amikacin, arbekacin, gentamicin, kanamycin, neomycin, netilmicin, paromomycin, rhodostreptomycin, streptomycin, tobramycin, and apramycin.

Aminoglycosides that are derived from bacteria of the *Streptomyces* genus are named with the *suffix-mycin*, while those which are derived from *Micromonospora* are named with the *suffix-micin*.

This nomenclature system is not specific for aminoglycosides. For example vancomycin is a glycopeptide antibiotic and erythromycin, which is produced from a species of *Saccharopolyspora* (which was previously misclassified as *Streptomyces*) along with its synthetic derivatives clarithromycin and azithromycin are macrolides—all of which differ in their mechanisms of action.

Aminoglycosides have several potential antibiotic mechanisms, some as protein synthesis inhibitors, although their exact mechanism of action is not fully known:

1. They interfere with the proofreading process, causing increased rate of error in synthesis with premature termination.
2. Also, there is evidence of inhibition of ribosomal translocation where the peptidyl-tRNA moves from the A-site to the P-site.
3. They can also disrupt the integrity of bacterial cell membrane.

They bind to the bacterial 30S ribosomal subunit (some work by binding to the 50S subunit). There is a significant relationship between the dose administered and the resultant plasma level in blood.

TDM, therapeutic drug monitoring, is necessary to obtain the correct dose. These agents exhibit a post-antibiotic effect in which there is no or very little drug levels detectable in blood, but there still seems to be inhibition of bacterial re-growth. This is due to strong, irreversible binding to the ribosome, and remains intracellular long after plasma levels drop. This allows a prolonged dosage interval. Depending on their concentration they act as bacteriostatic or bactericidal agents.

The protein synthesis inhibition of aminoglycosides does not usually produce a bactericidal effect, let alone a rapid one as is frequently observed on susceptible Gram-negative bacilli. Aminoglycosides competitively displace cell biofilm-associated Mg^{2+} and Ca^{2+} that link the polysaccharides of adjacent lipopolysaccharide molecules. 'The result is shedding of cell membrane blebs, with formation of transient holes in the cell wall and disruption of the normal permeability of the cell wall. This action alone may be sufficient to kill most susceptible Gram-negative bacteria before the aminoglycoside has a chance to reach the 30S ribosome'.

Traditionally, the antibacterial properties of aminoglycosides were believed to result from inhibition of bacterial protein synthesis through irreversible binding to the 30S bacterial ribosome. This explanation, however, does not account for the potent bactericidal properties of these agents, since other antibiotics that inhibit the synthesis of proteins (such as tetracycline) are not bactericidal. Recent experimental studies show that the initial site of action is the outer bacterial membrane. The cationic antibiotic molecules create fissures in the outer cell membrane, resulting in leakage of intracellular contents and enhanced antibiotic uptake. This rapid action at the outer membrane probably accounts for most of the bactericidal activity. Energy is needed for aminoglycoside uptake into the bacterial cell. Anaerobes have less energy available for this uptake, so aminoglycosides are less active against anaerobes. Aminoglycosides are useful primarily in infections involving aerobic, Gram-negative bacteria, such as *Pseudomonas*, *Acinetobacter*, and *Enterobacter*. In addition, some *Mycobacteria*, including the bacteria that cause tuberculosis, are susceptible to aminoglycosides. The most frequent use of aminoglycosides is empiric therapy for serious infections such as septicemia, complicated intraabdominal infections, complicated urinary tract infections, and nosocomial respiratory tract infections. Usually, once cultures of the causal organism are grown and their susceptibilities tested, aminoglycosides are discontinued in favour of less toxic antibiotics.

Streptomycin was the first effective drug in the treatment of tuberculosis, though the role of aminoglycosides such as streptomycin and amikacin has been eclipsed (because of their toxicity and inconvenient route of administration) except for multiple drug resistant strains.

Infections caused by Gram-positive bacteria can also be treated with aminoglycosides, but other types of antibiotics are more potent and less damaging to the host. In the past the aminoglycosides have been used in conjunction with β-lactam antibiotics in streptococcal infections for their synergistic effects, particularly in endocarditis. One of the most frequent combinations is ampicillin (a β-lactam, or penicillin-related antibiotic) and gentamicin. Often, hospital staff refer to this combination as 'amp and gent' or more recently called 'pen and gent' for penicillin and gentamicin.

Aminoglycosides are mostly ineffective against anaerobic bacteria, fungi and viruses.

Routes of administration: Since they are not absorbed from the gut, they are administered intravenously and intramuscularly. Some are used in topical preparations for wounds. Oral administration can be used for gut decontamination (e.g. in hepatic encephalopathy). Tobramycin may be administered in a nebulised form.

Streptomycin: Streptomycin is an antibiotic drug, the first of a class of drugs called aminoglycosides to be discovered, and was the first antibiotic remedy for tuberculosis. It is derived from the actinobacterium

Streptomyces griseus. Streptomycin is a bactericidal antibiotic. Streptomycin cannot be given orally, but must be administered by regular intramuscular injections. An adverse effect of this medicine is ototoxicity, which can lead to hearing loss.

Streptomycin is a protein synthesis inhibitor. It binds to the S12 protein of the 30S subunit of the bacterial ribosome, interfering with the binding of formyl-methionyl-tRNA to the 30S subunit. This prevents initiation of protein synthesis and leads to death of microbial cells. Humans have structurally different ribosomes from bacteria, thereby allowing the selectivity of this antibiotic for bacteria. However at low concentrations Streptomycin only inhibits growth of the bacteria, this is done by inducing prokaryotic ribosomes to misread mRNA.

While streptomycin is traditionally given intramuscularly (indeed, in many countries it is only licensed to be used intramuscularly), the drug may also be administered intravenously.

When grown on medium containing streptomycin, bacteria such as *Escherichia coli* are dependent upon expression of the *aadA* gene in order to survive. Thus, a suitably engineered *E. coli* strain, can be combined with a streptomycin-doped medium to select only bacteria hosting a successful interaction in two-hybrid screening experiments and methods derivative of two-hybrid screening. Streptomycin is an antibiotic that inhibits both Gram-positive and Gram-negative bacteria, and is a therefore a useful broad spectrum antibiotic. Streptomycin is also used as a pesticide, to combat the growth of bacteria, fungi, and algae. Streptomycin controls bacterial and fungal diseases of certain fruit, vegetables, seed, and ornamental crops, and controls algae in ornamental ponds and aquaria. A major use is in the control of fireblight on apple and pear trees. As in medical applications, extensive use can be associated with the development of resistant strains.

Streptomycin was the first aminoglycoside used for antibiotic therapy. Its activity against *Mycobacterium tuberculosis* initiated the widespread introduction of antibiotic treatment to combat tuberculosis. Aminoglycosides are potent antibiotics and have activity against both Gram-positive and Gram-negative bacteria as well as against mycobacteria. Unfortunately they can have nephro-(kidney) and oto-toxicities (hearing) and care has to be taken in their use in treatment of serious infections.

Aminoglycosides are bacterial and work by binding to the 30S ribosome subunit which prevents protein synthesis. There are many aminoglycosides in medical use and are all derived from *actinomyces* spp. For example; streptomycin (*S. griseus*), gentamicin (*Micromonospora purpurea*), tobramycin (*S. tenebrarius*), kanamycin (*S. kanamyceticus*), sisomicin (*M. inyoesis*). Some have been modified chemically to produce derivatives with resistance to clinical isolates with acquired resistance to earlier aminoglycoside types. Of particular interest is the use of the hydroxy-γ-aminobutyryl side-chain to give anti-pseudomonal activity to amikacin. This side chain occurs naturally in the aminoglycoside butirosin produced by *Bacillus circulans*. Netilmicin is chemically derived from sisomicin.

Bacterial resistance occurs by enzymic modification, e.g. acylation, phosphylation or adenylation of the various amine and hydroxyl groups (Fig. 13.11).

All ring structures of these antibiotics are derived from glucose, synthesised separately and then assembled into the final molecule. Most of the biosynthetic enzymes and their associated genes have been identified. Many similarities in biosynthesis have been seen across the wide variety of aminoglycosides. In addition one culture can produce a variety of molecules, e.g. kanamycin A, B, C or gentamicin C_1, C_2, C_{1a}, C_{2a}, A. Recombinant DNA techniques have been used to produce hybrid aminoglycosides (mutasynthesis) and many novel structures have been produced, however, none has been found to be superior to existing structures. Strain improvement programs have been successful in increasing fermentation titres to 15–20 mg ml^{-1}.

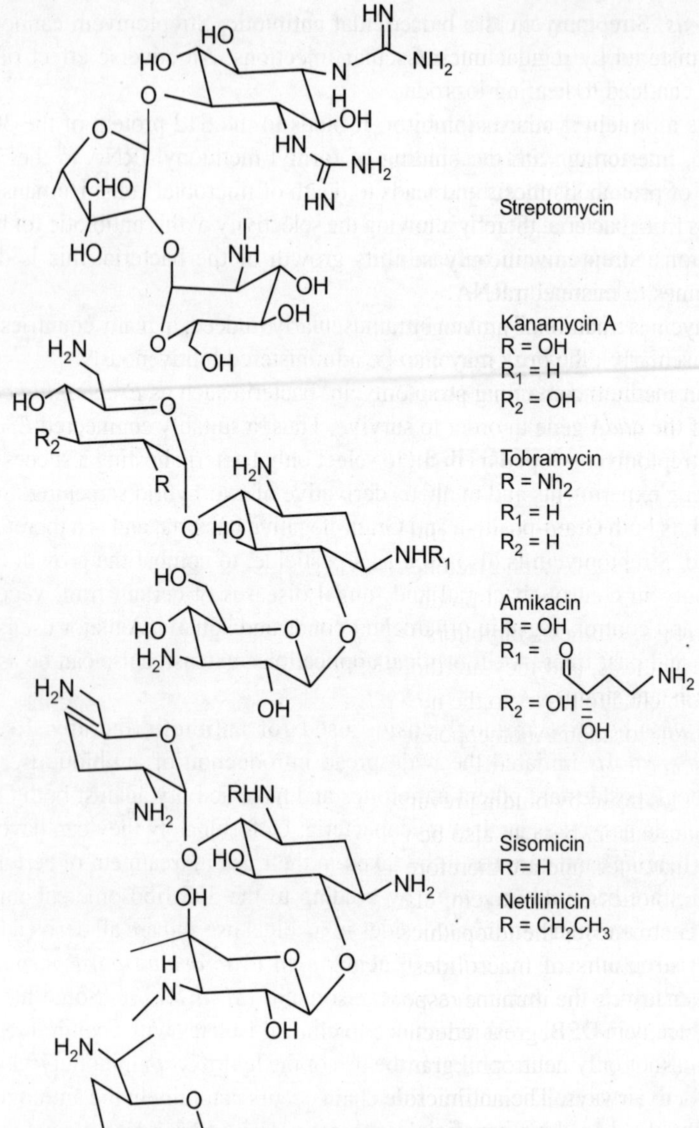

Fig. 13.11. Aminoglycosides.

Additional challenges have been to either reduce the production of unwanted products, e.g. kanamycin C or maintain the required ratios, e.g. gentamicin C_1, C_2, C_{1a}. Large-scale fermentations of aminoglycosides have several similar features. Use of soya products is common, e.g. soya flour or soya meal. Antibiotic synthesis is sensitive to feedback repression by glucose, ammonia and phosphate.

For these reasons ammonium and phosphate salts are not used in the starting batch. Nitrogen is obtained from the slow metabolism of the soya proteins and the necessary phosphate is obtained from organic sources such as phytic acid. Starch is commonly used in the starting batch as streptomyces have poor amylase activities and the enzymic release of glucose is slow and rate limiting. Alternatively, corn syrups can be fed at pre-determined rates.

Due to the general basic nature of aminoglycosides, they are generally recovered by a combination of resin column treatments, e.g. weak cationic IRC50, non-ionicXAD or alumina. Activated carbon treatment is often necessary and the final product can be precipitated as the sulphate salt.

MACROCYCLIC LACTONE ANTIBIOTICS

The macrolides are a group of drugs (typically antibiotics) whose activity stems from the presence of a macrolide ring, a large macrocyclic lactone ring to which one or more deoxy sugars, usually cladinose and desosamine, may be attached. The lactone rings are usually 14, 15 or 16-membered. Macrolides belong to the polyketide class of natural products.

Macrolides are protein synthesis inhibitors. The mechanism of action of macrolides is inhibition of bacterial protein biosynthesis, and they are thought to do this by preventing peptidyltransferase from adding the peptidyl attached to tRNA to the next amino acid (similarily to chloramphenicol) as well as inhibiting ribosomal translocation. Another potential mechanism is premature dissociation of the peptidyl-tRNA from the ribosome.

Macrolide antibiotics do so by binding reversibly to the subunit 50S of the bacterial ribosome. This action is mainly bacteriostatic, but can also be bactericidal in high concentrations. Macrolides tend to accumulate within leukocytes, and are therefore actually transported into the site of infection.

The macrolide antibiotics erythromycin, clarithromycin and roxithromycin have proven to be an effective long-term treatment for the idiopathic, Asian-prevalent lung disease diffuse panbronchiolitis (DPB). The successful results of macrolides in DPB stems from controlling symptoms through immunomodulation (adjusting the immune response), with the added benefit of low-dose requirements.

With macrolide therapy in DPB, great reduction in bronchiolar inflammation and damage is achieved through suppression of not only neutrophil granulocyte proliferation, but also lymphocyte activity and obstructive secretions in airways. The antimicrobial and antibiotic effects of macrolides, however, are not believed to be involved in their beneficial effects toward treating DPB. This is evident, as the treatment dosage is much too low to fight infection, and in DPB cases with the occurrence of the macrolide-resistant bacterium *Pseudomonas aeruginosa*, macrolide therapy still produces substantial anti-inflammatory results.

Ansamycins: Ansamycins is a family of secondary metabolites that show antimicrobial activity against many Gram-positive and some Gram-negative bacteria and includes various compounds among which: streptovaricins and rifamycins. In addition, these compounds demonstrate antiviral activity towards bacteriophages and poxviruses.

They are named ansamycins—*ansa* from the Latin for handle—because of their unique structure which comprises an aromatic moiety bridged by an aliphatic chain. The main difference between various derivatives of ansamycins is the aromatic moiety, which can be a naphthalene ring or a naphthoquinone ring as in rifamycin and naphthomycin. Another variation comprises benzene or a benzoquinone ring system as in geldanamycin or ansamitocin. Ansamycins were first discovered in 1959 by Sensi from *Amycolatopsis mediterranei*, an Actinomycete. Rifamycins are a subclass of ansamycins with high potency against mycobacterial activity. This resulted in their wide use in the treatment of tuberculosis, leprosy, and AIDS-related mycobacterial infections. Since then various analogues have been isolated from other prokaryotes.

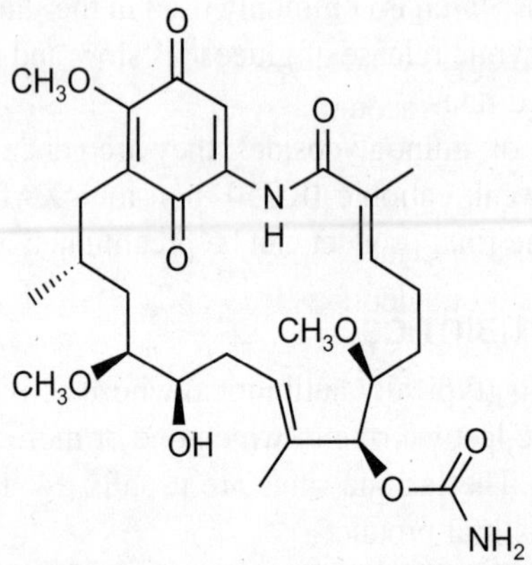

Structure of geldanamycin, one of the benzoquinone ansamycins

Rifamycins: Rifamycins are a group of antibiotics which are synthesised either naturally by the bacterium *Amycolatopsis mediterranei*, or artificially. They are a subclass of the larger family, Ansamycin. Rifamycins are particularly effective against mycobacteria, and are therefore used to treat tuberculosis, leprosy, and mycobacterium avium complex (MAC) infections. The rifamycin group includes the 'classic' rifamycin drugs as well as the rifamycin derivatives rifampicin (or rifampin), rifabutin and rifapentine.

Rifamycins were first isolated in 1957 from a fermentation culture of *Streptomyces mediterranei*. Eventually around seven rifamycins were discovered, named Rifamycin A, B, C, D, E, S and SV. Structure of rifamycin is shown below.

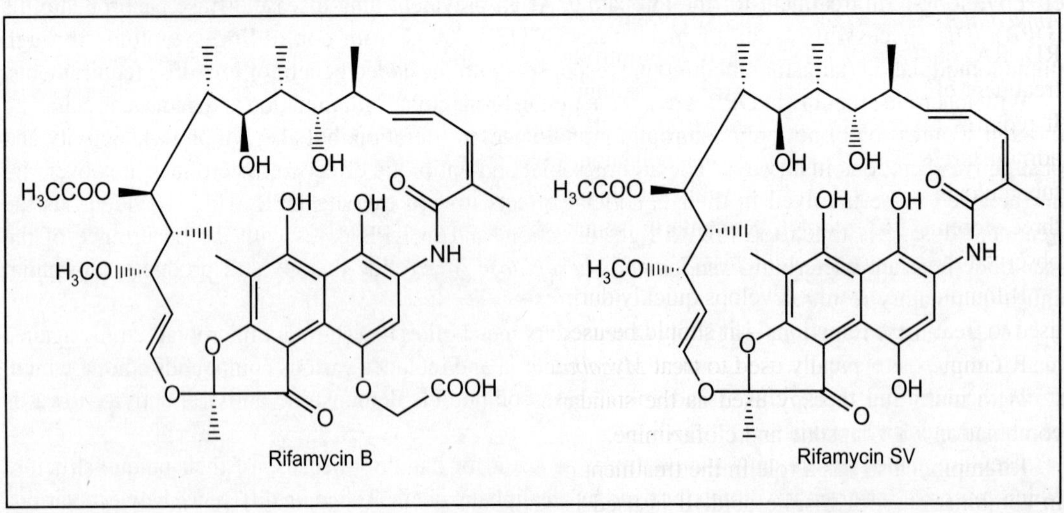

Structures of rifamycin

Of the various rifamycins rifamycin B was first introduced commercially. The drug is widely regarded as having helped conquer the issue of drug-resistant tuberculosis in the 1960s.

Rifamycins have been used for the treatment of many diseases, most importantly HIV-related Tuberculosis. Due to the large number of available analogues and derivatives, rifamycins have been widely utilised in the elimination of pathogenic bacteria that have become resistant to commonly used antibiotics. For instance, rifampicin is known for its potent effect and ability to prevent drug resistance. It rapidly kills fast-dividing bacilli strains as well as 'persisters' cells, which remain biologically inactive for long periods of time that allow them to evade antibiotic activity. In addition, rifabutin and rifapentine have both been used against tuberculosis acquired in HIV-positive patients.

The biological activity of rifamycins relies on the inhibition of DNA-dependent RNA synthesis. This is due to the high affinity of rifamycins to prokaryotic RNA polymerase. Crystal structure data of the antibiotic bound to RNA polymerase indicates that rifamycin blocks synthesis by causing strong steric clashes with the growing oligonucleotide. If rifamycin binds the polymerase after the chain elongation process has started, no effect is observed on the biosynthesis, which is consistent with a model that suggests rifamycin physically blocks chain elongation. In addition, rifamycins showed potency towards HIV. This is due to their inhibition of the enzyme reverse transcriptase, which is essential for tumor persistence. However, rifamycin's potency proved to be mild and this never lead to their introduction to clinical trials.

Despite the fact that rifamycin B is a mild antibacterial compound, it is known to be the precursor of various other clinically-utilised potent derivatives. The general scheme of biosynthesis starts with the uncommon starting unit, 3-amino-5-dihydroxybenzoic acid (AHBA), via type I polyketide pathway (PKS I) in which chain extension is performed using two acetate and eight propionate units. AHBA is believed to have originated from the Shikimate pathway, however this was not incorporated into the biosynthetic mechanism. This is due to the observation that three amino-acid analogues converted into AHBA in cell-free extracts of *A. mediterranei*.

Rifampicin: Rifampicin or rifampin (USAN) is a bactericidal antibiotic drug of the rifamycin group. It is a semisynthetic compound derived from *Amycolatopsis rifamycinica* (formerly known as *Amycolatopsis mediterranei* and *Streptomyces mediterranei*). Rifampicin may be abbreviated R, RMP, RD, RA, or RIF (US). Rifampicin was introduced in 1967, as a major addition to the cocktail-drug treatment of tuberculosis and inactive meningitis, along with isoniazid, ethambutol, and streptomycin. It requires a prescription in industrial North America, but is not a controlled substance. It must be administered regularly daily for several months without break otherwise, the risk of drug-resistant tuberculosis is greatly increased. In fact, this is the primary reason that it is used in tandem with the three aforementioned drugs, particularly isoniazid. This is also the primary motivation behind directly observed therapy for tuberculosis.

Rifampicin resistance develops quickly during treatment and rifampicin monotherapy should not be used to treat these infections—it should be used in combination with other antibiotics.

Rifampicin is typically used to treat *Mycobacterium* infections, including tuberculosis and leprosy.

With multidrug therapy used as the standard treatment of leprosy, rifampicin is always used in combination with dapsone and clofazimine.

Rifampicin also has a role in the treatment of methicillin-resistant *Staphylococcus aureus* (MRSA) in combination with fusidic acid. It is used in prophylactic therapy against *Neisseria meningitidis* (meningococcal) infection.

It is also used to treat infection by *Listeria* species, *Neisseria gonorrhoeae*, *Haemophilus influenzae* and *Legionella pneumophila*. For these non-standard indications, sensitivity testing should be done (if possible) before starting rifampicin therapy.

The *Enterobacteriaceae*, *Acinetobacter*, and *Pseudomonas* species are intrinsically resistant to rifampicin. Further, it has been used with amphotericin B in largely unsuccessful attempts to treat primary amoebic meningoencephalitis caused by *Naegleria fowleri*.

Rifampicin has some effectiveness against vaccinia virus.

Rifampicin inhibits DNA-dependent RNA polymerase in bacterial cells by binding its β-subunit, thus preventing transcription to RNA and subsequent translation to proteins. Its lipophilic nature makes it a good candidate to treat the meningitis form of tuberculosis, which requires distribution to the central nervous system and penetration through the blood-brain barrier.

Rifampin-resistant bacteria produce RNA Polymerases with subtly different β-subunit structures which are not readily inhibited by the drug. In molecular biology research, plasmids containing rifampicin-resistant genes are often used for colony screening. Many plasmids containing these resistant genes are commercially available to researchers.

The most serious adverse effect is related to rifampicin's hepatotoxicity, and patients receiving rifampicin often undergo baseline and frequent liver function tests to detect liver damage.

In short, adverse effects include:

1. Hepatotoxic—Hepatitis, jaundice, liver failure in severe cases.
2. Respiratory—breathlessness.
3. Cutaneous—flushing, pruritus, rash, redness and watering of eyes.
4. Abdominal—nausea, vomiting, abdominal cramps with or without diarrhea.
5. Flu-like symptoms—with chills, fever, headache, arthralgia, and malaise.

TETRACYCLINES AND ANTHRACYCLINES

Tetracycline is a broad-spectrum polyketide antibiotic produced by the *Streptomyces* genus of Actinobacteria, indicated for use against many bacterial infections. It is a protein synthesis inhibitor. It is commonly used to treat acne today, and, more recently, rosacea, and played a historical role in stamping out cholera in the developed world. It is sold under the brand names Sumycin, Terramycin, Tetracyn, and Panmycin, among others. Actisite is a thread-like fibre form, used in dental applications. It is also used to produce several semi-synthetic derivatives, which together are known as the tetracycline antibiotics.

Tetracyclines work by binding the 30S ribosomal subunit, and, through an interaction with 16S rRNA, they prevent the docking of amino-acylated tRNA.

Resistance to tetracyclines can arise through drug efflux, ribosomal protection proteins, 16S rRNA mutation, and drug inactivation through the action of a monooxygenase

Tetracycline sparked the development of many chemically altered antibiotics and in doing so has proved to be one of the most important discoveries made in the field of antibiotics. It is used to treat many Gram-positive and Gram-negative bacteria and some protozoa. It like some other antibiotics, is also used in the treatment of acne.

Since tetracycline is absorbed into bone, it is used as a marker of bone growth for biopsies in humans, and as a biomarker in wildlife to detect consumption of medicine or vaccine-containing baits. The presence of tetracycline in bone is detected by its fluorescence.

In genetic engineering, tetracycline is used in transcriptional activation. Tetracycline is also one of the antibiotics used to treat ulcers caused by bacterial infections. In cancer research at Harvard Medical School, tetracycline has been used to reliably cause regression of advanced stages of leukemia in mice, by placing it in their drinking water.

Tetracycline is used in cell biology as selective agent in cell culture systems. It is toxic to prokaryotic and eukaryotic cells and selects for cells harboring the bacterial *tet*^{r} gene, which encodes a 399-amino acid membrane-associated protein. This protein actively exports tetracycline out of the cell, rendering cells harboring this gene more resistant to the drug. The yellow crystalline powder can be dissolved in water (20 mg/ml) or ethanol (5 mg/ml), and is routinely used at 10 mg/l in cell culture. In cell culture at 37°C it is stable for 4 days. The basic structure of tetracycline consists of a naphthacene ring. Clinically important tetracyclines, produced by fermentation or semi-synthesis, vary with respect to specific ring substituents. Chlortetracycline and oxytetracycline are the major tetracyclines produced by *Streptomyces* whereas tetracycline is normally only formed in minor amounts. *Streptomyces aureofaciens* strains, mutated to block the chlorination reaction, excrete tetracycline as the major product.

Tetracyclines were the first group of antibiotics recognised to have broad spectrum activity. They act by preventing protein synthesis at the 30S ribosome interaction with tRNA. They are used for urinary tract infections, chronic bronchitis, rickettsial and chlamydial infections. They also have broad applications in veterinary use, despite the concern and known relationship of widespread use with resistance build-up. Novel applications include activity against *Helicobacter pylori* to combat stomach ulcers and as a prophylactic against malaria. Chlorotetracycline and tetracycline are produced by *S. rimosus*. Chlorotetracycline production is stimulated by chloride ions and tetracycline by bromide ions. The chlorination gene can be deleted making the bacterium produce only tetracycline. Tetracyclines have been modified chemically to produce products with improved activity and stability. These include doxycycline and minocycline (Fig. 13.12).

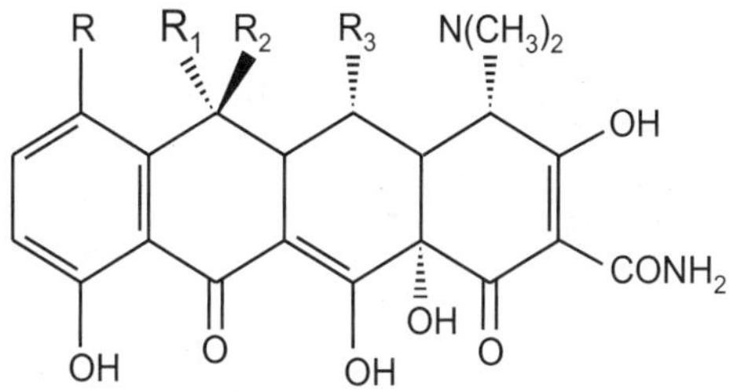

	R	R_1	R_2	R_3
Chlorotetracycline	Cl	CH_3	OH	H
Oxytetracycline	H	CH_3	OH	OH
Tetracycline	H	CH_3	OH	H
Doxycycline	H	CH_3	H	OH
Minocycline	$N(CH_3)_2$	H	H	H

Fig. 13.12. Tetracyclines.

The biosynthesis and genetics of tetracyclines have been well described. The starting polyketide chain is first cyclised into the four ring structure that is then sequentially modified in a specific order. From cloning studies of the biosynthetic enzymes the phenomenon of gene clustering was first recognised. Knowledge of the genetics of the producing organism has been a great asset to strain improvement. Tetracycline resistance genes have been identified and mapped and have played an important role in the build up of product resistance in the producing culture (which itself has 30S ribosomes). Little detain exists for the industrial fermentation of tetracyclines. In common with other streptomycete fermentations, soya flour, peanut meal or corn steep liquor are the main supply of nitrogen in the initial batch medium. Corn syrups are used as carbon feeds throughout the fermentation to maintain a balanced control of growth and product synthesis. Ammonium and phosphate have to be maintained at low concentrations to achieve successful fermentations. Various methods have been described for the recovery of tetracyclines. The antibiotic can be extracted into *n*-butanol or methylisobutyl ketone at acid or alkaline conditions or in the presence of quaternary ammonium compounds or adsorbed on to active carbon for subsequent selective elution.

Macrolides

Macrolides are a diverse class of antibiotics, produced by actinomyces. Macrolides with antibacterial properties have in common a 12, 14 or 16 carbon macrocyclic lactone ring, substituted with sugar molecules. Larger ring macrolides, the polyene macrolides, can have lactone rings of 26–38 carbons. These polyenes are mainly antifungal, e.g. nystatin and amphotericin. The non-polyene macrolides are bacteriostatic. They inhibit protein synthesis by reversibly binding to the 50S portion of the ribosome.

Erythromycin and clarithromycin (chemical derivative of erythromycin) are the most prescribed macrolides (Fig. 13.13). They have a similar activity spectrum to the penicillins and are used by penicillin-sensitive people to combat Gram-positive bacteria and in addition, are used against *Mycoplasma, Compylobacter, Bordetella* and *Legionella*. Clarithromycin is currently prescribed to combat *Helicobacter pylori*. Erythromycin is a 14 carbon macrolide produced by *S. erythreus*. Other 14 carbon macrolides include oleandomycin from *S. antibioticus*, pikromycin from *S. felleus*, megalomicin from *Micromonospora inositola*. Tylosin is a 16 carbon macrolide produced by *S. fradiae* and is produced industrially for animal use.

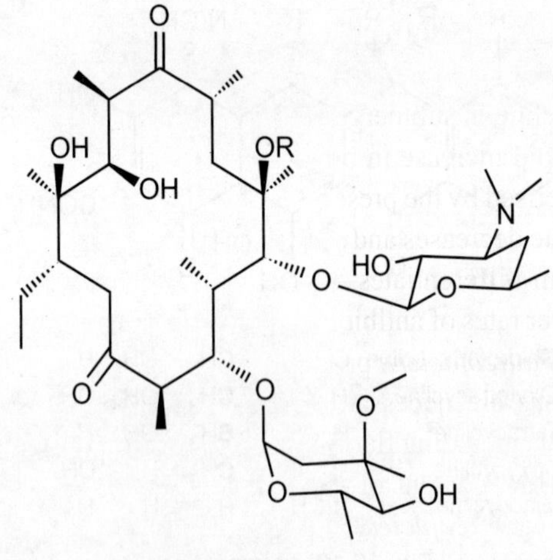

Erythromycin R = H
Clarithromycin R = CH_3

Fig. 13.13. Macrolides.

The general biosynthetic pathways and their associated enzymes and genes have been identified for many of the macrolides. Acetate, propionate and butyrate are the buildings blocks of the lactone ring and glucose is the precursor of the sugar units. Many commonalities have been recognised and considerable research has been focused on the creation of hybrid macrolides using recombinant techniques.

Tetracycline Biosynthetic Pathway

Chlortetracycline synthesis is a complex metabolic pathway having 72 intermediates and involving more than 300 genes. Initial stages involve formation of malonamoyl CoA bound to the enzyme complex anthracene synthase. Malonamoyl CoA condenses with 8 molecules of malonyl CoA and cyclisation occurs with eventual formation of chlortetracycline. High-yielding tetracycline strains are characterised by a lower rate of glycolysis and chlortetracycline production may be enhanced by use of the glycolysis inhibitor benzylthiocyanate. Under these conditions activity of the pentose phosphate cycle increases. The rate-limiting enzyme in chlortetracycline biosynthesis may be anhydrotetracycline oxygenase, the second last enzyme in the biosynthetic pathway. Its activity appears to be proportional to the rate of antibiotic synthesis. Synthesis of this enzyme is repressed by phosphate and stimulated by benzylthiocyanate. There is also an inverse relationship between the level of adenylates and activity of this enzyme. The ATP level or the total adenylate level appears to act as the metabolic effectors in catabolite regulation of tetracycline biosynthesis.

Production of Chlortetracycline

Current industrial yields of tetracyclines are around 20,000 μg ml^{-1}. Because of the complexity of the biosynthetic pathway, strain yield improvement has depended solely on mutation/selection techniques. Selection of strains resistant to the produced antibiotic is another method which has been applied to improving production capacity. Typical fermentation production media contain sucrose, corn steep, ammonium phosphate and salts with pH and temperature maintained at 5.8–6.0 and 28°C, respectively. High aeration rates are necessary particularly in the biomass growth stages. If glucose is used, continuous feeding is necessary. Because of phosphate repression, tetracycline fermentations are run under phosphate-limited conditions.

Production of chlortetracycline in submerged culture may be subdivided into three phases. The first phase is characterised by a rapid increase in biomass and rapid consumption of nutrients. During this phase the mycelium is characterised by the presence of thick basophilic hyphae with a high RNA content. In the second phase, growth rate decreases and sometimes ceases, maximum rates of antibiotic synthesis are observed and the organism differentiates. Hyphal filaments appear then and contain a low RNA content. In the third phase, lower rates of antibiotic production are observed and mycelium fragmentation and lysis occurs. *Streptomyces aureofaciens* produces a certain proportion of tetracycline in addition to chlortetracycline. Chloride ions are necessary for chlortetracycline formation, particularly in high producing mutants. Other agents including fluoride ions, copper, methionine and 5-fluorouracil suppress production of tetracycline. A low content of chloride ions in the medium is the basic condition for production of tetracycline by *S. aureofaciens*. Oxytetracycline is produced by *Streptomyces rimosus* under phosphate-limited conditions.

Anthracycline

Anthracyclines (or anthracycline antibiotics) are a class of drugs used in cancer chemotherapy derived from Streptomyces bacteria (more specifically, *Streptomyces peucetius* var. *caesius*). These compounds are used to treat a wide range of cancers, including leukemias, lymphomas, and breast, uterine, ovarian, and lung cancers. The anthracyclines are some of the most effective anticancer treatments ever developed and are effective against more types of cancer than any other class of chemotherapy agents. Their main adverse effects are heart damage (cardiotoxicity), which considerably limits their usefulness, and vomiting.

The first anthracycline discovered was daunorubicin (trade name Daunomycin), which is produced naturally by *Streptomyces peucetius*, a species of actinobacteria. Doxorubicin (Adriamycin) was developed shortly after, and many other related compounds have followed, although few are in clinical use. Anthracycline has three mechanisms of action:

1. Inhibits DNA and RNA synthesis by intercalating between base pairs of the DNA/RNA strand, thus preventing the replication of rapidly-growing cancer cells.
2. Inhibits topoiosomerase II enzyme, preventing the relaxing of supercoiled DNA and thus blocking DNA transcription and replication.
3. Creates iron-mediated free oxygen radicals that damage the DNA and cell membranes.

As well as many of the expected adverse reactions of chemotherapeutic agents, anthracyclines are notorious for causing cardiotoxicity. This cardiotoxicity may be caused by many factors, which may include interference with the ryanodine receptors of the sarcoplasmic reticulum in the heart muscle cells, free radical formation in the heart or from buildup of metabolic products of the anthracycline in the heart.

Daunorubicin or daunomycin (daunomycin cerubidine) is chemotherapy of the anthracycline family that is given as a treatment for some types of cancer. It is most commonly used to treat specific types of leukaemia (acute myeloid leukemia and acute lymphocytic leukemia). It was initially isolated from *Streptomyces peucetius*.

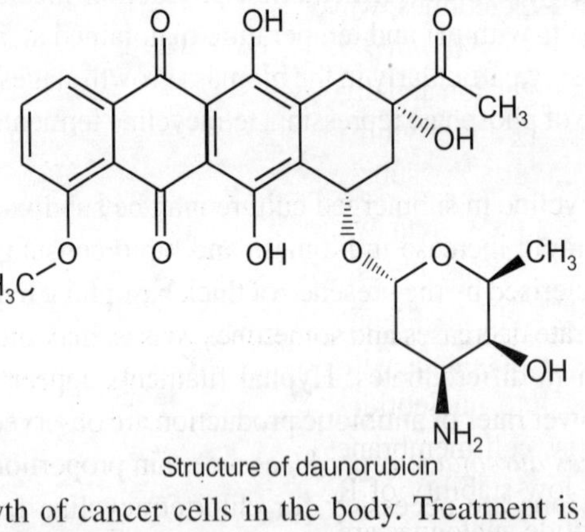

Structure of daunorubicin

It slows or stops the growth of cancer cells in the body. Treatment is usually together with other chemotherapy drugs (such as cytarabine), and its administration depends on the type of tumor and the degree of response. In addition to its major use in treating AML, daunorubicin is also used to treat neuroblastoma. Daunorubicin has been used with other chemotherapy agents to treat the blastic phase of chronic myelogenous leukemia.

Daunorubicin is also used as the starting material for semi-synthetic manufacturing of doxorubicin, epirubicin and idarubicin. Daunorubicin should only be administered in a rapid intravenous infusion. It should not be administered intramuscularly or subcutaneously, since it may cause extensive tissue necrosis. It should also never be administered intrathecally (into the spinal canal), as this will cause extensive damage to the nervous system and may lead to death.

NUCLEOSIDE ANTIBIOTICS

Nucleosides are glycosylamines consisting of a nucleobase (often referred to simply *base*) bound to a ribose or deoxyribose sugar. Examples of these include cytidine, uridine, adenosine, guanosine, thymidine and inosine.

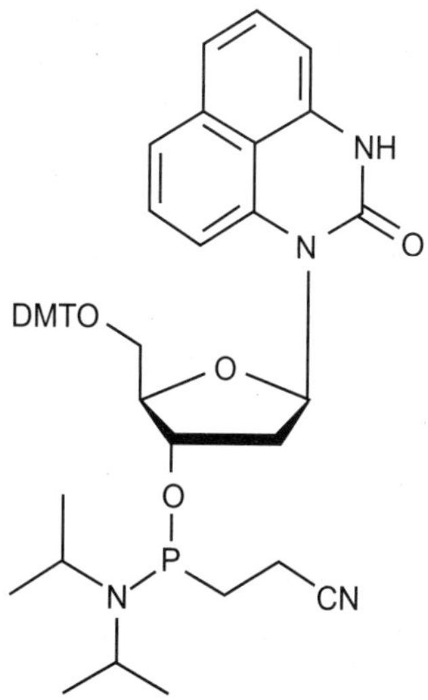

Structure of nucleoside

Nucleosides can be phosphorylated by specific kinases in the cell on the sugar's primary alcohol group (—CH_2·OH), producing nucleotides, which are the molecular building blocks of DNA and RNA.

Nucleosides can be produced by *de novo* synthesis pathways, particularly in the liver, but they are more abundantly supplied via ingestion and digestion of nucleic acids in the diet, whereby nucleotidases break down nucleotides (such as the thymine nucleotide) into nucleosides (such as thymidine) and phosphate. The nucleosides, in turn, are subsequently broken down:

1. In the lumen of the digestive system by nucleosidases into nucleobases and ribose or deoxyribose.
2. Inside the cell into nitrogenous bases, and ribose-1-phosphate or deoxyribose-1-phosphate.

In medicine several nucleoside analogues are used as antiviral or anticancer agents. The viral polymerase incorporates these compounds with non-canonical bases. These compounds are activated in the cells by being converted into nucleotides, they are administered as nucleosides since charged nucleotides cannot easily cross cell membranes. In molecular biology several analogues of the sugar back bone exist. Due to the low stability of RNA, which is prone to hydrolysis, several more stable alternative nucleoside/nucleotide analogues are used which correctly bind to RNA. This is achieved by using a different backbone sugar. These analogues include LNA, morpholino, PNA.

In sequencing dideoxynucleotides are used. These nucleotides possess the non-canon sugar dideoxyribose, which lacks 3′-hydroxyl group (which accepts the phosphate) and therefore cannot bond with the next base, terminating the chain as DNA polymerases mistake it for a regular deoxyribonucleotide.

Puromycin: Puromycin is an antibiotic that is a protein synthesis inhibitor by inhibiting prokaryotic translation. Puromycin is an aminonucleoside antibiotic, derived from the *Streptomyces alboniger* bacterium, that causes premature chain termination during translation taking place in the ribosome. Part of the molecule resembles the 3′ end of the aminoacylated tRNA. It enters the A site and transfers to the growing chain, causing premature chain release. The exact mechanism of action is unknown at this time, but, the 3′ position contains an amide linkage instead of the normal ester linkage of tRNA, the amide bond makes the molecule much more resistant to hydrolysis and thus causes the ribosome to become stopped.

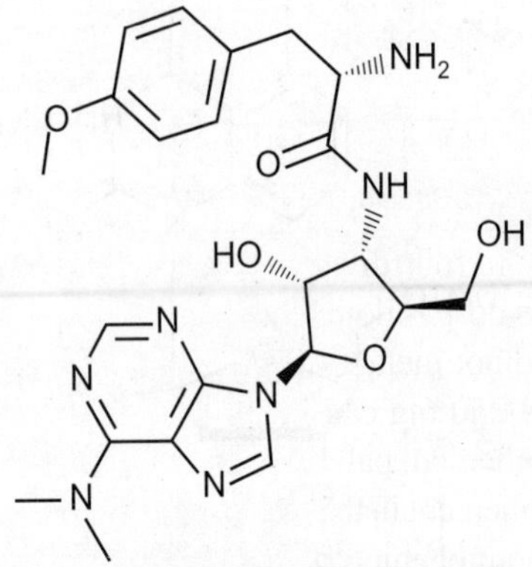

Structure of puromycin

It is not selective for either prokaryotes or eukaryotes. Also of note, puromycin is critical in mRNA display as it allows the growing peptide chain to be covalently bonded to its own mRNA template.

Puromycin is used in cell biology as selective agent in cell culture systems. It is toxic to prokaryotic and eukaryotic cells. Resistance to puromycin is conferred by the Pac gene encoding a puromycin *N*-acetyl-transferase (PAC) that was found in a *Streptomyces* producer strain. Puromycin is soluble in water (50 mg/ml) as colourless solution at 10 mg/ml. Puromycin is stable for one year as solution when stored at –20°C. The recommended dose as a selection agent in cell cultures is within a range of 1–10 μg/ml, although it can be toxic to eukaryotic cells at concentrations as low as 1 μg/ml. It acts quickly and can kill up to 99 per cent of nonresistant cells within 2 days.

AROMATIC ANTIBIOTICS

These are the heterogeneous group of antibiotics with aromatic rings in the molecule some of the aromatic antibiotics are discussed in this section.

Chloramphenicol

Chloramphenicol is a bacteriostatic antimicrobial originally derived from the bacterium *Streptomyces venezuelae*, isolated by David Gottlieb, and introduced into clinical practice in 1949. It was the first antibiotic to be manufactured synthetically on a large scale, and alongside the tetracyclines, is considered the prototypical broad-spectrum antibiotic.

Chloramphenicol is effective against a wide variety of Gram-positive and Gram-negative bacteria, including most anaerobic organisms. Due to resistance and safety concerns, it is no longer a first-line agent for any indication in developed nations and has been replaced by newer drugs in this setting, although it is sometimes used topically for eye infections. In low-income countries, chloramphenicol is still widely used because it is exceedingly inexpensive and readily available.

The most serious adverse effect associated with chloramphenicol use is bone marrow toxicity, which may occur in two distinct forms: bone marrow suppression, which is a direct toxic effect of the drug and is usually reversible, and aplastic anemia, which is idiosyncratic (rare, unpredictable, and unrelated to dose) and generally fatal.

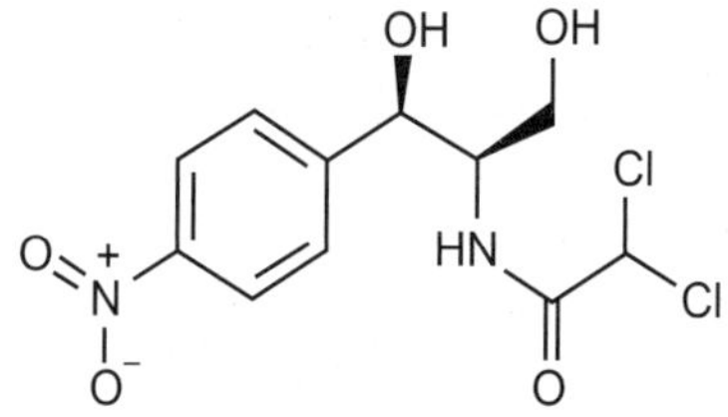

Structure of chloramphenicol

The usual dose is 50 mg/kg/day in four divided doses: the usual dose in an adult male is therefore around 750 mg four times daily; this dose is doubled in severe illness. Half the dose is used in premature babies or neonates, because they do not metabolise the drug as effectively.

Chloramphenicol is available as 250 mg capsules or as a liquid (125 mg/5 ml). In some countries, chloramphenicol is sold as chloramphenicol palmitate ester. Chloramphenicol palmitate ester is inactive, and is hydrolysed to active chloramphenicol in the small intestine. There is no difference in bioavailability between chloramphenicol and chloramphenicol palmitate.

The intravenous (IV) preparation of chloramphenicol is the succinate ester, because pure chloramphenicol does not dissolve in water. This creates a problem: chloramphenicol succinate ester is an inactive prodrug and must first be hydrolysed to chloramphenicol; the hydrolysis process is incomplete and 30 per cent of the dose is lost unchanged in the urine, therefore serum concentrations of chloramphenicol are only 70 per cent of those achieved when chloramphenicol is given orally. For this reason, the chloramphenicol dose needs to be increased to 75 mg/kg/day when administered IV in order to achieve levels equivalent to the oral dose. The oral route is therefore preferred to the intravenous route. Chloramphenicol is metabolised by the liver to chloramphenicol glucuronate (which is inactive). In liver impairment, the dose of chloramphenicol must therefore be reduced. There is no standard dose reduction for chloramphenicol in liver impairment, and the dose should be adjusted according to measured plasma concentrations.

The majority of the chloramphenicol dose is excreted by the kidneys as the inactive metabolite, chloramphenicol glucuronate. Only a tiny fraction of the chloramphenicol is excreted by the kidneys unchanged. It is suggested that plasma levels be monitored in patients with renal impairment, but this is not mandatory. Chloramphenicol succinate ester (the inactive intravenous form of the drug) is readily excreted unchanged by the kidneys, more so than chloramphenicol base, and this is the major reason why levels of chloramphenicol in the blood are much lower when given intravenously than orally.

Because it functions by inhibiting bacterial protein synthesis, chloramphenicol has a very broad spectrum of activity: it is active against Gram-positive bacteria (including most strains of MRSA), Gram-negative bacteria and anaerobes. It is not active against *Pseudomonas aeruginosa*, *Chlamydiae*, or *Enterobacter* species. It has some activity against *Burkholderia pseudomallei*, but is no longer routinely used to treat infections caused by this organism (it has been superseded by ceftazidime and meropenem). In the West, chloramphenicol is mostly restricted to topical uses because of the worries about the risk of aplastic anaemia. The most serious side effect of chloramphenicol treatment is aplastic anaemia. This effect is rare and is generally fatal: there is no treatment and there is no way of predicting who may or may not get this side effect. The effect usually occurs weeks or months after chloramphenicol treatment has been stopped and there may be a genetic predisposition.

Griseofulvin

Griseofulvin (also known as Grisovin, a proprietary name of Glaxo Laboratories) is an antifungal drug. It is used both in animals and in humans, to treat fungal infections of the skin (commonly known as ringworm) and nails. It is derived from the mould *Penicillium griseofulvum*. It is administered orally.

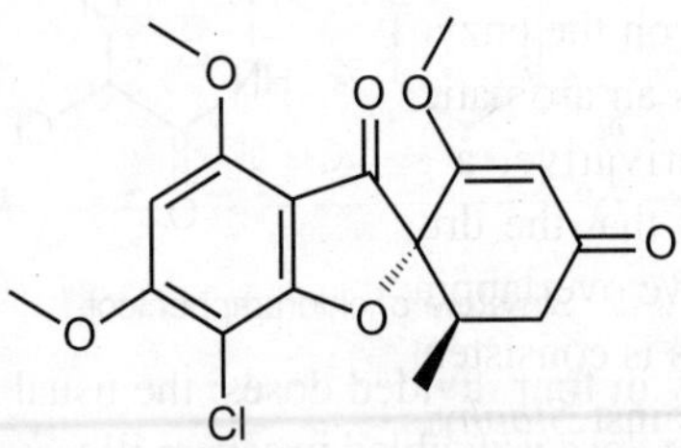

Structure of griseofulvin

The drug binds to tubulin, interfering with microtubule function, thus inhibiting mitosis. It binds to keratin in keratin precursor cells and makes them resistant to fungal infections. It is only when hair or skin is replaced by the keratin-griseofulvin complex that the drug reaches its site of action. Griseofulvin will then enter the dermatophyte through energy dependent transport processes and bind to fungal microtubules. This alters the processing for mitosis and also underlying information for deposition of fungal cell walls.

Griseofulvin is used to treat the following fungal infections:

1. Tinea capitis (ringworm of the scalp).
2. Tinea corporis (ringworm of the body).
3. Tinea pedis (athlete's foot).
4. Tinea unguium (onychomycosis).
5. Tinea cruris (ringworm of the thigh).
6. Tinea barbae (barber's itch).

Novobiocin

Novobiocin, also known as albamycin or cathomycin, is an aminocoumarin antibiotic that is produced by the actinomycete *Streptomyces niveus*, which has recently been identified as a subjective synonym for *S. spheroides* a member of the order Actinobacteria. Other aminocoumarin antibiotics include clorobiocin and coumermycin A1. Novobiocin was first reported in the mid-1950s (then called streptonivicin).

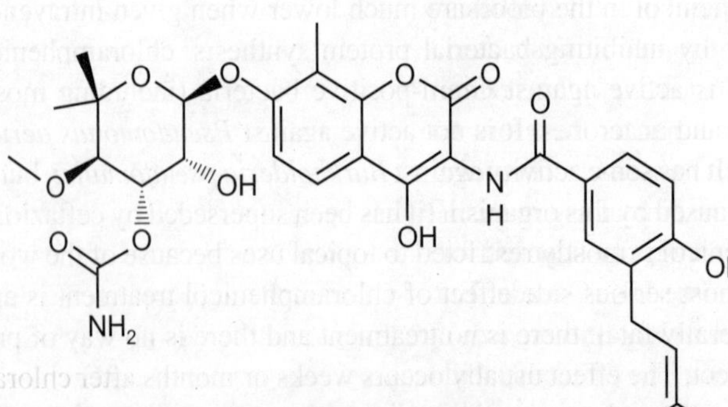

Structure of novobiocin

The molecular basis of action of novobiocin, and other related drugs clorobiocin and coumermycin A1 has been examined. Aminocoumarins are very potent inhibitors of bacterial DNA gyrase and work by targeting the GyrB subunit of the enzyme involved in energy transduction. Novobiocin as well as the other aminocoumarin antibiotics act as competitive inhibitors of the ATPase reaction catalysed by GyrB. The potency of novobiocin is considerably higher than that of the fluoroquinolones that also target DNA gyrase, but at a different site on the enzyme. The GyrA subunit is involved in the DNA nicking and ligation activity. Novobiocin is an aromatic ether compound. Novobiocin may be divided up into three entities; a benzoic acid derivative, a coumarin residue, and the sugar novobiose. X-ray crytallographic studies have found that the drug-receptor complex of Novobiocin and DNA Gyrase shows that ATP and Novobiocin have overlapping binding sites on the gyrase molecule. The overlap of the coumarin and ATP-binding sites is consistent with aminocoumarins being competitive inhib tors of the ATPase activity. It is active against *Staphylococcus epidermidis* and may be used to diffeɩentiate from the other coagulase-negative *Staphylococcus saprophyticus*, which is resistant to novobiocin, in culture.

OTHER COMMERCIALLY PRODUCED ANTIBIOTICS

Some of the antibiotics discussed here are used in thearapy or have other uses.

Fusidic Acid

Fusidic acid is a bacteriostatic antibiotic that is often used topically in creams and eyedrops, but may also be given systemically as tablets or injections.

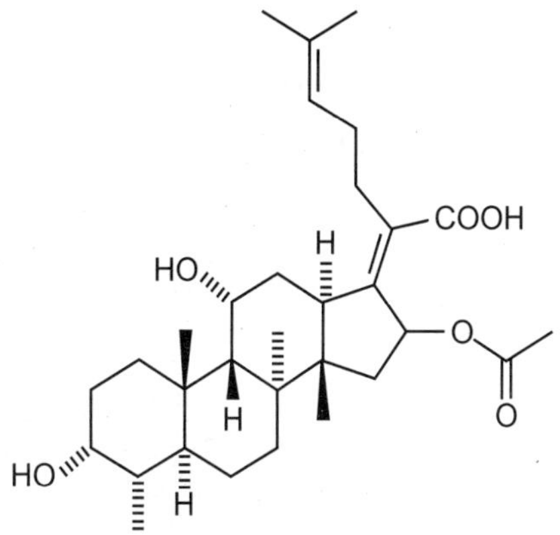

Structure of fusidic acid

Fusidic acid is a bacterial protein synthesis inhibitor by preventing the turnover of elongation factor G (EF-G) from the ribosome. Fusidic acid is only effective on Gram-positive bacteria such as *Staphylococcus* species and *Corynebacterium* species. Fusidic acid inhibits bacterial replication and does not kill the bacteria, and is therefore termed 'bacteriostatic'. Fusidic acid is a true antibiotic, derived from the fungus *Fusidium coccineum* and was developed by Leo Laboratories in Ballerup, Denmark and released for clinical use in the 1960s. It has also been isolated from *Mucor ramannianus* and *Isaria kogana*. The drug is not licensed for use in the US, but, as sodium fusidate, it is approved for use under prescription in the UK, Canada, Europe, Israel, Australia and New Zealand. Fusidic acid is active *in vitro* against *Staphylococcus aureus*, most coagulase-negative staphylococci, *Corynebacterium* species, most *clostridium* species. Fusidic acid has no useful activity against enterococci or most Gram-negative bacteria (except Neisseria, Moraxella, *Legionella pneumophila* and *Bacteroides fragilis*).

Mitomycin

The mitomycins are a family of aziridine-containing natural products isolated from *Streptomyces caespitosus* or *Streptomyces lavendulae.* One of these compounds, mitomycin C, finds use as a chemotherapeutic agent by virtue of its antitumour antibiotic activity. It is given intravenously to treat upper gastro-intestinal (e.g. esophageal carcinoma) and breast cancers, as well as by bladder instillation for superficial bladder tumours. It causes delayed bone marrow toxicity and therefore it is usually administered at 6-weekly intervals. Prolonged use may result in permanent bone-marrow damage. It may also cause lung fibrosis and renal damage. Mitomycin C has also been used topically rather than intravenously in several areas. The first is cancers, particularly bladder cancers and intraperitoneal tumours.It is now well known that a single instillation of this agent within 6 hrs of bladder tumor resection can prevent recurrence. The second is in eye surgery and the third is in esophageal and tracheal stenosis where application of mitomycin C onto the mucosa immediately following dilatation will decrease re-stenosis by decreasing the production of fibroblasts and scar tissue.

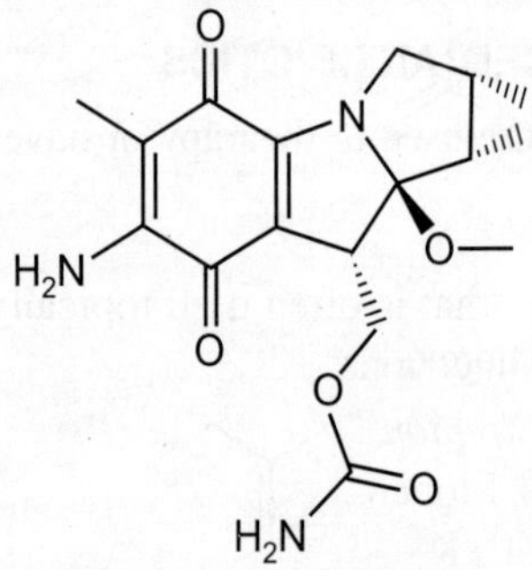

Structure of mitomycin

Mitomycin C is a potent DNA cross-linker. A single cross-link per genome has shown to be effective in killing bacteria. This is accomplished by reductive activation followed by two *N*-alkylations. Both alkylations are sequence specific for a guanine nucleoside in the sequence 5′-CpG-3′. Potential bis-alkylating heterocylic quinones were synthetised in order to explore their antitumoral activities by bioreductive alkylation.

In general the biosynthesis of all mitomycins proceed via combination of 3-amino-5-hydroxybenzoic acid (AHBA), D-glucosamine, and carbamoyl phosphate, to form the mitosane core, followed by specific tailoring steps. The key intermediate, AHBA, is a common precursor to other anticancer drugs, such as rifamycin and ansamycin.

Specifically, the biosynthesis begins with the addition of phosphoenolpyruvate (PEP) to erythrose-4-phosphate (E4P) with a yet undiscovered enzyme, which is then ammoniated to give 4-amino-3-deoxy-D-arabino heptulosonic acid-7-phosphate (aminoDHAP).

Next, DHQ synthase catalyses a ring closure to give 4-amino3-dehydroquinate (aminoDHQ), which is then undergoes a double oxidation via aminoDHQ dehydratase to give 4-amino-dehydroshikimate (aminoDHS). The key intermediate, 3-amino-5-hydroxybenzoic acid (AHBA), is made via aromatisation by AHBA synthase.

The mitosane core is synthesised via condensation of AHBA and D-glucosamine, although no specific enzyme has been characterised that mediates this transformation. Once this condensation has occurred, the mitosane core is tailored by a variety of enzymes. Unfortunately, both the sequence and the identity of these steps are yet to be determined.

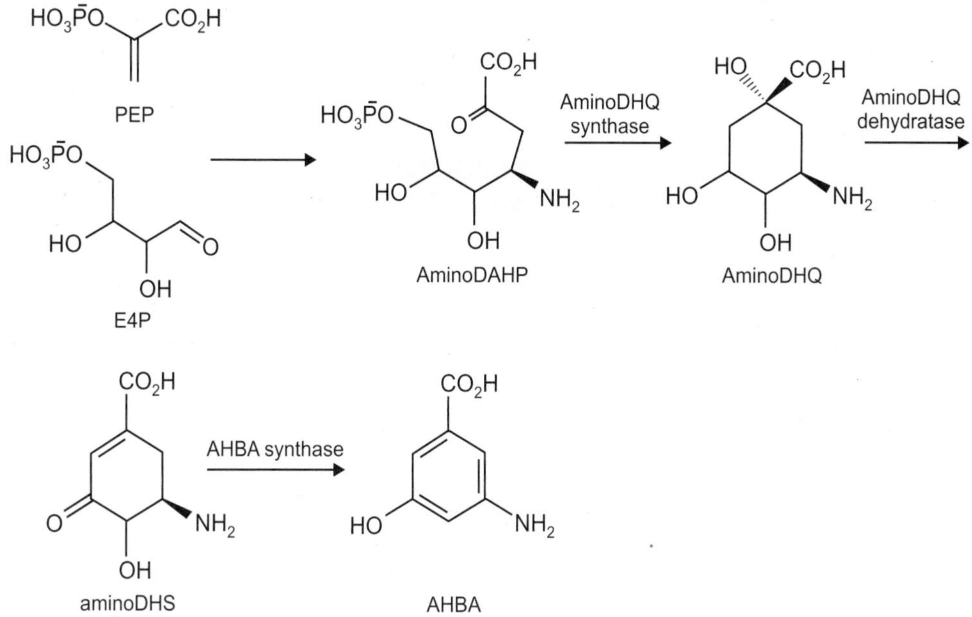

Synthesis of the key intermediate, 3-amino-5-hydroxy-benzoic acid.

Monensin: Monensin isolated from *Streptomyces cinnamonensis*, is a well-known representative of naturally polyether ionophore antibiotics. Monensin A exhibits significant preference to form complexes with monovalent cations such as: Li^+, Na^+, K^+, Rb^+, Ag^+ and Tl^+. Monensin A is able to transport these cations across lipid membranes of cells, playing an important role as an Na^+/H^+ antiporter. It blocks intracellular protein transport, and exhibits antibiotic, antimalarial, and other biological activities. The antibacterial properties of monensin and its derivatives are a result of their ability to transport metal cations through cellular and subcellular membranes.

Monensin is used extensively in the beef and dairy industries to prevent coccidiosis, increase the production of propionic acid and prevent bloat. Furthermore monensin, but also its derivatives monensin methyl ester (MME), and particularly monensin decyl ester (MDE) are widely used in ion selective electrodes.

Salinomycin: Salinomycin is an antibacterial and coccidiostat ionophore therapeutic drug. Salinomycin can kill breast cancer stem cells at least 100 times more effectively than another popular anti-cancer drug (paclitaxel) in mice. The study screened 16,000 different chemical compounds and found that only a small subset, including salinomycin, targeted cancer stem cells responsible for metastasis and relapse.

Chapter 14

Ergot Alkaloids and their Synthesis

INTRODUCTION

Alkaloids are naturally occurring chemical compounds containing basic nitrogen atoms. The name derives from the word alkaline and was used to describe any nitrogen-containing base. Alkaloids are produced by a large variety of organisms, including bacteria, fungi, plants, and animals and are part of the group of natural products (also called secondary metabolites). Many alkaloids can be purified from crude extracts by acid-base extraction.

Many alkaloids are toxic to other organisms. They often have pharmacological effects and are used as medications, as recreational drugs, or in entheogenic rituals. Examples are the local anesthetic and stimulant cocaine, the stimulant caffeine, nicotine, the analgesic morphine, or the antimalarial drug quinine. Some alkaloids have a bitter taste. The chemical structure of phenethylamine alkaloid and caffeine is given below.

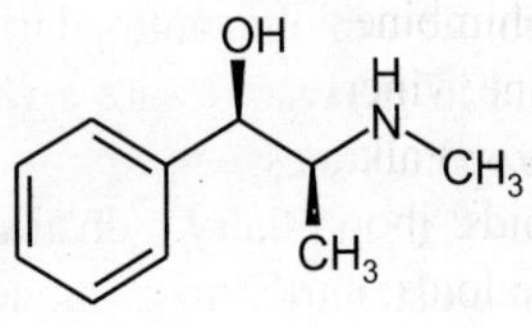

Chemical structure of ephedrine, a phenethylamine alkaloid

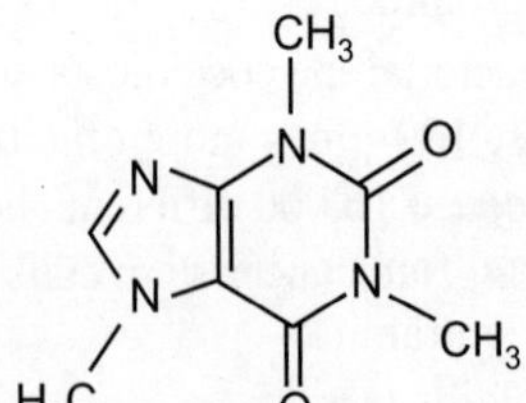

Caffeine

CLASSIFICATION OF ALKALOIDS

Generally speaking, alkaloids are categorised under three main categories,depending on their biogenic origin. For those containing at least a nitrogen atom in a ring system derived from amino acids (i.e. alkaloids derived from phenylalanine are not grouped in this category), they are true alkaloids. The alkaloids derived from phenylalanine are categorised as protoalkaloids. While the remaining ones, such as steroidal alkaloids and purine alkaloids, are classified as pseudoalkaloids.

Alkaloids are usually classified by their common molecular precursors, based on the metabolic pathway used to construct the molecule. When not much was known about the biosynthesis of alkaloids, they were grouped under the names of known compounds, even some non-nitrogenous ones (since those molecules' structures appear in the finished product; the opium alkaloids are sometimes called 'phenanthrenes', for example), or by the plants or animals they were isolated from.

When more is learned about a certain alkaloid, the grouping is changed to reflect the new knowledge, usually taking the name of a biologically-important amine that stands out in the synthesis process.

1. Pyridine group: piperine, coniine, trigonelline, arecoline, arecaidine, guvacine, cytisine, lobeline, nicotine, anabasine, sparteine, pelletierine.
2. Pyrrolidine group: hygrine, cuscohygrine, nicotine.
3. Tropane group: atropine, cocaine, ecgonine, scopolamine, catuabine.
4. Indolizidine group: senecionine, swainsonine.
5. Quinoline group: quinine, quinidine, dihydroquinine, dihydroquinidine, strychnine, brucine, veratrine, cevadine.
6. Isoquinoline group: opium alkaloids (papaverine, narcotine, narceine, morphine, codeine, heroine), sanguinarine, hydrastine, berberine, emetine, berbamine, oxyacanthine.
7. Phenanthrene alkaloids: opium alkaloids (morphine, codeine, thebaine).
8. Phenethylamine group: mescaline, ephedrine, dopamine.
9. Indole group:
    (a) Tryptamines: serotonin, DMT, 5-MeO-DMT, bufotenine, psilocybin.
    (b) Ergolines (the ergot alkaloids): ergine, ergotamine, lysergic acid.
    (c) Beta-carbolines: harmine, harmaline, tetrahydroharmine.
    (d) Yohimbans: reserpine, yohimbine.
    (e) Vinca alkaloids: vinblastine, vincristine.
    (f) Kratom (*Mitragyna speciosa*) alkaloids: mitragynine, 7-hydroxymitragynine.
    (g) *Tabernanthe iboga* alkaloids: ibogaine, voacangine, coronaridine.
    (h) *Strychnos nux-vomica* alkaloids: strychnine, brucine.
10. Purine group:
    (a) Xanthines: caffeine, theobromine, theophylline.
11. Terpenoid group:
    (a) *Aconitum* alkaloids: aconitine.
    (b) Steroid alkaloids (containing a steroid skeleton in a nitrogen containing structure):
        (i) *Solanum* (e.g. potato and tomato) alkaloids (solanidine, solanine, chaconine).
        (ii) *Veratrum* alkaloids (veratramine, cyclopamine, cycloposine, jervine, muldamine).
        (iii) Fire Salamander alkaloids (samandarin).
        (iv) Others: conessine.
12. Quaternary ammonium compounds: muscarine, choline, neurine.
13. Miscellaneous: capsaicin, cynarin, phytolaccine, phytolaccotoxin.

Physico-chemical Properties

Low-molecular weight alkaloids without hydrogen bond donors such as hydroxy groups are often liquid at room temperature, examples are nicotine, sparteine, coniine, and phenethylamine. The basicity of alkaloids depends on the lone pairs of electrons on their nitrogen atoms. As organic bases, alkaloids form salts with mineral acids such as hydrochloric acid and sulphuric acid and organic acids such as tartaric acid or maleic acid. These salts are usually more water-soluble than their free base form.

BIOLOGY AND MOLECULAR BIOLOGY OF ERGOT ALKALOIDS

Ergot alkaloid (EA) have been a major benefit, and a major detriment, to humans since early in recorded history. Their medicinal properties have been used, and continue to be used, to aid in childbirth, with new uses being found in the treatment of neurological and cardiovascular disorders. The surprisingly broad range of pharmaceutical uses for EA stems from their affinities for multiple receptors for three distinct neurotransmitters (serotonin, dopamine, and adrenaline), from the great structural diversity of natural EA, and from the application of chemical techniques that further expand that structural diversity. The dangers posed by EA to humans and their livestock stem from the ubiquity of ergot fungi (*Claviceps* species) as parasites of cereals, and of related grass endophytes (*Epichloë*, *Neotyphodium*, and *Balansia* species) that may inhabit pasture grasses and produce toxic levels of EA. Further concerns stem from saprophytic EA producers in the genera *Aspergillus* and *Penicillium*, especially *A. fumigatus*, an opportunistic pathogen of humans. Numerous fungal species produce EA with a wide variety of structures and properties. These alkaloids are associated with plants in the families Poaceae, Cyperaceae, and Convolvulaceae, apparently because these plants can have symbiotic fungi that produce EA. Pharmacological activities of EA relate to their specific structures. Known as potent vasoconstrictors, the ergopeptines include a lysergic acid substituent with an amide linkage to a complex cyclol-lactam ring structure generated from three amino acids. Simpler lysergyl amides and clavines are more apt to have oxytonic or psychotropic activities. One of the lysergyl amides is LSD, the most potent hallucinogen known. The EA biosynthetic pathway in *Claviceps* species has been studied extensively for many decades, and recent studies have also employed epichloës and *A. fumigatus*. The early pathway, shared among these fungi, begins with the action of an aromatic prenyl transferase, DMATrp synthase, which links a dimethylallyl chain to L-tryptophan. When the dmaW gene encoding DMATrp synthase was cloned and sequenced, the predicted product bore no identifiable resemblance to other known prenyl transferases.

The dmaW genes of *Claviceps* species are present in clusters of genes, several of which also have demonstrated roles in EA biosynthesis. In many other fungi, *dmaW* homologues are identifiable in otherwise very different gene clusters. The roles of DMATrp synthase homologues in these other fungi are probably quite variable. One of them is thought to prenylate the phenolic oxygen of L-tyrosine, and another catalyses the unusual reverse prenylation reaction in the biosynthesis of fumigaclavine C, an EA characteristic of *A. fumigatus*. The second step of the EA pathway is *N*-methylation of DMATrp, which is then subjected to a series of oxidation/oxygenation and reduction reactions to generate, in order, chanoclavine-I, agroclavine, and elymoclavine. Shunt reactions generate a wide variety of other clavines.

For example, *C. purpurea* P1 produces two distinct ergopeptines (ergotamine and ergocryptine), each of which is believed to be generated by multiple LPS 1 subunits encoded by separate, but related, genes (lpsA1 and lpsA2). The main ecological roles of EA in nature are probably to protect the fungi from consumption by vertebrate and invertebrate animals. The EA produced by plant-symbiotic fungi (such as epichloë endophytes) may protect the fungus by protecting the health and productivity of the host, which may otherwise suffer excessive grazing by animals. The EA, at levels typical of plants bearing these symbionts, can negatively affect the health of large mammals as well herbivorous insects. Some clavines have substantial anti-bacterial properties, which might protect the fungus and, in some cases, their host plants from infection. However, the fact that a large number of epichloë, and even several Claviceps species, produce no detectable EA indicates that the selection for their production is not universal. An unfortunate fact for many livestock producers is that some of the most popular forage grasses tend to possess EA-producing epichloë endophytes. Such endophytes are easily eliminated, but confer such fitness enhancements to their hosts that their presence is often preferred, despite the toxic EA. The future looks promising for continued interest in EA. Research continues into their pharmacological properties, medicinal uses, and structure-function relationships. New clavines and lysergic acid derivatives are identified regularly from new sources, such as marine animals. Also, programs are well underway to modify or replace epichloë endophytes of forage grasses in order to produce new grass cultivars that lack these toxins.

ERGOT

Ergot refers to a group of fungi of the genus *Claviceps* (ergot fungi). The most prominent member of this group is *Claviceps purpurea*. This fungus grows on rye and related plants, and can cause ergotism in humans and other mammals consuming seeds contaminated with the fruiting structure of this fungus, called an ergot sclerotium. There are about 50 known species of *Claviceps*, most of them in the tropical regions. Economically important species are *C. purpurea* (parasitic on grasses and cereals), *C. fusiformis* (on pearl millet, buffel grass), *C. paspali* (on dallis grass), and *C. africana*(on sorghum). *C. purpurea* most commonly affects outcrossing species such as rye (its most common host), as well as triticale, wheat and barley. It affects oats only rarely. Chemical structure of ergot is given below.

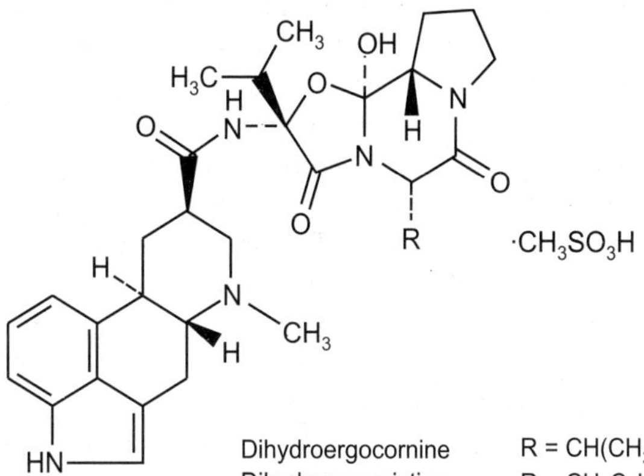

Dihydroergocornine	R = $CH(CH_3)_2$
Dihydroergocristine	R = $CH_2C_6H_5$
Dihydro-α-ergocryptine	R = $CH_2CH(CH_3)_2$
Dihydro-β-ergocryptine	R = $CH(CH_3)CH_2CH_3$

Chemical structure of ergot

There are at least three races or varieties of *C. purpurea*, differing in their host specificity:

1. G1—land grasses of open meadows and fields.
2. G2—grasses from moist, forest, and mountain habitats.
3. G3 (*C. purpurea* var. *spartinae*)—salt marsh grasses (*Spartina*, *Distichlis*).

Life Cycle

Claviceps purpurea

Claviceps purpurea is a fungus that grows on the ears of rye and related cereal and forage plants. Consumption of grains or seeds contaminated with the fruiting structure of this fungus, the ergot sclerotium, can cause ergotism in humans and other mammals. *C. purpurea* most commonly affects outcrossing species such as rye (its most common host), as well as triticale, wheat and barley. It affects oats only rarely.

Claviceps purpurea has been known to mankind for a long time, and its appearance has been linked to extremely cold winters that were followed by rainy summers.

The sclerotial stage of *C. purpurea* conspicuous on the heads of ryes and other such grains is known as ergot. Favourable temperatures for growth are in the range of 18°–30°C, while temperatures above 37°C will cause rapid germination of conidia. Sunlight has a chromogenic effect on the mycelium with intense colouration. Cereal mashes and sprouted rye are suitable substrates for growth of the fungus in the laboratory.

Claviceps africana

Claviceps africana infects sorghum and was first observed in south Texas in 1997. It only infects unfertilised ovaries, so self-pollination and fertilisation can decrease the presence of the disease, but male-sterile lines are extremely vulnerable to infection by this fungus. Symptoms of infection by *C. africana* include the secretion of honeydew (a fluid with high concentrates of sugar and conidia), which attracts insects like flies, beetles, and wasps that feed on it. This in turn contributes to spread of the fungus to uninfected plants.

C. africana caused ergot disease resulting in a famine in 1903–1906 in Northern Cameroon, West Africa, and also occurs in eastern and southern Africa, especially Zimbabwe and South Africa. Male sterile sorghums (also referred to as A-lines) are especially susceptible to infection, first recognised in the 1960s, and massive losses in seed yield have been noted. Infection is associated with cold night temperatures that are below twelve degrees Celsius occurring two to three weeks before flowering.

An ergot kernel called a sclerotium develops when a floret of flowering grass or cereal is infected by a spore of fungal species of the genus *Claviceps*. The infection process mimics a pollen grain growing into an ovary during fertilisation. Because infection requires access of the fungal spore to the stigma, plants infected by *Claviceps* are mainly outcrossing species with open flowers, such as rye (*Secale cereale*) and ryegrasses (genus *Lolium*). The proliferating fungal mycelium then destroys the plant ovary and connects with the vascular bundle originally intended for seed nutrition. The first stage of ergot infection manifests itself as a white soft tissue (known as *sphacelia*) producing sugary honeydew, which often drops out of the infected grass florets. This honeydew contains millions of asexual spores (conidia) which are dispersed to other florets by insects. Later, the sphacelia convert into a hard dry sclerotium inside the husk of the floret. At this stage, alkaloids and lipids accumulate in the sclerotium.

Claviceps species from tropic and subtropic regions produce macro- and microconidia in their honeydew. Macroconidia differ in shape and size between the species, whereas microconidia are rather uniform, oval to globose (5 × 3 μm). Macroconidia are able to produce secondary conidia. A germ tube emerges from a macroconidium through the surface of a honeydew drop and a secondary conidium of the oval to pearlike shape is formed to which the contents of the original macroconidium migrates. Secondary conidia form white frost-like surface on honeydew drops and are spread by wind. No such process occurs in *Claviceps purpurea*, *Claviceps grohii*, *Claviceps nigricans*, and *Claviceps zizaniae*, all from 'Northern temperate regions'.

When a mature sclerotium drops to the ground, the fungus remains dormant until proper conditions trigger its fruiting phase (onset of spring, rain period, etc.). It germinates, forming one or several fruiting bodies with head and stipe, variously coloured (resembling a tiny mushroom). In the head, threadlike sexual spores are formed, which are ejected simultaneously, when suitable grass hosts are flowering. Ergot infection causes a reduction in the yield and quality of grain and hay produced, and if infected grain or hay is fed to livestock it may cause a disease called ergotism. Black and protruding sclerotia of *C. purpurea* are well known.

However, many tropical ergots have brown or greyish sclerotia, mimicking the shape of the host seed. For this reason, the infection is often overlooked.

Insects, including flies and moths, have been shown to carry conidia of *Claviceps* species, but if insects play a role in spreading the fungus from infected to healthy plants is unknown.

Effects on Humans and Other Mammals

The ergot sclerotium contains high concentrations (up to 2 per cent of dry mass) of the alkaloid ergotamine, a complex molecule consisting of a tripeptide-derived cyclol-lactam ring connected via amide linkage to a lysergic acid (ergoline) moiety, and other alkaloids of the ergoline group that are biosynthesised by the fungus. Ergot alkaloids have a wide range of biological activities including effects on circulation and neurotransmission.

Ergotism is the name for sometimes severe pathological syndromes affecting humans or animals that have ingested ergot alkaloid-containing plant material, such as ergot-contaminated grains. Monks of the order of St. Anthony the Great specialised in treating ergotism victims with balms containing tranquilising and circulation-stimulating plant extracts; they were also skilled in amputations. The common name for ergotism is 'St. Anthony's fire', in reference to monks who cared for victims as well as symptoms, such as severe burning sensations in the limbs. These are caused by effects of ergot alkaloids on the vascular system due to vasoconstriction of blood vessels, sometimes leading to gangrene and loss of limbs due to severely restricted blood circulation.

The neurotropic activities of the ergot alkaloids may also cause hallucinations and attendant irrational behaviour, convulsions, and even death. Other symptoms include strong uterine contractions, nausea, seizures, and unconsciousness. Since the middle ages, controlled doses of ergot were used to induce abortions and to stop maternal bleeding after childbirth. Ergot alkaloids are also used in products such as Cafergot (containing caffeine and ergotamine or ergoline) to treat migraine headaches. Ergot extract is no longer used as a pharmaceutical preparation. In addition to ergot alkaloids, *Claviceps paspali* also produces tremorgens (paspalitrem) causing 'paspalum staggers' in cattle. Ergot alkaloids are also produced by fungi of the genera *Penicillium* and *Aspergillus*, notably by some isolates of the human pathogen *Aspergillus fumigatus*, and have been isolated from plants in the family Convolvulaceae, of which morning glory is best known.

Ergot contains no lysergic acid diethylamide (LSD) but instead contains ergotamine, which is used to synthesise lysergic acid, an analog of and precursor for synthesis of LSD. Moreover, ergot sclerotia naturally contain some amounts of lysergic acid.

ERGOT ALKALOIDS

One of the pharmacologically most important groups of indole alkaloids is the ergoline, or ergot, alkaloids. These alkaloids are isolated from the dried sclerotium of the fungus *Claviceps purpurea* (Hypocreaceae) (ergot). This fungus is a parasite on rye and wheat and other grains. Ingestion of contaminated grain, most often after the grain has been made into bread, causes ergotism, also known as the 'Devil's curse' or 'St. Anthony's fire', and has been a problem for centuries. It has been noted in writings from China as early as 1100 BC and in Assyria in 600 BC, and Julius Caesar's legions suffered an epidemic of ergotism during one of the campaigns in Gaul. In 994 AD, an epidemic in France killed between 20,000 and 50,000 people, and in 1926, at least 11,000 cases of ergotism occurred in Russia. Structure of ergot alkaloid family is given below.

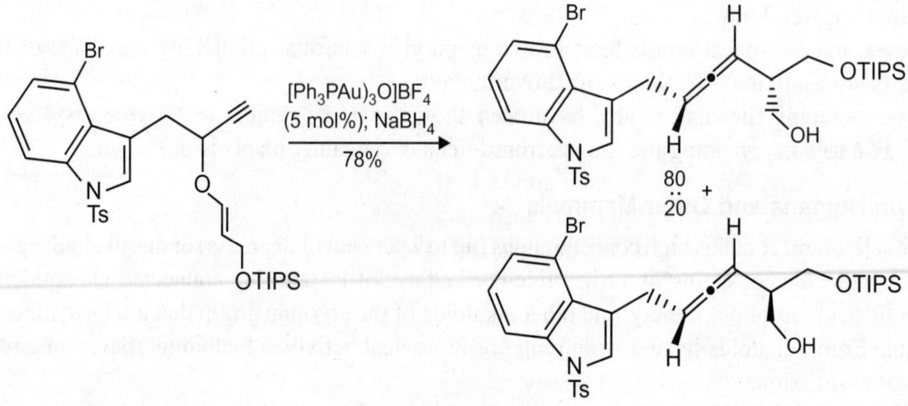

Ergot alkaloids family

Ergotism can cause convulsions, nausea, and diarrhea in mild forms, and there is some thought that an outbreak of ergotism may have been the cause of the 'bewitchings' which led to the Salem witch trials in the United States in 1691. Ergotism may also have caused some of the extreme destruction associated with the French Revolution. In the Middle Ages, ergotism was described as causing victims to dies 'miserably, their limbs eaten up by the holy fire that blacked like charcoal'. People turned to the church for help, assuming that the disease was retribution for their sins. In particular, they prayed to St. Anthony for deliverance, giving rise to the name for the disease. Ergotism takes two forms, gangrenous ergotism, in which tingling effects were felt in fingers and toes followed in many cases by dry gangrene of the limbs and finally loss of the limbs, and convulsive ergotism, in which the tingling was followed by hallucinations and delerium and epileptic-type seizures. In both cases, death was slow and painful. Ergotism has now been recognised as a result of infection by a mycotoxin, and the ergotism plagues have been eliminated.

However, the alkaloids derived from ergot have assumed new importance for their pharmacological properties, and ergot is produced commercially for the preparation of these alkaloids. There are three main groups of ergot alkaloids, the clavine type, the water-soluble lysergic acid type, and the water-insoluble lysergic acid type or peptide ergot alkaloids. The clavine type of alkaloids, such as agroclavine and elymoclavine, are generally regarded as precursors to the other groups of ergot alkaloids in the biogenetic pathway. These alkaloids are among several of the ergot alkaloids also isolated from higher plants, particularly the seeds of *Ipomoea violacea* and *Rivea corymbosa* ('ololiuqui', the Mexican morning glory), both members of the Convolvulaceae family. These alkaloids are not used pharmacologically, but agroclavine is a powerful uterine stimulant, and many of the ergot alkaloids are prolactin release inhibitors. The water-soluble lysergic acid derivatives are most often amide derivatives. Among the most important of these are ergonovine and methysergide. Ergonovine has potent uterine contraction activity and is used in treating postpartum hemorrhages. It has low vasoconstrictor action. Methysergide is used as a cranial vasodilator in the treatment of migraine headaches. Perhaps the most infamous of the semi-synthetic derivatives of the ergot alkaloids is lysergic acid diethyl amide (LSD). It was first synthesised by Albert Hofmann of Sandoz AG in 1938, but its hallucinogenic properties were not known until 1943. For a while following the discovery of the pharmacological effects of LSD, it was used in psychiatry, particularly in the treatment of alcoholic schizophrenia. In recent years, evidence has come out that it was used in the 1950s on military 'volunteers' to study its effect, presumably as a chemical warfare weapon.

In the early 1960s, proponents of its use as a way to achieve a state of nirvana, such as Dr. Timothy Leary, began using it heavily and distributing it throughout the high school and college age population, often adsorbed into a sugar cube. This attitude that LSD was a 'good drug' was fostered by popular songs, like the Jefferson Airplane's 'white rabbit' and the Beatles' 'Lucy in the sky with diamonds', which supposedly described the effects of good LSD trips. Unfortunately, many LSD trips turned out to be bad trips, and many heavy users of LSD experienced bad flashback trips at a later time. (One sidelight to the hallucinogenic effects of LSD is the result of finding ergot alkaloids in Mexican morning glory seeds.

Some seekers of nirvana through hallucinogens began ingesting large quantities of morning glory seeds. Rather than hallucinogenic activity, these foolish people experienced primarily toxic reactions thus obtaining nirvana in an unintended way. After a relatively brief time, the popularity of LSD as a hallucinogen diminished, and it became somewhat of a historical relic. However, it has started to make a comeback in the drug underground in recent years. This time, the target seems to be grade school age children, and the pushing of LSD to these children may be viewed by some as a means of hooking them on hard drugs like cocaine and heroin as they grow older.

The water-insoluble lysergic acid derivatives are primarily peptide ergot alkaloids like ergotamine. This compound was first isolated in 1918, but its structure was not determined until 1951. Ergotamine, as its tartrate salt, is a analgesic specifically used for treatment of severe migraine headaches. It is often used in conjunction with caffeine, which constricts cerebral blood vessels, and a dose of 2 mg taken orally often results in quick relief. 2-Bromo-á-ergocryptine, a semisynthetic derivative, has reduced toxicity and is now commercially available to be used in the reduction of lactation in women. It has also been used in treatment of sexual disorders, and has been shown to enhance sexual libidos in both men and women.

Toxicon–Lactam Ergot Alkaloids

Four major alkaloids in the extracts from sclerotia of *Claviceps purpurea*, picked from wild grasses, have been identified as lactam (non-cyclol) ergot alkaloids. The structural information was obtained from ion trap MS and NMR spectroscopy. The data for one of the lactam ergot alkaloids were coinciding with ergocristam [*N*-(lysergyl-valyl)-cyclo(phenylalanyl-prolyl)]. The structural information of two further lactam alkaloids was suggestive of either α- or β-ergocryptam [*N*-(lysergyl-valyl)-cyclo(leucyl-prolyl) or *N*-(lysergyl-valyl)-cyclo(isoleucyl-prolyl)] and ergoannam [*N*-(lysergyl-leucyl)-cyclo(leucyl-prolyl) or *N*-(lysergyl-isoleucyl)-cyclo(isoleucyl-prolyl)]. The constitution of the fourth lactam ergot alkaloid corresponded to *N*-(lysergyl-isoleucyl)-cyclo(phenylalanyl-prolyl), a new ergopeptam, which has not been described before. Additionally, the cyclol-analogue of the new ergopeptam was detected in the extracts and has been identified on the basis of its product ion spectrum from fragmentation of $[M + H]^+$. It has been found that lactam ergot alkaloids may not only be minor products of ergopeptine biosynthesis, as has been suggested hitherto, but may be major biosynthetic endproducts for some ergot strains. Thus, it demonstrats the production of an ergot alkaloid that contains isoleucine as the second amino acid, i.e. the *N*-(lysergyl-isoleucyl)-moiety, by parasitic, naturally growing *C. purpurea*. This unusual type of ergot alkaloid has so far only been found in saprophytic cultures of *C. purpurea*.

Dihydroergosine: A New Naturally Occurring Alkaloid from the Sclerotia of *Sphacelia sorghi* (McRae)

Infection of the florets of *Sorghum vulgare* by *Sphacelia sorghi* results initially in the characteristic sphacelial stage, but proceeds to replace the ovary by a sclerotium, off-white flecked with red, which protrudes only slightly from the floral cavity. Although the perfect stage of this organism has not been observed it is probably related to the genus *Claviceps*, and may therefore be regarded as an ergot fungus.

TOXIN BIOSYNTHESIS GENES IN ERGOT ALKALOID-PRODUCING FUNGI

Ergot alkaloids are a complex family of mycotoxins produced by several fungi including some *Neotyphodium* species, which grow as symbiotic endophytes (contained within the plant and with no symptoms of infection) within pasture grasses, and *Aspergillus fumigatus*, a common mould and opportunistic human pathogen. In US agriculture, ergot alkaloids produced by endophytic fungi in grasses are associated with problems in grazing animals that result in annual losses of hundreds of millions of dollars. Many ergot alkaloids are toxic to animals but some alkaloids may be necessary for maintaining desirable agronomic properties also associated with endophyte-infection of grasses. We are characterising the biochemical pathway by which simple ergot alkaloids are made into progressively more complex ergot alkaloids, by cloning the genes involved and inactivating them by a process called gene knockout. By truncating the pathway at different points, we can alter the ergot alkaloid profile of fungi in a controlled way. A thorough understanding of the ergot alkaloid biochemical pathway will allow for control of the types of ergot alkaloids that accumulate in endophyte-infected grasses. Ideally, toxicity may be minimised but agronomic quality maintained.

Objectives

Ergot alkaloid biosynthesis is characterise pathways of fungi for the purpose of facilitating truncation of the pathway at different points for basic studies and future applications. This overall goal is addressed in the following four objectives: (i) analysis of genomic DNA surrounding ergot alkaloid biosynthesis genes in *Neotyphodium* sp. Lp1, for the purpose of identifying additional ergot alkaloid-associated genes, and comparison to ergot gene clusters of other fungi; (ii) systematic functional analysis of genes controlling the shared steps in the ergot alkaloid pathway by gene knockout in *Aspergillus fumigatus*; (iii) complementation of knockouts and tests of congruence of critical shared pathway steps; and (iv) gene knockout analysis of fungal species-specific pathway steps.

Approach

Potential ergot alkaloid pathway genes in the grass endophyte *Neotyphodium* sp. Lp1 will be identified by DNA sequence analysis of genes clustered with known ergot alkaloid genes and by comparison to previously characterised ergot alkaloid gene clusters of *Claviceps purpurea* and *Aspergillus fumigatus*. Biochemical functions of genes in the core part of the pathway (shared among all ergot alkaloid-producing fungi) will be determined by gene knockout in Aspergillus fumigatus and subsequent biochemical characterisation of the gene-knockout mutants. The functions of individual genes will be confirmed, and congruence in the pathways of different fungi demonstrated, by complementation of the *A. fumigatus* gene-knockout mutants with wild-type *A. fumigatus* genes and with homologous genes from *Neotyphodium* sp. Lp1. The biochemical function of critical genes in the parts of the pathway that are unique to *A. fumigatus* or *Neotyphodium* spp. will be determined by gene knockout and biochemical characterisation of gene knockout mutants.

Outputs

Genetic and biochemical experiments were conducted to elucidate the functions of several genes in the pathway that fungi use to make toxic ergot alkaloids. An understanding of the genes in the ergot alkaloid pathway and the biochemical steps that they control provides the basis for blocking the ergot alkaloid pathway to alter the spectrum of alkaloids that will accumulate in agriculturally important grasses that contain ergot alkaloid producing fungi. One of the primary goals of this project is elucidate the functions of several genes (ergot alkaloid synthesis-'eas'- genes) that are shared among different fungi that produce ergot alkaloids. These genes are hypothesised to control early steps in the biochemical pathway to toxic ergot alkaloids. Chemical analyses of *Aspergillus fumigatus* strains containing a mutated copy of the gene named easF demonstrated that the mutant strains accumulated dimethylallyltryptophan, which is the first intermediate in the ergot alkaloid pathway, but not other alkaloids from the ergot pathway. Complementation of the easF mutant with a wild-type copy of easF restored production of typical ergot alkaloids. The formation of the ergoline ring proceeds from mevalonic acid to tryptophan shown below:

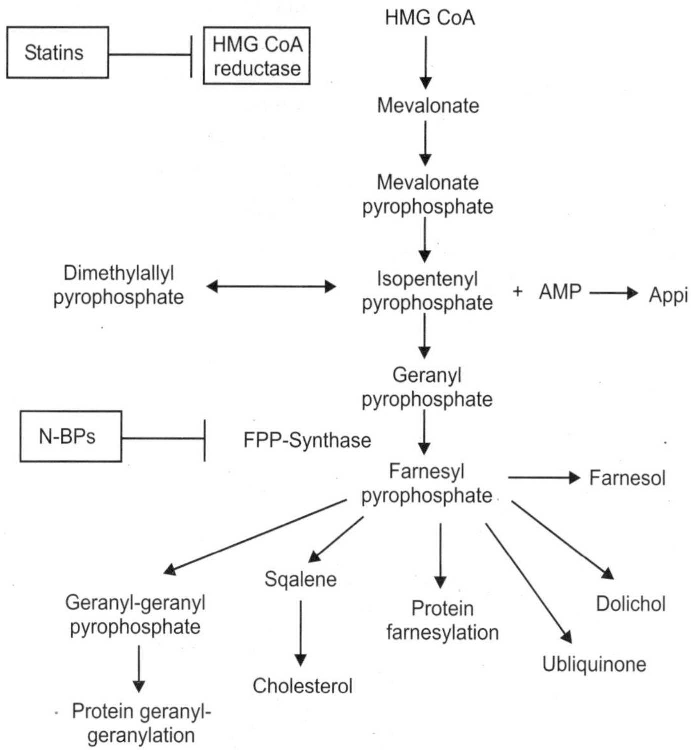

Mevalonate pathway

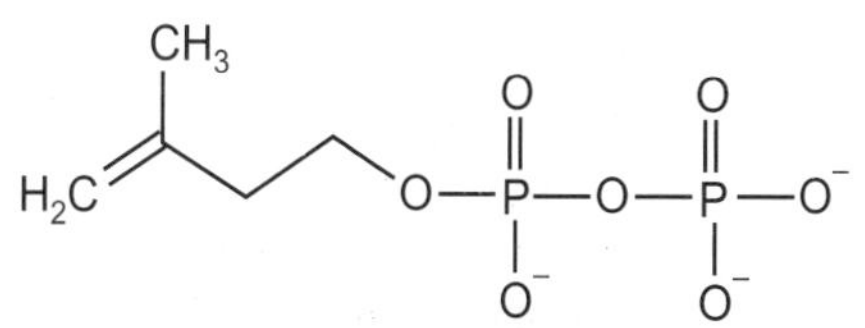

Isopentenyl pyrophosphate

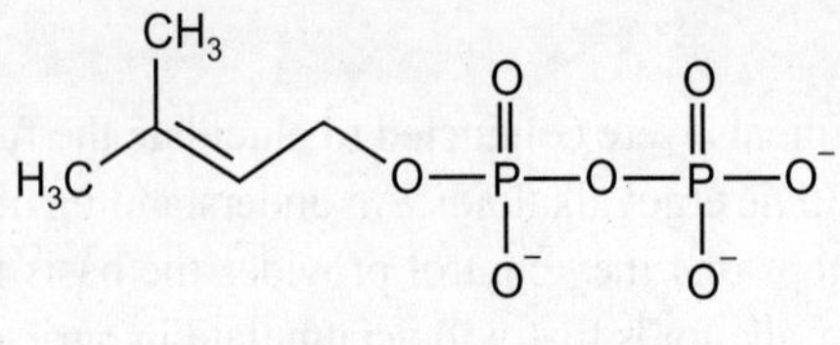

Dimethylallyl-pyrophosphate

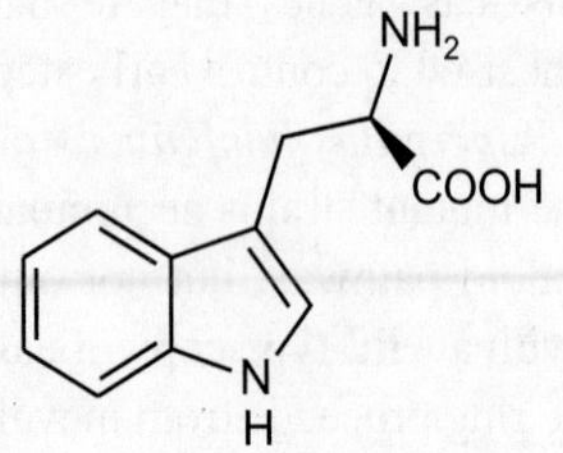

Tryptophan

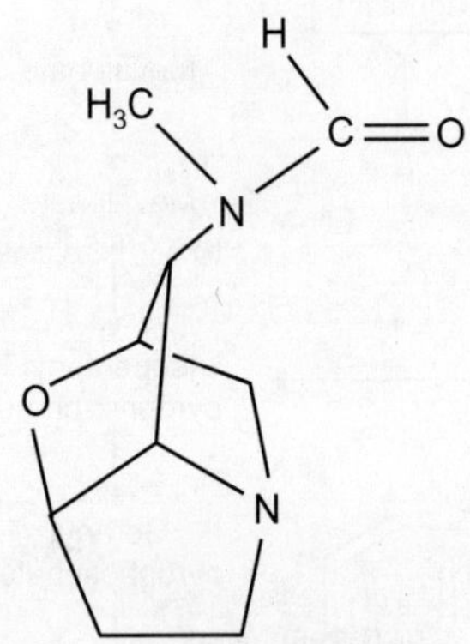

Dimethylallyl-tryptophan

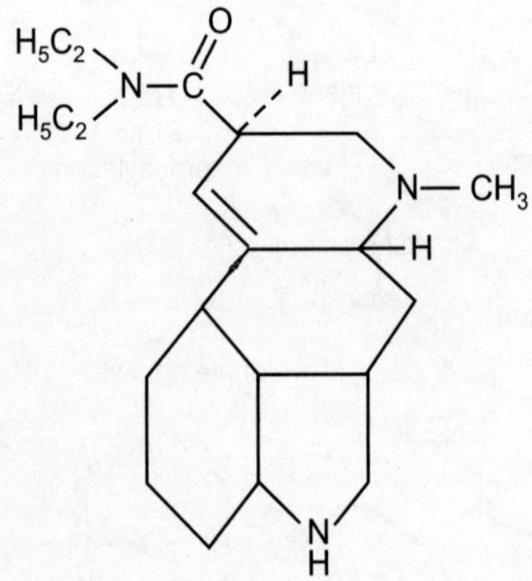

Chenoclovine

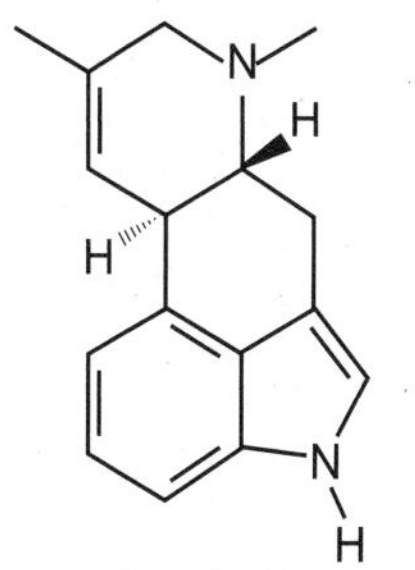

Agroclavine

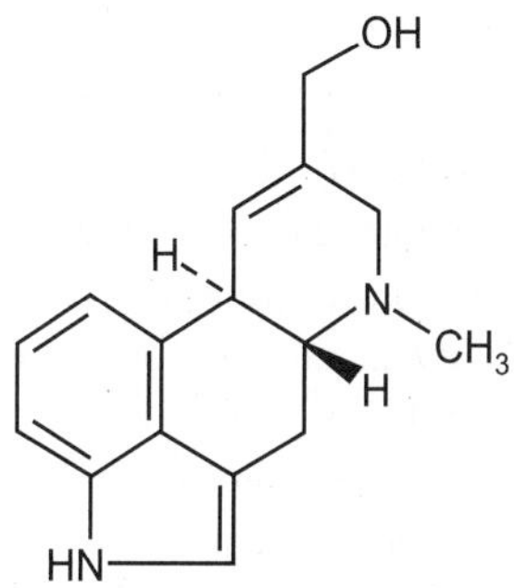

Elymoclavine

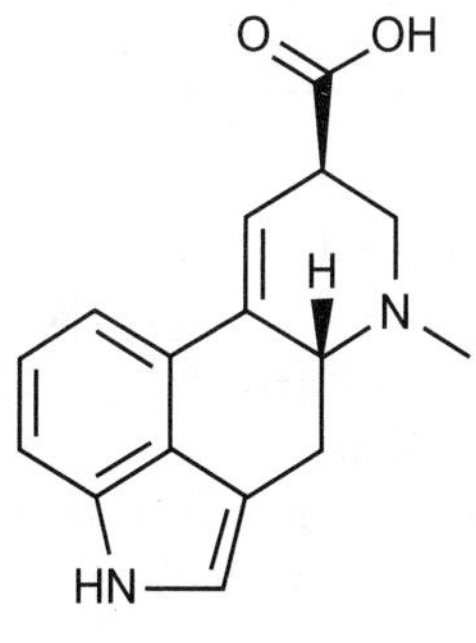

Lysergic acid

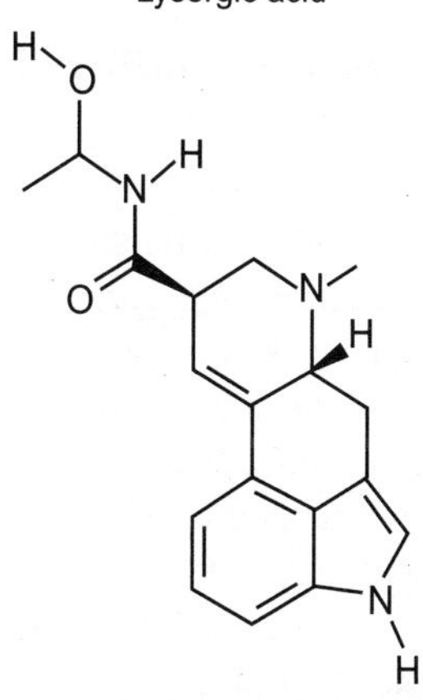

Hydroxyethyllysergamide

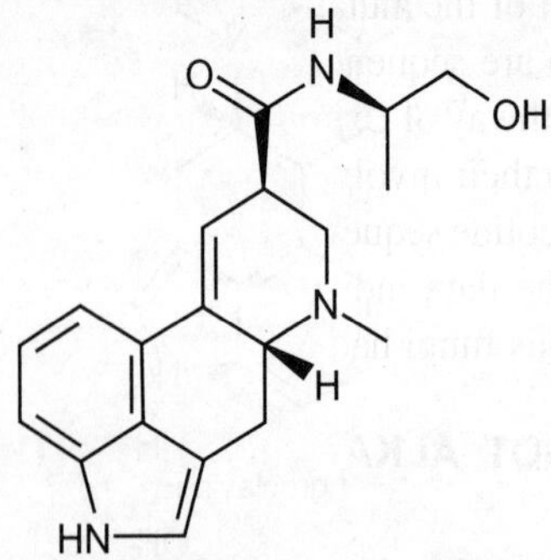

Ergometrine

These data demonstrate the association of easF with a methylation step in the pathway that immediately follows formation of dimethylallyltryptophan. Similar mutational analyses were carried out with the gene easE, which has structural properties indicating the capacity to encode an oxidoreductase.

Strains in which easE was functionally inactivated by recombination with an altered copy of the gene no longer produced the ergot alkaloids typically observed in the wild-type fungus. Instead these strain accumulated a product that was characterised to be *N*-methyl-dimethylallyltryptophan. These data indicate that the easE-encoded oxidoreductase catalyses the first oxidation step in the ergot alkaloid pathway and acts immediately after the product of the easF gene.

Strains in which the gene named easA was inactivated accumulated the ergot pathway intermediate chanoclavine aldehyde, demonstrating that it controls a later step in the pathway.

Impact

The data provided insight into genes controlling biochemical reaction that occur early in the ergot alkaloid pathway. Manipulation of these genes alters ergot alkaloid production in the producing fungus. The data have shown that ergot alkaloids from different stages of the pathway confer different properties that affect mammals versus insects differentially.

Elucidation of the genes in the ergot alkaloid pathway and the ability to manipulate them may facilitate alteration of the ergot alkaloid profile in such a way that anti-mammalian effects are minimised but anti-insect effects are retained.

ERGOT ALKALOID BIOSYNTHESIS GENE AND CLUSTERED HYPOTHETICAL GENES FROM *ASPERGILLUS FUMIGATUS*

The ergot alkaloids are a family of indole-derived mycotoxins with a variety of significant biological activities. *Aspergillus fumigatus*, a common airborne fungus and opportunistic human pathogen, and several fungi in the relatively distant taxon *Clavicipitaceae* (clavicipitaceous fungi) produce different sets of ergot alkaloids. The ergot alkaloids of these divergent fungi share a four-member ergoline ring but differ in the number, type, and position of the side chains. Several genes required for ergot alkaloid production are known in the clavicipitaceous fungi, and these genes are clustered in the genome of the ergot fungus *Claviceps purpurea*. Smith investigated whether the ergot alkaloids of *A. fumigatus* have a common biosynthetic and genetic origin with those of the clavicipitaceous fungi. A homolog of *dmaW*, the gene controlling the determinant step in the ergot alkaloid pathway of clavicipitaceous fungi, was identified in the *A. fumigatus* genome. Knockout of *dmaW* eliminated all known ergot alkaloids from

A. fumigatus, and complementation of the mutation restored ergot alkaloid production. Clustered with *dmaW* in the *A. fumigatus* genome are sequences corresponding to five genes previously proposed to encode steps in the ergot alkaloid pathway of *C. purpurea*, as well as additional sequences whose deduced protein products are consistent with their involvement in the ergot alkaloid pathway. The corresponding genes have similarities in their nucleotide sequences, but the orientations and positions within the cluster of several of these genes differ. The data indicate that the ergot alkaloid biosynthetic capabilities in *A. fumigatus* and the clavicipitaceous fungi had a common origin.

DETERMINANT STEP IN ERGOT ALKALOID BIOSYNTHESIS BY AN ENDOPHYTE OF PERENNIAL RYEGRASS

Many cool-season grasses harbor fungal endophytes in the genus Neotyphodium, which enhance host fitness, but some also produce metabolites—such as ergovaline—believed to cause livestock toxicoses. In *Claviceps* species the first step in ergot alkaloid biosynthesis is thought to be dimethylallyltryptophan (DMAT) synthase, encoded by *dmaW*, previously cloned from *Claviceps fusiformis*. Here Smith reported the cloning and characterisation of dmaW from *Neotyphodium* sp. isolate Lp1, an endophyte of perennial ryegrass (*Lolium perenne*). The gene was then disrupted, and the mutant failed to produce any detectable ergovaline or simpler ergot and clavine alkaloids. The disruption was complemented with the *C. fusiformis* gene, which restored ergovaline production. Thus, the biosynthetic role of DMAT synthase was confirmed, and a mutant was generated for future studies of the ecological and agricultural importance of ergot alkaloids in endophytes of grasses.

ALKALOID BIOSYNTHESIS—THE BASIS FOR METABOLIC ENGINEERING OF MEDICINAL PLANTS

Alkaloids—the term is linguistically derived from the Arabic word *al-qali*, the plant from which soda was first obtained are nitrogenous compounds that constitute the pharmacologically active 'basic principles' of predominantly, although not exclusively, flowering plants. Since the identification of the first alkaloid, morphine, from the opium poppy, *Papaver somniferum*, by Sertürner in 1806, ~10,000 alkaloids have been isolated and their structures elucidated. Historically, the use of alkaloid-containing plant extracts as potions, medicines, and poisons can be traced back almost to the start of civilisation. Famed examples include Socrates' death in 399 BC by consumption of coniine-containing hemlock (*Conium maculatum*) and Cleopatra's use during the last century BC of atropine-containing extracts of Egyptian henbane *(Hyoscyamus muficus)* to dilate her pupils and thereby appear more alluring. Medieval European women utilised extracts of deadly nightshade, *Atropa belladonna*, for the same purpose, hence the name belladonna. Although coniine is too toxic to find therapeutic use today outside of homeopathy, tropicamide, an anticholinergic that is a synthetic derivative of atropine, is routinely used in eye examinations to dilate the pupil. Tropicamide has also recently shown promise as an early diagnostic tool in the detection of Alzheimer's disease. A tonic prepared from the bark of *Cinchona officinalis* that contains the antimalarial drug quinine greatly facilitated European exploration and inhabitation of the tropics during the past two centuries.

In total, ~13,000 plant species are known to have been used as drugs throughout the world. Approximately 25 per cent of contemporary materia medica is derived from plants and used either as pure compounds (such as the narcotic analgesic morphine, the analgesic and antitussive codeine, and the chemotherapeutic agents vincristine and vinblastine; oras teas and extracts. Plant constituents have also served as models for modern synthetic drugs, such as atropine for tropicamide, quinine for chloroquine, and cocaine for procaine and tetracaine. In fact, active plant extract screening programs continue to result in new drug discoveries. The most recent examples of anticancer alkaloids are taxol from the western yew, *Taxus brevifolia*, and camptothecin (and derivatives currently in clinical trials) from the Chinese 'happy tree', *xi shu* (*Camptotheca acuminata*), both of which were originally isolated and assayed for biological activity in the 1960s in the laboratory of ME Wall. In other areas, there are intense searches for novel antivirals and antimalarials.

Smith also encounter alkaloids as the stimulants caffeine in coffee and tea and nicotine in cigarettes. Although a wealth of information is available on the pharmacological effects of these compounds, surprisingly little is known about how plants synthesise these substances, and almost nothing is known about how this synthesis is regulated. This is due, in part, to the complex chemical structures of many alkaloids, which contain multiple asymmetric centres. For example, although nicotine (one asymmetric center) was discovered in 1828, its structure was not known until it was synthesised in 1904, and the structure of morphine (five asymmetric centers) was not unequivocally elucidated until 1952, 146 years after its isolation. Beginning in the late 1950s, radiolabelled precursors were fed to plants and the resultant radioactive alkaloids were chemically degraded to identify the position of the label. This opened the field of alkaloid biosynthesis to experimentation. As analytical instrumentation became more sophisticated, precursors labelled with stable isotopes were fed to plants and the products analysed by nuclear magnetic resonance spectroscopy. No real progress was made in identifying alkaloid biosynthetic enzymes until the use of plant cell cultures as experimental systems was introduced in the 1970s. Since then, on the order of 80 new enzymes that catalyse steps in the biosynthesis of the indole, isoquinoline, tropane, pyrrolizidine, and purine classes of alkaloids have been discovered and partially characterised.

Alkaloids belong to the broad category of secondary metabolites. This class of molecule has historically been defined as naturally occurring substances that are not vital to the organism that produces them. Alkaloids have traditionally been of interest only due to their pronounced and various physiological activities in animals and humans.

A picture has now begun to emerge that alkaloids do have important ecochemical functions in the defense of the plant against pathogenic organisms and herbivores or, as in the case of pyrrolizidine alkaloids, as pro-toxins for insects, which further modifiy the alkaloids and then incorporate them into their own defense secretions. Alkaloids have now been isolated from such diverse organisms as frogs, ants (pheromones), butterflies (defense), marine bacteria, sponges, fungi, spiders (venom neurotoxin), beetles (defense), and mammals, although is not yet clear whether de novo alkaloid biosynthesis occurs in each organism.

With the introduction of molecular biology into the plant alkaloid field, induction of alkaloid biosynthesis in response to exposure to wounding or to elicitors can be analysed at the level of gene activation, and gene expression patterns in the plant can be determined and interpreted as a first indication of possible function. We also now have the capability to alter the pattern of alkaloid accumulation in plants for the purpose of studying the biological function of alkaloids, for engineering tailor-made plants that accumulate increased quantities of desired pharmaceuticals, or for producing foodstuff plants with lower alkaloid content (for example, coffee without caffeine).

Plants are some of nature's very best chemists, and sophisticated structures such as codeine, vinblastine, taxol, and camptothecin remain well beyond the reach of commercially feasible total chemical syntheses. With the ability to express alkaloid biosynthetic enzymes heterologously in organisms with better fermentation characteristics than plants, we can achieve unlimited quantities of these 'biocatalysts' for use in syntheses of important drugs.

Monoterpenoid Indole and Clavine Alkaloids

The monoterpenoid indole alkaloids comprise a large family of alkaloids, with over 1800 members of rich structural diversity. Many of these natural products are physiologically active in mammals. Among the monoterpenoid indole alkaloid pharmaceuticals that are still commercially isolated from plant material are the antimalarial drug quinine from *C. officinalis*, the antineoplastic drug camptothecin from *C. acuminata,* the rat poison and homeopathic drug strychnine from *Strychnos nuxvomica*, and the antineoplastic chemotherapeutic agents vincristine and vinblastine from *Catharanthus meus* (periwinkle). Total chemical syntheses of these complex alkaloids would be of academic interest but due to low yields are not likely to be applied commercially. To develop novel sources of these drugs, two options are available. The cDNAs for enzymes that catalyse those biosynthetic steps that are difficult to achieve by chemical means can be isolated and heterologously expressed for use in biornimetic syntheses. Alternatively, instead of single transformation steps, micro-organisms could be engineered to express short pathways, thus producing an end-product alkaloid of interest. Alkaloid biosynthetic pathways that are too long to be introduced into a single micro-organism could be modified in the parent plant using antisense or cosuppression technologies such that a desired alkaloid can be accumulated by blocking side pathways or catabolic steps.

All of these approaches require thorough knowledge of the alkaloid biosynthetic pathway and the enzymes that catalyse the individual transformation steps. Progress toward identifying the enzymes of monoterpenoid indole alkaloid biosynthesis has been made primarily in the laboratory of J. Stockigt using *Rauvolfia serpentina* cell suspension cultures in studies of the biosynthesis of the antiarrythmic drug ajmaline and in the laboratory of V. De Luca using *C. roseus* cell suspension cultures and plants to study the biosynthesis of vindoline, a precursor to the antineoplastics vincristine and vinblastine. The first successful cDNA cloning experiments in the alkaloid field were achieved with two cDNAs encoding enzymes that catalyse early steps in the biosynthetic pathway that leads to all monoterpenoid indole alkaloids, tryptophan decarboxylase and strictosidine synthase.

Increase in the indole alkaloid production and its excretion into the culture medium by calcium antagonists in *Catharanthus roseus* hairy roots: Treatment of *Catharanthus roseus* hairy roots with antagonists, like verapamil and $CdCl_2$, that block the Ca^{2+} flux across the plasma membrane enhanced the total alkaloid content by 25 per cent and their secretion 10 times. The specific Ca^{2+} chelator, EGTA, stimulated 90 per cent of the total alkaloid secretion. Treatment with inhibitors of intracellular Ca^{2+} movement, like TMB-8 and trapsigargin, enhanced the total alkaloid content by 74 per cent and their secretion into the culture media by 4- to 6-fold. The results suggest that an inhibition of external and internal Ca^{2+} fluxes induces an increase in the indole alkaloid accumulation and secretion in *C. roseus* hairy roots.

Alkaloids originally isolated from the ergot fungus *Claviceps purpurea* (Hypocreaceae). They include compounds that are structurally related to ergoline (ergolines) and ergotamine (ergotamines). Many of the ergot alkaloids act as alpha-adrenergic antagonists.

Recently the invention of fermentation process for the preparation of ergot alkaloids, primarily ergocornine and β-ergocryptine relates to a new process for the preparation of ergot alkaloids, primarily ergocornine and β-ergocryptine, by subjecting a *Claviceps purpurea* strain to fermentation under aerobic conditions in a culture medium which contains carbon and nitrogen sources, mineral salts and optionally other additives, too. According to the invention a *Claviceps purpurea* variant strain deposited under No. MNG 00186 is applied as alkaloid-producing strain.

ERGOTISM

Ergotism is the effect of long-term ergot poisoning, traditionally due to the ingestion of the alkaloids produced by the *Claviceps purpurea* fungus which infects rye and other cereals, and more recently by the action of a number of ergoline-based drugs. It is also known as ergotoxicosis, ergot poisoning and Saint Anthony's fire. Ergot poisoning is one of the explanations of bewitchment.

There is evidence of ergot poisoning serving a ritual purpose in the ritual killing of certain bog bodies. Found in peat swamps, Grauballe Man and Tollund Man have been preserved so well that large amounts of rotten cereals and weeds have been extracted from their stomachs, clearly showing force-feeding and primitive sedation.

When milled, the ergot is reduced to a red powder, obvious in lighter grasses but easy to miss in dark rye-flour. In less wealthy countries, ergotism still occurs; an outbreak in Ethiopia occurred in mid-2001 from contaminated barley. Whenever there is a combination of moist weather, cool temperatures, delayed harvest in lowland crops and rye consumption, an outbreak is possible. Russia has been particularly afflicted. Poisonings due to consumption of seeds treated with mercury compounds are sometimes misidentified as ergotism, possibly including the case of mass-poisoning in the French village Pont-Saint-Esprit in 1951.

Causes

The toxic ergoline derivatives are found in ergot-based drugs (such as methylergometrine, ergotamine or, previously, ergotoxine). The deleterious side-effects occur either under high dose or when moderate doses interact with potentiators such as azithromycin.

Traditionally, eating grain products contaminated with the fungus *Claviceps purpurea* also caused ergotism. Finally, the alkaloids can also pass through lactation from mother to child, causing ergotism in infants.

Symptoms

The symptoms can be roughly divided into convulsive symptoms and gangrenous symptoms.

Convulsive symptoms

Convulsive symptoms include painful seizures and spasms, diarrhea, paresthesias, itching, headaches, nausea and vomiting. Usually the gastrointestinal effects precede central nervous system effects. As well as seizures there can be hallucinations resembling those produced by lysergic acid diethylamide (LSD, to which the ergot alkaloid ergotamine is an immediate precursor and therefore shares some structural similarities), and mental effects including mania or psychosis. The convulsive symptoms are caused by clavine alkaloids.

Gangrenous symptoms

The dry gangrene is a result of vasoconstriction induced by the ergotamine-ergocristine alkaloids of the fungus. It affects the more poorly vascularised distal structures, such as the fingers and toes. Symptoms include desquamation, weak periphery pulse, loss of peripheral sensation, edema and ultimately the death and loss of affected tissues.

Chapter 15

Microbial Transformations

INTRODUCTION

Micro-organisms have the ability to convert one organic molecule into another. Many complex bioconversions that are achieved by micro-organisms, cannot be achieved by normal chemical means and are achieved with great difficulty by using organisms other than microbes. Microbial transformations of the molecules which have found industrial applications involve oxidation, reduction, isomerisation, hydrolysis, condensation, etc.

Transformation of organic compounds are achieved by using the spores, growing cultures, resting cells, enzymes, immobilised cells or enzymes derived from micro-organisms. Transformation reactions at large scale are carried out under sterile conditions in aerated and stirred fermenters.

Sterilisation is necessary because contamination can cause the production of undesired substances or can suppress the desired reaction. The end products are generally secreted outside the cell and they remain either dissolved or suspended in the fermentation broth. For further processing, if fungus is the micro-organism used, it is separated by filtration.

Recovery of the product is achieved by precipitation as calcium salt, by adsorption to ion exchangers, by extraction with appropriate solvents or by direct distillation from the medium (for volatile substances).

BIOCONVERSION

Bioconversion has two meanings is biotechnology. The first one is also known as biotransformation and is the use of micro-organisms to carry out a chemical reaction that is more costly or not feasible nonbiologically. The micro-organisms converts a substance to a chemically modified form. An example is the industrial production of cortisone. One step is the bioconversion of progesterone to 11-α-hydroxyprogesterone by *Rhizopus nigricans.*

The second is the conversion of organic materials, such as plant or animal waste, into usable products or energy sources by biological processes or agents, such as certain micro-organisms or enzymes.

New cellulosic ethanol conversion processes have enabled the variety and volume of feedstock that can be bioconverted to expand rapidly. Feedstock now includes materials derived from plant or animal waste such as paper, auto-fluff, tyres, fabric, construction materials, municipal solid waste (MSW), sludge, sewage, etc.

There are two principal processes for bioconversion:

1. Enzymatic hydrolysis: In enzymatic hydrolysis a single source of feedstock, switchgrass for example, is mixed with strong enzymes which convert a portion of cellulosic material into sugars which can then be fermented into ethanol.

2 Synthesis gas fermentation: In synthesis gas fermentation a blend of feedstock, not exceeding 30 per cent water, is gasified in a closed environment into a syngas containing mostly carbon monoxide and hydrogen. The cooled syngas is then converted into usable products through exposure to bacteria or other catalysts.

BIOTRANSFORMATION

When an organic compound is modified by simple chemically defined reactions, catalysed by enzymes present in cells, into a product that is recoverable, it is called biotransformation. Both the substrate and the product are not involved in the primary or secondary metabolism of the organism employed. This is in contrast to the various metabolites, e.g. organic acids, amino acids, antibiotics, etc. produced by the complex pathways of primary or secondary metabolism of the organisms. Biotransformations are performed by microbes, plant cells as well as animal cells. But microbial process is far more efficient and economical due to the rapid microbial growth and high metabolic rates of micro-organisms.

Inoculum

Vegetative cells, resting cells, dried cells, spores, or immobilised cells act as agents of biotransformation. The micro-organisms used for biotransformations are selected by primary screening of culture collections or from among isolates obtained from nature. The selected organisms are subjected to strain improvement, and mutants deficient in specific enzymes are often isolated to prevent undesirable modification of the product. The microbes used for biotransformation are eubacteria, streptomycetes and fungi; their cultures are usually maintained in frozen state at –20° to –170°C. The inoculum is developed in stages as with other fermentation operations.

Incubation

The incolum is transferred into a fermenter and conditions for optimal growth are provided. When the desired cell density is achieved, the substrate is added to the fermenter broth and incubated for 2–5 days. Substrates insoluble in water, e.g. steroids may be added as follows: (i) dissolved in a water-miscible organic solvent like ethanol, methanol, acetone etc. (solution added slowly with vigorous agitation), (ii) suspended in a surfactant, e.g. Tween 80, as fine particles, and (iii) as a fine powder.

Alternatively the cells may be separated from the broth by filtration or centrifugation, washed and resuspended in a suitable buffer. The substrate, e.g. a steroid, is added into the buffer; this simplifies recovery and purification of the product. Yet another approach uses cells/mycelium immobilised on a solid support which is then packed in a column. The substrate is passed through the column to allow biotransformation.

The recovery of products, which are released into the medium, is based on extraction and subsequent purification by column chromatography and other methods. For water insoluble products like steroids, solvents like methylene chloride, chloroform or ethyl acetate are used for extraction.

The chemical modifications occurring during biotransformation are: oxidation, reduction, hydrolysis, hydroxylation, isomerisation etc. Many of these reactions can be performed chemically. But biotransformation is highly specific and only a single isomer of the product is usually produced in very high yield (typically 90 per cent). Also biotransformation occurs at ambient temperature (20°–40°C), which reduces cost due to much lower energy needs in addition, the entire process leads to much lower pollution problems.

Commercial Examples

Conversion of wine into vinegar was practised in Babylon as early as 5000 BC; in 1864, Pasteur showed the role of microbes in the process. By 1900, several bioconversions, e.g. ethanol to acetic acid, glucose to gluconic acid, fructose to mannitol, tannin to gallic acid, etc. were known. But the first biotransformation to be commercialised was conversion of deoxycorticosterone into corticosterone by oxadrenals.

But this bioconversion is now achieved by much more efficient microbial process. The three examples described here briefly are: (i) conversion of D-sorbitol to L-sorbose, (ii) biotransformation of antibiotics, and (iii) biotransformation of steroids.

Biotransformation of D-sorbitol to L-sorbose

Ascorbic acid is commercially produced from glucose (Fig. 15.1). Glucose is first converted chemically into D-sorbitol, which is then biotransformed into L-sorbose by *Acetobacter suboxydans*. L-sorbose is converted into ascorbic acid by chemical process. Similarly, glycerol is biotransformed by *Gluconobacter melanogenus* or *A. suboxydans* into dihydroxyacetone which is mainly used as sustanning agent.

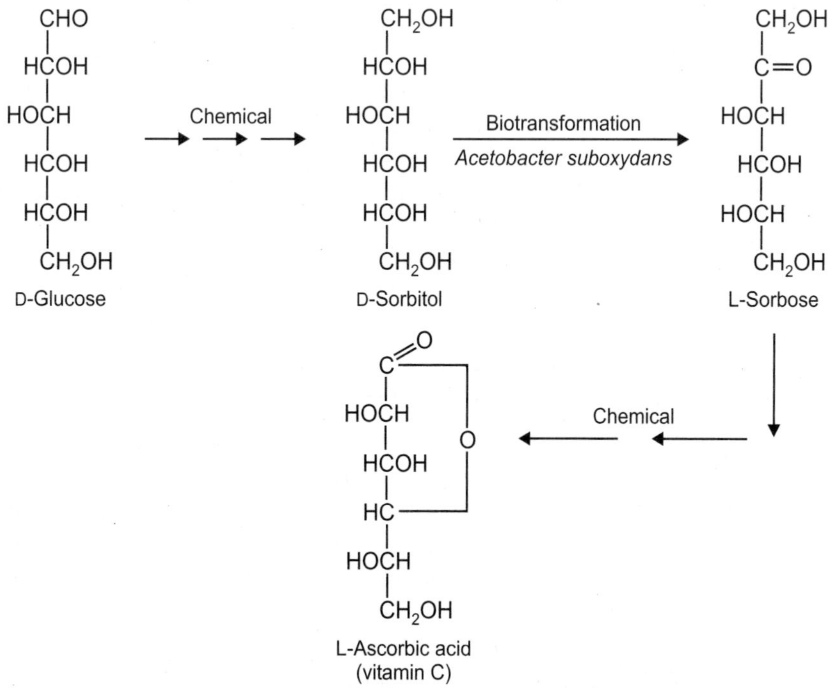

Fig. 15.1. Biotransformation of D-sorbitol into L-sorbose as a part of the process of vitamin C manufacture using glucose. The earlier and later steps in the process are achieved chemically.

For L-sorbose production, the medium is supplemented with 15–30 per cent D-sorbitol; *A. suboxydans* is inoculated and incubated at 30°C. Vigorous aeration and agitation is essential. In 1–2 days, 90–95 per cent conversion is obtained.

Cells are filtered out, and the filtrate is concentrated under vacuum; L-sorbose crystallises on cooling. About 65 per cent of L-sorbose is recovered in this manner.

Biotransformation of antibiotics

Biotransformation is routinely used for commercial production of several useful antibiotics. Semisynthetic antibiotics, e.g. semisynthetic penicillin and cephalosporins, are produced by chemical modification of the penicillin nucleus, e.g. 6-aminopenicillanic acid. The penicillin nucleus is produced by microbial deacylation of naturally produced penicillin G and penicillin V (Fig. 15.2).

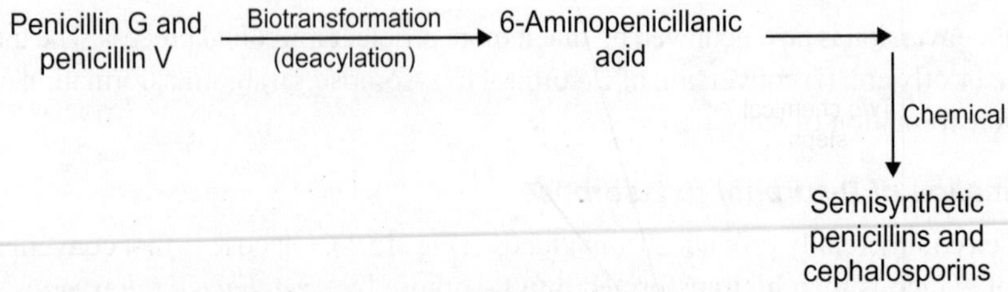

Fig. 15.2. Biotransformation of penicillin V (by penicillin-V-acylase) and penicillin G (by penicillin-G-acylase) to obtain the penicillin nucleus which is further modified chemically to yield semisynthetic penicillins and cephalosporins. The reactions is catalysed using immobilised enzymes.

Biotransformation of steroids

After antibiotics, steroid biotransformation is the most important microbial process yielding pharmaceuticals; it yields the adrenocrotical hormones like corticosterone, cortisone and hydrocortisone, and their therapeutically superior derivatives, e.g. prednisone, prednisolone, triamcinolone, etc. These conversion depend mainly on the addition of an oxygen atom at carbon 11 of the steroid molecule. The steroids are used as contraceptive (5 per cent of production) and as corticosteroids (95 per cent of production).

Steroids are produced from the following substrates: stigmasterol from soyabean, diosgenin from Dioscorea, β-sitosterol and campesterol from soyabean, cholesterol (natural substrate for steroid hormones in mammals) from wool grease, and hecogeni-*n* (used on a limited scale) from African sisal. Chemical and biotransformation procedures are combined in steroid production. For example chemical procedures are used to convert diosgenin (2 steps) and stigmasterol into progesterone which is then biotransformed as follows (on industrial scale) (Fig. 15.3).

1. α-Hydroxylation at carbon 11 by *Rhizopus nigricans* (85 per cent conversion) into 11-α-hydroxyprogesterone. A mutant of *Aspergillus ochraceus* is used on a limited scale and *R. arrhizus* is usually not preferred. α-Hydroxyprogesterone is converted chemically into hydrocortisone and cortisone, and into prednisolone and prednisone by a combination of chemical modification and biotransformation.
2. 16-α-Hydroxylation is due to *Streptomyces argenteolus*; the product 16-α-hydroxyprogesterone is used to produce triamcinolone by a combination of chemical modification and biotransformation.
3. 11-β-Hydroxylation by *Cunninghamella blakesleeana* or *Curvularia lunata* (Fig. 15.3) yields 11-β-hydroxyprogesterone, which is then converted by a combination of chemical and microbial modification into prednisolone.
4. C-1 dehydrogenation is effected preferably by *Arthrobacter simplex* or *Septomyxa affinis* to produce 1-dehydroprogesterone. C-1 dehydrogenation activity is also used in the production of triamcinolone.

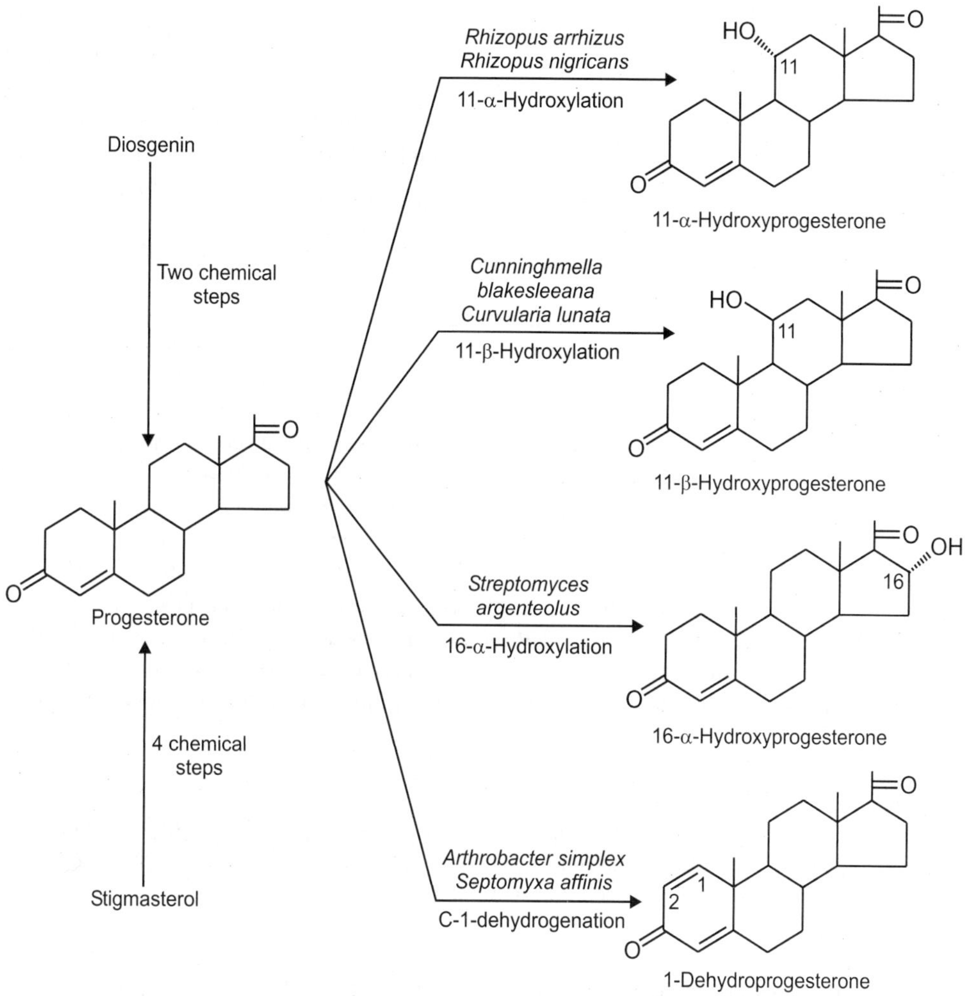

Fig. 15.3. Selected examples of biotransformation in commercial production of certain steroids of pharmaceutical value.

The steroid derivatives, e.g. prednisolone, triamcinolone etc. are more potent than cortisone or hydrocortisone. For example, prednisolone has 4–10 times more anti-inflamatory activity than cortisone and hydrocortisone. The 11-α-hydroxylation of steroids is achieved in a single step by biotransformation while it requires many chemical steps.

BIOINSECTICIDES

Micro-organisms used for insect control are often called bioinsecticides, while the term biopesticides is used for all biocontrol agents. Viruses, bacteria, fungi, protozoa and mites are employed to control a variety of insects attacking both plants and animals. Although a large number of micro-organisms attack insects, only a limited number of them have found commercial application. Recently the technology for production and application of biopesticides has been developed, and a *Bacillus thuringiensis* based insecticide is being commercially produced.

Bacillus thuringiensis is a spore forming bacterium, which produces a crystal protein as parasporic inclusion in the vegetative cell containing the spore. The crystal protein is a protoxin which is processed into a toxin by the proteases present in insect mid-gut. The toxin fragment binds to highly specific receptors present in the membranes of gut epithelium cells; this binding ultimately causes insect death. Several different types of crystal proteins are produced by different strains of the bacterium, each protein having a specific and ordinarily narrow insect host range. This highlights a very important feature of biopesticides: they are specific to a well-defined range of target species. This is in contrast to the chemical pesticides which usually affect a very wide range of target species. This property must be well appreciated by the user of a biopesticide.

B. thuringiensis cells are produced in fermenters, as are other bacteria and fungi. The commercial preparations usually contain a mixture of spores, crystal proteins and inert carriers. The formulation may be in the form of a water dispersible powder, wettable powder, emulsifiable concentrate, flowable concentrate, granules or dust. It may be mixed with chemical insecticides or fungicides, if required. The crystal protein activity disappears usually within 24–40 hrs after application, but the spores may persit for long periods and cause harm to nontarget useful insects, e.g. silkworm. This problem is sought to be overcome by developing mutants which produce the crystal protein but do not form spores. Such mutants have been isolated by various scientists, and their usefulness as biopesticide is being evaluated.

At least two insect viruses, called by the group name bacculoviruses, are being used/likely to be used on commercial scale for insect control. The nucleopolyhedrosis viruses (NPV) are used for insect control in USA for control of *Helicoverpa* (*Heliothis*) *armigera*. Another bacculovirus, the granulosis viruses (GV), is also promising and may achieve a commercial status for *Spodoptera litura* control in USA and India. Commercial production of these viruses is done in the larvae of the target insects, e.g. NPV in Heliothis larvae.

Bioherbicides

Fungal pathogens are attractive biocontrol agents for weed control in view of their host specificity and ease in production and inoculation in the field where, once established, they will spread on their own. At least two fungal pathogens have achieved commercial status in USA.

Disease control

Biocontrol of plant diseases employs both bacteria and fungi, and at least three commercial examples are available. In India, the technology for large scale production of *Trichoderma* and its use for control of soil-borne fungal pathogens like *Macrophomina phaseolona* etc. and for seed treatment has been developed; efforts are being made for its commercialisation.

Advantages and limitations

The interest in biopesticides is based in the disadvantages associated with chemical pesticides, some of which are as follows: (i) extensive pollution of the environment, (ii) serious health hazard due to the presence of their residues in food, fibre an fodder, and (iii) increasing cases of insects developing resistance, e.g. *Helicoverpa* (*Heliothis*) has become resistant to most of the insecticides.

In contrast, biopesticides: (i) do not leave harmful residues, (ii) do not destroy nontarget species, and (iii) are often cheaper than chemical pesticides. Their chief disadvantages are: (i) very high specificity which will require exact identification of the pest/pathogen and may require multiple pesticides to be used, and (ii) often variable efficacy due to the influences of various biotic and abiotic factors (since biopesticides are usually living organisms which bring about pest/pathogen control by multiplying within the target insect pest/pathogen) (Table 15.1).

Table 15.1. A comparison between biopesticides and chemical pesticides.

Biopesticides	*Chemical pesticides*
These do not harm nontarget species	Nontarget species are also harmed
They do not pollute the environment	Cause pollution; sometimes serious
No harmful residues remain in food, fodder and fibres	Harmful residues may often remain in food, fodder and fibres
Relatively cheaper	Relatively costlier
Insects are expected not to develop resistance to biopesticides	Insects may become resistant, e.g. *Heliathis* has become resistant to most insecticides
Since they are highly specific, correct identification of the pest is essential	It is often not critical
High specificity may often make the use of two or more biopesticides necessary	Often not required
Performance may be variable due to the influence of biotic and abiotic factors of the environment	This is not often the case

BIOFERTILISERS

Micro-organisms employed to enhance availability of nutrients, viz. nitrogen (by fixing atmospheric N_2) and phosphorus (by solublising soil phosphorus), to the crops are called biofertilisers. The various micro-organisms having realised/potential applications as biofertilisers are: bacteria (*Rhizobium* spp., *Azospirillum*, *Azotobacter*), fungi (mycorrhizae like *Glomus*), blue-green algae or cyanobacteria (*Anabaena*, *Nostoc*, etc.) and *Azolla* (a fern containing symbiotic *Anabaena azollae*) (Table 15.2).

Table 15.2. A list of some important micro-organisms with practical/potential application as biofertilizers.

Organism	*Activity*	*Association, if any*	*Used in crops*
Rhizobium (*leguminosarum*, *japonicum*, *phaseoli* etc.)	N_2-fixation	Symbiotic	Legumes (pulses, oilseeds, forage crops)
Azospirillum	N_2-fixation	Associative	Graminaceous crops like wheat, rice, sugarcane, jowar
Azotobacter	N_2-fixation	Asymbiotic	Wheat, rice, vegetables
Blue-green algae (*Anabaena*, *Nostoc*, *Plectonema* etc.)	N_2-fixation	Asymbiotic	Rice
Azolla-Anabaena complex	N_2-fixation	Symbiotic	Rice
Phosphate solublising bacteria (*Thiobacillus*, *Bacillus* etc.)	Phosphate solublisation	Asymbiotic	Many crops
Mycorrhiza (*Glomus*)	Phosphate solublisation	Associative	Many crops including pulses

Rhizobium spp.

These are Gram-negative soil bacteria capable of forming root nodules in most leguminous plants and some nonleguminous plants. In some cases, stem nodules are also produced. *Rhizobium* is divided into several species chiefly on the basis of the legume species they are able nodulate, e.g. *R. leguminosarum* (nodulates pea), *R. phaseoli* (*Phaseolus* sp.), *R. trifolii* (*Trifolium* sp.), *R. lupini* (lupins), *R. melilotii*

(*Melilotus* sp.) etc. Cowpea rhizobia are now classified as *Bradyrhizobium*. *Rhizobium* cells contain genes for nitrogen fixation (*nif* genes) on a megaplasmid. The bacteria enter the roots through root hairs, the interaction being highly specific and progressing through several steps; it ultimately results in nodule formation. Many genes of *Rhizobium* as well the host legume are involved in the process. Inside the nodule many bacterial cells change into nondividing bacterioids which produce nitrogenase, the enzyme which reduces atmospheric nitrogen into ammonia. Nitrogenase is highly sensitive to O_2; it is protected from O_2 by the pink pigment leghaemoglobin, which binds to O_2; produced by the legume and present in the nodules.

Different strains of a *Rhizobium* species differ in their ability to fix nitrogen; this trait is also affected by the genes of host legumes as well. Therefore, extensive screening for efficient N_2-fixers is undertaken (strain development) on the basis of N_2-fixation occurring in association with the host legumes. Some mutant, of *Rhizobium* are more efficient N_2-fixers than the wild type, e.g. nitrate reductase deficient mutants. Further, *hup*$^+$ strains of *Rhizobium* are more efficient since they are able to recycle the H_2 produced by bacterial cells, which otherwise is released as gas. The energy and carbon requirements of the bacteria are provided for by the host legume. In return, the ammonium produced by the bacterial cells is made available to the host.

The estimated N_2 fixed by *Rhizobium* species range from 50 to 150 kg/ha or even more, especially in case of clovers. Field trials suggest a 10–15 per cent yield increase in inoculated pigeonpea and chickpea over the uninoculated controls. *Rhizobium* spp. are attacked by viruses called rhizobiophages and the inoculated strains have to compete with the native rhizobia. It is therefore important that the selected strain should be tolerant to elevated temperatures (to survive the hot summer), have long shelf life when the inoculum is prepared and should out-compete (when inoculated) the native rhizobia present in soil.

Azotobacter and *Azospirillum*

Azotobacter uses the organic matter present in soil to fix nitrogen asymbiotically; it is capable of fixing up to 30 kg nitrogen/ha/yr. *Azospirillum* species occur in association with the roots of many plants of the grass family, e.g. jowar, wheat, bajra etc. These bacteria are capable of fixing over 30 kg N/ha/yr. Field inoculation of crops with *Azotobacter* or *Azospirillum* is estimated to save 15–25 kg N/ha.

Blue-Green Algae and Azolla

Blue-green algae (cyanobacteria) are photosynthetic, prokaryotic organisms which fix N_2 asymbiotically; some cyanobacteria are known to form symbiotic associations, e.g. Azolla (a fern)—*Anabaena azollae* (a blue-green alga). Examples of cyanobacteria are *Anabaena*, *Nostoc*, *Plectonema* etc. Usually, composite cultures containing two or more genera are used for field inoculation since they are often superior to single strain inoculations. Cyanobacteria produce nitrogenase and N_2-fixation occurs in specialised structures called heterocysts in which the *nif* region becomes reorganised (this is essential for N_2-fixation). In addition, heterocysts act as O_2-proof compartments which protect nitrogenase from O_2 inactivation.

Azolla owes its N_2-fixing capability to the symbiont *Anabaena azollae*. Azolla is widely used in Vietnam as biofertiliser for rice.

Cyanobacteria, in addition to N_2-fixation, accumulate biomass which improves the physical properties of soil, produce growth promoting substances and are useful in reclaimation of alkaline soils. They are used for rice, the inoculum being introduced in the field about 10 days after transplantation.

Phosphate Solublising Micro-organisms

Some bacteria, e.g. *Thiobacillus*, *Bacillus* etc. convert nonavailable inorganic phosphorus present in soil into an available form utilisable by crop plants. These bacteria also produce iron chelating substances, e.g. pseudobactin, called siderophores which chelate the iron present in the root zone; this iron becomes nonavailable to harmful micro-organisms and, in this manner, crop plants are protected from them. In addition, certain fungi, e.g. *Glomus*, form associations with plants roots; these are called mycorrhiza. The fungus may be located at the root surface (*ectomycorrhiza*) or it may be present inside the roots (*endomycorrhiza*). These fungi convert nonavailable phosphorus into an available form, produce growth promoting substance and also protect against soil pathogens.

These micro-organisms are yet to be exploited on a commercial scale. An example of commercial application of mycorrhizal fungi is found in *Citrus* in USA, seedlings are inoculated with the fungus in nurseries before being transplanted into the field.

Large Scale Production

Rhizobium inoculum is produced in shake-flasks or fermenters using appropriate growth medium. After a culture attains the desired cell density, contamination is checked before it is mixed with an appropriate inert and sterile carrier, e.g. peat, which is subsequently used as inoculum for coating the seeds. Alternative methods of *Rhizobium* inoculation have also been developed.

Blue-green algae are multiplied in troughs or small tanks. The algal biomass is harvested, dried and stored in gunny bags. The algal flakes are used at 10 kg/ha for inoculation in rice fields.

Advantages and Limitations of Biofertilisers

Advantages

The relevance of biofertilisers is increasing rapidly since chemical fertilisers (i) utilise petroleum (nitrogenous fertilisers), (ii) are costly, (iii) are short in supply, and (iv) damage the environment.

Disadvantages

In contrast, biofertilisers are (i) low cost inputs, (ii) lead to soil enrichment and (iii) are compatible with long-term sustainability. Further, (iv) they are eco-friendly and pose no danger to the environment. However, the acceptability of biofertilisers has been rather low chiefly because they do not produce quick and spectacular responses. In addition, the amount of nutrients provided by them is not enough to adequately meet the total needs of crops for high yields. Therefore, a pragmatic approach more likely to succeed will be to develop a rational and effective combination of biofertilisers and conventional fertilisers for optimum crop yields.

BIODEGRADATION

Biological degradation or simply biodegradation is generally considered as a phenomenon of biological transformation of organic compounds by living organisms, particularly microbes. The role of micro-organisms in the decomposition of sewage and other organic wastes is long known. It has been considered as a natural process in the microbial world as carbon and energy source for their growth and takes a pivotal role in the recycling of materials in the natural ecosystem. It brings about changes in the molecular structure of a compound ultimately yielding simpler (mineralisation) and comparatively harmless (non-toxic) products like CO_2, H_2O, NH_3, CH_4, H_2S or PO_3. When the compound is not fully broken, it is termed biotransformation. Many of the recalcitrant substances produced by biotransformation may sometimes be more toxic than the original compound. Such changes are brought about by the catabolic activities of bacteria or fungi by their intracellular or extracellular enzymes secreted in the medium. Biological fate of xenobiotic (novel to microbial system) compounds in the environment can be indicated as shown in Fig. 15.4.

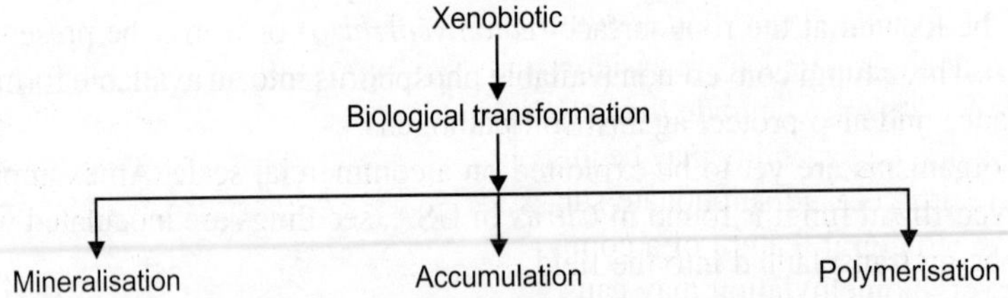

Fig. 15.4. Biological fate of xenobiotic compounds.

Perhaps biosynthetic abilities of the living world and the possibility of their degradation by catabolic enzymes evolved parallely but slowly in nature. This has apparently ensured that under suitable conditions all natural organics get decomposed and are not deposited in the environment with the exception of natural polymers like lignin and soil humus getting degraded very slowly. This has helped microbes to act as scavengers and reduce the pollution load in natural ecosystem. Bioremediation of polluted environment capitalises on the activities of aerobic or anaerobic heterotrophic microbes. The general scheme of such degradation may be represented as shown in Fig. 15.5.

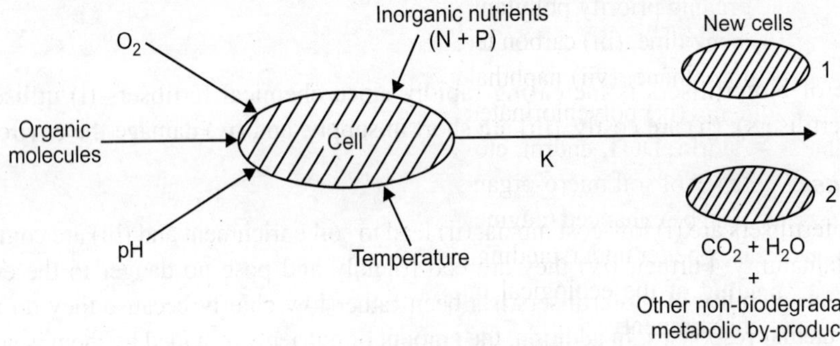

Fig. 15.5. Biodegradation process.

Here, K is the rate coefficient and is a function of biodegradability of organics. Many of the wastes are generally complex in nature. While a particular strain of a microbe may degrade only one type of compound or its related group, for some chemical substances the synergistic action of microbial communities or consortia in a polyculture, displaying wide range of degradative abilities rather than a monoculture, is desirable. Sometimes when the degrading material does not serve as a sole source of carbon and energy (non-growth substance) for the organism, but is associated with another growth substrate, then also it gets biotransformed and the phenomenon is termed as cometabolism. For example, toluene is primary substrate for *Pseudomonas putida* in breaking trichloroethylene. The insecticide parathion is decomposed by cometabolism of *Pseudomonas aeruginosa* and *P. stutzeri*. The controlled environment in a laboratory condition with one type of organism acting on a single substrate, however, does not always reflect the real outdoor situation.

The factors that affect biodegradation *in situ* are temperature, pH, redox potential, availability of nutrients, O_2 supply, biomass of the degrader, competition among microbial communities and the nature and concentration of the substrate as well.

The chemical nature of the compound has also great influence in the process, which is as follows:

1. Aliphatic compounds are degraded more easily than the aromatic ones. Algae and fungi cannot cleave aromatic rings, whereas bacteria can.
2. Recalcitrance of a compound increase with increased branching, polymerisation, and presence of polycyclic and heterocyclic residues.
3. Water soluble compounds are easier to degrade than insoluble forms.
4. Alkenes are easier to degrade than alkanes, while alkanes are more amenable than aromatics;
5. For aromatics, degradability may be influenced by molecular orientation, e.g. *ortho* > *para* > *meta*.
6. Halogen, nitrogen and sulphonate substitutions inhibit biodegradability.

Thus from the structural feature of a compound, one may get a rough idea about its biodegradable potential. However, biomethylation may cause problem in natural biodegradation process.

Realisation of this beneficial characteristic in microbes and the recent knowledge of the possibility of its genetic manipulation has given the biodegradation process a new dimension. This is particularly so with respect to various complex compounds getting injected daily into our environment through anthropogenic activities. Over the last few decades enormous quantities of synthetic chemicals have been released into the environment. They are posing serious problems being xenobiotic and highly hazardous, such as phenols, PCBs, hydrocarbons and other recalcitrant and persistent aromatics. These foreign substances of industrial origin are novel (unknown to nature earlier) to the normal microbial enzymatic degradative process. In solving this problem, biotechnology is likely to provide a new and safer approach in the early part of this millennium.

EPA list of some organic priority pollutants injected into the environment by human activities, are: (i) acenaphthene, (ii) benzidine, (iii) carbon tetrachloride (CCl_4), (iv) chlorinated phenols, (v) dichlorobenzene, (vi) hexachloroethane, (vii) naphthalene, (viii) polynucleated aromatic hydrocarbons (PAH), viz. benzopyrine, toluene, (ix) polychlorinated biphenyls (PCBs), (x) hexachlorocyclohexane—BHC, and (xi) pesticides—aldrin, DOT, endrin, etc.

The degrading potency of soil micro-organisms are being isolated these days from toxic waste sites. The inherent capacity can be enhanced (augmented) through genetic upgrading of the degradative genes. In order to achieve this, a better understanding of their catabolic pathways and genetic bases is essential. Moreover, understanding of the ecological interactions of these new microbial strains is needed for their success in the environment.

Till 2007 about 40–50 microbial strains with suitable degradative potentials for a variety of complex compounds have been isolated worldwide. Some of the compounds like organophosphate pesticides, parathion, DDT, 2,4-D, CCl_4, PCBs, toluene, biphenyls, heavy metals and their organic derivatives have been proved to be successful. Agricultural University at Wageningen (Netherlands) has developed a bacterial strain that can break down vienyl chloride (a carcinogen). Vinyl chloride is not easily amenable to degradation, which is being widely used for various purposes including the manufacture of furniture. A strictly anaerobic bacterium *Dehalococcoides ethenogenes* strain195 has shown its potentiality of degrading vinyl chloride as part of its energy metabolism, generating environmentally benign products like biomass, ethene and inorganic chloride through reductive dechlorination. Recently at the University of California at Davis, K. Scow and his group have identified a microbe (scavenger bug) named PM1, which can degrade MTBE (methyl tertiary butyl ether), a fuel-additive and a potential carcinogen present in the soil, in six days. Some of the microbes which can degrade various chemicals are given in Table 15.3.

Table 15.3. Microbes which can degrade various chemicals.

Chemicals	*Microbes*
Hydrocarbons	*Pseudomonas, Nocardia, Arthobacter, Mycobacterium*
PCBs	*Pseudomonas, Candida, Alcaligenes*
Phenolics	*Pseudomonas, Flavobacterium, Trichosporon, Bacillus, Candida, Aspergillus,*
Polycyclic aromatics	*Arthrobacter, Nocardia, Alcaligenes, Pseudomonas*
Naphthalene	*Pseudomonas, Nocardia*
Organophosphates	*Pseudomonas*
Benzene	*Mycobacterium, Alcaligenes*

Pseudomonas is a versatile bacterium for biodegradation of toxic wastes as it possesses various types of catabolic plasmids.

Microbial degradation constitute the basic principle of bioremediation of organic wastes in the environmental clean-up. It involves steps in the proper utilisation of indigenous microflora and/or development of 'enriched' microbes with genetic manipulation for better efficiency. Many biotech companies in the West now market the 'inocula' of such genetically improved microbes. Recent biotech clean-up of Alaskan beaches is the largest application of this emerging technology. The other biotech innovation in this area is the development of 'deep-shaft' fermentation system (with deep hole in the ground) by ICI limited, which is economical with respect to land use and man power in organic waste treatment. It produces much less sludge than the conventional system.

Aerobic vs Anaerobic Degradation

Microbial degradation or transformation of organic compounds may involve either of the two processes of aerobic (oxygen dependant) or anaerobic situation, while in some cases it may need both the conditions to detoxify some xenobiotic compounds.

Aerobic degradation

In the conventional aerobic system, the substrate is used as a source of carbon and energy. It serves as an electron donor resulting in bacterial growth. The extent of degradation is correlated with the rate of O_2 consumption, as also previous acclimation of the organism in the same substrate. Two enzymes primarily involved in the process are di- and mono-oxygenases. The latter enzyme can act on both aromatic and aliphatic compounds, while for the former, only aromatic compounds can act as substrates. Another class of enzymes involved in aerobic condition are peroxidases, which are receiving attention recently for their ability to degrade lignin.

Anaerobic degradation

This process is of widespread occurrence and relies on the metabolic versatility of mixed microbial populations present in soils or sediments when O_2 supply is limited. Growth yield of anaerobic bacteria is extremely low due to low energy yields. It has drawn attention these years due to the possibility of decomposition of extremely recalcitrant xenobiotics through this process.

Though the anaerobic process is slow, needs long retention time and produces H_2S gas, yet it is more advantageous than the aerobic one due to its non-dependence of O_2 supply. It thus saves the cost of energy for O_2 transfer. Materials like cellulose and fats, which remain unaffected by the aerobic process, breakdown under this situation. The overall process of anaerobic degradation of complex wastes is shown in Fig. 15.6.

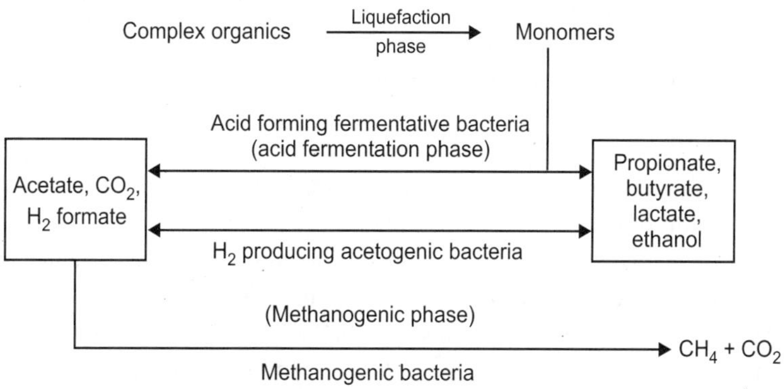

Fig. 15.6. Anaerobic degradation.

Three temperature ranges are used in anaerobic digestion:

1. Cold digestion at about 20°C.
2. Mesophylic digestion at 20°–40°C.
3. Thermophilic digestion at 40°–55°C.

Denitrification, sulphate reduction, dehalogenation and fermentation coupled to methanogenesis may occur concurrently in the same soil or sediment as different conditions exist at different microhabitats harbouring consortia of various degrading microbes and their interactions.

Some anaerobic microbial transformation reactions of organic compounds are:

Reaction type	*Examples*
Hydro-/dehydrogenations	Phenol, catechol, benzoate, fatty acids, unsaturated hydrocarbons
Carboxy-/decarboxylations	Cresol, toluene, benzoate, short hydrocarbons
Reductive dehalogenation	Polychlorinated aromatic compounds
Dechlorination	PCBs, phenols, chlorinated ethylenes
Methylation	Heavy metals

The anaerobic methods of waste-water treatment are considered safe, since few toxic chemicals can be stripped into the ambient air. Chlorinated xenobiotics need to be dehalogenated to make them harmless and biological treatment is an attractive proposition. They mostly need anaerobic situation for dehalogenation by bacterial genera like *Pseudomonas*, *Arthrobacter*, *Mycobacterium*, etc. Unlike aerobic condition, in an anaerobic degradation the chlorinated molecule is used as a direct source of electron. This is exemplified by 3-chlorobenzoate reduction.

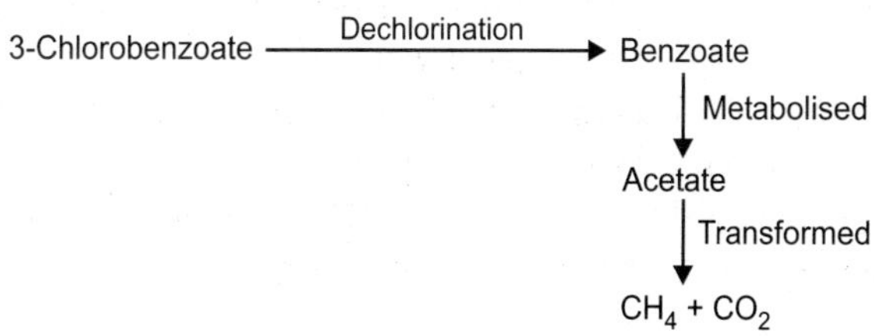

Reductive dechlorination involves successive 'shedding' of chlorine atoms under reduced anaerobic condition and is a common initial step in the biodegradation of chlorinated organics. The process is rather rapid for chemicals having a high number of chlorine substitution, viz. PCBs, hexachlorobenzene, trichloroethane, etc. Initial dechlorination is needed to make the compound less harmful for the microbes working under subsequent aero- and/or anaerobic degradation.

Sequential degradation

In many cases, both anaerobic and aerobic sequences are combined. This helps in the reduction of toxicity and mineralisation of compounds, which are otherwise recalcitrant. For example, tetrachloroethylene and tetrachloromethane may be mineralised in sequential steps of anaerobic and aerobic conditions, so that initially TCE and chloroform are formed, which, later in aerobic methanogenic stage, are converted into CO_2 and H_2O. In such sequential stages, the BOD reduction is also taken care of, as is being done in case of waste-water from pulp and paper industries.

For some xenobiotic chemicals, a single bacterium may transform it to another form, but cannot complete the breakdown. In such a situation a second group of microbe may act in a complementary fashion to have the total breakdown of the compound through a 'team work'. This sort of 'synergistic' action also prevents the build-up of toxic intermediates in the environment.

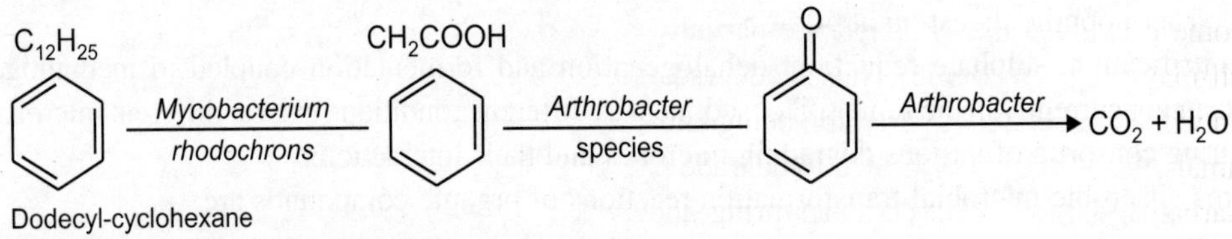

Microbial Basis of Biodegradation

Microbes in natural population may live in commensalism, which is an interactive association between two populations of different species. A sort of mutualistic interaction. For example, *Lactobacillus arabinosus* and *Streptococcus faecalis* association depends on nutrition requirements. The former makes folic acid required by the later, while the later species makes phenylalanine needed by the former for its growth and metabolism.

The other factor which works in the degradation phenomenon is the cometabolism. It is the ability of one organism to transform a non-growth substance so long as growth substance is present in the medium. During *in situ* biodegradation, the adopted organism's catabolic activity proceeds till some nutrients or electron acceptors reach a limiting concentration. Oxygen level often acts as a limiting one.

Biodegradation of herbicides and pesticides

With the advent of *Green Revolution*, there had been a quantum jump in the use of synthetic herbicides and pesticides throughout the world to sustain high-yielding crop varieties as they get easily attacked by pests. They have now become a part of modern agriculture.

Some of the common herbicides based on the chemical group are—Propham (carbamate); Dicamba (aromatic acid); Propanil (anilides); Simazine, Atrazine (triazines); Paraquat, Picloram (pyridines); Glyphosate (organophosphate); 2,4-D, MCPA, 2,4,5-T (Phenoxyacetic acid). Many of them are highly toxic, persistent and show recalcitrance to natural decomposition.

Apart from the chlorinated ones, some also have 'surfactants' (surface active agent) as adjuvants in their formulations for the purpose of retention by leaf surface.

A number of them are now found to contaminate the surface and ground waters through run-off from agricultural fields. Commonly used ones are triazine derivatives, carbamates, organophosphates, aldrin, parquat, diuran, parathion, malathion, etc. Some have even been found to be carcinogenic.

Herbicides can be partly decomposed in soils through chemical or photochemical reactions. However, the biodegradability is very much variable. Some of the pesticides may appear as recalcitrants, but can be degraded by the process of co-metabolism. A number of microbial cell-bound enzymes and extracellular ones can catalyse the breakage of bonds in herbicide molecules. Besides the enzymatic pathways, microbial activities may alter the pH of the soil and thus help in the detoxification of some of them.

Cometabolism of MCA and MCPA:

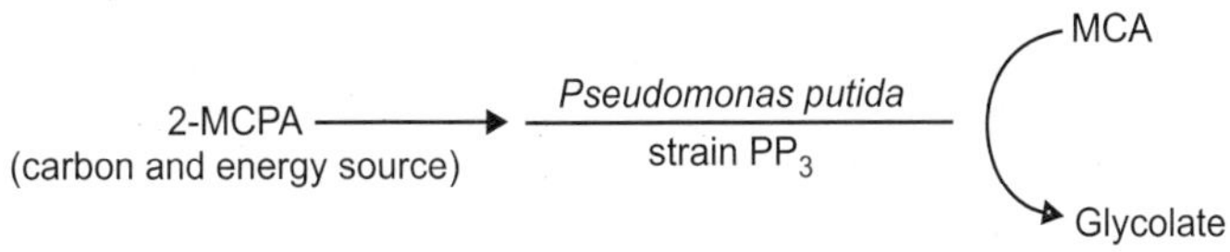

P. putida as such cannot metabolise MCA (monochloroacetate). But while metabolising MCPA (monochloropropionate), the organisms catalyse dehalogenation of MCA.

Some pesticides like propanil are partially biodegraded to form azocompounds (N_2-containing), which may be carcinogenic. Thus biodegradation of such a pesticide may lead to another problem. Again while fatal effect of parathion being known, its molecular modification malathion is less toxic to mammals. Some algae are known to degrade parathion.

Carbamate group of insecticides bring about their toxic effects by the inhibition of the enzyme acetylcholine esterase. Carbaryl is the most widely used in this group. It has low toxicity to mammals and gets hydrolysed to naphthol. Degradation of carbamates are brought about by *Pseudomonas*, *Achromobacter* and *Flavobacterium*.

Herbicides like atrazine (a highly toxic chlorinated triazine compound) are biodegraded sequentially through the removal of alkyl side chains followed by deamination, dehalogenation and ring cleavage.

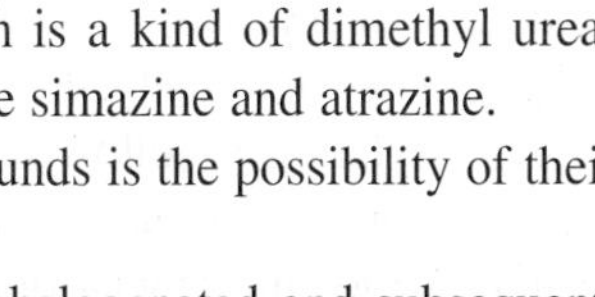

Atrazine

Other commonly used compounds are decamba (3-dichloro-*o*-anisic acid), MCPA (4-chloro-2-methyl phenoxy acetic acid), diuran (which is a kind of dimethyl urea) and glyphosate (N-phosphomethyl glycine), and triazine derivatives like simazine and atrazine.

Major advantage of these compounds is the possibility of their rapid biodegradation under suitable environmental conditions.

The chlorinated ones are first dehalogenated and subsequently broken down. The degradation is brought about by *Pseudonomas*, *Azotobacter*, *Bacillus*, *E. coli* and some others. The rate of the process would, however, depend on the density of the microbes and their direct contact; while mobility of the compounds in the soil layers may reduce the decomposition considerably.

Although DDT is now banned, for its effect on CNS (central nervous system), as an organochlorine pesticide, it has, however, been reported to be degraded to *p*-chlorophenyl acetic acid by a number of bacteria, algae and fungi. The filamentous basidiomycetous fungus *Phaenerocheate* (lignin degrader) is considered as a versatile one and can degrade the fungicide PCP (pentachlorophenol) and DDT as well with the help of a mixture of peroxidase enzymes. It is now widely used in the decontamination of chemically contaminated soil. Biodegradation steps of 2,4-D, propanil and diuran have been worked out. For 2,4-D, the steps are as follows:

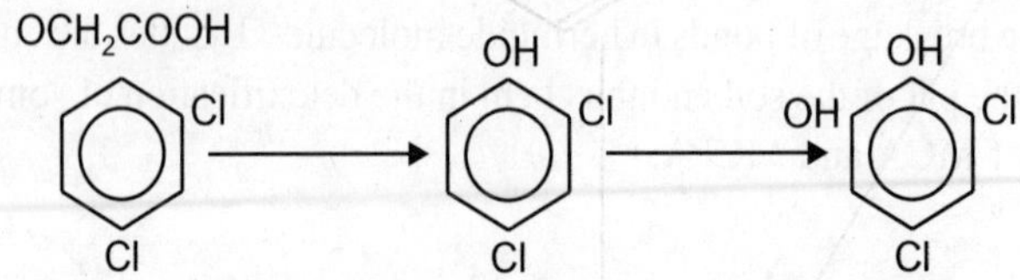

2,4-D (2,4-Dichloro-phenoxy acetic acdi) 2,4-Dichloro phenol 3,5-Dichloro catechol

Among the organic mercuric fungicides, the common one is dithiocarbamate. Some of them may exert toxicity as chelating agents (Thiram), while others (Maneb) decompose to isothiocyanates and creates another problem being a respiratory inhibitor.

Some herbicides may be degraded through the actions of different groups of microbes in sequential steps, e.g. Dalapon requires at least six types of organisms to act in complementary fashion. In order to cope with the newer herbicides released in the environment, microbes also undergo mutation, and in evolution, new enzymes are synthesised to deal with the novel substrates.

Bioremediation of contaminated soil by lindane, a chlorinated pesticide, was achieved by ITRC (Industrial Toxicological Research Centre), Lucknow, India, with the help of an isolated bacterial strain assigned as EL-l. The microbe could degrade all the major isomers (α, β, γ and δ) of lindane which was not possible earlier.

BIOCHEMICAL PATHWAYS OF BIODEGRADATION

The biodegradation of complex hydrocarbons, pesticides, herbicides and xenobiotics generally requires the concerted effort of a number of enzymes and in many cases more than one micro-organism. Hydrocarbons are stable reduced compounds and therefore degradation generally proceeds by oxidation under either aerobic or anaerobic conditions. Microbial degradation of monocyclic and polycyclic aromatic hydrocarbons has been studied extensively. Degradation of non-halogen xenobiotics is carried out by a range of enzymes to convert them into catechol or protocatechuate (Fig. 15.7).

The subsequent metabolism of catechol can take one of two pathways; ortho-cleavage yields *cis,cis*-muconate, whereas meta-cleavage yields 2-hydroxymuconic semialdehyde. Both pathways lead to the compounds pyruvate, acetaldehyde; succinate, and acetyl-CoA, which can enter the Krebs cycle (Fig. 15.8). In general aromatic ring hydroxylation is followed by ring cleavage, and both of these reactions are carried out by oxygenases. The incorporation of two oxygen molecules by dioxygenases introduces two hydroxyl groups that can undergo either meta- or ortho-cleavage. This process is found in bacteria and algae. Incorporation of a single oxygen molecule is catalysed by cytochrome P450 monooxygenases and is found in fungi and algae (Fig. 15.9). The white rot fungi have a different pathway, forming quinones before ring cleavage.

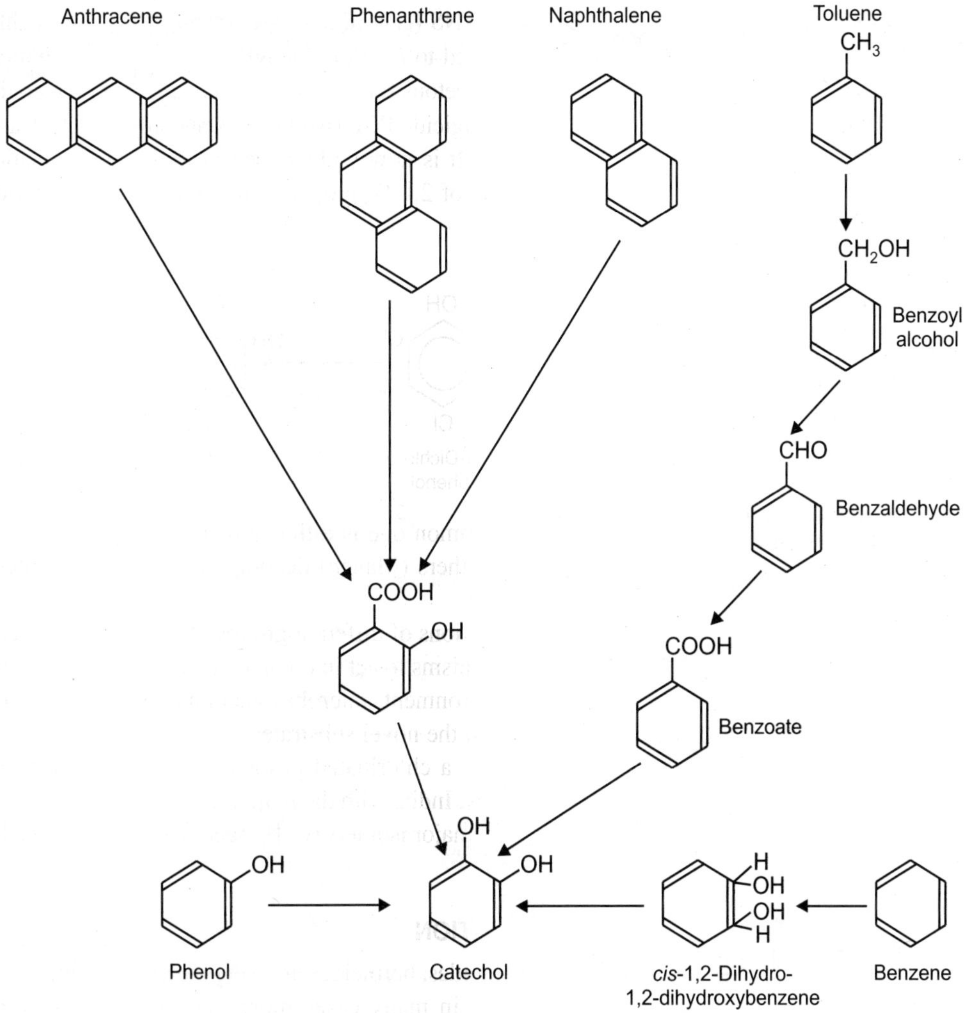

Fig. 15.7. The breakdown of toluene and PAHs. The compounds are degraded to catechol.

Halogenated xenobiotics are the main ingredients of herbicides and pesticides. Organisms capable of degrading these xenobiotics have been found in soil and sediment, particularly from contaminated sites, and include bacteria, fungi, and algae. The degradation pathways for some of the chloroaromatics have been determined under both aerobic and anaerobic conditions. The chloroaromatics are normally cleaved by monooxygenases and dioxygenases similar to those found with the degradation of PAHs. Superimposed on the degradation is the process of dehalogenation which can be by one of four mechanisms.

1. Oxidative dehalogenation; here the halogen is removed and replaced by two hydroxyl ions.
2. Eliminative dehalogenation; the simultaneous removal of the halogen and an adjacent hydrogen ion.
3. Hydrolytic (substitutive) dehalogenation; the substitution of the halogen with a hydroxyl ion.
4. Reductive dehalogenation; the halogen is replaced by a hydrogen ion.

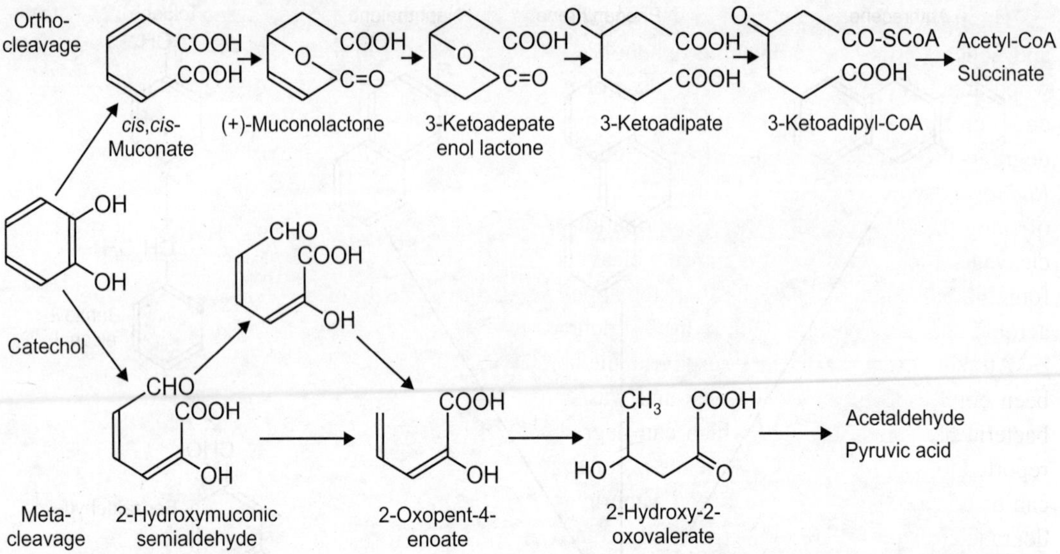

Fig. 15.8. The pathway for the degradation of catechol showing both *ortho-* and *meta-*cleavage pathways.

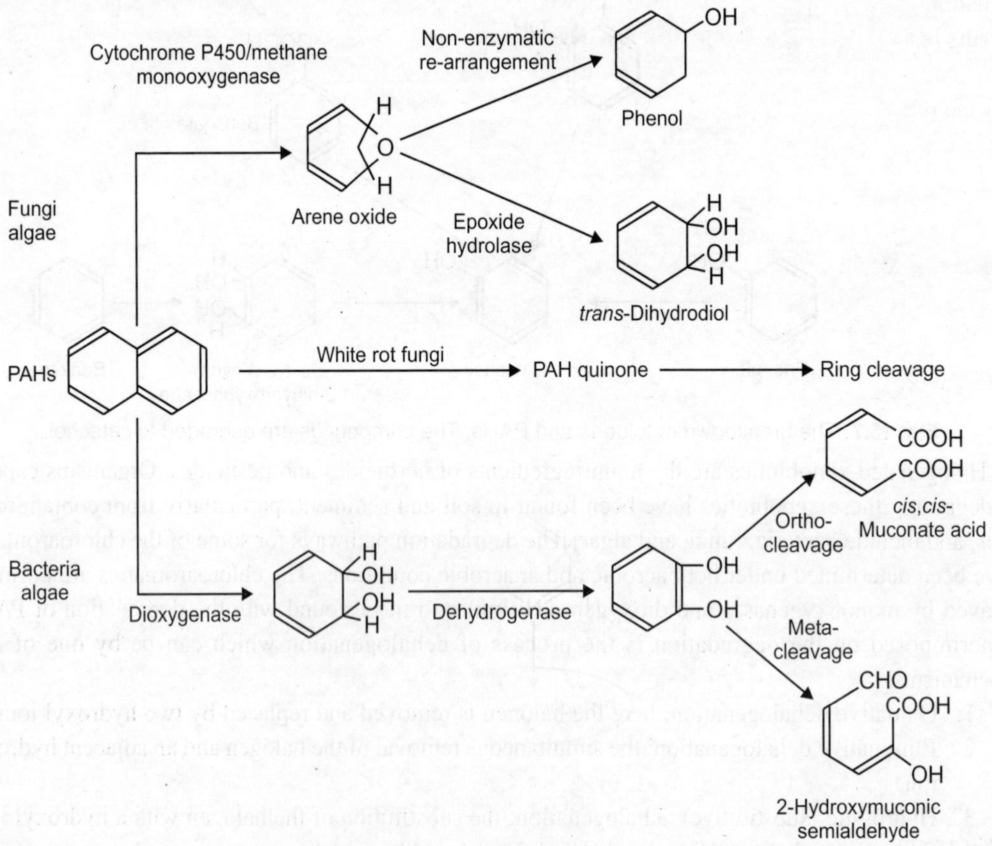

Fig. 15.9. The initial steps in the degradation of polycyclic aromatic hydrocarbons by fungi, bacteria, and algae.

The following are examples of the degradation of organochlorines illustrating individual pathways and general processes. Pentachlorophenol is a herbicide and fungicide used for the preservation of wood and is a priority pollutant. Because of its toxicity its manufacture has all but ceased in Europe, although treated wood is still imported. A number of micro-organisms have been isolated which can degrade PCP under aerobic and anaerobic conditions and include *Flavobacterium*, *Arthrobacter*, *Rhodococcus*, and the white rot fungus *Phanerochaete chryososporium*. Most pathways for the breakdown of chlorophenols consist of the dechlorination and hydroxylation of the aromatic ring followed by ring cleavage. Both hydroxylation and ring cleavage are catalysed by oxygenases similar in nature to those found in PAH metabolism. The first step appears to be the rate-limiting step in PCP degradation where aerobic degradation starts with reductive dehalogenation.

Atrazine is the most widely used triazine herbicide and is effective against broad-leaf weeds. It had been employed for some 40 years and was considered to be recalcitrant. However, pure cultures of bacteria have been isolated which can degrade atrazine although bacterial consortia have also been reported to degrade atrazine. Atrazine is converted to cyanuric acid in three steps and the cyanuric acid can be converted to CO_2 and NH_3. Cyanuric acid can also be metabolised by soil bacteria that cannot degrade atrazine. Three genes are involved in coding for the enzymes that convert atrazine to cyanuric acid and are located on a large plasmid. In *Pseudomonas* spp. these genes are found on a plasmid. The isolation of these three enzymes has allowed determination of the form of metabolism sharing that occurs in a bacterial consortium that can degrade atrazine. Figure 15.10 shows the contribution made by *Clavibacter* and *Pseudomonas* where *Clavibacter* are responsible for the first two steps and *Pseudomonas* for the remaining steps.

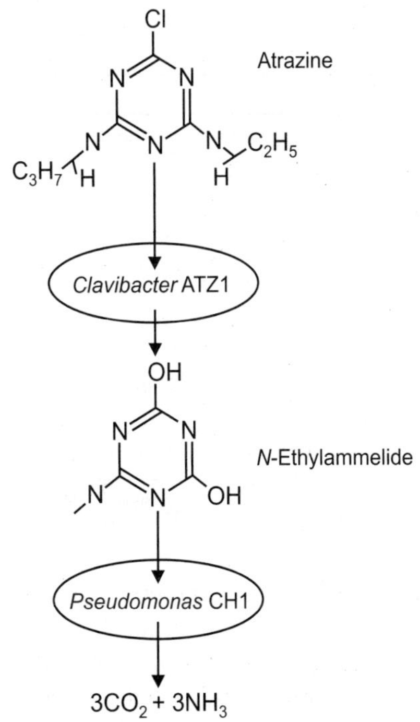

Fig. 15.10. The degradation of atrazine by a consortium of bacteria, *Clavibacter* ATZ1 and *Pseudomonas* CH1.

The bioremediation of hydrocarbon-contaminated soil cannot always be maintained under aerobic conditions due to waterlogging, the fine particle structure of the soil, and blocking of the soil pores with the biomass itself. However, aliphatic, monocyclic, and polycyclic aromatic hydrocarbons can be degraded anaerobically provided oxygen can be obtained from water under methanogenic conditions, from nitrate under nitrifying conditions, and sulphate under sulphur-reducing conditions. The hydrocarbons are converted to central metabolic intermediates by hydration, dehydration, reductive dehydroxylation, nitroreduction, and carboxylation. The central intermediates are benzoyl-CoA and sometimes resorcinol, which are reduced and hydrolysed and finally transformed to compounds which can enter the Krebs cycle. The only disadvantage with anaerobic degradation is that the process is much slower than the aerobic pathway.

MICROBIAL POLYMERS

Many micro-organisms are known to produce extracellular polysaccharides as part of a capsule or slime layer, or free in the medium. The production of a capsule and slime layer is related to resistance to drying and pathogens, and attachment to surfaces and formation of biofilms. Polysaccharides are used extensively in industry as adhesives, gums, thickeners, gelling agents, stabilisers, and binding agents. The polysaccharides used are extracted from plants and large marine algae and include starch, alginate, carrageenan, and agar. The only microbial polysaccharide currently produced on a large scale is xanthan but other microbial products have the potential to replace the various polymers.

Research into microbial polysaccharides started in the 1940s with the development of dextran as a blood plasma extender. Several polysaccharides are now used widely in a number of industries, of which the best known is xanthan (Table 15.4).

Table 15.4. Microbial polysaccharides.

Product	*Micro-organism*	*Molecular weight (Da)*
Curdlan	*Agrobacterium* spp. *Rhizobium* spp.	7.4×10^4
Dextran	*Leuconostoc mesenteroides*	$(4–5) \times 10^7$
Gellan	*Spingomonas paucimoblis*	5×10^5
Pullulan	*Aureobasidium pullulans*	$1 \times 10^3–3 \times 10^6$
Schleroglucan	*Sclerotium rolfsii, Sclerotium glucanium*	$1.3 \times 10^5–6 \times 10^6$
Xanthan	*Xanthomonas campestris*	$2 \times 10^6–1.5 \times 10^7$

Xanthan is a polymer produced by *Xanthomonas campestris* and consists of an alternating glucose backbone carrying side chains of D-mannose and D-glucuronic acid. *X. campestris* is a Gram-negative rod, a plant pathogen causing black rot in brassicas. Solutions of xanthan are pseudoplastic, being able to regain their viscosity after shearing. The pseudoplastic property is of particular value in drilling muds that act as a seal and lubricant during the drilling of oil wells. The mud needs to be viscous to form a good seal but flow when the drill bit is rotated.

Curdlan is a neutral gel-forming 1,3-β-D-glucan (74000 Da) which can be formed by a number of bacteria such as *Agrobacterium* and *Rhizobium*. It forms a gel, which is elastic and does not melt when heated. The property of forming a gel upon acidification has also suggested its use in oil wells. Scheroglucan is a glucose homopolymer produced by a number of micro-organisms including *Sclerotium glucanium*. Scleroglucan is produced commercially and has been developed as an alternative to xanthan for enhanced oil recovery (EOR) by Elf Aquitane. However, the yields are lower than of xanthan, the process takes longer with a lower final concentration. Dextrans are different from the other microbial polysaccharides in that they are produced outside the cell. The substrate sucrose is converted by a extracellular enzyme dextransucrase into the branched α-D-glucan dextran.

The processes for the production of microbial polymers are similar to those for the production of antibiotics apart from the viscosity of the culture, which means that the impeller design has to change. In large stirred-tank bioreactors viscous pseudoplastic materials mean that the culture thins when sheared and thus the viscosity will be lowest near the impeller. This leads to the division of the bioreactor into well-mixed and aerated areas near the impeller and the remainder being poorly aerated and mixed. Poor mixing and low aeration will reduce growth and yield. A number of modifications to the impeller in the normal stirred-tank bioreactor have been proposed in a number of cases and one, the helical ribbon-screw design, did give a significant improvement. Microbial polysaccharides have not replaced plant or algal polysaccharides but have established a market of their own, as they possess a number of advantages over conventional rivals.

MICROBIAL PLASTICS

Plastics are used for packaging and this is dominated by expanded polystyrene made from naphtha, a fraction of non-renewable crude oil. Although plastics are being used increasingly because of their durability, ease of moulding, and resistance to biodegradation it is this last property that is causing concern. Plastic wastes in rivers, in lakes, and on land do not degrade and threaten the environment. Much of the plastic waste that is collected is disposed of in landfill sites which is another problem due to its high volume/weight ratio and resistance to degradation. The possibilities of recycling plastics are limited and incineration yields toxic compounds.

Biodegradable plastic would reduce pressure on landfill sites and littering and also contribute to a more sustainable society as they would be produced from renewable resources. Degradable plastics can be biodegradable or photodegradable. Photodegradable plastics will break down into smaller fragments and therefore lose their structure, but the smaller fragments are not normally degradable. In contrast biodegradable plastic will be metabolised by micro-organisms. A compromise is semidegradable plastics which contain starch, cellulose, and polyethylene, but to achieve complete degradation a 50 per cent mix is required, which compromises the structural properties of the plastic.

However, development is under way of a number of biodegradable plastics including polyalkanoates (PHAs), polyactides, aliphatic polyesters, polysaccharides, and blends of these. One of the most promising is the PHAs, one of the best known of which is polyhydroxybutyrate (PHB). PHB is an intracellular microbial plastic, which is produced by a number of bacteria and was first discovered in 1926 as a component of *Bacillus megaterium*. There are over 80 different types of PHA but they are mainly formed from 3-hydroxyalkanoate acid monomers of 3–14 carbons in length and the polymer has a molecular weight of between 2×10^5 and 3×10^6 Da, containing 100–3000 monomers depending on the growth conditions and micro-organism. There are over 90 genera of bacteria (300 species) that have been shown to produce PHA (Table 15.5) and it appears to be produced as an energy store under conditions of limited nutrients. When the limitation is lifted the PHA is broken down. PHA appears in the microbial cells as granules of 0.2–0.5 μm in diameter, which are refractile.

The commercial value of a biodegradable plastic was clear and ICI used *Alcaligenes eutrophus* (now known as *Ralstonia eutropha*) to produce PHB (poly(3-hydroxybuyrate)). However, the application of PHB was limited as the polymer had a low thermal stability and was brittle. Later a copolymer, P(3HB-co-3HV) or poly(3-hydroxybutyrate-co-3-hydroxyvalerate), was produced by adding propionate to the culture and the polymer was more flexible and tougher.

Table 15.5. Production of poly (3-hydroxybutyrate) by various micro-organisms.

Organism	*Carbon source*	*PHB content (%)*	*Productivity (g/l/hr.)*
*Ralstonia eutropha**	Glucose	76	2.42
*Ralstonia eutropha**	CO_2	68	1.55
*Ralstonia eutropha**	Tabioca hydrolyate	58	1.03
Alcaligenes latus	Sucrose	50	3.97
Azotobacter vinelandii	Glucose	80	0.68
Azotobacter chroococcum	Starch	74	0.01
Haloferax mediterrenei	Starch	60	–
Methylobacterium organophilum	Methanol	52	1.86
Methylobacterium sp. ZP24	Whey	60	0.12
Protomonas extorquens	Methanol	64	0.88
Pseudomonas cepacia	Lactose	56	0.02
Pseudomonas cepacia	Xylose	60	0.03
Recombinant *E. coli*	Glucose	80	2.08
Recombinant *Klebsiella aerogenes*	Molasses	65	0.75

*New name for *Alcaligenes eutrophus*.

The biosynthesis of PHB is the best-studied pathway although there have been four different pathways found to date. PHB in *R. eutropha* is formed by three enzymatic reactions, starting from acetyl-CoA that can be supplied by the Krebs cycle (Fig. 15.11). In the first stage two molecules of acetyl CoA are condensed to acetoacetyl-CoA by the enzyme 3-ketothiolase. In the second stage acetoacetyl-CoA is reduced by an NADPH-dependent acetoacetyl-CoA reductase to 3-hydroxybutryl-CoA. This is then polymerised by the enzyme PHB polymerase. The other pathway differs only in minor ways.

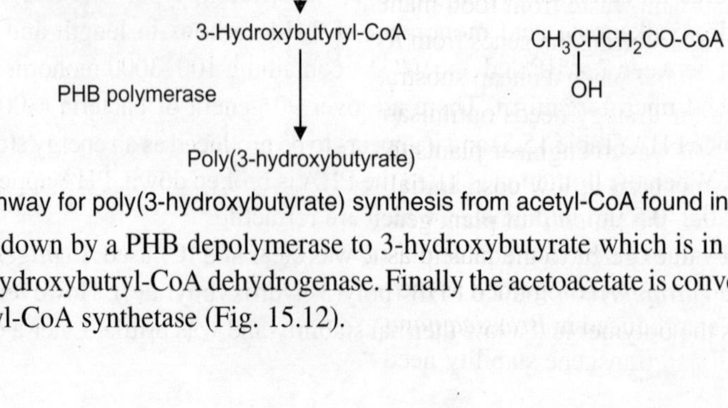

Fig. 15.11. The pathway for poly(3-hydroxybutyrate) synthesis from acetyl-CoA found in *Ralstonia eutropha*.

PHB is broken down by a PHB depolymerase to 3-hydroxybutyrate which is in turn converted into acetoacetate by 3-hydroxybutryl-CoA dehydrogenase. Finally the acetoacetate is converted to acetoacetyl-CoA by acetoacetyl-CoA synthetase (Fig. 15.12).

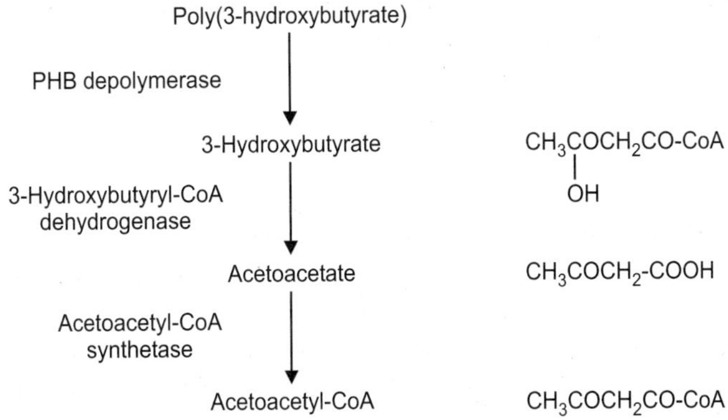

Fig. 15.12. The degradation pathway for poly(3-hydroxyburyrate).

A large number of bacteria and fungi have been shown to degrade PHB and a 1-mm section of P(3HB-co-3HV) was completely degraded in 6, 75, and 351 weeks in anaerobic sewage, soil, and sea water respectively.

In most bacteria PHA is synthesised and accumulated inside the cells under conditions of growth where some nutrient is limiting, for example nitrogen, phosphate, oxygen, and magnesium. Therefore, to maximise PHA production the culture strategy has to provide these conditions while having as high a cell density as possible.

Growth Strategy

The bacteria that can accumulate PHA can be divided into two groups; those that require some form of nutrient depletion to trigger accumulation and others that can accumulate PHA during growth. The first group is represented by *R. eutropha*, *Protomonas extorquens*, and *Pseudomonas oleovorans* and the second by *Alcaligenes latus* and *Azotobacter vinelandii*. In the cultivation of *R. eutropha* a glucose-salts medium is used containing a limited amount of phosphate which runs out after 60 hrs and the eventual PHA concentration reaches 45–80 per cent dry weight. If propionic acid is fed during the accumulation phase P(3HB-co-3HV) is produced. In this group a fed-batch process can also be used where the cells are grown to a high cell concentration in the first stage and in the second stage nutrient limitation is applied. An example of the second group, *Al. latus*, can be grown on glucose in fed-batch and continuous cultures.

Despite the high yields of PHA the cost of the product still requires reducing and a number of strategies have been adopted to reduce the cost. One of the main costs of production is the substrate: a number of carbon sources have been proposed including sugars, oils, alcohols, acids, starch, whey, organic waste-water, and waste from food manufacture. Genetic manipulation of *E. coli* has been used to produce PHA by transferring the genes from *R. eutropha*. Accumulation of PHA reached 80–90 per cent of the cell dry weight but when a cheap substrate, molasses, was used the yield was 45 per cent and so clearly the cultivation strategy needs optimisation.

Another approach was to engineer plants to produce PHA, as crop plants are capable of producing large quantities of material at low cost. Initial research used the genes from *R. eutropha* in the plant *Arabidopsis thaliana*, the '*E. coli*' of plant genetic engineering. In the second generation of engineered plants the genes were targeted to the plastid as the levels of acetyl-CoA were much higher than in the cytoplasm. Oilseed crops were regarded as the best since both oil and PHA are synthesised from acetyl-CoA. PHA has been produced in *Brassica napus* cotton, and maize but problems of phenotype changes, low product yield, and transgene stability need to be solved before the production is economical. The problem of the growth of transgenic plants containing an antibiotic-selection gene has been solved by the development of a non-antibiotic selection method.

EXPLOSIVES

The manufacture, use, and disposal of explosives from military operations have resulted in extensive contamination of soils and groundwater. Figure 15.13 illustrates the most commonly used military explosives: TNT, RDX, and HMX. Treatment of waste-waters from the manufacture of explosives is a long-standing challenge. Interest in decommissioning military bases for other uses has led to an intensified need to remediate explosives-contaminated environments. While incineration is one of the options commonly considered for soils, it is impossible for waste-waters and expensive in general. Research on the potential for bioremediation is an ongoing effort that is yielding some positive results.

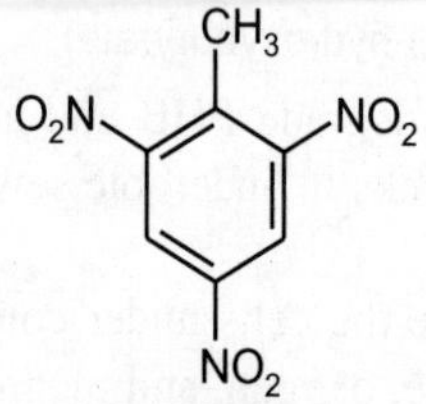

TNT
(2,4,6-Trinitrotoluene)

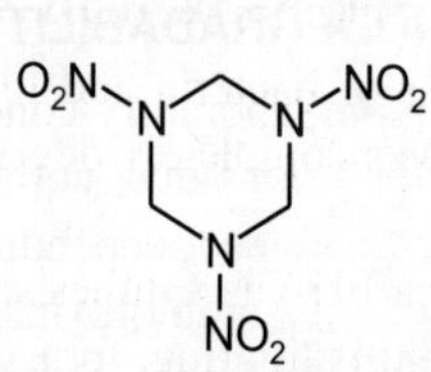

RDX
(Hexahydro-1,3,5-trinitro-1,3,5-triazine)

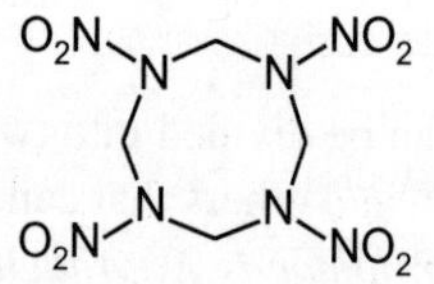

HMX
(Octahydro-1,3,5,7-tetra-nitro-1,3,5,7-tetrazocine)

Fig. 15.13. Common explosives of environmental concern.

A characteristic of explosives is the presence of nitro ($-NO_2$) groups on the molecule (Fig 15.13). Transformation of TNT generally results in the reduction of a nitro group to form an amino group. Thus, commonly found in TNT-contaminated soils are intermediate products such as 4-amino-2,6-dinitrotoluene (4ADNT) and 2-amino-4,6-dinitrotoluene (2ADNT). Less frequently found are the further reduction products, 2,4-diamino-6-nitrotoluene (2,4DANT), and 2,6-diamino-4-nitrotoluene (2,6DANT). The further reduction product 2,4,6-triaminotoluene (TAT) is found in the laboratory, but has not been reported in the environment. The reductions can be enzymatically catalysed or, under proper reducing conditions, abiotic.

The amino transformation products are subject to further interactions with each other and with components of soils, forming stable complexes through covalent bonding. Resulting products are large and insoluble. Thus, disappearance of TNT is often observed in biologically active environments, but mineralisation does not commonly occur. Whether or not the products formed from transformation and interactions are harmful to humans or the environment is not known. Slow mineralisation of TNT and its transformation products does appear to occur in soil in some instances.

While TNT biodegradation is problematic, toluene, benzene, and phenols with one or two nitro groups on the molecule are more subject to biodegradation and mineralisation. Initial monooxygenation or dioxygenation reactions can lead to the release of $-NO_2$ groups and substitution of –OH. This can lead to mineralisation. In some cases, the nitro group is reduced to an amino group, which accumulate as dead-end products.

RDX is more readily mineralised than TNT and is not subject to partial transformation and sequestration processes. However, the rate at which RDX disappears usually is slower than for TNT. HMX biotransformation occurs primarily under anaerobic conditions, leading to the formation of mono- and di-nitroso intermediates. Its biotransformation is slower than that of RDX. The explosives are also transformed, at least partially, by plants, leading to some interest in the potential for phytoremediation. Because of the complexity of the transformation processes involved, the unknown pathways for transformation, and the unknown environmental and health hazards of degradation products, much must be learned about the potential effectiveness of bioremediation for explosives.

TESTING FOR BIODEGRADABILITY

It is usually done in the laboratory conditions, as it is often difficult to conduct the same in the field, though laboratory conditions may not always simulate the field situations. However, a variety of test methods have been developed, depending on the purpose.

Biodegradability of a xenobiotic may depend on or be influenced by a host of factors.

1. Chemical structure of the compound and various substitutions,
2. Environmental factors such as the presence or absence of oxygen, pH and temperature of the medium.
3. Whether the substrate can provide carbon and energy for the metabolism and growth of the degrading organism.

So a successful bioremediation of a novel substrate necessitates considerations and attempts from various angles. In many cases, time and efforts may even go waste. So a preliminary idea about the biodegradable potential of the chemical concerned may be obtained through some simple techniques and procedures.

1. If structural features suggests its recalcitrance with respect to a particular microbe, it is advisable to check the need of synergistic effects of microbial community.
2. By comparing the disappearance of the compound from a biologically active or poisoned (with 1 per cent $HgC1_2$) growth medium.
3. Decreased UV absorption at 280 nm may help monitoring the degradation of aromatic compounds.
4. For specific chemical monitoring of certain compounds like aromatic amines and phenols, diazotisation reaction or Folin-Cricalteau reaction may be followed.
5. In certain cases, radiolabelling with C^{14} may indicate the complete mineralisation to $^{14}CO_2$ as an indication of biodegradability.
6. Use of indicator reagents in the growth media may also help detecting (i) conversion of annilines to azobenzene by spraying *p*-anisidine, whereby red-brown discolouration of microbial colonies may occur, and (ii) for dechlorination-acid-base indicators are used such as eosin-methylene blue or bromo-cresol purple. The change of colour indicates the release of chlorine.

Most non-specific analytical methods monitor one of the three parameters. These are O_2 uptake in an oxidation test (BOD), CO_2 production from complete mineralisation of organics and loss of dissolved organic carbon (DOC). The simplest is the determination of BOD_5 and its comparison with COD. BOD_5 determines the amount of O_2 required for microbial decomposition in a 5-day test at 20°C, while

COD indicates the amount of O_2 necessary for chemical oxidation. If BOD/COD is more than 0.6, the organic substance is easily biodegradable; if it is between 0.3 to 0.6, then it points to the possibility of biodegradation, and when the ratio is less than 0.3 it is not bioamenable. Non-specific methods are, however, less sensitive than the chemical specific methods mentioned above.

Test guidelines, primarily for screening, have been suggested by OECD (Organisation for Economic Cooperation and Development) and are generally adopted by EPAs (Environmental Protection Agencies) in different countries.

It is clear that the use of bioremediation for the removal of pollutants from contaminated sites or from process wastes will continue to increase. Increasing legislation will help with the use of bioremediation as a viable alternative to chemical treatment. There are biological processes that can be used for all forms of contamination including soil, water, and gaseous. The processes offer *in situ* treatment or after-excavation treatment on or off site. Whatever process is used the main aim is to encourage the aerobic growth of the indigenous micro-organisms by providing air, nutrients, and the correct pH. The addition of specific micro-organisms tailored for the contaminant in a process known as bioaugmentation has given mixed results. One positive aspect of bioaugmentation is that it probably supplies the plasmids containing the relevant genes, which can be spread through the population even if the culture itself does not survive. Micro-organisms are not the only organism used for remediation. Plants are being used to degrade xenobiotics, sequester metals, and treat waste-waters. Plants offer a slow but cheap method of remediation that is well regarded by the public.

Chapter 16

Single Cell Protein

INTRODUCTION

Food chains describe the eating relationships between species within an ecosystem or a particular living place. Many types of food chains or webs are applicable depending on habitat or environmental factors. Every known food chain begins with a type of autotroph, whether it be a plant or some kind of unicellular organism.

Energy enters the food chain from the sun. Some energy and/or biomass is lost at each stage of the food chain as; faeces (solid waste), movement energy and heat energy (especially by warm-blooded creatures). Therefore, only a small amount of energy and biomass is incorporated into the consumer's body and transferred to the next feeding level, thus showing a pyramid of biomass.

Primary producers, commonly forming autotrophs, produce complex organic substances (essentially food) from an energy source and materials. These organisms are typically photosynthetic plants, which use sunlight as their energy source. A few, such as those organisms forming the base of deep-sea vent food webs, are chemotrophic, using chemical energy instead. Organisms that get their energy by organic substances are called heterotrophs. Heterotrophs include herbivores, which obtain their energy by consuming live plants; carnivores, which obtain energy from eating live animals. Ultimately detritivores, scavengers and decomposers may predate living or consume dead biomass.

A food chain is the flow of energy from one organism to the next and to the next and so on. Organisms in a food chain are grouped into trophic levels, based on how many links they are removed from the primary producers. Trophic levels may contain either a single species or a group of species that are presumed to share both predators and prey, and usually start with a plant and end with a carnivore.

It is often the case that the biomass of each trophic level decreases from the base of the chain to the top. This is because energy is lost to the environment with each transfer. On average, only 10 per cent of the organism's energy is passed on to its predator. The other 90 per cent is used for the organism's life processes or is lost as heat to the environment. Graphic representations of the biomass or productivity at each tropic level are called trophic pyramids.

Some producers, especially phytoplankton, are so productive and have such a high turnover rate that they can actually support a larger biomass of grazers. This is called an inverted pyramid, and can occur when consumers live longer and grow more slowly than the organisms they consume.

A pyramid of numbers shows the number of consumers at each level drops significantly, so that a single top consumer (e.g. a polar bear) will be supported by literally millions of separate producers (e.g. phytoplankton).

Food web: Food chains are overly simplistic as representatives of what typically happens in nature. The food chain shows only one pathway of energy and material transfer. Most consumers feed on multiple species and are, in turn, fed upon by multiple other species. The relations of detritivores and parasites are seldom adequately characterised in such chains as well.

A food web is a set of interconnected food chains by which energy and materials circulate within an ecosystem. The food web is divided into two broad categories: the grazing web, which typically begins with green plants, algae, or photosynthesising plankton, and the detrital web, which begins with organic debris. These webs are made up of individual food chains. In a grazing web, materials typically pass from plants to plant eaters to flesh eaters. In a detrital web, materials pass from plant and animal matter to bacteria and fungi (decomposers), then to detritivores, and then to their predators (carnivores) as shown in Fig. 16.1.

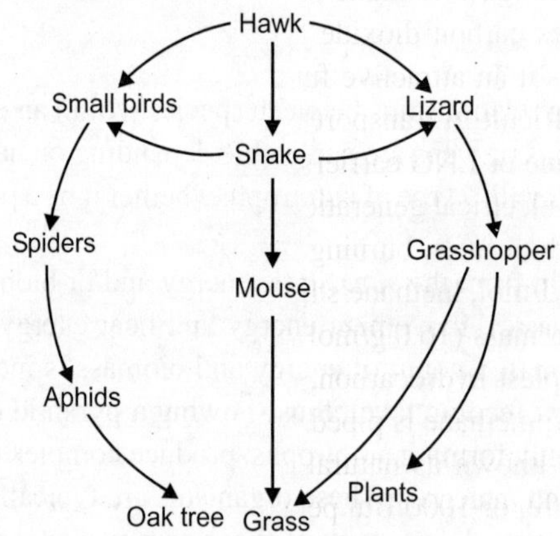

Fig. 16.1. Food web.

Generally, many interconnections exist within food webs. For example, the fungi that decompose matter in a detrital web may sprout mushrooms that are consumed by squirrels, mice, and deer in a grazing web. Robins are omnivores (consumers of both plants and animals) and thus are in both detrital and grazing webs. Robins typically feed on earthworms, which are detritivores that feed upon decaying leaves.

Herbivores belong to the second trophic level. Carnivores, predators feeding upon the herbivores, belong to the third. Omnivores belong to both the second and third. Secondary carnivores, which are predators that feed on other predators, belong to the fourth trophic level. As the trophic levels rise, the predators become fewer, larger, fiercer and more agile. At the second and higher levels, decomposers of the available materials function as herbivores or carnivores depending on whether their food is plant or animal material.

The feeding of one organism upon another in a sequence of food transfers is known as a food chain. Another definition is the chain of energy transfer from one organism to another.

Energy flow: Through these series of steps of consuming and being consumed, energy flows from one trophic level to another. Green plants, or other photosynthesising organisms, use light energy from the sun to manufacture carbohydrates for their own needs. Most of this chemical energy is processed in metabolism and dissipated as heat in respiration. Plants convert the remaining energy to biomass, both above ground as woody and herbaceous tissue and below ground as roots. Ultimately, this material, which is stored energy, is transferred to the second trophic level, which comprises grazing herbivores, decomposers, and detrital feeders. Most of the energy assimilated at the second trophic level is again lost as heat in respiration; a fraction becomes new biomass. Organisms in each trophic level pass on as biomass much less energy than they receive. Thus, the more steps between producer and final consumer, the less energy remains available. Seldom are there more than four links, or five levels, in a food web. Eventually, all energy flowing through the trophic levels is dissipated as heat. The process whereby energy loses its capacity to do work is called entropy. The biomass of one plant will affect another and therefore the cycle will continue.

METHANE

Methane is a chemical compound with the chemical formula CH_4. It is the simplest alkane, and the principal component of natural gas. Methane's bond angles are 109.5 degrees. Burning methane in the presence of oxygen produces carbon dioxide and water. The relative abundance of methane and its clean burning process makes it an attractive fuel. However, because it is a gas at normal temperature and pressure, methane is difficult to transport from its source. In its natural gas form, it is generally transported in bulk by pipeline or LNG carriers; few countries transport it by truck.

Methane is important for electrical generation by burning it as a fuel in a gas turbine or steam boiler. Compared to other hydrocarbon fuels, burning methane produces less carbon dioxide for each unit of heat released. At about 891 kJ/mol, methane's heat of combustion is lower than any other hydrocarbon; but a ratio with the molecular mass (16.0 g/mol) divided by the heat of combustion (891 kJ/mol) shows that methane, being the simplest hydrocarbon, produces more heat per mass unit than other complex hydrocarbons. In many cities, methane is piped into homes for domestic heating and cooking purposes. In this context it is usually known as natural gas, and is considered to have an energy content of 39 megajoules per cubic metre, or 1000 Btu per standard cubic foot.

Methane in the form of compressed natural gas is used as a vehicle fuel, and is claimed to be more environmentally friendly than other fossil fuels such as gasoline/petrol and diesel.

Methane is used in industrial chemical processes and may be transported as a refrigerated liquid (liquefied natural gas, or LNG). While leaks from a refrigerated liquid container are initially heavier than air due to the increased density of the cold gas, the gas at ambient temperature is lighter than air. Gas pipelines distribute large amounts of natural gas, of which methane is the principal component.

Sources of Methane

Natural gas fields

The major source of methane is extraction from geological deposits known as natural gas fields. It is associated with other hydrocarbon fuels and sometimes accompanied by helium and nitrogen. The gas at shallow levels (low pressure) is formed by anaerobic decay of organic matter and reworked methane from deep under the earth's surface. In general, sediments buried deeper and at higher temperatures than those which give oil generate natural gas. Methane is also produced in considerable quantities from the decaying organic wastes of solid waste landfills.

Alternative sources

Apart from gas fields, an alternative method of obtaining methane is via biogas generated by the fermentation of organic matter including manure, waste-water sludge, municipal solid waste (including landfills), or any other biodegradable feedstock, under anaerobic conditions. Methane hydrates/clathrates (icelike combinations of methane and water on the sea floor, found in vast quantities) are a potential future source of methane. Cattle belch methane accounts for 16 per cent of the world's annual methane emissions to the atmosphere. The livestock sector in general (primarily cattle, chickens, and pigs) produces 37 per cent of all human-induced methane. However, animals 'that put their energies into making gas are less efficient at producing milk and meat'. Early research has found a number of medical treatments and dietary adjustments that help limit the production of methane in ruminants.

SINGLE CELL PROTEIN

Single cell protein (SCP) typically refers to sources of mixed protein extracted from pure or mixed cultures of algae, yeasts, fungi or bacteria (grown on agricultural wastes) used as a substitute for protein-rich foods, in human and animal feeds.

Single cell proteins develop when microbes ferment waste materials (including wood, straw, cannery and food processing wastes, residues from alcohol production, hydrocarbons, or human and animal excreta). The problem with extracting single cell proteins from the wastes is the dilution and cost. They are found in very low concentrations, usually less than 5 per cent. Engineers have developed ways to increase the concentrations including centrifugation, flotation, precipitation, coagulation and filtration, or the use of semi-permeable membranes.

The single cell protein needs to be dehydrated to approximately 10 per cent moisture content and/or acidified to aid in storage and prevent spoilage. The methods to increase the concentrations to adequate levels, and de-watering process require equipment that is expensive and not always suitable for small-scale operations. It is economically prudent to feed the product locally and shortly after it is produced.

Examples: Microbes employed include yeasts [*Saccharomyces cerevisiae*, *Candida utilis* = *Torulopsis* and *Geotrichum candidum* (=*Oidium lactis*)], other fungi (*Aspergillus oryzae*, *Sclerotium rolfsii*, *Polyporus* and *Trichoderma*), bacteria (*Rhodopseudomonas capsulata*), and algae (*Chorella* and *Spirulina*). Typical yields of 43 to 56 per cent, with protein contents of 44 to 60 per cent.

The fungus *Scytalidium acidophilum* grows at below pH 1, offering advantages of (i) low-cost aseptic conditions, (ii) avoiding over 100-fold dilution of the acidic hydrolysates to pH values needed for other microbes, and (iii) after the biomass is harvested, the acids can be reused. Commercial production of SCP (*Spirulina*) includes Cyanotech in Hawaii and Earthrise in California.

Micro-organisms

Algae, fungi (filamentous), yeast and bacteria are used for SCP production (Table 16.1). The micro-organisms used for SCP production must be: (i) nonpathogenic to plants, animals and man, (ii) of good nutritional value, (iii) easily and cheaply produced on scale, (iv) toxin-free, (v) fast-growing, and (vi) easy to separate from medium and to dry, etc. The salient features of different micro-organisms are summarised in Table 16.2.

Algae

Members of the genera *Chlorella*, *Scenedesmus* and *Spirulina* are generally grown in ponds/tanks. They use CO_2 and sunlight as substrate which are without any cost. Generally, the limiting factor in their large scale production is illumination. Algal SCP has about 60 per cent crude protein, which is generally good in amino acid composition except for some deficiency in sulphur-containing amino acids (Table 16.2).

They are suitable for animal feed as protein-rich supplement. The disadvantages of algal SCP are: (i) rich chlorophyll content which is not suitable for human use, (ii) serious problems when *Chlorella* and *Scenedesmus* are used in human diet (*Spirulina* is more suited for human use), (iii) low cell density, e.g. 1–2 g dry weight/l, (iv) serious risk of contamination, and (v) costly recovery methods for unicellular algae (*Spirulina* harvested by filtration or simply by skimming).

Table 16.1. Some important micro-organisms used for SCP production.

Micro-organism	*Substrate*	*Used as*	*Countries using commercially*
Algae			
Chlorella sp.	CO_2 + sunlight	Feed	Japan and Taiwan
Scenedesmus acutus	CO_2 + sunlight	–	–
Spirulina maxima	CO_2 + sunlight	Feed	Mexico
Yeasts			
Candida utilis (Torula yeast)	Confectionery effluents	–	(UK
	Ethanol	Food	USA
	Sulphite liquor	–	Europe, USA, Russia
C. intermedia	Whey	–	Vienna
C. krusei (+ *Lactobacillus bulgaricus*)	Whey	–	–
C. lipolytica	*n*-Alkanes (C_{10}–C_{23}) + ammonia	–	Russia
Kluyveromyces fragilis	Whey	Food	France
Saccharomyces cerevisiae (Baker's yeast)	Molasses	(Food)*	–
Fungi			
Chaetomium cellulolyticum	Cellulosic wastes	–	–
Fusarium graminearum	Starch hydrolysate	Food	UK
Paecilomyces varioti	Sulphite liquor	Feed	Finland
Bacteria			
Brevibacterium sp.	C_1–C_4 hydrocarbons	–	–
Methylophilus methylotrophus	Methanol	Feed	UK

*Mainly used for fermentation of dough in bakeries; thus eaten indirectly as components of food.

Filamentous fungi

Filamentous fungi have been used to produce SCP mainly from polysaccharide hydrolysates, e.g. starch hydrolysates, sulphite liquor from wood pulp industries etc. These are usually grown as submerged cultures in which they grow as yeast-like cells, in filamentous form or in pellets. They have crude protein content of 50–55 per cent; the protein is low in S containing amino acids, but otherwise is excellent in amino acid composition (Table 16.2). The recovery of filamentous and pellet-forms is rather easy by filtration.

Table 16.2. Some important features of different micro-organisms and the SCP produced from them.

Feature	*Algae*	*Bacteria*	*Yeast*	*Filamentous fungi*
Growth rate	Low	Highest	Quite high	Lower than bacteria and yeast
Substrate	Light, inorganic carbon sources, e.g. CO_2 (preferably)	A wide range of substrates	Most substrates, except, hydrocarbons and CO_2	Limited substrates (mostly starchy and cellulosic materials)
pH range	Upto 11	5–7	5–7	3–8
Cultivated in	Open ponds, tanks; in sunlight	Bioreactors	Bioreactors	Bioreactors
Risk of contamination	Serious	High; precautions necessary	Low	Low if grown below pH 5
Biomass recovery	Difficult and costly with unicellular algae	Problematic; improved methods are needed	Easy; by centrifugation	Easy for filamentous or pellet forms
Protein content (crude)	Upto 60%	80% or more	55–60%	50–55%
Amino acid profile	Generally good; low in S-containing amino acids	Generally good; a small deficit in S-containing amino acids	Generally good; deficit in S-containing amino acids	Low in S-containing amino acids
Nucleic acid content	–	Very high (20% RNA)	High (15% RNA)	High (15% RNA)
Removal of nucleic acids	–	Necessary	Necessary	Necessary
Toxins	–	Geam bacteria may produce endotoxins	–	Many species produce mycotoxins
Other features	Low yield (1–2 g dry wt./l) High chlorophyll content; unsuitable for humans	–	High B-vitamin content	Chitin may contain a significant proportion of N-content which is unavailable

The problems associated with fungi are: (i) slower growth rates than bacteria and yeast, but some microfungi may be comparable to yeast, (ii) contamination by yeast may be frequent if sterility is not maintained, while that by bacteria can be minimised by keeping the pH of broth below 5, (iii) they have high nucleic acid content (upto 15 per cent RNA) which must be reduced, and (iv) the strains have to be thoroughly evaluated for mycotoxin production. However, a substantial part of the N may be present in chitin, a component of the cell wall.

Yeasts

Members of *Saccharomyces*, *Candida* and *Torulopsis* have been widely studied for SCP production and those of the first two genera are used for some commercial processes (Table 16.1) using various substrates.

The SCP has 55–60 per cent crude protein which has good amino acid balance except for a deficiency in S-containing amino acids.

It is usually very rich in B-group of vitamins. The SCP is used both for human food and animal feed supplementation. The risk of bacterial contamination is low and recovery by continuous centrifugation is easy (Table 16.2).

The difficulties in use of yeasts as SCP are: (i) slower growth rates than fastest growing bacteria, (ii) high nucleic acid content (upto 15 per cent) which needs to be reduced, and (iii) methionine supplementation may be done to overcome S-containing amino acid deficiency of its proteins.

Bacteria

A large number of bacterial species have been evaluated for SCP production: (i) using a wide variety of substrates (Table 16.3), (ii) some of these are used for production at commercial scales, e.g. *Methylophilus methylotrophus* using methanol (Table 16.1). SCP has very high crude protein (over 80 per cent) of good amino acid composition, although in some cases a small deficit of sulphur-containing amino acids may be encountered.

Table 16.3. Various substrates used for SCP production using different micro-organisms.

Substrate	*Micro-organisms*	*Status*
Fossil carbon sources		
Liquid hydrocarbons (C_{16}–C_{23} *n*-alkanes + ammonia)	*Candida lipolytica*	Product 'Toprina' released and approved, commercial production only in erstwhile Soviet Union (ca. 2,00,000 tons/yr)
Gaseous hydrocarbons		
Methane (CH_4)	*Preudomonas methanica, Methanomonas methanica, Methylococcus capsulatus, Pseudomonas methanitrificans*	Only upto pilot scale; commercial production presents many technical problems
C_1 to C_4 hydrocarbons	*Brevibacterium* sp.	Process developed
Methanol (produced by chemical conversion of C_4, etc.)	Bacteria: *Methylophilus methylotrophus*	Grows at 35°–40°C; commercial production in UK; marketed as 'Pruteen'; 71% protein
	Other bacteria: *Methylomonas methanolica, Streptomyces* sp., *P. utilis, P. extroquens*	
	Yeast: *Torulopsis glabrata, Candida boidinii*, etc.	
Ethanol (by catalytic addition of water to ethylene)	*Candida utilis* (Torula yeast)	Commercial production in USA; used as human food supplement
Renewable carbon sources		
Carbon dioxide (CO_2)	*Spirulina*	Grows in alkaline lakes (upto pH 11); commercial production in Mexico in 1 m deep ponds; yield, 50 tons/ha/yr
	Chlorella	Commercial production in Japan and Taiwans

(Contd ...)

Substrate	Micro-organisms	Status
Molasses	*Saccharomyces cerevisiae* (baker's yeast)	Over 2,00,000 tons (dry wt) 1 yr; produced as by product of ethanol production; used in dough fermentation
Whey (World production, 74 million tons/yr in 1973)	*Kluyveromyces fragilis*	23,000 tons/yr SCP; used as human food in France for 2 decades
	Lactobacillus bulgaricus + *Candida krusei*	–
	Candida intermedia	–
Starch hydrolysates	*Fusarium graminearum*	Meat like product approved for human use; marketed as 'Mycoprotein' in UK
Industrial effluents		
Confectionery effluents	*Candida utilis*	Commercial production in UK
Sulphite-liquor (from wood pulp mills)	*Paecilomyces varioti*	Commercial production in Finland; used as animal feed
—	*Candida utilis*	Used in USA, Europe and erstwhile Soviet Union
Cellulosic wastes, e.g. straw, bagasse, sawdust (thermally and chemically pretreated)	*Chaetomium cellulolyticum* *Mushrooms*	Semisolid fermentation; rapid degradation of cellulose (in 4 hr or so)

The disadvantages of using bacteria for SCP production are: (i) high nucleic acid, especially, RNA content (ca. 20 per cent) which must be reduced, (ii) sterility must be maintained during the production process since pH of cultures is kept between 5 and 7, (iii) risk of contamination by pathogenic bacteria is considerable, (iv) recovery of cells is problematic, and (v) careful evaluation for endotoxin production is essential particularly when Gram-negative bacteria are used.

Nucleic acids and their removal

Degration of nucleic acids produces uric acid which may accumulate to damaging levels in humans since they do not possess uricase activity. It is therefore, necessary that nucleic acids be reduced to acceptably low levels especially in SCP intended for human use. In the rapidly proliferating microbial cells used for SCP, the bulk of nucleic acids are RNAs. The level of RNA in SCP is reduced using one of the following methods: (i) activation of endogenous RNases by, usually, a brief heat treatment, e.g. 20 mins at 64°C reduces RNA from 10 to 1 per cent of dry weight in *Fusarium graminearum* SCP, (ii) alkaline hydrolysis, (iii) chemical extraction, and (iv) suitable manipulation of growth and physiology of the microbial cells.

Substrates

A variety of substrates, ranging from inorganic carbon, e.g. CO_2 (no cost for substrate), through industrial effluents, e.g. confectionery effluents, whey, etc. (utilisation of the substrate helps effluent disposal), and low cost organic materials, e.g. cellulosic wastes like straw, etc. to high cost materials like starch hydrolysate are used for SCP production (Tables 16.1 and 16.3). These substrates can be divided into two broad groups: (i) fossil carbon sources (non-renewable), and (ii) renewable carbon sources. But substrates like methanol and ethanol can be produced from both renewable and non-renewable carbon sources.

Fossil carbon sources

The various substrates in this group are: (i) gaseous hydrocarbons, (ii) liquid hydrocarbons, (iii) methanol and (iv) ethanol; use of these for SCP production is briefly described below.

Gaseous hydrocarbons

C_1 to C_4 gaseous hydrocarbons have been used for SCP production. Methane (C_1) has been extensively studied, and is highly desirable as it is available in high purity from natural gas which in some parts of the world is burnt at the oil wells. In addition, it is readily removed from the fermentation medium and supports high productivity in continuous processes. Methane is utilised by certain bacteria, e.g. *Pseudomonas methanica*, *Methanomonas methanica*, the thermophilic *Methylococcus capsulatus*, and *Pseudomonas methanitrificans*; the last bacterium utilises methane as well as fixes atmospheric nitrogen.

But use of methane for SCP production presents 3 important problems: (i) substrate limitation either due to methane or oxygen, (ii) heat generation necessitating efficient cooling, and (iii) hazards of explosion when over 12 per cent O_2 (v/v) is used. These and some other associated problems have limited the activities only upto pilot plant scale.

Liquid hydrocarbons

Saturated, straight-chain hydrocarbons, called *n*-alkanes, constitute 0–30 per cent of crude oil. C_5 to C_8 *n*-alkanes are liquid at room temperature, but are generally toxic to cells due to their solvent action. Generally, *n*-alkanes having 9 to 18 carbons (C_9–C_{18}) are used as SCP substrates. These *n*-alkanes are insoluble in water; a micro-emulsion is formed due to the combined actions of agitation in the bioreactor and the surfactants produced by micro-organisms. The hydrocarbons are taken up by cells and oxidised; thus the process requires very high O_2. Many bacteria, actinomycetes, yeast and moulds are able to use liquid hydrocarbons. Gas oil, in combination with ammonia and mineral nutrients, was used in airlift fermenters to produce *Candida*.

Pure *n*-alkanes (>97.5 per cent), having 10 to 23 carbons, was used by British Petroleum to culture *Candida lipolytica* in a continuous process; the medium consisted of *n*-alkanes, minerals and ammonia. The *n*-alkanes were almost completely utilised, biomass recovery was easier and cheaper, and the SCP having about 60 per cent crude protein was marketed as 'Toprina'.

The SCP was comparable to soyabean meal and fishmeal, and was cleared after nutritional and toxicological tests. However, large scale production was discontinued mainly due to legal difficulties and/or public protests. The process was being reportedly used in the erstwhile Soviet Union for industrial SCP production.

Ethanol

Ethanol can be obtained from ethylene by catalytic addition of a water molecule, or from organic substrates by alcoholic fermentation. It offers advantages similar to methanol, and is even more acceptable as a substrate for SCP production for human use. Several bacteria, yeasts and mycelial fungi utilise ethanol. Amco foods, USA produces food grade SCP by growing *Candida utilis*.

Renewable carbon sources

The renewable carbon source include: (i) CO_2, (ii) molasses, (iii) whey, (iv) cellulose hydrolysates, (v) starch hydrolysates, (vi) industrial effluents, and (vii) cellulosic wastes. SCP production using these substrates is briefly described below.

Carbon dioxide (CO_2)

This substrate is utilised by algae which derive the required energy from sunlight. *Spirulina* is grown on a commercial scale by Sosa Texcoco Co., Mexico, producing upto 5 tons dry biomass per day. The blue-green alga is cultivated in 1 m deep ponds and is harvested by filtration, followed by vacuum filtration and is dried using drums. *Chlorella* is commercially grown in Japan and Taiwan to get about 15,000 tons/yr dry biomass from many small production units.

The algae cultivation is more or less like crop cultivation, and offers the following advantages over the latter: (i) year-round cultivation, (ii) more efficient harvesting of sunlight, (iii) optimal utilisation of mineral nutrients, (iv) cultivation in seawater and alkaline lakes, and (v) very high yields, e.g. 50 tons/ha/yr *Spirulina* biomass. The disadvantages are: (i) dependence on climate, including sunlight, (ii) utilises considerable area of land, and (iii) is attacked by animals, weeds, and pathogens.

Molasses

This substrate is used for alcoholic fermentation from which yeast biomass is obtained as a by-product. Traditionally, the baker's yeast (*Saccharomyces cerevisiae*) and Torula yeast (*Candida utilis*) are produced on this substrate. About 200,000 tons dry weight of *S. cerevisiae* is produced each year which is used as inoculum for dough fermentation in bakeries.

Whey

Whey is the liquid portion of milk remaining after the curd is separated during cheese production. On dry weight basis whey contains 70 per cent lactose, 9–14 per cent protein and 9 per cent ash. About 90 million tons of whey is produced each year of which over 50 per cent is released as effluent which necessitates treatments to avoid pollution. Yeasts (*Kluyveromyces fragilis*, *Candida krusei* in combination with *Lactobacillus bulgaricus*, and *Candida intermedia*) are the most widely used for commercial scale production of SCP from whey. For example, Fromageries Le Bel, France, produces about 2300 tons of SCP/yr using *K. fragilis*; the SCP is being used as human food supplement for about two decades.

Cellulose hydrolysate

Cellulase obtained from fungi like *Trichoderma viride* has been used to hydrolyse cellulose and to produce glucose. The main advantage of this approach is the large quantities of cheap substrate. But the chief problems are: (i) difficult and expensive process for complete hydrolysis of lignocellulose, (ii) the amount of pretreatment necessary, and (iii) the extent of sugar destruction and by-product formation during chemical hydrolysis. The best utilisation of cellulosic substrates appears to be for mushroom production and for semisolid fermentation systems.

Starch hydrolysate

This rather expensive substrate is used for SCP production using *Fusarium graminearum*. The SCP has a fibrous, meat-like texture, is marketed as 'Mycoprotein', is approved for human use in UK, and food products having Mycoprotein are being marketed. The fungus grows at 30°C on a variety of mono- and oligosaccharides, is recovered by vacuum filtration, and the recovered biomass is held at 60°C for 20 mins to activate the native RNase (this reduces the RNA content of SCP from 10 per cent to about 1 per cent of dry weight).

Industrial effluents

Effluents from many industries, e.g. breweries, distilleries, confectionery industries, potato and canning industries, sulphite liquor from wood pulp mills, etc. contain large amounts of carbohydrates and other organic compounds which can be used for SCP production. Some of these are already being used for the purpose.

Cellulosic wastes

A new approach to use agricultural and forestry cellulosic wastes, e.g. straw, bagasse, sawdust, etc. far SCP production uses semisolid fermentation. The cellulosic material is pretreated thermally and chemically, and is then fermented with the fungus *Chaetomium cellulolyticum*. The fungus is reported to degrade the cellulosic wastes in very short time, e.g. 4 hrs. This promising approach may facilitate an efficient and economic utilisation of cellulosic wastes far SCP production.

Production of SCP

The following basic steps are involved in the SCP production.

1. Provision of a carbon source; it may need physical and/or chemical pretreatments.
2. Addition, to the carbon source, of sources of nitrogen, phosphorus and other nutrients needed to support optimal growth of the selected micro-organism.
3. Prevention of contamination by maintaining sterile or hygienic conditions. The medium components may be heated or sterilised by filtration and fermentation equipments may be sterilised.
4. The selected micro-organism is inoculated in a pure state.
5. SCP processes are highly aerobic (except those using algae). Therefore, adequate aeration must be provided. In addition, cooling is necessary as considerable heat is generated.
6. The microbial biomass is recovered from the medium.
7. Processing of the biomass far enhancing its usefulness and/or storability.

Biomass production is ordinarily carried out in the continuous mode to maximise yields and economic returns.

Biomass recovery

The general approaches for the recovery of biomass are as follows: (i) bacteria—flocculation and floatation combined with centrifugation, (ii) yeast-centrifugation, (iii) filamentous organisms—filtration. It is important to reduce water as much as possible to reduce drying costs. Sun-drying is cheap but it reduces the quality of SCP. Since recovery processes are not done under sterile conditions, adequate hygiene must be maintained and care should be taken to prevent growth of undesirable micro-organisms. Heat-treatments are used during the final stages of harvesting to inactivate heat-sensitive organisms (including undesirable contaminants) and to reduce RNA content. The cell walls must be broken to enhance the nutritional value of SCP. The biomass may be further processed or even the protein may be isolated and purified.

Large scale release of either living or dead micro-organisms into the environment must be prevented. If the spent medium has high BOD, it must be treated prior to discharge or, better still, may be recycled.

Nutritional and safety evaluations

1. The SCP chemical composition must be characterised in terms of protein, amino acid, nucleic acid, lipid, vitamin, etc. contents.
2. Analysis of substrate residues and toxic substances, e.g. heavy metals, mycotoxins, palycyclic hydrocarbons, etc. must be done.

3. Physical properties like density, particle size, texture, colour, storage, etc. and properties should be determined.
4. Microbiological description, e.g. species, strain, should be provided, and information on contamination be also given.
5. The nutritional value should be evaluated on the target species, and other species should also be included. The products for human use will, of necessity, be evaluated over a longer period using a multistaged process.
6. Possible toxic or carcinogenic compounds must be assayed for. These compounds may have been present in the substrate, may be synthesised by the organism or produced during the processing of SCP.

Advantages of SCP

The SCP processes and products offer several advantages as listed below.

1. The SCP is rich in high quality protein and is rather poor in fats, which is rather desirable (Table 16.4).
2. They can be produced all the year round and are not dependent of the climate (except the algal processes).
3. The microbes are very fast growing and produce large quantities of SCP from relatively very small area of land.
4. They use low cast substrates and, in same cases, such substrates which are being wasted and causing pollution to the environment.
5. When the substrate used for SCP process is a source of pollution. SCP production helps reduce pollution.
6. Strains having high biomass yields and a desirable amino acid composition can be easily selected or produced by genetic engineering.
7. Some SCPs are good sources of vitamins, particularly B-group of vitamins, as well, e.g. yeasts and mushrooms.
8. Mushrooms are considered as delicacy in the human diet.
9. At present, SCP appears to be the only feasible approach to bridge the gap between requirement and supply of proteins.

Table 16.4. The nutritional status of SCP obtained from certain micro-organisms.

Nutrient	*Micro-organism*			
	Paecilomyces varioti	*Candida utilis*	*Methylophilus methylotrophus*	*Spirulina maxima*
Protein	55	55	83	62
Fat	1	5	7	3
Ash	6	8	9	2

Product Safety and Quality

Some contaminants can produce mycotoxins. Some bacterial SCP have amino acid profiles different from animal proteins. Yeast and fungal proteins tend to be deficient in methionine.

Microbial biomass has a high nucleic acid content, and levels need to be limited in the diets of monogastric animals to <50 g per day. Ingestion of purine compounds arising from RNA breakdown, leads to increased plasma levels of uric acid, which can cause gout and kidney stones. Uric acid can be converted to allantoin, which is excreted in urine. Nucleic acid removal is not necessary from animal feeds but is from human foods. A temperature hold at 64°C inactivates fungal proteases and allows RNases to hydrolyse RNA with release of nucleotides from cell to culture broth.

METHANOL

A very efficient chemical conversion of methane yields methanol; it can also be produced from coal, gas oil, wood, naphtha, etc. Methanol is fully water soluble, is used by many bacteria (Table 16.3), and there is little danger of explosion (compare with methane). A highly successful process by Imperial Chemical Industries Ltd. (ICI) uses the bacterium *Methylophilus methylotrophus* for a continuous production process at 35°–40°C in a specially designed 'pressure cycle' fermenter (the fermenter is designed to achieve very efficient substrate mixing and O_2 transfer).

The cells are harvested by flocculation and floatation to give about 10 per cent solids; the biomass is then subjected to centrifugation to remove water and is then air-dried. The SCP contains 71 per cent protein, is marketed as 'Pruteen', and is used as a milk substitute in calf feeding. The bacterium *M. methylotrophus* had its NH_3-assimilation modified by genetic engineering; this has significantly increased its biomass production.

Methanol, also known as methyl alcohol, wood alcohol, wood naphtha or wood spirits, is a chemical with formula CH_3OH (often abbreviated MeOH). It is toxic: drinking 10 ml will cause blindness, and as little as 100 ml will cause death. It is the simplest alcohol, and is a light, volatile, colourless, flammable, liquid with a distinctive odour that is very similar to but slightly sweeter than ethanol (drinking alcohol). At room temperature it is a polar liquid and is used as an antifreeze, solvent, fuel, and as a denaturant for ethanol. It is also used for producing biodiesel via transesterification reaction.

Methanol is produced naturally in the anaerobic metabolism of many varieties of bacteria, and is ubiquitous in the environment. As a result, there is a small fraction of methanol vapour in the atmosphere. Over the course of several days, atmospheric methanol is oxidised by oxygen with the help of sunlight to carbon dioxide and water. Methanol is therefore biodegradable. Methanol burns in air forming carbon dioxide and water:

$$2CH_3OH + 3O_2 \rightarrow 2CO_2 + 4H_2O$$

A methanol flame is almost colourless in bright sunlight conditions, causing an additional safety hazard around open methanol flames. A methanol fire can, however, be extinguished with water.

Because of its toxic properties, methanol is frequently used as a denaturant additive for ethanol manufactured for industrial uses—this addition of methanol exempts industrial ethanol from liquor excise taxation. Methanol is often called wood alcohol because it was once produced chiefly as a by-product of the destructive distillation of wood.

Fuel for Vehicles

Methanol is used on a limited basis to fuel internal combustion engines, mainly by virtue of the fact that it is not nearly as flammable as gasoline. Methanol is harder to ignite than gasoline and produces just one-eighth of the heat upon burning. Pure methanol is required by rule to be used in Champcars, Monster Trucks, USAC sprint cars (as well as midgets, modifieds, etc.), and other dirt track series such as World of Outlaws, and Motorcycle Speedway. Methanol is also used, as the primary fuel ingredient since the late 1940s, in the powerplants for radio control, control line and free flight airplanes (as methanol is required in the 'glow-plug' engines that primarily power them), cars and trucks, from such an engine's use of a platinum filament glow plug being able to ignite the methanol vapour through a catalytic reaction. Drag racers and mud racers also use methanol as their primary fuel source. Methanol is required with a supercharged engine in a Top Alcohol Dragster and, until the end of the 2006 season, all vehicles in the Indianapolis 500 had to run methanol. Mud racers have mixed methanol with gasoline and nitrous oxide to produce more power than gasoline and nitrous oxide alone.

One of the drawbacks of methanol as a fuel is its corrosivity to some metals, including aluminium. Methanol, although a weak acid, attacks the oxide coating that normally protects the aluminium from corrosion:

$$6CH_3OH + Al_2O_3 \rightarrow 2Al(OCH_3)_3 + 3H_2O$$

The resulting methoxide salts are soluble in methanol, resulting in clean aluminium surface, which is readily oxidised by some dissolved oxygen. Also the methanol can act as an oxidiser:

$$6CH_3OH + 2Al \rightarrow 2Al(OCH_3)_3 + 3H_2$$

This reciprocal process effectively fuels corrosion until either the metal is eaten away or the concentration of CH_3OH is negligible. Concerns with methanol's corrosivity have been addressed by using methanol compatible materials, and fuel additives that serve as corrosion inhibitors.

When produced from wood or other organic materials, the resulting organic methanol (bioalcohol) has been suggested as renewable alternative to petroleum-based hydrocarbons. Low levels of methanol can be used in existing vehicles, with the use of proper cosolvents and corrosion inhibitors. The European fuel quality directive allows up to 3 per cent methanol with an equal amount of cosolvent to be blending in gasoline sold in Europe. Today, China uses more than one billion gallons of methanol per year as a transportation fuel in both low level blends used in existing vehicles, and as high level blends in vehicles designed to accommodate the use of methanol fuels.

Health and Safety

Toxicity

Methanol has a high toxicity in humans. If ingested, as little as 10 ml can cause permanent blindness by destruction of the optic nerve and 30 ml is potentially fatal. Although the usual fatal dose is typically 100–125 ml (4 fl oz). Toxic effects take hours to start and effective antidotes can often prevent permanent damage. Because of its similarities to ethanol (the alcohol in beverages), it is difficult to differentiate between the two (such is the case with denatured alcohol).

Methanol is toxic by two mechanisms. Firstly, methanol (whether it enters the body by ingestion, inhalation, or absorption through the skin) can be fatal due to its CNS depressant properties in the same manner as ethanol poisoning. Secondly, in a process of toxication, it is metabolised to formic acid (which is present as the formate ion) via formaldehyde in a process initiated by the enzyme alcohol dehydrogenase in the liver. The reaction to formate proceeds completely, with no detectable formaldehyde remaining. Fetal tissue will not tolerate methanol. Methanol poisoning can be treated with the antidotes ethanol or fomepizole. Both of these drugs act to reduce the action of alcohol dehydrogenase on methanol by means of competitive inhibition, so that it is excreted by the kidneys rather than being transformed into toxic metabolites. Further treatment may include giving sodium bicarbonate for metabolic acidosis and haemodialysis or haemodiafiltration can be used to remove methanol and formate from the blood. Folinic acid or folic acid is also administered to enhance the metabolism of formate.

The initial symptoms of methanol intoxication include central nervous system depression, headache, dizziness, nausea, lack of coordination, confusion, and with sufficiently large doses, unconsciousness and death. The initial symptoms of methanol exposure are usually less severe than the symptoms resulting from the ingestion of a similar quantity of ethanol. Once the initial symptoms have passed, a second set of symptoms arises, 10 to as many as 30 hrs after the initial exposure to methanol, including blurring or complete loss of vision and acidosis. These symptoms result from the accumulation of toxic levels of formate in the bloodstream, and may progress to death by respiratory failure. The ester derivatives of methanol do not share this toxicity. Ethanol is sometimes denatured (adulterated), and thus made undrinkable, by the addition of methanol. The result is known as methylated spirit or meths. The latter should not be confused with meth, a common US abbreviation for methamphetamine.

Safety in automotive fuels

Pure methanol has been used in open wheel auto racing since the mid-1960s. Unlike petroleum fires, methanol fires can be extinguished with plain water. A methanol-based fire burns invisibly, unlike gasoline, which burns with a visible flame. If a fire occurs on the track, there is no flame or smoke to obstruct the view of fast approaching drivers, but this can also delay visual detection of the fire and the initiation of fire suppression actions. The decision to permanently switch to methanol in American IndyCar racing was a result of the devastating crash and explosion at the 1964 Indianapolis 500 which killed drivers Eddie Sachs and Dave MacDonald. However, in 2007 IndyCars switched back to ethanol.

Methanol is readily biodegradable in both aerobic (oxygen present) and anaerobic (oxygen absent) environments. Methanol will not persist in the environment. The 'half-life' for methanol in groundwater is just one to seven days, while many common gasoline components have half-lives in the hundreds of days (such as benzene at 10–730 days). Since methanol is miscible with water and biodegradable, methanol is unlikely to accumulate in groundwater, surface water, air or soil.

Methanol Fuel

Methanol has been proposed as a fuel for internal combustion and other engines, mainly in combination with gasoline. Methanol fuel has received less attention than ethanol fuel as an alternative to petroleum based fuels. However, in 2005 Nobel prize winner George A. Olah advocated an entire methanol economy based on energy storage in synthetically produced methanol in an essay.

History and production

Historically, methanol was first produced from pyrolysis of wood, resulting in its common English name of wood alcohol. Presently, methanol is usually produced using methane (the chief constituent of natural gas) as a raw material.

It may also be produced by pyrolysis of many organic materials or by Fischer Tropsch from synthetic gas, so be called biomethanol. Production of methanol from synthesis gas using biomass-to-liquid can offer methanol production from biomass at efficiencies up to 75 per cent. Widespread production by this route has a postulated potential to offer methanol fuel at a low cost and with benefits to the environment. These production methods, however, are not suitable for small scale production.

Use as internal combustion engine fuel

Both methanol and ethanol burn at lower temperatures than gasoline, and both are more volatile, making engine starting in cold weather more difficult. Using methanol as a fuel in spark ignition engines can offer an increased thermal efficiency and increased power output (as compared to gasoline) due to its high octane rating and high heat of vapourisation. However, its low energy content of 19.7 MJ/kg and stoichiometric air fuel ratio of 6.42:1 mean that fuel consumption (on volume or mass basis) will be higher than hydrocarbon fuels. The extra water produced also makes the charge rather wet (similar to hydrogen/oxygen combustion engines)and combined with the formation of acidic products during combustion, the wearing of valves, valveseats and cylinder might be higher than with hydrocarbon burning. Certain additives may be added to motor oil in order to neutralise these acids.

Methanol, just like ethanol, contains soluble and insoluble contaminents. These soluble contaminants, halide ions such as cloride ions, have a large effect on the corrosivity of alcohol fuels. Halide ions increase corrosion in two ways; they chemically attack passivating oxide films on several metals causing pitting corrosion, and they increase the conductivity of the fuel. Increased electrical conductivity promotes electric, galvanic, and ordinary corrosion in the fuel system. Soluble contaminents, such as aluminum hydroxide, itself a product of corrosion by halide ions, clog the fuel system over time.

Methanol is hygroscopic, meaning it will absorb water vapour directly from the atmosphere. Because absorbed water dilutes the fuel value of the methanol (although, it suppresses engine knock), and may cause phase separation of methanol-gasoline blends, containers of methanol fuels must be kept tightly sealed.

Toxicity

Methanol is poisonous; ingestion of only 10ml can cause blindness and 60ml can be fatal, and it does not have to be swallowed to be dangerous since the liquid can be absorbed through the skin, and the vapours through the lungs. US maximum allowed exposure in air (40 hr/week) is 1900 mg/m^3 for ethanol, 900 mg/m^3 for gasoline, and 1260 mg/m^3 for methanol. However, it is less volatile than gasoline, and therefore decreases evaporative emissions. Use of methanol, like ethanol, significantly reduces the emissions of certain hydrocarbon-related toxins such as benzene and 1,3 butadiene. But as gasoline and ethanol are already quite toxic, safety protocol is the same.

Safety

Since methanol vapour is heavier than air, it will linger close to the ground or in a pit unless there is good ventilation, and if the concentration of methanol is above 6.7 per cent in air it can be lit by a spark, and will explode above 54°F/12°C. Once ablaze, the flames give out very little light making it very hard to see the fire or even estimate its size, especially in bright daylight. If you are unlucky enough to be exposed to the poisonous substance through your respiratory system, its pungent odour should give you some warning of its presence. However, it is difficult to smell methanol in the air at less than 2000 ppm (0.2 per cent), and it can be dangerous at lower concentrations than that.

METHANOL ECONOMY

The methanol economy is a suggested future economy in which methanol replaces fossil fuels as a means of energy storage, fuel and raw material for synthetic hydrocarbons and their products. It offers an alternative to the proposed hydrogen economy or ethanol economy.

Methanol is a fuel for heat engines and fuel cells. Due to its high octane rating it can be used directly as a fuel in flex-fuel cars (including hybrid and plug-in hybrid vehicles) using existing internal combustion engines (ICE). Methanol can also be used as a fuel in fuel cells, either directly in direct methanol fuel cells (DMFC) or indirectly after conversion into hydrogen by reforming.

Methanol is a liquid under normal conditions, allowing it to be stored, transported and dispensed easily, much like gasoline and diesel fuel is currently. It can also be readily transformed by dehydration into dimethyl ether, a diesel fuel substitute with a cetane number of 55.

Methanol is already used today on a large scale (about 40 million tonnes per year) as a raw material to produce numerous chemical products and materials. In addition, it can be readily converted in the methanol to olefin (MTO) process into ethylene and propylene, which can be used to produce synthetic hydrocarbons and their products, currently obtained from oil and natural gas.

Methanol can be efficiently produced from a wide variety of sources including still abundant fossil fuels (natural gas, coal, oil shale, tar sands, etc.), but also agricultural products and municipal waste, wood and varied biomass. More importantly, it can also be made from chemical recycling of carbon dioxide. Initially the major source will be the CO_2 rich flue gases of fossil fuel burning power plants or exhaust of cement and other factories. In the longer range however, considering diminishing fossil fuel resources and the effect of their utilisation on earth's atmosphere, even the low concentration of atmospheric CO_2 itself could be captured and recycled via methanol, thus supplementing nature's own photosynthetic cycle.

Efficient new absorbents to capture atmospheric CO_2 are being developed, mimicking plant life's ability. Chemical recycling of CO_2 to new fuels and materials could thus become feasible, making them renewable on the human timescale. One m^3 of methanol at ambient pressure and temperature contains 1.660 Nm^3 of hydrogen (H_2) compared to liquid hydrogen; one m^3 of liquid hydrogen (LH_2) at –253°C contains 788 Nm^3 of hydrogen (H_2).

Uses of Methanol in a Methanol Economy

Fuel uses

In an economy based on methanol, methanol could be used as a fuel:

1. In internal combustion engines (ICEs): Methanol has a high octane rating (RON of 107 and MON of 92), which makes it a suitable gasoline substitute. It has a higher flame speed than gasoline, leading to higher efficiency as well as a higher latent heat of vapourisation (3.7 times higher than gasoline), meaning that the heat generated by the engine can be removed more effectively, making it possible to use air cooled engines. Besides this methanol burns cleaner than gasoline and is safer in the case of a fire. However, methanol has only half the volumetric energy content of gasoline (8600 Btu/lb).
2. In compression ignition engines (diesel engine): Methanol itself is not a good substitute for diesel fuels. Methanol can, however, be converted by dehydration to dimethyl ether, which is a good diesel fuel with a cetane number of 55–60 as compared to 45–55 for regular diesel fuel. Compared to diesel fuel, DME has much lower emissions of particulate matter, NO_x and CO and does not emit any SO_x. Methanol can also be used, and is in fact already used, to produce biodiesel via transesterification of vegetable oil (SVO).
3. In advanced methanol powered vehicles: The use of methanol and dimethyl ether can be combined with hybrid and plug-in vehicle technologies allowing higher gas mileage and lower emissions. These fuels can also be used in fuel cells either via onboard reforming to hydrogen or directly in direct methanol fuel cells (DMFC).
4. For electricity production: Methanol and DME can be used in existing gas turbines to generate electricity. Fuel cells (PAFC, MCFC, SOFC) can also be used for electricity generation.

5. As a domestic fuel: Methanol and DME can be used in commercial buildings and homes to generate heat and/or electricity. DME can be used in a commercial gas stove without modifications. In developing countries methanol could also be used as a cooking fuel, burning much cleaner than wood, thus mitigating indoor air quality problems.

Raw material for chemicals

Methanol is already used today on a large scale as raw material to produce a variety of chemicals and products. Through the methanol to gasoline (MTG) process, it can be transformed into gasoline. Using the methanol to olefin (MTO) process, methanol can also be converted to ethylene and propylene, the two largest chemicals produced by the petrochemical industry. These are important building blocks for the production of essential polymers (LDPE, HDPE, PP) and other chemical intermediates are currently produced mainly from petroleum feedstock. Their production from methanol could therefore reduce our dependency on petroleum. It would also make it possible to still produce these chemicals when fossil fuels reserves will be depleted.

Methanol production for a methanol economy

The methanol needed in the methanol economy can be synthesised from a wide array of carbon sources including still available fossil fuels and biomass but also CO_2 emitted from fossil fuel burning power plants and other industries and eventually even the CO_2 contained in the air.

Today methanol is produced exclusively from methane through syngas. Although conventional natural gas resources are currently the preferred feedstock for the production of methanol, unconventional gas resources such as coalbed methane, tight sand gas and eventually the very large methane hydrate resources present under the continental shelves of the seas and Siberian and Canadian tundra could also be used. Besides methane all other conventional or unconventional (tar sands, oil shale, etc.) fossil fuels could be utilised to produce methanol. Besides the conventional route to methanol from methane passing through syngas generation by steam reforming combined (or not) with partial oxidation, new and more efficient ways to produce methanol from methane are being developed. These include:

1. Methane oxidation with homogeneous catalysts in sulphuric acid media.
2. Methane bromination followed by hydrolysis of the obtained bromomethane.
3. Direct oxidation of methane with oxygen.
4. Microbial or photochemical conversion of methane.

The use of methane or another fossil fuel for the production of methanol using all the above mentioned synthetic routes has a potential drawback: the emission of the greenhouse gas carbon dioxide CO_2. To mitigate this, methanol can be made through ways minimising the emission of CO_2. One solution is to produce it from syngas obtained by biomass gasification. For this purpose any biomass can be used including wood, wood wastes, grass, agricultural crops and their by-products, animal waste, aquatic plants and municipal waste. There is no need to use food crops as in the case of ethanol from corn, sugar cane and wheat.

$$\text{Biomass} \rightarrow \text{Syngas } (CO, CO_2, H_2) \rightarrow CH_3OH$$

More importantly, methanol can also be produced from CO_2 by catalytic hydrogenation of CO_2 with H_2 where the hydrogen has been obtained from water electrolysis. Methanol may also be produced through CO_2 electrochemical reduction, if electrical power is available. The energy needed for these reactions in order to be carbon neutral would come form renewable energy sources such as wind, hydroelectricity and solar as well as nuclear power. In effect, all of them allow free energy to be stored in easily transportable methanol, which is made immediately from hydrogen and carbon dioxide, rather than attempting to store energy in free hydrogen.

$$CO_2 + 3H_2 \rightarrow CH_3OH + H_2O$$

$$CO_2 + 2H_2O + \text{electrons} \rightarrow CO + 2H_2 \ (+ 3/2O_2) \rightarrow CH_3OH$$

The necessary CO_2 would be captured from fossil fuel burning power plants and other industrial flue gases including cement factories. With diminishing fossil fuel resources and therefore CO_2 emissions, the CO_2 content in the air could also be used.

Considering the low concentration of CO_2 in air (0.037 per cent) improved and economically viable technologies to absorb CO_2 will have to be developed. This would allow the chemical recycling of CO_2, thus mimicking nature's photosynthesis.

Advantages over other energy storage media

In the process of photosynthesis, green plants use the energy of sunlight to split water into free oxygen (which is released) and free hydrogen. Rather than attempt to store the hydrogen, plants immediately capture carbon dioxide from the air to allow the hydrogen to reduce it to storable fuels such as hydrocarbons (plant oils and terpenes) and polyalcohols (glycerol, sugars and starches). In the methanol economy, any process which similarly produces free hydrogen, proposes to immediately use it 'captively' to reduce carbon dioxide into methanol, which, like plant products from photosynthesis, has great advantages in storage and transport over free hydrogen itself.

Advantages over hydrogen

Methanol economy advantages compared to a hydrogen economy:

1. Efficient energy storage (by volume) and also by weight as compared with compressed hydrogen, when hydrogen pressure-confinement vessel is taken into account. The volumetric energy density of methanol is considerably higher than liquid hydrogen, in part because of the low density of liquid hydrogen of 71 g/l. Hence there is actually more hydrogen in a litre of methanol (99 g/l) than in a litre of liquid hydrogen, and methanol needs no cryogenic container maintained at a temperature of –253°C.
2. Required hydrogen infrastructure would be prohibitively expensive. Methanol can use existing gasoline infrastructure with only limited modifications.
3. Can be blended with gasoline (for example in M85, a mixture containing 85 per cent methanol and 15 per cent gasoline).
4. User friendly. Hydrogen is volatile and requires high pressure or cryogenic system confinement.

Methanol economy advantages compared to ethanol

1. Can be made from any organic material using the proven Fischer Tropsch method going through syngas. There is no need to use food crops and compete with food production. The amount of methanol that can be generated from biomass is much greater than ethanol.
2. Can compete with and complement ethanol in a diversified energy marketplace. Methanol obtained from fossil fuels has a lower price than ethanol.
3. Can be blended in gasoline like ethanol. In 2007, China blended more than 1 billion gallons of methanol into fuel and will introduce methanol fuel standard by 2011. M85, a mixture of 85 per cent methanol and 15 per cent gasoline can be used much like E85 sold in some gas stations today.

Methanol economy disadvantages

1. High energy costs associated with generating hydrogen (when needed to synthesise methanol).
2. Depending on the feedstock the generation in itself can be not clean.
3. Presently generated from syngas still dependent on fossil fuels (although in theory any energy source can be used).
4. Energy density (by weight or volume) one half of that of gasoline and 24 per cent less than ethanol.
5. Corrosive to some metals including aluminium, zinc and manganese. Parts of the engine fuel-intake systems are made from aluminum. Similar to ethanol, compatible material for fuel tanks, gasket and engine intake have to be used.
6. Hydrophilic: attracts water: In mixture with gasoline this could lead to phase separation and to difficulty starting the engine or making it run smoothly.
7. As with similarly corrosive and hydrophilic ethanol, existing pipelines designed for petroleum products cannot handle methanol. Thus methanol requires shipment at higher energy cost in trucks and trains, until a whole new pipeline infrastructure can be built.
8. Methanol, as an alcohol, increases the permeability of some plastics to fuel vapours (e.g. high-density polyethylene). This property of methanol has the possibility of increasing emissions of volatile organic compounds (VOCs) from fuel, which contributes to increased tropospheric ozone and possibly human exposure.
9. Low volatility in cold weather: Pure methanol-fuelled engines can be difficult to start, and they run inefficiently until warmed up. This is why a mixture containing 85 per cent methanol and 15 per cent gasoline called M85 is generally used in ICEs. The gasoline allows the engine to start even at lower temperatures.
10. Methanol is generally considered toxic. Methanol is in fact toxic and eventually lethal when ingested in larger amounts (30 to 100 ml). But so are most motor fuels, including gasoline (120 to 300 ml) and diesel fuel. Gasoline also contains many compounds known to be carcinogenic (e.g. benzene). Methanol is not a carcinogen, nor does it contain any carcinogens. However, methanol may be metabolised in the body to formaldehyde, which is both toxic and carcinogenic.
11. Methanol is a liquid: This creates a greater fire risk compared to hydrogen in open spaces. Methanol leaks do not dissipate. A methanol-based fire burns invisibly unlike gasoline. Compared to gasoline, however, methanol is much safer. It is more difficult to ignite and releases less heat when it burns. Methanol fires can be extinguished with plain water, whereas gasoline floats on water and continues to burn. The EPA has estimated that switching fuels from gasoline to methanol would reduce the incidence of fuel related fires by 90 per cent.
12. Methanol accidentally released from leaking underground fuel storage tanks may undergo relatively rapid groundwater transport and contaminate well water, although this risk has not been thoroughly studied. The history of the fuel additive methyl *t*-butyl ether (MTBE) as a groundwater contaminant has highlighted the importance of assessing the potential impacts of fuel and fuel additives on multiple environmental media. An accidental release of methanol in the environment would, however, cause much less damage than a comparable gasoline or crude oil spill. Unlike these fuels, methanol, being totally soluble in water, would be rapidly diluted to a concentration low enough for micro-organism to start biodegradation. Methanol is in fact used for denitrification in water treatment plant as a nutrient for bacterias.

METHANOGENESIS

Methanogenesis or biomethanation is the formation of methane by microbes known as methanogens. Organisms capable of producing methane have been identified only from the kingdom Archaea, a group phylogenetically distinct from both eukaryotes and bacteria, although many live in close association with anaerobic bacteria. The production of methane is an important and widespread form of microbial metabolism. In most environments, it is the final step in the decomposition of biomass.

Recently, some experiments have suggested that leaf tissues of living plants emit methane. Other research has indicated that the plants are not actually generating methane; they are just absorbing methane from the soil and the emitting it through their leaf tissues. There may still be some unknown mechanism by which plants produce methane, but that is by no means certain.

Biochemistry of Methanogenesis

Methanogenesis in microbes is a form of anaerobic respiration. Methanogens do not use oxygen to breathe; in fact, oxygen inhibits the growth of methanogens. The terminal electron acceptor in methanogenesis is not oxygen, but carbon. The carbon can occur in a small number of organic compounds, all with low molecular weights. The two best described pathways involve the use of carbon dioxide and acetic acid as terminal electron acceptors:

$$CO_2 + 4H_2 \rightarrow CH_4 + 2H_2O$$

$$CH_3COOH \rightarrow CH_4 + CO_2$$

However, methanogenesis has been shown to use carbon from other small organic compounds, such as formic acid (formate), methanol, methylamines, dimethyl sulphide, and methanethiol.

The biochemistry of methanogenesis is relatively complex, involving the following coenzymes and cofactors: F430, coenzyme B, coenzyme M, methanofuran, and methanopterin.

Importance in carbon cycle

Methanogenesis is the final step in the decay of organic matter. During the decay process, electron acceptors (such as oxygen, ferric iron, sulphate, nitrate, and manganese) become depleted, while hydrogen (H_2) and carbon dioxide accumulate. Light organics produced by fermentation also accumulate. During advanced stages of organic decay, all electron acceptors become depleted except carbon dioxide. Carbon dioxide is a product of most catabolic processes, so it is not depleted like other potential electron acceptors.

Only methanogenesis and fermentation can occur in the absence of electron acceptors other than carbon. Fermentation only allows the breakdown of larger organic compounds, and produces small organic compounds.

Methanogenesis effectively removes the semi-final products of decay: hydrogen, small organics, and carbon dioxide. Without methanogenesis, a great deal of carbon (in the form of fermentation products) would accumulate in anaerobic environments.

In ruminants

Methanogenesis occurs in the guts of humans and other animals, especially ruminants. In the rumen, anaerobic organisms including methanogens digest cellulose into forms usable by the animal, without them, livestock such as cattle would not be able to graze grass. The useful products of methanogenesis are absorbed by the gut, but the methane is released from the animal mainly by belching (eructation). The average cow emits around 250 litres of methane per day.

In humans

Some humans produce flatus that contains methane. In one study of the faeces of nine adults, only five of the samples contained archaea capable of producing methane. Similar results are found in samples of gas obtained from within the rectum.

Even among humans whose flatus does contain methane, the amount is only 0–10 per cent of the total amount of gas.

Role in global warming

Methane in the earth's atmosphere is an important greenhouse gas with a global warming potential 21 times greater than carbon dioxide (averaged over 100 years), and methanogenesis in livestock and the decay of organic material is thus a considerable contributor to global warming. It may not be a net contributor in the sense that it works on organic material which used up atmospheric carbon dioxide when it was created, but its overall effect is to convert the carbon dioxide into methane which is a much more potent greenhouse gas.

Methanogenesis can also be beneficially exploited, to treat organic waste, to produce useful compounds, and the methane can be collected and used as biogas, a fuel.

Factors Affecting Biomethanation

Biomethanation as an industrial process is gaining importance and its success is dependent upon the control over a diverse microbial population and its biological activity. Besides microbial population, the environmental factors, which govern the process of methanogenesis, are anaerobiosis, temperature, pH, substrate composition, carbon:nitrogen (C:N) ratio, micronutrient, the presence or absence of toxic materials in the substrate biomass and accumulation of inhibitory by-products or intermediates of the process.

Anaerobiosis

Methanogenesis is a strict anaerobic microbial process. The major trophic groups and methanogens in this process die in the presence of oxygen. Though the major microbial activity in methanogenesis is anaerobic, the presence of aerobic and facultative anaerobic bacteria has been reported by many workers. The number of facultative anaerobic bacteria in piggery waste digesters was found to be the same as that of anaerobes. These bacteria do not have any role in the main degradative reactions of digestion but play some role in general sugar fermentation. Most probably these bacteria have a role in the anaerobic digestion as scavengers of oxygen reducing the system to be suitable for the growth of the methanogens. Therefore, the requirement for anaerobiosis in digesters can be formulated in a dynamic way by minimising the diffusion of air into the digestion liquor. Most of the digesters are designed in such a way that the ecosystem is closed to atmospheric air so that the microflora in the digester is not exposed to oxygen tension.

Temperature

Temperature in all microbial systems is known to enhance the microbial activity to a certain optimal level. Methanogenesis occurs practically in an optimal way in two ranges of temperatures. Mesophilic methanogenesis occurs at a range of 30°–37°C and thermophilic methanogenesis between 50° and 65°C. Adoption of mesophilic digester to thermophilic conditions is possible (in about 10–20 days), as it always contains about 10 per cent of thermophilic micro-organisms. Thermophilic digestion processes offer a number of advantages. The rapid metabolic activity in a thermophilic digester helps in reducing the retention time and increasing the loading rates and hence a small digester volume. In addition to increased rates of methane production, which is reported to be 1.5 times faster than mesophilic digestion, thermophilic digestion has improved dewaterability and enhanced, if not complete, killing of pathogenic organisms. However, changes in the overall process efficiency due to increased metabolic activity are balanced by a corresponding increase in the rates of microbial inactivation. Hence, the optimal process temperature represents the best compromise between these two factors. The required optimal digestion temperature may also depend on the waste being treated. For example, higher conversion of protein is observed at 30°–50°C and conversely lower temperatures were found suitable for lipid digestion. Low temperature waste digestion is slower than mesophilic and thermophilic digestion.

pH

A pH value between 6.5 and 7.7 has been found to be optimum for hydrogen-utilising bacterium, *Methanobacterium ruminantium*, seen in digesters and rumen. Smith used a pH of 7.2 for optimum running of the methanogenic stage of a two-stage laboratory digester and Cohen used 7.8 for the same. Maintenance of near neutral pH is due to the conversion of acid end-product to methane by the combined effect of acetogenic and methanogenic bacteria. It should be recalled that the methanogenic process is alkali. The major controlling buffer in this pH range is the carbonate-bicarbonate system. Volatile fatty acids (VFAs) and ammonia also have been known to regulate the buffering capacity of the digester.

The sensitivity of anaerobic digestion to pH levels has been due to the critically pH-sensitive methanogenic population. However, it is evident that acid-tolerant methanogens do exist in peat bog environment. Methanogenesis beyond the range of 6.5–8.0 pH has been found to be less yielding.

Substrate composition

The biomass substrate used in methanogenesis is comprised of various carbon sources such as cellulose, pectin, hemicellulose, protein, lipid, starch, lignin and other organic materials at varied concentration. For a pure substrate, it is possible to calculate the amount of gas, which may be produced on the basis of a simple carbon balance.

Theoretical yields of biogas (m^3/kg VS destroyed) from various components of organic matter are 0.886 (carbohydrates), 1.535 (fat) and 0.587 (proteins) with a methane content of 50, 70 and 84 per cent respectively.

C:N ratio

Extra attention has to be paid to the nitrogen content of the substrate biomass for two reasons. On the one hand, nitrogen is required for cell growth, a phenomenon necessarily linked to methane production; on the other, the nitrogenous compounds contribute to the buffering of the digesting liquor by the release of ammonium cation. Investigations by various authors have shown that C:N ratio of 16–19 is required for maximum methanogenic performance. Stevens and Brune used glucose and urea to adjust the C:N ratio of swine manure between 20 and 25. They have reported low methane yields at low C:N ratios because of ammonia inhibition. A concentration above 3000 mg/l of ammonia nitrogen is supposed to be toxic at all pH levels.

Maximum methane yields have been reported at C:N ratios of between 16 and 19. Decreasing methane yields have also been reported when the C:N ratio increased above 20, probably because of low nitrogen availability. The findings of Hashimoto have also endorsed the above result while studying anaerobic digestion of cattle manure and molasses.

Micronutrients

The bacteria in the anaerobic digestion process also require phosphorous (P) and metal ions for optimum growth. Murray and Berg have reported that nickel and cobalt ions are necessary at concentrations of 100 nM and 50 nM, respectively. Molybdenum ions at a concentration of 50 nM may enhance the joint effect of nickel and cobalt. Iron cations up to 2 nM and copper ions at a concentration of 4 nM are necessary to enhance the performance of aceticlastic methanogenic archaebacteria. An increase of up to three-fold can be achieved by incorporating the above minerals to a digester deficient in these components. The requirement of P has also been studied and a C:P ratio of 100–200 is said to be optimum. The required optimum C:N:P ratio for enhanced yield of methane was found to be 100:2.5:0.5.

Toxins and inhibitors

It has been observed that metal ions exert a toxic effect when they exceed the required concentration. Mosey and Hughes proposed the critical value of pS > 14.0 as threshold value indicating the inhibitory presence of heavy metal ions such as Zn^{2+}, Cd^{2+}, Fe^{2+} or Cu^{2+}. Hayes and Theis described the concentrations inhibitory to anaerobic digestion. Accumulation of NH_4^+, SO_2, HSO_3^- and SO_3^{2-} was reported to inhibit a methanogenic bacterial community at a concentration beyond 1 mg/l of digesting liquor. The presence of organic compounds (amino derivatives, higher alcohols (5–12 carbon) and ketones (5–8 carbon) is inhibitory only when their concentration exceeds their solubilities in water. Petrochemicals are toxic when they are present as derivatives of –Cl, $-NH_2$, –C = O, di–COOH, though accumulation is possible with severely increased time requirements. However, chloroform ($CHCl_3$), an analogue of methane, is a very potent inhibitor of the methanogenic archaebacteria.

Intermediates

The VFAs form the intermediates of the anaerobic digestion process. The C_2–C_5 acids such as acetate, propionate and butyrate may be inhibitory to methanogenesis at higher levels though they are also the substrates for the reaction. However, acetic or butyric acid inhibition is not significant. It has been observed that a concentration of 10,000 mg/l of acetic and butyric acid had no inhibitory effect on *M. formicicum.* However, Andrews observed that propionic acid was inhibitory to laboratory digestion at concentrations as low as 100 mg/l and can also inhibit growth of *M. formicicum*. The composition of volatile fatty acid (VFA) in cattle manure biogas digester is about 87–88 per cent of branched-chain fatty acids comprising isobutyrate, isovalarate and were found to produce an inhibitory effect. Isobutyric acid, in particular, was inhibitory at concentrations as low as 50 ppm. The toxic effect of VFAS at high concentrations is attributed either to the toxicity of the VFAS themselves or to the decrease in pH brought about by them. However, inhibition due to VFAS in anaerobic digestion is varied.

Hydraulic retention time

Hydraulic retention time (HRT) expresses the volume of fluids in the reactor per volume of fluid passing into and out of the reactor on a daily basis. It indicates the contact time allowed between the substrate and micro-organisms in the system. Maintaining optimum retention time is an important factor to cause efficient conversion of organic matter to methane. Use of very long retention times would result in inefficient use of digester capacity as it may allow the bacterial population to proceed beyond the phase of exponential multiplication. At too short HRT, the rate of bacterial multiplication may not be sufficient to compensate for the bacteria discharged into the effluent from the digester resulting in an unstable bacterial population. The HRT maintained in conventional anaerobic digesters utilising raw sewage solids and organic refuse ranges from 15 to 30 days. For cellulosic material, an HRT of 10 days was reported as optimum and the lignaceous material was non-degradable even at 30 days HRT.

Loading rate

The loading rate is the amount of raw material introduced into the digester per unit time. It concerns the interaction between substrate and micro-organisms and directly influences the equilibrium between different physiological groups of bacteria involved in the process. With certain substances, adverse effects of overloading are immediate and drastic (for example, sewage sludge) and the effects are less drastic when the substrate is heterogeneous in nature (for example, organic refuse). High loading brings about unfavourable pH, leading to low methane production. An appropriate loading rate is required to maintain the balanced activity of different physiological groups leading to a stable digestion process. The loading rate of organic matter is expressed in terms of kilogram (kg) of volatile solids (VS) per cubic meter of digester or percentage of VS.

The reported loading rate for various substrates ranges from 0.6 to 1.6 kg VS/m^3 for sewage sludge 1.27–4.80 kg VS/m^3 for organic refuse 2.56–3.50 kg VS/m^3 for piggery waste, 1.12–27.24 kg VS/m^3 for cattle manure, 2.88 kg VS/m^3 for manure and 3.50–4.25 kg VS/m^3 for fruit and vegetable waste.

Reactor Design

The commercial methane-generating continuous fermenters use systems, which have arrangements for retaining biomass within the reactor and usually, the contents are not completely homogeneous. Two distinct approaches have been followed to retain the biomass. In one approach, bacterial population, which is flocculent in nature is used so that its loss during high feeding rates could be minimised. In the second approach microbial flora, critical to the treatment process, are immobilised on different types of granular material so that the loss of biomass is restricted to the maximum possible extent. The different types of modern-day approaches to the reactor models are described below.

Upflow anaerobic sludge blanket or upflow anaerobic sludge bed

Upflow anaerobic sludge blankets (UASBS) are very widely employed in anaerobic waste-water treatment in breweries and sugar industries. The unique feature of the reactor is that it allows the flocculation of active microbial cells within the system and the size of the microbial flocculent particles is enough to sediment at a very fast rate. The sedimentation potential is very good; as such the effluent is not allowed to flow through the reactor without disturbing the sedimented microbes. The cell-biomass, therefore, is not lost through the reactor exit along with the treated effluent. The organic matter content is reduced to a greater extent. The present UASB reactors are the result of extensive work carried out by Lettinga and other Dutch scientists in the last decade. UASB avoids plugging by reducing the volume of the packing material and uses a gas collector to encourage settling of sludge. High microbial populations are needed in addition to proper development of a granular sludge to promote settling. It has been shown that for wastes with an organic matter concentration of the order of 500 mg/l, most of which are in solution, the utilisation is of the order of approximately 90 per cent with an HRT of less than one day, which is an extraordinarily efficient performance. But this requires careful adoption of the system. These results, of course, do show a small degree of fluctuation and more knowledge of the physiology and biochemistry of the performance of the microbial populations involved is required.

A key factor in the system is the ability to maintain suitable sedimentation in the floc. Establishing an upflow sludge blanket system depends on suitable microbial population. For this purpose, sewage sludge is a very good source, provided the selection leading to the loss of non-flocculent material and the retention of particles efficiently is established.

The characteristics of suitable sludge blanket bed methanogenesis, as described by Lettinga are mentioned below:

1. It should contain good amounts of suitable sludge that sediments well and gives a sludge bed with concentration of total solids of 100–150 g/l.
2. The flocculent occupies the upper part of the blanket. Its upward movement and penetration, to some degree, into the precise arrangement of particles within the reactor are not constant enough for the production and evolution of gas, which serves to mix the contents to some extent. This is needed to prevent more compact sediment which in turn will reduce the contact of the effluent with the microbial blanket on the bed.
3. Extreme agitation should not be done as this would disturb the microbial blanket and result in the loss of biomass.

The success of UASB reactor depends on the design of the separating system through which the treated effluent leaves the digester. Inevitably some biomass leaves the digester when it reaches the turbulent area. Here, retention of biomass is essential.

Anaerobic contact processor or anaerobic contact reactor

Anaerobic contact reactor is different from the UASB reactor in that it does not require sedimentation of biomass within the primary vessel. However, it must be capable of removing the biomass from the bioreactor in some way. Collection by sedimentation is the most economic way since one requires an associative population which may not be large enough to facilitate increased concentration of biomass in the digester but must be rapid enough to leave suspension when the effluent reaches a separating device. The rector, in fact, uses sludge, which sediments to the bottom and contacts the raw waste. Thus, the settling ability and the mixing of sludge with the waste are important for efficient bioconversion. This type of bioreactor works well for particulate wastes, which settle easily and are completely biodegradable. Though this process is attractive, difficulties such as the failure of organisms leaving the reactor to sediment properly in the separator have been experienced. In such cases, the biomass may be collected by centrifuge and returned. However, this is relatively costlier and shear forces may damage the organisms so that activity is reduced. Flocculation or floatation chemicals may be used to facilitate the collection of biomass, although it would be necessary to ensure that they are free from toxicity. The anaerobic contact process may be regarded as a method for overcoming inadequacies in the natural retraction of microbes within the fermenter and would not be the method of choice if a natural sedimentation could be guaranteed.

Anaerobic attached film reactor or downflow stationary fixed film reactor

This reactor can accommodate a wide variety of wastes. Both the UASB and anaerobic contact processes exploit the ability of microbes to attach to others, yielding sedimented particles in the digester. The particulate material may in turn attach to the microbial cells so as to increase the size and effective density of the floc.

Better degradation/breakdown of the substrates in thus facilitated. The downflow stationary fixed film reactor is an alternative to the above systems. Here, the approach is to deliberately attach the cells to solids support, which will invariably remain within the reactor because of its density, together with its microbial load. The solid may be static or mobile as in fluidised bed systems.

Static system or anaerobic filter reactor

In static systems, the population of microbes is allowed to attach to large-sized supports. They may be rock, stone, pumice or bamboo cuttings. This retains suspended bacteria and waste in a packing material or solid support, where degradation occurs. This is advantageous to treat dilute soluble wastes. However, it is easily plugged by suspended particulate waste. This system could be operated not only as an upflow device but also as a horizontal system or down-flow system. The passage of the effluent leads to methane generation. The suspended solids are retarded in their flow and retained by allowing increased time for their breakdown. The applicability of this system to wastes with a BOD of 1000–2000 mg/l was also studied.

Expanded or fluidised bed reactors

Reactors of this type have been a recent addition to the already existing bioreactors. The potential of this system could be used to the fullest only in anaerobic systems and these reactors could be exploited very efficiently for dilute as well as concentrated wastes. The concentration of biomass in the reactor can approach 30 kg/m^3 while the bulk density of the film is in excess of 100 kg VS/m^3. This high density allows an 80 per cent reduction in COD (from an initial value of 200 mg/l) from settled sewage in treatment times of no more than one hour.

Additionally, the unit removes suspended solids with high efficiency from the waste stream. The expanded bed systems are resistant to shock-loading and are able to operate at ambient temperatures even when these fall close to zero.

Two-Phase Anaerobic Digestion

Considering the diverse nature of the requirement of trophic microbial population in anaerobic digestion, the interspecies electron flow and the apparent limitation of methanogenic process have led to the proposal of phase separation. In phase 1, organic matter is hydrolysed and acidified and in phase 2 the same is subjected to methanation. It would be advantageous as the process of methanogenesis is step-wise. The environmental and operational parameters may be optimised separately to improve the overall kinetics of the process. This would not only protect the limiting microbial population from inhibitory effects of intermediates but also enhance the overall performance of anaerobic digestion.

The process is separated into acid and methane forming phases generally by selective inhibition of methanogens in the first phase digestion either by manipulation of operating parameters or by addition of some chemical inhibitors. The first phase product (acidified substrate) is then transferred to methanogenic phase either by selective diffusion or by manipulation of the flow rate. Two-phase anaerobic digestion has been demonstrated for liquid and particulate substrate by many scientists.

Phase separation has certainly proved better as it can handle very high loading rates in half the residence time required in conventional digestion system. A comparison of the two-phase and conventional anaerobic digestion of soft drink waste water has indicated that in two-phase system, loading rate could be increased up to 4.8 kg VS/m^3/day, while the high rate system could sustain only 0.64 kg/m^3/day. The rate of gas production has also increased to 2.74 volumes compared to 0.4 volumes per liquid volume per day obtained in conventional biomethanation. Two-phase system has also recorded 96 per cent of COD removal whereas it was 84 per cent in conventional system.

Substrates for methanogenesis

Recently, due to widespread use of anaerobic digestion for waste treatment as well as for biogas production a wide variety of substrates have been tried and commercially exploited. The range of biomass materials that can be broken down by anaerobic digestion is very broad. More than 30 types of wastes with potential for methane generation have been listed in Table 16.5. Domestic wastes like cattle manure, sewage sludge, domestic garbage, night soil, agricultural and industrial wastes such as piggery, poultry, crop residues and distillery waste have been successfully used for biomethanation in the past. Till recently, biogas substrates were those which were either natural habitat for the bacteria involved in anaerobic digestion (namely, cattle manure, sewage sludge and fuel wastes) or those which were capable of supporting anaerobic digestion on their own as they contained all the necessary nutrient components.

Table 16.5. Potential organic matter for methane generation.

Crop wastes	Sugarcane trash, weeds, corn and related crop stubble, straw, spoiled fodder
Wastes of animal origin	Cattle shed wastes (dung, urine, litter), poultry litter, sheep and goat droppings, slaughter house wastes (blood and meat), fishery, leather, wool wastes
Wastes of human origin	Faeces, urine, refuse
By-products and wastes from agriculture-based industries	Oil cakes, bagasse, rice bran, tobacco wastes and seeds, wastes from fruit and vegetable processing, press-mud from sugar factories, tea wastes, cotton dust from textile factories
Forest litter	Twigs, bark, branches, leaves
Waste from aquatic plant	Marine algae, sea weeds, water hyacinths

Chapter 17

Newer Sewage and Sludge Treatment Processes

INTRODUCTION

Sewage treatment, or domestic waste-water treatment, is the process of removing contaminants from waste-water and household sewage, both run-off (effluents) and domestic. It includes physical, chemical, and biological processes to remove physical, chemical and biological contaminants. Its objective is to produce a waste stream (or treated effluent) and a solid waste or sludge suitable for discharge or reuse back into the environment. This material is often inadvertently contaminated with many toxic organic and inorganic compounds.

Sewage is created by residences, institutions, hospitals and commercial and industrial establishments. Raw influent (sewage) includes household waste liquid from toilets, baths, showers, kitchens, sinks, and so forth that is disposed of via sewers. In many areas, sewage also includes liquid waste from industry and commerce.

The separation and draining of household waste into greywater and blackwater is becoming more common in the developed world, with greywater being permitted to be used for watering plants or recycled for flushing toilets. A lot of sewage also includes some surface water from roofs or hard-standing areas. Municipal waste-water therefore includes residential, commercial, and industrial liquid waste discharges, and may include stormwater runoff.

Sewage systems capable of handling stormwater are known as combined systems or combined sewers. Such systems are usually avoided since they complicate and thereby reduce the efficiency of sewage treatment plants owing to their seasonality. The variability in flow also leads to often larger than necessary, and subsequently more expensive, treatment facilities.

In addition, heavy storms that contribute more flows than the treatment plant can handle may overwhelm the sewage treatment system, causing a spill or overflow (called a combined sewer overflow, or CSO, in the United States). It is preferable to have a separate storm drain system for stormwater in areas that are developed with sewer systems.

As rainfall runs over the surface of roofs and the ground, it may pick up various contaminants including soil particles and other sediment, heavy metals, organic compounds, animal waste, and oil and grease. Some jurisdictions require stormwater to receive some level of treatment before being discharged directly into waterways. Examples of treatment processes used for stormwater include sedimentation basins, wetlands, buried concrete vaults with various kinds of filters, and vortex separators (to remove coarse solids).

PROCESS OVERVIEW

Sewage can be treated close to where it is created (in septic tanks, biofilters or aerobic treatment systems), or collected and transported via a network of pipes and pump stations to a municipal treatment plant. Sewage collection and treatment is typically subject to state and central regulations and standards. Industrial sources of waste-water often require specialised treatment processes.

Conventional sewage treatment may involve three stages, called primary, secondary and tertiary treatment. Primary treatment consists of temporarily holding the sewage in a quiescent basin where heavy solids can settle to the bottom while oil, grease and lighter solids float to the surface. The settled and floating materials are removed and the remaining liquid may be discharged or subjected to secondary treatment. Secondary treatment removes dissolved and suspended biological matter. Secondary treatment is typically performed by indigenous, water-borne micro-organisms in a managed habitat. Secondary treatment may require a separation process to remove the micro-organisms from the treated water prior to discharge or tertiary treatment. Tertiary treatment is sometimes defined as anything more than primary and secondary treatment. Treated water is sometimes disinfected chemically or physically (for example by lagoons and microfiltration) prior to discharge into a stream, river, bay, lagoon or wetland, or it can be used for the irrigation of a golf course, green way or park. If it is sufficiently clean, it can also be used for groundwater recharge or agricultural purposes (Fig. 17.1).

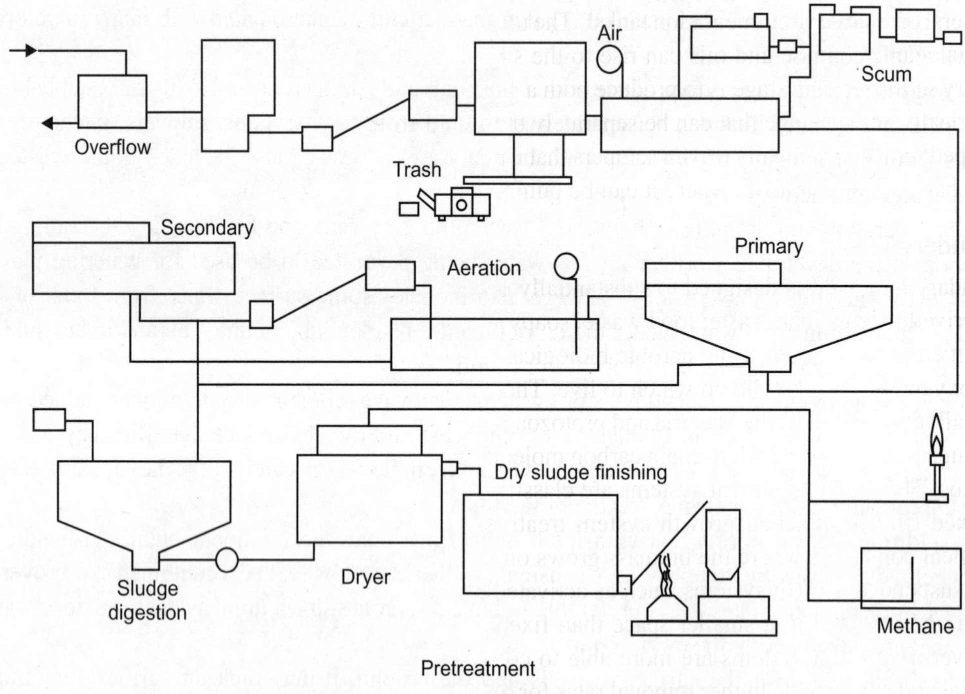

Fig. 17.1. Process flow diagram for a typical large-scale treatment plant.

Pretreatment

Pretreatment removes materials that can be easily collected from the raw waste-water before they damage or clog the pumps and skimmers of primary treatment clarifiers.

Screening

The influent sewage water is strained to remove all large objects carried in the sewage stream, such as rags, sticks, tampons, cans, fruit, etc. This is most commonly done with an automated mechanically raked bar screen in modern plants serving large populations, whilst in smaller or less modern plants manually cleaned screen may be used. The raking action of a mechanical bar screen is typically paced according to the accumulation on the bar screens and/or flow rate. The solids are collected and later disposed in a landfill or incinerated.

Grit removal

Pretreatment may include a sand or grit channel or chamber (sometimes called a degritter) where the velocity of the incoming waste-water is carefully controlled to allow sand, grit and stones to settle, while keeping the majority of the suspended organic material in the water column. Sometimes there is a sand washer (grit classifier) followed by a conveyor that transports the sand to a container for disposal. The contents from the sand catcher may be fed into the incinerator in a sludge processing plant, but in many cases, the sand and grit is sent to a landfill.

Primary Treatment

In the primary sedimentation stage, sewage flows through large tanks, commonly called 'primary clarifiers' or 'primary sedimentation tanks'. The tanks are large enough that sludge can settle and floating material such as grease and oils can rise to the surface and be skimmed off. The main purpose of the primary sedimentation stage is to produce both a generally homogeneous liquid capable of being treated biologically and a sludge that can be separately treated or processed. Primary settling tanks are usually equipped with mechanically driven scrapers that continually drive the collected sludge towards a hopper in the base of the tank from where it can be pumped to further sludge treatment stages.

Secondary Treatment

Secondary treatment is designed to substantially degrade the biological content of the sewage such as are derived from human waste, food waste, soaps and detergent. The majority of municipal plants treat the settled sewage liquor using aerobic biological processes. For this to be effective, the biota require both oxygen and a substrate on which to live. There are a number of ways in which this is done.

In all these methods, the bacteria and protozoa consume biodegradable soluble organic contaminants (e.g. sugars, fats, organic short-chain carbon molecules, etc.) and bind much of the less soluble fractions into floc. Secondary treatment systems are classified as: (i) fixed-film, and (ii) suspended-growth.

Fixed-film or attached growth system treatment process including trickling filter and rotating biological contactors where the biomass grows on media and the sewage passes over its surface.

In suspended-growth systems, such as activated sludge, the biomass is well mixed with the sewage and can be operated in a smaller space than fixed-film systems that treat the same amount of water. However, fixed-film systems are more able to cope with drastic changes in the amount of biological material and can provide higher removal rates for organic material and suspended solids than suspended growth systems.

Roughing filters are intended to treat particularly strong or variable organic loads, typically industrial, to allow them to be treated by conventional secondary treatment processes. Characteristics include typically tall, circular filters filled with open synthetic filter media to which waste-water is applied at a relatively high rate. They are designed to allow high hydraulic loading and a high flow-through of air.

On larger installations, air is forced through the media using blowers. The resultant waste-water is usually within the normal range for conventional treatment processes.

ACTIVATED SLUDGE

In general, activated sludge plants encompass a variety of mechanisms and processes that use dissolved oxygen to promote the growth of biological floc that substantially removes organic material.

The process traps particulate material and can, under ideal conditions, convert ammonia to nitrite and nitrate and ultimately to nitrogen gas (Fig. 17.2).

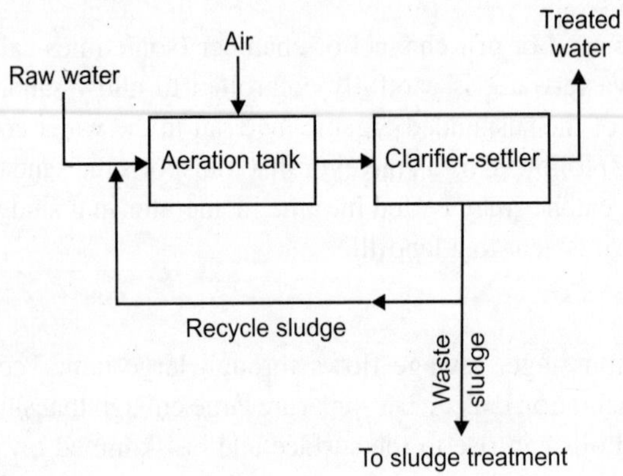

Fig. 17.2. Schematic diagram of an activated sludge process.

Surface-Aerated Basins

Most biological oxidation processes for treating industrial waste-waters have in common the use of oxygen (or air) and microbial action. Surface-aerated basins achieve 80 to 90 per cent removal of (BOD) biochemical oxygen demand with retention times of 1 to 10 days. The basins may range in depth from 1.5 to 5.0 metres and use motor-driven aerators floating on the surface of the waste-water (Fig. 17.3).

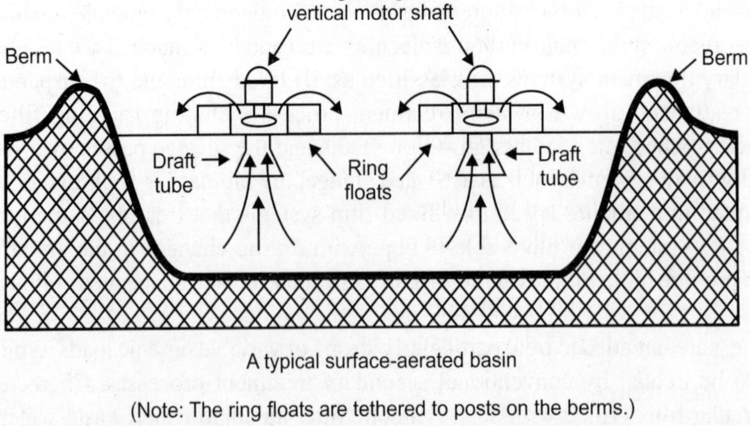

A typical surface-aerated basin

(Note: The ring floats are tethered to posts on the berms.)

Fig. 17.3. A typical surface-aerated basin (using motor-driven floating aerators).

In an aerated basin system, the aerators provide two functions: they transfer air into the basins required by the biological oxidation reactions, and they provide the mixing required for dispersing the air and for contacting the reactants (that is, oxygen, waste-water and microbes). Typically, the floating surface aerators are rated to deliver the amount of air equivalent to 1.8 to 2.7 kg O_2/kW·h. However, they do not provide as good mixing as is normally achieved in activated sludge systems and therefore aerated basins do not achieve the same performance level as activated sludge units.

Filter Beds (Oxidising Beds)

In older plants and plants receiving more variable loads, trickling filter beds are used where the settled sewage liquor is spread onto the surface of a deep bed made up of coke (carbonised coal), limestone chips or specially fabricated plastic media. Such media must have high surface areas to support the biofilms that form. The liquor is distributed through perforated rotating arms radiating from a central pivot. The distributed liquor trickles through this bed and is collected in drains at the base. These drains also provide a source of air which percolates up through the bed, keeping it aerobic. Biological films of bacteria, protozoa and fungi form on the media's surfaces and eat or otherwise reduce the organic content. This biofilm is grazed by insect larvae and worms which help maintain an optimal thickness. Overloading of beds increases the thickness of the film leading to clogging of the filter media and ponding on the surface.

Biological Aerated Filters

Biological aerated (or anoxic) filter (BAF) or biofilters combine filtration with biological carbon reduction, nitrification or denitrification. BAF usually includes a reactor filled with a filter media. The media is either in suspension or supported by a gravel layer at the foot of the filter. The dual purpose of this media is to support highly active biomass that is attached to it and to filter suspended solids. Carbon reduction and ammonia conversion occurs in aerobic mode and sometime achieved in a single reactor while nitrate conversion occurs in anoxic mode. BAF is operated either in upflow or downflow configuration depending on design specified by manufacturer.

Membrane Bioreactors

Membrane bioreactors (MBR) combines activated sludge treatment with a membrane liquid-solid separation process. The membrane component uses low pressure microfiltration or ultra filtration membranes and eliminates the need for clarification and tertiary filtration. The membranes are typically immersed in the aeration tank (however, some applications utilise a separate membrane tank). One of the key benefits of a membrane bioreactor system is that it effectively overcomes the limitations associated with poor settling of sludge in conventional activated sludge (CAS) processes. The technology permits bioreactor operation with considerably higher mixed liquor suspended solids (MLSS) concentration than CAS systems, which are limited by sludge settling. The process is typically operated at MLSS in the range of 8000–12,000 mg/l, while CAS are operated in the range of 2000–3000 mg/l. The elevated biomass concentration in the membrane bioreactor process allows for very effective removal of both soluble and particulate biodegradable materials at higher loading rates. Thus increased sludge retention times (SRTs)—usually exceeding 15 days—ensure complete nitrification even in extremely cold weather.

Secondary Sedimentation

The final step in the secondary treatment stage is to settle out the biological floc or filter material and produce sewage water containing very low levels of organic material and suspended matter.

Rotating Biological Contactors

Rotating biological contactors (RBCs) are mechanical secondary treatment systems, which are robust and capable of withstanding surges in organic load. RBCs were first installed in Germany in 1960 and have since been developed and refined into a reliable operating unit. The rotating disks support the growth of bacteria and micro-organisms present in the sewage, which breakdown and stabilise organic pollutants. To be successful, micro-organisms need both oxygen to live and food to grow. Oxygen is obtained from the atmosphere as the disks rotate. As the micro-organisms grow, they build up on the media until they are sloughed off due to shear forces provided by the rotating discs in the sewage. Effluent from the RBC is then passed through final clarifiers where the micro-organisms in suspension settle as a sludge. The sludge is withdrawn from the clarifier for further treatment (Fig. 17.4).

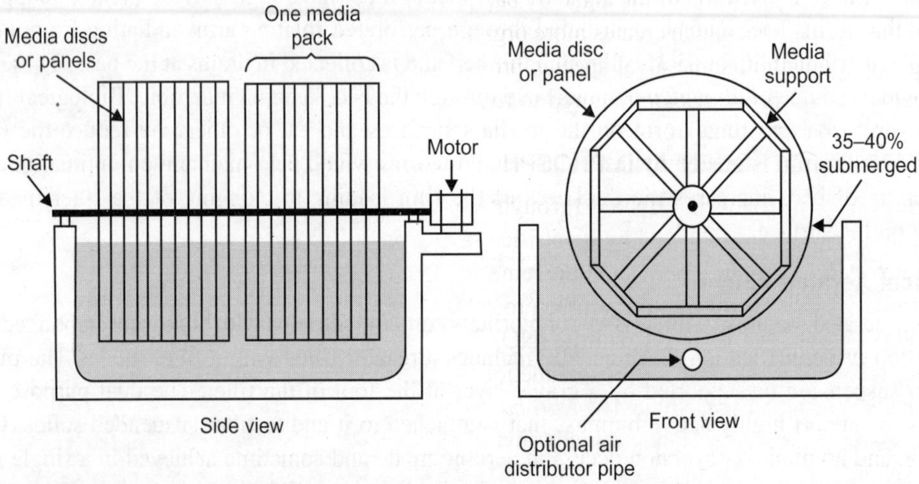

Fig. 17.4. Schematic diagram of a typical rotating biological contactor (RBC). The treated effluent clarifier/settler is not included in the diagram.

Tertiary Treatment

The purpose of tertiary treatment is to provide a final treatment stage to raise the effluent quality before it is discharged to the receiving environment (sea, river, lake, ground, etc.). More than one tertiary treatment process may be used at any treatment plant. If disinfection is practiced, it is always the final process. It is also called 'effluent polishing'.

Filtration

Sand filtration removes much of the residual suspended matter. Filtration over activated carbon removes residual toxins.

Lagooning

Lagooning provides settlement and further biological improvement through storage in large man-made ponds or lagoons. These lagoons are highly aerobic and colonisation by native macrophytes, especially reeds, is often encouraged. Small filter feeding invertebrates such as Daphnia and species of Rotifera greatly assist in treatment by removing fine particulates.

Constructed wetlands

Constructed wetlands include engineered reedbeds and a range of similar methodologies, all of which provide a high degree of aerobic biological improvement and can often be used instead of secondary treatment for small communities.

Nutrient removal

Waste-water may contain high levels of the nutrients nitrogen and phosphorus. Excessive release to the environment can lead to a build up of nutrients, called eutrophication, which can in turn encourage the overgrowth of weeds, algae, and cyanobacteria (blue-green algae). This may cause an algal bloom, a rapid growth in the population of algae. The algae numbers are unsustainable and eventually most of them die. The decomposition of the algae by bacteria uses up so much of oxygen in the water that most or all of the animals die, which creates more organic matter for the bacteria to decompose. In addition to causing deoxygenation, some algal species produce toxins that contaminate drinking water supplies. Different treatment processes are required to remove nitrogen and phosphorus.

REMOVAL OF NITROGEN AND PHOSPHORUS

The removal of nitrogen is effected through the biological oxidation of nitrogen from ammonia (nitrification) to nitrate, followed by denitrification, the reduction of nitrate to nitrogen gas. Nitrogen gas is released to the atmosphere and thus removed from the water.

Nitrification itself is a two-step aerobic process, each step facilitated by a different type of bacteria. The oxidation of ammonia (NH_3) to nitrite (NO_2^-) is most often facilitated by *Nitrosomonas* spp. (nitroso referring to the formation of a nitroso functional group). Nitrite oxidation to nitrate (NO_3^-), though traditionally believed to be facilitated by *Nitrobacter* spp. (nitro referring the formation of a nitro functional group), is now known to be facilitated in the environment almost exclusively by *Nitrospira* spp.

Phosphorus removal is important as it is a limiting nutrient for algae growth in many fresh water systems. It is also particularly important for water reuse systems where high phosphorus concentrations may lead to fouling of downstream equipment such as reverse osmosis.

Phosphorus can be removed biologically in a process called enhanced biological phosphorus removal. In this process, specific bacteria, called polyphosphate accumulating organisms (PAOs), are selectively enriched and accumulate large quantities of phosphorus within their cells (up to 20 per cent of their mass). When the biomass enriched in these bacteria is separated from the treated water, these biosolids have a high fertiliser value.

Nitrification

Ammonia is oxidised rapidly to nitrate in the environment and in waste-water-treatment systems in a process known as nitrification. The conversion is carried out by two groups of chemoautotrophic bacteria, which use the oxidation of ammonia as a source of energy. The first stage of ammonia oxidation is carried out mainly by the genera *Nitrosomonas, Nitrosococcus, Nitrosospira, Nitrocystis,* and *Nitrosogloea*. The reaction is as follows, although the oxidation of ammonia is more complex than given in the Eq. 17.1.

$$2NH_4^+ + 3O_2 \longrightarrow 2NO_2^- + 4H^+ + 2H_2O + (\text{energy } 480\text{–}700 \text{ kJ}) \qquad \text{... (17.1)}$$

The energy released is used by the organisms to synthesise cell components from inorganic sources. The release of hydrogen ions can cause a drop in pH and it is clear that a good supply of oxygen is required. The growth of nitrifying bacteria is very slow (μ_{max} 0.1–1 day^{-1}) compared with that of heterotrophic bacteria (μ_{max} 0.46–2.2 day^{-1}).

The nitrite formed is converted to nitrate by the genera *Nitrobacter, Nitrocystis, Nitrosococcus,* and *Nitrosocystis*, but *Nitrobacter* has been the most studied. The reactions is given in Eq. 17.2.

$$2NO_2^- + O_2 \longrightarrow 2NO_3^- + (\text{energy } 130\text{–}180 \text{ kJ}) \qquad \text{... (17.2)}$$

As the oxidation of nitrite to nitrate yields less energy than the oxidation of ammonia the cell yield of *Nitrobacter* is less than *Nitrosomonas* and the growth rates are also slow with a μ_{max} of 0.28–1.44 day^{-1}. The characteristics of the organisms involved in nitrification affect the waste-water treatment as follows:

1. The growth rate is slower than for heterotrophic organisms so that the organic load has to be balanced to their slower growth rate, otherwise the organisms will be washed out.
2. There is a low cell yield per unit of ammonium oxidised.
3. The organisms require significant amounts of oxygen, 4.2 g/g of NH_4^+ converted.
4. The system may need some form of buffering due to the acid conditions produced by the hydrogen ions.

If nitrification is not required in the sewage process then a higher rate of flow can be used. Recently bioaugmentation with nitrifying bacteria has been shown to be effective in maintaining nitrification in stress situations such as low temperatures and high SRT values.

Denitrification

Nitrification in treatment plants and soils combined with the nitrates from agricultural run-off can give rise to high nitrate levels (above 50 mg/l) in waterways, which are used to supply drinking water. High nitrate levels are associated with one disease, methaemoglobinaemia, which affects children below the age of 6 months. The children have an incomplete digestive system and the intake of nitrate leads to the accumulation of nitrite ions, which enter the blood system and block oxygen transport by haemoglobin. Thus the EU have set an absolute limit of 50 mg/l for nitrate, and a recommended limit of 25 mg/l, although a number of UK water-treatment systems are working at 80 mg/l.

Nitrate can be converted to nitrite in the human stomach and nitrite has been shown to be converted to carcinogenic nitrosamines and leads to concern over the development of stomach cancer on consumption of high nitrate water.

Nitrate removal, ion exchange, or biological processes can carry out denitrification. The ion-exchange process depends on the resin's affinity, which on a conventional anion-exchange resin is given as under.

$$SO_4^{2-} >> NO_3^- > Cl^- \geq HCO_3$$

Any sulphate in the waste will bind in preference to nitrate, but once this has occurred nitrate will exchange with chloride. Once the resin is exhausted it will require regeneration with excess sodium chloride which yields a solution Containing high concentrations of sodium sulphate, sodium nitrate, and sodium chloride which will need disposal. Adding these high-salt solutions to waterways is unacceptable and in practice it is passed on to sewage works for treatment.

The biological conversion of nitrate to nitrite and eventually nitrogen occurs under conditions where oxygen is very low or absent. The process of oxidation involves the loss of electrons and in normal conditions oxygen acts as an electron acceptor but when oxygen levels are low inorganic ions such as nitrate, phosphate, and sulphate can act as electron acceptors. In waste-water where nitrification has occurred, combined with nitrate from agricultural run-off the concentration of nitrate will be higher than sulphate or phosphate. A number of facultative heterotrophic micro-organisms occur in sewage-treatment systems which are capable of converting nitrate to nitrogen provided an electron donor is present.

$$3NO_3^- + 6H^+ \longrightarrow 3NO_2^- + 3H_2O \qquad \text{... (17.3)}$$

$$2NO_2^- + 8H^+ + 6e^- \longrightarrow N_2 + 4H_2O \qquad \text{... (17.4)}$$

The electron donor is usually an organic compound and in some cases methanol which has been used to supplement the normal organic source. The reactions with methanol are as follows:

$$3NO_3^- + CH_3OH \longrightarrow 3NO_2^- + CO_2 + 2H_2O \qquad \text{... (17.5)}$$

$$2NO_2^- + CH_3OH \longrightarrow N_2 + CO_2 + H_2O + 2OH^- \qquad \text{... (17.6)}$$

The process of denitrification requires low oxygen levels (anaerobic), an organic carbon energy source, a level of nitrate of 2 mg/l or above, and a pH of 6.5–7.5.

Nitrification and Denitrification Processes

In order to meet increasing European standards, the removal of ammonia and other nitrogenous compounds from waste-water has been developed. Within the sewage system the processes of biological nitrification and denitrification can be organised in a number of ways. In general, the first step of the removal of ammonia by nitrification can be carried out in parallel with the removal of organic material, provided that the hydraulic retention time is not too short. Denitrification, in contrast, requires a change in growth conditions from aerobic to anaerobic and an organic carbon source. Both nitrification and denitrification can be achieved by partitioning the sewage-treatment system (Fig. 17.5) or providing separate reactors which can be used with both suspended and fixed-film cultures (Fig. 17.6). In the single vessel the anoxic zone is situated at the start of the aeration tank where the carbon levels are high, and the anoxic conditions are achieved by stopping aeration. The process using separate vessels is much easier to use and control.

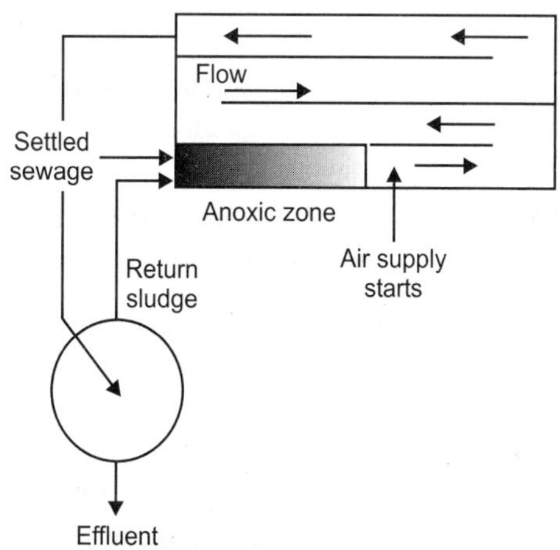

Fig. 17.5. An activated-sludge process with the provision of an anoxic zone (anaerobic) at the start in order to achieve denitrification. The anoxic zone is formed by stopping aeration at the first stage where the organic material is highest.

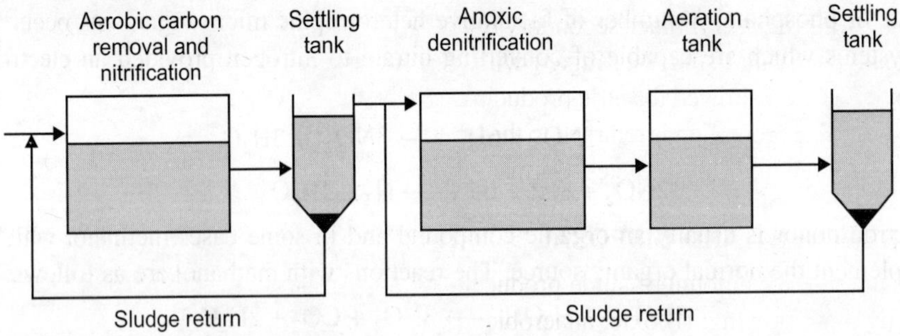

Fig. 17.6. An outline of a two-stage or sequencing process for both nitrification and denitrification. In the first stage the normal activated-sludge process occurs with the removal of organic materials and nitrification. In the second stage anoxic conditions cause denitrification which is followed by an aeration tank to strip out the nitrogen formed to ensure precipitation in the settling tank.

Another system for combined nitrification and denitrification is the sequencing batch reactor (SBR) where a single vessel is used but a programmed sequence of operations is applied, which can be feeding, anaerobic conditions, aerobic conditions, sludge settling, and effluent removal. This type of operation has been used for the treatment of a number of wastes such as agricultural run-off and landfill leachates but it has the potential for the combining nitrification and denitrification. The ammonia levels drop during the initial anaerobic phase and the subsequent aerobic phase. Nitrate concentration in contrast is low at the start but rises due to nitrification in the aerobic phase. Both nitrate and nitrite are denitrified during the anoxic phase. The sequencing mode can also be applied to anaerobic reactors and in two-stage processes where the second stage is the anoxic phase. Some new systems have been developed based on partial nitrification of ammonia to nitrite in anaerobic ammonium oxidation. The systems are based on the following reaction carried out by planctomyces-like bacteria.

$$NH_4^+ + NO_2^- \longrightarrow N_2 + 2H_2O \quad \text{... (17.7)}$$

Disinfection

The purpose of disinfection in the treatment of waste-water is to substantially reduce the number of micro-organisms in the water to be discharged back into the environment. The effectiveness of disinfection depends on the quality of the water being treated (e.g., cloudiness, pH, etc.), the type of disinfection being used, the disinfectant dosage (concentration and time), and other environmental variables. Cloudy water will be treated less successfully since solid matter can shield organisms, especially from ultraviolet light or if contact times are low. Generally, short contact times, low doses and high flows all militate against effective disinfection. Common methods of disinfection include ozone, chlorine, or ultraviolet light. Chloramine, which is used for drinking water, is not used in waste-water treatment because of its persistence.

Package plants and batch reactors

In order to use less space, treat difficult waste, deal with intermittent flow or achieve higher environmental standards, a number of designs of hybrid treatment plants have been produced. Such plants often combine all or at least two stages of the three main treatment stages into one combined stage. In the UK, where a large number of sewage treatment plants serve small populations, package plants are a viable alternative to building discrete structures for each process stage.

One type of system that combines secondary treatment and settlement is the sequencing batch reactor (SBR). Typically, activated sludge is mixed with raw incoming sewage and mixed and aerated. The resultant mixture is then allowed to settle producing a high quality effluent. The settled sludge is run-off and re-aerated before a proportion is returned to the headworks. SBR plants are now being deployed in many parts of the world including United States, Germany.

SLUDGE TREATMENT AND DISPOSAL

One solution to the large amount of slude produced is to reduce the levels of sludge produced. this can be achieved by the following: (i) oxic, anaerobic digester, (ii) high-dissolved oxygen, (iii) uncouplers, and (iv) ozonation.

The sludges accumulated in a waste-water treatment process must be treated and disposed of in a safe and effective manner. The purpose of digestion is to reduce the amount of organic matter and the number of disease-causing micro-organisms present in the solids. The most common treatment options include anaerobic digestion, aerobic digestion, and composting. Incineration is also used albeit to a much lesser degree.

Choice of a waste-water solid treatment method depends on the amount of solids generated and other site-specific conditions. However, in general, composting is most often applied to smaller-scale applications followed by aerobic digestion and then lastly anaerobic digestion for the larger-scale municipal applications.

Anaerobic Digestion

Anaerobic digestion is a bacterial process that is carried out in the absence of oxygen. The process can either be thermophilic digestion, in which sludge is fermented in tanks at a temperature of 55°C, or mesophilic, at a temperature of around 36°C. Though allowing shorter retention time (and thus smaller tanks), thermophilic digestion is more expensive in terms of energy consumption for heating the sludge. Flow diagram of conventional anaerobic-sludge process shown in Fig. 17.7.

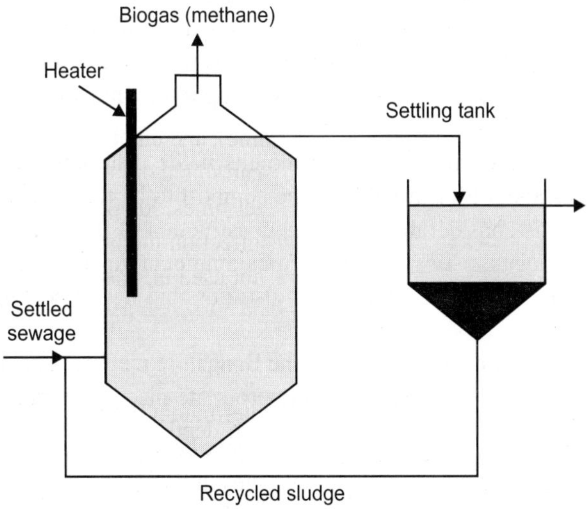

Fig. 17.7. Conventional anaerobic-sludge process. The vessel is heated and sealed so that the biogas can be collected.

Aerobic Digestion

Aerobic digestion is a bacterial process occurring in the presence of oxygen. Under aerobic conditions, bacteria rapidly consume organic matter and convert it into carbon dioxide. The operating costs are characteristically much greater for aerobic digestion because of the energy costs needed to add oxygen to the process.

Composting

Composting is also an aerobic process that involves mixing the sludge with sources of carbon such as sawdust, straw or wood chips. In the presence of oxygen, bacteria digest both the waste-water solids and the added carbon source and, in doing so, produce a large amount of heat.

Composting can be defined as the aerobic, thermophilic, biodegradation of organic wastes at 40–70 per cent moisture content to a relatively stable, humus-like material. This material is used primarily as a soil conditioner. The process can be expressed as follows:

$$\underset{\text{Matter}}{\text{Organic}} + \underset{\text{Oxygen}}{O_2} + \text{micro-organisms} \longrightarrow \underset{\substack{\text{Carbon}\\\text{dioxide}}}{CO_2} + \underset{\text{Water}}{H_2O} + \text{heat}$$

Composting was originally developed in Asian countries where it is still in widespread use because of a scarcity of fertiliser and labour is inexpensive.

Methods of composting

Basically, any system or design that ensure efficient decomposition of organic matter can constitute a composting methods. However, some methods, worked out long ago, have been in vogue over the past several years with a little or no modifications. These methods are based upon simple manipulations of the raw materials and conditions of composting and are being followed in most practical situations. Conventionally, two methods of composting are known in India. A brief description of these methods is given below:

Indore method: The Indore method is an aerobic process and hence precludes adequate supply of oxygen during the decomposition process. Waste organic materials such as straw, garbage, leaves and plant clippings are laid in a heap or pit in alternate layers with animal manure and soil. The layers are repeated until the heap reaches a height of 1.5–2.0 m. The final layer should be of the compostable material covered by a thin layer of soil, about 60 mm. To provide optimum moisture (60–70 per cent), water is sprinkled over each layer.

The finished compost is ready in about three months if the materials are properly shredded and layered. Generally, it takes about four months for the compost to be ready. The Indore method has the disadvantage that is demands considerable labour in construction of the heap, turning of material and maintenance of adequate moisture. Loss of nitrogen as ammonia gas also takes place, which can be considerably reduced. Economy in use of water is also possible if composting is carried out in pits instead of heaps.

Bangalore method: To carry out composting by the Bangalore method, trenches of about 1 m depth, 1.5–2.5 m width and of any length are made at an appropriate place, generally on the outskirts of the city. If material is limited, one can go in for pits of 1 m depth, 1.5 m width and 3–4 m length. The compostable refuse is dumped into the trench or pit and spread out with rakes or forked shovels to make a layer of about 1.5 cm thickness. Night soil or dung is then placed over the refuse in a layer of about 5 cm.

The process is repeated until the trench of pit is filled up to about 30 cm above the ground level and a final layer of compostable material is placed on the top. At each layering, water is sprinkled over the material to make it optimally moist. The above-ground material is made into a dome shape and covered with about 2.5 cm mud-plaster. If all operations are properly carried out, the compost is ready in about five to six months, a period of about one-and-a-half times longer than that for aerobic composting by the Indore method.

Beccari method: The Beccari method, which was developed to increase the efficiency of composting. This system allows for the recovery of gaseous products of the process, a thoroughness of the organic matter decay and a more stable end-product (humus).

The process operates as follows:

1. Initially, all the air intake and outflow valves are closed and the system is allowed to go anaerobic with temperatures rising to about 150°F with some organic matter volume reduction.
2. The vents and air valves are opened, allowing the gas products to escape. The mass is left to dry and air is then admitted. The process now goes aerobic.
3. A continuous operation results and air flows freely in and out.
4. The detention time for the organic matter in the container is about 45 days.

The efficiency of the process can be improved through.

1. Mixing.
2. Adding additional oxygen.
3. Forcing air through the mass up the bottom ports.
4. Shredding the material to effect a greater surface area to be biologically attacked.

Leaves can be composted almost anywhere, however, the process can be accelerated by constructing the leaves into long, high piles. Initially, biological degradation is aerobic in nature with its characteristics products of carbon dioxide (CO_2), wastes and nitrates. However, if the piles are not mixed and aerated, the oxygen in the windrow is depleted and facultative and anaerobic micro-organisms begin to grow.

The products of anaerobic decomposition are methane (CH_4), carbon dioxide (CO_2), water (H_2O), some organic acids, nitrogen (N_2), ammonia (NH_3), ammonium salts (NH_4) and sulphides of iron, manganese and hydrogen. Gases are given off until decomposition is complete and must be monitored because they can: (i) escape to the atmosphere, (ii) react with the soil, and (iii) react with water.

Usually, in leaf composting, the major concern is from the sulphides which cause odour problems. Odour complaints occur upon the emission of sulphide gases when the leaves are turned after the system has gone anaerobic. On the other hand, if leaves are continually mixed, the system stays aerobic, or, if the leaves are left standing, they might keep the gases sealed within the windrow.

Since, composting involves the biological degradation of organic matter by a variety of naturally-occurring micro-organisms, the micro-organisms are essential to the composting process. Commercial composting must provide an environment for these micro-organisms to perform most efficiently. Efficiencies are measured in time required for stabilisation. As stated, the most important parameters of the composting process are temperature, available oxygen and the nutrients and moisture content of the waste. The pH of the waste is also an important factor of composting.

All composting operations undergo five processes: (i) preparation, (ii) digestion, (iii) curing, (iv) finishing, and (v) storage or disposal.

Composting is dependent on micro-organisms utilising organic matter as nutrients. These micro-organisms are readily available in sufficient quantities in all materials to be composted, yet they may lack the nutrients necessary to sustain the process. These nutrients are supplied through the use of sewage sludge or other additives.

Factors controlling microbiological decomposition

Present methods of composting are mainly of aerobic nature. Therefore, to understand the principles and practice of composting, one must have an insight into the physical conditions necessary for successful composting. The success in preparing good quality compost depends upon the creation of a favourable environment for the growth and activities of the micro-organisms in the system. A compost heap is a miniature ecosystem where interactions between biological and abiological factors bring about the desired changes.

Chemical nature of raw material: The main constituents of plant residues are the carbonaceous compounds such as cellulose, hemicellulose and lignin, occurring in that order of abundance. Nitrogenous constituents (proteins) occur to a lesser extent. Protein constituents, cellulose and hemicellulose decompose easily. Lignin, being a complex aromatic polymer, is resistant to microbial attack to a considerable extent. That is how most components of lignin reach the finally produced humus in the compost. It is for this reason that most microflora, which develop on naturally decomposing plant materials, are largely cellulolytic bacteria and fungi. Substantial cellulase activity by such fungi on waste organic substrates has been reported by several investigators. Hence, good raw material for composting should be predominantly of cellulosic nature. The inclusion of weeds, clippings of green grass, bushes and green leguminous plants, which have high weeds, cellulose and low lignin contents, helps greatly in the rapid composting of mature organic residues. At the same time, lignin is not entirely recalcitrant to microbial decomposition. It also suffers slow degradation following the flush of cellulosic degradation. A number of fungi, particularly those belonging to the *Basidiomycetes* groups, are well known for their lignin decomposing property. Some bacteria and actinomycetes also possess lignolytic characteristics.

Carbon:Nitrogen ratio: The abundance of carbon and protein in organic matter and the carbon:nitrogen (C:N) ratio play a very significant role in microbial food supply. Generally, C:N ratios ranging from 30 to 50 have been used for composting. Higher ratios slow down decomposition and hence composting takes a longer time. On the other hand, ratios lower than 30 cause rapid decomposition and may result in an appreciable loss of nitrogen as ammonia. In fact, a higher C:N ratio causes greater conservation of the nitrogen originally present in the compostable material; but this can only be gained at the cost of composting time. The C:N ratio may be adjusted by incorporating materials such as leguminous hays, trimmings of green bushes, night soil or sewage sludge. Addition of inorganic nitrogen should generally be avoided, as it increases costs of composting. Addition of inorganic nitrogen does stimulate decomposition of nitrogen-poor residues. This recourse may be taken only when other means of C:N ratio adjustments are not available and the compost is required in a short time.

Moisture: Provision of optimum moisture in the substrate material is essential for the metabolism of micro-organisms. Moisture content of 60–70 per cent is generally considered optimum to start with. At later stages of decomposition, it may be 50–60 per cent. Excess moisture creates anaerobic conditions in the heap and brings about putrefaction. Putrefaction, the decomposition by anaerobic micro-organisms, produces disagreeable odours and undesirable products. If the material has insufficient water, the growth and proliferation of micro-organisms as well as the rate of decomposition of the organic material are slowed down or even stopped.

Aeration: Aeration controls the internal environment of the compost heaps, which in turn regulates the microbiological activity during composting. Excessive watering, use of large quantities of fine and green materials for composting and heaps of large dimensions, all these contribute to aeration. If aeration is poor, anaerobic conditions predominate. As a result, the rate of decomposition is slow. The nitrates formed during composting are converted to gaseous form of nitrogen. These escape into the air, lowering the mineral value of the compost. In all traditional methods, good aeration is achieved by giving frequent turnings to the material, as in the case of the Indore method. There is a slight loss of heat during turning, but the temperature again rises to its original level in two to three days.

Temperature: Under optimum conditions of air, temperature and food balance, there is a rapid rise in temperature of the heap or pit as the decomposition proceeds. This is because of the metabolic activity of microbes present in it. In the aerobic system, the temperature rises to 50°–60°C in just a few days and goes even up to 70°C in two to three weeks. The high temperature rise in the compost heap destroys weed seeds, pathogenic micro-organisms, maggots and worms and prevents fly breeding. When the compost is prepared in pits or trenches, the rise in temperature will not be as high as in heaps, but the rise is maintained over a longer period. The considerable rise in temperature during the early stages of decomposition is supported by several experimental evidence.

Reaction: Reaction or pH is the least significant factor in general composting processes. The optimum pH for most micro-organisms is between 6.5 and 7.5. The pH of the waste materials usually available for composting under rural conditions falls in this narrow range. Hence, the problem of pH control does not arise. No doubt, organic acids are produced during decomposition of the organic matter, yet in an active composting system, their existence is only transitory. Problems may arise if the material obtained for composting has undergone putrefaction, as appreciable amounts of troublesome organic acids are produced during anaerobic decomposition. In such a situation, addition of materials like lime or wood ash helps in neutralising excessive acidity. However, a rise in pH beyond 7.5 may make the environment alkaline, which may cause loss of nitrogen as ammonia.

Incineration

Incineration of sludge is less common due to air emissions concerns and the supplemental fuel (typically natural gas or fuel oil) required to burn the low calorific value sludge and vapourise residual water. Stepped multiple hearth incinerators with high residence time as well as fluidised bed incinerators are the most common systems used to combust waste-water sludge. Co-firing in municipal waste-to-energy plants is occasionally done, this option being less expensive assuming the facilities already exist for solid waste as well as no need for auxiliary fuel. An outline of the process of sewage sludge incineration in shown in Fig. 17.8.

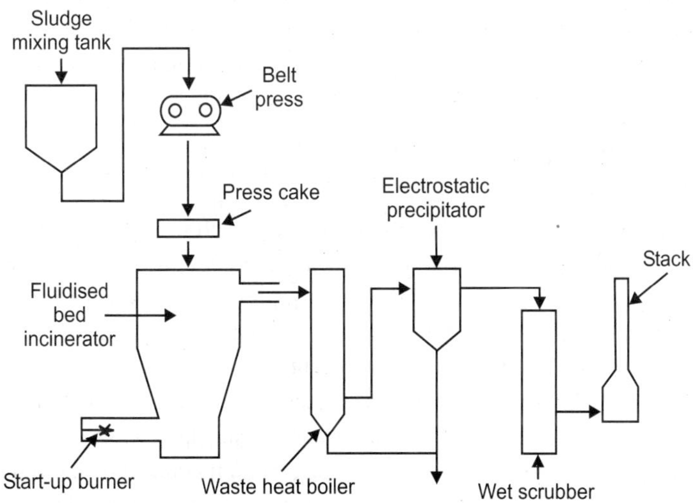

Fig. 17.8. An outline of the process of autocalorific sewage sludge incineration, including an electrostatic precipitator and wet scrubber to clean the flue gases.

Sludge disposal

When a liquid sludge is produced, further treatment may be required to make it suitable for final disposal. Typically, sludges are thickened (dewatered) to reduce the volumes transported off-site for disposal. There is no process which completely eliminates the need to dispose of biosolids. There is, however, an additional step some cities are taking to superheat the waste-water sludge and convert it into small pelletised granules that are high in nitrogen and other organic materials. In New York City, for example, several sewage treatment plants have dewatering facilities that use large centrifuges along with the addition of chemicals such as polymer to further remove liquid from the sludge. The removed fluid, called centrate, is typically reintroduced into the waste-water process. The product which is left is called 'cake' and that is picked up by companies which turn it into fertiliser pellets. This product is then sold to local farmers and turf farms as a soil amendment or fertiliser, reducing the amount of space required to dispose of sludge in landfills.

Treatment in the Receiving Environment

Many processes in a waste-water treatment plant are designed to mimic the natural treatment processes that occur in the environment, whether that environment is a natural water body or the ground. If not overloaded, bacteria in the environment will consume organic contaminants, although this will reduce the levels of oxygen in the water and may significantly change the overall ecology of the receiving water. Native bacterial populations feed on the organic contaminants, and the numbers of disease-causing micro-organisms are reduced by natural environmental conditions such as predation or exposure to ultraviolet radiation. Consequently, in cases where the receiving environment provides a high level of dilution, a high degree of waste-water treatment may not be required. However, recent evidence has demonstrated that very low levels of certain contaminants in waste-water, including hormones (from animal husbandry and residue from human hormonal contraception methods) and synthetic materials such as phthalates that mimic hormones in their action, can have an unpredictable adverse impact on the natural biota and potentially on humans if the water is reused for drinking water. In the US and EU, uncontrolled discharges of waste-water to the environment are not permitted under law, and strict water quality requirements are to be met. A significant threat in the coming decades will be the increasing uncontrolled discharges of waste-water within rapidly developing countries.

Sewage Treatment in Developing Countries

There are few reliable figures on the share of the waste-water collected in sewers that is being treated in the world. In many developing countries the bulk of domestic and industrial waste-water is discharged without any treatment or after primary treatment only. In Latin America about 15 per cent of collected waste-water passes through treatment plants (with varying levels of actual treatment). In Venezuela, a below average country in South America with respect to waste-water treatment, 97 per cent of the country's sewage is discharged raw into the environment.

In a relatively developed Middle Eastern country such as Iran, Tehran's majority of population has totally untreated sewage injected to the city's groundwater. Israel has also aggressively pursued the use of treated sewer water for irrigation. In 2008, agriculture in Israel consumed 500 million cubic metres of potable water and an equal amount of treated sewer water. The country plans to provide a further 200 million cubic metres of recycled sewer water and build more desalination plants to supply even more water. Most of sub-Saharan Africa is without waste-water treatment.

SEWAGE TREATMENT USING MICROBIAL SYSTEMS—DEGRADATION OF SEWAGE

The role of micro-organisms in the decomposition of sewage and other waste materials has long been recognised. Conventional sewage treatment involves the use of micro-organisms, which develop naturally within the sewage treatment system.

In some newer approaches, however, the sewage is inoculated with a specific micro-organism, which has been specially selected for that particular sewage treatment process. Such organisms might be called 'starter cultures'. The use of starter cultures increases the efficiency of sewage degradation. Following are some of the examples of sewage treatment by using starter cultures:

1. Bacteria have been developed, which degrade alkanes and aromatic compounds at 0°–15°C, in saline habitats. These could be useful in the degradation of oil spills in the ocean.
2. Starter cultures of mixed micro-organisms are available, which metabolise DDT and polychlorinated diphenols or phenols and which possess high protease, lipase or cellulase activity.
3. Starter cultures have also been used to deodourise animal excrements.
4. A strain of *Pseudomonas putida* containing plasmids has been developed, which can degrade octane, xylene, metaxylene and camphor.

The degradation of sewage by micro-organisms requires large amount of oxygen, so that in order to provide room for oxygen most sewage treatment plants are bulky.

For improving the efficiency, tubular loop reactors and air lift fermenters have been developed. They both have better oxygen transfer efficiency. Aeration by pure oxygen has also been tried in USA and West Germany. Commercial sewage treatment plants that use pure oxygen, are marketed by several companies.

ADDITION OF PURE OXYGEN

The aeration of the activated sludge can be improved by the addition of pure oxygen to a closed system or to open tanks. The closed system has the advantage that the oxygen is not lost to the atmosphere but the presence of high oxygen concentrations does constitute an explosive hazard requiring strict safety precautions. The high aeration also causes an accumulation of carbon dioxide that can reduce the pH and thus reduce nitrification (Fig. 17.9). A system like this was marketed in the UK as the Unox system. The tanks are divided into a series of compartments, each of which is mixed by a surface aerator. This type of system can be used to sustain a higher biomass (increased MLSS), with lower sludge production and double the loading rate.

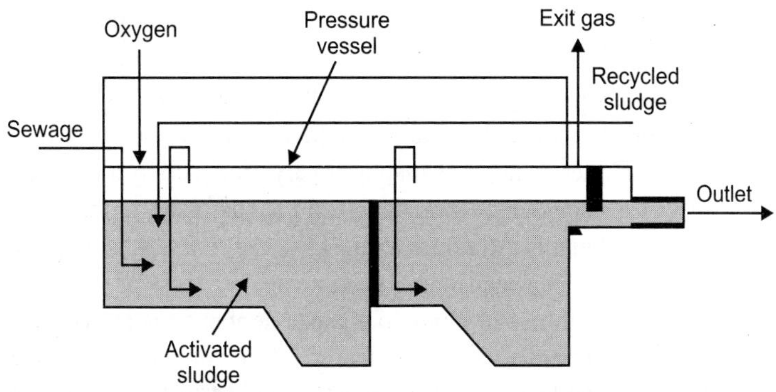

Fig. 17.9. An activated-sludge process using pure oxygen to improve waste removal. Oxygen has a low solubility in water (7 mg/l) so that the application of pure oxygen will improve oxygen transfer in this system. An example is the Unox system, which is sealed to avoid the loss of expensive oxygen and to reduce the fire hazard of pure oxygen. Oxygen applied under pressure is forced in the activated sludge in a number of chambers arranged in a cascade. The carbon dioxide and residual oxygen is collected and treated at 1000°C.

Captor Process

In order to maintain a high biomass at the start of the plug flow process of waste treatment by activated sludge a modification has been used. Here the activated-sludge biomass is immobilised in reticulated plastic pads of 25 mm × 25 mm × 12 mm in dimension, similar in nature to washing-up pads. The activated sludge micro-organisms from aggregates readily colonise these pads.

The pads are retained in the early part of the aeration tank by screens and have been shown to give higher biomass levels of 6–8 g/l (Fig. 17.10). To maintain an active biomass some of the pads are stripped of excess biomass by a system that removes the pads, squeezes out the sludge, and returns the empty pads to the tank.

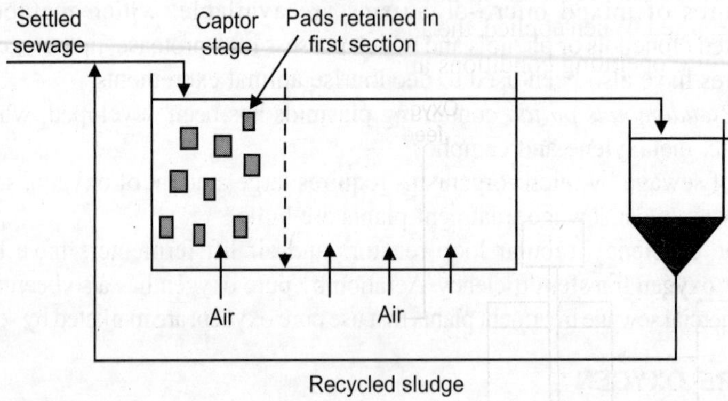

Fig. 17.10. The Captor process. A high biomass is retained at the start of the activated-sludge process where the organic material is highest by immobilising the biomass is plastic pads. The pads are small (10 cm) plastic meshes that form a large area on which micro-organisms.

UNOX SYSTEM

The UNOX System has been developed to improve upon the conventional activated sludge process. The use of enriched oxygen in a simple and economical multistage gas liquid contacting device allows oxygen to be transferred to waste-water at increased rates with significant decreases in power requirements over those required when using air as the oxygen supply media. The elimination of the mass transfer restriction allows operation at solids levels of 4500–7000 mg/l while maintaining a dissolved oxygen level of 8–10 mg/l in the mixed liquor. Retention times for the process can be correspondingly decreased to 1–2 hrs. A highly flocculant sludge is obtained which has excellent settling and dewatering characteristics and is produced in less quantities than normally produced by a conventional air activated sludge process. The process has been demonstrated in a 2.5 mgd activated sludge plant at Batavia, New York. During the Federal Water Quality Administration (FWQA) contract the UNOX process was able to demonstrate consistent BOD and suspended solids removals in excess of 90 per cent.

A number of pilot plant programmes in municipal waste applications are continuing to verify and confirm the excellent treatment effectiveness and decreased power requirements achieved with the system. Field tests are also being conducted on the treatment of industrial wastes. A pilot plant programme is successfully underway treating a mixed petrochemical waste in one of the Union Carbide large petrochemical plants. Plans are being made to pilot plant the UNOX System on a pulp and paper mill waste stream during 1971.

UNOX-BNR System

System description

The UNOX-BNR system from Lotepro provides an effective solution to the problem of phosphorus and nitrogen removal from waste-water. Depending upon the type of nutrient to be removed, the UNOX-BNR system combines the high purity oxygen activated sludge process with an anaerobic and/or anoxic process. The basic elements of a UNOX-BNR system are shown in Fig. 17.11.

The UNOX-BNR system configuration presented in Fig. 17.11 combines the UNOX System for carbonaceous BOD removal and nitrification, with an anaerobic stage for biological phosphorus removal and an anoxic stage for denitrification. Variations of the UNOX-BNR system include elimination of the anaerobic stage if phosphorus removal is not required or the elimination of the anoxic stage if denitrification is not required. When applied, the anaerobic and anoxic stage typically consists of multiple compartments to optimise operating conditions and assure achievement of design effluent levels.

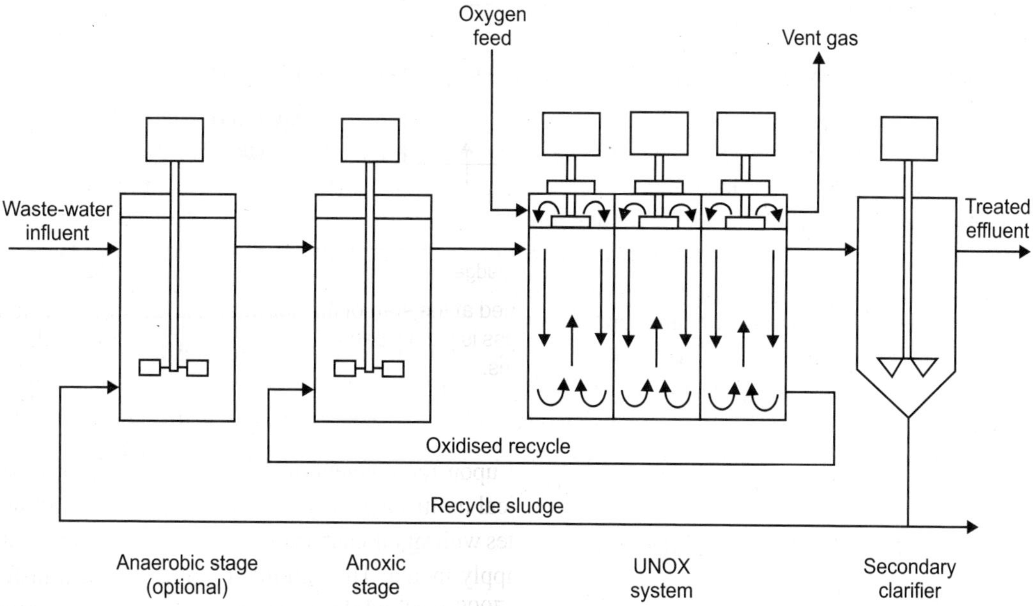

Fig. 17.11. Schematic diagram of UNOX-BNR system.

Typically an anaerobic process is combined with the UNOX system to provide a competitive advantage for the polyphosphate storing micro-organisms that are responsible for biological phosphorus removal by absorbing BOD and accumulating large quantities of phosphorus within their cell structure. The accumulated phosphorus is removed from the plant with discharge of the waste sludge. The anoxic process allows metabolism to proceed under conditions where oxygen is absent and nitrate is present. Under anoxic conditions the denitrifying micro-organisms utilise the oxygen from nitrate as an oxidising agent for stabilisation of BOD. The anoxic process would follow the anaerobic process if phosphorus removal was required in addition to denitrification.

System advantages

The key features and advantages for the UNOX system are applicable to the UNOX-BNR system. Combining the selector technology with the UNOX system results in a biological nutrient removal process that achieves the nitrogen and phosphorus removal levels required by imposed discharge permits. The UNOX-BNR system provides economical and operational advantages for both new plants and upgrading existing plants. The UNOX-BNR system operates at high dissolved oxygen levels for optimum nitrification and the improved sludge settling characteristics allows high MLSS levels to be maintained resulting in increased sludge age in smaller reactors. The process eliminates the need for a supplemental carbon source for denitrification and has lower oxygen and alkalinity requirements as a result of recovery of these materials during denitrification. Additionally, the rate of denitrification is increased as a result of the utilisation of sorbed BOD_5 in the anoxic stage.

Reference plants

The combining of the high purity oxygen process with the selector technology has been used by a number of waste-water treatment facilities. This experience together with the wide application of the UNOX system for carbonaceous BOD removal and for a combination of BOD removal and nitrification provides a solid basis for the UNOX-BNR system.

PURE OXYGEN AERATION SYSTEM FOR WASTE-WATER TREATMENT

Disclosed is a pure oxygen aeration system for waste-water treatment, which biologically treats waste-water using microbes of activated sludge in an aeration tank. The pure oxygen aeration system comprises a pure oxygen supply device including a pure oxygen generator and at least one oxygen supply pipe extending from the pure oxygen generator and directed toward the internal space of the aeration tank, a high-speed jet injection device installed in the aeration tank, a mixed liquor circulation device for circulating and introducing the mixed liquor in which microbes of activated sludge, waste-water and pure oxygen are mixed, into the aeration tank through the high-speed jet injection device, and an oxygen suction pipe for sucking in the oxygen remaining in the headspace of the aeration tank and reintroducing the sucked oxygen into the water in the aeration tank. The present invention provides a pure oxygen aeration system for waste-water treatment which is economical, increases an oxygen utilisation efficiency, secures easily the required land, saves the expense and can maintain an optimum level of dissolved oxygen and discharge smoothly and rapidly harmful gases.

BIOTECHNOLOGY AND DEODOURISATION OF WASTE-WATER

Conventional sewage treatment involves the use of micro-organisms which develop naturally within the sewage treatment systems. In more advanced approaches, sewage is inoculated with a specific micro-organism which has been specifically selected for the particular sewage treatment process.

These starter cultures as they are called enhance the efficiency of sewage degradation. These starter cultures have been used to deodourise animal excrements.

METHANE PRODUCTION

Methane fermentation is a versatile biotechnology capable of converting almost all types of polymeric materials to methane and carbon dioxide under anaerobic conditions. This is achieved as a result of the consecutive biochemical breakdown of polymers to methane and carbon dioxide in an environment in which a variety of micro-organisms which include fermentative microbes (acidogens); hydrogen-producing, acetate-forming microbes (acetogens); and methane-producing microbes (methanogens) harmoniously grow and produce reduced end-products. Anaerobes play important roles in establishing a stable environment at various stages of methane fermentation.

Methane fermentation offers an effective means of pollution reduction, superior to that achieved via conventional aerobic processes. Although practiced for decades, interest in anaerobic fermentation has only recently focused on its use in the economic recovery of fuel gas from industrial and agricultural surpluses.

The biochemistry and microbiology of the anaerobic breakdown of polymeric materials to methane and the roles of the various micro-organisms involved, are discussed here. Recent progress in the molecular biology of methanogens is reviewed, new digesters are described and improvements in the operation of various types of bioreactors are also discussed.

Microbial Consortia and Biological Aspects of Methane Fermentation

Methane fermentation is the consequence of a series of metabolic interactions among various groups of micro-organisms. A description of micro-organisms involved in methane fermentation, based on an analysis of bacteria isolated from sewage sludge digesters and from the rumen of some animals, is summarised in Fig. 17.12. The first group of micro-organisms secrete enzymes which hydrolyse polymeric materials to monomers such as glucose and amino acids, which are subsequently converted to higher volatile fatty acids, H_2 and acetic acid (Fig. 17.12; stage 1). In the second stage, hydrogen-producing acetogenic bacteria convert the higher volatile fatty acids, e.g. propionic and butyric acids, produced, to H_2, CO_2, and acetic acid. Finally, the third group, methanogenic bacteria convert H_2, CO_2, and acetate, to CH_4 and CO_2.

Hydrolysis and acidogenesis

Polymeric materials such as lipids, proteins, and carbohydrates are primarily hydrolysed by extracellular, hydrolases, excreted by microbes present in Stage 1 (Fig. 17.12). Hydrolytic enzymes, (lipases, proteases, cellulases, amylases, etc.) hydrolyse their respective polymers into smaller molecules, primarily monomeric units, which are then consumed by microbes. In methane fermentation of waste-waters containing high concentrations of organic polymers, the hydrolytic activity relevant to each polymer is of paramount significance, in that polymer hydrolysis may become a rate-limiting step for the production of simpler bacterial substrates to be used in subsequent degradation steps.

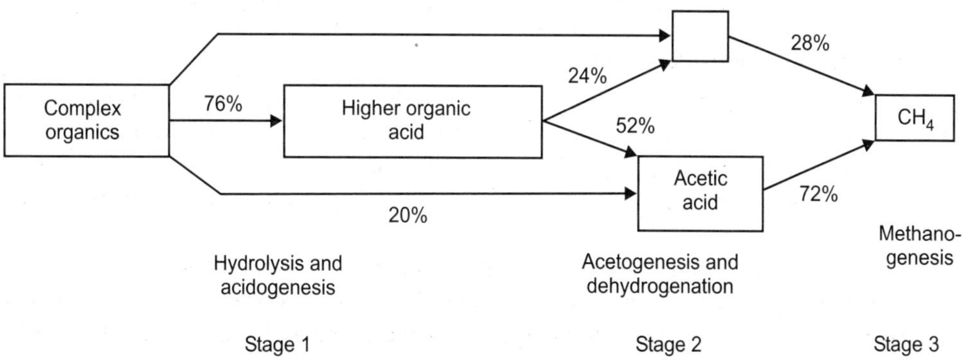

Fig. 17.12. Stages of methane fermentation.

Lipases convert lipids to long-chain fatty acids. A population density of 10^4–10^5 lipolytic bacteria per ml of digester fluid has been reported. Clostridia and the micrococci appear to be responsible for most of the extracellular lipase producers. The long-chain fatty acids produced are further degraded by p-oxidation to produce acetyl CoA.

Proteins are generally hydrolysed to amino acids by proteases, secreted by *Bacteroides, Butyrivibrio, Clostridium, Fusobacterium, Selenomonas,* and *Streptococcus.* The amino acids produced are then degraded to fatty acids such as acetate, propionate, and butyrate, and to ammonia as found in *Clostridium, Peptococcus, Selenomonas, Campylobacter,* and *Bacteroides.*

Polysaccharides such as cellulose, starch, and pectin are hydrolysed by cellulases, amylases, and pectinases. The majority of microbial cellulases are composed of three species: (i) endo-(3-1,4-glucanases, (ii) exo-*p*-1,4-glucanases, (iii) cellobiase or *p*-glucosidase. These three enzymes act synergistically on cellulose effectively hydrolysing its crystal structure, to produce glucose. Microbial hydrolysis of raw starch to glucose requires amylolytic activity, which consist of 5 amylase species: (i) α-amylases that endocleave α ± 1–4 bonds, (ii) *p*-amylases that exocleave α±1–4 bonds, (iii) amyloglucosidases that exocleave α ± 1–4 and α ± 1–6 bonds, and (iv) debranching enzymes that act on α ± 1–6 bonds; (v) maltase that acts on maltose liberating glucose. Pectins are degraded by pectinases, including pectinesterases and depolymerases. Xylans are degraded with α^2-endo-xylanase and α^2-xylosidase to produce xylose.

Hexoses and pentoses are generally converted to C_2 and C_3 intermediates and to reduced electron carriers (e.g. NADH) via common pathways. Most anaerobic bacteria undergo hexose metabolism via the Emden-Meyerhof-Parnas pathway (EMP) which produces pyruvate as an intermediate along with NADH.

The pyruvate and NADH thus generated, are transformed into fermentation end products such as lactate, propionate, acetate, and ethanol by other enzymatic activities which vary tremendously with microbial species.

Thus, in hydrolysis and acidogenesis (Fig. 17.12; Stage 1), sugars, amino acids, and fatty acids produced by microbial degradation of biopolymers are successively metabolised by fermentation end products such as lactate, propionate, acetate, and ethanol by other enzymatic activities which vary tremendously with microbial species.

Thus, in hydrolysis and acidogenesis (Fig. 17.12; Stage 1), sugars, ammo acids, and fatty acids produced by microbial degradation of biopolymers are successively metabolised by groups of bacteria and are primarily fermented to acetate, propionate, butyrate, lactate, ethanol, carbon dioxide, and hydrogen.

Acetogenesis and dehydrogenation

Although some acetate (20 per cent) and H_2 (4 per cent) are directly produced by acidogenic fermentation of sugars, and amino acids, both products are primarily derived from the acetogenesis and dehydrogenation of higher volatile fatty acids (Fig. 17.12; Stage 2).

Obligate H_2-producing acetogenic bacteria are capable of producing acetate and H_2 from higher fatty acids.

Only *Syntrophobacter wolinii,* a propionate decomposer and *Sytrophomonos wolfei,* a butyrate decomposer have thus far been isolated due to technical difficulties involved in the isolation of pure strains, since H_2 produced, severely inhibits the growth of these strains. The use of co-culture techniques incorporating H_2 consumers such as methanogens and sulphate-reducing bacteria may therefore facilitate elucidation of the biochemical breakdown of fatty acids.

Methane Production Using Microbial Systems

Methane is produced during anaerobic decomposition of sewage and other organic wastes by bacteria. The methane thus produced is collected and used as fuel in many countries.

The methogenic bacteria are able to utilise acetate, methanol, formate, and $H_2 + CO_2$ of the organic wastes for the production of methane gas.

The starting materials in the wastes are, however, complex organic molecules such as cellulose, starch, fats and proteins. These are first broken down to simpler substrates such as acetate and $H_2 + CO_2$ by other micro-organisms.

In the sewage treatment system, the degradation of macromolecules (cellulose, protein, fats, etc.) to smaller molecules is called liquefaction phase. A large number of micro-organisms can perform this degradation. However, only a restricted group of organisms are able to produce methane from the degraded products.

The composition and the amount of the gas (called biogas), produced from the degradation of biological products, is related to the quality of the substrate and the temperature. In anaerobic sludge digestion process, upto 600 litres of methane is produced per kg dry organic matter.

The biogas generated in such a process contains 60–70 per cent methane, 25–30 per cent CO_2 and the rest is $H_2 + N_2$. In many country, small and inexpensive installations have been set up in the villages to convert cow dung into biogas. About 200 to 1000 litres of methane per kg dry cow dung is produced at 55°C, in such installations.

In most systems, the amount of biogas generated increases with the increase in temperature. The time for the generation of gas is also reduced at high temperatures.

Method for Sewage Treatment with Bacteria

A process for the treatment of sewage wherein the sludge is innoculated with a bacteria, *L. plantarum*, and a carbohydrate such as lactose is admixed therewith. The addition of the bacteria and the carbohydrate without more, drops the pH of the sludge to below 4.0.

This results in the elimination of pathogenic bacteria and renders the sludge suitable for use as a soil extender without any further environmental constraints.

BULKING AND OTHER SLUDGE SETTLING PROBLEMS

The successful application of the activated sludge process requires that the sludge floc settles and compacts well in the settling tank. It must settle well so that the effluent, has little suspended solids. High suspended solids in the effluent are the main cause of unsatisfactory effluent quality, and they also make proper SRT control difficult. The sludge must compact well so that the sludge can be returned successfully to the aeration basin. A highly compacted sludge also reduces the costs for dewatering and disposal of the wasted solids.

The major problem in activated sludge operation is the development of poor settling sludge. When this occurs, suspended solids pass to the effluent in concentrations that often exceed regulatory standards for SS, the desired SRT necessary cannot be maintained, and effluent BOD limitations are often exceeded. Many sludge settling problems can occur, and their causes are not at all the same. The cause of a particular sludge-settling problem must be recognised before a satisfactory solution can be found. The cause may be related to the particular configuration of the treatment plant, to the loading applied, to the environmental conditions (such as temperature, pH, and dissolved oxygen concentration), or to the presence of particular waste-water constituents or by the absence of other.

A summary of the different solids-separation problems that can develop in activated sludge operation. is given in Table 17.1.

Table 17.1. Biosolids separation problems encountered in activated sludge operation.

Biosolids separation problem	*Cause of problem*	*Effect of problem*
Bulking	Filamentous organisms extend from flocs into the bulk solution and interfere with compaction and settling	High sludge volume index with clear supernatant. Overflow of sludge blanket can occur. Solids handling processes become hydraulically overloaded
Viscous bulking or non-filamentous bulking	Micro-organisms present in large amounts of exocellular slime. In severe cases, the slime imparts a jelly-like consistency	Reduced settling and compaction rates. Can result in overflow of sludge blanket from secondary clarifier or formation of a viscous foam
Dispersed growth	Micro-organisms do not form flocs, but are dispersed, forming only small clumps or single cells	Turbid effluent. No zone settling of sludge
Pin floc or pinpoint floc	Small, compact, weak, roughly spherical flocs. Larger aggregates settle rapidly, smaller ones slowly	Low sludge volume index and cloudy turbid effluent
Foaming/scum formation	Caused by (i) nondegradable surfactants, or (ii) the presence of *Norcardia* sp. and/or *Microthrix parvicella*	Foams float large amounts of biosolids to surface of treatment units. Micro-organism caused foams are persistent and difficult to break. Causes solids overflow into secondary effluent and onto walkways. Anaerobic digestion foaming can also result
Blanket rising	Denitrification in settler releases poorly soluble N_2 gas, which attaches to activated sludge flocs and floats them to the clarifier surface	'Chunks' of activated sludge collect on the surface of the settler and may result in turbid effluent

Bulking Sludge

Bulking sludge is the most pervasive and difficult problem. Bulking is the term used to describe the formation of activated sludge floc that settles slowly and compacts poorly. Bulking makes it difficult to remove the activated sludge from the settling tank for return to the aeration basin. Higher and higher recycle rates are required because the concentration of the settled sludge is so low. If the sludge cannot be removed fast enough from the underflow, then the sludge blanket rises until it fills the settler. Then, the activated sludge solids discharge into the effluent. This causes a massive loss of biomass, which decreases the SRT in an uncontrolled manner and destroys effluent SS and BOD quality. A prolonged siege of sludge bulking can lead to the total failure of the activated sludge process.

A compact and good-settling activated sludge results from a floc microstructure consisting of a backbone of filamentous bacteria to which zoogleal micro-organisms attach to form a strong and compact macrostructure. If there are not sufficient filamentous bacteria, the floc is weak and subject to break up into smaller particles.

On the other hand, too many filamentous bacteria are the cause of sludge bulking. The problem becomes serious when the filaments extend outside the compact floc. These extended filaments create bridges between the flocs. The bridging causes two serious and negative effects. First, the bridges prevent the flocs from coming close together, or compacting. Second, the bridges trap water within and between the flocs. As the sludge flocs try to move downward and compact in the settler, the water they displace must move upward. The bridging prevents the movement of the water through and away from the flocs. These two effects of extended filaments cause the bulking sludge to settle slowly and compact very poorly.

Reduced-sulphur bulking occurs whenever reduced sulphur compounds, mostly sulphides, enter the activated sludge unit. Sulphur-oxidising species, such as Thiothrix and 021N, are filament formers that gain a competitive advantage from the chemolithotrophic electron donor. So far, the only reliable means to eliminate reduced sulphur bulking is to eliminate all inputs of reduced sulphur. If the source cannot be eliminated, the reduced sulphur can be chemically oxidised before the waste-water enters the activated sludge process. Although a number of chemical oxidants can be used, the easiest one is hydrogen peroxide, which oxidises sulphides according to

$$4H_2O_2 + HS^- \longrightarrow SO_4^{2-} + 4H_2O + H^+$$

Foaming and Scum Control

Another very common problem in activated sludge systems is the formation of foam or scum on the surface of aeration tanks. Foam and scum cause many problems in plant operation, including excessive suspended solids in the effluent, unsightly and dangerous conditions, such as slippery walkways around them, and great difficulties in making a sludge inventory. In addition, the causative organisms can create foaming problems in anaerobic digesters that receive waste activated sludge.

While the many factors causing foaming and scum formation are not well known, the problem often is associated with long SRT and high waste-water temperatures. This suggests that the causative organisms are slow growers. In some cases, chlorination of return activated sludge helps control this problem. Since the organisms cause problems by accumulating as foam and scum on the surfaces of the tanks, another control strategy — and probably the most promising — is to implement a vigorous program to remove scum and foam. The concept is to severely reduce the SRT of the nuisance micro-organisms by wasting them very aggressively from where they preferentially accumulate.

Rising Sludge

Rising sludge is a problem sometimes occurring in the settling tank of activated sludge plants in which ammonium is nitrified to nitrate. If denitrification of the nitrate to N_2 takes place in the sludge blanket of the settler, gas bubbles can attach to the settled sludge particles. 'Chunks' of sludge then become buoyant and rise to the surface of the settler, where they collect. These pieces of the sludge blanket are unsightly, and they can cause a very large increase in effluent suspended solids if they escape to the effluent.

The most effective cure for rising sludge is to stop nitrification in the activated sludge. This is accomplished by reducing the SRT and washing out the slow-growing nitrifiers. If no nitrate is formed by nitrification, then none can be denitrified to N_2 gas. Another way to prevent rising sludge is to promote denitrification as part of the activated sludge process. If the nitrate is removed before the mixed liquor enters the settler, then denitrification cannot occur in the settler.

Another way to reduce denitrification in the settler is by improving settler design. The idea is to prevent the sludge from 'sitting' in the sludge blanket for too long. Circular settlers with vacuum-type sludge removal devices work particularly well here. Rectangular clarifiers can be effective if sludge-removal scrapers remove the sludge with sufficient speed and if accumulation of sludge in quiescent corners is prevented.

Chapter 18

Bioleaching

INTRODUCTION

In nature sulphidic ores are decayed by weathering under the influence of oxygen and water. Microbiological investigations reveal that certain bacteria are the main agent in this process. Several bacteria, especially Thiobacilli, are able to solubilise heavy metal minerals by oxidising ferrous to ferric iron as well as elemental sulphur, sulphide and other sulphur compounds to sulphate. So they enhance leaching of heavy metals from sulphidic ores under aerobic conditions about 104-fold or more compared with weathering without bacteria.

The principal bacterium in ore leaching is *Thiobacillus ferrooxidans*, which is capable of oxidising ferrous iron as well as sulphur and sulphur compounds. But there are some other bacteria which may also be involved. For example the *Thermophilic sulpholobus* plays a role in leaching at elevated temperatures. *Thiobacillus thiooxidans*, which oxidises merely sulphur and sulphur compounds but not iron, and *Leptospirillum ferrooxidans*, which contrarily oxidises only ferrous iron, may play a role if they work together or with other bacteria.

Bacterial ore leaching can be applied to extract heavy metals from low grade ores, industrial wastes and other materials on an industrial scale by different procedures: dump leaching, *in situ* leaching, tank leaching, leaching in suspension. Sulphidic copper and uranium ores are the principle ores leached in several countries. So 20 to 25 per cent of the copper production in the USA and about 5 per cent of the world copper production is obtained by bacterial leaching. This process is a very slow one and needs a long time (years) for good recovery, but its main advantages are low investment costs and low operating costs.

Current investigations deal with the leaching of ores other than those mentioned, leaching industrial wastes to recover metals, desulphurising of coal, developing methods for *in situ* leaching and using other micro-organisms than those used until now. Basic microbiological research focuses on the biochemistry, physiology and genetics of the involved micro-organisms and on the complex interrelationships in the microbial community of leaching biotopes.

It is a fact that resources of metal ores are limited and that sooner or later these resources will be exhausted. But how great are our resources in naturally occurring deposits? Before we answer this question we have to define what a metal ore deposit is. A metal ore deposit is a naturally occurring concentration of a metal or some metals from which this metal can be obtained in an economic way. So, whether or not a deposit of metal ore can be considered are source or not depends on the costs we have to pay for extracting the metal from the ore and on the price we can get for the pure metal on the market.

In other words if the price of a metal rises—as is to be expected with depletion of the resources—and the costs of extraction are lowered, the amount of resources in the world rises.

Microbial leaching of ores depends primarily on bacterial processes which are the essential causes of natural weathering of sulphidic minerals. If sulphidic heavy metal minerals come into contact with air and water they begin to decay with the formation of sulphate, sometimes sulphuric acid, and water soluble heavy metal cations.

Weathering of an ore body results in a typical picture.

1. An upper oxidation zone, being in contact with atmospheric oxygen and rain water, which contains secondary minerals formed by oxidation of the primary ore minerals and in most cases a remarkable enrichment of ferric iron minerals (limonite and others).
2. An underlying cementation zone just below the groundwater level, in which minerals, formed by the reaction of primary ore minerals with the constituents of the leaching solution descending from the oxidation zone, are accumulated.
3. A zone in which the primary ore minerals are unchanged.

So we have to look at these phenomena to understand what exactly happens in this process and to get an idea of how to apply these natural processes to ore leaching on an industrial scale (Fig. 18.1).

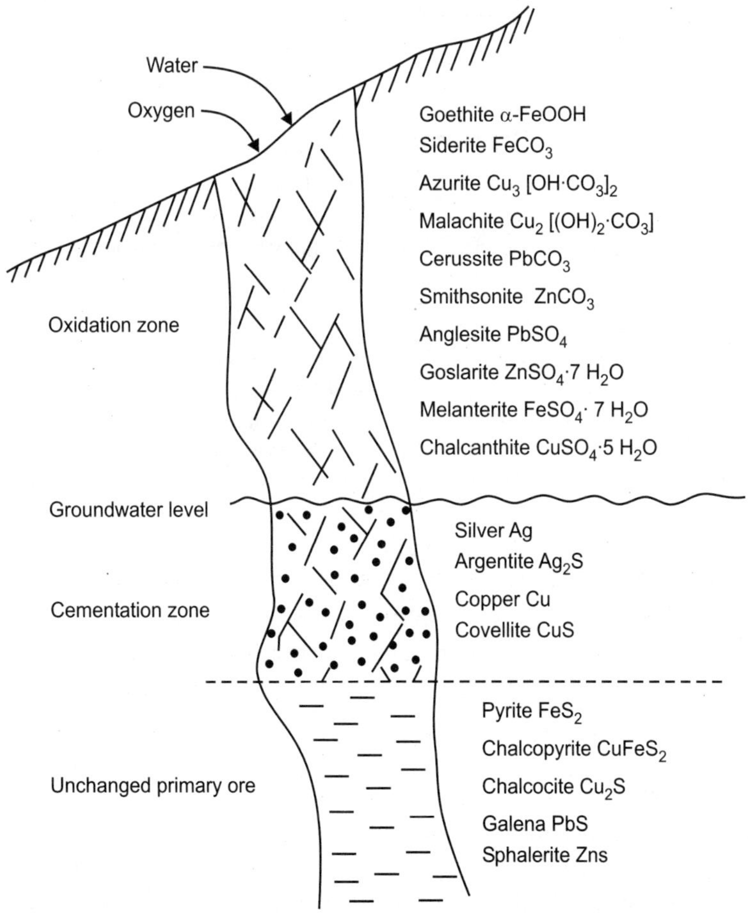

Fig. 18.1. Weathering of an ore body.

MICROBIOLOGY OF ORE LEACHING

Microbiological investigations revealed that certain bacteria are the main agent in natural weathering of sulphidic heavy metal minerals.

Thiobacilli

The principal bacteria which play the most important role in solubilising sulphidic metal minerals at moderate temperatures are species of the genus *Thiobacillus*. They are Gram-negative rods, either polarly or nonflagellated. Most species are acidotolerant, some even extremely acidotolerant and acidophilic. Some grow best at pH 2 and may grow at pH 1 or even at pH 0.5. Most species are tolerant against heavy metal toxicity.

Thiobacilli are chemolithoautotrophs, that means CO_2 may be the only source of carbon and they derive their energy from a chemical transformation of inorganic matter. All Thiobacilli oxidise sulphur or sulphur compounds to sulphate or sulphuric acid (Figs 18.2 and 18.3).

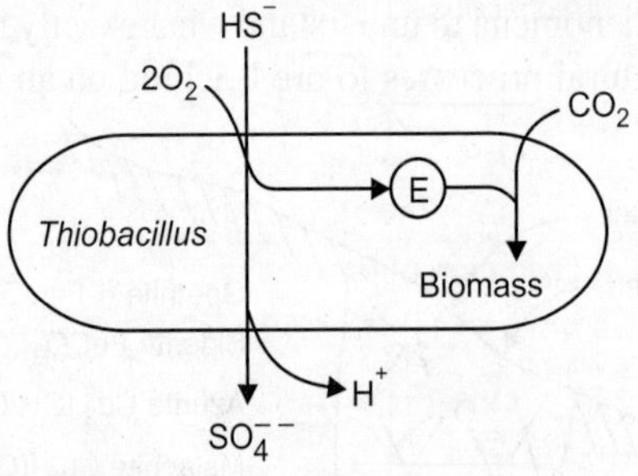

Fig. 18.2. Oxidation of hydrogen sulphide by thiobacilli.

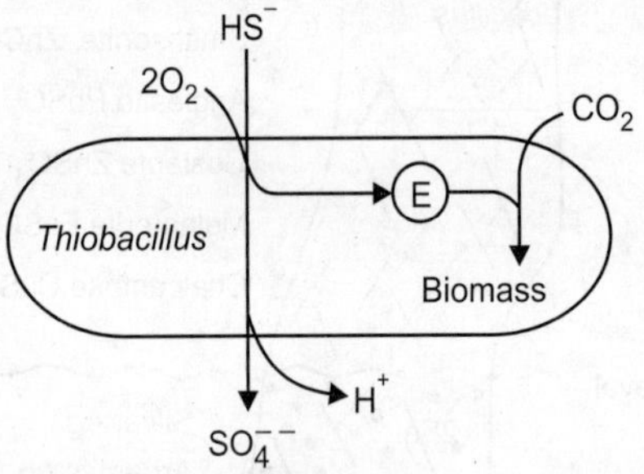

Fig. 18.3. Oxidation of elemental sulphur by thiobacilli.

If they oxidise hydrogen sulphide, thiosulphate, polythionates or elemental sulphur they produce hydrogen ions and so they lower the pH of the medium, often below pH 2, in some cases below pH 1.

$$HS^- + 2O_2 \longrightarrow SO_4^{--} + H^+ \qquad \text{... (18.1)}$$

$$S^\circ + H_2O + 1\tfrac{1}{2}O_2 \longrightarrow SO_4^{--} + 2\,H^+ \qquad \text{... (18.2)}$$

Thiobacillus ferrooxidans

In addition to the oxidation of sulphur and sulphur compounds *Thiobacillus ferrooxidans* is able to oxidise ferrous to ferric iron and so derive its energy from this exergonic reaction. In this reaction hydrogen ions are consumed and so the pH of the medium should rise. But at pH values higher than 2 the ferric iron precipitates as ferric hydroxide, jarosites or similar compounds and this results in the formation of hydrogen ions, so that the pH of the medium is lowered as is the case with oxidation of sulphur compounds:

$$2Fe^{++} + 2H^+ + 1/2O_2 \longrightarrow 2Fe^{+++} + H_2O \quad \text{... (18.3)}$$

$$2Fe^{+++} + 6H_2O \longrightarrow 2Fe(OH)_3 + 6H^+ \quad \text{... (18.4)}$$

$$2Fe^{++} + 5H_2O + 1/2O_2 \longrightarrow 2Fe(OH)_3 + 4H^+ \quad \text{... (18.5)}$$

Oxidation of ferrous iron by *T. ferrooxidans* and oxidation of ferrous iron by *T. ferrooxidans* with subsequent precipitation of ferric hydroxide are shown in Figs 18.4 and 18.5.

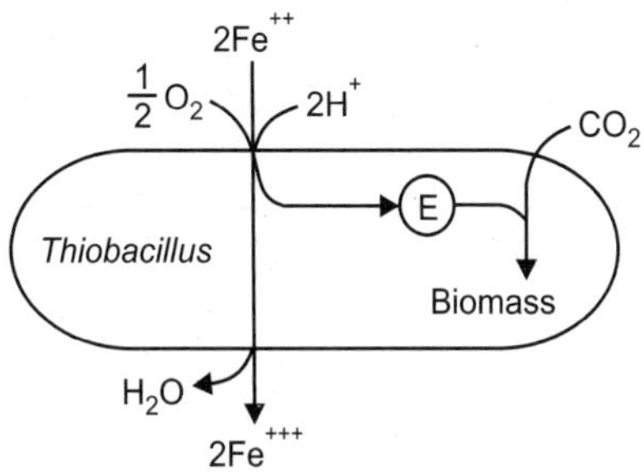

Fig. 18.4. Oxidation of ferrous iron by *T. ferrooxidans.*

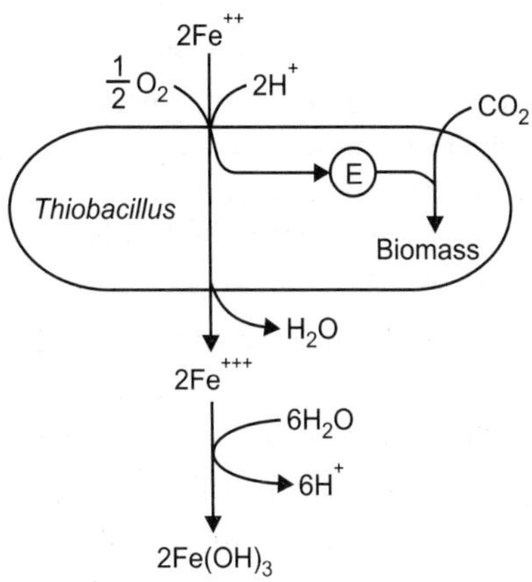

Fig. 18.5. Oxidation of ferrous iron by *T. ferrooxidans* with subsequent precipitation of ferric hydroxide.

As will be shown later, owing to its ability to oxidise ferrous iron, *T. ferrooxidans* is the principal agent of bacterial ore leaching at moderate temperatures.

Thiobacilli and Sulphidic Minerals

Some Thiobacilli, especially *T. ferrooxidans*, are able to oxidise sulphide and some heavy metals—mainly iron but also copper, zinc, molybdenum and presumable some other metals—in the form of sulphidic heavy metal minerals which are of very low solubility in water, practically insoluble. These oxidations result in a solubilisation of the minerals. This is often seen in the case of pyrite or marcasite, both FeS_2, minerals which are oxidised very easily by Thiobacilli:

$$FeS_2 + H_2O + 3\tfrac{1}{2}O_2 \longrightarrow Fe^{++} + 2SO_4^{--} + 2H^+ \quad \text{... (18.6)}$$

But also in the case of other minerals. Oxidation of the sulphide of a divalent metal:

$$Me^{II}S + 2O_2 \longrightarrow Me^{++} + SO_4^{--} \qquad \text{... (18.7)}$$

Direct solubilisation of sulphidic heavy metal minerals by Thiobacilli

In the solubilisation of sulphidic minerals there are several reactions involved which are not fully understood in all details nor in their relative importance. But some mechanisms are clear:

1. The oxidation of sulphide ions and of metal ions disturb the solubility equilibrium and so the sulphide mineral may dissolve slowly (Fig. 18.6).
2. Hydrogen ions formed in connection with sulphide and ferrous iron oxidation by the bacteria attack the mineral and release metal ions and hydrogen sulphide or elemental sulphur:

$$NiS + 2H^+ \longrightarrow Ni^{++} + H_2S \qquad \text{... (18.8)}$$

$$FeS_2 + 2H^+ \longrightarrow Fe^{++} + H_2S + S^\circ \qquad \text{... (18.9)}$$

Hydrogen sulphide and elemental sulphur are then oxidised by the bacteria to sulphuric acid, which gives rise to more hydrogen ions.

3. The combination of hydrogen ion attack and oxidation with oxygen releases metal ions and elemental sulphur:

$$Me^{II}S + 2H^+ + \tfrac{1}{2}O_2 \longrightarrow Me^{++} + H_2O + S^\circ \qquad \text{... (18.10)}$$

In the case of chalcocite (Cu_2S) it forms covellite (CuS), and copper ions:

$$Cu_2S + 2H^+ + \tfrac{1}{2}O_2 \longrightarrow CuS + Cu^{++} + H_2O \qquad \text{... (18.11)}$$

These processes are called the direct mechanisms of bacterial mineral solubilisation to distinguish them from an indirect mechanism:

4. Ferric ions, formed by oxidation of ferrous iron by *T. ferrooxidans*, are a strong oxidant and may oxidise sulphidic bound metals so that soluble metal cations are formed:

$$Me^{II}S + 2F^{+++} \longrightarrow Me^{++} + 2F^{++} + S^\circ \qquad \text{... (18.12)}$$

The iron is thereby reduced to ferrous iron which is oxidised to ferric iron again by *Thiobacillus ferrooxidans*:

$$2Fe^{++} + 2H^+ + \tfrac{1}{2}O_2 \longrightarrow 2F^{+++} + H_2O \qquad \text{... (18.13)}$$

The elemental sulphur may be oxidised by Thiobacilli to sulphuric acid which supports the dissolution of the mineral according to equations (18.8) to (18.11):

$$S^\circ + H_2O + 1\tfrac{1}{2}O_2 \longrightarrow SO_4^{--} + 2H^+ \qquad \text{... (18.14)}$$

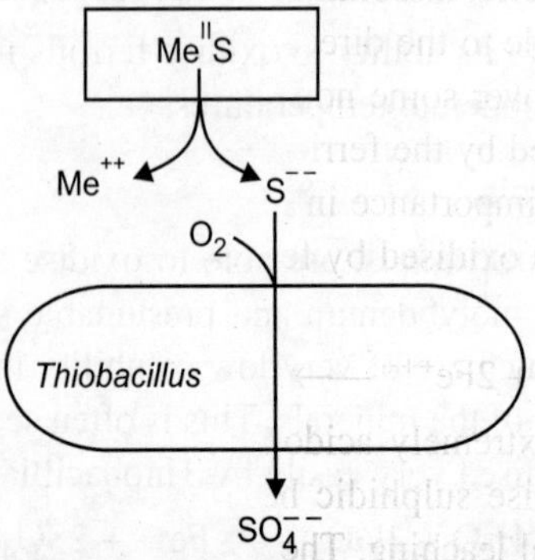

Fig. 18.6. Direct solubilisation of sulphidic heavy metal minerals by *Thiobacilli*.

Indirect solubilisation of sulphidic heavy metal minerals by *Thiobacillus ferrooxidans* and indirect solubilisation of uraninite by *Thiobacillus ferrooxidans* shown in Figs 18.7 and 18.8.

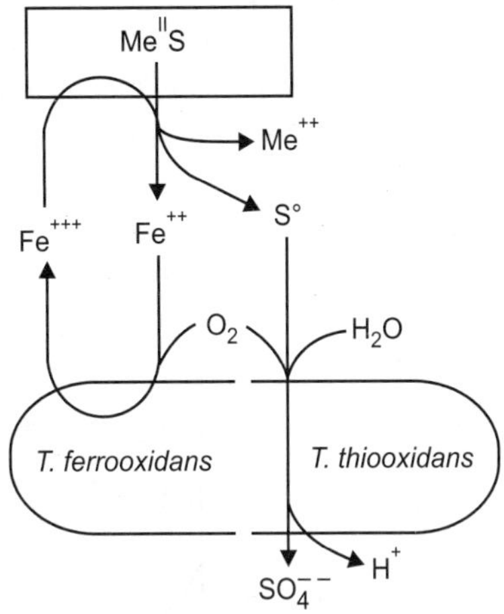

Fig. 18.7. Indirect solubilisation of sulphidic heavy metal minerals by *Thiobacillus ferrooxidans*.

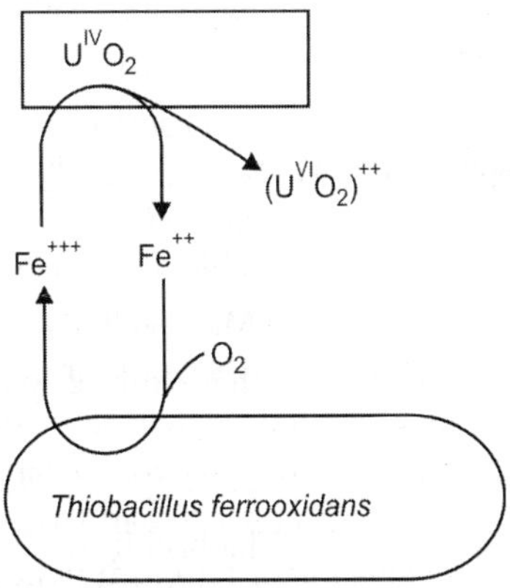

Fig. 18.8. Indirect solubilisation of uraninite by Thiobacillus ferrooxidans.

By this indirect mechanism of bacterial dissolution of sulphidic minerals also heavy metal minerals can be attacked which are not accessible to the direct mechanisms, especially whose metals which cannot be oxidised by the bacteria. Moreover some non-sulphidic heavy metal minerals can be brought into solution through oxidation mediated by the ferric/ferrous iron system.

This latter fact is of particular importance in leaching uranium ores: uranium (IV) for example as uranium dioxide UO_2, uraninite, is oxidised by ferric iron to uranium (VI) and so soluble uranyl ions UO_2 are formed:

$$UO_2 + 2Fe^{+++} \longrightarrow (UO_2)^{++} + 2Fe^{++} \quad \text{... (18.15)}$$

Thiobacillus thiooxidans, an extremely acidophilic but not ferrous iron oxidising species of the Thiobacilli, is not able to solubilise sulphidic heavy metal minerals in pure culture. Nevertheless *T. thiooxidans* plays a role in metal leaching. The solubilisation of sulphidic minerals by *Thiobacillus ferrooxidans* is increased by cooperation with *T. thiooxidans* as compared with the effect of *T. ferrooxidans* alone. We can assume that the cause of this enhancement is the oxidation of elemental sulphur and hydrogen sulphide which is formed as a result of the oxidation by ferric iron according to Eq. 18.12, for this oxidation produces hydrogen ions which in turn attack the minerals according to Eqs 18.8 and 18.9.

Direct solubilisation of pyrite or marcasite by *Thiobacillus ferrooxidans* is shown in Fig. 18.9.

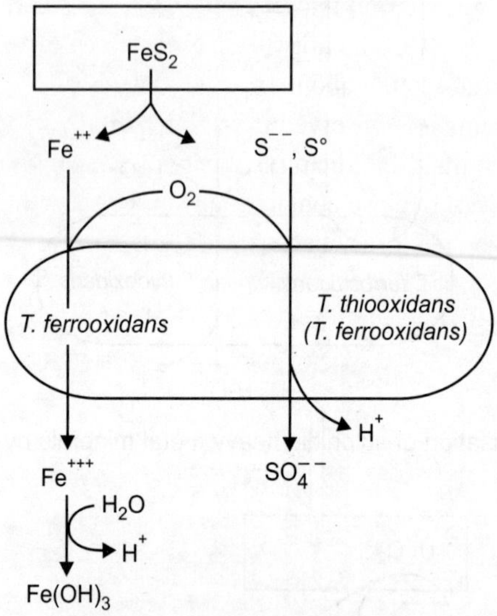

Fig. 18.9. Direct solubilisation of pyrite or marcasite by *Thiobacillus ferrooxidans.*

OTHER BACTERIA

In addition to Thiobacilli there are some other bacteria known to be effective in solubilising sulphidic minerals. In hot biotopes containing sulphur or oxidisable sulphur compounds, such as hydrothermal vents and self heating brown coal dumps, one can find an archaebacterium named *Sulfolobus*. This is a bacterium without a rigid cell wall, round shaped, about 0.8 to 1.0 cm in diameter.

Like Thiobacilli it is acidophilic, chemolithoautotroph and derives its energy from oxidation of sulphur and sulphur compounds and from oxidation of ferrous iron like *Thiobacillus ferrooxidans*. Its pH-range of growth is pH 1.0–6.0 and its optimum at about pH 2. A salient characteristic is its thermophily: its growth range is 45°–85°C, its optimum 70°–75°C. Species of this genus, especially *S. brierleyi* seem to be the main agent in metal leaching at high temperatures.

Leptospirillum

Often one can see in acid metal leaching biotopes spirilloid bacteria. They belong to the species *Leptospirillum ferrooxidans*, a Gram-negative spirillum, facultatively chemolithoautotroph, deriving its energy from oxidising ferrous iron like *Thiobacillus ferrooxidans*. But in contrast to this latter bacterium it cannot oxidise sulphur or sulphur compounds and is incapable of utilising the iron of sulphidic minerals.

Leptospirillum ferrooxidans alone cannot solubilise sulphidic ferrous iron containing minerals. But incooperation with *Thiobacillus thiooxidans*, which, for its part alone, is also unable to dissolve sulphidic minerals, it can; both bacteria together disintegrate sulphidic ferrous iron containing minerals by oxidation and bringing them into solution.

Bacterial Leaching versus Abiotic Leaching

Simple laboratory experiments can show, that chemical reactions catalysed by bacteria are the essential processes which lead to decay of sulphidic heavy metal minerals and some other minerals and that abiotic reactions play a negligible role. If sulphidic ores are percolated with simple water or diluted salt solutions under aeration in laboratory percolators in parallel sets, one set not sterilised or inoculated with natural acid mine effluent, another set under sterile conditions, it can be seen that disintegration of ore and leaching of metals proceeds in the not sterilised or inoculated percolators very much quicker than in the sterilised ones, the ratio being about 10^4 or higher.

In such percolator experiments it is observed that almost all the bacteria adhere to the pieces of ore and especially to the surfaces of the sulphidic minerals. Only a small amount of bacteria is floating free in the medium. So the bacteria are in close contact to the almost insoluble substrate which they oxidise to yield energy. This seems to be necessary because we can assume, that solubilisation of the minerals by some direct mechanisms requires direct contact (Fig. 18.10).

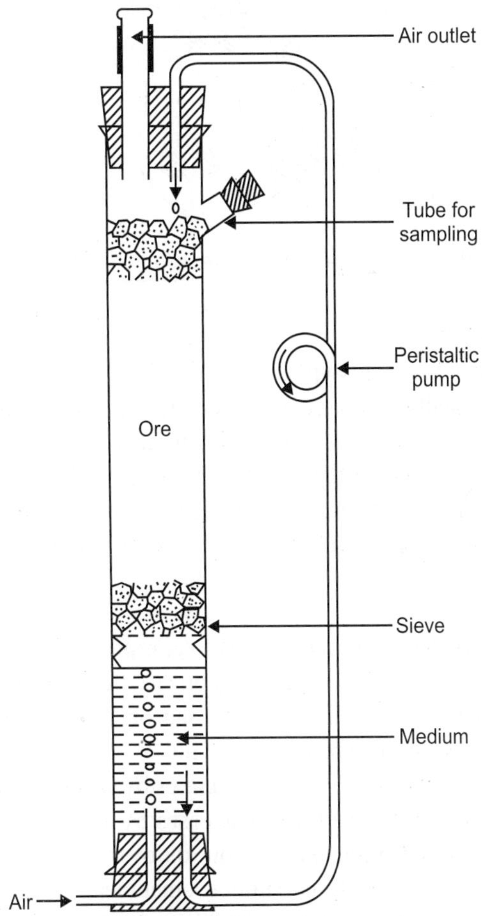

Fig. 18.10. Percolator experiments.

The rate of dissolution of the metal minerals is essentially limited by the accessible surface of the minerals and can be enhanced by grinding the minerals or the pieces of ore responsible to smaller grains. If the sulphidic minerals are not freely exposed, but are embedded in rock, as is normally the case with heavy metal ores, the rate of leaching is limited above all by the diffusion rates of solutes through fissures. Oxygen, ferric ions and hydrogen ions have to diffuse from the outside of the piece of ore, to the metal minerals inside and, conversely, metal, sulphate and hydrogen ions have to diffuse out to the surrounding medium, regardless of whether the bacteria are within the fissures on the sulphidic minerals or on the outside of the piece of ore.

Bacterial leaching of a piece of ore with imbedded sulphidic ore minerals is shown in Fig. 18.11.

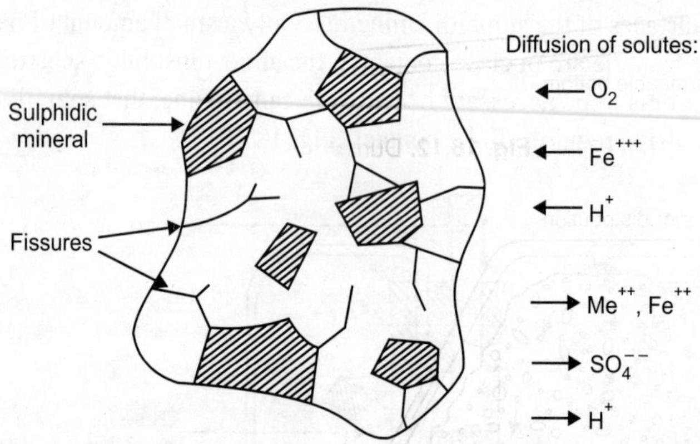

Fig. 18.11. Bacterial leaching of a piece of ore with imbedded sulphidic ore minerals.

TECHNICAL APPLICATIONS

Bacterial disintegration of ores has been applied on a technical scale for many years, almost solely to leach copper and uranium. Actually it was used for extracting copper from sulphidic ores long ago and long before bacteria were recognised as the cause of natural weathering. In some places ore leaching was operated some centuries ago, for instance at Rio Tinto in Spain. In the last decades bacterial ore leaching was carried out in many countries: Canada, USA, Mexico, Australia, India, erstwile USSR, Turkey, Yugoslavia, Romania, Hungary, Spain and some other countries.

Dump Leaching

The most commonly applied method is that of the percolator principle. Big dumps of ore are set up on an impermeable ground. The grain size has to be so that on the one hand the leaching liquor can percolate through the dump and air may enter from the sides, and on the other hand the distances for mass diffusion inside the grains are as short as possible.

The leaching liquor is distributed on the top of the dumps by sprinklers or by intermittent flooding of ponds. At the bottom the liquor is collected, in some cases by a drainage system, and conducted to a collecting reservoir from which it is pumped back on top of the dump. Before pumping back to the dump the whole liquor or a part of it may be conditioned, that means extracting the dissolved metal (for instance copper by cementation with iron scrap), addition of sulphuric acid if the pH is too high and addition of nutrient salts if desired (Figs 18.12 and 18.13).

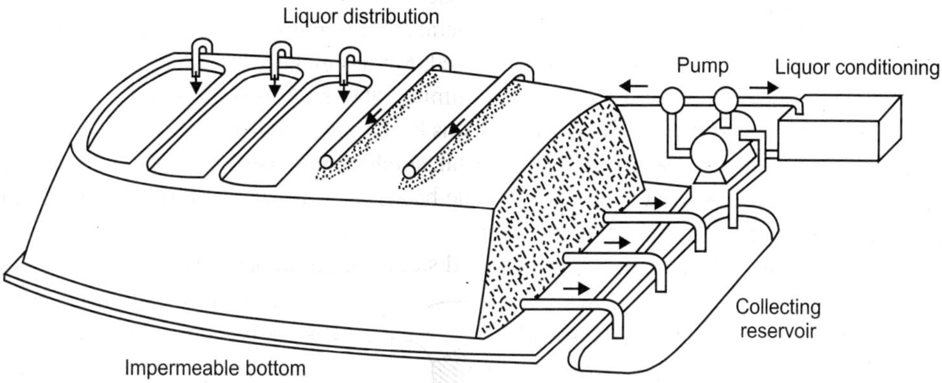

Fig. 18.12. Dump ore leaching.

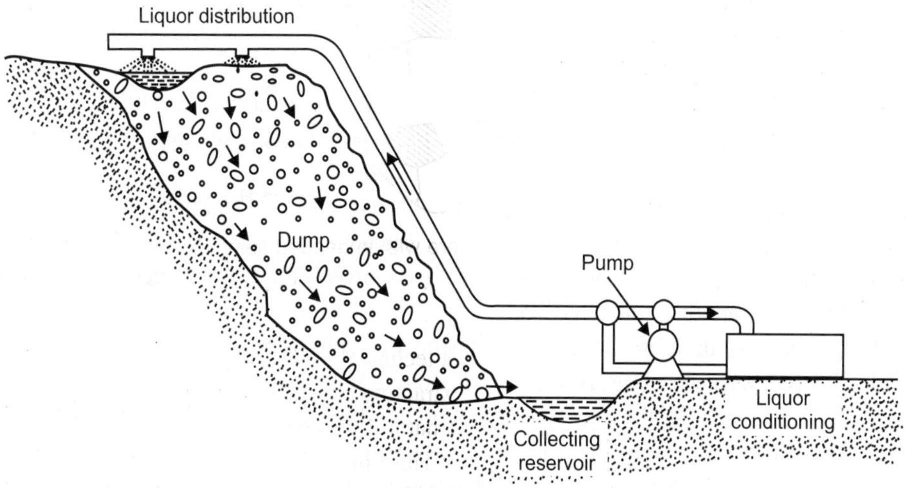

Fig. 18.13. Dump ore leaching on a slope.

Copper from ores which contain sulphides are leached on the whole by dump leaching. Chiefly copper ores of the porphyric type (disseminated copper ores) with low concentrations of copper (below 0.6 per cent Cu) are leached in this way. For instance in some states of the USA at some open pit mines, in which low grade copper ores are excavated, big dump leaching facilities are operated. The height of the dumps ranges from 20 m to about 200 m and they may contain up to 10^9 T of ore at one mine. The grain size is up to 1 m^3, the copper concentration is 0.1 to 0.6 per cent.

The pH of the circulating liquor is about 2.0–3.5, its iron concentration about 35–60 mmol/l. In the on-flowing liquor the iron is almost completely ferrous iron, whereas in the outflow only 108 to 40 per cent, sometimes 70 per cent, of the iron is ferrous iron. So we can conclude, that iron is oxidised by the bacteria almost exclusively inside the dump. This fits with the observation, that almost all bacteria adhere on the ore and only a small amount is free in the fluid as mentioned above. Therefore a good aeration of the dumps is necessary, but this occurs unaided at least in their outer and upper parts by thermic air buoyancy for the temperatures in the dumps are elevated by the reaction heat up to

30°– 40°C, and in some spots temperatures near 60°C were measured. By the way: out-streaming air at the top of, the dumps contains much less oxygen than does normal air.

In most cases the addition of nutrients is not necessary because Thiobacilli are lithoautotrophs and need only some inorganic nutrients besides an energy source. The required inorganic nutrients may be taken from the ore. The nitrogen source may be an exception for ores usually contain only small amounts of nitrogen compounds. But it has been found that strains of *Thiobacillus ferrooxidans* are able to reduce molecular nitrogen and so meet their demand for nitrogen.

Operating big dumps the circulation rate is about 5000 m^3 of liquor per hour (20–30 $1 \times m^{-2} \times hr^{-1}$). The copper concentration of the out-flowing liquor is about 8 mmol/l (500 g/m^3). In the USA 2,00,000 to 2,50,000 T of copper are produced annually by bacterial leaching, equivalent to 20–25 per cent of the total copper production. In the whole world about 5 per cent of the total copper production is obtained by bacterial leaching.

Bacterial leaching is a very slow process. Around 3 to 10 per cent of the copper content is leached out of a low grade sulphidic copper ore per year. So dumps may be operated 10 to 20 years. But on the other hand dump leaching is a simple and cheap method. It needs only a little capital investment, has low operating costs, requiring little labour, and is well-suited to low grade ores if they contain the metal in sulphidic minerals or if sulphides are contained in addition. A certain amount of pyrite in the ore is favourable because oxidation of pyrite by Thiobacilli releases enough hydrogen ions to lower the pH value and enough ferric iron for the indirect oxidation mechanism.

Besides copper uranium is leached by bacteria from its ores on a technical scale. This leaching depends wholly on indirect oxidation by means of the ferric/ferrous iron system according to equation (18.15). So the leaching of uranium ores which contain pyrite as an iron source is most economical. Otherwise one has to add pyrite or another source of iron.

The technical set-up of uranium ore leaching may be the dump method, but sometimes a variation of this, so-called heap or basin leaching is applied. The ore is set up in basins. The mode of operation is preferably a two stage leaching: the out-flowing liquor, in which the iron is largely in the ferrous form, is treated in an oxidation pond. In this the liquor is aerated to enable *Thiobacillus ferrooxidans* to oxidise ferrous iron and to obtain the ferric iron required for oxidation of uranium. The oxidised liquor is then pumped back to the dump or basin.

In situ Leaching

In a few cases it has been attempted to leach ores by means of bacteria without excavating the ore prior to leaching. At first sight it seems advantageously to leach ores on the spot were they are, for excavating costs can be saved. But difficulties arise if the ore body is impermeable or if there are only a few channels through which the leaching liquor would stream downwards without percolating the ore body entirely. In such cases the ore body has to be cracked by explosions.

In situ ore leaching from injection wells to producing wells

Moreover there may be some difficulties connected with the geological situation because it is necessary to collect the liquor after it has passed through the ore body. Unsuitable siting may lead to large amounts of the leaching fluid escaping underground. To my knowledge bacterial leaching *in situ*, in a strict sense, has not yet been performed. In the USA uranium deposits were leached *in situ* underground as shown in the Fig. 18.14.

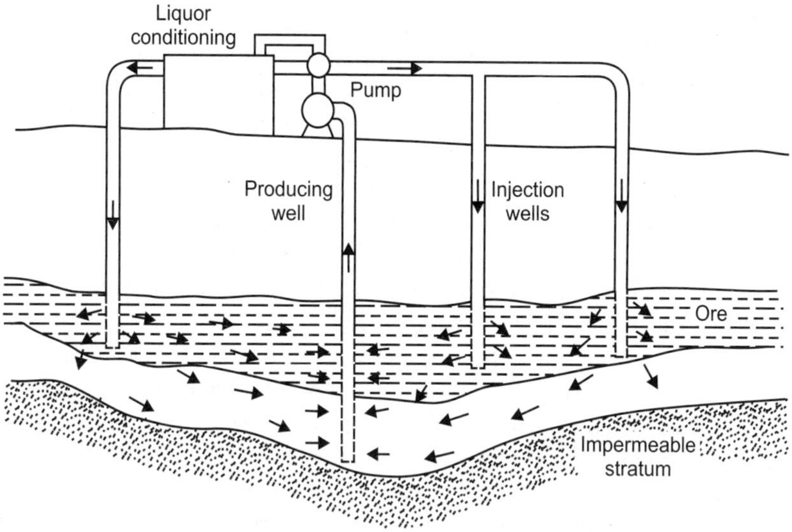

Fig. 18.14. *In situ* ore leaching from injection wells to producing wells.

But these leachings were done abiotic without using bacteria. There are some bacterial leaching set-ups which in a broader sense can be called *in situ* leaching. To this belongs the percolating of a worked-out mine with residues of ore as is schematically shown.In Canadian uranium mines after they were worked-out the walls, roofs and floors were hosed down at intervals of several months.The water was collected and the uranium extracted (Fig. 18.15).

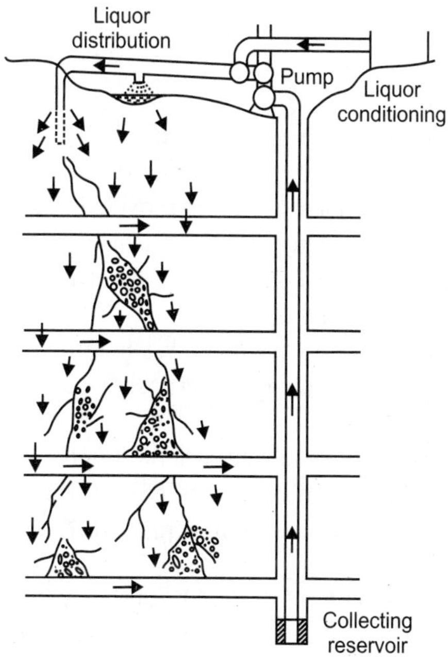

Fig. 18.15. *In situ* leaching in a mine.

Other Types of Bacterial Leaching Plants

Some other types of bacterial ore leaching arrangements were set up on a laboratory scale as well as on a semi-technical scale. Big tanks may be filled with pieces of ore like a laboratory percolator and then the ore may be percolated. Such an pilot plant has been set up at the John D. Sullivan Centre for *In situ* Mining Research in Socorro, New Mexico. The advantage of such a tank leaching is that the process can be easily controlled and regulated. The ore can be heated simply by insulating the walls of the tank, so the reaction heat of the oxidation is used for heating. Of course leaching in tanks is more expensive than dump leaching and could therefore be applied only to special purposes.

Leaching ground ore in suspension: A very interesting method is leaching ground ore in suspension. Grinding ore down to particle size of below 0.1 mm increases considerably the specific surface area and so increases the leaching rate substantially. But ore which is ground to low particle size cannot be percolated, it has to be treated in suspension. Therefore a reactor is required in which the suspension can be agitated and aerated. The pulp may contain 10 per cent to 20 per cent solids in suspension (pulp density). Suspension leaching is a very effective method and has the advantage that it can easily be controlled and regulated. So it may be possible to chose a favourable temperature and to add phosphate, ammonia, carbon dioxide, sulphuric acid, iron or other additives in order to accelerate the leaching process. But on the other hand it is expensive and its application is restricted to special purposes, for instance to the leaching of concentrates Suspension leaching on a laboratory scale in agitated flasks is a convenient tool to investigate the leachability of an ore and to reveal the optimal leaching conditions.

PROBLEMS

There are many possibilities for disturbing bacterial leaching. Lack of iron can be met in most cases by adding iron in some form, preferably as pyrite because by oxidation of pyrite not only iron ions are formed but also hydrogen ions. Therefore addition of pyrite is well suited if it is necessary to lower the pH. For this latter purpose also elemental sulphur may be added instead of pyrite, Thiobacilli will then oxidise sulphur to sulphuric acid.

A large amount of carbonates may cause serious disruption because Thiobacilli and other bacteria concerned with ore leaching are acidophilic. They are inactive and do not grow in a neutral or alkaline milieu. If enough hydrogen ions are formed by bacterial oxidation activity alkali of earth carbonates may be neutralised and decomposed. But then another problem arises: the alkali of earth ions precipitate as sulphates and these may disturb the leaching by plugging and by covering the surfaces of the ore minerals.

In ponds on the top of dumps operated by the pond system ferric iron compounds often precipitate. This hinders the infiltration of the liquor by plugging the upper layer of the ore dump. From time to time the precipitates have to be scraped off.

Toxic substances in the ore may inhibit or kill the bacteria. *Thiobacilli*, *Sulfolobus* and *Leptospirillum ferrooxidans* are very tolerant against dissolved heavy metals. The following limits of heavy metal tolerance of *T. ferrooxidans* were observed is shown in Table 18.1.

Arsenic, molybdenum, silver and mercury may be toxic to *Thiobacilli*. Noteworthy is the higher tolerance against molybdenum of *Sulfolobus brierleyi*: this bacterium metabolises without inhibition at a molybdenum concentration of 20 mmol/l or higher, whereas *Thiobacilli* tolerate molybdenum only up to about 1 mmol/l. In some cases the tolerance of leaching bacteria against toxic substances may be developed by adaptation.

Table 18.1. Limits of heavy metals tolerance of *T. ferrooxidans*.

Cu	0.87 mol/l	55 g/l
Zn	1.83 mol/l	120 g/l
Ni	0.85 mol/l	150 g/l
U	0.004 mol/l	1 g/l without adaptation
		12 g/l after adaptation
	0.05 mol/l	
Mo	0.0008 mol/l	0.08 g/l

FURTHER INVESTIGATIONS

Many factors influence bacterial ore leaching are discussed below.

Properties of the micro-organisms mineral species including accompanying minerals surface area of the minerals, particle size water availability temperature pH redox potential oxygen supply carbon dioxide supply, supply of other nutrients e.g. nitrogen compounds, phosphate toxic substances light formation of secondary minerals

Much work has been done on the influence of these factors, qualitatively and quantitatively. Further effort is needed to understand fully all dependencies in all cases of bacterial leaching. Much work has to be done in order to find new applications and new methods. Many research groups in several countries work in this field. An interesting approach is genetical manipulation of leaching bacteria.

But here we should confine ourself to report what is done in the Federal Republic of Germany:

1. Dr. Bosecker in his laboratory at the Bundesanstalt für Geowissenschaften in Hanover investigates the application of bacterial leaching to new ores. In particular he has tried to leach copper from copper bearing black shale, nickel out of gabbro and other basic plutonic rocks, and zinc out of old dumps which were left by miners some centuries ago in Germany.
2. Bacterial leaching of industrial waste materials is done on an laboratory scale and in pilot plants by the group of Prof. Onken in Dortmund and the German mining company Preussag at the Harz Mountains (Goslar). Tailings from flotation plants, metal containing drosses and similar materials are leached, mainly as suspensions in different bioreactors.
3. Coal often contains considerable amounts of pyrite which on combustion is oxidised to sulphur dioxide. To minimise the emission of this toxic and acid forming gas Dr. Ebner and his co-workers in a laboratory of the Bergbau-Forschung in Essen try to desulphurise coal by bacterial leaching.
4. As was mentioned above *in situ* leaching is problematic. The mining company Preussag set up an underground pilot plant in an old mine in which a complex sulphidic ore has been excavated for more than a thousand years. In about 2 years it will be worked-out. In the upper and older parts of the mine the miners of past centuries left a lot of ore which is now slowly leached by natural bacterial oxidation and which cannot be excavated economically. The company is investigating whether or not these residues can be recovered economically by bacterial leaching.
5. Prof. Stetter and his group at the University of Regensburg have isolated about 350 new strains of bacteria, some of these, from hot springs and other hot biotopes in areas with volcanic activity in Italy and Iceland, are thermophilic, some extremely so. Prof. Stetter and his co-workers have examined these new strains in respect of their ability to leach metals out of ore minerals. Some

170 out of the 350 isolates are active in metal leaching. It may be that some of the thermophilic strains are suited to leaching at high temperatures with high rates on a technical scale.

6. The mining company 'Uranerzbergbau' did some work on leaching special uranium ores and investigated conditions for leaching with rates which make this operation economical.
7. In one laboratory in Braunschweig Smith investigated the ecology of the microbial communities which develop in acid ore leaching biotopes. The main object of the work is an old mine in the Harz Mountains in which the mining company has installed an underground *in situ* leaching pilot plant. The company isolated a lot of micro-organisms from this mine biotope in which the pH is between 2 and 3.

We have been able to show that many neutrophilic, heterotrophic bacteria, isolated from the mine, become acidotolerant growing in a spent culture medium of Thiobacilli. This fact explains why so many neutrophilic bacteria are found in acid leaching biotopes.

We wondered why so many heterotrophic bacteria live in an inorganic biotope which is thought to be free of organic matter, for heterotrophic bacteria need organic matter as an energy source and nutrient. The answer is that the autotrophic Thiobacilli are primary producers of organic matter, because they can synthesise biomass from carbon dioxide. It was found that many of the heterotrophic bacteria isolated from this biotope can grow at the expense of organic matter produced and partly excreted by the Thiobacilli.

Growth of heterotrophic bacteria at the expense of organic matter, able to reduce molecular nitrogen and bacterial leaching of heavy metals from sulphidic ore minerals are shown in Figs 18.16 and 18.17.

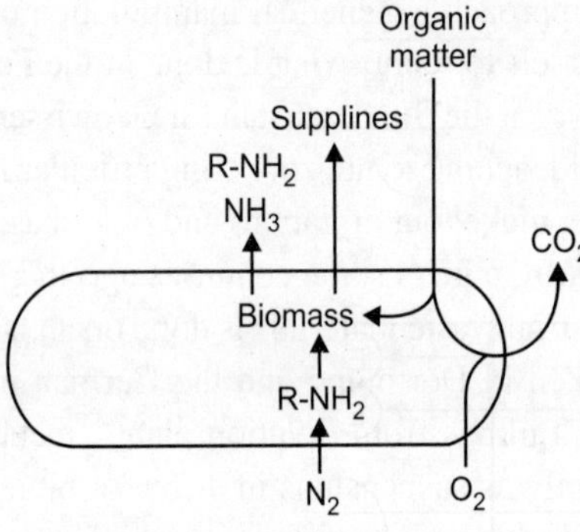

Fig. 18.16. Growth of heterotrophic bacteria at the expense of organic matter, able to reduce molecular nitrogen.

It is known that the metabolic activity and the growth of Thiobacilli are inhibited by some organic compounds. It was found that this is also the case with organic matter which is excreted by the Thiobacilli themselves. The consumption of this organic matter by heterotrophic bacteria therefore supports the metabolic activity and the growth of Thiobacilli, as we were able to show.

Bacterial leaching of heavy metals from sulphidic ore minerals. Almost all of the isolated strains of *Thiobacillus ferrooxidans*, but none of the *T. thiooxidans* strains, are able to utilise molecular nitrogen as a nitrogen source.

The *T. thiooxidans* strains cannot grow without addition of nitrogen compounds. Among the isolated heterotrophic bacteria were some strains able to reduce molecular nitrogen. In mixed cultures with these heterotrophic strains the *T. thiooxidans* strains grow and leach metals from ore without addition of nitrogen compounds. So we know that Thiobacilli may be provided with an utilisable nitrogen source by heterotrophic nitrogen reducing bacteria (Fig. 18.18).

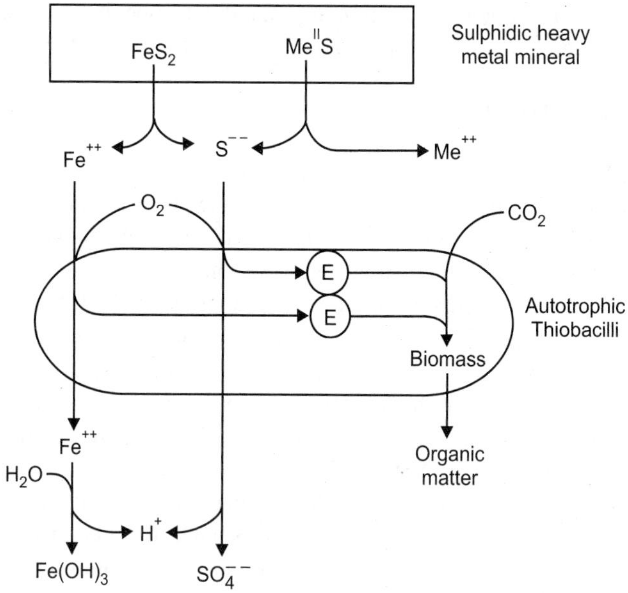

Fig. 18.17. Bacterial leaching of heavy metals from sulphidic ore minerals.

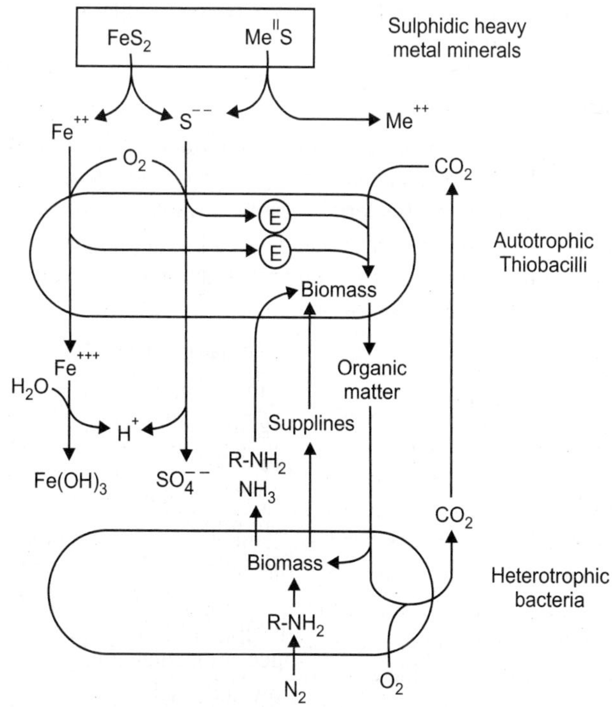

Fig. 18.18. Bacterial leaching of heavy metals.

In summary there are several interactions between the autotrophic Thiobacilli and their heterotrophic companions, and we do not know yet all of them. So we hope to learn more about the interrelationships in these interesting biocenoses of acid leaching biotopes, and we hope that more detailed knowledge in this field can help influence bacterial leaching methods towards greater efficiency.

BIOLEACHING OF VARIOUS METALS

Bioleaching of Copper

The physical configurations of bioleaching operations worldwide for copper are mostly uniform. Typically, copper ore mined from open pits is segregated, higher-grade material is concentrated to produce feed for smelting, while the lower-grade ore is subjected to leaching. The ore is piled on an impermeable surface until a dump of suitable dimension forms. After the top is levelled, leach solution is flooded or sprayed onto the dump. A copper dump is a complex and heterogenous microbiological habitat. It contains solids ranging in size from boulders to fine sand and includes material of complex mineralogy. Bacterial colonisation occurs mainly in the top one metre or so. The temperature may reach 900°C in the interior of the dump and supports a range of thermophillic micro-organism, which are often anaerobic, or microaerophilic. In these regions indirect leaching by ferric sulphate also prevails. The exterior of the dump is at ambient temperature and undergoes changes in temperature reflecting seasonal and diurnal fluctuations. Many different micro-organisms have been isolated from copper dumps, some of which have been studied in the laboratory. These include a variety of mesophilic, aerobic iron and sulphur oxidising micro-organisms; thermophilic iron and sulphur oxidising micro-organisms; and anaerobic sulphate reducing bacteria. Some are heterotrophic bacteria, which indirectly affect metal solubilisation by affecting the growth and activity of metal solubilising bacteria. Others are protozoa, which interact with and prey on different types of bacteria.

Bioleaching of Uranium

Uranium leaching proceeds by the indirect mechanism as *Acidithiobacillus ferrooxidans* does not directly interact with uranium minerals. The role of *Acidithiobacillus ferrooxidans* in uranium leaching is the best example of the indirect mechanism. Bacterial activity is limited to oxidation of pyrite and ferrous iron. The process involves periodic spraying or flooding of worked-out stopes and tunnels of underground mines with lixiviant. Another method in use for uranium extraction is vat leaching. Bioleaching has also been used succesfully to obtain uranium from waste gold ore.

Bioliberation of Gold

Iron- and sulphur-oxidising acidophilic bacteria are able to oxidise certain sulphidic ores containing encapsulated particles of elemental gold, resulting in improved accessibility of gold to complexation by leaching agents such as cyanide. Bio-oxidation of gold ores is a less costly, less polluting alternative to other oxidative pretreatments such as roasting and pressure oxidation.

Recently, bio-oxidation of gold ores has been implemented as a commercial process, and is under study worldwide for further application to refractory gold ores. Technology developed by K. A. Natarajan and co-workers at the Indian Institute of Science is being applied at the Hutti Gold Mines, Karnataka, India for extraction of gold. Bio-oxidation involves treatment with *Acidithiobacillus ferrooxidans* to oxidise the sulphide matrix prior to cyanide extraction. Commercial exploitation has made use of heap leaching technology for refractory gold ores. Refractory sulphidic gold ores contain mainly two types of sulphides: pyrite and arsenopyrite. Since gold is usually finely disseminated in the sulphide matrix, the objective of biooxidation of refractory gold ores is to break the sulphide matrix by dissolution of pyrite and arsenopyrite.

GREEN CHEMISTRY APPROACH FOR LEACHING GOLD

Over the past fifteen years there has been significant interest in the concepts of sustainable development and mining. Many governments, mining companies and non-government organisations have embraced the concepts of sustainable developnlent. One concept generated from the ideas stemnling from sustainable mining is that of 'green gold'. Similar to sustainable mining, green gold's meaning is defined by the originator.

Built into the definition of green gold are the diverse concerns for human well-being, community robustness, environmental integrity, and overall social and ethical responsibility. This section concentrates on the concerns of environmental integrity. The potential for an alternative to cyanide leaching of gold ores is examined in terms of green chemistiy. Also discussed is the potential of cyanide leaching becoming more environmentally friendly. The leaching reagents cyanide, thiocyanate, thiosulphate, and thiourea are compared using some of the principles of green chemistiy, specifically:

1. Prevention.
2. Atom economy.
3. Safer auxiliaries.
4. Real-time analysis for pollution prevention.

Sustainable Development

Sustainable development is a growing concern expressed by many businesses, organisations and individuals. Yet, no workable quantifiable definition of sustainability is available for evaluation of specific projects or operations. The lack of such a definition inherently gives engineers difficulty. Sustainability of human activities (predominantly production and consumption) is a growing concern among businesses, customers, governments, international bodies and non-governmental organisations. These concerns are often linked to energy efficiency, reduction of environmentally harmful emissionse, cosystem reservation and other 'save the earth' efforts.

They are becoming a part of the 'triple bottom line' for business accounting: financial, social and environmental. Despite its increasing importance, current definitions of 'sustainability' are somewhat vacuous. The most commonly accepted description was provided by the World Commission on Environinent and Development in the 'Brundtland report'. According to this report, the goal of sustainability is to 'meet the needs of the present generation without compromising the ability of future generations to meet their own needs'.

Sustainable Mining

Obviously, sustainable mining is part of the more general notion of sustainable development. But what is it? The definitions and the perceived concepts associated with the term, 'sustainable mining' are as diverse and numerous as the special interest groups that use the terminology. How do we recognise sustainable development and mining? When the concepts of sustainable development are applied to mineral resources, there is little agreement on what is to be sustained, and by what means. Without such agreement, sustainable development will remain little more than a slogan of little practical value to public-policy makers.

Part of what is missing from most discussion is an understanding of the nature of mineral resources and the dynamics of their development. Without this understanding the use of the sustainable development and mining sustainability becomes phrases that often confuse sustainability with environmental protection and other lofty goals that, strictly speaking, are not required for sustainable operations.

The corporate leaders of the minerals industry have responded to the call for sustainable development. The best source of their response is found in the report of the Mining, Minerals and Sustainable Development Project. In the report they state, 'One of the greatest challenges facing the world today is integrating economic activity with environmental integrity, social concerns, and effective governance systems. The goal of that integration can be seen as 'sustainable development'. In the context of the minerals sector, the goal should be to maximise the contribution to the well-being of the current generation in a way that ensures an equitable distribution of its costs and benefits, without reducing the potential for future generations to meet their own needs'. Sustainable development and mining sustainability are supported by four 'pillars'; economic sphere, social sphere, environmental sphere, and governance sphere. Each sphere has a set of guiding principles as indicated in Table 18.2.

Table 18.2. Sustainable development principles.

Economic sphere
Maximise human well-being
Ensure efficient use of all resources, natural and otherwise, by maximising rents
Seek to identify and internalise environmental and social costs
Maintain and enhance the conditions for viable enterprise
Social sphere
Ensure a fair distribution of the costs and benefits of development for all those alive today
Respect and reinforce the fundamental rights of human beings, including civil and political liberties, cultural autonomy, social and economic freedoms, and personal security
See to sustain improvements over time; ensure that depletion of natural resources will not deprive future generations through replacement with other forms of capital
Environmental sphere
Promote responsible stewardship of natural resources and the environment, including remediation of past damage
Minimise waste and environmental damage along the whole of the supply chain
Exercise prudence where impacts are unknown or uncertain
Operate within ecological limits and protect critical natural capital
Governance sphere
Support representative democracy, including participatory decision-making
Encourage free enterprise within a system of clear and fair rules and incentives
Avoid excessive concentration of power through appropriate checks and balances
Ensure transparency through providing all stakeholders with access to relevant and accurate information
Ensure accountability for decisions and actions, which are based on comprehensive and reliable analysis
Encourage cooperation in order to build trust and shared goals and values
Ensure that decisions are made at the appropriate level, adhering to the principle of subsidiarity where possible.

The principles of sustainable mining are well-defined by various groups, but the challenges for meeting the principles are great and complex. Many of the challenges facing sustainable mining are challenges that public opinion makers and government officials will have major roles in overcoming them. As technical leaders and educators we do have influence on how the challenge of mining, minerals beneficiation can be less threatening to the local environment and ecology. We can certainly aid in meeting the challenge of developing an integrated approach to using minerals.

One of the challenges is the large amounts of material involved in large-scale mining and minerals extraction. Although Korte and others have written about the effects of the cyanide leaching gold recovery process on the environment and the ramifications of it in terms of sustainable mining, some of their conclusions can be easily extrapolated to all large mining operations. Without question their point that the problems arising from the change of the chemistry of million of tons of natural ore during the grinding procedure, producing changes in the bioavailabilities of metals and other substances are not well understood. In the case of phosphate production not only are there large quantities of ore, but large quantities of mineral product that is chemically treated. Their conclusion is that when living in a time where the word 'sustainability' has a certain weight, can such operations ever be considered sustainable.

Another is challenge of the significant altering of material by the use of chemicals in the treatment of the ores. The American Chemical Society (ACS) supports the use of the twelve Green Chemistry principles. Bucknam points out that four of these principles:

1. Prevention.
2. Atom economy.
3. Safer auxiliaries.
4. Real-time analysis for pollution prevention provide guidance for sustainable use of chemicals in metallurgical processing.

Mining produces very large volumes of waste, so decisions about where and how to dispose of it are often virtually irreversible. Because decisions about waste handling and other aspects of operations are often so difficult an expensive to reverse, they need to be made correctly in the first place through mine closure planning.

A third challenge is the environmental legacy left by mining. The environmental issues of current mining operations are daunting enough. But in many ways far more troubling are some of the continuing effects of past mining and smelting. The development of an environmental management system (EMS) in which an environmental impact assessment (EIA) is an integral part would permit the mining company and the regulating authority a structure method for having an awareness and control of the performance of a project that could be applied at all stages of the mining life cycle. In the times of sustainable development the EIA is insufficient. Now and in the future there is a greater need for integrated impact assessments. The minimum loss of biodiversity is the other great challenge of mining sustainability. The loss of biodiversity is an irreversible loss. Conservation practices that guarantee a minimum impact on biodiversity must be adopted and implemented.

As important as the methods of mining and beneficiation is how the minerals are used in efforts to develop sustainable development. An integrated approach for the use of minerals must be utilised. The use and downstream supply of mineral products has implications for sustainable development and must be considered along with mining and processing of minerals. Current patterns of minerals use raise concerns about efficiency and the need for more equitable access to resources world-wide. The environniental and health impacts of different mineral products in use need to be carefully managed. Where the risks associated with use are deemed unacceptable or are not known, the costs associated with using certain minerals may outweigh the benefits.

Gold Hydrometallurgy and Cyanide

The cyanidation process is remarkably robust and the gold-cyanide complex ion is very stable and gives very few operation problems. In addition to valuable results from research and development, industrial operation experiences have made the cyanidation process a great success. However, the toxicity of cyanide is a concern. With increasing regulations on the use of cyanide and the lowering of acceptable cyanide discharge levels, the gold industry faces more challenges than ever before. Cyanide is unable to effectively extract gold from certain difficult-to-treat ores leading to poor gold recoveiy from carbonaceous refractory ores and lack of selectivity in the treatment of copper bearing gold ores. In these cases, cyanide consumption is generally high and tailings discharge more difficult.

Alternative Lixiviants

Alternative reagents that have been seriously considered include the halide system (chlorine, bromine and iodine), the thiosystem (thiosulphate, thiocyanate and thiourea), and polysulphide system (S_x^{2-}), and the ammonia system (ammonia and ammonium copper-cyanide).

In the alkaline copper-ammonia-thiosulphate system most common sulphide minerals which exist in refractory gold ores become unstable in thiosulphate solution under gold leaching conditions. The presence of certain sulphide minerals accelerates thiosulphate degradation, consumes thiosulphate, and retards gold dissolution. Leaching experimental results confirm that thiosulphate is much more stable in the leaching of oxide ores compared to its stability in the leaching of sulphidic and/or carbonaceous ores.

Thiocyanate is an effective lixiviant for gold leaching in an acidic environment. Thiocyanate is a pseudohalide. The pseudohalide behaviour of thiocyanate results in the formation of insoluble salts when the thiocyanate reacts with silver, mercury, lead and copper ion under certain conditions depending on the thiocyanate concentration. At low thiocyanate concentrations, the formation of insoluble compounds such as AgSCN and CuSCN may precipitate on the gold surface and retard gold dissolution.

For the ferric-thiourea system, a high reagent consumption and gold passivation make the system impractical for commercial operation. However, the copper-ammonia-thiosulphate process has a number of significant problems which have impeded its implementation industrially. These include:

1. High rates of thiosulphate oxidation in the presence of oxygen.
2. Generation of polythionates which hinder the resin adsorption process.
3. The requirement for relatively high concentrations of ammonia, and hence the environmental impact of ammonia/ammonium discharges needs to be considered.

On possible innovation to the thiosulphate process is the application of the ferric-oxalatethiosulphate process. Studies have been conducted in which the ferric ions stabilised using oxalate was used as the oxidant in the system, and thiourea was used as an additive to improve the leaching kinetics. It was shown that the gold leach rate and solution stability were dependent on the ferric to oxalate ratio, and for a ratio of Fe(1II):oxalate of 1:3, the solution E_H and the gold leach rate were stable with time. At lower Fe(III):oxalate, some of the Fe(III) is reactive towards thiosulphate.

Barbosa-Filho and Monhemius showed that thiocyanate can leach gold in the pH range of 1–3 at higher temperatures (up to 85°C). The low pH and the higher temperatures would indicate that high capital costs would be required for a leach plant. The higher temperatures would also mean that higher operating costs would be required compared to cyanidation. The availability of thiocyanate may also be a restriction and if thiocyanate had to be detoxified, a considerable oxygen demand would be necessary, which would even further increase the operating costs.

This lixiviant, however, may be suitable for most ore types and the recyclability could be possible if the temperature is not too high to decompose to a considerable extent. In practice, high temperature around 85°C is, unfortunately, necessary to achieve satisfactory leach performance. An ultimate oxidation cyanate and sulphate is also possible. These costs, however, as already mentioned, would be considerable, since five moles of active oxygen are required per mole of thiocyanate to oxidise it to sulphate and cyanate.

Compared to cyanide, the use of thiocyanate as an alternative lixiviant is probably not economically viable, because of the high temperatures required, even if detoxification costs are not considered. Although specific cases of viability may be possible with high grade concentrates, for example, the operating costs would be the limiting factor for universal applicability.

After examining the chemical characteristics and the health and safety profiles of the many proposed alternative lixiviants for cyanide it is not suprising that cyanide is still the most common, if not only universally applicable lixiviant for gold bearing ores. The application of green chemistry for the leaching of gold ores becomes more of an engineering problem than a problem directly associated with chemistry. Atom economy, or the use of the minimum cyanide required is one aspect that must be determined if the cyanide pollution problem is to be satisfactorily abated. Atom economy cannot be achieved without real-time analysis for pollution prevention.

Even with the acceptance of cyanide or some other lixiviant for gold extraction the public debate over the social value of any gold mining will not abate. The contentions that gold mining has no social value are not answered by improved use of lixiviants nor are the answers within the scope of this paper.

Chapter 19

Extracellular Polysaccharides and their Derivatives

INTRODUCTION

Polysaccharides are polymeric carbohydrate structures, formed of repeating units (either mono- or di-saccharides) joined together by glycosidic bonds. These structures are often linear, but may contain various degrees of branching. Polysaccharides are often quite heterogeneous, containing slight modifications of the repeating unit. Depending on the structure, these macromolecules can have distinct properties from their monosaccharide building blocks. They may be amorphous or even insoluble in water.

When all the monosaccharides in a polysaccharide are the same type the polysaccharide is called a homopolysaccharide, but when more than one type of monosaccharide is present they are called heteropolysaccharides.

Examples include storage polysaccharides such as starch and glycogen, and structural polysaccharides such as cellulose and chitin.

Polysaccharides have a general formula of $C_x(H_2O)_y$ where, x is usually a large number between 200 and 2500. Considering that the repeating units in the polymer backbone are often six-carbon mono-saccharides, the general formula can also be represented as $(C_6H_{10}O_5)_n$ where $40 = n = 3000$.

Exopolysaccharide: Exopolysaccharides are high-molecular-weight polymers that are composed of sugar residues and are secreted by a micro-organism into the surrounding environment. Micro-organisms synthesise a wide spectrum of multifunctional polysaccharides including intracellular polysaccharides, structural polysaccharides and extracellular polysaccharides or exopolysaccharides (EPS). Exo-polysaccharides generally constitute of monosaccharides and some non-carbohydrate substituents (such as acetate, pyruvate, succinate, and phosphate). Owing to the wide diversity in composition, exopolysaccharides have found multifarious applications in various food and pharmaceutical industries. Many microbial EPS provide properties that are almost identical to the gums currently in use. With innovative approaches, efforts are underway to supersede the traditionally used plant and algal gums by their microbial counterparts. Moreover, considerable progress has been made in discovering and developing new microbial EPS that possess novel industrial significance.

The sensory benefits of the exopolysaccharides of lactic acid bacteria are well established and there is evidence for the health properties that are attributable to exopolysaccharides from lactic acid bacteria.

Capsular exopolysaccharides can protect pathogenic bacteria and contribute to their pathogenicity. Attachment of nitrogen-fixing bacteria to plant roots and soil particles, which is important for colonisation of rhizosphere and roots and for infection of the plant, can be mediated by exopolysaccharides. An example for industrial use of exopolysaccharides is the application of dextran in panettone and other breads in the bakery industry.

BACTERIAL POLYSACCHARIDES

Bacterial polysaccharides represent a diverse range of macromolecules that include peptidoglycan, lipopolysaccharides, capsules and exopolysaccharides, compounds whose functions range from structural cell-wall components (e.g. peptidoglycan), and important virulence factors (e.g. poly-*N*-acetyl-glucosamine in *S. aureus*), to permitting the bacterium to survive in harsh environments (e.g. *Pseudomonas aeruginosa* in the human lung). Polysaccharide biosynthesis is a tightly regulated, energy intensive process and understanding the subtle interplay between the regulation and energy conservation, polymer modification and synthesis, and the external ecological functions is a huge area of research. The potential benefits are enormous and should enable for example the development of novel antibacterial strategies (e.g. new antibiotics and vaccines) and the commercial exploitation to develop novel applications.

Pathogenic bacteria commonly produce a thick, mucous-like, layer of polysaccharide. This 'capsule' cloaks antigenic proteins on the bacterial surface that would otherwise provoke an immune response and thereby lead to the destruction of the bacteria. Capsular polysaccharides are water soluble, commonly acidic, and have molecular weights on the order of 100–1000 kDa. They are linear and consist of regularly repeating subunits of one to six monosaccharides. There is enormous structural diversity; nearly two hundred different polysaccharides are produced by *E. coli* alone. Mixtures of capsular polysaccharides, either conjugated or native are used as vaccines.

Bacteria and many other microbes, including fungi and algae, often secrete polysaccharides as an evolutionary adaptation to help them adhere to surfaces and to prevent them from drying out. Humans have developed some of these polysaccharides into useful products, including xanthan gum, dextran, gellan gum, and pullulan.

Cell-surface polysaccharides play diverse roles in bacterial ecology and physiology. They serve as a barrier between the cell wall and the environment, mediate host-pathogen interactions, and form structural components of biofilms. These polysaccharides are synthesised from nucleotide-activated precursors (called nucleotide sugars) and, in most cases, all the enzymes necessary for biosynthesis, assembly and transport of the completed polymer are encoded by genes organised in dedicated clusters within the genome of the organism. Lipopolysaccharide is one of the most important cell-surface polysaccharides, as it plays a key structural role in outer membrane integrity, as well as being an important mediator of host-pathogen interactions.

EXTRACELLULAR POLYSACCHARIDES

Microbial extracellular polysaccharide (EPS) are the key components for the aggregation of micro-organisms in biofilms, flocs and sludge. They are composed of polysaccharides, proteins, nucleic acids, lipids and other biological macromolecules. Cell free EPS have been separated from the cells which have produced them.

Acidithiobacillus ferrooxidans EPS consists mainly of neutral sugars and lipids produced from UDP-glucose, UDP-galactose, and dTDP-rhamnose precursors. Extracellular polysaccharides(ECPS) are widely used in the oil, printing and dyeing , papermaking, and pesticide industries. They serve the function of coating and preserving material. Besides having pseudoplastic rheological properties and excellent pH stability, it is compatible with cationic dyes without the precipitation factor.

Unlike other microbial polysaccharides, the viscosity of the polymer keeps constant below 65°C, and decreases sharply in a narrow range of temperature between 70°–75°C.

XANTHAN

Natural gums occur in all life forms. Many types of gums can be obtained from plant sources. Their collection is costly and requires skilled labourers. Moreover, seasonal variations affect the quality and quantity of plant gums. Gum is the common term for hydrocolloidal gels—polysaccharides that have an affinity for water and exhibit binding properties with water and other organic/inorganic materials. Traditionally, gums have been derived from a wide variety of plants. Chemically, gums are carbohydrate polymers or polysaccharides (however, gelatine is a protein).

Polysaccharides are present in all life forms. They have a number of unique chemical and physical properties. They serve as structural material to the plant kingdom, as energy reserves, adhesives and also information-transfer agents. Microbial polysaccharides are composed of regular repeating units of simple sugars like glucose, mannose, fructose etc. These polysaccharides are sometimes termed as slime or exopolysaccharides. Naturally occurring polysaccharides from seaweeds have been in use from many decades in immense quantities. Dextran, discovered in early 1940s, was the first microbial polysaccharide to be commercialised. The second microbial polysaccharide commercialised was xanthan.

Xanthan gum is a water-soluble exopolysaccharide. It is produced industrially from carbon sources by fermentation using the Gram-negative bacterium *xanthomonas*. There have been various attempts to produce xanthan gum by fermentation method using bacteria and yeast by using various cheap raw materials. Xanthan gum is widely used in various industries such as food, agriculture, oil, paint and various chemical industries. Xanthan gum is a natural polysaccharide and an important industrial biopolymer. The polysaccharide, B-1459 or xanthan gum, produced by the bacterium *Xanthomonas campestris* NRRL B-1459, was extensively studied because of its properties that would allow it to supplement other known natural and synthetic water-soluble gums. Extensive research was carried out in several industrial laboratories during the 1960s, culminating in semi-commercial production as Kelzan by Kelco. Substantial commercial production began in early 1964. Today, the major producers of xanthan are Merck and Pfizer in the United States, Rhone Poulenc and Sanofi-Elf in France, and Jungbunzlauer in Austria.

The toxicological and safety properties of xanthan gum for food and pharmaceutical applications have been extensively researched. Xanthan is nontoxic and does not inhibit growth. It is nonsensitising and does not cause skin or eye irritation. On this basis, xanthan has been approved by the United States Food and Drug Administration (FDA) for use as food additive without any specific quantity limitations. In 1980, The Canadian Governor-in-Council also approved general use of xanthan in food. Xanthan is on Annex-I of the European Economic Community emulsifier/stabiliser list as item E-415.

The Joint Expert Committee of the Food and Agriculture Organisation/World Health Organisation of the United Nations (FAO/WHO) have issued an acceptable daily intake (ADI) of xanthan. In addition, many other countries have approved xanthan for various food uses.

The excellent rheological properties of xanthan gum contribute to its wide-range of applications as a suspending, stabilising, and/or thickening agent in the food industry and its use as an emulsifier, lubricant, thickening agent, and/or mobility control agent to enhance oil recovery.

The petrochemical industry uses other plant-derived polysaccharides and synthetic polymers instead of xanthan gum based on the relative costs of xanthan gum to the other polymers. In the United States, the only commercially available xanthan gum is food grade. Commercially available xanthan gum is relatively expensive due to glucose or sucrose being used as the sole carbon source and the very stringent purity standards of the FDA for foods. For food-grade xanthan gum, up to 50 per cent of the production costs are related to downstream purification steps, many of which would not be necessary for non-food applications. Another cost reduction could be achieved by using less expensive substrates, such as waste agricultural products.

Structure of Xanthan Gum

Xanthan gum is a heteropolysaccharide with a primary structure consisting of repeated pentasaccharide units formed by two glucose units, two mannose units, and one glucuronic acid unit, in the molar ratio 2.8:2.0:2.0 (Fig. 19.1).

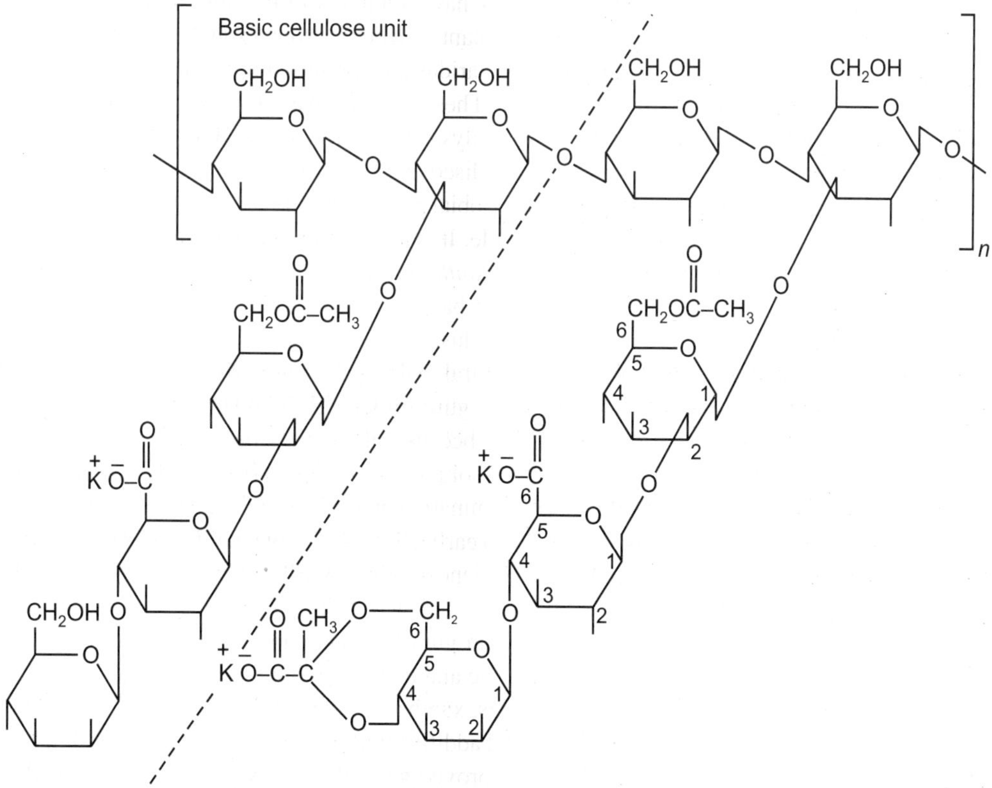

Fig. 19.1. Cellulose-like backbone and trisaccharide side chains.

The chemical structure of the main chain is identical to that of cellulose. Trisaccharide side chains contain a D-glucuronic acid unit between two D-mannose units linked at the O-3 position of every other glucose residue in the main chain. Approximately one-half of the terminals D-mannose contain a pyruvic acid residue linked via keto group to the 4 and 6 positions, with an unknown distribution. D-Mannose unit linked to the main chain contains an acetyl group at position O-6. The presence of acetic and pyruvic acids produces an anionic polysaccharide type. The trisaccharide branches appear to be closely aligned with the polymer backbone. The resulting stiff chain may exist as a single, double, or triple helix, which interacts with other polymer molecules to form a complex.

The molecular weight distribution ranges from 2×10^6 to 20×10^6 Da. This molecular weight distribution depends on the association between chains, forming aggregates of several individual chains. The variations of the fermentation conditions used in production are factors that can influence the molecular weight of xanthan.

Application of Xanthan Gum in the Food Industry

The major food applications of xanthan gum stem from the fact that when it is dispersed either in hot or cold water, the resultant aqueous dispersions are thixotropic.

The weak gel-like structure formed results in an unusually high 'low shear rate viscosity' at low polymer concentrations, which can be used to thicken aqueous samples and permits stabilisation of emulsions, foams and particulate suspensions. Finally, the reversible shear thinning behaviour allows manipulation and control of processes such as spreading, pumping, pouring and spraying.

The major applications of xanthan gum are in food industry as a suspending and thickening agent for fruit pulp and chocolates. US-FDA have approved xanthan on the basis of toxicology tests for use in human food. Many of today's foods require unique texturisation, viscosity, flavour release, appearance and water-control properties.

Xanthan gum improves all these properties and additionally controls the rheology of the final food product. It exhibits pseudoplastic properties in solutions, and has less 'gummy' mouth feel than gums with more Newtonian characteristics.

Bakery products

In the bakery, xanthan gum is used to increase water binding during baking and storage and extends the shelf-life of baked goods and refrigerated doughs. In soft-baked goods xanthan gum can also be used as an egg replacer, in particular the egg-white content can be reduced without affecting appearance and taste. Xanthan inhibits syneresis and prevents the filling from being absorbed by the pastry. In baking goods and refrigerated dough, it increases moisture retention and inhibits retrogradation, thus extending shelf-life. Xanthan gum also contributes smoothness, air incorporation and retention, and recipe tolerance to batters for cakes, muffins, biscuits and bread mixes. Baked goods have increased volume and moisture, higher crumb strength, less crumbling and greater resistance to shipping damage.

Batters

In wet prepared batters, xanthan gum reduces flour sedimentation; improves gas retention; imparts enzyme, shear and freeze thaw stability; and provides uniform coating and good cling. In predusts, xanthan gum improves adhesion and controls moisture migration during frying. In pancake batters, xanthan gum improves spread control, volume and air retention.

Bakery and pie fillings

Adding xanthan gum to either coldor hot processed bakery and fruit pie fillings improves texture and flavour release. The added benefits incream and fruit fillings are extended shelf stability, freeze-thaw stability and syneresis control.

Beverages

Xanthan is used as a bodying agent in beverages and squashes. When these drinks contain particles of fruit pulp, inclusion of xanthan helps in maintaining the suspension, resulting in good product appearance and texture. Xanthan contributes to pleasing mouth-feel, rapid and complete solubility at low pH with excellent suspension of insolubles and compatibility with most components. Xanthan gum in dry-mix beverage bases provides enhanced body and quality to the reconstituted drink, along with rapid viscosity development.

Dairy

Blends of xanthan gum, carrageenan, guar, locust bean gum (LBG), and galactomannans are excellent stabiliser for ice cream, ice milk, sherbet, milk shakes and water ices. Xanthan with methyl carboxymethyl cellulose works for frozen dairy and with carboxymethyl cellulose for directly acidified yogurts.

Similar blends are used for dessert puddings, acidified milk gels and others.These economical blends provide optimal viscosity, long-term stability, improved heat transfer during processing, enhanced flavour release, heat-shock protection and ice-crystal control. The xanthan, guar and LBG blend is vital to sliceability, firm body and flavour release of cream cheese. Also, xanthan works with wheat flour and soya flour as a matrix for spices and flavouring. Xanthan thickens cottage cheese dressings by providing good drainage control. Xanthan improves consistency, body and syneresis in sour cream.

Dressings

Xanthan is widely used in pourable salad dressings (alone or in combination with propylene glycol alginate or pectin), imparting the clean mouth-feel due to its high pseudoplasticity that also helps keep the dressing on top of the salad. Xanthan gum's stability to acid and salt, effectiveness at low concentrations and highly pseudoplastic rheology make it the ideal stabiliser for pourable no-oil, low-oil and regular salad dressings. Dressings with xanthan gum have excellent long-term emulsion stability and a relatively constant viscosity over a wide temperature range; they pour easily, but cling well to the salad. As a partial replacement for starch in regular and reduced calorie spoonable dressings, xanthan gum imparts desirable body, texture and freeze-thaw stability, as well as improved flavour release and eating sensation.

Pet food

Xanthan, along with LBG or guar produces a homogeneous gelled product for blood chunks or semi-moist pet foods. In liquid milk replacers for calves and piglets, xanthan gum is used to stabilise the suspension of insoluble substances. Xanthan is often used in combination with LBG and guar gum as stabiliser and binder in the formulation of canned gravy based pet food.

Syrups and topping

The outstanding solution properties of xanthan are utilised in syrups and toppings. Buttered syrups and chocolate toppings containing xanthan have excellent consistency and flow properties and because of their high at-rest viscosities, appear thick and appetising on products such as pancakes, ice cream and cooked meats. Xanthan is an ideal thickener for these products because of its high acid stability and the pseudoplastic flow properties it imparts. Frozen nondairy whipped topping concentrates have firm texture, high overrun and excellent freeze-thaw stability.

Relish

The addition of xanthan gum to relish improves the drained weight and virtually eliminates the loss of liquor during handling. In portion pack relish, xanthan gum keeps the relish and liquor uniformly distributed during filling and prevents spattering.

Sauces and gravies

Low levels of xanthan gum provide high viscosity in sauces and gravies at both acid and neutral pH. Viscosity is also extremely stable to temperature change and is maintained under a variety of long-term storage conditions. These sauces and gravies cling to hot foods and have excellent flavour release and appearance. There are still several issues that need to be addressed in order to produce xanthan gum biotechnologically within the targeted cost, such as the development of high-performance xanthan gum-producing micro-organisms and the lowering of the costs of raw materials and fermentation processes. The biotechnological processes for the production of xanthan gum from cheap raw materials should be improved further to make them effective.

Allergy

Some people are allergic to xanthan gum, with symptoms of intestinal gripes, diarrhea, temporary high blood pressure, and migraine headaches. Evaluation of workers exposed to xanthan gum dust found little evidence that respiratory symptoms were associated with exposure to xanthan gum dust. Since xanthan gum is sometimes produced by a bacterium that is fed corn to grow, some people allergic to corn may also react to it. Yellow Phrygian Husk is a common source of bacterium in which xanthan gum is created. However, some xanthan gum is not corn-derived.

ALGINATE

Alginic acid, also called algin or alginate, is a viscous gum that is abundant in the cell walls of brown algae. It ranges from white to yellowish-brown, and takes filamentous, granular and powdered forms. Alginates are produced by brown seaweeds (*Phaeophyceae*, mainly Laminaria) and have the structure as shown in Fig. 19.2.

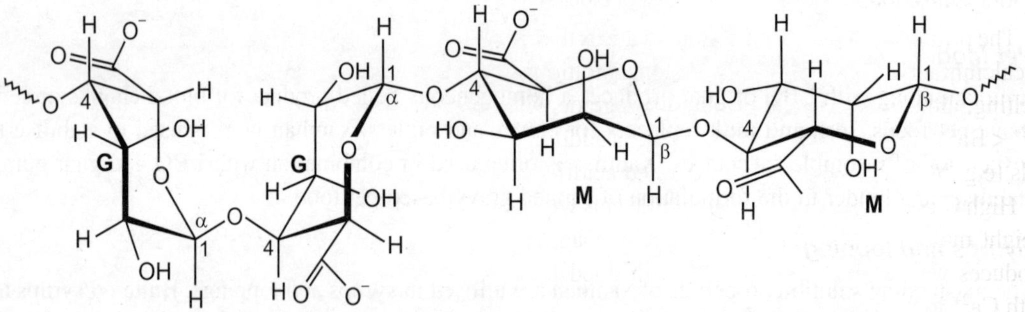

Fig. 19.2. Structure of alginate.

They are linear unbranched polymers containing D-mannuronic acid (**M**) and L-guluronic acid (**G**) residues. Alginates are not random copolymers but, according to the source algae, consist of blocks of similar and strictly alternating residues (i.e. **MMMMMM, GGGGGG** and **GMGMGMGM**), each of which have different conformational preferences and behaviour. As examples, the **M/G** ratio of alginate from Macrocystis pyrifera is about 1.6, whereas that from Laminaria hyperborea is about 0.45. Alginates may be prepared with a wide range of average molecular weights (50–1,00,000 residues) to suit the application (Fig. 19.3).

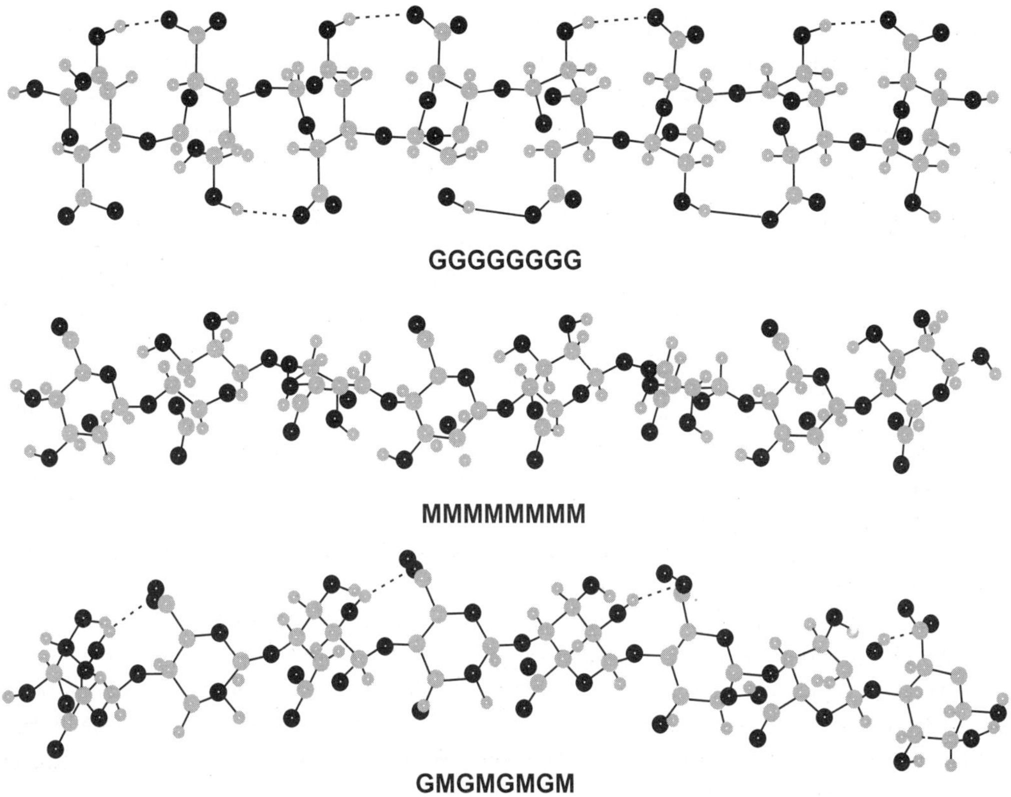

Fig. 19.3. Alginates consist of blocks of similar and strictly alternating residues.

The primary function of the alginates are as thermally stable cold setting gelling agents in the presence of calcium ions; gelling at far lower concentrations than gelatine. Such gels can be heat treated without melting, although they may eventually degrade. Gelling depends on the ion binding (Mg^{2+} << Ca^{2+} < Sr^{2+} < Ba^{2+}) with the control of the dication addition being important for the production of homogeneous gels (e.g. by ionic diffusion or controlled acidification of $CaCO_3$).

High **G** content produces strong brittle gels with good heat stability (except if present in low molecular weight molecules), but prone to water weepage (syneresis) on freeze-thaw, whereas high **M** content produces weaker more-elastic gels with good freeze-thaw behaviour and high **MGMG** content zips with Ca^{2+} ions to reduce shear.

However, at low or very high Ca^{2+} concentrations high **M** alginates produce the stronger gels. So long as the average chain lengths are not particularly short, the gelling properties correlate with average **G** block length (optimum block size ~12) and not necessarily with the **M/G** ratio, which may be primarily due to alternating **MGMG** chains. The future prospects are excellent as recombinant epimerases with different specificities may be used to produce novel designer alginates.

Alginate's solubility and water-holding capacity depend on pH (precipitating below about pH 3.5), molecular weight (lower molecular weight calcium alginate chains with less than 500 residues showing increasing water binding with increasing size), ionic strength (low ionic strength increasing the extended nature of the chains) and the nature of the ions present.

Generally alginates show high water absorption and may be used as low viscosity emulsifiers and shear-thinning thickeners. They can be used to stabilise phase separation in low fat, fat-substitutes, e.g. as alginate/caseinate blends in starch three-phase systems. Alginate is used in a wide variety of foodstuff such as pet food chunks, onion rings, stuffed olives, low fat spreads, sauces and pie fillings.

Sodium Alginate

The chemical compound sodium alginate is the sodium salt of alginic acid. Its empirical formula is $NaC_6H_7O_6$. It form as a flavourless gum, when extracted from the cell walls of brown algae, is used by the foods industry to increase viscosity and as an emulsifier. It is also used in indigestion tablets and the preparation of dental impressions. Another major use of sodium alginate is reactive dye printing, where it is used in the textile industry.

A major application for sodium alginate is as thickener for reactive dyestuffs (such as the Procion cotton-reactive dyes) in textile screen-printing and carpet jet-printing. Alginates do not react with these dyes and wash out easily, unlike starch-based thickeners. Sodium alginate is a good chelator for pulling radioactive toxins such as iodine-131 and strontium-90 from the body which have taken the place of their non-radioactive counterparts. It is also used in immobilising enzymes by inclusion.

As a food additive, sodium alginate is used especially in the production of gel-like foods. For example, bakers', chellies are often gelled alginate jam.

Also, the pimento stuffing in prepared cocktail olives is usually injected as a slurry at the same time that the stone is ejected; the slurry is subsequently set by immersing the olive in a solution of a calcium salt which causes rapid gelation by electrostatic cross-linking. A similar process can be used to make 'chunks' of everything from cat food through 'reformed' ham or fish to 'fruit' pieces for pies. It has the E-number 401. Now-a-days it is also used in the biological experiments for the immobilisation of cells to obtain important products like alcohols, organic acids, etc.

In recent years sodium alginate has been used in molecular gastronomy at some of the best restaurants in the world. Ferran Adria pioneered the technique and it has since been used by chefs such as Grant Achatz and Heston Blumenthal. Sodium alginate is combined with calcium lactate or similar compound to create spheres of liquid surrounded by a thin jelly membrane.

CURDLAN

Curdlan, an insoluble microbial exopolymer is composed almost exclusively of β-(1,3)-glucosidic linkages. One of the unique features of curdlan is that aqueous suspensions can be thermally induced to produce high-set gels, which will not return to the liquid state upon reheating, and this has attracted the attention of the food industry.

In addition to this, curdlan offers many health benefits, as the β-glucan family is well known among the scientific community to have immunestimulatory effects.

Structure

Both chemical and enzymatic analyses have confirmed that curdlan is a homopolymer of D-glucose linked in β-(1,3) fashion (Fig. 19.4). Curdlan has an average degree of polymerisation (DP) of approximately 450, and is unbranched. Nakata reported that the average molecular weight of curdlan in 0.3 N NaOH in the range of 5.3×10^4 to 2.0×10^6 daltons. Within the class of polysaccharides classified as β-(1,3)-glucans, there are a number of structural variants. The sources of glucans and their structural differences are listed in Table 19.1. Mycelial fungi are an abundant source of β-(1,3)-glucans; grifolan, which is produced from *Grifola frondosa* and stimulates cytokine production from macrophages, is a β-(1,3)-glucan with a molecular weight of >4.5 × 10^5 daltons, while lentinan, from *Lentinus edodes*, has a molecular weight of 5 × 10^5 daltons and two glucose branches for every five β-(1,3)-glucosyl units in the backbone.

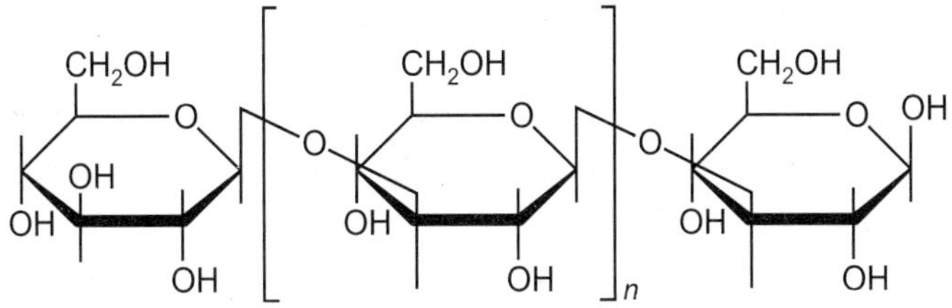

Fig. 19.4. Structure of curdlan.

Table. 19.1. A variety of glucans having β-1,3 linkage in their backbones.

Source	*Branch*	M_w
Bacteria		
Curdlan (*Agrobacterium* sp. *Alcaligens* sp.)	Exclusively β-(1,3)-glucosidic linkages	5.3 × 10^4 – 2.0 × 10^6
Fungi		
Grifolan (*Grifola frondosa*)	Branched β-1,3-gulcan	4.5 × 10^5
Lentinan (*Lentinus eeodes*)	Two glucose branches for every five glucose unit	5 × 10^5
Schizophyllan (*Schizophyllum commune*)	One glucose branch for every third glucose unit	4.5 × 10^5
Scleroglucan (*Sclerotium glucanum*)	One glucose branch for every third glucose unit	1.6–5.0 × 10^6
SSG (*Sclerotinia sclerotiorum*)	Highly branched β-1,3-glucan	2 × 10^5–2 × 10^6
Pachyman (*Poria cocos*)	Several β-1,6-linked branch points per molecule	2.06 × 10^4, 8.93 × 10^4
Krestin (*Coriolus versicolor*)	β-1,3-glucan	1.0 × 10^6
Yeast (*Saccharomyces cerevisiae*)		
Soluble glucan	β(1,6) linkage to β-(1,3) backbone	2 × 10^5–2 × 10^6
Insoluble glucan	β-(1,6) linkage to β-(1,3) backbone	3.53 × 10^4, 4.57 × 10^6
Brown algea		
Laminarin (*Laminaria digitata*)	β-1,3-glucan and β-1,6-glucan	–

The structure of schizophyllan is very similar to that of lentinan, but it has one glucose branch for every third glucose in the β-1,3-backbone and its molecular weight is 4.5 × 10^5 daltons. Scleroglucan from *Sclerotium rolfsii* has one glucose branch for every third glucose unit, whereas SSG from *Sclerotinia sclerotiorum* is a highly branched β-(1,3)-glucan. Pachyman, from *Poria cocos*, has an average of 3.2

branch points per molecule of β-(1,3)-glucan, whilst krestin is a protein-linked β-glucan with a molecular weight of c. 1,00,000 daltons, which can be extracted from the mycelia of *Coriolus versicolor*. β-(1,3)-Glucan is also present in the inner cell wall of the bakers' yeast *Saccharomyces cerevisiae* to support the structural strength of its cell wall. Unlike lentinan, schizophyllan and scleroglucan, the side branches of yeast β-(1,3)-glucans are chains of glucose molecules, and not single glucose residues. Depending on the extraction procedure used and their subsequent treatment, the yeast glucans may be either particulate water-insoluble or water-soluble macromolecules.

Conformation in Solution

Many researchers have investigated the molecular structures of curdlan in aqueous system. Three conformers of soluble curdlan have been reported, including single-helix, triple-helix, and random coil. Ogawa studied the conformational behaviour of curdlan in alkaline solution by measuring the optical rotatory dispersion, intrinsic viscosity and flow birefringence. At low concentrations of sodium hydroxide, curdlan has a helical (ordered) conformation, but a significant conformational change occurs at a NaOH concentration of 0.19–0.24 N NaOH.

In alkaline solution >0.2 N NaOH, curdlan is completely soluble and exists as random coils, but upon neutralisation the polymer adopts an 'ordered state', which is composed of a mixture of single and triple helices. A ^{13}C-NMR study supported this finding. Increasing the salt concentration shifts the point of conformational transition to a higher alkali concentration, and addition of nonsolvents such as 2-chloroethanol, dioxan or water to dimethylsulphoxide (DMSO) solution also changes the conformation of curdlan to a rigid, ordered structure. These workers also showed that the optical rotation was dependent upon the DP of the curdlan in 0.1 N sodium hydroxide, and concluded that the content of the ordered form increases with DP until becoming constant at DP values of about 200. Electron microscopic comparison of the molecular structures of curdlan with different DPs showed that only curdlan with higher DP can form a gel when heated.

The conversion between triple-helix and single-helix conformers is mediated by different chemical or physical treatments. Treatment of the triple-helix schizophyllan with NaOH has been used to prepare single helix-rich forms. Young proposed a transition mechanism after an investigation using fluorescence resonance energy transfer spectroscopy, which showed that a partially opened triple-helix conformer was formed on treatment with NaOH, and that increasing degrees of strand opening were associated with increasing concentrations of NaOH. After neutralising the NaOH, the partially opened conformers gradually reverted to the triple-helix.

Gel Structure

Some clarification of the fine structure of dispersed molecules and networks is necessary to understand the viscoelastic properties of curdlan, which forms two distinct types of gel. For both gels—described as low-set and high-set gels—transmission electron microscopy (TEM) showed them to be composed of three curdlan molecules that are associated to form a triple helix. Tada proposed a mechanism of formation of the low-set gel using static light-scattering measurements.

The molecular associates are formed at a NaOH concentration of 0.01–0.1 N at 25°C, and this association progresses with as the NaOH concentration decreases. Consequently, the average molecular weight for the molecular associate in 0.01 N NaOH is higher than that in 0.1 N NaOH aqueous solution. The molecular associates consist of a dense core and hydrophilic surface at low NaOH concentrations. By contrast, Kasai and Harada proposed an annealing model to form a highset, resilient gel upon heating to >80°C. This annealing is associated with the irreversible loss of water, and resulted in a more tightly coiled triple helix. The structure crystallises as a triplex of right-handed, sixfold helical chains in a hexagonal unit cell with a fiber repeating length of 18.78 Å. Further removal of water from this structure, by drying under vacuum, results in further tightening of the six-fold triple helix and a decrease in the fiber period to only 5.87Å.

Occurrence

β-(1-3)-D-Glucans are present in a variety of living systems, including fungi, yeasts, algae, bacteria and higher plants. However, until now only bacteria belonging to the *Alcaligenes* and *Agrobacterium* species have been reported to produce the linear β-(1,3)-glucan type of homopolymer, curdlan. Figure 19.5 illustrates the lineage of the curdlan-producing strains since Harada first isolated *Alcaligenes faecalis* var. *myxogenes* 10C3 during the screening of soil bacteria capable of metabolising various petroleum fractions.

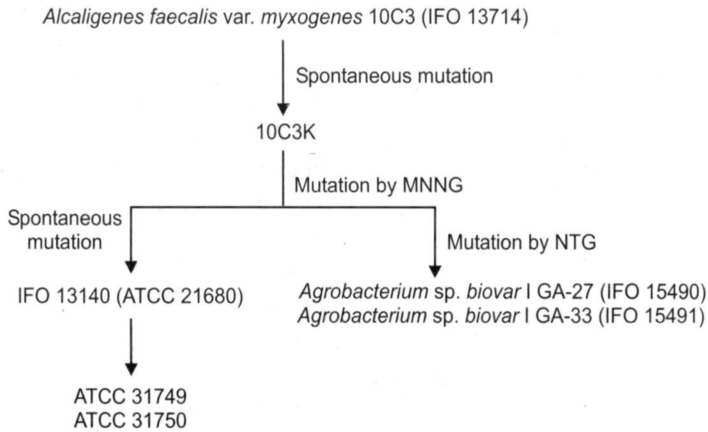

Fig. 19.5. Lineage of representative strains for curdlan production.

The parent strain produced two different types of exopolysaccharide; a water insoluble neutral homoglucan called 'curdlan', and a water-soluble acidic heteroglucan containing about 10 per cent succinic acid, and referred to as 'succinoglucan'. Moreover, a mutant strain 10C3K was isolated from a stock culture of 10C3, which produced only curdlan. Strain 10C3K is a spontaneous mutant with a stable ability to produce exocellular polysaccharide. By inducing mutagenesis with N-methyl-N-nitro-N-nitrosoguanidine (MNNG), Takeda Chemical Industries Ltd. in United States later isolated a uracil auxotrophic mutant from strain 10C3K, which was named *Alcaligenes faecalis* var. *myxogenes* IFO 13140 (ATCC 21680) and had improved gelforming β-(1,3)-glucan-producing ability. Phillips and Lawford isolated a mutant strain from strain ATCC 21680 in a nitrogen-limited chemostat culture (the accession number was ATCC 31749). Unlike its auxotrophic parent strain, ATCC 31749 does not require uracil for its growth, and is not a revertant as it can be distinguished from 10C3K by its inability to hydrolyse starch and its ability to grow on citric acid as sole carbon source. ATCC 31750, which arose as a spontaneous variant of the parent ATCC 31749, produces only the waterinsoluble curdlan-type glucan, while the parent strain produces both soluble and insoluble polysaccharides. All of these strains, which were formerly regarded as *Alcaligenes* species, have now been taxonomically reclassified as *Agrobacterium* species. Takeda Chemical Industries Ltd. derived further mutant strains from strain

10C3K, which reduced the activity of the enzyme, phosphoenol pyruvic acid carboxykinase. Naganishi examined the occurrence of curdlan-type polysaccharides in micro-organisms by using the water-soluble dye aniline blue, with which curdlan forms a blue complex. It was also shown that the rate of colour complex formation was dependent both on the polymer concentration and DP; hence these findings provided an excellent tool for the screening of curdlan-producing bacteria. Naganishi tested 687 strains of different genus of bacteria using the aniline blue staining technique. Among those examined, some strains of *Alcaligenes* and *Agrobacterium* species turned blue on agar plates containing aniline blue, and these have been used widely in the production of curdlan-type polysaccharides. Some strains of *Bacillus* formed blue complexes with aniline blue, but their polymeric constitution has not yet been studied.

Biosynthesis

Sutherland generalised the biosynthesis of extracellular polysaccharides into three major steps: (i) substrate uptake, (ii) intracellular formation of polysaccharide, and (iii) extrusion from cell. A metabolic pathway for exopolysaccharide biosynthesis is shown schematically in Fig. 19.6. First, a carbohydrate substrate enters the cell by active transport and group translocation involving substrate phosphorylation. The substrate is then directed along either catabolic pathways, or those leading to polysaccharide synthesis. UDP-glucose, a key precursor, is synthesised by the UDPglucose pyrophosphorylase-induced conversion of glucose-1-phosphate to UDP-glucose. Subsequently, polymer construction occurs together with the transfer of monosaccharides from UDP-glucose to a carrier lipid. After further chain elongation, the polymer is extruded from the cells.

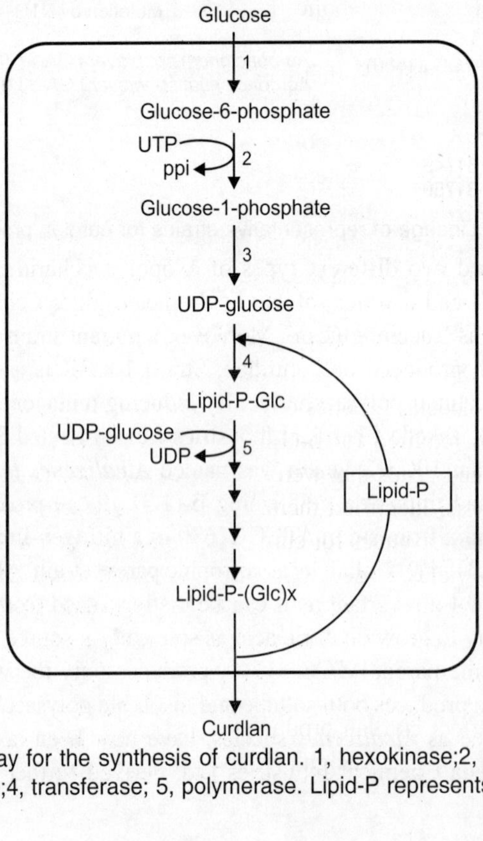

Fig. 19.6. Metabolic pathway for the synthesis of curdlan. 1, hexokinase;2, phosphoglucomutase; 3, UDP-glucose pyrophosphorylase;4, transferase; 5, polymerase. Lipid-P represents isoprenoid lipid phosphate.

Nitrogen-limited culture has also been employed for the production of curdlan, and has been generally explained by the roles of the carrier lipids. The availability of isoprenoid lipid may provide a way of regulating polysaccharide synthesis. Since curdlan biosynthesis takes place most extensively after cell growth has been stopped due to nitrogen exhaustion, isoprenoid lipids would be more available for carrying oligosaccharides instead of cellular liposaccharide and peptidoglycan.

The synthesis of precursor molecules is also of considerable importance in polysaccharide synthesis, in terms of the metabolic driving force. UDP-glucose serves as an activated precursor for glycosyl moieties in the synthesis of curdlan, a homopolysaccharide composed exclusively of β-1,3-linked glucose residues. In addition, cellular nucleotides not only play an important role in the synthesis of sugar nucleotides, but also have a widespread regulatory potential in cellular metabolism. Kim examined the change of intracellular nucleotide levels and their stimulatory effects on curdlan synthesis in *Agrobacterium* species under different culture conditions. Under nitrogen-limited conditions where curdlan synthesis was stimulated, intracellular levels of UMP and AMP were at least twice as high as those occurring under nitrogen-sufficient conditions, though UDP-glucose levels were similar. The time profiles of curdlan synthesis and cellular nucleotide levels showed that curdlan synthesis is positively related with intracellular levels of UMP and AMP.

In vitro enzyme reactions involved in the synthesis of UDP-glucose showed that a higher UMP concentration promotes the synthesis of UDP-glucose, while AMP neither inhibits nor facilitates the activity of UDP-glucose pyrophosphorylase. The addition of UMP to the medium also increased curdlan synthesis. From these results, these workers concluded that the higher intracellular UMP levels caused by nitrogen limitation enhance the metabolic flux of curdlan synthesis by promoting cellular UDP-glucose synthesis.

Kai reported a study of the biosynthetic pathway of curdlan using ^{13}C-labelled glucose in *Agrobacterium* sp. ATCC 31749. By analysing the labelled products, the biosynthesis of curdlan was interpreted as involving five routes: direct polymerisation from glucose; rearrangement; isomerisation of cleaved trioses; from fructose-6-phosphate; and from fructose fragments produced in various pathways of glycolysis. However, it was noted that more than 60 per cent of curdlan is synthesised by direct polymerisation, and that curdlan biosynthesis via glycolysis is comparatively low. This analysis also indicated that glycolysis occurs mainly via the pentose cycle and the Entner-Doudoroff pathway rather than the Embden-Meyerhof pathway.

Molecular Genetics

Little is known about the molecular genetics of bacterial curdlan biosynthesis, while there is growing information about the genes required for β-(1,3)-glucan synthesis in yeasts and filamentous fungi. Recently, Stasinopoulos cloned genes that were essential for the production of curdlan and, by using comparative sequence analysis, identified them as putative curdlan synthase genes. Further genetic investigations will open up new avenues for curdlan synthesis, and these will doubtlessly be exploited to produce curdlan in higher yields.

Production

Carbon source

Many crucial factors that affect curdlan production, including carbon, nitrogen, phosphate, oxygen supply, and pH have been investigated. High productivity using cheap carbon sources is important for the industrial production of curdlan. Lee, reported that maltose and sucrose were efficient carbon sources for the production of curdlan by a strain of *Agrobacterium* species, with maximal production (60 g l^{-1}) being obtained from sucrose, with a productivity of 0.5 g l^{-1} h^{-1} when nitrogen was limited at a cell concentration of 16 g l^{-1}. Molasses, which contains large amounts of sucrose, might also be the substrate of choice, with up to 42 g l^{-1} of curdlan, at a yield of 0.35 g curdlan per gram total sugar, being obtained in 5-day cultivation. Sucrose is a less expensive substrate than glucose, and as sugar beet or sugar cane molasses are cheap byproducts widely available from the sugar industry, they are very attractive carbon sources from an economic point of view.

Nitrogen effect

As previously described, relatively few strains of *Agrobacterium* and *Alcaligenes* species are known to produce curdlan. In such strains, curdlan production is associated with the poststationary phase of nitrogen depletion; thus, the operation involves an initial production of biomass, which is followed by curdlan production. Therefore, it is important to determine the initial concentration of the nitrogen source because it provides the limiting factor for cell growth during batch fermentation. Kim, reported that the cell growth rate decreased as the ammonium concentration increased. However, since higher cell concentrations produce more curdlan, an optimal ammonium concentration should be determined to provide an appropriate cell concentration while minimising the inhibitory effect of the ammonium ion.

Oxygen supply

Curdlan-producing stains are highly aerobic, and an adequate oxygen supply is therefore a key factor in production. Since curdlan is insoluble in water, the fermentation broth is of relatively low viscosity, and there is little resistance to oxygen transfer from gas to the liquid. However, a layer of insoluble exopolymer surrounding the cell mass offers resistance to oxygen transfer from the liquid into the cell, and therefore a high dissolved oxygen concentration is required for maximal productivity. Shake-flask fermentation results have shown that the specific production rate decreases as the volume of medium is increased, indicating that these cultures are limited by the relatively low oxygen transfer capacity of the system. Several investigations have been made into developing the process for curdlan production, especially with respect to reactor design. These workers employed two different types of impeller: a radial-flow, flat-blade impeller; and an axial-flow impeller.

The radial-flow impeller was effective at providing high oxygen transfer rates to increase the production of curdlan, but the high shear characteristics of this design yielded a product of inferior quality in terms of tensile strength of the thermally induced gel. An axial-flow impeller typically produces less shear and more pumping. High volumetric oxygen transfer can be achieved by low shear designs equipped with sparging devices, which consist of microporous materials through which oxygen-enriched air is dispersed. The maximal specific production rate was 90 mg per g cells h^{-1} when 30 per cent oxygen-enriched air was supplied in the low shear system.

Phosphate effect

Phosphate concentration must also be considered because it significantly influences cell growth and product formation. The production of rhamnose-containing polysaccharide by a *Klebsiella* strain was enhanced by a reduction in the phosphate content of the medium. In contrast, a sufficiency of phosphate resulted in good alginate yields in *Pseudomonas* strains and showed growth-associated production.

Thus, the effect of phosphate on the production of polysaccharides is variable. Kim, investigated the influence of inorganic phosphate concentration on the production of curdlan by *Agrobacterium* species. Under nitrogen-limited conditions which allow curdlan production, the concentration of phosphate remains constant as it is not further utilised for cell growth. The optimal residual phosphate concentration for curdlan production was in the range 0.1–0.5 g l^{-1}. Relatively low concentrations appeared to be optimal for curdlan production, although without phosphate, curdlan production was extremely low. However, on increasing the cell phosphate concentration from 0.42 to 1.68 g l^{-1}, curdlan production increased from 0.44 to 2.80 g l^{-1}. Moreover, the optimal phosphate concentration range was not dependent upon cell concentration, and the specific production rate was about 70 mg curdlan per g cells h^{-1}, irrespective of cell concentration.

pH effect

The pH of the culture is one of the most important factors because it significantly influences rates of both cell growth and product formation. High viscosity of the culture broth is often a critical problem in polysaccharide production. However, the viscosity problem can be obviated by operating the fermentation at a slightly acidic pH, since curdlan is insoluble under these conditions. Moreover, there appears to be more than one single optimal pH because fermentation of the culture for curdlan production is divided into two phases—the cell growth phase and the curdlan production phase.

Lee, sought an optimal pH profile to maximise curdlan production in a batch fermentation of *Agrobacterium* species. The cell growth rate was maximal at pH7.0, while curdlan production was maximal at pH 5.5. The pH profile provided a strategy to shift the culture pH from the optimal growth condition (pH 7.0) to the optimal production one (pH5.5) at the time of ammonium exhaustion. By adopting the optimal pH profile in a batch process, these workers obtained a significant improvement in curdlan production (64 g l^{-1}) compared with that obtained using a constant-pH operation (36 g l^{-1}).

SCLEROGLUCAN

Among different neutral polysaccharides from natural sources, scleroglucan from *Sclerotium glucanicum* significantly inhibits the replication of herpes simplex virus type 1 on Vero cells. Scleroglucan belongs to a class of exopolymers, expressed by members of genus *Sclerotium* and consists of a linear β-1,3-linked glucopyranose with side chains of single glucopyranose residues linked through β-1,6 glycosidic bonds. The effective antiviral concentration of this polysaccharide is far from the cytotoxicity threshold and consequently this natural product possesses a good selectivity index. Results obtained in experiments carried out in order to clarify the mechanism of action of this carbohydrate indicate that the block of infection occurs during the very early phases of the viral mutliplication cycle since the highest inhibitory effect took place when it was added during the attachment step. The antiviral effect of scleroglucan seems to be related to its binding with membrane glycoproteins of HSV-1 particles which impedes the complex interactions of the virus with the cell plasma membrane. Figure 19.7 shows the structure of scleroglucan.

Scleroglucan is a water soluble, natural polysaccharide produced by fermentation with the filamentous fungus *Sclerotium rolfsii*. It is a highly versatile ingredient, which improves the sensory characteristic of personal care products.

Scleroglucan has rheological properties, and unlike most natural and synthetic gums has high thermal stability, is resistant to hydrolysis and retains skin moisture.

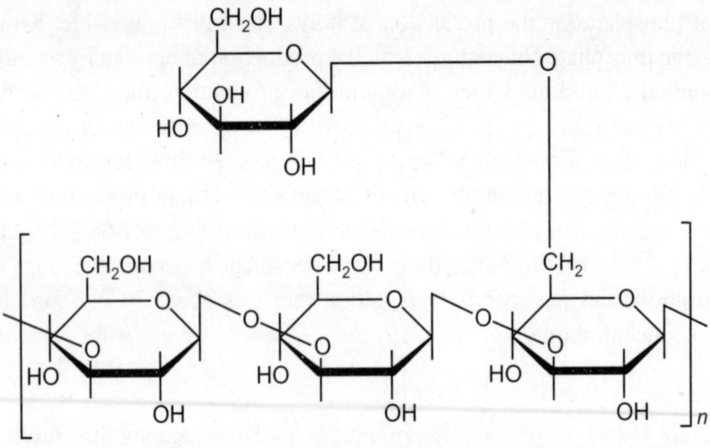

Fig. 19.7. Structure of scleroglucan.

Scleroglucan is a water soluble natural polymer produced by fermentation of the filamentous fungi *Sclerotium rofsii*. Its remarkable rheological properties and stability over a wide range of pH's, salinities, and temperatures make this neutral non-ionic polysaccharide suitable for a number of varied technical applications. Scleroglucan can be used in creams and lotions in hair care, skin care, sun care and bath and body products and colour cosmetics. As a—thickening agent, sparkling clarity, electrolyte tolerant, forms fluid gels with very high and unique suspension properties, sparkling clarity, process flexibility and tolerance, excellent and exceptional suspension for insoluble solids and oil droplets, highly effective at low concentrations, shear reversible behaviour, excellent emulsion and foam stabilisation, and excellent stability in extremely high conditions. Unlike most natural and synthetic gums, high temperature has little effect on the viscosity of a solution of scleroglucan. Below 10°C (50°F), the solutions have a semi-gelled appearance that disappears on shaking or heating. Solutions of scleroglucan may be sterilised by heating at 121°C (250°F) for 20 hrs without affecting their viscosity.

PULLULAN

Pullulan is a polysaccharide polymer consisting of maltotriose units, also known as α-1,4-; α-1,6-glucan. Three glucose units in maltotriose are connected by an α-1,4-glycosidic bond, whereas consecutive maltotriose units are connected to each other by an α-1,6-glycosidic bond. Pullulan is produced from starch by the fungus *Aureobasidium pullulans*. Structure of pullulan is shown in Fig. 19.8. As an edible, mostly tasteless polymer, the chief commercial use of pullulan is in the manufacture of edible films that are used in various breath freshener or oral hygiene products such as Listerine Cool Mint PocketPaks. As a food additive, it is known by the E number E1204.

Pullulanase is a specific kind of glucanase, an amylolytic exoenzyme, that degrades pullulan. It is produced as an extracellular, cell surface-anchored lipoprotein by Gram-negative bacteria of the genus *Klebsiella*. Type I pullulanases specifically attack α-1,6 linkages, while type II pullulanases are also able to hydrolyse α-1,4 linkages. It is also produced by some other bacteria and archaea. Pullulanase is used as a detergent in biotechnology. Pullulanase (EC 3.2.1.41) is also known as pullulan-6-glucanohydrolase (Debranching enzyme). Its substrate, pullulan, is regarded as a chain of maltotriose units linked by α-1,6-glycosidic bonds. Pullulanase will hydrolytically cleave pullulan (alpha-glucan polysaccharides). NPcaps® capsules are an all-natural, two-piece non-animal capsule suitable for addressing a variety of cultural and dietary requirements, including those of vegetarians, diabetics and patients with restricted diets. NPcaps capsules are made from pullulan, a water-soluble polysaccharide produced through a fermentation process. Widely used in Japan, pullulan is very stable and well-characterised, and has achieved broad regulatory acceptance around the world with its proven safety record. It has been in commercial production for more than 25 years, having numerous uses in the food and pharmaceutical industries.

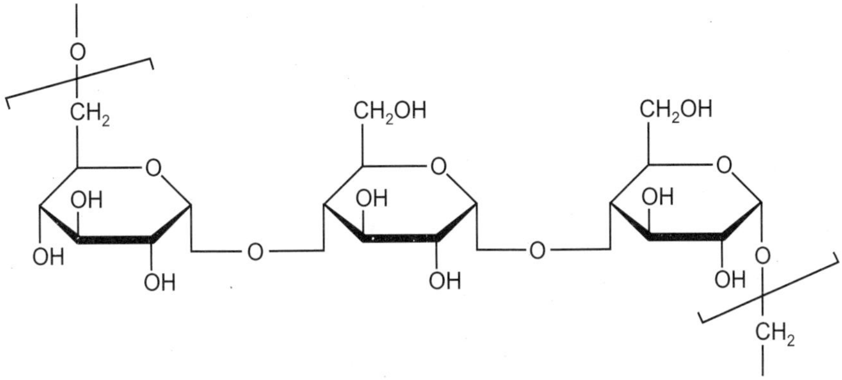

Fig. 19.8. Structure of pullulan.

Natural Balance between Health and Technology

With the trend moving towards healthier lifestyles, consumers are becoming more and more selective. That's why Capsugel is one step ahead of the development curve with NPcaps pullulan capsules, a highly effective encapsulation option that helps differentiate your product in a 100 per cent natural, crystal-clear capsule material.

NPcaps pullulan capsules are odorless, tasteless, and completely biodegradable two-piece capsules made from this completely natural, vegetable-derived polysaccharide. Because pullulan capsules are impermeable to oxygen transmission, NPcaps capsules are recommended for encapsulating oxidation-sensitive ingredients to provide enhanced protection.

Characteristics of Pullulan-based Edible Films

Pullulan is an extracellular bacterial polysaccharide produced from starch by *Aureobasidium pullulans*. It is a linear polysaccharide made up of α–1,6-linked maltotriose residues. As an odourless white coloured powder, pullulan is easily soluble in water to make clear and viscous solution. This polymer also has high adhesion, sticking, lubrication, and film forming abilities.

The main objectives of this study were to prepare pullulan-based edible films with or without gelling agents such as carrageenan, alginate, gellan, and agar and to determine some selected properties like tensile strength (TS), elongation at break (E), water vapour permeability (WVP), and water solubility (WS) of the films.

Five grams pullulan was dissolved into 150 ml distilled water to make a clear solution. The film solution was casted onto a levelled Teflon-coated glass plate and dried at room temperature for 24 hr. In addition, pullulan films were prepared with gelling agents with 5–10 per cent of pullulan by weight. TS and E of the films were measured using an Instron Universal Testing Machine and WVP was measured by the gravitational method using a water vapour permeability cup. Transparent pullulan films with or without gelling agents were prepared. WVP, WS, TS and E of pullulan films were 0.69×10^{-9} g.m/m^2.s.Pa, 92.4 per cent, 45.0 Mpa and 4.0 per cent, respectively. Physical properties of pullulan films were greatly affected by adding gelling agents. As a whole, TS increased by adding gelling agents without significant change in E. Though WS of the films decreased by adding gelling agents, WVP increased. Among the gelling agents tested agar was the most effective to increase TS and gellan was the most effective to reduce WS of the pullulan-based films.

Free standing films were prepared with pullulan and their properties could be properly modified by adding various gelling agents.

DEXTRAN

Dextran is a complex, branched glucan (polysaccharide made of many glucose molecules) composed of chains of varying lengths (from 10 to 150 kilodaltons). It is used medicinally as an antithrombotic (anti-platelet), to reduce blood viscosity, and as a volume expander in anemia. Figure 19.9 shows the structure of dextran. The straight chain consists of α-1,6-glycosidic linkages between glucose molecules, while branches begin from α-1,4 linkages (and in some cases, α-1,2 and α-1,3-linkages as well). Dextran is synthesised from sucrose by certain lactic-acid bacteria, the best-known being *Leuconostoc mesenteroides* and *Streptococcus mutans*. Dental plaque is rich in dextrans. Dextran is also formed by the lactic acid bacterium *Lactobacillus brevis* to create the crystals of tibicos, or water kefir fermented beverage which supposedly has some health benefits.

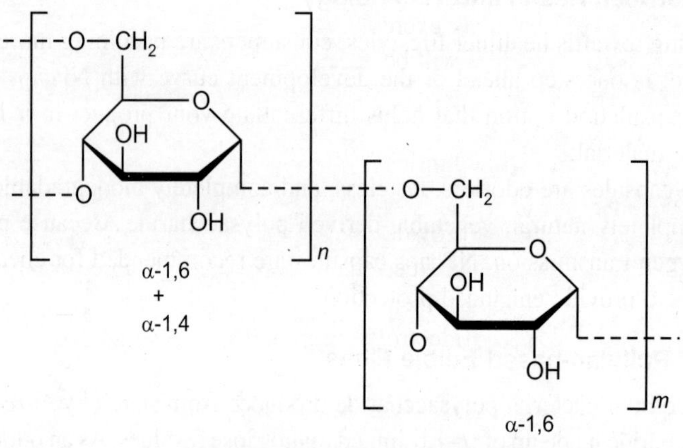

Fig. 19.9. Structure of dextran.

Dextran Characteristics

Dextran fractions are characterised by their average molecular weights and molecular weight distributions. Dextran is used in various fields such as pharmaceutical, photographic, agricultural, and food industries.

The versatile use of Dextran products relates to the favourable properties:

1. Dextran is neutral and water soluble.
2. Dextran is easily filtered.
3. Dextran is biocompatible.
4. Dextran is biodegradable.
5. Dextran is stable for more than 5 years.

Uses

Microsurgery uses

These agents are used commonly by microsurgeons to decrease vascular thrombosis. The antithrombotic effect of dextran is mediated through its binding of erythrocytes, platelets, and vascular endothelium, increasing their electronegativity and thus reducing erythrocyte aggregation and platelet adhesiveness. Dextrans also reduce factor VIII-Ag Von Willebrand factor, thereby decreasing platelet function. Clots formed after administration of dextrans are more easily lysed due to an altered thrombus structure (more evenly distributed platelets with coarser fibrin). By inhibiting α-2 antiplasmin, dextran serves as a plasminogen activator and therefore possesses thrombolytic features.

Outside from these features, larger dextrans, which do not pass out of the vessels, are potent osmotic agents, and thus have been used urgently to treat hypovolemia. The hemodilution caused by volume expansion with dextran use improves blood flow, thus further improving patency of microanastomoses and reducing thrombosis.

Still, no difference has been detected in antithrombotic effectiveness in comparison of intraarterial and intravenous administration of dextran.

Dextrans are available in multiple molecular weights ranging from 10,000 Da to 1,50,000 Da. The larger dextrans are excreted poorly from the kidney and therefore remain in the blood for as long as weeks until they are metabolised. Subsequently, they have prolonged antithrombotic and colloidal effects. In this family, dextran-40 (MW: 40,000 Da), has been the most popular member for anticoagulation therapy.

Close to 70 per cent of dextran-40 is excreted in urine within the first 24 hrs after intravenous infusion while the remaining 30 per cent will be retained for several more days.

Other medical uses

1. It is used in some eye drops as a lubricant, and in certain intravenous fluids to solubilise other factors, e.g. iron (iron dextran).
2. Dextran in intravenous solution provides an osmotically neutral fluid that once in the body is digested by cells into glucose and free water. It is occasionally used to replace lost blood in emergency situations, when replacement blood is not available, but must be used with caution as it does not provide necessary electrolytes and can cause hyponatremia or other electrolyte disturbances.
3. It also increases blood sugar levels.

Laboratory uses

1. Dextran is used in the osmotic stress technique for applying osmotic pressure to biological molecules.
2. It is also used in some size-exclusion chromatography matrices; an example is Sephadex.
3. Dextran has also been used in bead form to aid in bioreactor applications.
4. Dextran has been used in immobilisation in biosensors.
5. Dextran preferentially binds to early endosomes; fluorescently-labelled dextran can be used to visualise these endosomes under a fluorescent microscope.
6. Dextran can be used as a stabilising coating to protect metal nanoparticles from oxidation and improve biocompatibility.

7. Dextran coupled with a fluorescent molecule (such as FITC) can be used to create concentration gradients of diffusible molecules for imaging and allow subsequent characterisation of gradient slope.

Side Effects

Although there are relatively few side-effects associated with dextran use, these side-effects can be very serious. These include anaphylaxis, volume overload, pulmonary edema, cerebral edema, or platelet dysfunction. An uncommon but significant complication of dextran osmotic effect is acute renal failure. The pathogenesis of this renal failure is the subject of many debates with direct toxic effect on tubules and glomerulus versus intraluminal hyperviscosity being some of the proposed mechanisms. Patients with history of diabetes mellitus, renal insufficiency, or vascular disorders are most at risk. Brooks and others recommend the avoidance of dextran therapy in patients with chronic renal insufficiency and CrCl<40 cc per minute.

Chapter 20

Other Fermentation Processes and Future Prospects

INTRODUCTION

Fermentation is the process of deriving energy from the oxidation of organic compounds, such as carbohydrates, and using an endogenous electron acceptor, which is usually an organic compound, as opposed to respiration where electrons are donated to an exogenous electron acceptor, such as oxygen, via an electron transport chain. Fermentation does not necessarily have to be carried out in an anaerobic environment. For example, even in the presence of abundant oxygen, yeast cells greatly prefer fermentation to oxidative phosphorylation, as long as sugars are readily available for consumption.

This chapter aims at conveying several fermentations that do not fit into the categories established in previous chapters, including which are not of commercial use but may become significant in coming future.

GIBBERELLIN

Gibberellins (GAs) are plant hormones that regulate growth and influence various developmental processes, including stem elongation, germination, dormancy, flowering, sex expression, enzyme induction and leaf and fruit senescence. Gibberellin was first recognised in 1926 by a Japanese scientist, Eiichi Kurosawa, studying *bakanae*, the 'foolish seedling' disease in rice. It was first isolated in 1935 by Teijiro Yabuta, from fungal strains (*Gibberella fujikuroi*) provided by Kurosawa. Yabuta called the isolate gibberellin. All known gibberellins are diterpenoid acids that are synthesised by the terpenoid pathway in plastids and then modified in the endoplasmic reticulum and cytosol until they reach their biologically-active form. All gibberellins are derived from the *ent*-gibberellane skeleton, but are synthesised via *ent*-kaurene. The gibberellins are named GA1....GA*n* in order of discovery. Gibberellic acid, which was the first gibberellin to be structurally characterised, is GA3.

As of 2003, there were 126 GAs identified from plants, fungi, and bacteria. Gibberellins are tetracyclic diterpene acids. There are two classes based on the presence of 19 carbons or 20 carbons. The 19-carbon gibberellins, such as gibberellic acid, have lost carbon 20 and, in place, possess a five-member lactone bridge that links carbons 4 and 10. The 19-carbon forms are, in general, the biologically active forms of gibberellins. Hydroxylation also has a great effect on the biological activity of the gibberellin.. In general, the most biologically active compounds are dihydroxylated gibberellins, which possess hydroxyl groups on both carbon 3 and carbon 13. Gibberellic acid is a dihydroxylated gibberellin.

Gibberellins are involved in the natural process of breaking dormancy and various other aspects of germination. Before the photosynthetic apparatus develops sufficiently in the early stages of germination, the stored energy reserves of starch nourish the seedling. Usually in germination, the breakdown of starch to glucose in the endosperm begins shortly after the seed is exposed to water. It is believed that gibberellins in the seed embryo signal starch hydrolysis through inducing the synthesis of the enzyme α-amylase in the aleurone cells. In the model for gibberellin-induced production of α-amylase, it is demonstrated that gibberellins (denoted by GA) produced in the scutellum diffuse to the aleurone cells where they stimulate the secretion α-amylase. α-Amylase then hydrolyses starch, which is abundant in many seeds, into glucose that can be utilised in cellular respiration to produce energy for the seed embryo. Studies of this process have indicated that gibberellins cause higher levels of transcription of the gene coding for the α-amylase enzyme, in order to stimulate the synthesis of α-amylase.

Gibberellins are produced in greater mass when the plant is exposed to cold temperatures. They stimulate cell elongation, breaking and budding, seedless fruits, and seed germination. They do the last by breaking the seed's dormancy and acting as a chemical messenger. Its hormone binds to a receptor, and Ca^{2+} activates a protein, calmodulin, and the complex binds to DNA, producing an enzyme to stimulate growth in the embryo. *Gibberella fujikuroi* is a fungal plant pathogen. It causes bakanae disease in rice seedlings, by overloading them with the phytohormone gibberellin as its own metabolic by-product.

ZEARALENONE

Zearalenone (ZEA), also known as RAL and F-2 mycotoxin, is a potent estrogenic metabolite produced by some Giberella species.

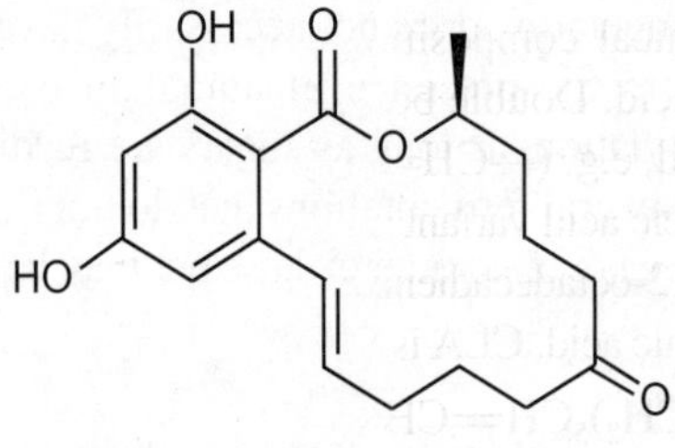

Zearalenone

Several *Fusarium* species produce toxic substances of considerable concern to livestock and poultry producers: namely, deoxynivalenol, T-2 toxin, HT-2 toxin, diacetoxyscirpenol (DAS) and zearalenone.

Zearalenone is the primary toxin causing infertility, abortion or other breeding problems, especially in swine. Zearalenone is heat-stable and is found worldwide in a number of cereal crops, such as maize, barley, oats, wheat, rice, and sorghum and also in bread.

Chemical and Physical Properties

Zearalenone is a white crystalline solid. It exhibits blue-green fluorescence when excited by long wavelength UV light (360 nm) and a more intense green fluorescence when excited with short wavelength UV light (260 nm). In methanol, UV absorption maxima occur at 236 (e = 29,700), 274 (e = 13,909) and 316 (e = 6020). Maximum fluorescence in ethanol occurs with irradiation at 314 nm and with emission at 450 nm. Solubility in water is about 0.002 g/100 ml. It is slightly soluble in hexane and progressively more so in benzene, acetonitrile, methylene chloride, methanol, ethanol and acetone. It is also soluble in aqueous alkali.

Sampling and Analysis

In common with other mycotoxins sampling food commodities for zearalenone must be carried out to obtain samples representative of the consignment under test. Commonly used extraction solvents are aqueous mixtures of methanol, acetonitrile or ethyl acetate followed by a range of different clean-up procedures that depend in part on the food and on the detection method in use. TLC methods and HPLC are commonly used. HPLC alone in not sufficient as it may often yield false positive results. The TLC method for zearalenone is normal phase silica gel plates, the eluent: 90 per cent dichloromethane, 10 per cent v/v acetone; or reverse phase C18 silica plates; the eluent: 90 per cent v/v methanol, 10 per cent water. Zearalenone gives unmistakable blue luminiscence under UV.

FATTY ACIDS AND TRIGLYCERIDES

Lipids consist of numerous fat-like chemical compounds that are insoluble in water but soluble in organic solvents. Lipid compounds include monoglycerides, diglycerides, triglycerides, phosphatides, cerebrosides, sterols, terpenes, fatty alcohols, and fatty acids. Dietary fats supply energy, carry fat-soluble vitamins (A, D, E, K), and are a source of antioxidants and bioactive compounds. Fats are also incorporated as structural components of the brain and cell membranes. Fatty acids consist of the elements carbon (C), hydrogen (H) and oxygen (O) arranged as a carbon chain skeleton with a carboxyl group (—COOH) at one end. Saturated fatty acids (SFAs) have all the hydrogen that the carbon atoms can hold, and therefore, have no double bonds between the carbons. Monounsaturated fatty acids (MUFAs) have only one double bond. Polyunsaturated fatty acids (PUFAs) have more than one double bond. Fatty acids are frequently represented by a notation such as C18:2 that indicates that the fatty acid consists of an 18-carbon chain and 2 double bonds. Although this could refer to any of several possible fatty acid isomers with this chemical composition, it implies the naturally-occurring fatty acid with these characteristics, i.e. linoleic acid. Double bonds are said to be conjugated when they are separated from each other by one single bond, e.g. (—CH=CH—CH=CH—). The term 'conjugated linoleic acid' (CLA) refers to several C18:2 linoleic acid variants such as 9,11-CLA and 10,12-CLA which correspond to 9,11-octadecadienoic acid and 10,12-octadecadienoic acid. The principal dietary isomer of CLA is *cis*-9, *trans*-11 CLA, also known as rumenic acid. CLA is found naturally in meats, eggs, cheese, milk and yogurt.

$$CH_3(CH_2)_5CH{=}CH{-}CH{=}CH(CH_2)_7COOH$$

9,11-Conjugated linoleic acid

TRIGLYCERIDE

Triacylglyceride (TAG) is a glyceride in which the glycerol is esterified with three fatty acids. It is the main constituent of vegetable oil and animal fats.

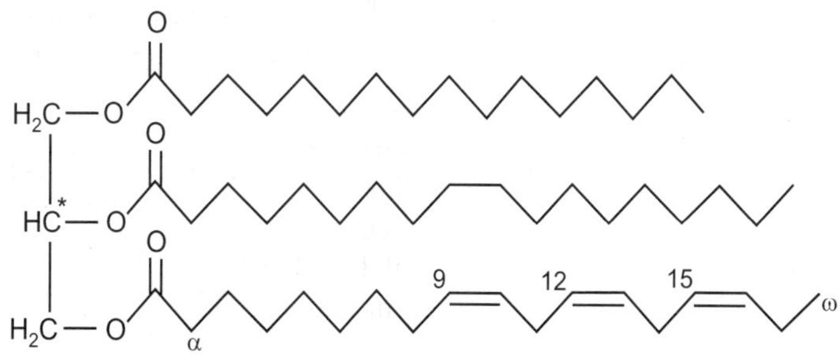

Example of an unsaturated fat triglyceride. Left part: glycerol, right part from top to bottom: palmitic acid, oleic acid, α-linolenic acid, chemical formula: $C_{55}H_{98}O_6$.

Chemical Structure

Triglycerides are formed from a single molecule of glycerol, combined with three fatty acids on each of the OH groups, and make up most of fats digested by humans. Ester bonds form between each fatty acid and the glycerol molecule. This is where the enzyme pancreatic lipase acts, hydrolysing the bond and 'releasing' the fatty acid. In triglyceride form, lipids cannot be absorbed by the duodenum. Fatty acids, monoglycerides (one glycerol, one fatty acid) and some diglycerides are absorbed by the duodenum, once the triglycerides have been broken down. The chemical formula is RCOO—CH_2CH (—OOCR′) CH_2—OOCR′, where R, R′, and R″ are longer alkyl chains. The three fatty acids RCOOH, R′COOH and R″COOH can be all different, all the same, or only two the same.

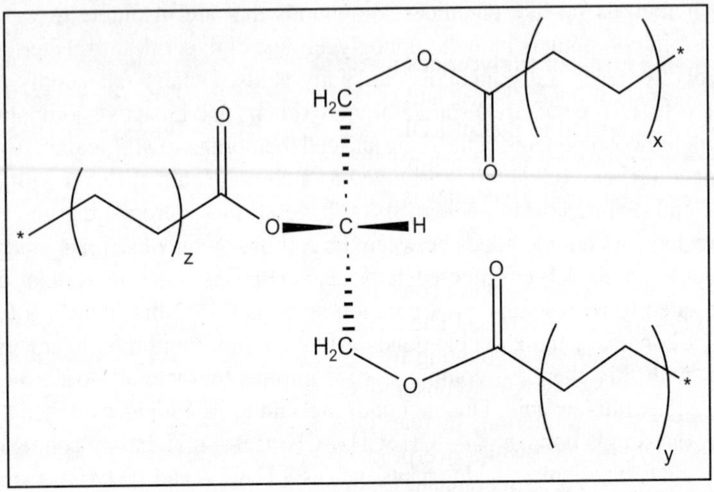

General structure of a triglyceride

Chain lengths of the fatty acids in naturally occurring triglycerides can be of varying lengths, but 16, 18 and 20 carbons are the most common. Natural fatty acids found in plants and animals are typically composed only of even numbers of carbon atoms due to the way they are biosynthesised from acetyl CoA. Bacteria, however, possess the ability to synthesise odd- and branched-chain fatty acids. Consequently, ruminant animal fat contains odd numbered fatty acids, such as 15, due to the action of bacteria in the rumen. Most natural fats contain a complex mixture of individual triglycerides. Because of this, they melt over a broad range of temperatures. Cocoa butter is unusual in that it is composed of only a few triglycerides, one of which contains palmitic, oleic and stearic acids in that order. This gives rise to a fairly sharp melting point, causing chocolate to melt in the mouth without feeling greasy.

Metabolism

Triglycerides, as major components of very low density lipoprotein (VLDL) and chylomicrons, play an important role in metabolism as energy sources and transporters of dietary fat. They contain more than twice as much energy (9 kcal/g) as carbohydrates and proteins. In the intestine, triglycerides are split into monoacylglycerol and free fatty acids in a process called lipolysis, with the secretion of lipases and bile, which are subsequently moved to absorptive enterocytes, cells lining the intestines. The triglycerides are rebuilt in the enterocytes from their fragments and packaged together with cholesterol and proteins to form chylomicrons. These are excreted from the cells and collected by the lymph system and transported to the large vessels near the heart before being mixed into the blood. Various tissues can capture the chylomicrons, releasing the triglycerides to be used as a source of energy. Fat and liver cells can synthesise and store triglycerides. When the body requires fatty acids as an energy source, the hormone glucagon signals the breakdown of the triglycerides by hormone-sensitive lipase to release free fatty acids. As the brain cannot utilise fatty acids as an energy source (unless converted to a ketone), the glycerol component of triglycerides can be converted into glucose, via gluconeogenesis, for brain fuel when it is broken down. Fat cells may also be broken down for that reason, if the brain's needs ever outweigh the body's.

Triglycerides cannot pass through cell membranes freely. Special enzymes on the walls of blood vessels called lipoprotein lipases must break down triglycerides into free fatty acids and glycerol. Fatty acids can then be taken up by cells via the fatty acid transporter (FAT).

Role in Disease

In the human body, high levels of triglycerides in the bloodstream have been linked to atherosclerosis, and, by extension, the risk of heart disease and stroke. However, the relative negative impact of raised levels of triglycerides compared to that of LDL:HDL ratios is as yet unknown. The risk can be partly accounted for by a strong inverse relationship between triglyceride level and HDL-cholesterol level. Another disease caused by high triglycerides is pancreatitis.

Reducing triglyceride levels

To lower triglyceride levels, one may reduce consumption of fats, alcohol and carbohydrates, particularly in rice, and engage in aerobic exercise. The American Heart Association notes that diets high in carbohydrates, with carbohydrates accounting for more than 60 per ent of the total caloric intake, can increase triglyceride levels. Carbohydrate consumption increases insulin production, which is associated with increased triglyceride production. Increased exercise and reduced carbohydrate consumption ameliorate one potential cause of insulin overproduction to help maintain sensible triglyceride levels. Triglyceride levels are also reduced by omega-3 fatty acids from fish, flax seed oil and other sources.

Industrial Uses

Triglycerides are also split into their components via transesterification during the manufacture of biodiesel. The fatty acid monoalkyl ester can be used as fuel in diesel engines. Glycerine has many uses, such as in the manufacture of food and in the production of pharmaceuticals. Other examples are the triglyceride process in the decaffeination of coffee beans. Triglycerides are also a major feedstock source for biodiesel.

Staining

Staining for fatty acids, triglycerides, lipoproteins, and other lipids is done through the use of lysochromes (fat-soluble dyes). These dyes can allow the qualification of a certain fat of interest by staining the material a specific colour. Some examples: Sudan IV, oil red O, and Sudan black B.

ENZYME INHIBITORS

Enzyme inhibitors are molecules that bind to enzymes and decrease their activity. Since blocking an enzyme's activity can kill a pathogen or correct a metabolic imbalance, many drugs are enzyme inhibitors. They are also used as herbicides and pesticides. Not all molecules that bind to enzymes are inhibitors; enzyme activators bind to enzymes and increase their enzymatic activity.

The binding of an inhibitor can stop a substrate from entering the enzyme's active site and/or hinder the enzyme from catalysing its reaction. Inhibitor binding is either reversible or irreversible. Irreversible inhibitors usually react with the enzyme and change it chemically. These inhibitors modify key amino acid residues needed for enzymatic activity. In contrast, reversible inhibitors bind non-covalently and different types of inhibition are produced depending on whether these inhibitors bind the enzyme, the enzyme-substrate complex, or both. Many drug molecules are enzyme inhibitors, so their discovery and improvement is an active area of research in biochemistry and pharmacology. A medicinal enzyme inhibitor is often judged by its specificity (its lack of binding to other proteins) and its potency (its dissociation constant, which indicates the concentration needed to inhibit the enzyme). A high specificity and potency ensure that a drug will have few side effects and thus low toxicity.

Enzyme inhibitors also occur naturally and are involved in the regulation of metabolism. For example, enzymes in a metabolic pathway can be inhibited by downstream products. This type of negative feedback slows flux through a pathway when the products begin to build up and is an important way to maintain homeostasis in a cell. Other cellular enzyme inhibitors are proteins that specifically bind to and inhibit an enzyme target. This can help control enzymes that may be damaging to a cell, such as proteases or nucleases; a well-characterised example is the ribonuclease inhibitor, which binds to ribonucleases in one of the tightest known protein–protein interactions. Natural enzyme inhibitors can also be poisons and are used as defenses against predators or as ways of killing prey. Figure 20.1 shows the peptide structure of the substance antipain an inhibitor of propane.

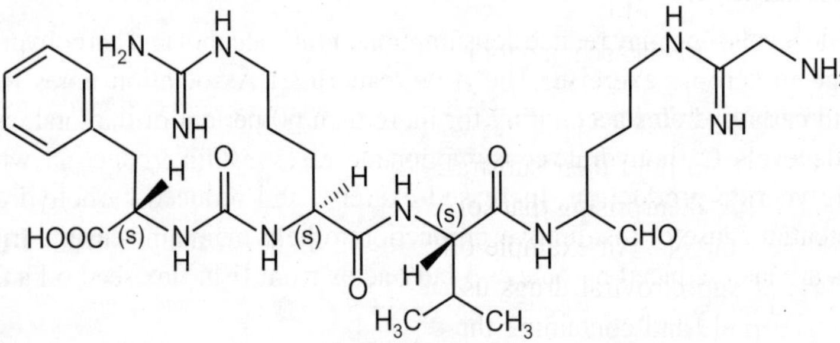

Fig. 20.1. Peptide structure of the substance antipain an inhibitor of propane.

Reversible Inhibitors

Reversible inhibitors bind to enzymes with non-covalent interactions such as hydrogen bonds, hydrophobic interactions and ionic bonds. Multiple weak bonds between the inhibitor and the active site combine to produce strong and specific binding. In contrast to substrates and irreversible inhibitors, reversible inhibitors generally do not undergo chemical reactions when bound to the enzyme and can be easily removed by dilution or dialysis.

Quantitative description of reversible inhibition

Reversible inhibition can be described quantitatively in terms of the inhibitor's binding to the enzyme and to the enzyme-substrate complex, and its effects on the kinetic constants of the enzyme. In the classic Michaelis-Menten scheme below, an enzyme (E) binds to its substrate (S) to form the enzyme-substrate complex ES. Upon catalysis, this complex breaks down to release product P and free enzyme. The inhibitor (I) can bind to either E or ES with the dissociation constants K_i or K_i', respectively.

1. Competitive inhibitors can bind to E, but not to ES. Competitive inhibition increases K_m (i.e. the inhibitor interferes with substrate binding), but does not affect V_{max} (the inhibitor does not hamper catalysis in ES because it cannot bind to ES).
2. Non-competitive inhibitors have identical affinities for E and ES ($K_i = K_i'$). Non-competitive inhibition does not change K_m (i.e. it does not affect substrate binding) but decreases V_{max} (i.e. inhibitor binding hampers catalysis).

3. Mixed-type inhibitors bind to both E and ES, but their affinities for these two forms of the enzyme are different ($K_i \neq K_i'$). Thus, mixed-type inhibitors interfere with substrate binding (increase K_m) and hamper catalysis in the ES complex (decrease V_{max}).

When an enzyme has multiple substrates, inhibitors can show different types of inhibition depending on which substrate is considered. This results from the active site containing two different binding sites within the active site, one for each substrate. For example, an inhibitor might compete with substrate A for the first binding site, but be a non-competitive inhibitor with respect to substrate B in the second binding site.

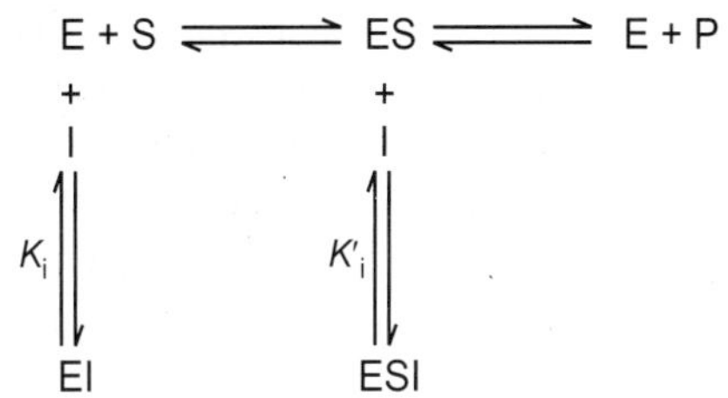

Kinetic scheme for reversible enzyme inhibitors.

Examples of reversible inhibitors

As enzymes have evolved to bind their substrates tightly, and most reversible inhibitors bind in the active site of enzymes, it is unsurprising that some of these inhibitors are strikingly similar in structure to the substrates of their targets. An example of these substrate mimics are the protease inhibitors, a very successful class of antiretroviral drugs used to treat HIV. The structure of ritonavir, a protease inhibitor based on a peptide and containing three peptide bonds, is shown on the right. As this drug resembles the protein that is the substrate of the HIV protease, it competes with this substrate in the enzyme's active site.

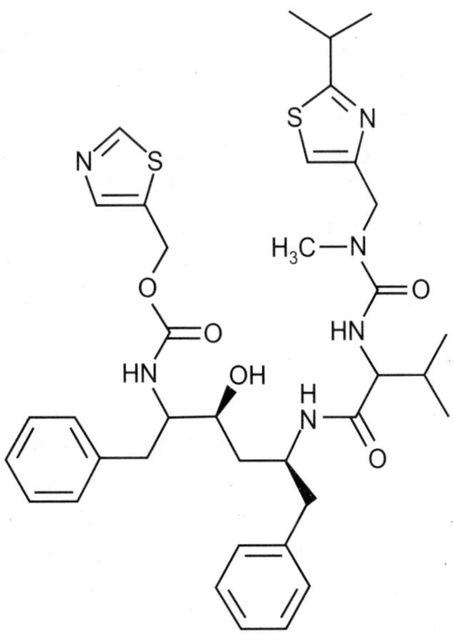

Peptide-based protease inhibitor ritonavir.

Enzyme inhibitors are often designed to mimic the transition state or intermediate of an enzyme-catalysed reaction. This ensures that the inhibitor exploits the transition state stabilising effect of the enzyme, resulting in a better binding affinity (lower K_i) than substrate-based designs. An example of such a transition state inhibitor is the antiviral drug oseltamivir; this drug mimics the planar nature of the ring oxonium ion in the reaction of the viral enzyme neuraminidase.

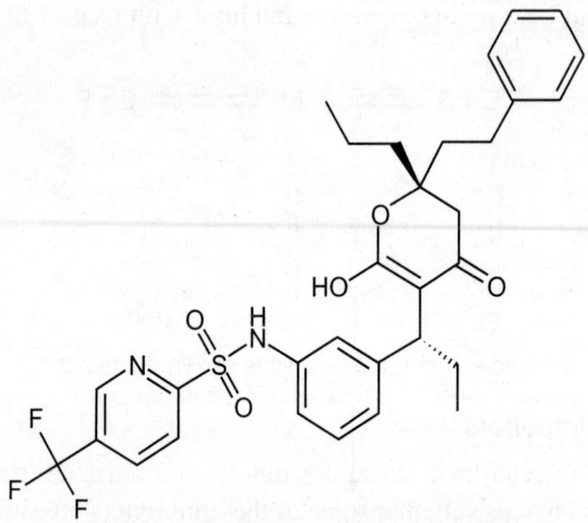

Nonpeptidic protease inhibitor tipranavir.

However, not all inhibitors are based on the structures of substrates. For example, the structure of another HIV protease inhibitor tipranavir is shown on the left. This molecule is not based on a peptide and has no obvious structural similarity to a protein substrate. These non-peptide inhibitors can be more stable than inhibitors containing peptide bonds, because they will not be substrates for peptidases and are less likely to be degraded.

In drug design it is important to consider the concentrations of substrates to which the target enzymes are exposed. For example, some protein kinase inhibitors have chemical structures that are similar to adenosine triphosphate, one of the substrates of these enzymes. However, drugs that are simple competitive inhibitors will have to compete with the high concentrations of ATP in the cell. Protein kinases can also be inhibited by competition at the binding sites where the kinases interact with their substrate proteins, and most proteins are present inside cells at concentrations much lower than the concentration of ATP. As a consequence, if two protein kinase inhibitors both bind in the active site with similar affinity, but only one has to compete with ATP, then the competitive inhibitor at the protein-binding site will inhibit the enzyme more effectively.

Irreversible Inhibitors

Irreversible inhibitors usually covalently modify an enzyme, and inhibition cannot therefore be reversed. Irreversible inhibitors often contain reactive functional groups such as nitrogen mustards, aldehydes, haloalkanes, alkenes, Michael acceptors, phenyl sulphonates, or fluorophosphonates. These electrophilic groups react with amino acid side chains to form covalent adducts. The residues modified are those with side chains containing nucleophiles such as hydroxyl or sulphhydryl groups; these include the amino acids serine (as in DFP, right), cysteine, threonine or tyrosine.

Irreversible inhibition is different from irreversible enzyme inactivation. Irreversible inhibitors are generally specific for one class of enzyme and do not inactivate all proteins; they do not function by destroying protein structure but by specifically altering the active site of their target. For example, extremes of pH or temperature usually cause denaturation of all protein structure, but this is a non-specific effect. Similarly, some non-specific chemical treatments destroy protein structure: for example, heating in concentrated hydrochloric acid will hydrolyse the peptide bonds holding proteins together, releasing free amino acids. Figure 20.2 shows the reaction of the irreversible inhibitor diisopropylfluorophosphate (DFP) with a serine protease

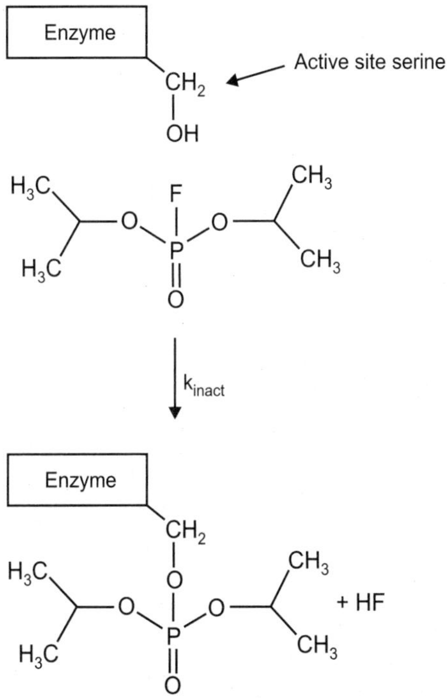

Fig. 20.2. Reaction of the irreversible inhibitor diisopropylfluorophosphate (DFP) with a serine protease.

Irreversible inhibitors display time-dependent inhibition and their potency therefore cannot be characterised by an IC_{50} value. This is because the amount of active enzyme at a given concentration of irreversible inhibitor will be different depending on how long the inhibitor is preincubated with the enzyme. Instead, $k_{obs}/[I]$ values are used, where, k_{obs} is the observed pseudo-first order rate of inactivation (obtained by plotting the log of per cent activity vs. time) and $[I]$ is the concentration of inhibitor. The $k_{obs}/[I]$ parameter is valid as long as the inhibitor does not saturate binding with the enzyme (in which case $k_{obs} = k_{inact}$).

Examples of irreversible inhibitors

Diisopropylfluorophosphate (DFP) is shown as an example of an irreversible protease inhibitor in the figure above right. The enzyme hydrolyses the phosphorus–fluorine bond, but the phosphate residue remains bound to the serine in the active site, deactivating it. Similarly, DFP also reacts with the active site of acetylcholine esterase in the synapses of neurons, and consequently is a potent neurotoxin, with a lethal dose of less than 100 mg.

Suicide inhibition is an unusual type of irreversible inhibition where the enzyme converts the inhibitor into a reactive form in its active site. An example is the inhibitor of polyamine biosynthesis, α-difluoromethylornithine or DFMO, which is an analogue of the amino acid ornithine, and is used to treat African trypanosomiasis (sleeping sickness). Ornithine decarboxylase can catalyse the decarboxylation of DFMO instead of ornithine. However, this decarboxylation reaction is followed by the elimination of a fluorine atom, which converts this catalytic intermediate into a conjugated imine, a highly electrophilic species.

This reactive form of DFMO then reacts with either a cysteine or lysine residue in the active site to irreversibly inactivate the enzyme.

Uses of Inhibitors

Enzyme inhibitors are found in nature and are also designed and produced as part of pharmacology and biochemistry. Natural poisons are often enzyme inhibitors that have evolved to defend a plant or animal against predators. These natural toxins include some of the most poisonous compounds known. Artificial inhibitors are often used as drugs, but can also be insecticides such as malathion, herbicides such as glyphosate, or disinfectants such as triclosan.

Chemotherapy

The most common uses for enzyme inhibitors are as drugs to treat disease. Many of these inhibitors target a human enzyme and aim to correct a pathological condition. However, not all drugs are enzyme inhibitors. Some, such as anti-epileptic drugs, alter enzyme activity by causing more or less of the enzyme to be produced. These effects are called enzyme induction and inhibition and are alterations in gene expression, which is unrelated to the type of enzyme inhibition. Other drugs interact with cellular targets that are not enzymes, such as ion channels or membrane receptors.

Metabolic control

Enzyme inhibitors are also important in metabolic control. Many metabolic pathways in the cell are inhibited by metabolites that control enzyme activity through allosteric regulation or substrate inhibition.

Physiological enzyme inhibition can also be produced by specific protein inhibitors. This mechanism occurs in the pancreas, which synthesises many digestive precursor enzymes known as zymogens.

Acetylcholinesterase inhibitors

Acetylcholinesterase (AChE) is an enzyme found in animals from insects to humans. It is essential to nerve cell function through its mechanism of breaking down the neurotransmitter acetylcholine into its constituents, acetate and choline.

α-Glucosidase Inhibitor

α-Glucosidase inhibitors are oral anti-diabetic drugs used for diabetes mellitus type 2 that work by preventing the digestion of carbohydrates (such as starch and table sugar). Carbohydrates are normally converted into simple sugars (monosaccharides), which can be absorbed through the intestine. Hence, α-glucosidase inhibitors reduce the impact of carbohydrates on blood sugar. Examples of α-glucosidase inhibitors include: (i) acarbose-precose, (ii) miglitol—glyset, and (iii) voglibose.

Even though the drugs have a similar mechanism of action, there are subtle differences between acarbose and miglitol. Acarbose is an oligosaccharide, whereas miglitol resembles a monosaccharide. Miglitol is fairly well-absorbed by the body, as opposed to acarbose. Moreover, acarbose inhibits pancreatic alpha-amylase in addition to α-glucosidase.

Alpha-glucosidase inhibitors are used to establish greater glycemic control over hyperglycemia in diabetes mellitus type 2, particularly with regard to postprandial hyperglycemia. They may be used as monotherapy in conjunction with an appropriate diabetic diet and exercise, or they may be used in conjunction with other antidiabetic drugs.

α-Glucosidase inhibitors may also be useful in patients with diabetes mellitus type 1; however, this use has not been officially approved by the Food and Drug Administration.

Mechanism of action

α-Glucosidase inhibitors are saccharides that act as competitive inhibitors of enzymes needed to digest carbohydrates: specifically α-glucosidase enzymes in the brush border of the small intestines. The membrane-bound intestinal α-glucosidases hydrolyse oligosaccharides, trisaccharides, and disaccharides to glucose and other monosaccharides in the small intestine.

Acarbose also blocks pancreatic α-amylase in addition to inhibiting membrane-bound α-glucosidases. Pancreatic α-amylase hydrolyses complex starches to oligosaccharides in the lumen of the small intestine. Inhibition of these enzyme systems reduces the rate of digestion of carbohydrates. Less glucose is absorbed because the carbohydrates are not broken down into glucose molecules. In diabetic patients, the short-term effect of these drugs therapies is to decrease current blood glucose levels: the long term effect is a small reduction in hemoglobin level.

Dosing

Since α-glucosidase inhibitors are competitive inhibitors of the digestive enzymes, they must be taken at the start of main meals to have maximal effect. Their effects on blood sugar levels following meals will depend on the amount of complex carbohydrates in the meal.

Side effects and precautions

Since α-glucosidase inhibitors prevent the degradation of complex carbohydrates into glucose, the carbohydrates will remain in the intestine. In the colon, bacteria will digest the complex carbohydrates, thereby causing gastrointestinal side effects such as flatulence and diarrhea. Since these effects are dose-related, it is generally advised to start with a low dose and gradually increase the dose to the desired amount. Voglibose, in contrast to acarbose, has less of these side effects, and is hence preferred lately. It is also more economical compared to acarbose.

If a patient using an α-glucosidase inhibitor suffers from an episode of hypoglycemia, the patient should eat something containing monosaccharides, such as glucose tablets. Since the drug will prevent the digestion of polysaccharides (or non-monosaccharides), nonmonosaccharide foods may not effectively reverse a hypoglycemic episode in a patient taking an α-glucosidase inhibitor.

MICROBIAL INSECTICIDES

Synthetic chemical insecticides provide many benefits to food production and human health, but they also pose some hazards. In many instances, alternative methods of insect management offer adequate levels of pest control and pose fewer hazards. One such alternative is the use of microbial insecticides that contain micro-organisms or their by-products.

Microbial insecticides are especially valuable because their toxicity to non-target animals and humans is extremely low. Compared to other commonly used insecticides, they are safe for both the pesticide user and consumers of treated crops. Microbial insecticides also are known as biological pathogens, and biological control agents.

Microbial insecticides are comprised of microscopic living organisms (viruses, bacteria, fungi, protozoa, or nematodes) or the toxins produced by these organisms. They are formulated to be applied as conventional insecticidal sprays, dusts, liquid drenches, liquid concentrates, wettable powders, or granules. Each product's specific properties determine the ways in which it can be used most effectively.

Advantages of Microbial Insecticides

Individual products differ in important ways, but the following list of beneficial characteristics applies to microbial insecticides in general.

1. The organisms used in microbial insecticides are essentially nontoxic and nonpathogenic to wildlife, humans, and other organisms not closely related to the target pest. The safety offered by microbial insecticides is their greatest strength.
2. The toxic action of microbial insecticides is often specific to a single group or species of insects, and this specificity means that most microbial insecticides do not directly affect beneficial insects (including predators or parasites of pests) in treated areas.
3. If necessary, most microbial insecticides can be used in conjunction with synthetic chemical insecticides because in most cases the microbial product is not deactivated or damaged by residues of conventional insecticides. (Follow label directions concerning any limitations).
4. Because their residues present no hazards to humans or other animals, microbial insecticides can be applied even when a crop is almost ready for harvest.
5. In some cases, the pathogenic micro-organisms can become established in a pest population or its habitat and provide control during subsequent pest generations or seasons.

Disadvantages of Microbial Insecticides

The limitations or disadvantages listed below do not prevent the successful use of microbial insecticides. Understanding how these limitations affect specific micro-organisms will help users to choose effective products and take necessary steps to achieve successful results.

1. Because a single microbial insecticide is toxic to only a specific species or group of insects, each application may control only a portion of the pests present in a field, garden, or lawn. If other types of pests are present in the treated area, they will survive and may continue to cause damage. Conventional insecticides are subject to similar limitations because they too are not equally effective against all pests. Nonetheless, the negative aspect of selectivity is often more noticeable for microbials.
2. Heat, desiccation (drying out), or exposure to ultraviolet radiation reduces the effectiveness of several types of microbial insecticides. Consequently, proper timing and application procedures are especially important for some products.
3. Special formulation and storage procedures are necessary for some microbial pesticides. Although these procedures may complicate the production and distribution of certain products, storage requirements do not seriously limit the handling of microbial insecticides that are widely available. (Store all pesticides, including microbial insecticides, according to label directions.)
4. Because several microbial insecticides are pest-specific, the potential market for these products may be limited. Their development, registration, and production costs cannot be spread over a wide range of pest control sales. Consequently, some products are not widely available or are relatively expensive (several insect viruses, for example).

Bacteria

Bacterial pathogens used for insect control are spore-forming, rod-shaped bacteria in the genus *Bacillus*. They occur commonly in soils, and most insecticidal strains have been isolated from soil samples. Bacterial insecticides must be eaten to be effective; they are not contact poisons. Insecticidal products comprised of a single *Bacillus* species may be active against an entire order of insects, or they may be effective against only one or a few species. For example, products containing *Bacillus thuringiensis* var. *kurstaki* kill the caterpillar stage of a wide array of butterflies and moths. In contrast, *Bacillus popillae* (milky spore disease) kills Japanese beetle larvae but is not effective against the closely related annual white grubs (masked chafers in the genus *Cyclocephala*) that commonly infest lawns. The microbial insecticides most widely used in the United States since the 1960s are preparations of the bacterium *Bacillus thuringiensis* (abbreviated as *Bt*). *Bt* products are produced commercially in large industrial fermentation tanks. As the bacteria live and multiply in the right conditions, each cell produces (internally) a spore and a crystalline protein toxin called an endotoxin. Most commercial *Bt* products contain the protein toxin and spores, but some are cultured in a manner that yields only the toxin component.

Bt formulations that kill caterpillars

The best-known and most widely used *Bt* insecticides are formulated from *Bacillus thuringiensis* var. *kurstaki* isolates that are pathogenic and toxic only to larvae of the butterflies and moths. Many such *Bt* products have been registered with the United States Environmental Protection Agency. The most common trade names for these commercial products include Dipel®, Javelin®, Thuricide®, Worm Attack®, Caterpillar Killer®, Bactospeine®, and SOK-*Bt*®, but many small companies sell similar products under a variety of trade names. These products are commercially successful and widely available as liquid concentrates, wettable powders, and ready-to-use dusts and granules. They are used to control many common leaf-feeding caterpillars, including caterpillar pests on vegetables (especially the 'worms' that attack cabbage, broccoli, cauliflower, and Brussels sprouts), bagworms and tent caterpillars on trees and shrubs, larvae of the gypsy moth and other forest caterpillars, and European corn borer larvae in field corn. Several of these products are used to control Indianmeal moth larvae in stored grain. One product with a very specific target is Certan®, formulated from *Bacillus thuringiensis* var. *aizawai*, and used exclusively for the control of wax moth larvae in honeybee hives.

Common caterpillar pests that are controlled effectively with *Bacillus thuringiensis* var. *kurstaki* (*Bt*) include: (i) European corn borer in corn, (ii) Indianmeal moth in stored grain, (iii) cabbage looper, (iv) imported cabbageworm, (v) diamondback moth, (vi) tomato/tobacco hornworm, (vii) gypsy moth, (viii) spruce budworm, (ix) tent caterpillars, (x) fall webworm, (xi) mimosa webworm, (xii) bagworms, (xiii) spring and fall cankerworm, and (xiv) common caterpillar pests that are 'not' controlled by normal applications of *Bt* include: (a) corn earworm (on corn), (b) codling moth, (c) peach tree borer, and (d) squash vine borer.

Using Bt insecticides

Insecticides containing *Bt* can be very effective for insect control in a variety of situations. Reviewing a few key facts about these products can help users obtain the best results possible. Each *Bt* insecticide controls only certain types of insects; therefore, it is essential to identify the target pest and to confirm that the *Bt* insecticide controls only certain types of insects; therefore, it is essential to identify the target pest and to confirm that the *Bt* product label states that the insecticide is effective against that pest.

Separate stages of insects differ in their susceptibility to *Bt*; isolates that are effective against larval stages of butterflies, moths, or mosquitos will not kill adults. Because susceptible insects must consume *Bt* to be poisoned, treatments must be directed to the plant parts or other material that the target pest will eat. Where this is not possible (for example, where pests bore into plant tissue without feeding much on the surface foliage or fruits), *Bt* is usually not very effective. *Bt* does not kill susceptible insects immediately. Poisoned insects normally remain on plants for a day or two after treatment, but they do not continue feeding and will die soon.

Where *Bt* is applied to plant surfaces or other sites exposed to sunlight, it is deactivated rapidly by direct ultraviolet radiation. To maximise the effectiveness of *Bt* treatments, sprays should thoroughly cover all plant surfaces, including the undersides of leaves. Treating in the late afternoon or evening can be helpful because the insecticide remains effective on foliage overnight before being inactivated by exposure to intense sunlight the following day. Treating on cloudy (but not rainy) days provides a similar result. Production processes that encapsulate *Bt* spores or toxins in a granular matrix (such as starch) or within killed cells of other bacteria also provide protection from ultraviolet radiation. Registration and sale of products containing encapsulated *Bt* are forthcoming.

Some *Bt* isolates (not those used in currently available insecticides) produce significant amounts of an additional toxin called thuringiensin, an exotoxin that is released outside the bacterial cell wall. Research is underway to develop commercial insecticides containing this toxin. Although thuringiensin might be lauded as 'natural' because it is produced by living organisms, it is nonetheless toxic to a wide range of animal species and humans.

As thuringiensin insecticides are registered and become available, users should recognise the difference between thuringiensin and other *Bt* products. Thuringiensin is much more toxic and should be handled much more cautiously.

Although the issue of thuringiensin's toxicity to mammals is a unique characteristic that does not detract from the overall safety of registered microbial insecticides, users are advised to handle all microbial insecticides cautiously. Bacterial spores, mould spores, and virus particles become 'foreign proteins' if they are inhaled or rubbed into the skin. As such, they can cause serious allergic reactions. The dusts or liquids used to dilute and carry these micro-organisms also can act as allergens or irritants. These problems do not prevent the safe use of microbial insecticides, but they do mean that users should not breathe dusts or mists of microbial insecticides. Users should wear gloves, long sleeves, and long trousers during application and wash thoroughly after completing the application. These are common sense precautions that will help to prevent unexpected reactions and minimise any effects from unknown toxicity.

Recent advances in biotechnology have resulted not only in improved prospects for developing new *Bt* insecticides but also in an ability to place *Bt* toxins within crop plants in a variety of ways. For example, genes directing the production of *Bt* toxins can be incorporated into certain plant-dwelling bacteria. When these altered bacteria grow and multiply within an inoculated host plant, the *Bt* toxin is produced within the plant. Efforts are underway to test this type of *Bt* 'application' in corn to control the European corn borer.

Although the development of this technology may seem ideal, the season-long, high-level control it would provide would also pose a great risk for the development of insect resistance to the *Bt* toxin. As genes for production of insecticidal compounds are added to crop plants, developers must also devise methods of preventing or managing insecticide resistance in target pests.

Other bacterial insecticides

Viruses

The larvae of many insect species are vulnerable to devastating epidemics of viral diseases. The viruses that cause these outbreaks are very specific, usually acting against only a single insect genus or even a single species. Most of the viruses that are nuclear polyhedrosis viruses (NPV's), in which numerous virus particles are 'packaged' together in a crystalline envelope within insect cell nuclei, or granulosis viruses (GV's), in which one or two virus particles are surrounded by a granular or capsule-like protein crystal found in the host cell nucleus. These groups of viruses infect caterpillars and the larval stages of sawflies.

Fungi

Fungi, like viruses, often act as important natural control agents that limit insect populations. Most of the species that cause insect diseases spread by means of asexual spores called conidia. Although conidia of different fungi vary greatly in ability to survive adverse environmental conditions, desiccation and ultraviolet radiation are important causes of mortality in many species. Where viable conidia reach a susceptible host, free water or very high humidity is usually required for germination. Unlike bacterial spores or virus particles, fungal conidia can germinate on the insect cuticle and produce specialised structures that allow the fungus to penetrate the cuticle and enter the insect's body. Fungi do not have to be ingested to cause infections. In most instances, as fungal infections progress, infected insects are killed by fungal toxins, not by the chronic effects of parasitism.

Fungi used as insecticides include the following:

1. *Beauveria bassiana*: This common soil fungus has a broad host range that includes many beetles and fire ants. It infects both larvae and adults of many species. Understanding the interactions between *Beauveria bassiana* and other soil micro-organisms may be the key to successful use of this fungus.
2. *Nomuraea rileyi*: In soyabeans (especially in the southeastern United States), naturally occurring epidemic infections of *Nomuraea rileyi* cause dramatic reductions in populations of foliage-feeding caterpillars. Research directed at predicting disease outbreaks caused by this fungus may help in determining the need for application of insecticides.
3. *Vericillium lecanii:* This fungus (once sold under the trade name Vertelec®) has been used in greenhouses in Great Britain to control aphids and whiteflies.
4. *Lagenidium giganteum*: This aquatic fungus is highly infectious to larvae of several mosquito genera. It cycles effectively in the aquatic environment (spores produced in infected larvae persist and infect larvae of subsequent generations), even when mosquito density is low. Its effectiveness is limited by high temperatures.
5. *Hirsutella thompsonii*: Although preparations of this pathogen were once registered by the US EPA and marketed under the trade name Mycar®, it is no longer available commercially. *Hirsutella thompsonii* is a pathogen of the citrus rust mite.

Protozoa

Protozoan pathogens naturally infect a wide range of insect hosts. Although these pathogens can kill their insect hosts, many are more important for their chronic, debilitating effects. One important and common consequence of protozoan infection is a reduction in the number of offspring produced by infected insects. Although protozoan pathogens play a significant role in the natural limitation of insect populations, few appear to be suited for development as insecticides.

Nematodes

To be accurate, nematodes are not microbial agents. Instead, they are multicellular roundworms. Nematodes used in insecticidal products are, however, nearly microscopic in size, and they are used much like the truly microbial products. Nematodes used for insect control infect only insects or related arthropods; they are called entomogenous nematodes. The entomogenous nematodes *Steinernema feltiae* (sometimes identified as *Neoaplectana carpocapsae*), *S. scapteriscae*, *S. riobravis*, *S. carpocapsae* and *Heterorhabditis heliothidis* are the species most commonly used in insecticidal preparations. Within each of these species, different strains exhibit differences in their abilities to infect and kill specific insects. In general, however, these nematodes infect a wide range of insects. Microbial insecticides offer effective alternatives for the control of many insect pests. Their greatest strength is their safety, as they are essentially nontoxic and nonpathogenic to animals and humans. Although not every pest problem can be controlled by the use of a microbial insecticide, these products can be used successfully in place of more toxic insecticides to control many lawn and garden pests and several important field crop and forest insects. Because most microbial insecticides are effective against only a narrow range of pests and because these insecticides are vulnerable to rapid inactivation in the environment, users must properly identify target pests and plan the most effective application. But these same qualities mean that microbial insecticides can be used without undue risks of human injury or environmental damage. Consequently, microbial insecticides are likely to become increasingly important tools in insect management. Table 20.1 listed the summary of products and their uses in microbial insecticides.

Table 20.1. Microbial insecticides: A summary of products and their uses.

Pathogen	*Product name*	*Host range*	*Uses and comments*
		Bacteria	
Bacillus thuringiensis var. *kurstaki* (*Bt*)	Bactur®, Bactospeine®, Bioworm®, Caterpillar killer®, Dipel®, Futura®, Javelin®, SOK-*Bt*®, Thuricide®, Topside®, Tribactur®, Worthy Attack®	Caterpillars (larvae of moths and butterflies)	Effective for foliage-feeding caterpillars (and Indian meal moth in stored grain). Deactivated rapidly in sunlight; apply in the evening or on overcast days and direct some spray to lower surfaces or leaves. Does not cycle extensively in the environment. Available as liquid concentrates, wettable powders, and ready to use dusts and granules. Active only if ingested.
Bacillus thuringiensis var. *israelensis* (*Bt*)	Aquabee®, Bactimos®, Gnatrol®, LarvX®, Mosquito Attack®, Skeetal®, Teknar®, Vectobac®	Larvae of *Aedes* and *Psorophora* mosquitoes, black flies, and fungus gnats	Effective against larvae only. Active only if ingested. Culex and Anopheles mosquitoes are not controlled at normal application rates. Activity is reduced in highly turbid or polluted water. Does not cycle extensively in the environment. Applications generally made over wide areas by mosquito and blackfly abatement districts.

(Contd ...)

Pathogen	Product name	Host range	Uses and comments
Bacillus thuringiensis var. *tenebrinos*	Foil® M-One® M-Track®, Novardo® Trident®	Larvae of Colorado, potato beetle, elm leaf beetle adults	Effective against Colorado potato beetle larvae and the elm leaf beetle. Like other *Bt*s, it must be ingested. It is subject to breakdown in ultraviolet light and does not cycle extensively in the environment.
Bacillus thuringiensis var. aizawai	Certan®	Wax moth caterpillars	Used only for control of wasmoth infestations in honeybee hives.
Bacillus popilliae and *Bacillus lentimorbus*	Doom¨, Japidemic¨,® Milky Spore Disease, Grub Attack®	Larvae (grubs) of Japanese beetle	The main Illinois lawn grub (the annual white grub, *Cyclocephala* sp.) Is 'not' susceptible to milky spore disease. The disease is very effective against Japanese beetle grubs (not a major pest in Illinois) and cycles effectively for years in the soil.
Bacillus sphaericus	Vectolex CG®, Vectolex WDG®	Larvae of *Culex, Psorophora* and *Culiseta* mosquitos, larvae of some *Aedes* spp.	Active only if ingested, for use against *Culex, Psorophora,* and *Culiseta* species; also effective against *Aedes vexans*. Remains effective in stagnant or turbid water. Commercial formulations will not cycle to infect subsequent generations.
		Fungi	
Beauveria bassiana	Botanigard®, Mycotrol®, Naturalis®	Aphids, fungus gnats, mealy bugs, mites, thrips, whiteflies	Effective against several pests. High moisture requirements, lack of storage longevity, and competition with other soil micro-organisms are problems that remain to be solved.
Lagenidium giganteum	Laginex®	Larvae of most pest mosquito species	Effective against larvae of most pest mosquito species; remains infective in the environment through dry periods. A main drawback is its inability to survive high summertime temperatures.
		Protozoa	
Nosema locustae	NOLO Bait®, Grasshopper attack®	European cornborer caterpillars, grasshoppers and mormon crickets	Useful for rangeland grasshopper control. Active only if ingested. Not recommended for use on a small scale, such as backyard gardens, because the disease is slow acting and grasshoppers are very mobile. Also effective against caterpillars.
		Viruses	
Gypsy moth nuclear plyhedrosis (NPV)	Gypchek® virus	Gypsy moth caterpillars	All of the viral insecticides used for control of forest pests are produced and used exclusively by the US Forest Service.

Contd ...

Pathogen	*Product name*	*Host range*	*Uses and comments*
Tussock moth NPV	TM Biocontrol-1®	Tussock moth caterpillars	
Pine sawfly NPV	Neochek-S®	Pine sawfly larvae	
Codling moth granulosis virus (GV)	(see comments)	Codling moth caterpillars	Commercially produced and marketed briefly, but no longer registered or available. Future re-registration is possible. Active only if ingested. Subject to rapid breakdown in ultraviolet light.
		Entomogenous nematodes	
Steinernema feltiae (=*Neoaplectana carpocapsae*) *S. riobravis*, *S. carpocapsae* and other *Steinernema* species	Biosafe®, Ecomask®, Scanmask®, also sold generically (wholesale and retail), Vector®	Larvae of a wide variety of soil-dwelling and boring insects	*Steinernema riobravis* is the main nematode species marketed retail in the US because of moisture requirements, it is effective primarily against insects in moist soils or inside plant tissues. Prolonged storage or extreme temperatures before use may kill or debilitate the nematodes. Effective in cool temperatures.
Heterorhabditis heliothidis	Currently available on a wholesale basis for large scale operations	Larvae of a wide variety of soil-dwelling and boring insects	Not commonly available by retail in the US; this species is used more extensively in Europe. Available by wholesale or special order for research or large-scale commercial uses. Similar in use to *Steinernema* species but with some differences in host range, infectivity, and temperature requirements.
		Pathogen	
Steinernema scapterisci	Nematac®S late nymph and adult stages of mole crickets		*S. scapterisci* is the main nematode species marketed to target the tawny and southern mole cricket. Best applied where irrigation is available. Irrigate after application.

Botanical Insecticides

Botanical insecticides are naturally occurring chemicals extracted from plants. Natural pesticidal products are available as an alternative to synthetic chemical formulations but they are not necessarily less toxic to humans. Some of the most deadly, fast acting toxins and potent carcinogens occur naturally. Some of the botanical pesticides are very toxic to fish and other cold-blooded creatures and should be treated with care. Protective clothing should be worn when spraying, even though their toxicity is normally low to warm-blooded animals. Botanical insecticides break down readily in soil and are not stored in plant or animal tissue. Often their effects are not as long lasting as those of synthetic pesticides and some of these products may be very difficult to find.

Citrus oil (limonene, linalool) are extracts from citrus peels primarily used as flea dips, but have been combined with soaps as contact poisons against aphids and mites. They evaporate quickly after application and provide no residual control. Nicotine concentrate is very poisonous if inhaled. It is derived from tobacco and is commonly sold as a 40 per cent nicotine sulphate concentrate. Nicotine is a fast acting contact killer for soft bodied insects, but does not kill most chewing insects. It is less effective when applied during cool weather. Do not spray within 7 days of harvest.

Pyrethrin is a fast acting contact poison derived from the pyrethrum daisy. It is very toxic to cold blooded animals. Some people and most cats have allergic reactions to it. Pyrethrin is effective on most insects, but does not control mites. It rapidly breaks down in sunlight, air and water.

Rotenone is derived from the roots of over 68 plant species, and is very toxic to fish, pigs, and cool-blooded animals. It has a short residual. Rotenone is a broad spectrum poison mainly used to control leaf-eating caterpillars and beetles. Direct contact may cause skin and mucous membrane irritation. It is more toxic when inhaled. Ryania is a slow acting stomach poison. It has a longer residual than most botanicals. Toxicity to mammals is moderate.

Sabadilla is derived from the seeds of South American lilies. It is a broad spectrum contact poison, but has some activity as a stomach poison. It is most effective against true bugs such as harlequin bugs and squash bugs. Sabadilla degrades rapidly in air and sunlight, and has little residual toxicity. It is very toxic to honey bees. The least toxic botanical to humans.

Neem is a relatively new product on the market. It is derived from the neem tree that grows in arid tropical regions. Extracts from the neem tree have been reported to control over 200 types of insects, mites, and nematodes. The neem spray solution should not be exposed to sunlight and must be prepared with water having a temperature between 50°F and 90°F. The solution is effective for only 8 hrs after mixing. Neem is most effective under humid conditions or when the insect and plants are damp. It has a low toxicity to mammals.

Insecticide	*Use against*
Pyrethrum	Pickleworms, aphids, leafhoppers, spider mites, harlequin bugs, cabbage worms.
Neem	Cutworms, armyworms, sodworms.
Rotenone	Spittlebugs, aphids, potato beetles, harlequin bugs, chinch bugs, spider mites, carpenter ants
Ryania	Codling moths, Japanese beetles, squash bugs, potato aphids, onion thrips, corn earworms, silkworms
Sabadilla	Grasshoppers, codling moths, moths, armyworms, aphids, cabbage loopers, blister beetles, squash bugs, harlequin bugs
Nicotine	Aphids, thrips, caterpillars.

CYCLOSPORINE

Immunosuppressive medications play a large part of the management of many pediatric illnesses. Cyclosporine is the primary tool used to prevent rejection following solid organ and bone marrow transplantation. It has been estimated that cyclosporine is given to more than 90 per cent of children who have received a kidney transplant in the United States.

In addition, the ability of cyclosporine to inhibit T-cell activation has been shown to have a role in the treatment of diseases such as nephrotic syndrome, refractory Crohn's disease and ulcerative colitis, biliary cirrhosis, aplastic anemia, rheumatoid arthritis, myasthenia gravis, and dermatomyositis.

Though powerful, it causes very few serious reactions when used for short periods of time (days, weeks, or a few months). It is considered a short-term treatment for psoriasis and is safest when given for less than one year. Problems with kidney function, including kidney failure, and hypertension can occur in cases where the drug is used for over one year.

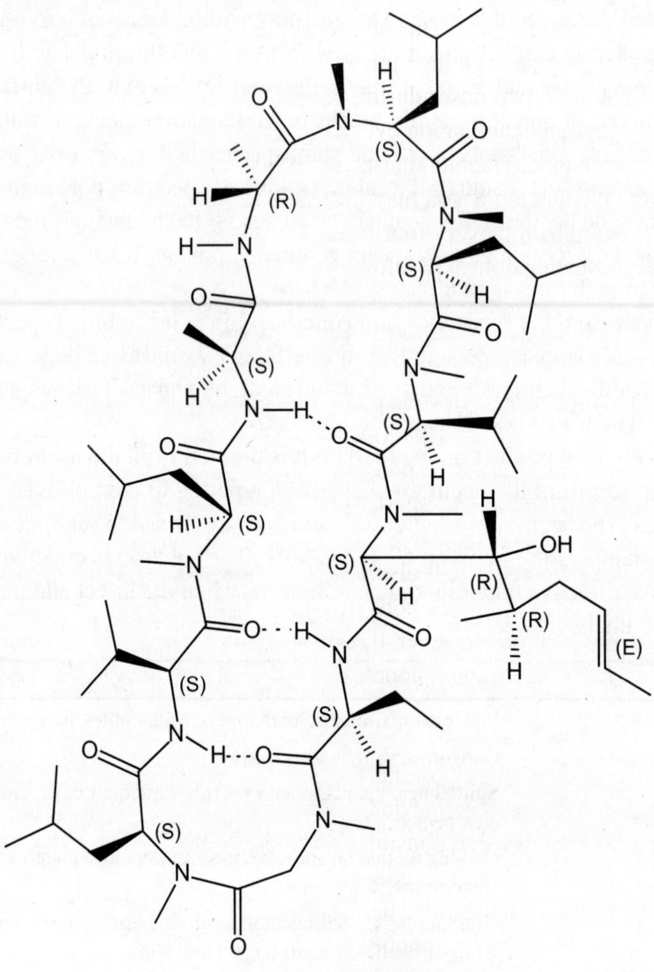

Cyclosporin structure

Short-term side effects of cyclosporine may include headache, nausea, tingling in the fingers and toes, aches in joints, growth of hair where it is not desired, and swelling of the gums. While these side effects sound awful, most people do not experience them at lower dosages. The medication need not be stopped in the case of these side effects, and they many go away with continued use of the medication.

Short-term use of cyclosporine can also cause elevation of the blood pressure, decrease in kidney function, as well as elevation of cholesterol and triglycerides. These functions must be monitored and the dosage lowered if they occur. Sometimes the drug must be stopped at least for a while.

To maximise safety, it is important to follow these rules:

1. All physicians, dentists, pharmacists you deal with must know you are on cyclosporine.
2. Do not take any new medications while on cyclosporine unless your doctor has checked to make sure it is safe with cyclosporine.

3. Medications, which could cause trouble with cyclosporine, include some anti-inflammatory medications, certain antibiotics (nafcillin, trimethoprim-sulpha, erythromycin, certain fungus medications, and others).
4. Avoid drinking grapefruit juice. This one type of juice has the ability to raise the blood levels of cyclosporine to high levels. Orange juice is ok.
5. Take the medication in two doses during the day, rather than all at once.
6. Get blood tests and an examination every two weeks for the first month and again two weeks after any dosage change. The blood tests should be obtained in the morning, without breakfast.
7. Do not take the medication if you may be pregnant, or are breastfeeding.
8. Have a dental examination every four months while on cyclosporine.

Following these recommendations will allow using this very effective medication with the greatest safety.

Mode of Action

Cyclosporin is thought to bind to the cytosolic protein cyclophilin (immunophilin) of immunocompetent lymphocytes, especially T-lymphocytes. This complex of cyclosporin and cyclophilin inhibits calcineurin, which, under normal circumstances, is responsible for activating the transcription of interleukin 2. It also inhibits lymphokine production and interleukin release and, therefore, leads to a reduced function of effector T-cells.

It does not affect cytostatic activity. It also has an effect on mitochondria. Cyclosporin A prevents the mitochondrial permeability transition pore from opening, thus inhibiting cytochrome c release, a potent apoptotic stimulation factor. However, this is not the primary mode of action for clinical use, but rather an important effect for research on apoptosis.

Biosynthesis

Cyclosporine A is synthesised by a nonribosomal peptide synthetase, cyclosporine synthetase. The enzyme contains an adenylation domain, a thiolation domain, a condensation domain, and an *N*-methyltransferase domain. The adenylation domain is responsible for substrate recognition and activation, whereas the thiolation domain covalently binds the adenylated amino acids to phosphopantetheine and the condensation domain elongates the peptide chain. Cyclosporine synthetase substrates include L-Valine, L-Leucine, L-Alanine, L-Glycine, 2-aminobutyric acid, 4-methylthreonine, and D-Alanine. With the adenylation domain, cyclosporine synthetase generates the acyl-adenylated amino acids, then covalently binds the amino acid to phosphopantetheine through a thioester linkage. Some of the amino acid substrates become N-methylated by S-Adenosyl methionine. The cyclisation step releases cyclosporine A from the enzyme. Amino acids such as D-Ala and butenyl-methyl-L-threonine indicates that cyclosporine synthetase requires the action of other enzymes such as a D-Alanine racemase. The racemisation of L-Ala to D-Ala is pyridoxal phosphate-dependent. The formation of butenyl-methyl-L-threonine is performed by a butenyl-methyl-L-threonine polyketide synthase that utilises acetate/malonate as its starting material.

Adverse Effects and Interactions

Treatment may be associated with a number of potentially serious adverse drug reactions (ADRs) and adverse drug interactions. Cyclosporine interacts with a wide variety of other drugs and other substances including grapefruit juice. There have been studies into the use of grapefruit juice to increase the blood level of cyclosporin. ADRs can include gum hyperplasia, convulsions, peptic ulcers, pancreatitis, fever, vomiting, diarrhea, confusion, breathing difficulties, numbness and tingling, pruritus, high blood pressure, potassium retention, and possibly hyperkalemia, kidney and liver dysfunction (nephrotoxicity and hepatotoxicity), and an increased vulnerability to opportunistic fungal and viral infections.

An alternate form of the drug, cyclosporine G (OG37-324), has been found to be much less nephrotoxic than the standard cyclosporine A. Cyclosporine G (lol. mass 1217) differs from cyclosporine A in the amino acid 2 position, where an L-nor-valine replaces the α-aminobutyric acid.

Formulations

The drug exhibits very poor solubility in water, and, as a consequence, suspension and emulsion forms of the drug have been developed for oral administration and for injection. Cyclosporine was originally brought to market by Sandoz, now Novartis, under the brand name of Sandimmune, which is available as soft-gelatine capsules, as an oral solution, and as a formulation for intravenous administration. These are all nonaqueous compositions Sandimmune-Novartis. A newer microemulsion orally-administered formulation Neoral Neoral-Novartis is available as a solution and as soft gelatine capsules. The Neoral compositions are designed to form microemulsions in contact with water. Generic cyclosporine preparations have been marketed under various trade names including Cicloral (Sandoz/Hexal) and Gengraf (Abbott). Since 2002, a topical emulsion of cyclosporine for treating keratoconjunctivitis sicca has been marketed under the trade name Restasis. Inhaled cyclosporine formulations are in clinical development, and include a solution in propylene glycol and liposome dispersions. The drug is also available in a dog preparation manufactured by Novartis called Atopica. Atopica is indicated for the treatment of atopic dermatitis in dogs. Unlike the human form of the drug, the lower doses used in dogs mean the drug acts as an immuno-modulator and has fewer side-effects than in man. The benefits of using this product include the reduced need for concurrent therapies to bring the condition under control.

BIOCHIPS

The development of biochips is a major thrust of the rapidly growing biotechnology industry, which encompasses a very diverse range of research efforts including genomics, proteomics, and pharmaceuticals, among other activities. Advances in these areas are giving scientists new methods for unraveling the complex biochemical processes occurring inside cells, with the larger goal of understanding and treating human diseases. At the same time, the semiconductor industry has been steadily perfecting the science of microminiaturisation. The merging of these two fields in recent years has enabled biotechnologists to begin packing their traditionally bulky sensing tools into smaller and smaller spaces, onto so-called biochips. These chips are essentially miniaturised laboratories that can perform hundreds or thousands of simultaneous biochemical reactions. Biochips enable researchers to quickly screen large numbers of biological analytes for a variety of purposes, from disease diagnosis to detection of bioterrorism agents. A biochip is a collection of miniaturised test sites (microarrays) arranged on a solid substrate that permits many tests to be performed at the same time in order to achieve higher output and speed. Biochips can also be used to perform techniques such as electrophoresis or PCR using microfluidics technology.

Microarray Fabrication

The microarray—the dense, two-dimensional grid of biosensors—is the critical component of a biochip platform. Typically, the sensors are deposited on a flat substrate, which may either be passive (e.g.

silicon or glass) or active, the latter consisting of integrated electronics or micromechanical devices that perform or assist signal transduction. Surface chemistry is used to covalently bind the sensor molecules to the substrate medium. The fabrication of microarrays is non-trivial and is a major economic and technological hurdle that may ultimately decide the success of future biochip platforms. The primary manufacturing challenge is the process of placing each sensor at a specific position (typically on a Cartesian grid) on the substrate.

Various means exist to achieve the placement, but typically robotic micro-pipetting or micro-printing systems are used to place tiny spots of sensor material on the chip surface. Because each sensor is unique, only a few spots can be placed at a time. The low-throughput nature of this process results in high manufacturing costs.

Protein Biochip Array and Other Microarray Technologies

Microarrays are not limited to DNA analysis; protein microarrays, antibody microarray, chemical compound microarray can also be produced using biochips. Randox Laboratories Ltd. launched Evidence, the first protein biochip array technology analyser in 2003. In protein biochip array technology, the biochip replaces the ELISA plate or cuvette as the reaction platform. The biochip is used to simultaneously analyse a panel of related tests in a single sample, producing a patient profile. The patient profile can be used in disease screening, diagnosis, monitoring disease progression or monitoring treatment. Performing multiple analyses simultaneously, described as multiplexing, allows a significant reduction in processing time and the amount of patient sample required. Biochip array technology is a novel application of a familiar methodology, using sandwich, competitive and antibody-capture immunoassays. The difference from conventional immunoassays is that the capture ligands are covalently attached to the surface of the biochip in an ordered array rather than in solution.

FLAVOURING SUBSTANCES

Flavour or flavour (see spelling differences) is the sensory impression of a food or other substance, and is determined mainly by the chemical senses of taste and smell. The 'trigeminal senses', which detect chemical irritants in the mouth and throat, may also occasionally determine flavour. The flavour of the food, as such, can be altered with natural or artificial flavourants, which affect these senses.

Flavourant is defined as a substance that gives another substance flavour, altering the characteristics of the solute, causing it to become sweet, sour, tangy, etc.

Of the three chemical senses, smell is the main determinant of a food item's flavour. While the taste of food is limited to sweet, sour, bitter, salty, and savory (umami)—the basic tastes—the smells of a food are potentially limitless. A food's flavour, therefore, can be easily altered by changing its smell while keeping its taste similar. Nowhere is this better exemplified than in artificially flavoured jellies, soft drinks and candies, which, while made of bases with a similar taste, have dramatically different flavours due to the use of different scents or fragrances. The flavourings of commercially produced food products are typically created by flavourists.

Although the terms 'flavouring' or 'flavourant' in common language denote the combined chemical sensations of taste and smell, the same terms are usually used in the fragrance and flavours industry to refer to edible chemicals and extracts that alter the flavour of food and food products through the sense of smell. Due to the high cost or unavailability of natural flavour extracts, most commercial flavourants are nature-identical, which means that they are the chemical equivalent of natural flavours but chemically synthesised rather than being extracted from the source materials.

Flavourants or Flavourings

Flavourings are focused on altering or enhancing the flavours of natural food product such as meats and vegetables, or creating flavour for food products that do not have the desired flavours such as candies and other snacks. Most types of flavourings are focused on scent and taste. Few commercial products exist to stimulate the trigeminal senses, since these are sharp, astringent, and typically unpleasant flavours.

There are three principal types of flavourings used in foods, under definitions agreed in the EU and Australia:

1. Natural flavouring substances: Flavouring substances obtained from plant or animal raw materials, by physical, microbiological or enzymatic processes. They can be either used in their natural state or processed for human consumption, but cannot contain any nature-identical or artificial flavouring substances.
2. Nature-identical flavouring substances: Flavouring substances that are obtained by synthesis or isolated through chemical processes, which are chemically identical to flavouring substances naturally present in products intended for human consumption. They cannot contain any artificial flavouring substances.
3. Artificial flavouring substances: Flavouring substances not identified in a natural product intended for human consumption, whether or not the product is processed.

Regulations on natural flavouring

UK Food Law defines a natural flavour as:

> 'A flavouring substance (or flavouring substances) which is (or are) obtained, by physical, enzymatic or microbiological processes, from material of vegetable or animal origin which material is either raw or has been subjected to a process normally used in preparing food for human consumption and to no process other than one normally so used.'

The US Code of Federal Regulations describes a 'natural flavourant' as:

> 'The essential oil, oleoresin, essence or extractive, protein hydrolysate, distillate, or any product of roasting, heating or enzymolysis, which contains the flavouring constituents derived from a spice, fruit or fruit juice, vegetable or vegetable juice, edible yeast, herb, bark, bud, root, leaf or any other edible portions of a plant, meat, seafood, poultry, eggs, dairy products, or fermentation products thereof, whose primary function in food is flavouring rather than nutritional.'

The European Union's guidelines for natural flavourants are slightly different. Certain artificial flavourants are given an E number, which may be included on food labels.

Smell

Smell flavourants, or simply, flavourants, are engineered and composed in similar ways as with industrial fragrances and fine perfumes. To produce natural flavours, the flavourant must first be extracted from the source substance. The methods of extraction can involve solvent extraction, distillation, or using force to squeeze it out.

The extracts are then usually further purified and subsequently added to food products to flavour them. To begin producing artificial flavours, flavour manufacturers must either find out the individual naturally occurring aroma chemicals and mix them appropriately to produce a desired flavour or create a novel non-toxic artificial compound that gives a specific flavour.

Most artificial flavours are specific and often complex mixtures of singular naturally occurring flavour compounds combined together to either imitate or enhance a natural flavour. These mixtures are formulated by flavourist to give a food product a unique flavour and to maintain flavour consistency between different product batches or after recipe changes.

The list of known flavouring agents includes thousands of molecular compounds, and the flavour chemist (flavourist) can often mix these together to produce many of the common flavours. Many flavourants are esters.

Chemical	*Odour*
Diacetyl	Buttery
Isoamyl acetate	Banana
Benzaldehyde	Bitter almond
Cinnamic aldehyde	Cinnamon
Ethyl propionate	Fruity
Methyl anthranilate	Grape
Limonene	Orange
Ethyl- (E,Z)-2,4-decadienoate	Pear
Allyl hexanoate	Pineapple
Ethyl maltol	Sugar, cotton candy
Ethylvanillin	Vanilla
Methyl salicylate	Wintergreen

The compounds used to produce artificial flavours are almost identical to those that occur naturally, and a natural origin for a substance does not necessarily imply that it is safe to consume. In fact, artificial flavours may be safer to consume than natural flavours due to the standards of purity and mixture consistency that are enforced either by the company or by law. Natural flavours in contrast may contain toxins from their sources while artificial flavours are typically more pure and are required to undergo more testing before being sold for consumption.

Flavours from food products are usually the result of a combination of natural flavours, which set up the basic smell profile of a food product while artificial flavours modify the smell to accent it.

Taste

While salt and sugar can technically be considered flavourants that enhance salty and sweet tastes, usually only compounds that enhance umami, as well as other secondary flavours are considered taste flavourants. Artificial sweeteners are also technically flavourants.

Umami or 'savory' flavourants, more commonly called taste or flavour enhancers are largely based on amino acids and nucleotides. These are manufactured as sodium or calcium salts. Umami flavourants recognised and approved by the European Union include:

1. Glutamic acid salts: This amino acid's sodium salt, monosodium glutamate (MSG) is one of the most commonly used flavour enhancers in food processing. Mono and diglutamate salts are also commonly used.
2. Glycine salts: A simple amino acid that is usually used in conjunction with glutamic acid as a flavour enhancer.
3. Guanylic acid salts: Nucleotide salts that is usually used in conjunction with glutamic acid as a flavour enhancer.

4. Inosinic acid salts: Nucleotide salts created from the breakdown of AMP. Due to high costs of production, it is usually used in conjunction with glutamic acid as a flavour enhancer.
5. 5'-Ribonucleotides salts.

Certain organic acids can be used to enhance sour tastes, but like salt and sugar these are usually not considered and regulated as flavourants under law. Each acid imparts a slightly different sour or tart taste that alters the flavour of a food.

1. Acetic acid: gives vinegar its sour taste and distinctive smell.
2. Citric acid: found in citrus fruits and gives them their sour taste.
3. Lactic acid: found in various milk products and give them a rich tartness.
4. Malic acid: found in apples and gives them their sour/tart taste.
5. Tartaric acid: found in grapes and wines and gives them a tart taste.

Dietary Restrictions

Food manufacturers are sometimes reluctant about informing consumers about the source from where the flavour is obtained and whether it has been produced with the incorporation of substances such as animal by-products glycerine, gelatine, and the like, and the use of alcohol in the flavours. In many western countries, consumers rely on a Jewish Kosher Pareve certification mark to indicate that natural flavourings used in a food product are free of meat and dairy (although they can still contain fish). The Vegan Society's sunflower symbol (which is currently used by over 260 companies world wide) can also be used to see which products do not use any animal ingredients (including flavourings and colourings).

Flavour Creation

Most food and beverage companies do not create their own flavours but instead employ the services of a flavour company. Food and beverage companies may require flavours for new products, product line extensions (e.g. low fat versions of existing products) or due to changes in formula or processing for existing products.

The flavour creation is done by a specially trained scientist called a 'flavourist'. The flavourist's job combines extensive scientific knowledge of the chemical palette with artistic creativity to develop new and distinctive flavours. The flavour creation begins when the flavourist receives a brief from the client. In the brief the client will attempt to communicate exactly what type of flavour they seek, in what application it will be used, and any special requirements (e.g. must be all natural).

The communication barrier can be quite difficult to overcome since most people are not experienced at describing flavours. The flavourist will use his or her knowledge of the available chemical ingredients to create a formula and compound it on an electronic balance. The flavour will then be submitted to the client for testing. Several iterations, with feedback from the client, may be needed before the right flavour is found.

Additional work may also be done by the flavour company. For example, the flavour company may conduct sensory taste tests to test consumer acceptance of a flavour before it is sent to the client or to further investigate the 'sensory space'.

The flavour company may also employ application specialists who work to ensure the flavour will work in the application for which it is intended. This may require special flavour delivery technologies that are used to protect the flavour during processing or cooking so that the flavour is only released when eaten by the end consumer.

Few standards are available or being prepared for sensory analysis of flavours. In chemical analysis of flavours, solid phase extraction (SPE), solid phase microextraction (SPME), and headspace gas chromatography are applied to extract and separate the flavour compounds in the sample. The determination is typically done by various mass spectrometric techniques.

Microbial Biocatalysis in the Generation of Flavour and Fragrance Chemcials

Microbial biocatalysis is used in the commercial production of many flavour and fragrance chemicals. Bulk flavouring chemicals such as citric acid, high fructose corn syrup, and glutamic acid are produced in millions of pounds annually using microbial processes. In the past few years, biocatalysis has also begun to play an increasingly important role in the production of many flavour and fragrance aroma chemicals. Microbial processes have traditionally played an integral role in the development of complex mixtures of flavour and aroma chemicals since the discovery of beer, wine, cheese, and soya sauce thousands of years ago. Today, contemporary microbiological techniques are being increasingly applied to enhance the efficiency of many microbial biocatalysts for the production of specific flavour and fragrance chemicals. However, to ensure commercial implementation of these new microbial processes, much more needs to be learned about the basic biochemistry and genetics of these novel biocatalysts.

Flavour and texture are the most pronounced factors that influence the quality and acceptance of yogurt and related fermented milks. Many parameters affect flavour, body, and texture of yogurt such as the starter culture, incubation temperature, processing conditions (e.g. heat treatment, homogenisation) and compositional properties of hulk base. A beer flavoured post-fermentation with an isoalpha acid or isohumulone flavouring agent, selected from unreduced and reduce alpha acids and isoalpha acids, which is essentially odour-causing-impurity free and which therefore, has a more reproducible and acceptable flavouring and aroma said flavouring agent being produced by separation of the flavouring agent as an organic phase from an aqueous solution containing undesirable odour-forming impurities at a pH above 4 in the presence of a food-grade water-soluble salt.

BIOFILTERS

Biofilters are an odour treatment technology that utilises biological processes as the treatment mechanism. Biofilters are considered to be a 'green' approach to odour control, because they utilise micro-organisms in media to oxidise odour and air emission compounds to carbon dioxide, water, biomass, and other benign by-products such as chloride and sulphate. The by-products are either emitted in the outlet air, drained from the biofilter, or consumed by the micro-organisms. The biological activity in a biofilter is similar to the activities performed by the micro-organisms in activated-sludge secondary waste-water treatment processes. Typical biofilter work by routing odourous air through a porous filter media. The media represents the contact surface area, on which the micro-organisms live, where the biological oxidation described above can take place. The key to effective biofilter operation is maintaining a healthy environment for these micro-organisms to thrive in. The most important parameters for maintaining a healthy environment is moisture content and pH. The micro-organisms need water to remain active, and the presence of water effects the transfer of the contaminants from the air to the media. The desired micro-organisms thrive at neutral to moderately acidic pH levels. Figure 20.3 shows biofilter sections.

Biofilters are less control-dependent than chemical scrubbers, because the treatment system is more self-regulating. It is essentially a self-contained ecosystem, and is therefore likely to function longer, without excessive control. It is very important, however, to ensure moisture levels are controlled in the biofilter for it to function properly. Moisture content is assisted during dry periods by simple sprinkler systems and air humidification. The pH levels are often self-regulating within the ecosystem, and are assisted by proper choice of media. However, the reaction products in a biofilter treating hydrogen sulphide is sulphuric acid so preventing the formation of a very low pH is important and design should include corrosion protection on concrete and other materials.

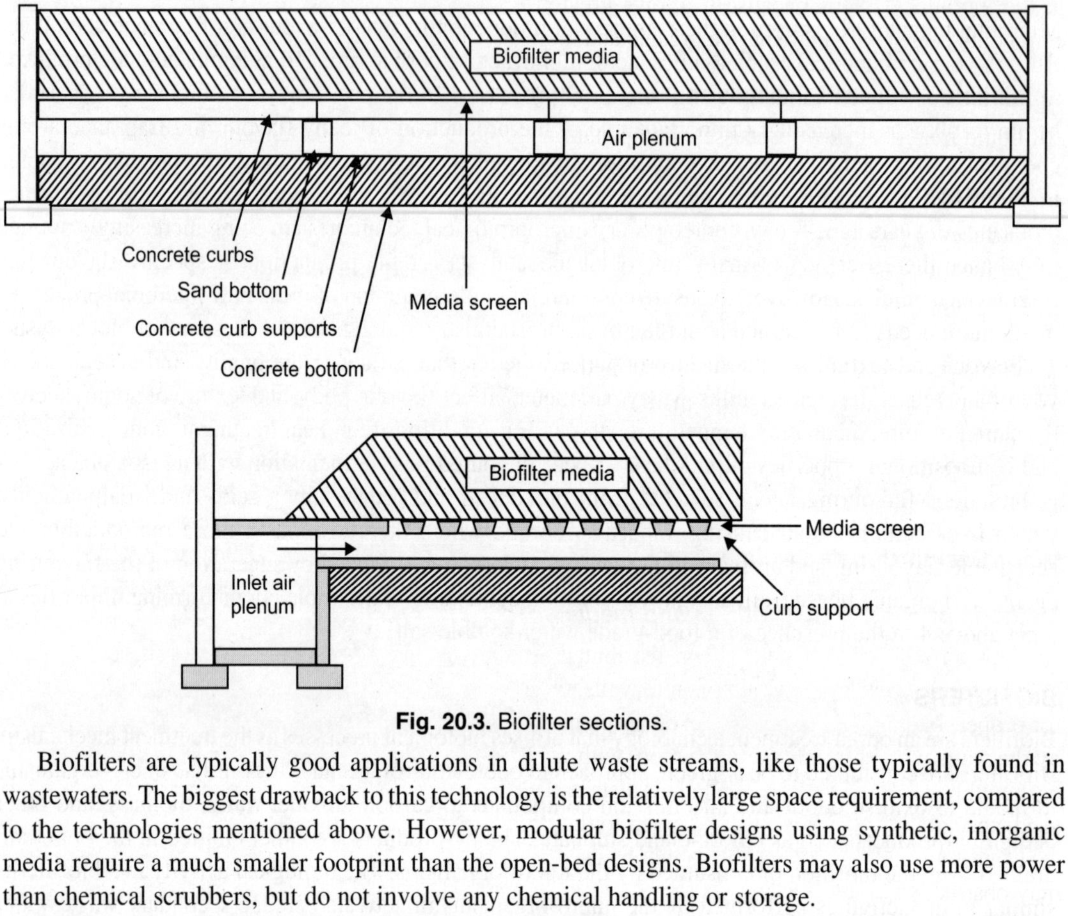

Fig. 20.3. Biofilter sections.

Biofilters are typically good applications in dilute waste streams, like those typically found in wastewaters. The biggest drawback to this technology is the relatively large space requirement, compared to the technologies mentioned above. However, modular biofilter designs using synthetic, inorganic media require a much smaller footprint than the open-bed designs. Biofilters may also use more power than chemical scrubbers, but do not involve any chemical handling or storage.

Air streams with high concentrations of reduced sulphur compounds (RSCs) such as mercaptans, dimethyl sulphide, dimethyl disulphide, diphenyl sulphide, carbonyl sulphide, and carbon disulphide can be treated in biofilters but the loading rates must be much lower than those required to treat H_2S.

Biofilter media types include wood-chip/bark media, soil media, and inorganic synthetic media. Wood-chip/bark media generally possess a large diversity and density of micro-organisms, accepts moisture relatively well, has low initial costs, and is readily available. The normal lifetime for wood-chip/bark media is 2–4 years.

Soil media is a blended mix of soils, primarily sand-based. The primary advantage of soil media over wood-chip/bark media are their life expectancy. Soil has an estimated lifetime of over 30 years as a filter media. Soil is denser than wood-chip/bark media and therefore resists compaction, it resists acidification because of it is inherent pH buffering properties, it is less difficult to rehydrate after drying out, and generally distributes the air more uniformly than wood-chip/bark media. The primary disadvantage is that it requires a smaller loading per square foot, and therefore may require a larger footprint and higher initial capital costs.

The inorganic synthetic media is newer to the market but well tested. It consists of strong, uniform sized gravel-like cores that do not compact as easily as organic media. This type of media may be used in the modular designs because it allows greater media depth and a smaller footprint. The cores are commonly coated with nutrient rich organic and inorganic adsorbents.

Applicable Treatment Processes

All liquid treatment plant processes, pump stations, sludge thickening, sludge dewatering.

Typical Design Criteria

Surface loading (wood-chip media)	3–4 cfm per sf media
Surface loading (soil media)	2 cfm per sf media
Inorganic, synthetic media	10–12 cfm/ft^2
Media bed depth (wood-chip/soil media)	3 ft.
Media bed depth (inorganic, synthetic media)	5 ft.
Detention time (wood-chip/soil media)	$\geq$ 60 seconds
Detention time (inorganic, synthetic media)	20–30 seconds
Pressure drop	6–12
H_2S removal efficiency	99 per cent

Major Design Considerations

Methods of air flow distribution and media support

Air flow through the biofilter may be distributed by several methods. The outer walls of the air plenum may be formed by earth berms, concrete walls, or other support mechanisms. A plenum lining provides for proper drainage of the biofilter. The air plenum below the media bed may be open air space formed by the walls with grating forming the top, with railroad ties forming the support and top of the plenum, or it could be formed with perforated air distribution piping buried in a coarse rock bed. If a rock bed is used, special consideration must be given to the type of rock.

Limestone and other soft rock cannot be used because it breaks down in the acidic environment and may obstruct air flow.

Media selection

Media may be purchased from manufacturers, or blended based on a recipe from locally available media constituents, such as wood-chips, bark, and various soil media constituents. Media replacement frequency is affected by media selection, as mentioned above.

Moisture control

Moisture control may be accomplished by pre-humidification of the air in a mist chamber with spray nozzles, with a packed tower humidification chamber, by keeping the media wet using soaker piping within the media bed, surface irrigation with spray nozzles, or a combination of these methods. Moisture sensors have not proven to be extremely reliable, therefore manual operator monitoring is typically used to ensure adequate moisture content.

Loadings

Loading of biofilters should be properly designed to prevent acid formation, corrosion problems, premature compaction of the media, short-circuiting the media bed, inadequate biological activity, and other problems which can result in sub-standard performance of the biofilter.

Corrosion protection

Due to the formation of sulphuric acid as a by-product in hydrogen sulphide treatment, the following corrosion protection should be included in the biofilter design:

1. Liners or protective coatings on concrete.
2. Installation of pH probes in drain water to measure pH.

References

Allen, K.L., *Basic Concepts of Microbiology*, Butterworths, London.

Arceivala, K.J., *Chemistry and Biology of Yeasts*, Marcel Dekker Inc., New York.

Batterman, S.A., *Yeasts and their Importance*, McGraw-Hill, Tokyo.

Benaim Pinto, C., *Microbial Biomass and Economic Microbiology*, Prentice-Hall, London.

Betina, V.K., *Dairy Microbiology*, Applied Science Publishers, London.

Bradley, R.S., *Pollution Prevention through Biotechnology*, Academic Press, London.

Brown, M.H., *Environmental Microbiology*, Cambridge University Press, Cambridge.

Budyko, M.I., *Bioactive Microbial Products,* D. Van Nostrand, New York.

Cambell, K.E. and Lemer, H.A., *Microbial Technology*, Academic Press, London.

Chang, J.C., *Biochemical Engineering and Biotechnology Handbook*, John Wiley & Sons, New York.

Cook, K., *Introduction to Environmental Microbiology*, John Wiley & Sons, New York.

Coolingwood, R.W., *Fermentation Technology and its Industrial Applications*, John Wiley & Sons, New York.

Downe, S.A., *Biochemistry of Industrial Micro-organisms*, John Wiley & Sons, New York.

Dugan, P.R., *Fungi, Food and Fermentation*, Plenum Publishing Corporation, London.

Goldman, M., *Yeast from Molasses Alcohol*, Gordon and Breach, Science Publishers, New York.

Gould, G.W., *Ecology of Micro-organisms*, D. Van Nostrand, New York.

Harding, G., *Acetic Acid Bacteria*, Prentice-Hall, London.

Hidy, G.M., *Biohazards,* Heinemann, London.

Huff, C.B., *Applied Environmental Microbiology*, Elsevier Scientific Publishing Co., Amsterdam.

Jackson, M.L., *Vinegar Fermentation*, Prentice-Hall, London.

Jackwerth, E., *Encyclopedia of Biotechnology and Industrial Microbiology*, Academic Press, London.

James, A. and Evison L., *Production of Industrial Enzymes*, John Wiley & Sons, New York.

Jencks, W.P., *Encyclopedia of Environmental Microbiology*, John Wiley & Sons, New York.

Kim, C.K., *Bioseparation Techniques*, Marcel Dekker, New York.

Lowman, R.L., *Industrial Processing with Membranes*, Chilton Book Co., USA.

Mason, R., *Industrial Aspects of Biochemistry and Microbiology,* McGraw-Hill, New York.

Odum, P.L., *Down Stream Processing in Biotechnology*, W.B. Saunders and Co., New York.

Wyatt, G.M., *Fermentation and Enzyme Technology*, Reston Publishing Co., Reston, Virginia.

William, T.M., *Bioreactors and Biotransformation*, Butterworths, London.

References

Index